APR 23 1991

KINGSTON PUBLIC LIBRARY

Decapod Crustacea
of the
Atlantic Coast of Canada

DEDICATION

For my grandson, Julian

Canadian Bulletin of Fisheries and Aquatic Sciences 221

Decapod Crustacea of the Atlantic Coast of Canada

Hubert J. Squires

*122 University Avenue
St. John's, Newfoundland A1B 1Z5*

Scientific Excellence
Resource Protection & Conservation
Benefits for Canadians

DEPARTMENT OF FISHERIES AND OCEANS
Ottawa 1990

The *Canadian Bulletins of Fisheries and Aquatic Sciences* are designed to interpret current knowledge in scientific fields pertinent to Canadian fisheries and aquatic environments.

The *Canadian Journal of Fisheries and Aquatic Sciences* is published in annual volumes of monthly issues. *Canadian Special Publications of Fisheries and Aquatic Sciences* are issued periodically. These series are available from authorized bookstore agents and other bookstores, or your may send your prepaid order to the Canadian Government Publishing Centre, Supply and Services Canada, Ottawa, Ont. K1A 0S9. Make cheques or money orders payable in Canadian funds to the Receiver General for Canada.

Communications Directorate

Director General: Nicole M. Deschênes
Director: John Camp
Editorial and Publishing Services: Gerald J. Neville

© Minister of Supply and Services Canada 1990

Available from authorized bookstore agents, other bookstores
or you may send your prepaid order to the
Canadian Government Publishing Centre
Supply and Services Canada, Ottawa, Ont. K1A 0S9.

Make cheques or money orders payable in Canadian funds
to the Receiver General for Canada.

A deposit copy of this publication is also available
for reference in public libraries across Canada.

Cat. No. Fs 94-221E   ISBN 0-660-13332-6
DFO/4309   ISSN 0706-6503

Printed in Canada

Think Recycling!

Pensez à recycler !

Printed on recycled paper

Correct citation for this publication:

SQUIRES, H. J. 1990. Decapod Crustacea of the Atlantic Coast of Canada. Can. Bull. Fish. Aquat. Sci. 221: 532 p.

CONTENTS

Abstract/Résumé	vi
Acknowledgements	vii
Introduction	1
Taxonomic Review of Species of Decapod Crustacea in the Area of Reference	11
Keys to Suborders, Infraorders and Families	15
Families, Genera and Species	
Suborder Dendrobranchiata	19
Aristeidae	20
Sergestidae	46
Luciferidae	57
Suborder Eukyphida	63
Oplophoridae	64
Nematocarcinidae	100
Pasiphaeaidae	110
Palaemonidae	126
Hippolytidae	147
Pandalidae	233
Crangonidae	261
Plates	313
Suborder Reptantia	325
Nephropidae	326
Axiidae	332
Callianassidae	343
Polychelidae	354
Paguridae	365
Lithodidae	391
Parapaguridae	402
Galatheidae	407
Majidae	428
Cancridae	450
Geryonidae	461
Portunidae	467
Xanthidae	488
Grapsidae	499
Extralimital Species	505
Glossary	507
References	511
Appendices	
Appendix 1. Summary of some distinctive features of mouthparts and antennae of selected families of decapods from the Atlantic coast of Canada	525
Appendix 2. Mythological references of names of some families, genera and species in the text	530
Index of Scientific Names	531

Abstract

SQUIRES, H. J. 1990. Decapod Crustacea of the Atlantic Coast of Canada. Can. Bull. Fish. Aquat. Sci. 221: 532 p.

Eighty-nine species of decapod crustaceans from the Atlantic coast of Canada are described (including details of mouthparts and other appendages) and original drawings provided for each. Distribution records for each species from 100° W Longitude (80° N latitude) in the Arctic (including Hudson Bay) to the northern edge of Georges Bank, and from 0 to 1000 m in depth are provided also. The Introduction has a brief history of collections of decapod Crustacea in the area, and the Appendix includes a list of mythological references to some taxonomic names used. Tables summarize details of features of appendages according to family. One new species, *Bythocaris spinipleura*, is described. A species of *Macrobrachium* trawled in deep water of the Fundian Channel, southwest of Nova Scotia is also described. Brief notes on the biology of and fishery for each species where applicable are included.

Résumé

SQUIRES, H. J. 1990. Decapod Crustacea of the Atlantic Coast of Canada. Can. Bull. Fish. Aquat. Sci. 221: 532 p.

L'auteur décrit 89 espèces de crustacés décapodes peuplant les eaux canadiennes de l'Atlantique, donne une description détaillée des parties buccales et d'autres appendices et présente des illustrations inédites de chacune. Il inclut aussi des données sur l'aire de répartition de chaque espèce, à partir de 100° de longitude ouest et 80° de latitude nord de l'Arctique (y compris la baie d'Hudson) jusqu'à la limite nord du banc Georges et de 0 à 1 000 m de profondeur. L'introduction comprend un bref historique des relevés de crustacés décapodes effectués dans cette région tandis que l'appendice inclut une liste des noms mythologiques d'où sont tirés des noms taxonomiques. Les caractéristiques des appendices sont résumées selon la famille sous forme de tableaux. L'auteur décrit une nouvelle espèce, *Bythocaris spinipleura*, ainsi qu'une espèce du genre *Macrobrachium* capturée au chalut dans les eaux profondes du chenal Fundian, au sud-ouest de la Nouvelle-Écosse. Il inclut aussi de brèves notes sur la biologie et l'exploitation, selon le cas, de chaque espèce.

ACKNOWLEDGEMENTS

I wish to express my appreciation for assistance from the following: Mr. D. G. Parsons of the Department of Fisheries and Oceans, Science Branch, Northwest Atlantic Fisheries Centre, White Hills, St. John's, for regular discussion of the work, provision of many services including specimens for drawings and coloured photographs of several species of shrimps, and for reading and commenting on preliminary and final drafts of the manuscript; Mr. E. L Rowe of St. John's for assistance with photography; other photo credits to Mr. P. J. Veitch and Dr. M. J. Dadswell; Mr. H. Mullett for drafting the map for Fig. 1; Dr. M. J. Dadswell formerly of the Atlantic Reference Centre, St. Andrews, N.B., for assisting with loans of specimens, and latterly of the Biology Department, Acadia University, for reading a final draft of the manuscript; Dr. G. Pohle of the Atlantic Reference Centre, Huntsman Marine Science Centre, St. Andrews, N. B., for providing loans of specimens and for valuable comments on part of the manuscript; Dr. R. W. Elner of the Department of Fisheries and Oceans, St. Andrews, N. B., for providing specimens of crabs; the Smithsonian Institution (Dr. R. B. Manning), the British Museum of Natural History (Dr. R. W. Ingle), the Rijksmuseum van Natuurlijke Historie (Dr. C. H. J. M. Fransen), Universitets Zoologiske Museum (Dr. Torben Wolff) and the National Museums of Canada (Natural History) (Dr. D. Laubitz) for loans of specimens difficult to obtain; Drs. F. A. Chace, Jr., A. B. Williams, L. B. Holthuis and R. W. Ingle for professional advice; Dr. G. Lilly for discussions of distributions, specimens, etc.; the American Geographical Society, New York, for provision of the map used for distributions formerly in their Folio 12 of the Serial Atlas of the Marine Environment.

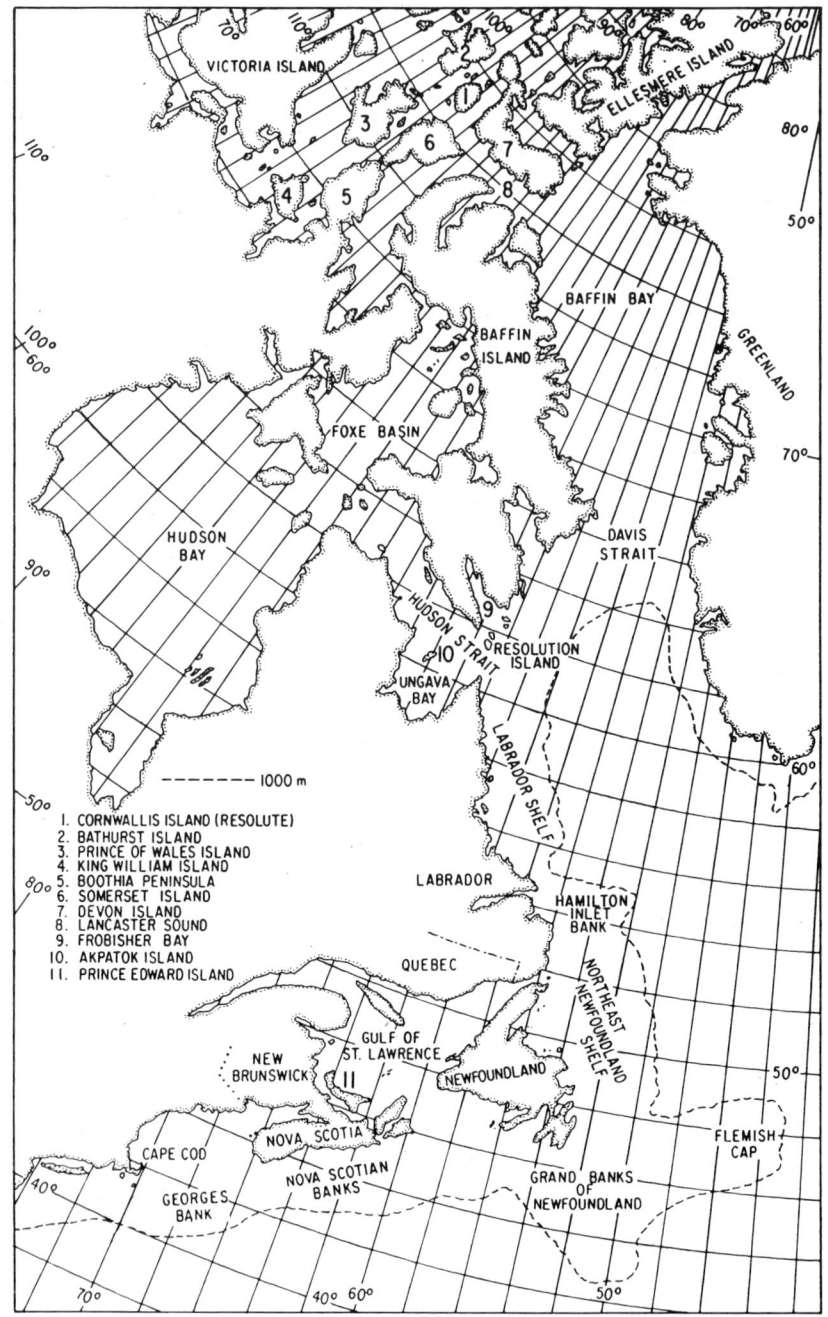

Fig. 1. Map of Area of reference and place names mentioned in text (original map designed by American Geographical Society for Folio 12, in Serial Atlas of Marine Environment).

INTRODUCTION

The purpose of this publication is to give a detailed description of the external morphology of species of decapod crustaceans found in the Canadian Arctic-boreal marine zoogeographic province of the Atlantic coast. The animals include shrimps, lobsters and crabs as well as shrimp-like or lobster-like and crab-like species of the Phylum Crustacea (Class Malacostraca: Order Decapoda (Schram 1986)). The area for this purpose is limited in the south by the northern part of Georges Bank or 42° N latitude and extends into the Canadian Arctic to include Hudson Bay and as far as Prince of Wales Island or 100° W longitude, and at depths not greater than 1 000 m throughout (Fig. 1). The decapod Crustacea in this region comprise only about 89 out of a world total of about 9 000 recent species (excluding fossil species; McLaughlin 1980). These crustaceans have many important commercial species throughout the world (Holthuis 1980) but only seven are exploited in commercial fisheries in the area of reference.

In this region the decapod crustaceans, although few in numbers of species, occur in considerable numbers, and many appear as prey species for marine mammals and fishes of major commercial importance. Apart from this the various species occupy ecological niches throughout the region, and they are so prevalent that any discrete marine area may be characterized by assemblages of them (Squires 1965a).

With respect to characteristic environmental assemblages of decapods the large area of Ungava Bay, for example, could be characterized by six decapod species that are prevalent throughout the bay (although 17 species were taken in all; Squires 1957). Also in a small local area in Bonavista Bay, Newfoundland, where specimens were obtained while Scuba diving, two species of shrimps (*Pandalus montagui* and *Eualus pusiolus*) were found to occupy a particular niche with a shallow rocky substrate (Ennis and Squires unpublished). On the other hand such assemblages can be predicted where well-defined conditions occurred. For example, where soft mud and detritus bottom occurs at more than 200 m deep (200–300 m) and where bottom temperatures are higher than 2° C there is almost certain to be a homogenous population of *Pandalus borealis* (Squires 1961). And where sandy mud and shallow eel-grass (*Zostera* sp.) beds occur in this region there will almost always be *Crangon septemspinosa* present.

To understand interaction between species of fishes and invertebrates in the ecosystem it is important to recognize prey species when found in stomachs. Often only fragments of decapod crustaceans are found, but with knowledge of detailed morphology of species, identification is often possible. Descriptions of species of decapods in the present work have therefore been presented in detail and with accompanying drawings so that even fragments may be recognizable to species on examination with a dissecting microscope.

Previous taxonomic studies of decapod crustaceans of this region are now out of print (Whiteaves 1901; Rathbun 1929; Squires 1965a) and out of date in at least some of the naming of species. Also the morphological details presented here can only be obtained by reference to many publications, or in some instances have not previously been published. All drawings in the present work have been made from actual specimens of the various species.

There is some overlap in distribution and treatment of the present species by the recent magnificent work of Williams (1984). Most of his species are from the Virginian and Carolinian zoogeographical provinces or farther south, and some of the more arctic or deepwater species are not included. However, some species included in the present work have their main centres of distribution more to the south and occur in the area of reference only ocasionally. An example of such a species is *Callinectes sapidus* which is found even in the Tropics but wanders into the Maritimes in some warm summers.

Review of Species of Decapod Crustacea in the Area of Reference

Keys to identify families, genera and species of decapod crustaceans in the region have, for the most part, been borrowed from authors as indicated but are usually modified to include only the species present. They have limitations for application to other areas, therefore, or to species not in the present work.

Synonymies included here are few and listed merely to indicate authors who provide descriptions and other data on the particular species which might be of further interest.

World distributions are mentioned as provided by authors referred to. Distribution in the area of reference is marked on a map for each species. Records for Greenland are generally not included as in Squires (1966), although a few, such as for *Pandalus borealis*, are put in to indicate a major fishing area or presence further north than our records. Distribution records for the southern part of the area of reference are from Williams and Wigley (1977).

Descriptions of Species

Equal treatment has been given as much as possible to the description of each species with a drawing provided for each part described. These drawings are representative in that they show proportions and shapes closely but cannot be said to depict the intrinsic beauty of form of the original with all its complexity. Descriptions follow the same pattern for each species and are prefaced by a statement of the distinguishing characteristics which summarizes features that may be used to identify the species. This is followed by a detailed description beginning with (a) the integument and colour of the whole animal; (b) the rostrum, carapace and abdomen; (c) the anterior sensory appendages, the antennules and antennae and the eyes; (d) the mouth parts used for manipulating items of food, including the paired mandibles, maxillules, maxillae and three pairs of maxillipeds; (e) the five pairs of legs or pereopods, the front pairs of which are generally chelate and used for the capture of food, and the others mainly used for locomotion; (f) the five pairs of swimmerets or pleopods in most species, used for locomotion but also modified for transferral of spermatophores in males and/or to carry eggs in females in some species.

Additional functions of the appendages include bailing water through the gills or branchia (including *podobranchs* on epipods, *arthrobranchs* at the membrane joining maxillipeds or legs to the thorax, and *pleurobranchs* on the walls of the thorax) by the exopod of the maxilla (the scaphognathite), also the action of grooming hooks of the epipods (Bauer 1981) in cleaning the setobranchs which stream thread-like setae over the gills to keep them free of fouling organisms. The somites or segments of the abdomen equipped with internal extensor and flexor muscles (not illustrated here) in shrimps or lobsters give powerful flexures of the abdomen, using the large tailfan of uropods and telson to propel the animal at considerable speed for sudden escape.

Morphology of Rostrum, Carapace, Abdomen and Appendages
(Figures 2 and 3)

The rostrum is an anterior extension of the median dorsal part of the carapace and may be very short or long and may have fixed or moveable spines dorsally and ventrally. Usually there is a lateral ridge to give it greater strength, and this is generally confluent with the orbit.

The carapace is a rigid covering of the head and thorax extending laterally to enclose the gills and origins of the appendages, usually a pair on each of the 14 somites. Anteriorly the orbits give protection to the eyes and there may be a series of spines: supraorbital, suborbital, antennal, branchiostegal and pterygostomian at or near the anterior edge. Other spines may be located on specific areas of the carapace such as the gastric, hepatic, cardiac and branchial. These areas may also exhibit ridges or grooves characteristic of a species.

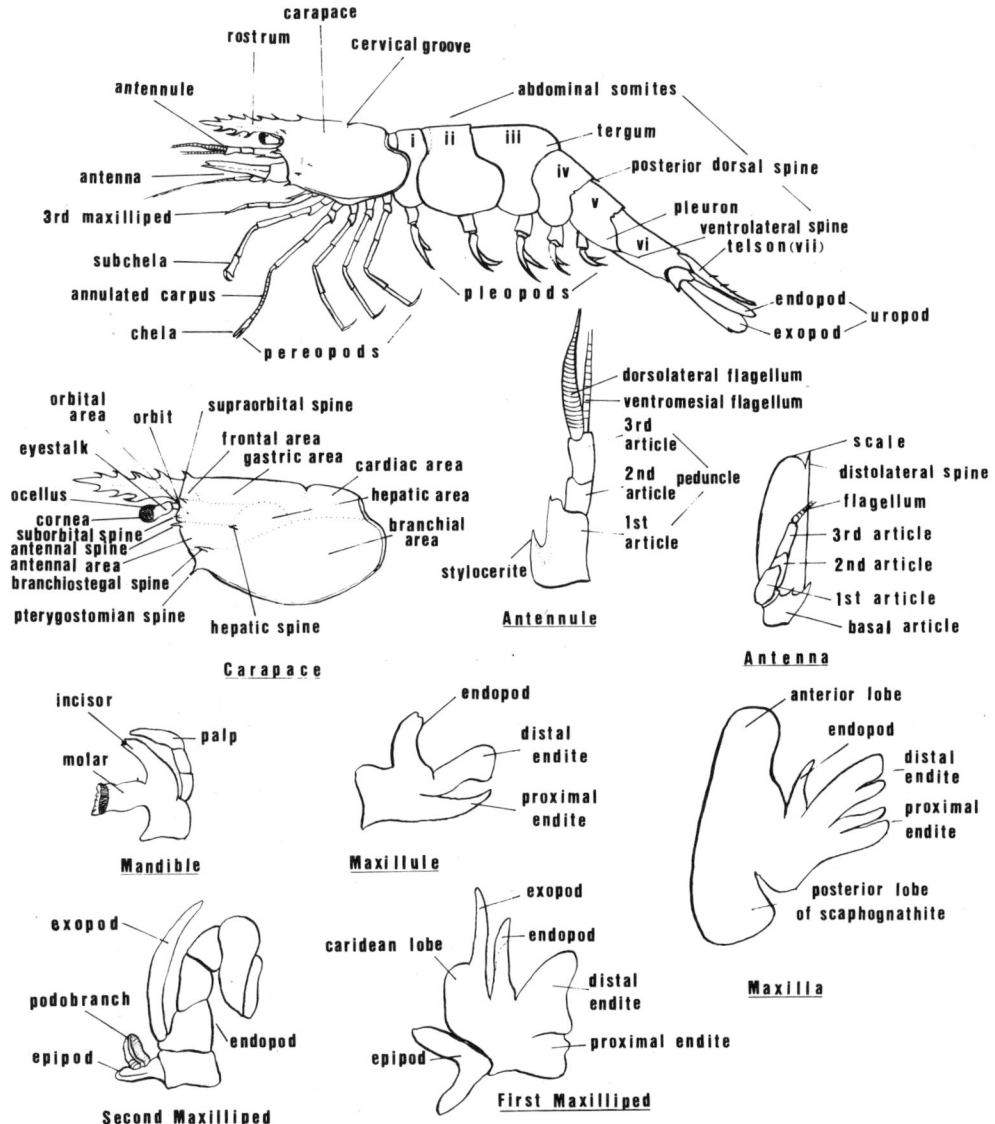

Fig. 2. Names of morphological features of decapod crustaceans referred to in the text.

The abdomen is equally a large part of the anatomy and more clearly shows body segmentation. Six of the seven somites each have a dorsolateral covering or tergum and a ventrolateral pleuron, the latter hanging down at each side and sometimes armed with spines. The seventh somite or telson is usually tapered and armed with dorsolateral and terminal spines, but is greatly varied according to species.

The eyes, when present, have a multi-faceted cornea, and sometimes an accessory ocellus. They are spherical or ovoid in shape from the front, of various colours with some translucency, and at the end of a somewhat conical stalk. The latter may sometimes have a tubercle, often characteristic of a species or genus.

The antennules (c) or first pair of antennae have a three-segmented peduncle, the first segment usually with a lateral wing-like expansion — the stylocerite — with sharp

3

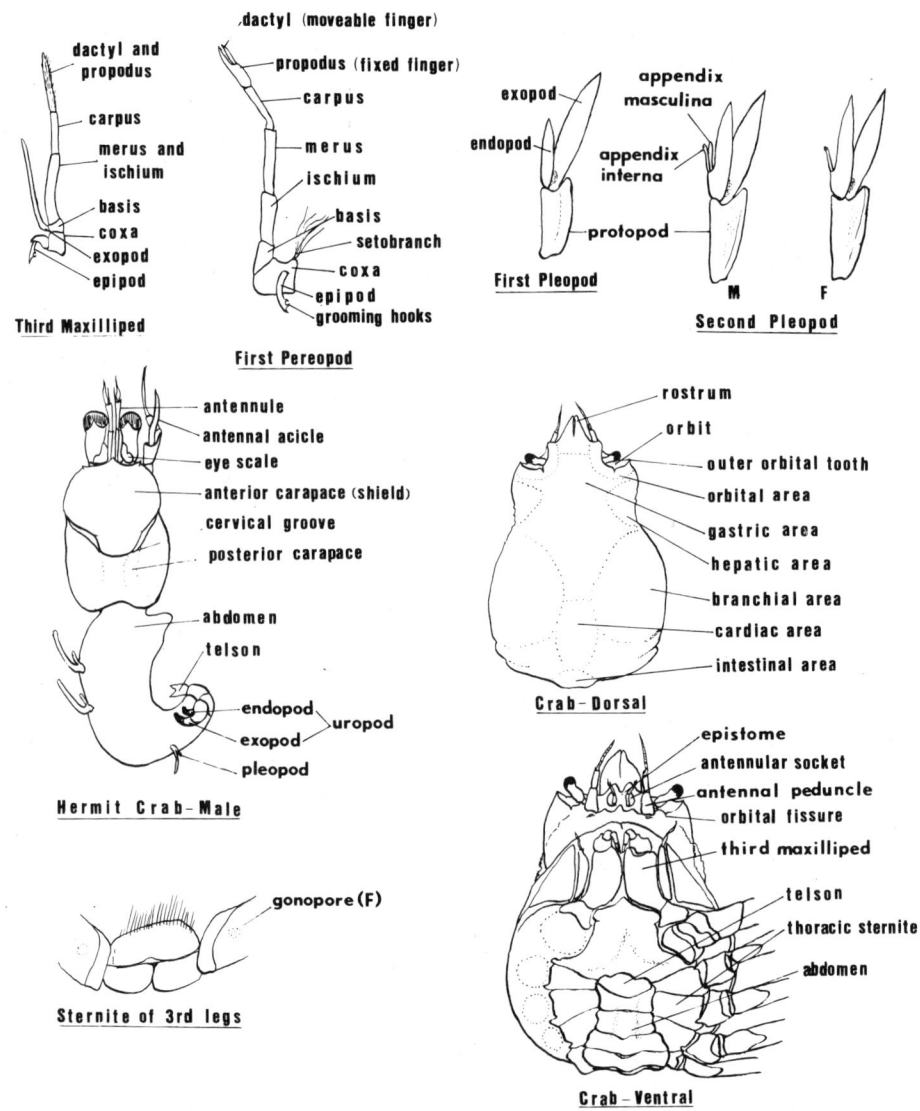

Fig. 3. Names of morphological features of decapod crustaceans referred to in the text.

tip, and the third with two flagella. The dorsolateral or outer flagellum is usually dilated proximally for part of its length, and the ventromesial or inner flagellum usually more slender and longer (the dorsolateral with two branches in the family Palaemonidae).

The antennae (d) or second pair of antennae usually have two basal segments and a three-segmented peduncle with a distal long flagellum. Distally the basal segments have a flat lamella or scale, the scaphocerite, in some species modified as a thorn.

The mandibles (e) or teeth may have a molar and incisor process and a segmented palp; in some species the molar or the incisor is present and the other parts may be missing. The right may be slightly different from the left. The right shears past outside the left.

The maxillules (f) are flat and slightly curved lying close to the mandibles and usually consisting of two endites fringed with spinous setae (stronger at the tips), an endopodal palp and an inconspicuous outer shoulder or exopod.

The maxillae (g) are also flat and slightly curved and lie close outside the maxillules with usually two pairs of fringed endites forming part of a setal basket to filter food particles. There is a thin endopod pointing anteriorly, and a two-lobed lateral scaphognathite strongly muscled at about the centre by which it is swung to and fro to act as a bailer of water through the gills.

The first maxillipeds (h), also thin and slightly curved, fit against the maxillae, and mostly consist of two large endites mesially, the proximal of which is triangular in cross-section with its thickened and setose edge at the centre, and a thin narrow endopod lying close to a foliaceous exopod which has a wide lobe in caridean shrimps. There is usually a bilobed epipod.

The second maxilliped (i) is somewhat leglike but small and flattened to fit against the 1st maxilliped: it sometimes has seven segments but usually fewer; a slender exopod is usually present and sometimes an epipod with attached podobranch.

The third or outer maxilliped (k) is large and extends forward anterior to and sometimes larger than the 1st leg but usually has fewer segments. Often it has a slender exopod and an epipod somewhat similar to the epipods of the legs and sometimes with an attached podobranch.

The five pairs of pereopods (the spelling follows Chace 1986, and Williams 1984, and other recent authors) are appendages of the thorax and all are quite similar: they each have seven segments, i.e. the *coxa* joined to the pre-coxa or body wall, and *basis*, *ischium*, *merus*, *carpus*, *propodus* and *dactyl* in order from the base to the tip. The first two legs are chelate in most caridean shrimps: the dactyl forming the moveable part or "finger" and an extension of the propodus forming the fixed finger, both usually with opposing rows of teeth. Other decapods may have one or more or no legs chelate.

The five pairs of pleopods consisting of a protopod and usually an endopod and exopod are ventrally on the first five abdominal somites. Often the anterior two pairs of pleopods are sexually modified, and in females of some species that carry eggs, the posterior pairs at least become covered with thread-like setae in "breeding dress" and exude a type of cement which causes the eggs to adhere to the setae. The 6th somite has a pair of uropods each with a protopod and an exopod and endopod. The 7th somite or telson (t; Schram 1986) has no appendages but dorsolateral and terminal spines.

In all species of decapod crustaceans all appendages and the integument in general are variously fringed or covered with setae which perform various functions, many sensory, but some perform a mechanical service such as the plumose setae fringing the scaphognathite which add to the stroking surface of the bailer. In a drawing only some of the numerous setae can be shown, and reference to them is generally omitted from descriptions in this work.

Measurements and Biological Assessment

Carapace length (cl) is taken as the standard measurement for shrimp-like or lobster-like decapods, i.e., from the posterior edge of the orbit to the posterior edge of the carapace in the midline (between parallels), and the length of the anterior part or calcified portion of hermit crabs, but carapace width (cw) is given additionally in crabs because of common usage. Total body length is mentioned where it is the only length given in the literature. Under "Biology" the greatest carapace length is given where available.

Also under "Biology" is discussed reproductive potential of the species where known, and food, where observations have been made and referred to in the literature. A brief note is also given on the fishery for the species if it is fished.

Only brief mention of parasites is made under species in question.

As is well known, all species of decapod crustaceans hatch from the egg as a larva which, as it grows, moults into a stage more like the adult than the former stage (four stages in most caridean shrimps). The post-larval stage can usually be identified to species without difficulty, although even in the juvenile form adult features may be somewhat modified (as in the juvenile form of *Eualus macilentus* called *E. stoneyi* by Rathbun 1902). Identifications of small juveniles in some species are therefore to be attempted with caution. Larval identifications are based on larvae hatched from eggs of identified adults (as for example, Haynes 1973, 1976, 1978a, b, 1979), although some have also been identified from the plankton (Frost 1936; Squires 1965b). Because of the length of the present study and lack of specimens of most species, larvae have not been included.

Determining Sex in Species

In penaeid shrimps the male has a *petasma* on the protopod of the first pleopod and the female has a *thelycum* formed from the sternites of the 3rd to 5th legs.

In caridean shrimps the male has an accesory appendage, the appendix masculina next to the appendix interna on the second pleopod.

In hermit crabs of the genus *Pagurus* the male has only 3 unpaired pleopods while the female has 4. In *Parapagurus* the male has paired large pleopods modified for transferral of spermatophores, and the female first pleopods are very small.

In lobsters and lobster-like species the first pleopods may be strongly modified in males for transferral of spermatophores while they are not so modified in females. However, there are exceptions such as in *Callianassa* where the male has no first pleopods. In true hermaphroditic species, such as *Calocaris templemani*, there are male openings of the vas deferens on the coxa of the 5th legs and of the female oviduct on the coxa of the 3rd legs, but the 1st pleopods are modified for transferral of spermatophores. In protandric hermaphrodites such as *Pandalus borealis* the male secondary characters change at the intersex phase and are lacking in the female.

In crabs there are only two pairs of strongly modified pleopods in males and the abdomen is narrow, sometimes with fused somites, while in females there are four pairs of setose pleopods (the first usually missing) and the abdomen is wide and without fused somites.

In decapods in general the genital openings are on body somite XII (of the third legs) in females and XIV (of the fifth legs) in males. There are also minor secondary characteristics in males and females that vary throughout the order. Some of these are referred to by Butler (1980); reference to them is made under species in the present work.

History of Collections of Decapods in the Area of Reference
(partly from Squires 1965a)

Sabine (1824) recorded five species of decapod crustaceans from Davis Strait as collected by the 1st Parry Expedition in search of a northwest passage to the Pacific during 1819–20. Two of these species, new to science, were named *Crangon septemcarinatus* (= *Sabinea septemcarinata*) and *Alphaeus polaris* (= *Lebbeus polaris*). The 2nd and 3rd Parry Expeditions collected 5 species near Igloolik in Foxe Basin (Ross 1826 and 1835). Pfeffer (1886) identified 2 species collected by the German Arctic Expedition of 1882–83 from Cumberland Sound. These were *Hippolyte groenlandicus* (= *Lebbeus groenlandicus*) and *Hippolyte amazo* (= *Lebbeus polaris*). Ohlin (1895) also recorded *Eualus fabricii* from the Cumberland Sound collection. These and other collections of decapod crustaceans from Davis Strait and West Greenland were reviewed by Hansen (1908), Hofsten (1916) and Stephensen (1935). Hansen also recorded the decapods collected in Davis Strait by the *Ingolf* in 1900 and Stephensen those of the *Godthaab*

Expedition in 1910. Stations of these expeditions and records of occurrence of species are reviewed by Squires (1965a, 1966). Also the deep-sea expeditions worked a few stations at depths less than 1 000 m on the continental slope: namely, the *Challenger* off Nova Scotia in 1873, and the *Hirondelle* and *Michael Sars* on the Grand Banks in 1877 and 1910, respectively. Bate (1888) listed 4 species, Milne-Edwards and Bouvier (1894) 4 species, and Sivertsen and Holthuis (1956) 9 species taken by these respective expeditions in the area of reference.

Whiteaves (1901) reviewed collections of invertebrates including crustaceans from eastern Canada during the mid and late 1800s. Those that led to publications about decapod crustaceans included collections by Stimpson (1854) at Grand Manan; Bell's Report of Progress of his survey of 1882–83–84 on the east coast of Ungava Bay and the north coast of Hudson Strait (Appendix iv by S. I. Smith, 1884); Packard's (1867) "View of the invertebrate fauna of Labrador"; Verrill in 1861 at Anticosti and Mingan Islands and Whiteaves 1871, 1872 and 1873 in the Gulf of St. Lawrence (Whiteaves 1901); U.S. Fish Commission 1872, 1877 and 1883 from Georges Bank to the coast of Nova Scotia (Smith 1895); Stearns 1882 in shallow water of the coast of Labrador (Smith 1895); Low in 1897, 1898 and 1899 in Ungava Bay and Hudson Bay including Richmond Gulf, and Wakeham in 1897 of the Hudson Bay Exploring Expedition in the *Diana* on the west coast of Hudson Bay (Rathbun 1919).

The *Diana* and *Neptune* expeditions in Hudson Bay in 1897 and 1903–04 collected decapod Crustacea that were reported on by Rathbun (1919). Further collections were made by the *Loubyrne* expedition in Hudson Bay in 1930 (Van Winkle and Schmitt 1936). Species collected were listed by Squires (1967a).

Outside of Hudson Bay the Canadian Arctic Expedition of 1910 collected decapods near the mouth of Hudson Bay and at Port Burwell in Ungava Bay (Rathbun 1919). Other collections from Hudson Strait and the coast of Labrador were reported by Rathbun (1919). Collections from Capt. Robert A. Bartlett's voyages into Foxe Basin in 1927 and south of Ellesmere Island in 1933 were included in Van Winkle and Schmitt's (1936) report.

Rathbun (1929) reviewed earlier collections from the Canadian Atlantic coast, revised the taxonomy and gave the ranges in distribution of 52 species of decapod crustaceans.

Other collections reported on were 7 species of hippolytids by Leim (1921) including a new species *Lebbeus zebra* (without comparing it with *L. microceros*); 6 species of decapod crustaceans from the Gulf of St. Lawrence by Boone (1930); 21 species collected in groundfish and plankton surveys off the Newfoundland coast by Frost and Thompson (Frost 1936), and 24 species taken in the St. Lawrence estuary by Prefontaine and Brunel (1962).

During 1947–61 the *Calanus* Expeditions of McGill University (Fisheries Research Board of Canada Eastern Arctic Investigations) collected fishes and invertebrates throughout the Arctic from Ungava Bay and Frobisher Bay to Foxe Basin and Hudson Bay. The collections of decapod crustaceans were reported on by Squires (1957, 1962, 1965a, 1966, 1967a). Extension of this arctic program was undertaken by the Arctic Biological Station, Ste-Anne-de-Bellevue, and the Freshwater Institute, Winnipeg, of the Fisheries Research Board of Canada (now Stations of the Science Branch of the Department of Fisheries and Oceans). Squires (1968a and 1969) reported on species of decapods from work done in 1960–65.

Extensive groundfish and shrimp surveys in the fishing area from the Nova Scotia Banks to Cape Dyer, Baffin Island, during 1946–60, by the Fisheries Research Board of Canada collected thousands of specimens comprising 48 species which were reported on by Squires (1965a and 1966).

More recent published accounts of decapod crustaceans from the area include studies in the St. Lawrence estuary and the Gulf of St. Lawrence by Couture and Trudel (1968 and 1969a, b) and Fréchette et al (1970).

Since 1960 fisheries surveys throughout the area by the Canadian Department of Fisheries and Oceans and provincial fisheries departments have been very comprehensive, identifying various stocks and leading to special fisheries for species of fishes as well as shrimps and crabs. Also some southerly surveys for the origin of squid spawning have collected pelagic and deep-sea species of decapods some of which are extra-limital to the present work.

Discovery of a Marine Species of *Macrobrachium*

A well-documented collection (three specimens including one male and two females) of a species of *Macrobrachium* from the Fundian Channel, off Nova Scotia (Lat. 43°28′ N, Long. 67°32′ W, on July 11, 1974, by the MV "*A. T. Cameron*", Trip 225, Station 18; collector: Dr. J. G. Scott) is a first for such a marine area (depth just over 200 m). The genus and all its species were formerly taken only in estuaries or completely freshwater locations, although it is well-known that some species effect migrations through marine areas. Unfortunately the large second legs of all three specimens were lost in their capture with fishing gear, so that naming the species is precluded since descriptions of species of this genus are dependant in part on the large legs. However, remaining features are described as for other species. The specimens, including 2 females 8 and 9 mm cl and 1 male 8 mm cl, (No. ARC 8958232) are in the collections of the Atlantic Reference Centre, Huntsman Marine Science Centre, St. Andrews, N. B.

Oceanography of the Area

A summary of hydrographic conditions in the region was given by Hachey (1961), reviewed for the northern part by Squires (1965a) and Templeman (1975) and by Petrie et al (1988) among others. Also regular overviews of conditions in the major part of the area are given such as those by Trites and Drinkwater (1984, 1985, 1986), and Drinkwater and Trites (1988). Conditions in the arctic were reviewed by Dunbar (1951).

The area is influenced greatly by the presence of arctic water of the Labrador Current even as far south as Georges Bank, and the volume of water with temperatures less than 0° C increases during November to April. Ice cover prevails in northern areas during these months (as long as 9 months in Hudson Bay and further north) and in some years reaches as far south as Nova Scotia. The cold water with sea ice in winter, and salinites a few parts per thousand (ppt) lower than oceanic water, is swept toward the east coast presumably by Coriolis force. In northern areas it tends to be deeper near the coast and becomes shallower offshore and is underlain by Atlantic water of higher density and temperature with which some mixing occurs especially at the edge of the continental shelf (Smith and Sandstrom 1988). As the arctic water reaches the Grand Banks off Newfoundland there is increased turbulence and greater mixing with Atlantic water which continues farther south. Throughout the region there are influences of stream runoff and inshore warming in summer and cooling with formation of shore ice especially in the north during winter. Also there are influences of the Gulf Stream Drift often in the form of giant eddies of warm core water (Smith and Sandstrom 1988) which contribute to the mixing and raising of temperature over the continental shelf and shallower banks. In the global sense the immense volume of warm water of the Gulf Stream and the northern edge of the Sargasso Sea lie only a few hundred kilometres to the southeast of the Grand Banks, and their moderating influence is felt to a geater or less extent from year to year depending upon weather.

In the Gulf of St. Lawrence there is penetration of warm (to 5° C) Atlantic water from the Laurentian Channel (Lauzier and Trites 1958), reaching as far as the Esquiman and Mingan channels as a bottom layer. Surface layers are contributed to by the flow of arctic water in through the Straits of Belle Isle and from the outflow of the St. Lawrence River. Surface currents are generally anticlockwise but otherwise deflected by winds as shown by the movement of ice cover in winter.

Over the Nova Scotian banks the surface layer is mostly under the influence of southbound arctic water of the Labrador Current and outflows from the Gulf of St. Lawrence. But there are intrusions of Atlantic water forming an intermediate cool layer and a warmer bottom layer (6–8° C), the latter occasioned by Gulf Stream incursions (Smith et al 1978). The Bay of Fundy area is influenced by extreme tides and southward flowing cool water from the surface layer of the Nova Scotian Banks mixing with upwelling and surface water from the Gulf of Maine (Scott and Scott 1988).

Seasonal changes greatly influence the life cycle of decapod crustaceans, the winter cold supressing reproductive as well as other development, and summer warmth allowing for growth through moulting and the development and release of sexual products. So propagation continues even in the far north and populations are maintained. However, pockets of cold arctic water such as in some deepwater bays in Newfoundland may influence negatively the capability of *Pandalus borealis* in such areas to be self-propagating (Squires 1968b), while other species such as *Eualus macilentus* appear to be more cold-adapted (Squires 1967b). Several species of decapods including *Homarus americanus*, *Crangon septemspinosa* and *Cancer irroratus* reach their furthest north (at the Straits of Belle Isle) in the area as a result of lack of cold adaptation, but many species, especially in the families Hippolytidae, Pandalidae and Crangonidae have adapted to the cold extremes of even the arctic part of the area.

Fisheries for Decapod Crustaceans in the Area of Reference

Market demand for the tasty meat of decapod crustaceans has led to their exploitation at least in the case of lobster since the late 19th century. However, fisheries for other species of decapods (mainly *Pandalus borealis* and *Chionoecetes opilio*) in Canadian waters did not begin until well after the middle of the 20th century. These fisheries are reviewed briefly under relevant species.

Fisheries research surveys conducted by the Fisheries Research Board of Canada (later the Science Branch of the Department of Fisheries and Oceans) from the late 1940s onward gave an indication of incidental occurrences of species such as those of decapod crustaceans and, to some extent, indicated their relative abundance throughout the area. In the Newfoundland area (for example) some follow-up was done in a directed survey for fishable quantities of *Pandalus borealis* in 1957 and 1958 with positive results (Squires 1961). Although well-publicised at the time, a fishery was not initiated until ten years later following further surveys (Frechet 1971) and demonstration fishing under the auspices of federal–provincial government departments (Brothers 1971). Other provincial surveys (Couture 1971; Frechet 1971; Legare 1971) have led to fisheries for *Pandalus borealis* in the Bay of Fundy, off Nova Scotia and in the Gulf of St. Lawrence. Fisheries to date for *Pandalus borealis* from Greenland to the Gulf of Maine have been reviewed by Parsons and Fréchette (1989).

Brunel (1963) reported an initial by-catch fishery for the Snow Crab (*Chionoecetes opilio*) off Gaspé, Quebec, in 1960 and 1961. The fishery for this species expanded rapidly after the mid-60s and reached a peak of 47 000 tons in 1982, fourth in landed value of Canadian fisheries species, i.e. cod, lobster, scallop and snow crab. Catches of other crabs are of much less significance and occur mainly as bycatches in fisheries for other species (Elner 1985).

Recent research surveys have located fishable quantities of *Pandalus montagui* which is now exploited commercially in Hudson Strait and Ungava Bay (D. G. Parsons, personal communication).

Research on Decapod Crustaceans

Apart from biological studies referred to under relevant species, present research is directed towards population assessments of lobster, northern or pink shrimp and snow crab in an effort to provide management advice for fisheries of these species. Also routine sampling from these populations provides opportunity for further biological studies related to growth, reproduction and other factors in their life history.

Relationships with other animal species in the area, principally fish and other large predators, are shown by the occurrence of decapod crustaceans in stomach contents taken incidentally to other studies of these species (Lilly, personal communication). Also their own stomach contents can indicate predation on species in their community in the ecosystem and their degree of competition with other species.

Studies on behaviour in aquaria and in their natural habitat have led to an understanding of the possibilities of aquaculture or "ranching" of some of the species.

TAXONOMIC REVIEW OF SPECIES OF DECAPOD CRUSTACEA IN THE AREA OF REFERENCE

Marine and arctic-boreal species of decapod crustaceans of the east and northeast coasts of Canada from the Queen Elizabeth Islands (100° W long.) to the northern edge of Georges Bank (42° N lat.). Depths from 0 to 1 000 m.

Phylum	CRUSTACEA Pennant, 1777
Class	MALACOSTRACA Latreille, 1806
Order	DECAPODA Latreille, 1803
Suborder	DENDROBRANCHIATA Bate, 1888
Infraorder	PENAEIDEA Boas, 1880
Superfamily	PENAEOIDEA Rafinesque, 1815
Family	ARISTEIDAE Alcock, 1901

Benthesicymus bartletti Smith, 1882
Gennadas elegans Smith, 1882
Gennadas valens Smith, 1884
Pleoticus robustus (Smith, 1885)
Plesiopenaeus edwardsianus (Johnson, 1867)

Superfamily	SERGESTOIDEA Dana, 1852
Family	SERGESTIDAE Dana, 1852

Sergestes arcticus Krøyer, 1859
Sergia robusta (Smith, 1882)

Family	LUCIFERIDAE Burkenroad, 1983

Lucifer faxoni Borradaile, 1915

Suborder	EUKYPHIDA Boas, 1880
Infraorder	CARIDEA Dana, 1952
Superfamily	OPLOPHOROIDEA Dana, 1852
Family	OPLOPHORIDAE Dana, 1852

Acanthephyra pelagica (Risso, 1816)
Acanthephyra purpurea A. Milne-Edwards, 1881
Hymenodora glacialis (Bucholz, 1874)
Notostomus elegans A. Milne-Edwards, 1881
Notostomus robustus Smith, 1884
Oplophorus spinosus (Brullé, 1839)
Systellaspis debilis (A. Milne-Edwards, 1881)

Family	NEMATOCARCINIDAE Smith, 1884

Nematocarcinus cursor A. Milne-Edwards, 1881
Nematocarcinus rotundus Crosnier et Forest, 1973

Superfamily	PASIPHAEOIDEA Dana, 1852

Family   PASIPHAEIDAE Dana, 1852
                  Parapasiphaea sulcatifrons (Smith, 1884)
                  Pasiphaea multidentata Esmark, 1866
                  Pasiphaea tarda Krøyer, 1845

Superfamily   PALAEMONOIDEA Rafinesque, 1815
Family   PALAEMONIDAE Rafinesque, 1815
                  Leander tenuicornis (Say, 1818)
                  Macrobrachium sp.
                  Palaemonetes pugio Holthuis, 1949
                  Palaemonetes vulgaris (Say, 1818)

Superfamily   ALPHEOIDEA Rafinesque, 1815
Family   HIPPOLYTIDAE Dana, 1852
                  Bythocaris gracilis Smith, 1885
                  Bythocaris payeri (Heller, 1875)
                  Bythocaris spinipleura, new species
                  Caridion gordoni (Bate, 1858)
                  Eualus fabricii (Krøyer, 1842)
                  Eualus gaimardi belcheri (Bell, 1855)
                  Eualus gaimardi gaimardi (H. Milne-Edwards, 1837)
                  Eualus macilentus (Krøyer, 1842)
                  Eualus pusiolus (Krøyer, 1842)
                  Hippolyte coerulescens (Fabricius, 1775)
                  Latreutes fucorum (Fabricius, 1798)
                  Lebbeus groenlandicus (Fabricius, 1775)
                  Lebbeus microceros (Krøyer, 1842)
                  Lebbeus polaris (Sabine, 1821)
                  Spirontocaris lilljeborgi (Danielssen, 1859)
                  Spirontocaris phippsi (Krøyer, 1842)
                  Spirontocaris spinus (Sowerby, 1805)

Superfamily   PANDALOIDEA Dana, 1852
Family   PANDALIDAE Dana, 1852
                  Dichelopandalus leptocerus (Smith, 1881)
                  Pandalus borealis Krøyer, 1838
                  Pandalus montagui Leach, 1814
                  Pandalus propinquus G. O. Sars, 1869
                  Stylopandalus richardi (Coutière, 1905)

Superfamily   CRANGONOIDEA H. Milne-Edwards, 1837
Family   CRANGONIDAE H. Milne-Edwards, 1837
                  Argis dentata (Rathbun, 1904)
                  Crangon septemspinosa Say, 1818
                  Metacrangon jacqueti agassizi (Smith, 1882)
                  Pontophilus brevirostris Smith, 1881
                  Pontophilus norvegicus (M. Sars, 1861)
                  Sabinea hystrix (A. Milne-Edwards, 1881)
                  Sabinea sarsi Smith, 1879
                  Sabinea septemcarinata (Sabine, 1824)
                  Sclerocrangon boreas (Phipps, 1774)
                  Sclerocrangon ferox (G. O. Sars, 1877)

Suborder	REPTANTIA Boas, 1880
Infraorder	ASTACIDEA Latreille, 1803
Family	NEPHROPIDAE Dana, 1852

    Homarus americanus H. Milne-Edwards, 1837

Infraorder	THALASSINIDEA Latreille, 1831
Family	AXIIDAE Huxley, 1879

    Axius serratus Stimpson, 1852
    Calocaris templemani Squires, 1965

Family	CALLIANASSIDAE Dana, 1852

    Callianassa atlantica Rathbun, 1926
    Callianassa biformis Biffar, 1971

Infraorder	PALINURA Latreille, 1803
Family	POLYCHELIDAE Wood-Mason, 1874

    Polycheles granulatus Faxon, 1893
    Stereomastis sculpta (Smith, 1880)

Infraorder	ANOMALA Boas, 1880
Family	PAGURIDAE Latreille, 1803

    Pagurus acadianus Benedict, 1901
    Pagurus arcuatus Squires, 1964
    Pagurus longicarpus Say, 1817
    Pagurus politus (Smith, 1881)
    Pagurus pubescens Krøyer, 1838

Family	LITHODIDAE Samouelle, 1819

    Lithodes maja (Linnaeus, 1758)
    Neolithodes grimaldii (A. Milne-Edwards and Bouvier, 1894)

Family	PARAPAGURIDAE Smith, 1882

    Parapagurus pilosimanus Smith, 1879

Family	GALATHEIDAE Samouelle, 1819

    Munida iris iris A. Milne-Edwards, 1880
    Munida tenuimana G. O. Sars, 1871
    Munida valida Smith, 1883
    Munidopsis curvirostra Whiteaves, 1874

Infraorder	BRACHYURA Latreille, 1803
Section	EUBRACHYURA de Saint Laurent, 1980
Subsection	HETEROTREMATA Guinot, 1977
Family	MAJIDAE Samouelle, 1819

    Chionoecetes opilio (O. Fabricius, 1780)
    Hyas araneus (Linnaeus, 1758)
    Hyas coarctatus coarctatus Leach, 1815
    Libinia emarginata Leach, 1815

Family   CANCRIDAE Latreille, 1803

Cancer borealis Stimpson, 1859
Cancer irroratus Say, 1817

Family   GERYONIDAE Colosi, 1923

Geryon quinquidens Smith, 1897

Family   PORTUNIDAE Rafinesque, 1815

Callinectes sapidus Rathbun, 1896
Carcinus maenas (Linnaeus, 1758)
Ovalipes ocellatus (Herbst, 1799)
Portunus sayi (Gibbes, 1850)

Family   XANTHIDAE MacLeay, 1838

Neopanope sayi (Smith, 1869)
Rhithropanopeus harrisi (Gould, 1841)

Subsection   THORACOTREMATA Guinot, 1977
Family   GRAPSIDAE MacLeay, 1838

Planes minutus (Linnaeus, 1758)

NOTE: Higher categories based on Schram (1986).

KEYS TO SUBORDERS, INFRAORDERS AND FAMILIES

N.B. The following keys refer only to species present in the area of reference and should be used with caution in other areas.

Key to the Suborders of the Decapoda
(From various authors)

1   Body usually flattened vertically (compressed); pereopods with the basis and ischium free; one fixed point in the carpo-propodal articulation 2

   Body usually flattened horizontally (depressed); pleopods almost never natatory; the anterior pereopod at least with the basis and ischium fused; two fixed points in the carpo-propodal articulation .... REPTANTIA

2   Third pereopods with a chela; pleura of 2nd abdominal somite not overlapping those of first and third; gills dendrobranchiate ....
   .... DENDROBRANCHIATA (Infraorder PENAEIDEA)

   Third pereopods without a chela; pleura of 2nd abdominal somite overlapping those of first and third; gills phyllobranchiate ....
   .... EUKYPHIDA (Infraorder CARIDEA)

Key to Families of Dendrobranchiata (Penaeidea)
(After Burkenroad, 1983)

1   Last two pairs of pereopods well developed; branchiae numerous ....
   .... ARISTEIDAE

   Last two pairs of pereopods reduced in length; not more than eight branchiae on either side 2

2   Carapace moderately compressed; lower antennular flagellum and gills present .... SERGESTIDAE

   Carapace highly compressed, anteriorly neck-like; lower flagellum and gills absent .... LUCIFERIDAE

Key to Families of Eukyphida (Caridea)
(After Holthuis, 1955)

1   First pair of pereopods chelate or simple 2

   First pair of pereopods subchelate .... CRANGONIDAE

2   Fingers of first pairs of chelae slender and long, their cutting edges pectinate
   .... PASIPHAEIDAE

   Cutting edges of fingers of chelae not all pectinate 3

3	Carpus of second pair of pereopods not annulated; first pair with well-developed chelae; some with exopods	4
	Carpus of second pair of pereopods with two or more annulations; pereopods without exopods	5
4	Last three pairs of pereopods not conspicuously lengthened; carpus of each shorter than propodus	OPLOPHORIDAE
	Last three pairs of pereopods enormously lengthened; carpus of these legs several times longer than propodus; ischio-meral joints of all legs larger than others	NEMATOCARCINIDAE
5	Chelae of first pair of pereopods distinct	6
	Chelae of first pair of pereopods microscopically small or indistinct	PANDALIDAE
6	First pair of chelae heavier than second	HIPPOLYTIDAE
	First pair of chelae usually more slender than or rarely subequal to second pair	PALAEMONIDAE

Key to the Infraorders of the Reptantia
(After various authors and Schram, 1986)

1	Sternite of the last thoracic somite free or modified from the others.	
	Telson asymmetrical or with sutures	ANOMALA
	Telson symmetrical and without sutures	THALASSINIDEA
	Sternite of last thoracic somite not free from others	2
2	Abdomen large, symmetrical; uropods present and developed into a tail fan.	
	First three pairs of pereopods chelate	ASTACIDEA
	First four or no pereopods chelate	PALINURA
	Abdomen small, dorsoventrally flattened, symmetrical, recurved to the sternal face of the enlarged cephalothorax and tail fan not developed; uropods rudimentary or none	BRACHYURA

Key to Families of the Anomala
(After Williams, 1984)

1	Uropods present; abdomen usually soft and with reduced pleura and terga	2
	No uropods; abdomen strongly calcified and pressed against sternum as in Brachyura	LITHODIDAE
2	Uropods and telson asymmetrical; uropods modified for holding body in shell.	
	Two genital openings; male without sexual pleopods	PAGURIDAE
	One genital opening in female; male with sexual pleopods	PARAPAGURIDAE
	Uropods symmetrical, unmodified; telson sutured	GALATHEIDAE

Key to Families of the Infraorder Thalassinidea
(After Williams, 1984)

1   Podobranchs on at least the first three pairs of pereopods; uropods with transverse suture on exopod .... AXIIDAE

      No podobranchs on pereopods; no transverse sutures on uropods .... CALLIANASSIDAE

Key to the Families of the Infraorder Brachyura
(After various authors)

1   Carapace usually conspicuously narrowed in front, with distinct rostrum sometimes forked .... MAJIDAE

      Carapace wide in front, rostrum reduced or absent 2

2   Fifth pair of legs with flattened paddle-like dactyl, adapted for swimming .... PORTUNIDAE

      Fifth pair of legs without flattened paddle-like dactyl 3

3   Front of carapace with 3 or more teeth, one median; carapace transversely oval .... CANCRIDAE

      Front of carapace without a single median tooth 4

4   Spines, not teeth, on anterolateral border of carapace, two median teeth at front .... GERYONIDAE

      Teeth, not spines, at anterolateral edge of carapace 5

5   Four teeth at anterolateral edge of carapace, carapace oval .... XANTHIDAE

      Only two, if any, modified teeth at anterolateral edge of carapace; carapace quadrate .... GRAPSIDAE

Suborder DENDROBRANCHIATA

Infraorder PENAEIDEA

Family ARISTEIDAE Wood-Mason, 1891
Burkenroad 1983: 281; Schram 1986: 252.

Second pleopods in male with appendices interna and masculina; third legs chelate; pleura of second abdominal somite does not overlap pleura of first; male with petasma on first pleopods, female with thelycum on posterior thoracic sterna; body laterally compressed, first somite of abdomen not shortened.

Key to species of the Aristeidae
(Partly after Burkenroad 1983)

1   Postorbital spine on carapace behind orbital margin; tip of telson flanked by conspicuous pair of fixed lateral spines; scale-like projection at base of eyestalk; endopod of second pleopod of male with spur as well as appendices .... *Pleoticus robustus*

    No postorbital spine; tip of telson flanked by spines that are moveable; no scale at base of eyestalk; endopod of second pleopod of male without a spur 2

2   Upper flagellum of antennule not much produced beyond its short, thickened basal part. Three or more rostral teeth .... *Plesiopenaeus edwardsianus*

    Both flagella of antennule long; rostral teeth usually two or less 3

3   Fifth somite of abdomen carinated in dorsal midline; exopod of 1st maxilliped distally segmented and narrow; a long moveable middorsal spine at middle of 5th somite .... *Benthesicymus bartletti*

    Fifth somite of abdomen not carinate; tip of exopod of 1st maxilliped not narrow and set off from rest 4

4   Carpus of 3rd leg shorter than merus; thelycum wider than long, ovate, no centre line .... *Gennadas elegans*

    Carpus of 3rd leg longer than merus; thelycum about as wide as long, almost circular, centre line .... *Gennadas valens*

Genus *Benthesicymus* Bate, 1881
Zariquiey Alvarez 1968: 33.

Rostrum short, reaching at most the the edge of the cornea, forming a crest above orbit with 1–3 spines; cervical groove conspicuous, post-cervical less well marked; eyestalk with small inner tubercle; upper antennular flagellum dilated in area as long as carapace; endopod of 1st maxilliped with distal segment much narrower than others; 4th and 5th pereopods very long; telson shorter than the 6th somite.

Benthesicymus bartletti Smith, 1882
Burkenroad 1936: 47; Crosnier et Forest 1973: 275, fig. 92 a–b; Zariquiey Alvarez 1968: 35.
(Figures 4 and 5)

DISTINGUISHING CHARACTERISTICS

A long slender middorsal spine at about the middle of the 5th abdominal somite extending over the 6th; rostrum short, crest over orbit with 2 spines; abdominal somites 5th and 6th carinate; cervical groove well marked on anterior third of carapace; telson shorter than 6th somite and uropods; eyestalks long, with inner tubercle.

DESCRIPTION

Integument thin, shiny, somewhat membraneous. Colour bright red.

Rostrum short, compressed, not exceeding eye, 2 teeth dorsally one behind or even with orbit, forming a crest at anterior edge of carapace and a median carina reaching the cervical groove.

Carapace with moderate branchiostegal spine beginning a prominent longitudinal carina which forms an arc over branchial region to posterior edge, also at slightly less than half carapace it joins short oblique hepatic carina to marginal carina below extending whole length of carapace; prominent cervical groove joins a short hepatic groove below rising towards orbit; a faint post-cervical groove also present.

Abdominal pleura rounded to straight ventrally, terga non-carinate except from middle of 5th to posterior 6th somites; at midline of 5th is a strong spinous process reaching over about 0.2 of 6th; abdomen thick at 1st to 3rd somites and tapering to very compressed 6th somite.

Telson (t) shorter than 6th somite and branches of uropod, outer branch much longer; telson rounded dorsally with more pronounced hump proximally, laterally scalloped proximally and tapering to blunt tip with a pair of small moveable spines and a few lateral plumose setae.

Eyestalk sinuous, depressed, with small inner tubercle; cornea large, oval.

Antennule (c) 2nd article longer than 3rd; 1st article hollowed dorsally, stylocerite about 0.8 times first article.

Antenna (d) scale about 0.6 times cl, width 0.3 times length; blade pointed distally greatly exceeding spine; peduncle about 0.3 times length of scale; basal article with bifurcate lobe and large curved inner dorsal spine distally.

Mandible (e) incisor with two sharp teeth, joined to small triangular molar with hollow crown; hollow area behind incisor. Palp large, two segments.

Maxillule (f) proximal endite oval, shorter than distal; endopod long with neck curved inward from thicker base.

Maxilla (g) both endites unequally bilobed, lobes narrow decreasing in length proximally; endopod stout sharp-pointed; anterior lobe of scaphognathite long, subrectilinear with rounded tip, posterior lobe short, axe-shaped.

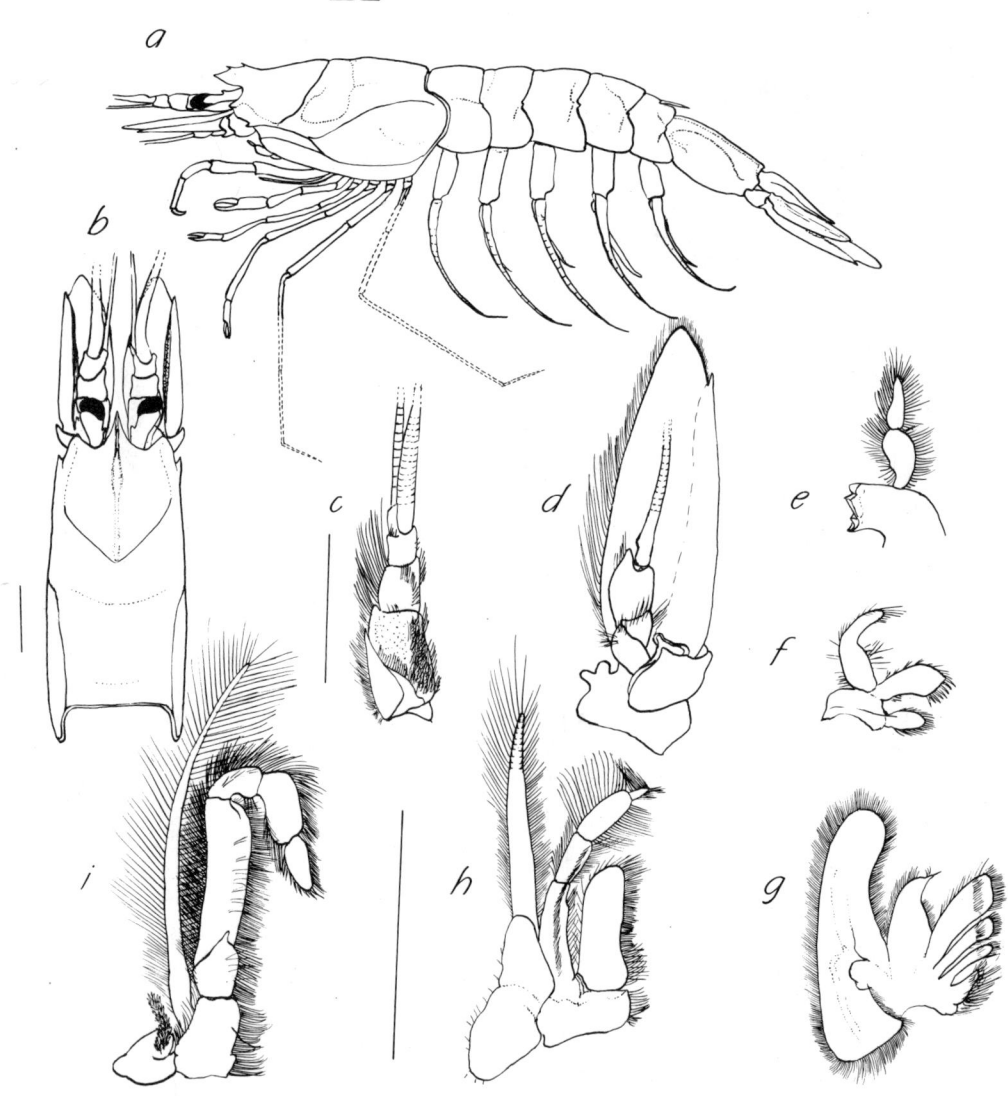

Fig. 4. *Benthesicymus bartletti*: *a*, whole shrimp from left side; *b*, carapace in dorsal aspect; *c*, antennule; *d*, antenna; *e*, mandible; *f*, maxillule; *g*, maxilla; *h*, first maxilliped; *i*, second maxilliped; solid line = scale of 10 mm.

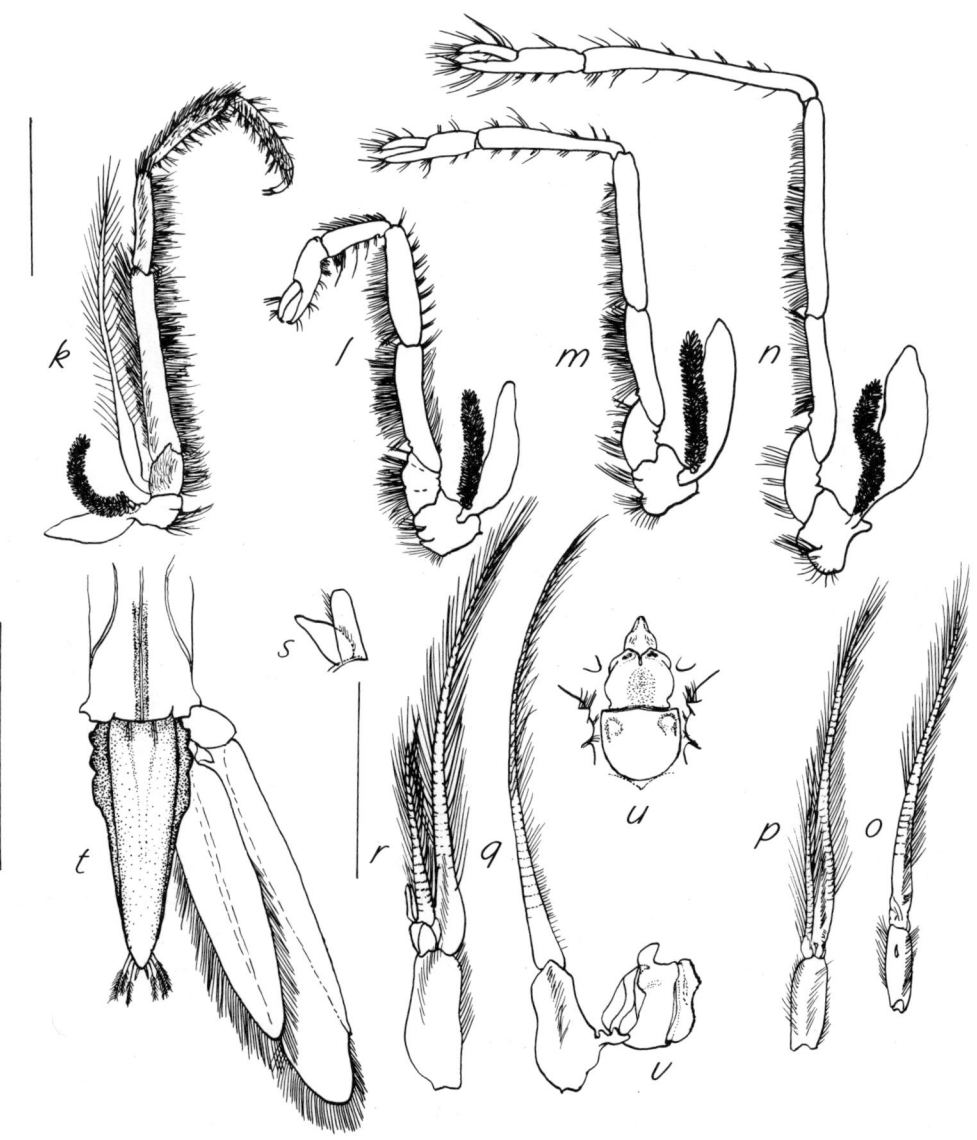

Fig. 5. *Benthesicymus bartletti*: k, third maxilliped; l, first pereopod; m, second pereopod; n, third pereopod; o, first pleopod (female); p, second pleopod (female); q first pleopod (male); r, second pleopod (male); s, appendix masculina; t, telson; u, thelycum; v, petasma; solid line = scale of 10 mm.

Maxilliped I (h) distal endite subrectilinear, long; proximal short; endopod long, first three segments stout and long but distal segment very narrow and short; exopod long and slender, with short flagellum; epipod bilobed, inflated.

Maxilliped II (i) leglike, only slightly compressed, distal segment longer than wide, with strong apical spines; exopod long and slender; epipod with podobranch.

Maxilliped III (k) leglike, with seven segments, the distal narrow and curved, with apical spines; exopod long and slender; epipod with podobranch.

Pereopods: I (l) chelate, moderately stout, shorter than others; large epipod with podobranch; II (m) chelate, slightly more slender and longer than I, epipod and podobranch; III (n) chelate, still more slender and longer than II, epipod and podobranch; IV and V much longer and slender than others, nonchelate, IV with epipod but no podobranch (both incomplete in specimens examined).

Pleopods: female I (o) one ramus only, about 4 times length of protopod, the latter with a moveable spine on distal third; female II (p), 2 rami, exopod about 3 times length of very slender endopod. Male I (q) with single ramus and petasma on proximal fourth of protopod; male II (r) endopod less than half length of exopod, with appendix interna sub-triangular or jar-shaped and appendix masculina (s) rectilinear, rounded at tip with a few setae subapically.

Thelycum (u) with sternal plate of 5th legs shield-shaped, anterior edge straight, with knobs or rounded protuberances at anterior corners; median sternal plate of 4th legs somewhat heart-shaped with narrow median cleft anteriorly and central hollow with a few setae at each side; sternite of 3rd legs with sub-triangular or dome-shaped protuberance with few short setae on raised surface.

Petasma (v) with disto-lateral projection curving outward from broad thin middle lobe; outer lobe with vertical fold and sinuous edge of hooks; inner lobe with sinuous vertical folds and attachment to protopod.

Range of Distribution

Western Atlantic from Nova Scotia to the West Indies and Gulf of Mexico, eastern Atlantic from the Azores to the West African coast off Gabon and the Congo; also from the Indo-Pacific from the Bay of Bengal, the Philippines and North Pacific (37°49′ N, 166°47′ W). Depths from 609 to 5 777 m (Crosnier et Forest 1973).

Records of occurrence in the area of reference are in Fig. 6.

Biology

Lengths in males to 37 mm cl and females 40 mm cl; total length in males 133 mm.

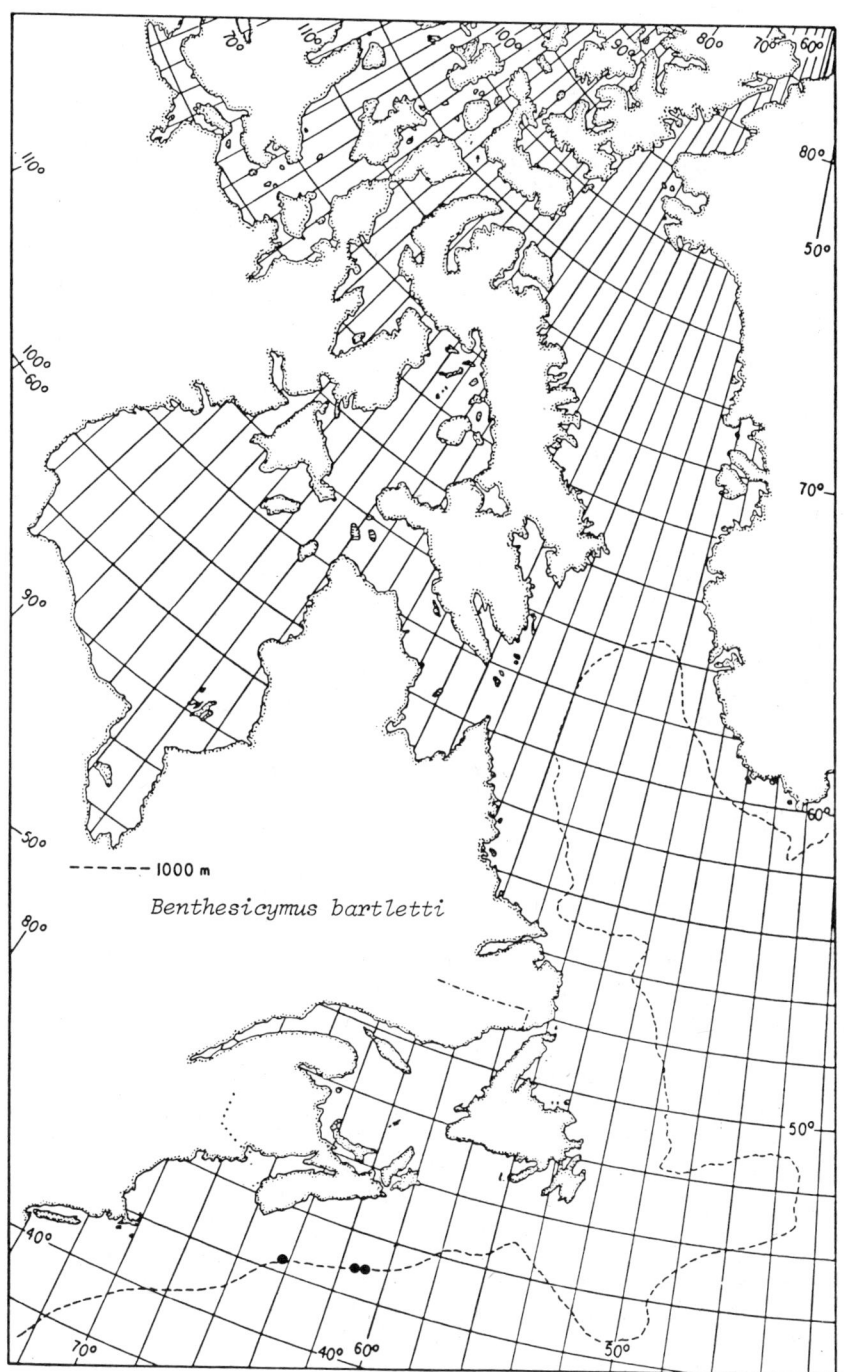

Fig. 6. *Benthesicymus bartletti*, distribution records in the area of reference.

Genus *Gennadas* Bate, 1881
Burkenroad 1936:
Zariquiey Alvarez 1968: 37.

Rostrum short with sharp point, rostral crest above orbit with one spine only; pronounced cervical and post-cervical grooves; antennular flagella long, proximal dilatation of upper shorter than peduncle; antennal flagellum with a kink not far from its origin; endopod of first maxilliped without terminal prolongation; proximal segments of second maxilliped with wide foliaceous lamina attached laterally; ischium and merus of third maxilliped dilated, dactyl spoon-shaped; without podobranchs on third maxilliped or pereopods; carinate on the 6th somite only.

Gennadas elegans (Smith, 1882)
Burkenroad 1936: 70, fig. 55;
Zariquiey Alvarez 1968: 38, fig. 20 a–d.
(Figures 7 and 8)

DISTINGUISHING CHARACTERISTICS

Thelycum a flattened oval, pointed laterally, and without a median division; merus of the 3rd leg equal to or longer than the carpus; infra-antennal angle at anterior edge of carapace wide and obtuse; tubercle on eyestalk at distal two-thirds; rostrum with sharp anterior point and crest over orbit with a single spine.

DESCRIPTION

Integument thin, shiny. Colour red to brownish red at front of carapace, lighter on abdomen, legs dark brownish red, with patches of deep blue. Eyes brownish.

Rostrum short, a sharp point only at the end of a sharp fold in the anterior middorsal edge of the carapace which forms a narrow crest with a single spine, and descends to form a low carina which almost reaches the posterior edge; where the crest descends is a small inconspicuous flat spine.

Carapace has a cervical and a post-cervical groove both of which descend to meet a horizontal carina, the latter begins near the orbital fissure curving upward to the base of the rostro-carapacial crest and extends laterally to reach the posterior edge where it turns downward a short distance; below this carina anteriorly a groove extends backward from the branchiostegal spine, turning downward at the middle of the carapace; ahead of the spine the antennal angle is wide and obtuse, and a gap is below at the anteroventral edge of the carapace; a submarginal carina extends the length of the carapace.

Abdomen has rounded terga except a mid-dorsal carina on the 6th somite; pleura are rounded but have a posterior expansion and lateral grooves on the 3rd to 5th somites.

Telson (t) is short with proximal scalloped edges, and tapers to a truncate end with a spine at each corner and a few setae in between; branches of the uropod are longer than the telson, the outer longest.

Eyestalk moderately long with strong inner projection on anterior two-thirds and a slight curve outward; cornea oval, brownish.

Antennule (c) 1st article thick, stylocerite reaching just past half the article; 2nd article about half as long as 3rd; dilation of upper flagellum less than length of peduncle.

Antenna (d) scale length 2.3 times width, tapering to narrow apex much exceeded by distolateral spine; peduncle less than one-half scale.

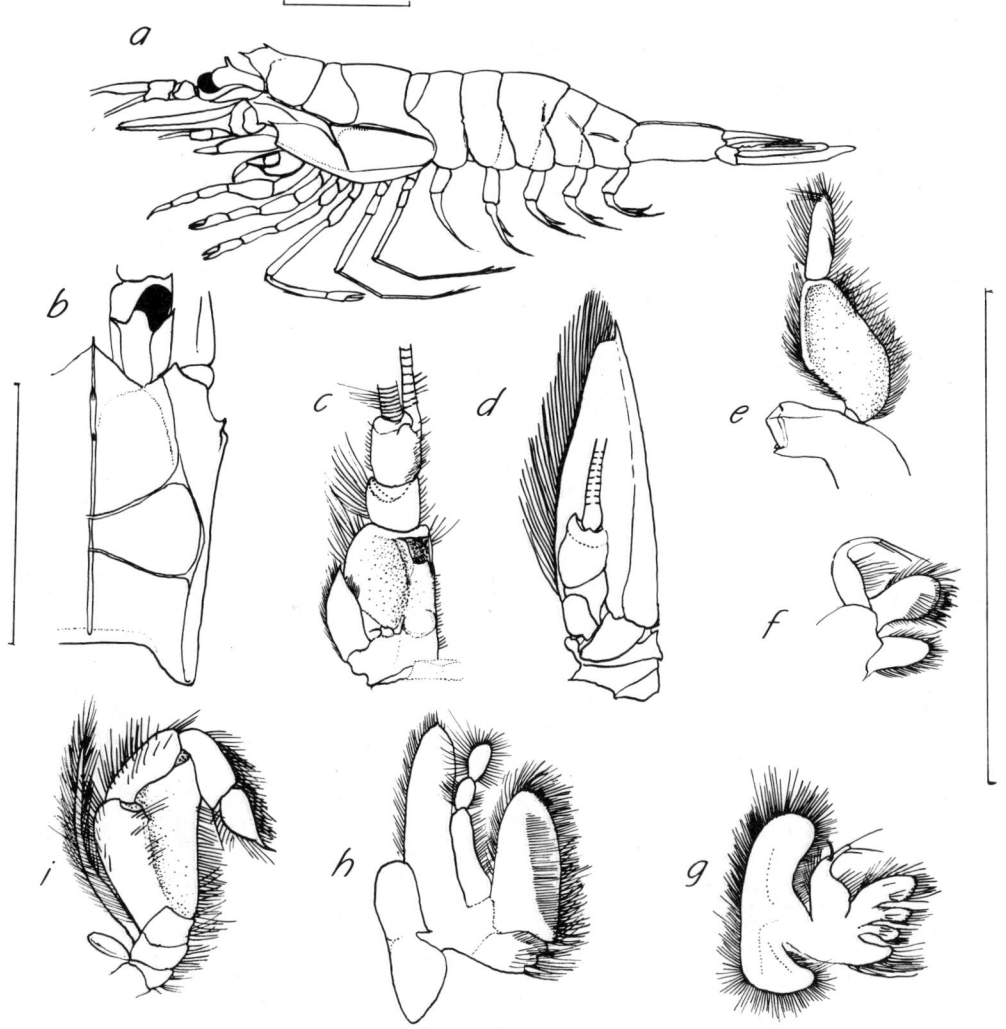

Fig. 7. *Gennadas elegans*: *a*, whole shrimp from left side; *b*, carapace in dorsal aspect; *c*, antennule; *d*, antenna; *e*, mandible; *f*, maxillule; *g*, maxilla; *h*, first maxilliped; *i*, second maxilliped; solid line = 10 mm.

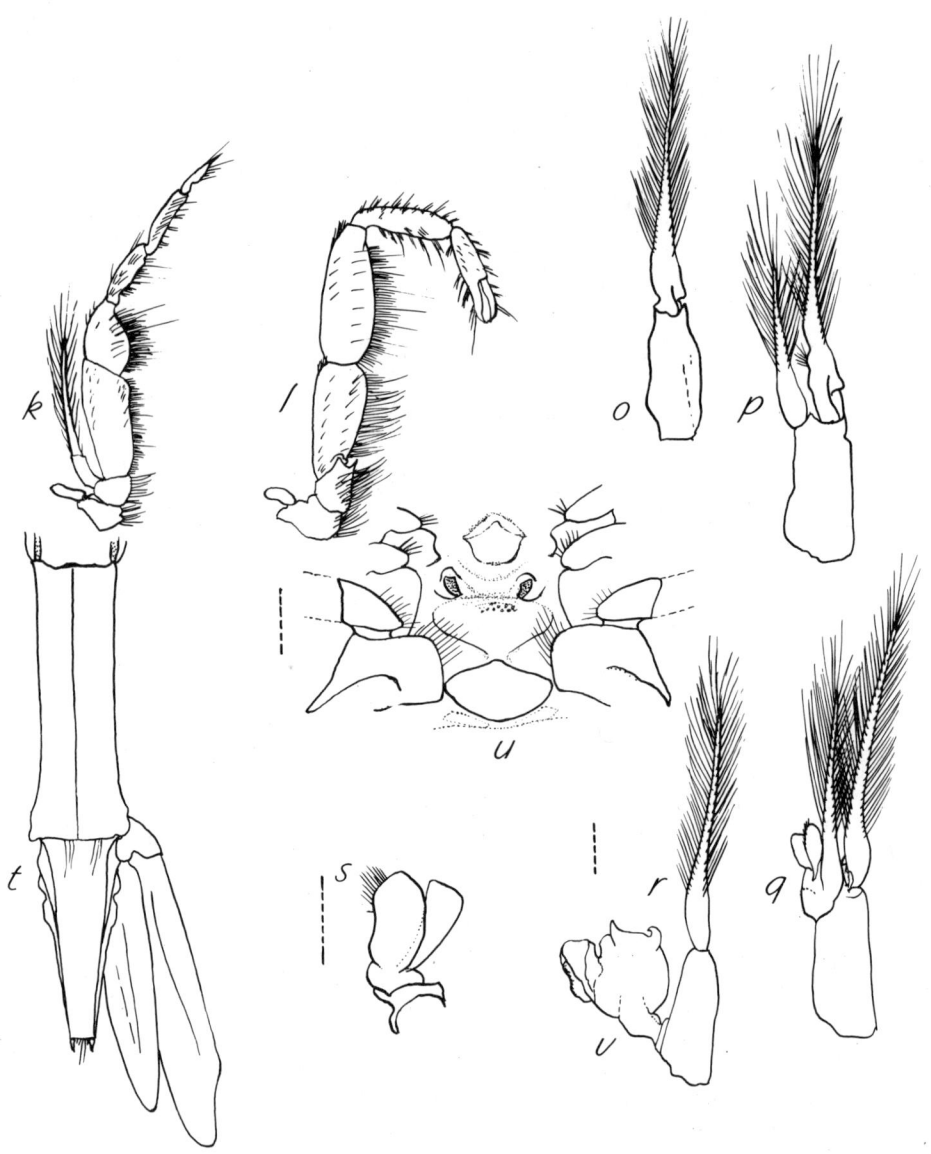

Fig. 8. *Gennadas elegans*: *k*, third maxilliped; *l*, first pereopod; *o*, F first pleopod; *p*, F second pleopod; *q*, M first pleopod; *r*, M second pleopod; *s*, appendix masculina; *t*, telson; *u*, thelycum; *v*, petasma; solid line = 10 mm, broken line = 1 mm.

Mandible (e) incisor with two pointed cusps at cutting edge, joined closely to molar (an oval flat surface behind incisor plus a large pointed cusp at posterior corner); palp with proximal segment about as wide as length of distal segment.

Maxillule (f) distal endite oval rounded, slightly longer than tapered proximal endite; endopod with narrow neck curved inward.

Maxilla (g) endites unequally bilobed, two middle lobes narrower; endopod with curved neck from wide base, 3 sharp curved spines (two small) proximally on neck; anterior lobe of scaphognathite sub-rectilineaar rounded at apex, longer than boot-shaped posterior lobe.

Maxilliped I (h) long distal endite, short proximal with scalloped edge; endopod long, three segments, distal ovate, almost as long as wide, slightly tapered exopod; epipod large, bilobed.

Maxilliped II (i) leglike, with 7 segments, longest or ischium with wide laminate lateral expansion; distal segment with long acuminate spinous tip; exopod slender; small epipod.

Maxilliped III (k) leglike, with 7 segments; ischium and merus much wider than others; dactyl shovel-like with strong apical spine; slender exopod; small epipod.

Pereopods: I (l) chelate, merus and ischium inflated, small epipod; II chelate, longer than I, not quite as stout; III chelate, more slender than I and II, with carpus longer than or equal to merus; IV and V as slender as III but shorter.

Pleopods: female I (o) single ramus about twice as long as protopod, tapered; female II (p) endopod about half exopod, no appendices; male I (q) single ramus, twice as long as protopod; petasma on proximal quarter of protopod; male II (r) endopod shorter than exopod, wide at base and with appendices interna and masculina (s) about equal in size, the latter with a few subapical setae.

Thelycum (u): between 5th legs sternal plate is flat, semicircular in outline posteriorly but forming an obtuse point anteriorly (point slightly projecting ventrally); between 4th legs the sternal plate is the reverse of the former, the lines forming an obtuse angle but not meeting at a point; between the 3rd legs (or slightly posterior) is an arcuate ridge directed posteriorly with a rounded orifice at each side and just anterior to it a muffin-shaped projecting area. All outlined in red.

Petasma (v) simpler in structure than in *G. valens*; main lobe is rounded below and has two small ear-like projections at the top; second lobe smaller and lateral to first, with funnel-like fold and outer edge with few rows of hooks.

RANGE OF DISTRIBUTION

West Atlantic from Davis Strait to 30° N, eastern Atlantic from British Isles to the Cape Verde Islands and in the Mediterranean as far as the Sea of Marmara and Egypt. Depths 400–3 000 m (Zariquiey Alvarez 1968).

Reported occurrences in the area of reference are in Fig. 9.

BIOLOGY

Lengths to 10 mm cl in males and 11 mm cl in females; body length to 40 mm.

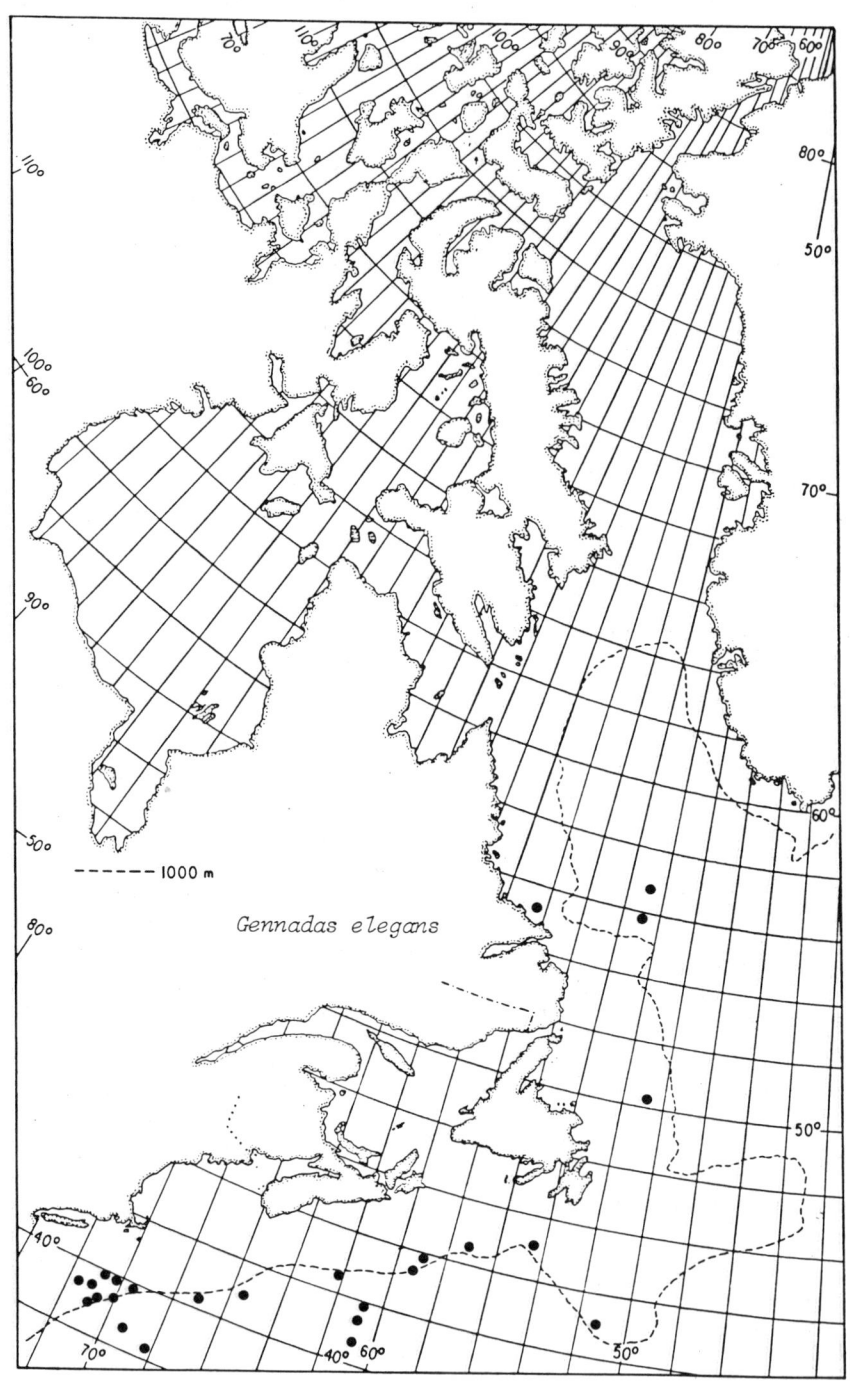

Fig. 9. *Gennadas elegans*, distribution records in the area of reference.

Gennadas valens (Smith, 1884)
Burkenroad 1936: 75;
Zariquiey Alvarez 1968: 38, fig. 20 f.
(Figures 10 and 11)

DISTINGUISHING CHARACTERISTICS

Thelycum almost circular with median division; merus of 3rd leg shorter than carpus; infra-antennal angle narrow and pointed; rostrum a sharp point at front of thin crest over the orbit with a forward-pointing spine at the highest point; cervical and post-cervical grooves conspicuous; lateral ridge turned downward at posterior edge of carapace more than in *G. elegans*; neck of endopod of maxilla with four hooked spines, only three in *G. elegans*.

DESCRIPTION

Integument thin, shiny, membraneous (somewhat wrinkled in preservative). Colour coral red with purplish spots.

Rostrum a sharp spine at the front of a thin blade-like high crest formed from the anterior edge of the carapace above the orbit, a sharp forward-directed spine is at the top of the crest and at its base a tiny flat spine.

Carapace with frontal crest continued as a median carina almost reaching the posterior margin with interruptions at the cervical and post-cervical grooves: these grooves form a loop laterally and a forward groove rising towards the orbit, they meet a lateral carina extending from the anterior edge to near the posterior edge where it turns downward; below the carina is a groove beginning near the branchiostegal spine and turning down at the hepatic region where it meets a short vertical carina joining the lateral carina above it; the infra-antennal angle is narrower than in *G. elegans* and the gap at the ventro-lateral edge rounder; a submarginal carina extends the length of the carapace.

Abdominal terga and pleura rounded except for lateral grooves on 3rd to 5th somites and a thin dorsal carina on the 6th.

Telson (t) proximally with lateral ventral scalloped edges, and lateral carinae tapering slightly to truncate tip with a spine at each corner and a few long setae in between. Branches of uropod longer than telson, the outer much longer.

Eye large, brownish, cornea globular; eyestalk not curved as in *G. elegans* and with prominent tubercle nearer the middle.

Antennule (c) 1st article about equal in length to other two, 3rd about twice as long as 2nd; stylocerite closely adherent to 1st article, triangular in cross-section, sharp point reaches slightly past half 1st article; dilated part of dorsolateral flagellum about as long as 3rd article.

Antenna (d) scale length 3.3 times width, tapering to narrow tip much exceeded by disto-lateral spine; peduncle less than one-half scale; flagellum about 5 times cl.

Mandible (e) incisor with sharp cutting edge flanked by two spines, joined to molar with oval almost flat grinding surface and shearing edge and a sharp cusp near join; distal segment of palp equal in length to width of proximal segment.

Maxillule (f) distal endite expanded, slightly curved and longer than proximal, the latter evenly rounded to a point; endopod proximally inflated, with short curved neck.

Maxilla (g) distal and proximal endites unequally bilobed, central lobes narrow; endopod moderate, from wide base with distal neck and four sharp hooked spines (two on each side) near apex; anterior lobe of scaphognathite sub-rectilinear but curved, longer than boot-shaped posterior lobe.

Maxilliped I (h) distal endite much longer than proximal; endopod three-segmented, almost as long as compressed club-shaped exopod; epipod with both lobes sub-triangular.

Maxilliped II (i) leglike, longest segment (ischium) with large laminate lateral expansion or wing; distal segment sub-triangular with strong curved apical spine; exopod very slender about as long as laminate expansion; epipod ovate.

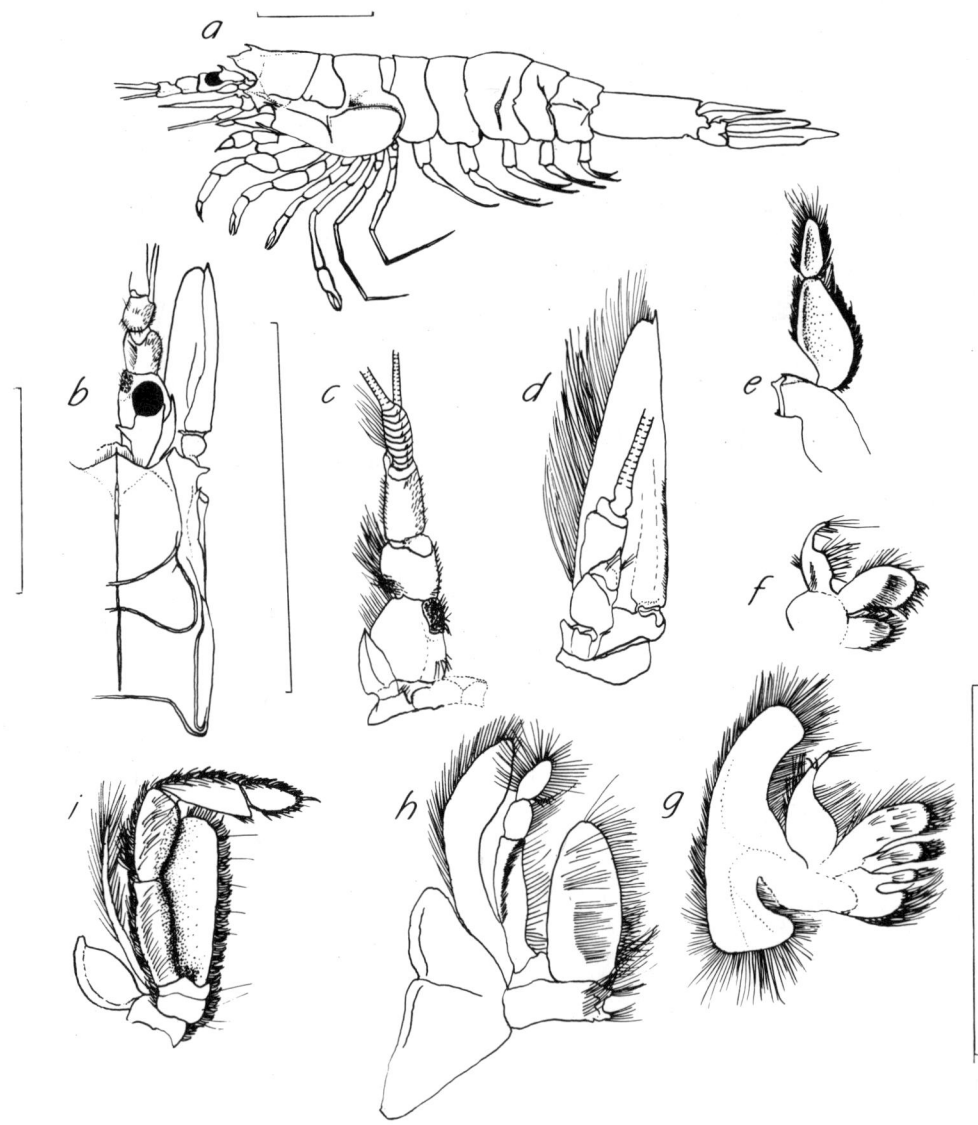

Fig. 10. *Gennadas valens*: *a*, whole shrimp from left side; *b* carapace in dorsal aspect; *c*, antennule; *d*, antenna; *e*, mandible; *f*, maxillule; *g*, maxilla; *h* first maxilliped; *i*, second maxilliped; solid line = 10 mm.

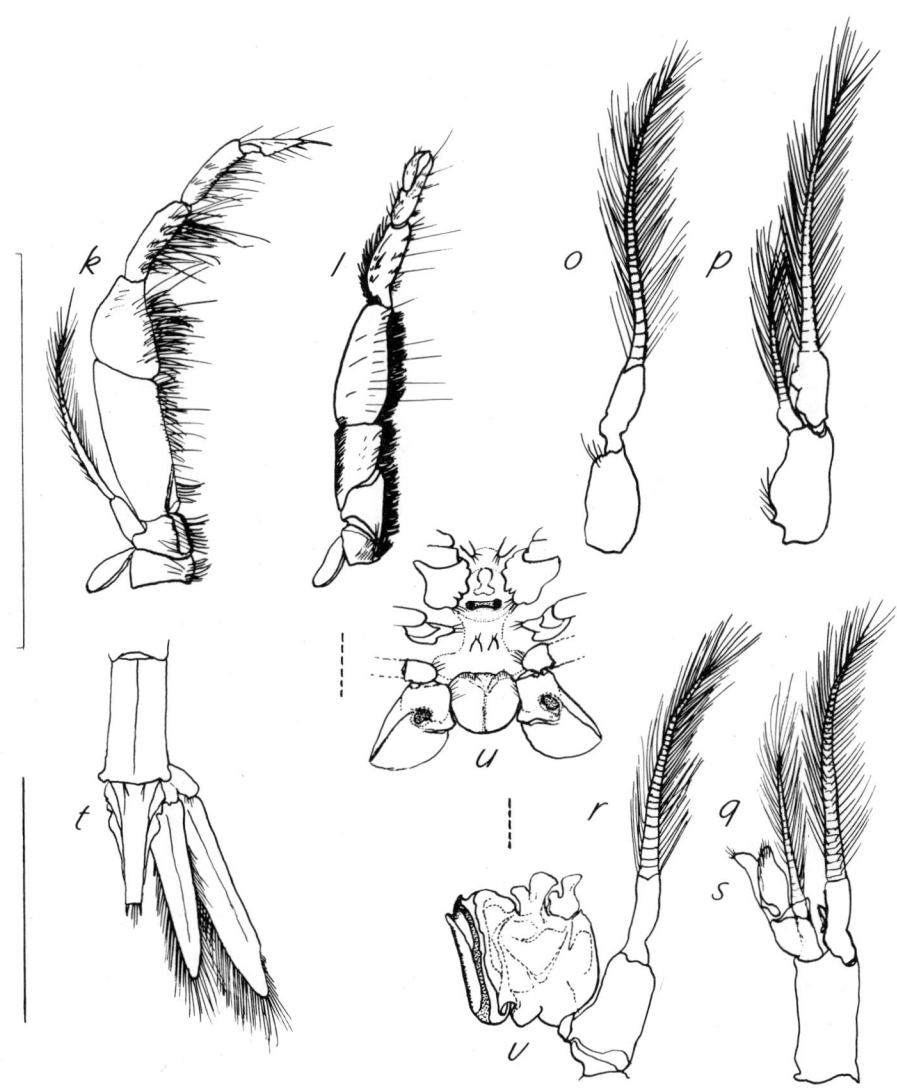

Fig. 11. *Gennadas valens*: *k*, third maxilliped; *l*, first pereopod; *o*, F first pleopod; *p*, F second pleopod; *q*, M first pleopod; *r*, M second pleopod; *s*, appendix masculina; *t*, telson; *u*, thelycum; *v*, petasma; solid line = 10 mm, broken line = 1 mm.

Maxilliped III (k) leglike, seven segments, ischium and merus compressed and wide; distal segment shovel-shaped and with a strong apical spine; exopod slender, long; epipod ovate.

Pereopods: I (l) stout, chelate, compressed, ischium and merus slightly wider than others, epipod small; II (m) chelate, about as long as I but more slender, epipod; III chelate, more slender and longer than II, carpus distinctly longer than merus, epipod; IV and V non-chelate, long, slender, IV only with epipod.

Pleopods: I (o) endopod uniramous much longer than protopod; II (p) female endopod longer than protopod but much shorter than exopod; I (q) male uniramous, protopod with attached petasma proximally; II (r) male endopod shorter than exopod, thick proximally with appendix interna about as long as appendix masculina (s) somewhat sinuous with moderate setae sub-apically.

Thelycum (u): between 5th legs sternal plate is projecting ventrally, almost circular in outline and with a dividing line at centre (not clear in some); immediately anterior between 4th legs are two small lappets pointed ventrally with two small apical spines: coxa of leg has a projection towards lappet with apical setae; between 3rd legs is a semicircular swollen area or thick lip with a round orifice at each end, and anterior to it a keyhole-shaped projection (not clear in some). All outlined in red.

Petasma (v) multilobular and not clearly divided into three major lobes, but the distomedian (Burkenroad, 1936: 61) or inner lobe has rows of hooks and a canal-like fold; a disto-lateral or middle lobe has an anterior (dorsomedian) small lobe of modified cones which may be palpatory or have suction; the other distoventral lobe has attachment to the protopod and anteriorly another small lobe similar to the above as well as other frills or lobules.

RANGE OF DISTRIBUTION

From the Gulf of Mexico north to 49° N in the western Atlantic, and from South Africa (35° S), the Azores and Sargasso Sea to the British Isles (51° N) in the eastern Atlantic as well as in the Mediterranean near Gibraltar. Depths 300–2 000 m (Zariquiey Alvarez 1968).

Records of occurrence in the area of reference are in Fig. 12.

BIOLOGY

Lengths to 48 mm body length; 9 mm cl in males and 11 mm cl in females examined.

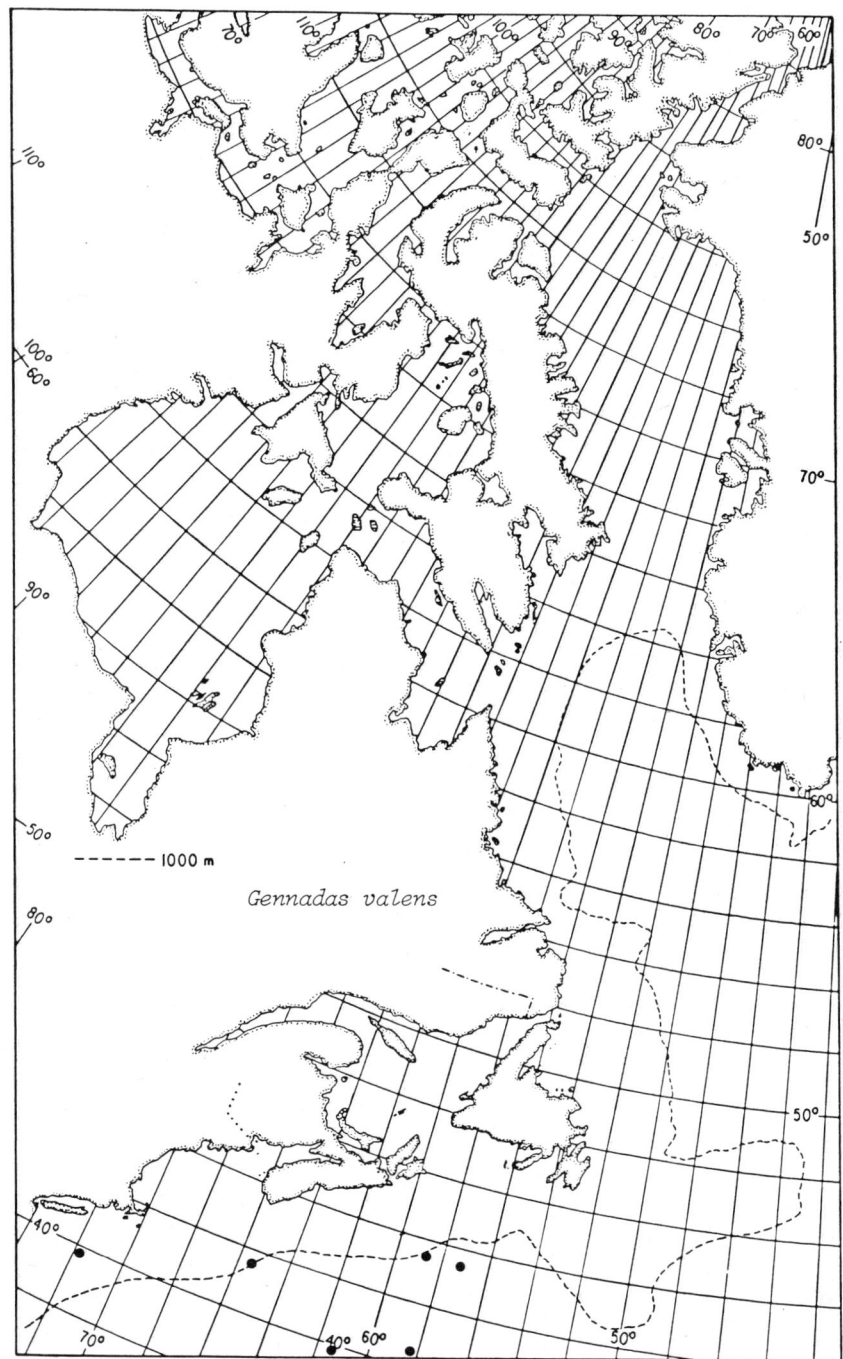

Fig. 12. *Gennadas valens*, distribution records in the area of reference.

Genus *Pleoticus* Bate, 1888
Perez Farfante 1977: 295.

Hymenopenaeus Smith, 1882
Burkenroad 1936: 102.

Epigastric tooth (posteriormost) separated from rostral teeth by an interval not much greater than that between first and second rostral teeth; rostrum not deep and with ventral margin straight and without spines, dorsally with 10–12 close-set spines; orbital, postorbital, antennal, and hepatic spines; both antennular flagella cylindrical in cross-section; with prosartema on 1st article of antennule; tip of telson tridentate (fixed spines on each side of terminal spine).

Pleoticus robustus (Smith, 1885)
Perez Farfante 1977: 297, figs. 9, 29–36.

Hymenopenaeus robustus, Burkenroad 1936: 118.
"Royal red shrimp" — "Salicoque royale rouge"
(Figures 13 and 14)

DISTINGUISHING CHARACTERISTICS

Rostrum moderate, not quite as long as antennal scale, slightly ascending, with close set teeth above and none below; cervical groove not interrupting middorsal carina which almost reaches the posterior margin; carapace with orbital, postorbital, antennal, hepatic and branchiostegal spines, the latter near the ventral margin; tip of telson is tridentate (a fixed spine on each side of the central sharp tip); three legs chelate; epipods of legs forked; body covered with short setae.

DESCRIPTION

Integument covered with short setae, otherwise smooth. Colour pink, salmon or orange or whitish, usually one colour predominates and there are bands or patches of the others (Perez Farfante 1977).

Rostrum exceeds eyes, is pointed slightly upwards or arched and has close set spines (getting closer distally) above and none below, intercalated with setae and with setae only below, about 7 spines on rostrum plus 4 on carapace; lateral carina confluent with orbit.

Carapace with postorbital spine dorsal to antennal, hepatic spine at lower end of cervical groove and branchiostegal spine behind anterior margin, with a groove that enters the hepatic area and rises over branchial area, also a branchiostegal carina that is submarginal with a membrane ventral to it. Median dorsal carina extending posteriorly to and with small tubercle at edge. A postorbital spine, and a tooth at corner of orbit.

Abdomen with pleura and terga rounded except dorsal carina on 5th and 6th somites, latter forming posterior spine, and also a posteroventral spine on 6th.

Telson (t) about as long as 6th somite; with dorsal sulcus and lateral ridges as far as sharp fixed spines lateral to sharp tip. Branches of uropod longer than telson, the outer longer and with a distolateral tooth.

Eye longer than stalk, cornea large ovate.

Antennule (c) 1st article longer than other two combined, 2nd longer than 3rd; prosartema inserted on proximal half of 1st, narrow, and about half as long as article; stylocerite sharp-pointed, small, slighlty less than one-half 1st article. Flagella heavily setose proximally (Perez Farfante 1977).

Antenna (d) scale length about 3 times width; disto-lateral spine about even with tip of blade; peduncle about half length of scale; ventral spine on basal article.

Mandible (e) incisor thin with 3 teeth, joined closely to molar. Palp large, two-segmented, setose.

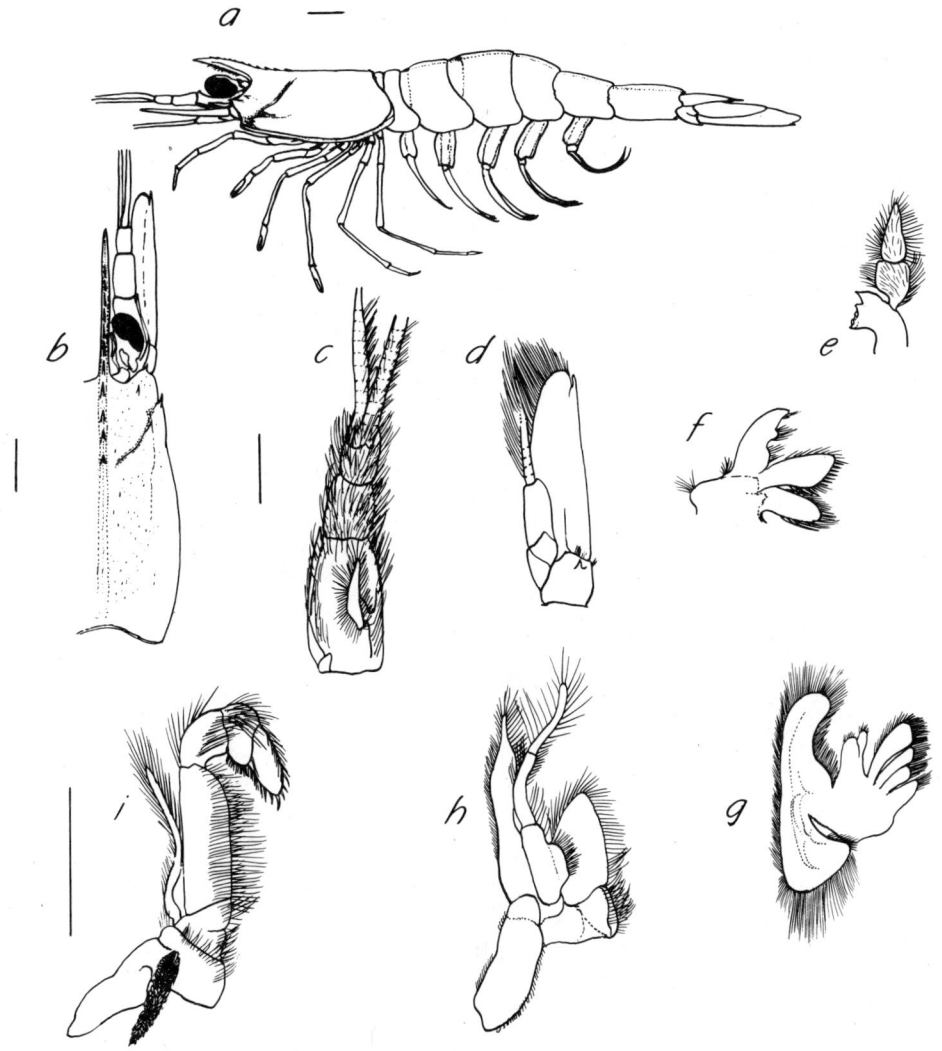

Fig. 13. *Pleoticus robustus*: *a*, whole shrimp from left side; *b*, carapace in dorsal aspect; *c*, antennule; *d*, antenna; *e*, mandible; *f*, maxillule; *g*, maxilla; *h*, first maxilliped; *i*, second maxilliped; solid line = 10 mm.

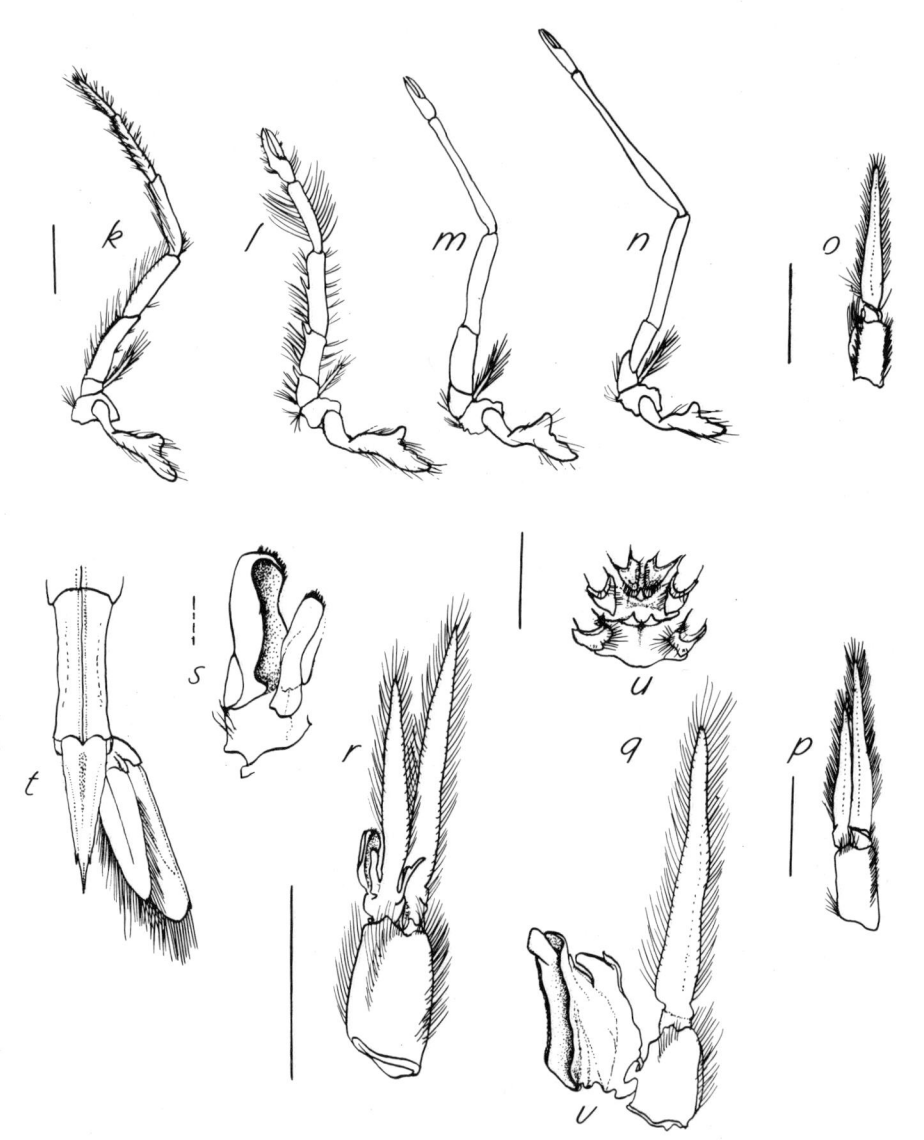

Fig. 14. *Pleoticus robustus*: *k*, third maxilliped; *l*, first pereopod; *m*, second pereopod; *n*, third pereopod; *o*, F first pleopod; *p*, F second pleopod; *q*, M first pleopod; *r*, M second pleopod; *s*, appendix masculina; *t*, telson; *u*, thelycum; *v*, petasma; solid line = 10 mm, broken line = 1 mm.

Maxillule (f) distal endite slightly expanded distally, proximal almost as long, narrow, tapering near tip; endopod with subapical projection indicating bifurcation.

Maxilla (g) endites unequally bilobed, two centre lobes narrow unequal in length; endopod short, rounded, bifurcate; anterior lobe of scaphognathite narrow, concave on inner edge, longer than axe-shaped posterior lobe.

Maxilliped I (h) distal endite subtriangular, longer than proximal, the latter with thickened edge and many setae; endopod three-segmented, sinuous, longer than exopod, proximal segment expanded laterally; exopod fusiform; epipod with anterior lobe very small, posterior sub-rectilinear.

Maxilliped II (i) leglike, compressed, distal segment longer than wide and with terminal spines; exopod slender, longer than longest segment; epipod with podobranch.

Maxilliped III (k) leglike, slender, longer than first leg; very short and slender exopod; epipod bifurcate.

Pereopods: I (l) chelate, stout, shorter than others; strong distal spine on basis and ischium and on merus just past middle; exopod very short and slender; epipod large bifurcate. II (m) chelate, more slender but longer than I, exopod very short and slender, epipod large, bifurcate; III (n) chelate, longer than II, very short exopod, epipod large, bifurcate; IV and V long slender, exopod but no epipod on V.

Pleopods: female I (o) uniramous; female II (p), endopod shorter than exopod, without appendices; male I (q) short thick protopod with attached petasma, near attachment a strong ovate projection; male II (r) endopod with small fingerlike appendix interna fitting into hollow side of thick and longer appendix masculina (s) both with short apical spines, on opposite side of endopod is a spur with elongate tip.

Thelycum (u) with a pair of triangular projections at front end of last thoracic somite (XIV), also a median pointed projection with wide rounded base on sternite XIII.

Petasma (v): edge of distoventral lobe has a short, rigid, free, distoventral projection; distolateral lobes have a very small projection from their lateral edge; distomedian lobe has rows of hooks along median edge nearly to distal end; distal margin of distoventral projection armed with series of short, stout, but minute spines; margin of distoventral flap bears minute spines (Burkenroad 1936).

Range of Distribution

Extending from Cape Cod to Brazil, including the Gulf of Mexico and the West Indies. Depths from 140 to 730 m (usually fishing concentrations at 250–475 m) (Perez Farfante 1977).

Records of distribution in the area of reference are in Fig. 15.

Biology

Body lengths 180 mm in males and 225 mm in females; 42 mm cl in males and 62 mm cl in females (Perez Farfante 1977).

Apparently digs grooves in substrate of blue/black mud, sand, muddy sand or white calcareous mud, but does not burrow. Feeds on small bottom-living organisms.

In the impregnated female a conjoined pair of large spermatophores cover the sternites of the 4th and 5th legs and project above the coxae of the legs (Burkenroad 1936).

Fishery

The most important fishing grounds are located off northeastern Florida. Total annual catch in the USA was 300 (1973), 181 (1974), 122 (1975), and 136 (1976) metric tonnes heads-on shrimps. They average 26 to 30 shrimps (heads-off) to the pound. (Holthuis 1980). Fishing is done by day as well as by night (Perez Farfante 1977).

The species has also been fished off the coast of Venezuela (Davant 1963).

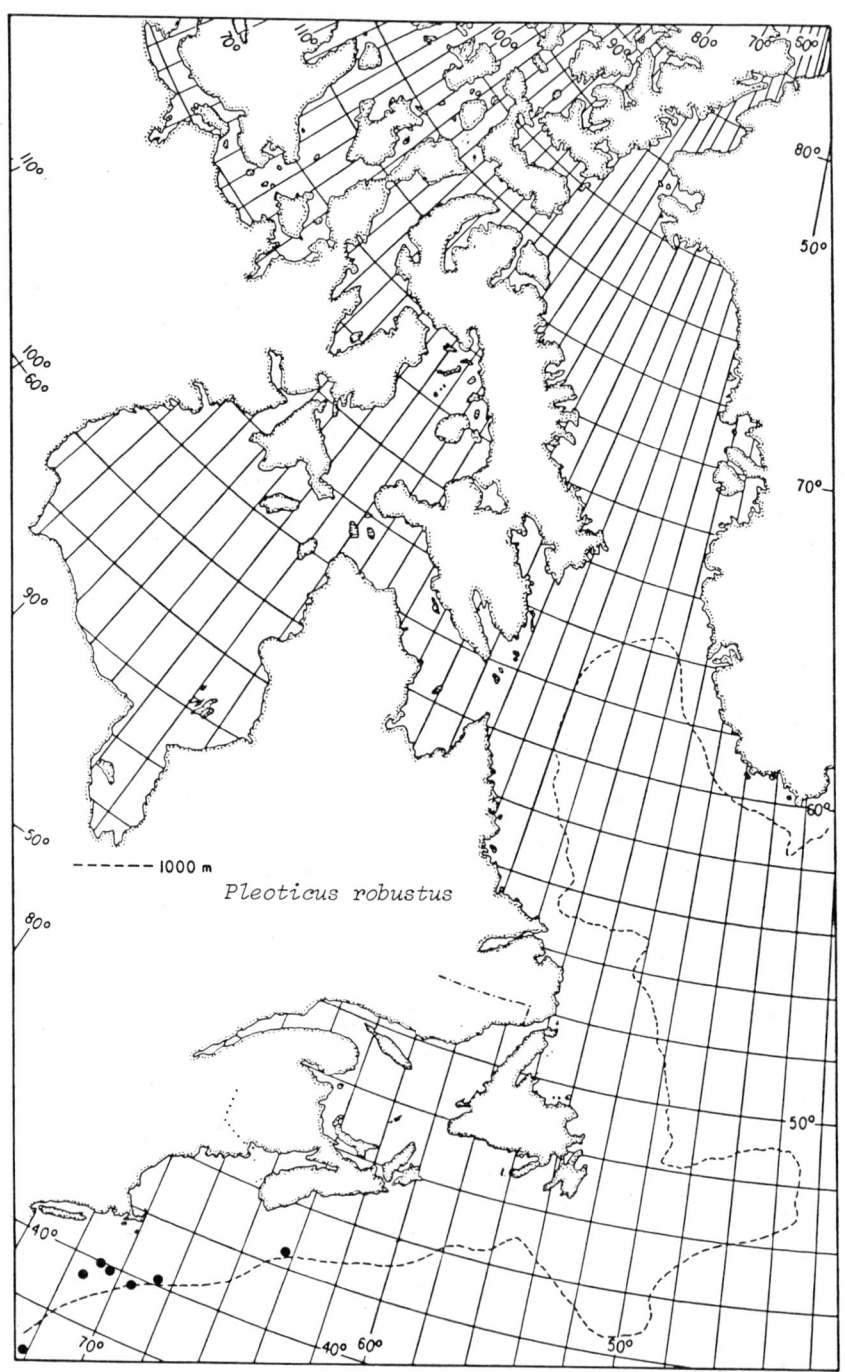

Fig. 15. *Pleoticus robustus*, distribution records in the area of reference.

Genus *Plesiopenaeus* Bate, 1881
Burkenroad 1936: 94; Zariquiey Alvarez 1968: 43.

Large epipod on somite XIII (4th legs), large podobranch on XII (3rd legs), epipods on all pereopods except 5th, no hepatic spine and a tridentate rostrum characterize the genus.

Plesiopenaeus edwardsianus (Johnson, 1867)
Crosnier et Forest 1973: 293, fig. 98;
Zariquiey Alvarez 1968: 44, fig. 21.
"Scarlet shrimp" — "Gambon ecarlat"
(Figures 16 and 17)

DISTINGUISHING CHARACTERISTICS

Intense brilliant crimson in colour with profuse fringes of gold setae; a large shrimp; rostrum moderate with 3 well-spaced dorsal teeth, the middle one over the orbit; well-marked cervical groove and a few lateral carinae; small antennal and strong branchiostegal spine; ventral edge of carapace with horizontal carinae and narrow folds; antennal scale with greatly extended narrow tip in males; second maxilliped with a very long profusely fringed exopod at ventro-anterior edge of carapace; abdomen carinate, each carina produced as a posterior spine on 3rd to 6th somites; telson short tapered to a sharp point and without dorsal spines.

DESCRIPTION

Integument heavy, smooth, shiny. Colour intense crimson fringed with gold setae.

Rostrum moderately long, about one-half carapace length. Spines of rostrum equidistant, the second over the orbit and the third behind it; lateral rostral carina ending above orbit, two ventral carinae confluent with orbit; rostral crest continued as a median carina, very sharp to the cervical groove, less afterwards and almost reaching the posterior edge of the carapace. Lateral carinae: gastro-orbital begins near the orbit and reaches the cervical groove; antennal from antennal spine for a short distance; hepatic from the branchiostegal spine, interrupted at the branchio-cardiac groove, then continues to posterior margin of the carapace. Ventrally is a strong even sulcus for the whole length of the carapace with a few submarginal carinae and folds posteriorly.

Abdomen compressed, lightly carinate on first somite but strongly on others and produced as a spine on each; lateral carinae on 4th and 5th somites and with scalloping on 6th; 6th also with ventro-posterior spinule.

Telson (t) shorter than 6th, tapering to sharp point, with a median sulcus but no dorso-lateral spines; inner branch of uropod longer than and outer branch about twice as long as telson, inner with central groove flanked by carinae and outer with two strong grooves and three carinae near outer edge.

Eye large, cornea globular.

Antennule (c) 1st article longer than other two combined, 3rd article very short, about one-third 2nd; stylocerite slender, sharp, about 0.7 length of 1st article; dorso-lateral flagellum short, compressed, ventromesial long.

Antenna (d) scale thin, wide, length 1.6 times width in females, but with a long distal extension in males (length about 4 times width); grooves and disto-lateral spine shorter than blade; peduncle short and slender; flagellum about 2.5 times body length.

Mandible (e) left incisor with two large saw teeth and a small one at edge and 5 cusps on molar, right with one low saw tooth and a fang at edge, the molar with 3 rounded and one pointed cusp; distal segment of palp with lateral expansion.

Maxillule (f) distal endite expanded from base with straight inner edge, proximal expanded from base but rounded; endopod short and curved.

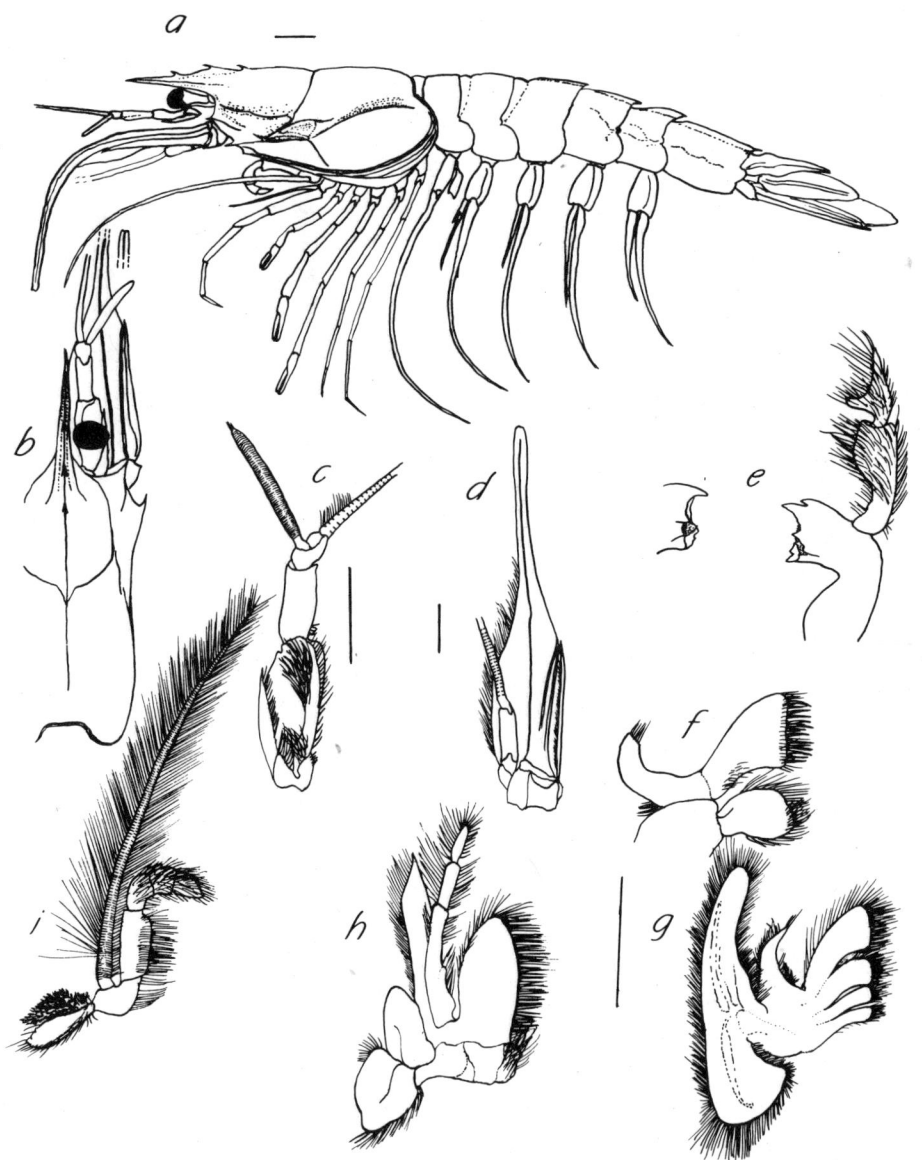

Fig. 16. *Plesiopenaeus edwardsianus*: *a*, whole shrimp from left side; *b*, carapace in dorsal aspect; *c*, antennule; *d*, antenna; *e*, mandible; *f*, maxillule; *g*, maxilla; *h*, first maxilliped; *i*, second maxilliped; solid line = 10 mm.

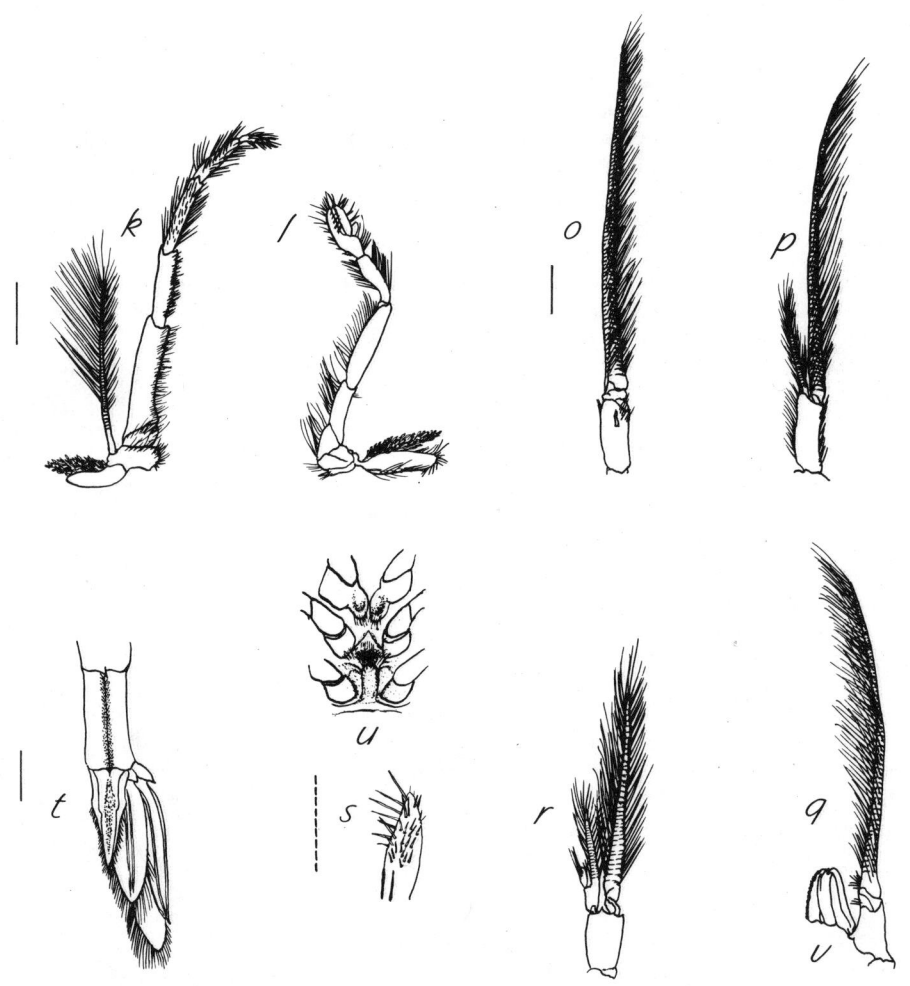

Fig. 17. *Plesiopenaeus edwardsianus*: k, third maxilliped; l, first pereopod; o, F first pleopod; p, F second pleopod; q, M first pleopod; r, M second pleopod; s, appendix masculina; t, telson; u, thelycum; v, petasma; solid line = 10 mm, broken line = 1 mm.

Maxilla (g) endites unequally bilobed; endopod with subapical strong hooked spines and apical setae; narrow anterior lobe of scaphognathite longer than wide axe-shaped but rounded posterior lobe.

Maxilliped I (h) long distal endite, short proximal; endopod three-segmented longer than exopod, the latter with short apical incipient lash; epipod bilobed, both thick, rounded.

Maxilliped II (i) leglike, compressed, exopod about twice as long as endopod, slender, annulated and fringed with long lateral setae; epipod subtriangular, with podobranch.

Maxilliped III (k) leglike, distal segment with 2 strong apical spines; exopod short; epipod small with straplike attachment and with podobranch.

Pereopods: I (l) chelate, moderately slender, shorter than leg II or maxilliped III, large epipod and podobranch; II chelate, slightly more slender than I but longer, epipod large with podobranch; III chelate, more slender and longer than II, epipod large with podobranch; IV and V non-chelate, about as long as III but more slender, IV only with epipod.

Pleopods: female I (o) uniramous, protopod only about one-seventh as long as exopod, on distal third of protopod a moveable spine; female II (p) endopod much shorter and more slender than exopod; male I (q) uniramous, very long exopod, protopod with petasma on inner half; male II (r) endopod less than half length of exopod, appendix interna slightly shorter than appendix masculina (s) the latter pointed slightly at apex and with a few moderate lateral spines and many short spines.

Thelycum (u) with median ventral protuberance on sternite XIV rounded posteriorly and trending forward with flat lateral plates meeting at each side and with many anterior setae meeting setae from median strong ventral protuberance on XIII (4th legs), the latter with sharp anterior projection almost reaching coxal saucer-like projections from 3rd legs, a deep declevity between both protuberances.

Petasma (v) distomedian lobe slightly pointed at apex, with rows of hooks along edge; distoventral lobe rectilinear, distolateral with pointed apex.

DISTRIBUTION

In the western Atlantic from south of Labrador to the Gulf of Mexico, West Indies and Surinam; in the eastern Atlantic from Portugal, the Azores to off South Africa. Also in the Indo-West-Pacific off the east coast of Africa, the Andaman Sea and Sumatra. Depths 274–1 850 m (Crosnier et Forest 1973).

Records of distribution in the area of reference are in Fig. 18.

BIOLOGY

Lengths to 197 mm total length (55 mm cl) in males (Squires 1965a) and to 334 mm total length (104 mm cl) in females (Crosnier et Forest 1973).

Stomach contents included remains of shrimps, euphausiids, amphipods, chaetognaths and polychaetes (Squires 1965a).

Sergestes arcticus and *Pasiphaea tarda* were in catches with this species (Squires 1965a).

FISHERY

Fished commercially by Spanish trawlers in the area of Senegal, Guinea and off Congo and Angola. The shrimps are frozen on board and mainly sold in Spanish markets (e.g. Barcelona), but also in France (Holthuis 1980).

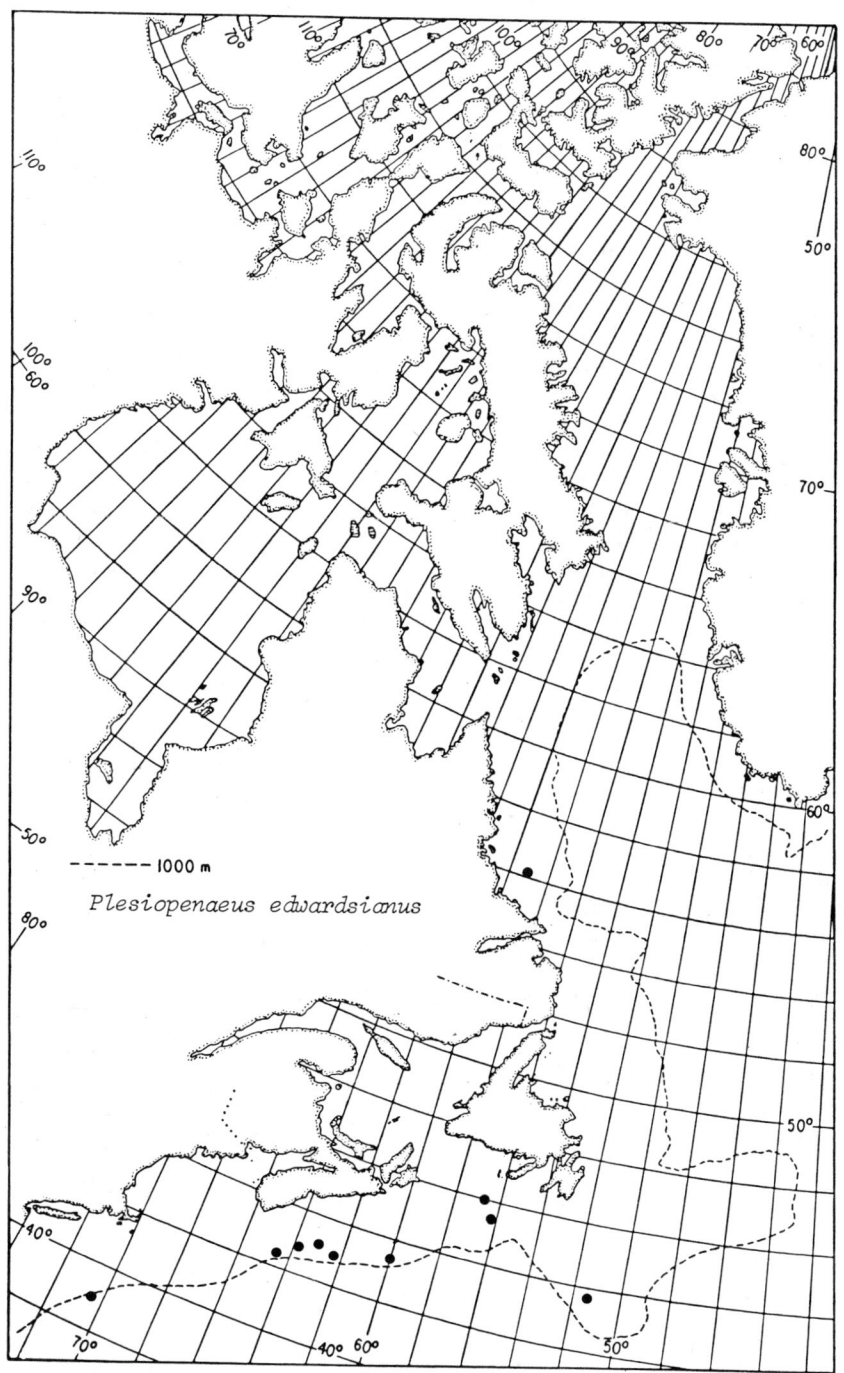

Fig. 18. *Plesiopenaeus edwardsianus*, distribution records in the area of reference.

Family SERGESTIDAE Dana, 1852
Butler 1980: 45; Zariquiey Alvarez 1968: 59.

Rostrum short; first legs without chela; fourth and fifth pairs of pereopods reduced, smaller than others; ventro-mesial flagellum of antennule modified as clasping organ in males; pereopods and third maxilliped without exopods.

Of the 53 species known only two are reported from the area of reference.

Key to the species of Sergestidae

1 Supraorbital spine well-developed; hepatic spine; 3rd antennular article very slender when viewed from above .... *Sergestes arcticus*

   No supraorbital or hepatic spine; 3rd antennular article robust as viewed from above .... *Sergia robusta*

Genus *Sergestes* H. Milne-Edwards, 1830
Burkenroad 1983: 283 (key); Butler 1980: 45;
Genthe 1969: 203; Yaldwin 1957: 7;
Zariquiey Alvarez 1968: 59.

Several luminescent organs in the cephalothorax, developed from parts of the gastrohepatic gland; ischium of 1st and 2nd pereopods with well-developed spine proximally; third article of antennular peduncle slender, about 3 times as long as its diameter; supraorbital and hepatic spines present.

Only one of the 26 known species is reported from the area of reference.

Sergestes arcticus Kroyer, 1855
Yaldwin 1957: 9, figs 1-5;
Zariquiey Alvarez 1968: 61.
(Figures 19 and 20)

DISTINGUISHING CHARACTERISTICS

Supraorbital and hepatic spines on carapace; antennular segments slender; rostrum very short; 4th and 5th legs shorter than others and more slender than in *Sergia robusta*; first leg without and 2nd and 3rd legs with chela, terminal segments of legs annulated; colour whitish translucent. Telson terminally rounded and with setae only.

DESCRIPTION

Integument thin, fragile, smooth and shiny. Colour whitish translucent with areas of red pigmentation, also luminescent organs (organs of Pesta) formed from parts of the hepatopancreas a little above the maxillae.

Rostrum is a short median frontal spine, pointing obliquely forward with a short supporting carina.

Carapace with supraorbital spine set back from edge of orbit with a short carina behind and in front; hepatic spine is low laterally at the end of a faint cervical groove and close to a lateral carina which extends the whole length of the carapace, a branch rising above the branchial area and a fainter branch towards the ventral margin obliquely; organ of Pesta slightly behind the hepatic spine.

Abdomen compressed, especially the 6th somite which is also deep and has a midventral and a ventro-lateral carina on each side; all pleura and terga rounded; faint arched carina on 5th laterally; small middorsal posterior spine on 6th.

Telson (t) with moderate ventro-lateral skirt on proximal third, tapering to narrow rounded tip with about eight strong terminal setae; dorso-lateral carinae and shallow central sulcus; both branches of uropod longer than telson, the outer much longer.

Eye with moderately long stalk and large globular cornea.

Antennule (c) exceeding in length the antennal scale; 1st article shorter than other two combined, the 2nd slightly longer than 3rd; stylocerite short about one-third 1st article; dorso-lateral flagellum with short dilated area and a length somewhat greater than the carapace, the ventro-mesial very short and slender in females and biramous and modified as a grasping organ in males.

Antenna (d) scale length about 4.5 times width, small disto-lateral spine exceeding blade; peduncle less than half scale; flagellum is about 1.5 times total body length.

Mandible (e) incisor with strong corner tooth and long cutting edge joined to ovalshaped molar; palp long and slender, the proximal segment about 1.8 times distal.

Maxillule (f) proximal endite ovate, distal endite twice as large, almost circular in outline; endopod short, rounded at tip with two sets of setae.

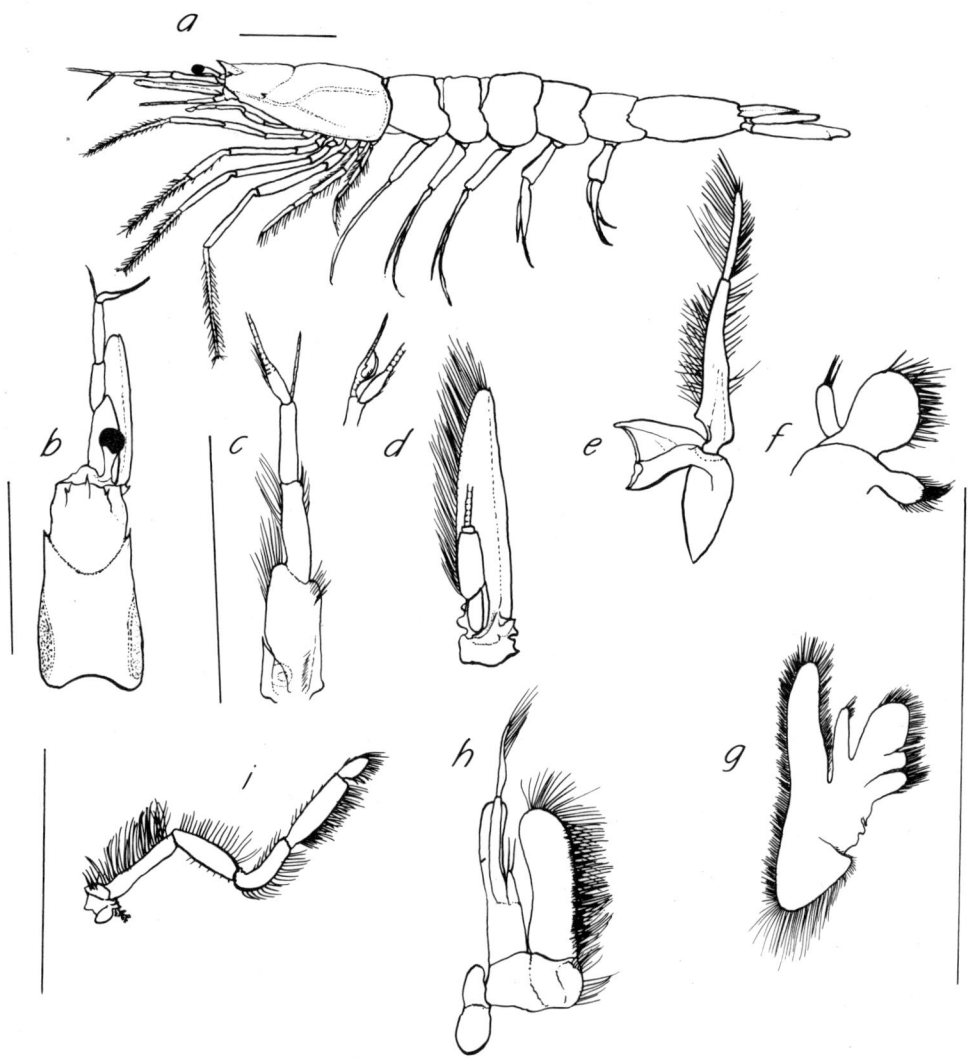

Fig. 19. *Sergestes arcticus*: *a*, whole shrimp from left side; *b*, carapace in dorsal aspect; *c*, antennule; *d*, antenna; *e*, mandible; *f*, maxillule; *g*, maxilla; *h*, first maxilliped; *i*, second maxilliped; solid line = 10 mm.

Fig. 20. *Sergestes arcticus*: *k*, third maxilliped; *l*, first pereopod; *m*, second pereopod; *o*, F first pleopod; *p*, second pleopod; *q*, M first pleopod; *r*, M second pleopod; *s*, appendix masculina; *t*, telson; *u*, thelycum; *v*, petasma; solid line = 10 mm, broken line = 1 mm.

Maxilla (g) proximal endite unilobed; distal endite bilobed, the distal lobe much the larger; endopod straight, short; anterior lobe of scaphognathite long and narrow from narrow base, much longer than wider, rounded posterior lobe.

Maxilliped I (h) endite long subrectilinear, slightly curved and rounded; proximal endite short; endopod with two segments, the proximal segment exceeding length of exopod (also a very short papilla at base of endopod); epipod bilobed, inflated.

Maxilliped II (i) elongate, leglike with 7 segments; carpus with strong curve; dactyl rounded apically; epipod with small podobranch.

Maxilliped III (k) leglike, slender, longer than 1st leg; terminal segment tapering and annulated.

Pereopods: I (l) non-chelate, very slender, propodus tapering and annulate, distally with strong setae; II (m) chelate, propodus very long, annulate, tapering, with strong seta at each annulation (on each side), small dactyl, tips of fingers with strong lateral setae; III chelate, longer than I or II, as in II the dactyl forming a small chela with tip of long propodus, fingers with tufts of setae; IV and V both shorter than others, V about half as long as IV, compressed but slender, terminally propodus only, no dactyl.

Pleopods: female I (o) uniramous, protopod triangular in cross-section, about half length of exopod; female II (p) biramous, endopod shorter than exopod; male I (q) uniramous, exopod very long, petasma attached to proximal fifth of protopod; male II (r) endopod shorter than exopod, appendix masculina (s) only, slightly pointed apically and with about 10 short spines and about 3 long spines on one side.

Thelycum (u) of the open type with an operculum, a transverse semicircular fold of the sternite of the 3rd legs, arching back over the atria, triangular openings to invaginated receptacles (Genthe 1969); posterior is a low ridge and a bracket-shaped suture across the sternite. Coxal nib and tooth of the 3rd legs are in the vicinity of and cover the oviducal openings.

Petasma (v) complex, short distomedian lobe subtriangular with edge of rows of hooks, disto-ventral largest with 3 apical variously armed (with retractile hooks) branches the central of which is biramous, the others tri- and uni-ramous, the distolateral with a single stem and a shield-shaped wing.

RANGE OF DISTRIBUTION

From 70° N off West Greenland and Norway in the North Atlantic to the Strait of Magellan and South Africa in the South Atlantic, also in the Mediterranean as far as the Adriatic, and in the Indo-Pacific including off New Zealand. Depths 250–4 500 m (Crosnier et Forest, 1973; Stephensen, 1935; Yaldwin, 1957).

Records of distribution in the area of reference are in Fig. 21.

BIOLOGY

Lengths reported are to 19 mm cl in males and 20 mm cl in females.

Temperatures where taken in the area of reference were 2–8° C, and other pelagic species such as *Pasiphaea tarda* and *Pasiphaea multidentata* and the euphausiid *Meganyctiphanes norvegica* were present in the catches (Squires 1965a).

Stomach contents were remains of calanoid copepods and chaetognaths.

Males were first mature at 11 mm cl in Hermitage Bay, Newfoundland (Squires 1965a).

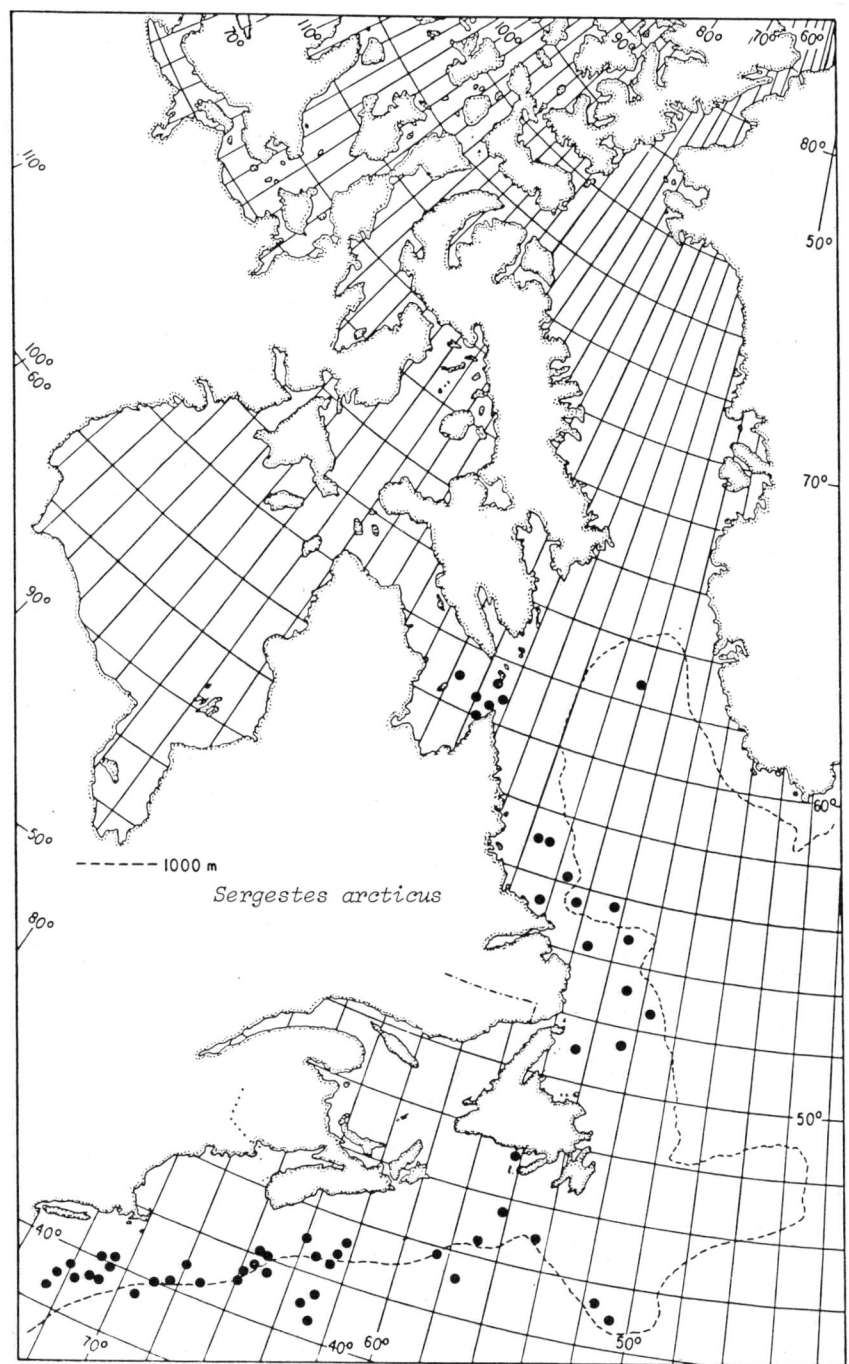

Fig. 21. *Sergestes arcticus*, distribution records in the area of reference.

Genus *Sergia* Stimpson, 1860
Burkenroad 1983: 283 (key);
Butler 1980: 45.

Luminescent organs, if present, cuticular; no supraorbital or hepatic spines; 2nd and 3rd articles of antennular peduncle stout; rostrum thin, with prolonged point and sometimes a denticle at the upper edge of its base; 2nd and 3rd legs with minute chela, 4th and 5th legs shorter than others but compressed and wider (wider than in *Sergestes*).

Sergia robusta (Smith, 1882)
Crosnier et Forest 1973: 327, figs. 111 d–i, 112 c–d;
Foxton 1970: 974, fig. 8.
(Figures 22 and 23)

Distinguishing Characteristics

Antennular peduncle stout; body robust; carapace without spines; rostrum short, with prolonged point; first three legs long, slender, 2nd and 3rd chelate, 4th and 5th shorter, compressed, moderately wide; bright reddish in colour; telson ending in a sharp point.

Description

Integument thin, smooth and shiny. Colour reddish to scarlet with some iridescence.

Rostrum short, sharp-pointed, slightly ascending, reaching just beyond anterior edge of carapace, in some specimens with one or two small teeth dorsally on short rostral carina.

Carapace with no spines; very low indeterminate horizontal carina from hepatic area rising over branchial area and meeting faint cervical and post cervical grooves and others towards anterior and ventral edges of carapace.

Abdomen with terga and pleura of first to fifth somites rounded but ventrally indented and with faint lateral ridges and grooves near points of articulation, most clear on 5th and 6th, the latter with a small spine at posterior edge dorsally.

Telson (t) with dorsolateral ridges and median sulcus tapering to short bevelled point with long lateral setae, also three pairs of tiny spines at distal quarter of lateral ridges; on proximal third is a pair of ventro-lateral lobes. Inner branch of uropod considerably longer than telson and outer branch much longer still.

Eye large, cornea globular much wider than tapered stalk.

Antennule (c) each article of peduncle only slightly shorter than one preceeding; stylocerite about half length of 1st article; dorsolateral flagellum compressed about equal to body length, ventromesial flagellum very slender, less than peduncle length, modified as a clasping organ with a fixed and a moveable ramus in male. Photophore on 1st article near point of stylocerite.

Antenna (d) scale length 2.8 times width; distolateral spine about even with blade; peduncle slightly longer than half scale.

Mandible (e) left incisor with two teeth at corner of sharp cutting edge, right with one tooth only; molar sub-triangular at surface, joined closely to incisor. Distal segment of palp slender, about half as long as thick proximal.

Maxillule (f) distal endite large, ovate; proximal narrow; endopod short with two strong apical setae.

Maxilla (g) proximal endite reduced: one lobe retained the other truncated; distal endite with two large lobes slightly unequal; endopod moderate, tapered, straight; anterior lobe of scaphognathite rectilinear, narrow, longer than axe-shaped posterior lobe.

Maxilliped I (h) distal endite as long as exopod, narrow, subrectilinear; proximal endite short with thickened edge; endopod with proximal shoulder with strong spines and tapered to narrow distal segment longer than exopod; epipod bilobed.

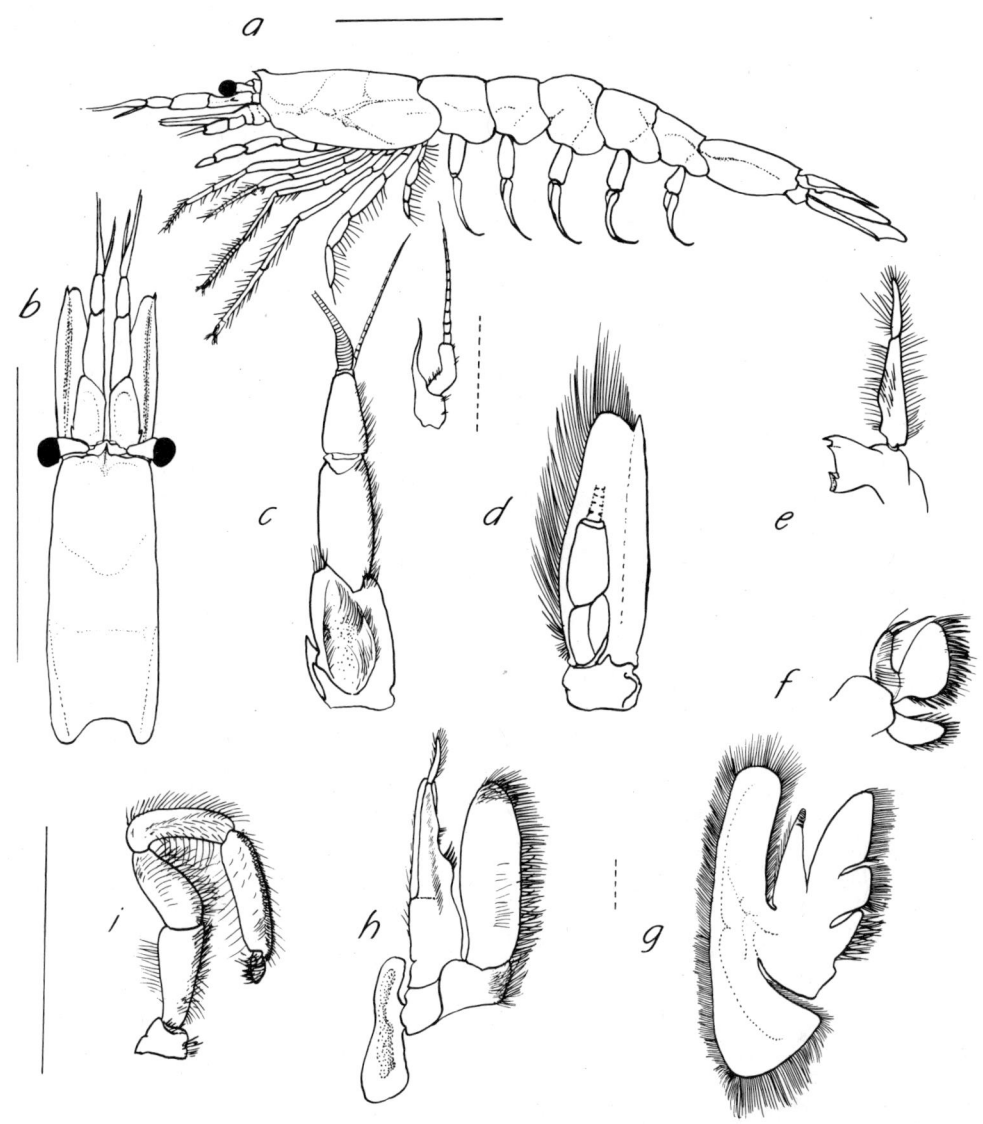

Fig. 22. *Sergia robusta*: *a*, whole shrimp from left side; *b*, carapace in dorsal aspect; *c*, antennule; *d*, antenna; *e*, mandible; *f*, maxillule; *g*, maxilla; *h*, first maxilliped; *i*, second maxilliped; solid line = 10 mm, broken line = 1 mm.

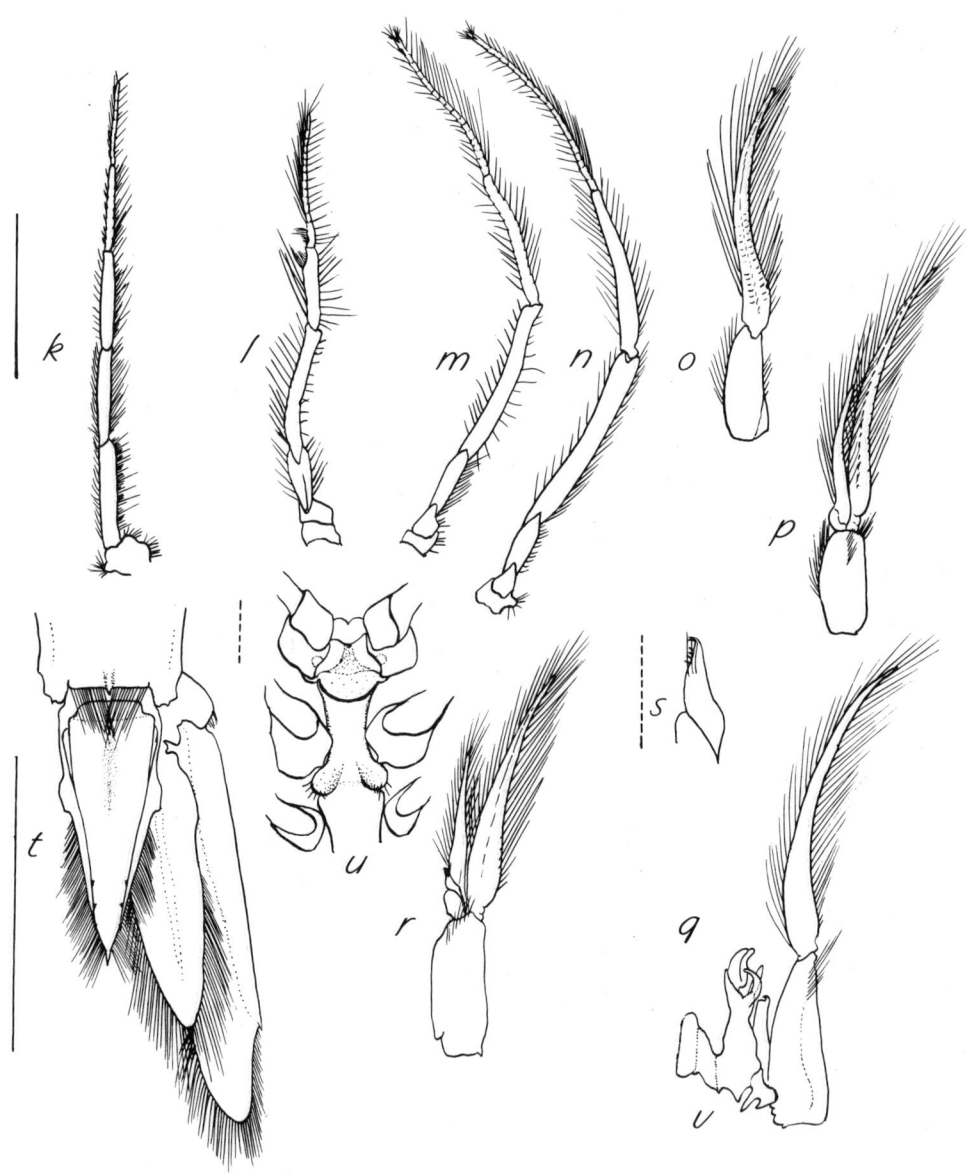

Fig. 23. *Sergia robusta*; *k*, third maxilliped; *l*, first pereopod; *m*, second pereopod; *n*, third pereopod; *o*, F first pleopod; *p*, F second pleopod; *q*, M first pleopod; *r*, M second pleopod; *s*, appendix masculina; *t*, telson; *u*, thelycum; *v*, petasma; solid line = 10 mm, broken line = 1 mm.

Maxilliped II (i) leglike, carpus with strong curve, distal segment small, longer than wide, with blunt tip.

Maxilliped III (k) leglike, slender, all segments about equal in length, except first, tapering to annulated distal segment. Slightly longer than first leg.

Pereopods: I (l) non-chelate, slender, propodus annulate, no dactyl; II (m) chelate, slender, tapering, propodus annulate, dactyl very short, fingers with distal tufts of setae; III (n) chelate, slender, longer and stouter than II, tapered with propodus annulate, dactyl short, fingers with distal tufts of setae; IV and V shorter than others, IV about twice as long as V, also compressed and wider than others, no dactyl.

Pleopods: female I (o) uniramous; female II (p) endopod about half length of exopod; male I (q) uniramous, protopod with petasma attached at proximal fifth; male II (r) appendix interna exceeded by appendix masculina (s) the latter tapered from rounded base, with two strong apical spines and four subapical spines and a few setae.

Thelycum (u) is the open type with modification of the sternite of third and fourth legs and coxae of the third legs. At the sternum of third legs a transverse semi-oval plate is between articulation points of the coxae. Anteriorly the plate descends and narrows to form a depression to where it meets a transverse suture; ahead of it is another suture with a faint spine at centre. Behind the thelycum the sternum of the 4th legs forms two lateral rounded lappets.

Petasma (v) distomedian lobe subrectilinear rounded at ends and with rows of hooks laterally, joined by fold to longer distoventral or central lobe which has apically a double connected small lobe (lobus connectans) with tips covered with rectractile hooks, and a curved lobe (lobus armatus) and straight lobe (lobus terminalis) both covered with small retractile hooks; the distolateral lobe consists of the uncifer process with terminal hook and connection to protopod: a subtriangular process with two lateral projections.

Range of Distribution

In the western Atlantic from Labrador to the West Indies and the Gulf of Mexico; in the eastern Atlantic from the Faroes to off Angola, and in the Mediterranean to the Aegean. Depths 0–5 000 m (Crosnier et Forest 1973).

Records of distribution in the area of reference are in Fig. 24.

Biology

Lengths to 9 mm cl in males (specimens examined) and 94 mm total length in females (Crosnier et Forest 1973).

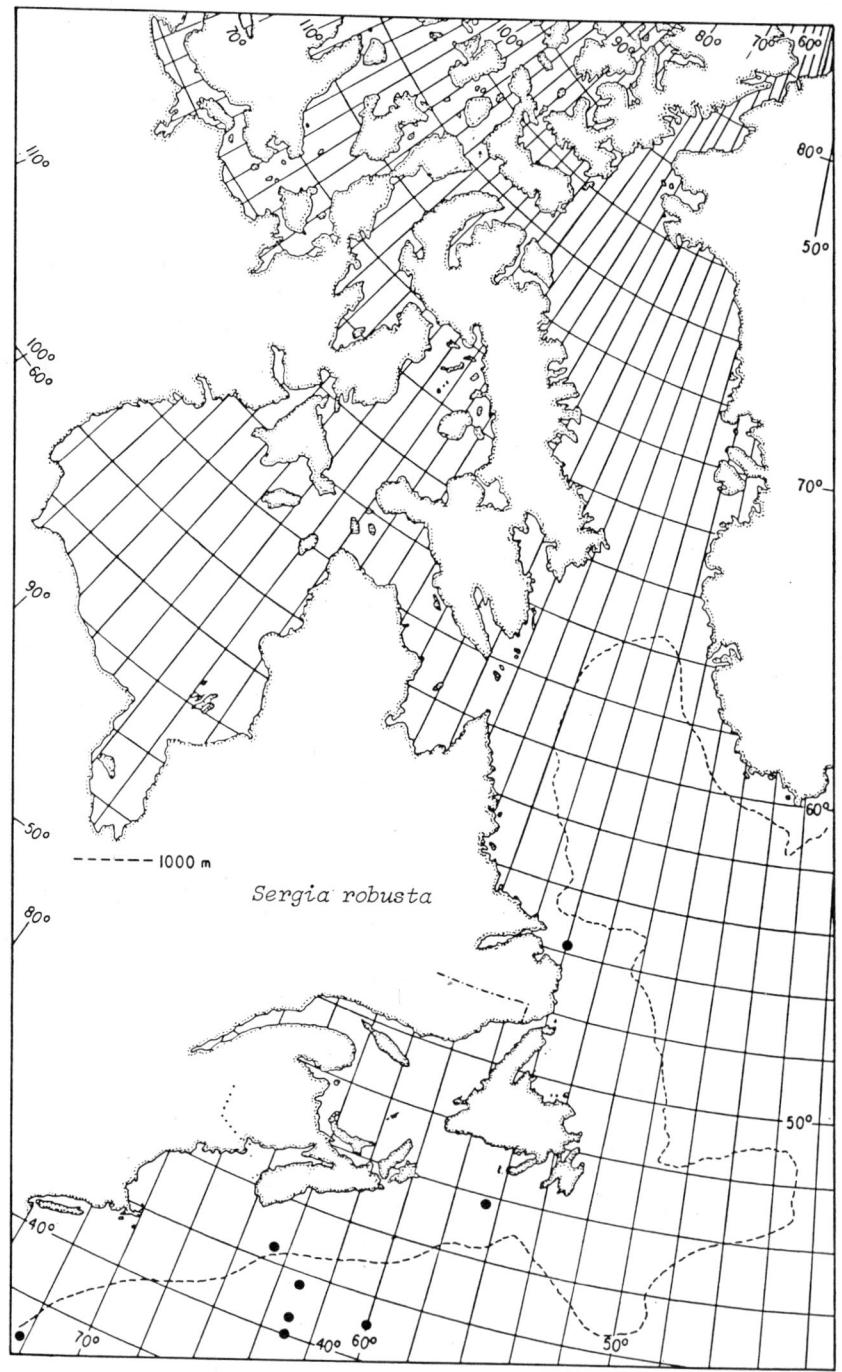

Fig. 24. *Sergia robusta*, distribution records in the area of reference.

Family LUCIFERIDAE Burkenroad, 1983
Williams 1984: 52; Zariquiey Alvarez 1968: 63.

Body extremely compressed, anterior part between orbit and mouth-field elongate forming a "neck"; antennule without ventro-mesial flagellum; third legs with tiny dactyl but no propodal finger; gills absent; mandible without palp; abdomen long, first 5 somites with ventral angle; 6th abdominal somite with ventral processes in male; telson of male with strong protuberance on ventral face; a single genital opening in both sexes.

Genus *Lucifer* Thompson, 1829
Williams 1984: 52; Zariquiey Alvarez 1968: 63.

Tiny, slender shrimps, anterior portion of carapace long, cylindrical, forming a neck in front of mouthparts; rostrum short; postorbital and antennal spine; antennule with only one flagellum, peduncle longer than antennal scale; mandible without palp; only three pairs of pereopods; abdomen long, the first five somites with ventral angle at insertion of pleopods; 6th somite with ventral processes in male.

Lucifer faxoni Borradaile, 1915
Williams 1984: 52, fig. 34;
Bowman 1967: figs. 2 c-d, 3 d, e.
(Figures 25 and 26)

DISTINGUISHING CHARACTERISTICS

Small shrimp with front of body elongate like a neck, rest of carapace short with only three pairs of legs, eye and eyestalk about one-third as long as neck; antennule with only one flagellum, peduncle longer than antennal scale, 1st article much longer than other two; antennal peduncle about half as long as scale; abdomen elongate, each of 1st to 5th somites with ventral angle where pleopods originate, 6th somite of male with 2 curved ventral spinous processes, also telson of male with ventral protuberance near tip; uropods much longer than telson.

DESCRIPTION

Integument thin, smooth, shiny. Colour almost completely translucent.

Carapace compressed, the anterior part like a long neck, cylindrical, with eyes, antennae, etc., at the front, and behind the neck the mouthparts, legs, etc. Rostrum short, a spinous projection of the dorsal edge of the carapace; at anterior edge also is a postorbital and an antennal spine, behind the neck is a hepatic spine and a faint cervical groove and the carapace expands posteriorly and ventrally in the form of a triangle; behind the legs under the carapace is an apparent globule of oil in female.

Abdomen extremely compressed, pleura of first five somites forming a central projection and spine ventrally where the pleopods are attached; 6th somite of the male longer than the 4th and 5th combined, with two prominent curved ventral spines, the posterior one larger, and a small posterior spine middorsally.

Telson (t) short, tapering from small proximal expansions to truncate concave tip with strong lateral spines and two pairs of small inner spines, also three pairs of dorsolateral spines; ventrally is a short stem-like projection near tip in males. Inner branch of uropod narrow about twice as long as telson; outer branch longer and wider than inner, outer edge slightly convex, distal notch and spine.

Eyestalk long and cornea large, globular, together about one-third as long as neck.

Antennule (c) 1st article slender, sinuous, longer than other two combined, 3rd less than half 2nd; no stylocerite or ventro-mesial flagellum.

Antenna (d) scale length about 7 times width, outer edge sinuous, tapering to fine point distally; peduncle slightly more than half scale.

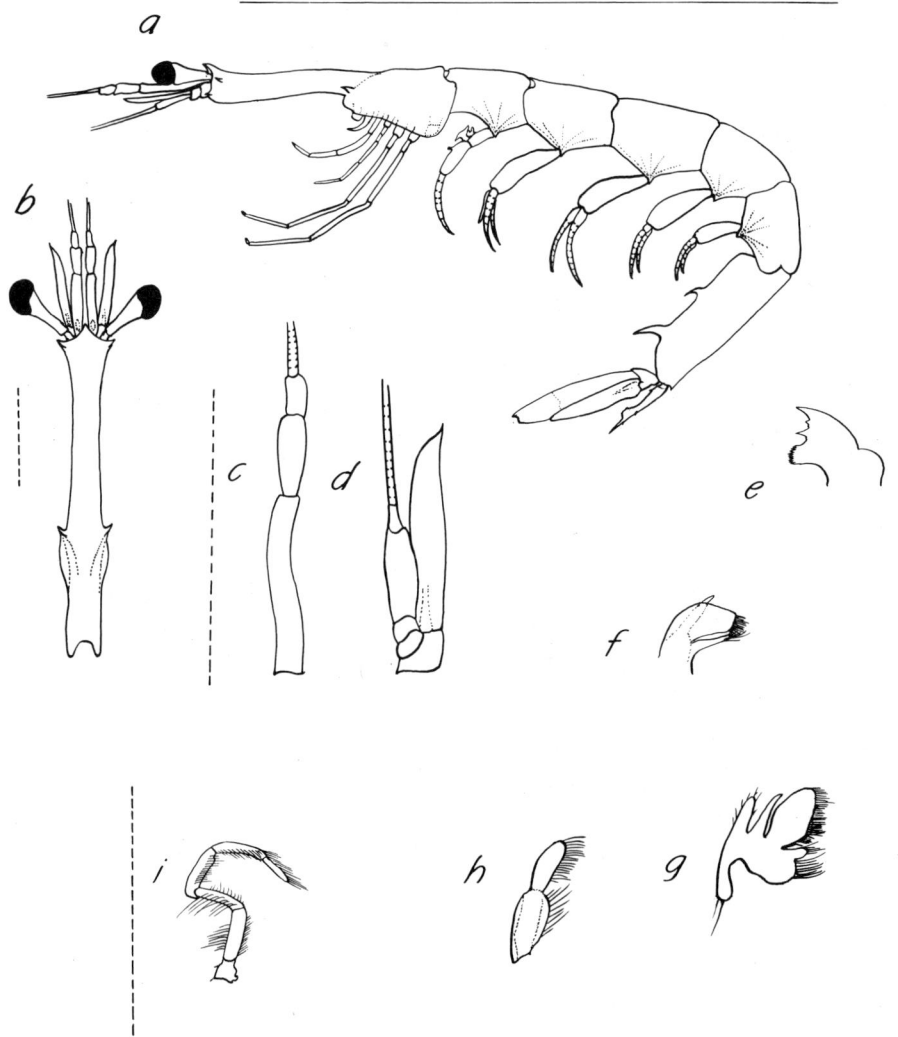

Fig. 25. *Lucifer faxoni*: *a*, whole shrimp from left side; *b*, carapace in dorsal aspect; *c*, antennule; *d*, antenna; *e*, mandible; *f*, maxillule; *g*, maxilla; *h*, first maxilliped; *i*, second maxilliped; solid line = 10 mm, broken line = 1 mm.

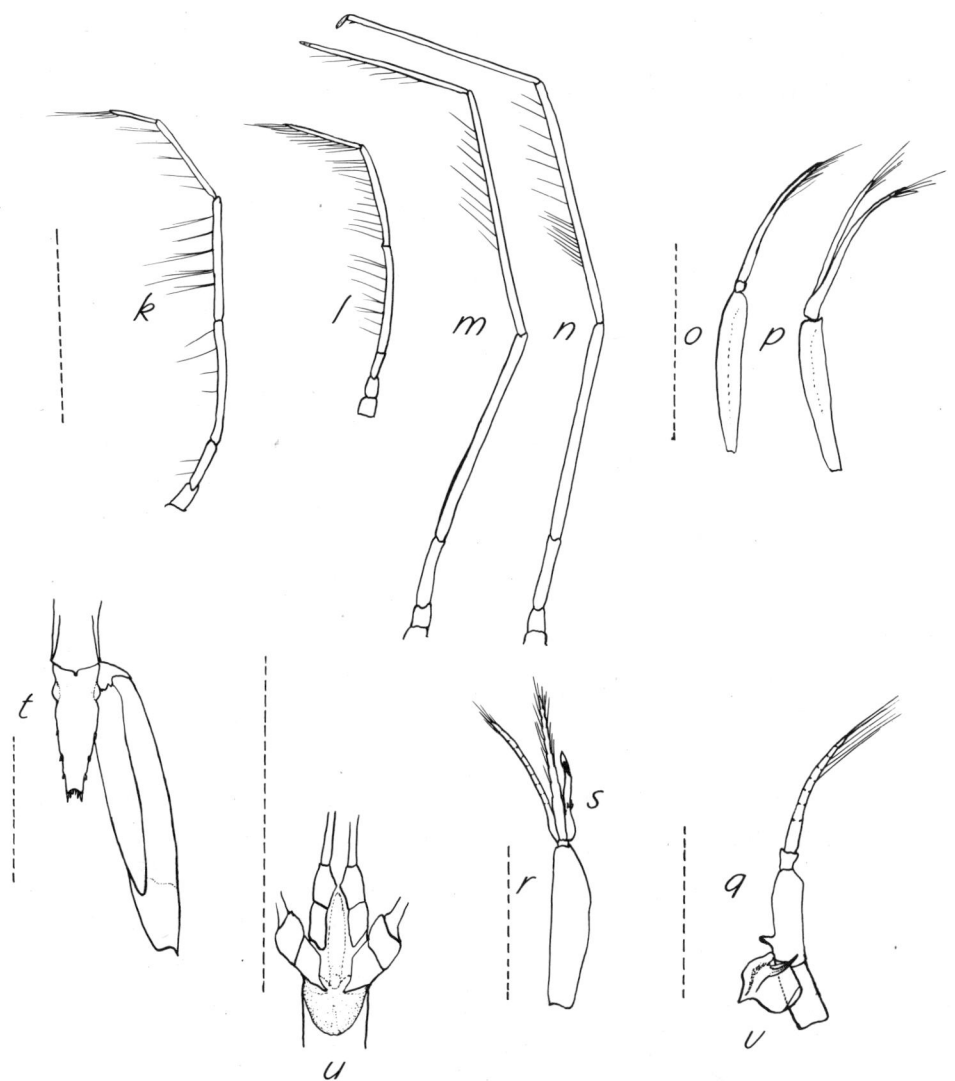

Fig. 26. *Lucifer faxoni*: k, third maxilliped; l, first pereopod; m, second pereopod; n, third pereopod; o, F first pleopod; p, F second pleopod; q, M first pleopod; r, second pleopod; s, appendix masculina; t, telson; u, thelycum; v, petasma; broken line = 1 mm.

Mandible (e) without palp; incisor with two major sharp teeth and a small one between them; molar has edge of small serrate teeth.

Maxillule (f) distal endite tapering from wide base to truncate spinous edge, much larger than narrow proximal pointed endite; endopod tiny.

Maxilla (g) distal endite a single wide lobe with setose inner edge, proximal with small papilla-like lobe and wide rounded lobe with long setae; endopod tapered distally, slightly longer than anterior lobe of scaphagnathite, the latter about equal to slightly expanded posterior lobe with a long distal seta.

Maxilliped I (h) appearing reduced to two segments possibly modified endites with other parts fused or reduced.

Maxilliped II (i) leglike, with six segments.

Maxilliped III (k) leglike, slender, with six segments, longer than 1st leg.

Pereopods: I (l) non-chelate, without dactyl, slender, shorter than others; II (m) non-chelate, dactyl small; III (n) non-chelate, slender, about the same length as II, propodus with distal spine, dactyl slightly longer than in II, rounded at tip. IV and V non-existant.

Pleopods: female I (o) uniramous, slender, endopod about as long as protopod; female II (p) biramous, slender, endopod about equal to exopod. Male I (q) uniramous, two-segmented, distal annulated; protopod also two-segmented, distal with a proximal inner truncate spine, proximal with petasma; male II (r) endopod about equal in length to but more slender than exopod, appendix masculina (s) inserted at base of endopod, long and slender with small apical projection and few subapical setae.

Thelycum (u) a rounded or conical mass behind and between 3rd legs; central groove obscure with a small circular pit at posterior end, the latter a single aperture for paired tubes from the seminal receptacle (Bowman 1967).

Petasma (v) distomedian lobe with edge rows of small hooks with small projections; distoventral lobe with apical needle-like projection; distolateral with small apical lobe and attachment along edge to protopod.

RANGE OF DISTRIBUTION

Open ocean from off the Grand Banks in waters of the Gulf Stream and southward to the West Indies, where it may be taken in shallow inshore areas (Chace 1972), also Gulf of Mexico and Caribbean to Brazil; in the eastern Atlantic off West Africa. Depths 0–90 m (Bowman and McCain 1967; Crosnier et Forest 1973; Williams 1984).

Records of distribution in the area of reference are in Fig. 27.

BIOLOGY

Length of body to 12 mm (Williams 1984).

First maturity appeared to be at 1.5 mm cl in males and in females at 1.8 mm cl in islands of the West Indies (Chace 1972). Unlike other penaeids the females are egg-bearing. Studies of estuarine populations have shown some females to be carrying eggs from spring to autumn and post-larvae to occur in the plankton during fall, winter and summer probably indicating production of several broods during the year. Williams (1984) reviews these studies.

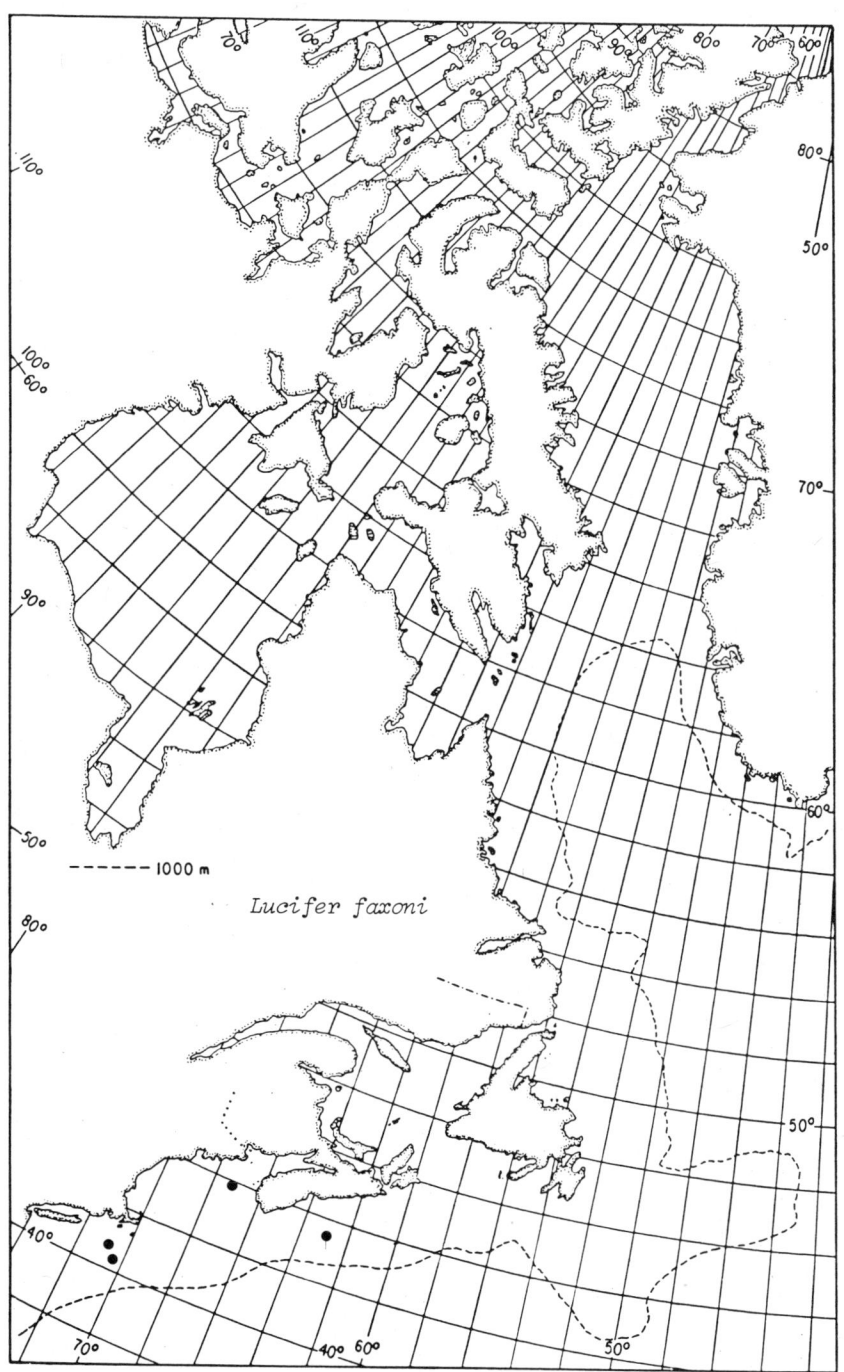

Fig. 27. *Lucifer faxoni*, distribution records in the area of reference.

Suborder   EUKYPHIDA

Infraorder CARIDEA

Family OPLOPHORIDAE Dana, 1852

Reddish deep-water pelagic shrimps with exopods on all pereopods, first two pairs of pereopods chelate, last three pairs moderately longer than first two; epipods straplike but with an end-part at right angles extending vertically among the gills and with mesial hooks at the angle on first three pereopods — very small or vestigial on fourth and none on fifth pereopod. Mandible with palp, molar and incisor processes separated by a small gap; appendix masculina on second pleopod compressed and distolaterally fringed with long fine spines in most species. Five of the 10 genera in this Family are represented in the area of reference (Chace 1986; Butler 1980).

Key to genera and species of family OPLOPHORIDAE
(Modified from Chace 1940).

1	Exopods of pereopods slender and flexible, telson subtruncate	2
	Exopods of third maxilliped and first pereopod wide and rigid; outer edge of antennal scale armed with series of spines; telson ending in sharp point .... *Oplophorus spinosus*	
2	Last 4 abdominal somites dorsally carinate	3
	Sixth abdominal somite not carinate	4
3	Carapace free of lateral carinae along lateral surface; rostrum with at least as many dorsal as ventral teeth .... *Acanthephyra*	
	Four pairs of dorsolateral spines on telson .... *A. purpurea*	
	About 10 pairs of dorsolateral spines on telson .... *A. pelagica*	
	Carapace with carinae along lateral surface from front to posterior margin .... *Notostomus*	
	Four lateral carinae on posterior half of carapace where second is short; rostrum short with one carina .... *N. robustus*	
	Five lateral carinae on posterior half of carapace where second is long; rostrum long with two carinae .... *N. elegans*	
4	Cornea of eye smaller than stalk and poorly pigmented; anterior margin of first abdominal somite without a lobe or tooth; telson truncate .... *Hymenodora glacialis*	
	Cornea of eye wider than stalk and well-pigmented; anterior margin of first abdominal somite with a lobe or tooth overlapping hind margin of carapace; telson ending in a sharp end-piece with lateral spines .... *Systellaspis debilis*	

Genus *Acanthephyra* A. Milne-Edwards, 1881
Chace 1986: 6, figs 2–14.

These are intense red deep-sea pelagic shrimps; body well compressed; rostrum mostly long with dorsal and ventral spines in species from this area; strong dorsal carina on carapace and second to sixth abdominal somites, carina of some somites projected posteriorly as a spine; all legs with exopods; epipods on first four legs; telson subtruncate, not acute; palp of 1st maxilliped with three divisions.

Acanthephyra pelagica (Risso, 1816)
Acanthephyra haeckeli (von Martens, 1868),
Chace 1940: 140, fig. 18.

Acanthephyra pelagica, Chace 1986:8 (key).
Sivertsen and Holthuis 1956: 7, figs. 3–7.

Acanthephyra purpurea, Rathbun 1929: 6, fig. 2.
(Figures 28 and 29, Plate 1a)

DISTINGUISHING CHARACTERISTICS

Rostrum about as long as carapace, sharp, pointing upward slightly, teeth about 8/5; telson with 7–11 dorsolateral spines on distal two-thirds; posterior median spine from carina on each of 3rd to 6th abdominal somites.

DESCRIPTION

Integument smooth but with minute pits and rugosities and firm but flexible. Colour brilliant scarlet red (Chace 1940).

Carapace. Rostrum almost equal to or as long as carapace and sharp, pointing upward slightly, with supporting lateral ridge widening and merging with carapace above orbit (a); with dorsal and ventral fixed teeth, ventral larger, first one on carina above eyestalk; a brush of ventral plumose setae at base over eye. Carapace with antennal and branchiostegal spines, the latter with a short but pronounced carina; low median dorsal carina on anterior two-thirds; two sub-marginal carinae along ventral edge. Oblique depression from antennal area to near end of carina from branchiostegal spine (a).

Abdomen compressed with prominent dorsal carina on all somites but first; carinae extended as long spine on third and as shorter spines on at least 5th and 6th, the latter over proximal telson. Pleura rounded and with setal fringes but no spines. Telson (t) long and narrow, 1.5 times 6th segment, with proximal sulcus and several pairs of lateral (sutured) spines: terminal spines with central long and curved downward, and with longer moveable spines on each side below a smaller distolateral pair. Outer branch of uropod longer than inner.

Eye globular, about 1.5 times average width of tapered stalk; inner tubercle on stalk distally.

Antennule (c). Peduncle about 0.3 antennal scale; stylocerite about 0.7 first article, short pointed tip; first article longer than other two together, the third slightly shorter than second; swollen base of upper flagellum about as long as peduncle.

Antenna (d). Scale almost as long or longer than carapace, narrow, tapering, with spine exceeding blade, length 4–6 times width. Basicerite with very strong fixed spine ventro-laterally; peduncle reaching about 0.3 length of scale.

Mandible (e). Left incisor with seven large pointed teeth from centre toward molar, and on other side of centre one large tooth and a low serrate edge — followed by large tooth in right; all chitinous tipped. Molar with hollow crown and curved triangular or squarish shape fringed with serrate chitin.

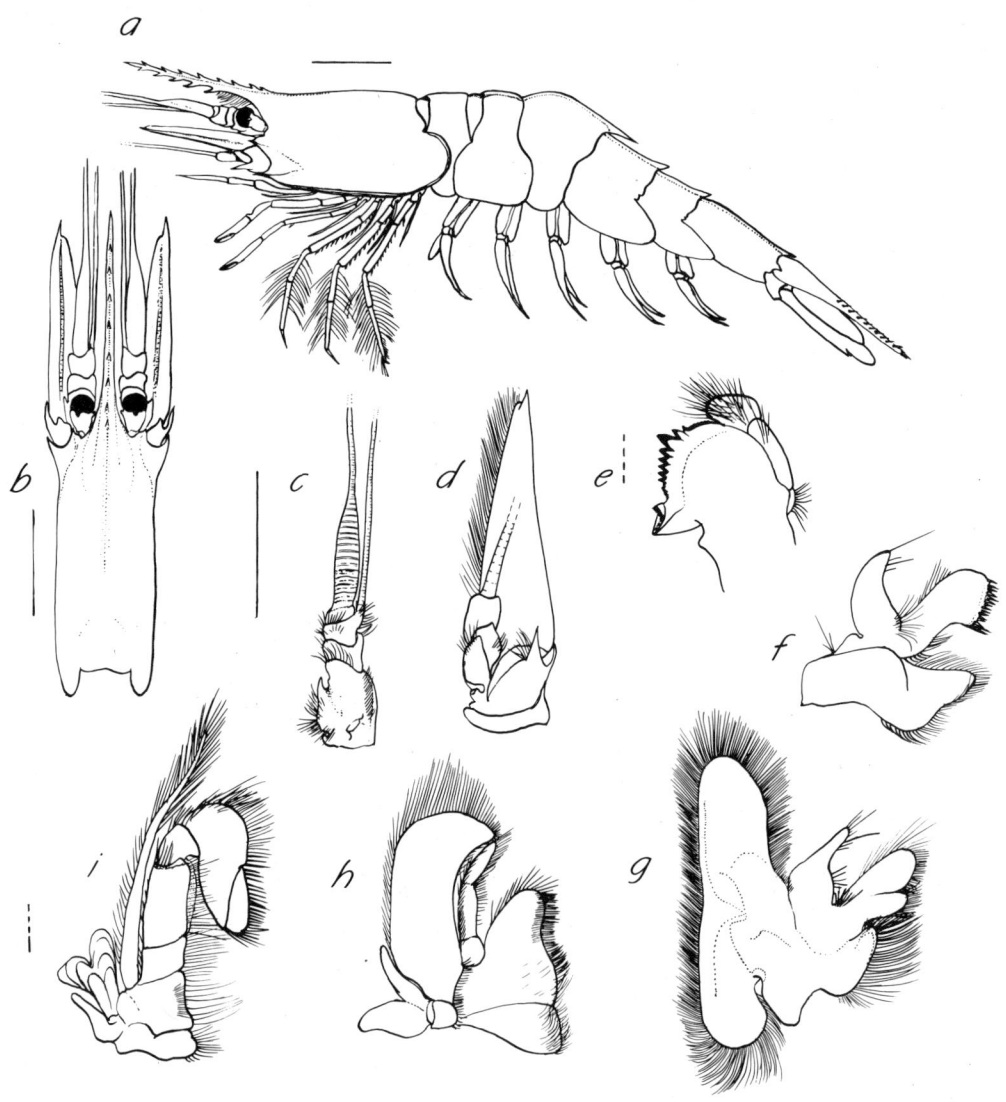

Fig. 28. *Acanthephyra pelagica*: *a*, whole shrimp from left side; *b*, carapace in dorsal aspect; *c*, antennule; *d*, antenna; *e*, mandible; *f*, maxillule; *g*, maxilla; *h*, first maxilliped; *i*, second maxilliped; solid line = 10 mm, broken line = 1 mm.

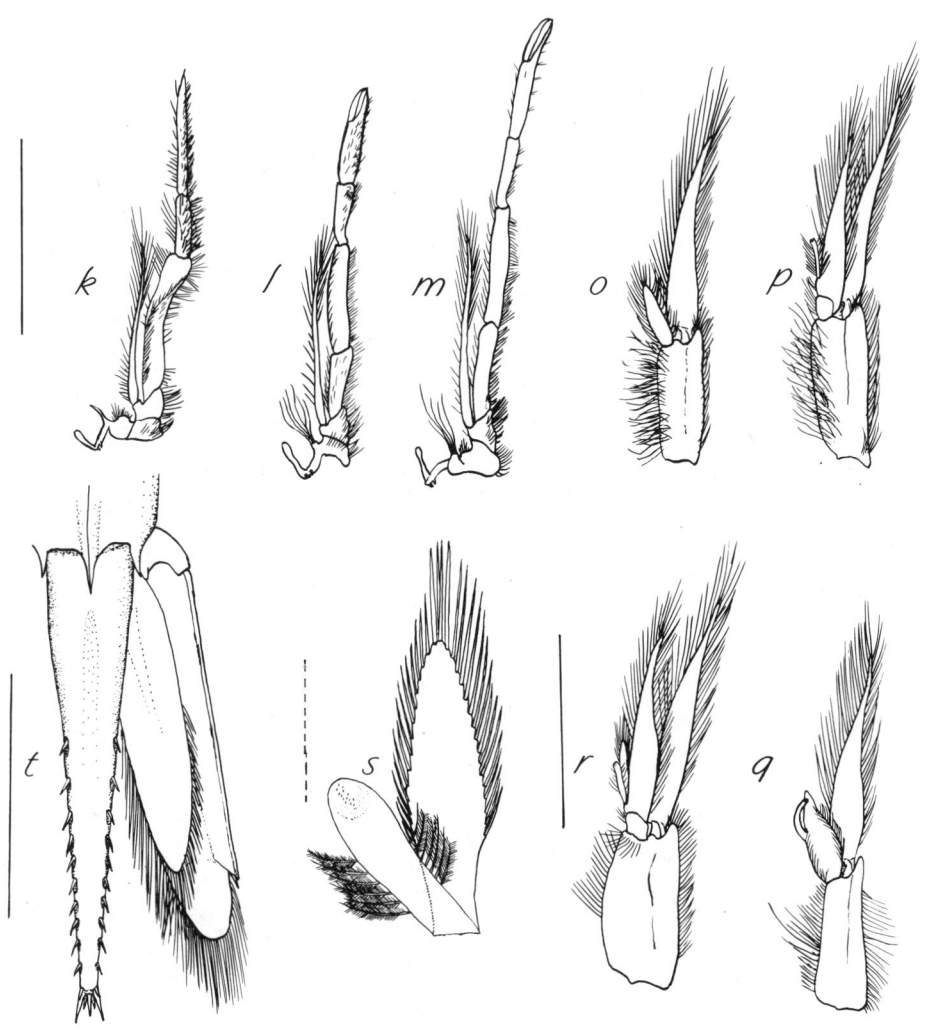

Fig. 29. *Acanthephyra pelagica*: *k*, third maxilliped; *l*, first pereopod; *m*, second pereopod; *o*, F first pleopod; *p*, second pleopod; *q*, M first pleopod; *r*, second pleopod; *s*, appendix masculina; *t*, telson; solid line = 10 mm, broken line = 1 mm.

Maxillule (f). Distal endite curved, with terminal rows of sharp teeth and setae; proximal subtriangular in outline. Endopod with a few subapical setae on inner side.

Maxilla (g). Scaphognathite with anterior lobe narrow and longer than posterior. Distal endite unequally bilobed, proximal reduced to single ovate lobe, but a lamina along edge outside has an upper lobe or papilla and fringe of setae. Endopod broad at base narrows distally and turns inward.

First maxilliped (h). Distal endite subtriangular, wide, edge concave; proximal much shorter thickened at edge. Endopodal palp or central lobe divided into three parts and twisted distally, lying close to rectilinear exopod which curves inward distally. Epipod unequally bilobed.

Second maxilliped (i). Leglike but compressed; distal segment joined diagonally along edge of penultimate segment; exopod slender about as long as endopod; epipod with podobranch.

Third maxilliped (k). Leglike, five segments, distal with strong apical spine; exopod slender, about 0.5 endopod; epipod narrow, straplike, with terminal part fleshy and at right angles, hooks at angle and sharp spine proximally.

Pereopods (l,m), all with exopods: I–III with epipods, podobranchs and setobranchs, IV with very short epipod and no podobranch, V without epipod; I and II chelate, II more slender, slightly longer, merus with distal spine; III–V longer than others, with series of spines on ischium and merus, dactyls of III and IV long, sharp; dactyl of V short, setose, with curved distal spine.

Pleopods: female I (o) with short setose endopod about 0.3 exopod; female II (p) with endopod longer and with appendix interna with a distal patch of hooks. Male I (q) endopod with curved appendix with apical patch of hooks; male II (r) with appendix interna with distal patch of hooks and masculina (s) with long slender spines on both sides.

RANGE OF DISTRIBUTION

From the North Atlantic as far north as Davis Strait and Iceland, southwards to about 13° S Latitude including the Mediterranean. Also in the Indian and Pacific oceans south of 32° S and 57° S, respectively. Depths of 500–1 650 m and temperatures of 3–12° C (Sivertsen and Holthuis 1956).

Distribution records in the area of reference are shown in Fig. 30.

BIOLOGY

Lengths to 24 mm cl in males and 31 mm cl in females.

Specimens from just south of the Grand Banks near the Gulf Stream had substantial numbers of females ovigerous and with eyed embryos in eggs during February. Also other northern collections had high percentages potentially ovigerous in autumn indicating that reproductive potential of the species is high in these areas. Dr. Chace's (1940) findings that reproductivity was comparatively low farther south and suggestion that this species was a more northerly form than *A. purpurea* is supported by our data. The *Michael Sars* Expedition collections also confirmed this opinion (Sivertsen and Holthuis 1956).

Stomach contents included fragments of chaetognaths and crustaceans (including euphausids and copepods) and small fishes (Squires 1965a; Sivertsen and Holthuis 1956).

Other decapod crustacean species taken in hauls with this species were *Sergestes arcticus, Pasiphaea tarda, Bythocaris payeri, Pandalus borealis, Pandalus propinquus, Pontophilus norvegicus, Munida tenuimana , Munidopsis curvirostra* (Squires 1965a) and other oplophorids such as *Systellaspis debilis* and *Oplophorus spinosus*.

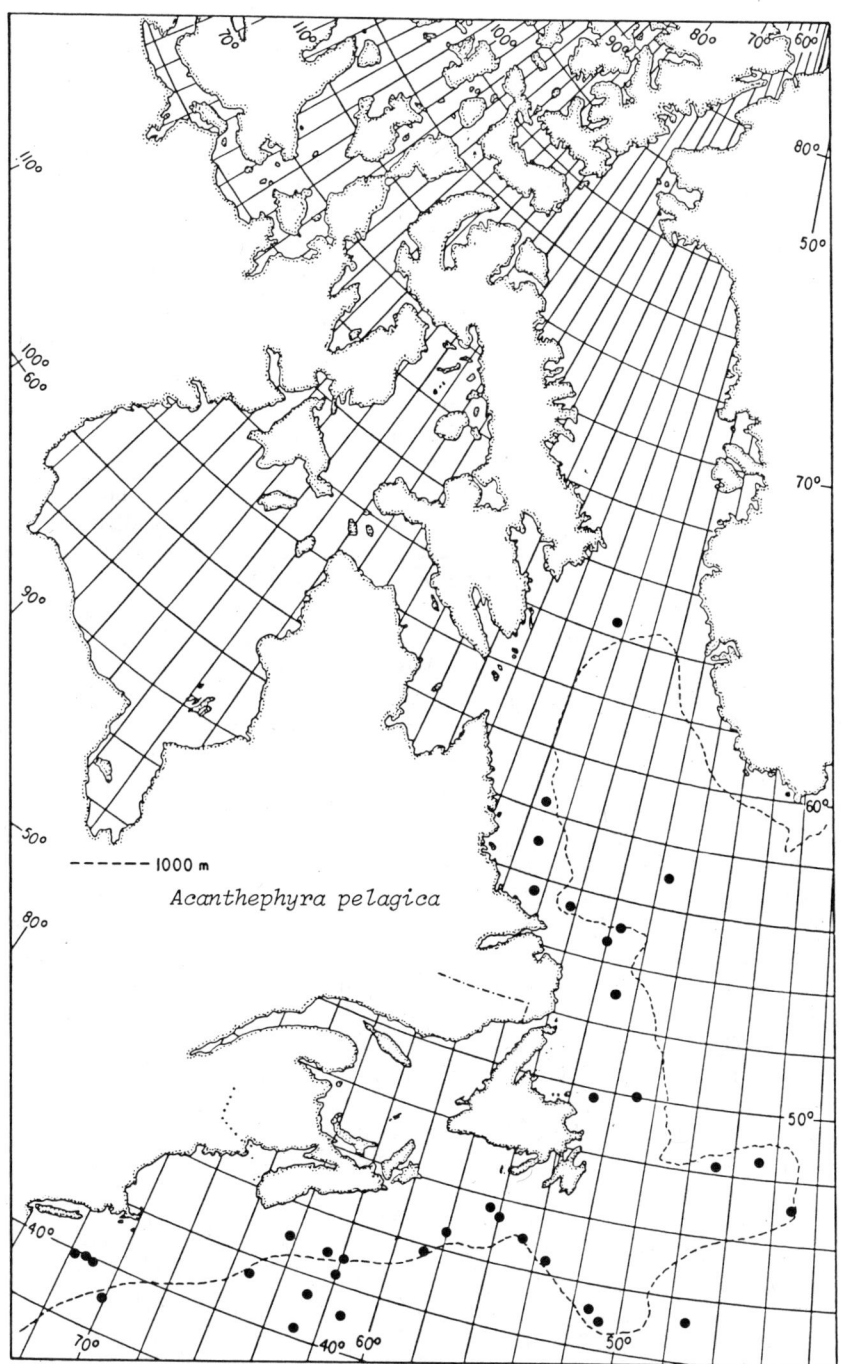

Fig. 30. *Acanthephyra pelagica*, distribution records in the area of reference.

Acanthephyra purpurea A. Milne-Edwards, 1881
Chace 1940: 134, figs. 11-17; 1986: 8;
Sivertsen and Holthuis 1956: 6, figs 2–6.
(Figures 31 and 32)

DISTINGUISHING CHARACTERISTICS

In comparison with *A. pelagica* this species is smaller, has only 4 pairs of dorsolateral spines on the telson and has a small or no tooth at the posterior median carina on the 4th abdominal somite. Its rostrum is proportionately longer, however, and teeth about 8/6, and the telson is only faintly sulcate above proximally. Appendix masculina with about 5 spines apically, 7 spines on one side and 18 on the other. Scale of antenna longer than carapace.

DESCRIPTION

Integument firm, apparently smooth but microscopically scaly in appearance with numerous pits, hollows and rugosities. Colour brilliant scarlet red (Chace 1940).

Rostrum longer than carapace and straight but rising slightly from carapace at an angle of about 5 degrees; dorsally about 7–10 strong teeth and ventrally about 5 similar in size, 7–10/5; laterally two carinae, the upper merging with median carapacial carina and the lower confluent with orbit; initial spine of rostral series smaller than others. Branchiostegal spine strong, with a short carina; a strong depression extends downward past the carina of the branchiostegal spine in the hepatic area. Two submarginal carinae along ventral edge of carapace.

Abdomen compressed, with middorsal carina on 2nd to 6th somites, projecting as a spine posteriorly on 2nd, 3rd, 5th and 6th somites; pleura rounded and with short reddish setal fringe; Telson (t) slightly rounded dorsally (deeply hollowed ventrally), with faint proximal sulcus and four pairs of dorsolateral spines, tapering to subtruncate tip with central spine and three pairs of superimposed spines laterally; outer branch of uropod longer than inner, shorter than telson, with axillary spine.

Eye moderate, with cornea larger than stalk, latter with small tubercle on inner edge.

Antennule (c) 1st article subequal to 2nd and 3rd combined, 3rd equal to 2nd; stylocerite short, spine not reaching distal end of 1st article.

Antenna (d) length of scale 4 times width, tapered to narrow distal blade, much exceeded by distolateral spine; 1.2 times carapace length; strong ventro-lateral spine on distal basal article.

Mandible (e) incisor wide with nine strong teeth, the 7th longest, plus a low series of small teeth; molar small, at corner of incisor; palp of three segments, middle longest.

Maxillule (f) distal endite narrow, expanding toward truncate tip; proximal wide, subtriangular; endopod slightly bent, uniramous.

Maxilla (g) distal endite equally bilobed; proximal reduced, ovate, with emarginate lamella outside, a rounded lobe and small papilla; endopod wide at base with narrow neck; anterior lobe of scaphognathite long with inner sloping edge, posterior axe-shaped, rounded.

Maxilliped I (h) distal endite large, subtriangular, slightly concave, proximal short with thickened edge; endopod three-segmented, not reaching curving inward corner of wide exopod (no lash); small slender bilobed epipod.

Maxilliped II (i) endopod with six segments, distal triangular-shaped with one very strong spine; exopod long slender; epipod with podobranch.

Maxilliped III (k) longer than 1st leg; five-segmented, distal long narrow with terminal sharp spine; exopod slender about as long as longest segment of endopod; epipod L-shaped with strong spine at one end and fleshy extension at other entering branchial chamber, grooming hooks at "elbow".

Pereopods: I chelate, slightly stouter and shorter than II; exopod long slender; epipod L-shaped; setobranch on coxa. II chelate, slender, palm longer than fingers; exopod long

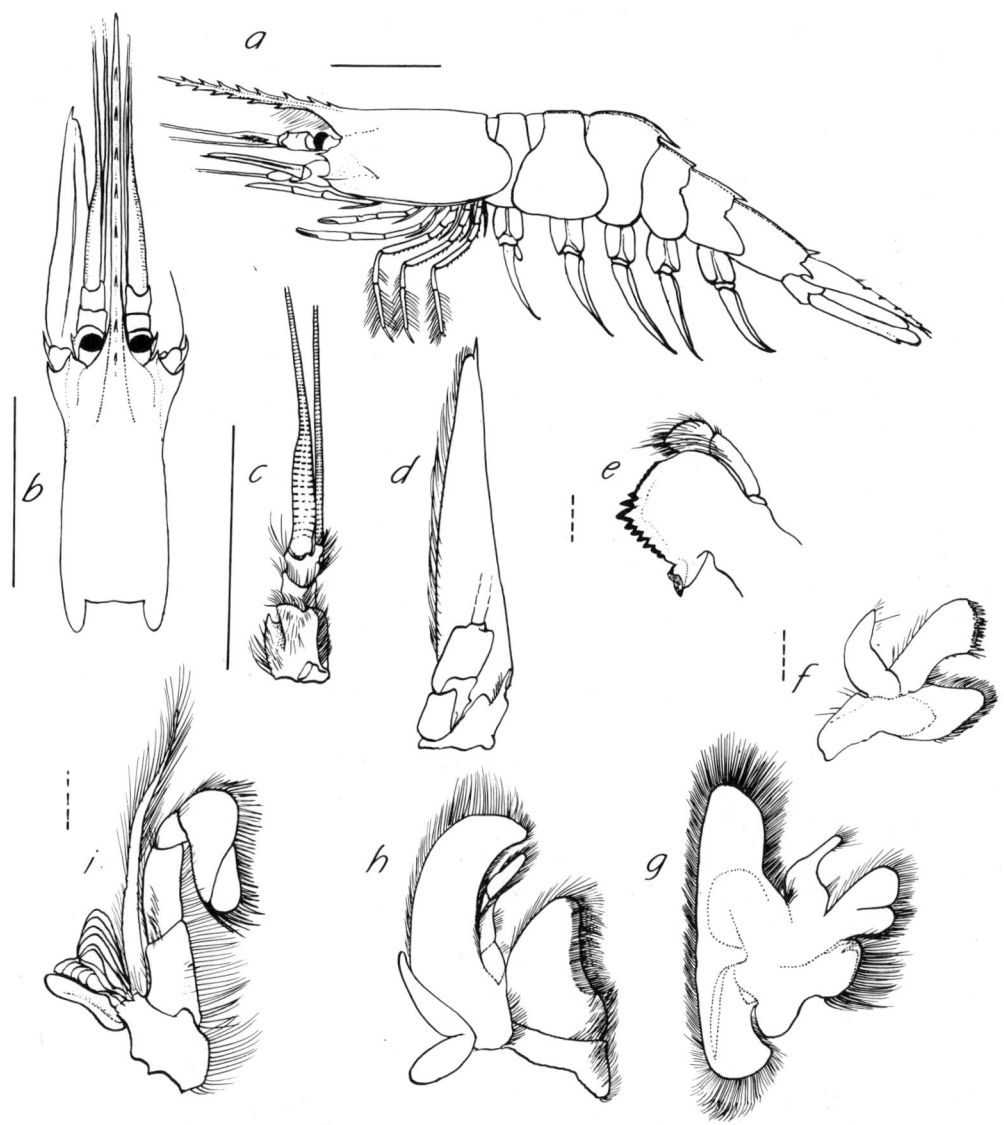

Fig. 31. *Acanthephyra purpurea*: *a*, whole shrimp from left side; *b*, carapace in dorsal aspect; *c*, antennule; *d*, antenna; *e*, mandible; *f*, maxillule; *g*, maxilla; *h*, first maxilliped; *i*, second maxilliped; solid line = 10 mm, broken line = 1 mm.

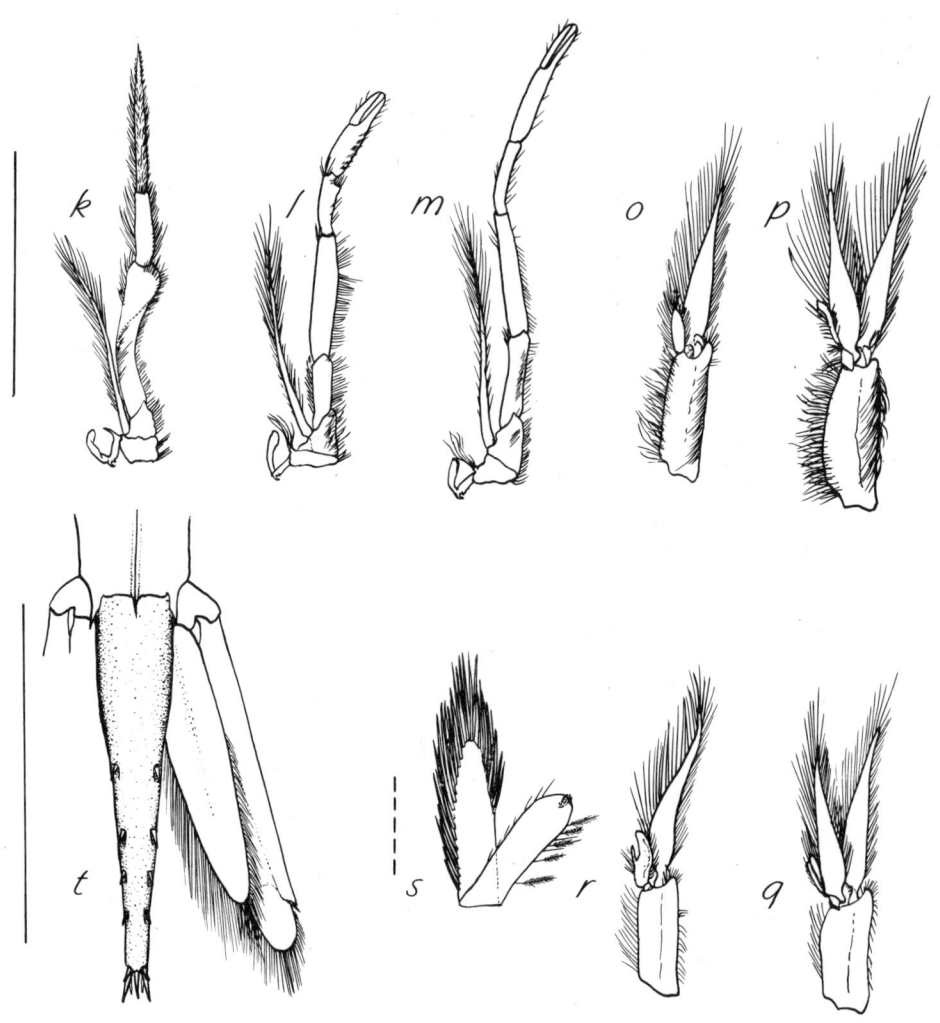

Fig. 32. *Acanthephyra purpurea*: *k*, third maxilliped; *l*, first pereopod; *m*, second pereopod; *o*, F first pleopod; *p*, F second pleopod; *q*, M first pleopod; *r*, M second pleopod; *s*, appendix masculina; *t*, telson; solid line = 10 mm, broken line = 1 mm.

slender; epipod L-shaped. III–V long and slender, series of spines on ischium and merus, all with exopods, no epipod on V; dactyls of III and IV slender, of V short with brush setae.

Pleopods: female I (o) endopod very much shorter than exopod and fringed with long plumose setae; female II (p) endopod and exopod subequal, appendix interna long and sinuous with apical hooks. Male I (q) endopod very short, with appendix interna at middle, laterally; male II (r) appendix interna slightly shorter than appendix masculina, latter (s) with about 5 long spines apically continuous with 7 laterally on one side and 18 spines on the other.

RANGE OF DISTRIBUTION

Recorded as far north as 53° N. Latitude and south to 20° N in the North Atlantic and not in the Mediterranean (Sivertsen and Holthuis 1956). Depths of 100–1 500 and temperatures of 4–18° C (Sivertsen and Holthuis 1956).

Distribution records in the area of reference are shown in Fig. 33.

BIOLOGY

Some specimens examined from south of the Grand Banks near the Gulf Stream were ovigerous in February. Chace (1940) noted that the smallest ovigerous females were only 10 mm cl in collections near Bermuda. In the southern area the proportion of young were fairly uniform from April to September possibly indicating that reproduction occurs throughout the summer at least or year round in the tropical or deep-sea stable environment (Dunbar 1960).

Stomach contents included crustaceans (from copepods to larger shrimp-like forms), pteropods, worms, radiolarians and fragments of blackish fish (Chace 1940).

Size to 23 mm cl in males and females combined (Sivertsen and Holthuis 1956); males to 19 mm, largest in specimens examined (Chace 1940).

Gennadas elegans, *Sergestes arcticus* and *Oplophorus spinosus* occured commonly in catches with this species.

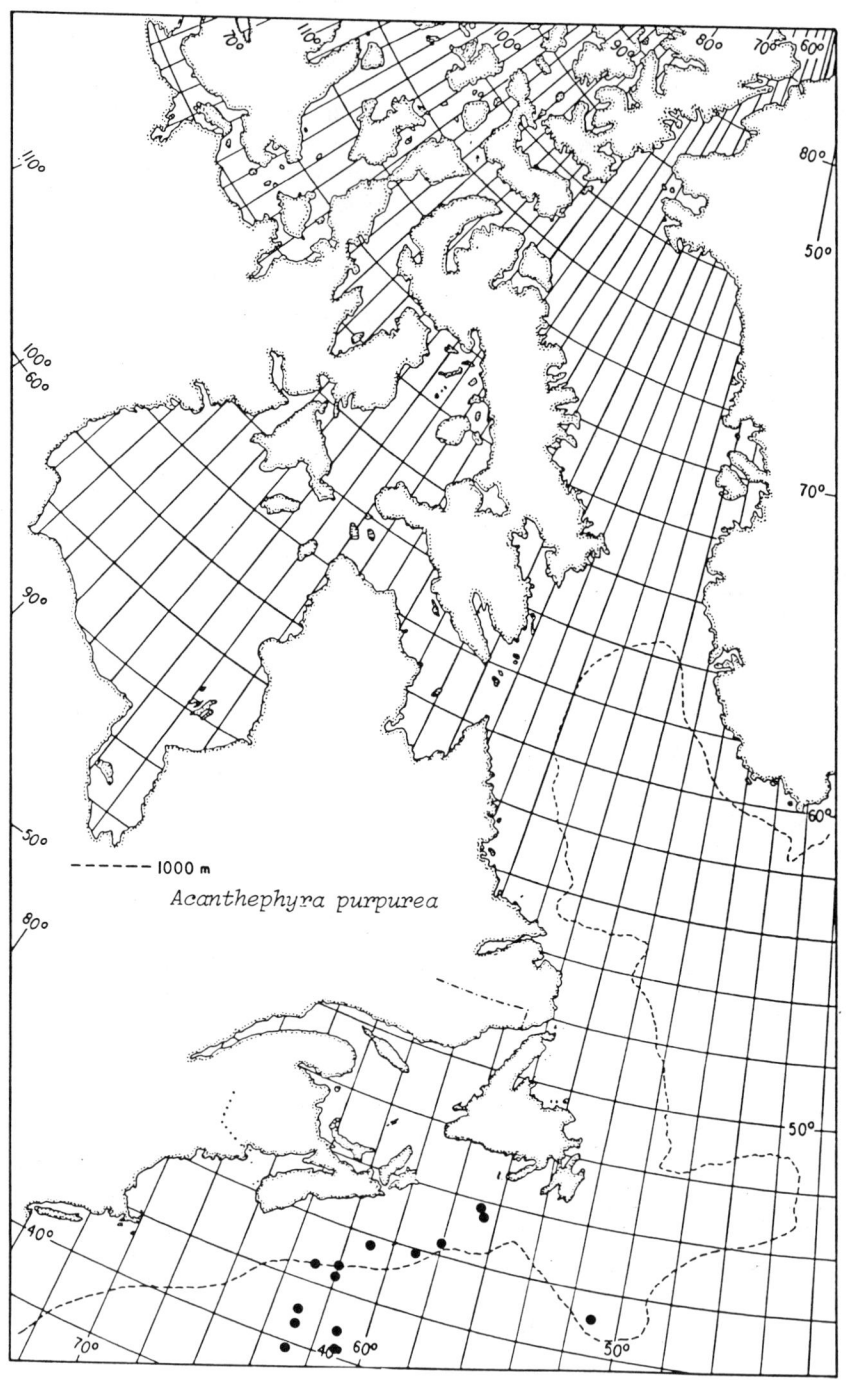

Fig. 33. *Acanthephyra purpurea*, distribution records in the area of reference.

Genus *Hymenodora* G. O. Sars, 1877
Butler 1980: 68.

Body rounded with thin soft integument microscopically pitted and scaly in appearance. Rostrum short and sharply pointed, triangular in cross-section, lateral carinae confluent with rim of orbits, and a dorsal carina continued on the carapace as a crest with 5–7 small spines. An orbito-gastric groove extends obliquely backward from the orbit to hepatic area. Antennal spine small, branchiostegal very small and without carina. Ventro-lateral edge of carapace with submarginal carina, minutely serrate anteriorly, and thin skirt.

Only one of the four species known has been recorded in the area of reference.

Hymenodora glacialis (Buchholz, 1874)
Pasiphae glacialis Buchholz, 1874.
Hymenodora glacialis, Rathbun 1929;
Butler 1980; Chace 1986.
(Figures 34 and 35)

DISTINGUISHING CHARACTERISTICS

A small shrimp with a soft, thin and fragile shell; carapace and abdomen rounded; short pointed rostrum with a dorsal carina continued on the carapace as a crest with 5–7 spines; lateral carina confluent with orbit; an orbito-gastric groove extending obliquely backward from the orbit to hepatic area branches toward gastric and branchial areas: between these branches is a short crescent-shaped groove forming the other side of a diamond laterally on the carapace. Antennal and branchiostegal spines small and without carinae. Eye with cornea smaller than stalk and poorly pigmented. Second maxilliped without podobranch.

DESCRIPTION

Integument thin, soft, and appears to be covered with small translucent imbricate scales. Colour blood red (Rathbun 1929). Rostrum short and sharply pointed, triangular in cross-section, lateral carina confluent with orbit, dorsal carina continued on the carapace as a crest with 5–7 small spines (3 behind orbit).

Carapace anteriorly pointed to meet the short rostrum; an orbito-gastric groove and gastro-orbital ridge extending backward from the orbit, the former joining a branchial groove below extending upward and also a short upward-trending gastric groove. Antennal and branchiostegal spines small, without carinae; ventro-lateral edge of carapace with submarginal carina minutely serrate anteriorly, and a thin skirt wider anteriorly.

Abdomen with all pleura and terga rounded; 6th somite twice as long as 5th but shorter than telson. Faint transverse sulcus dorsally on first somite and an antero-lateral point on the pleuron; a faint ridge on 4th and 5th pleura laterally at the level of the articulation points. Telson (t) moderately wide and with a median sulcus proximally, tapering to a subtruncate tip with about 5 pairs of very small dorsolateral spines, and terminally small corner spines above a long pair of lateral spines and a shorter central pair with a small central spine.

Eye small with rounded cornea but little pigment and comparatively large surface cells (ommatidia), an inner tubercle distally on eyestalk which is wider than cornea.

Antennule (c): peduncle reaches less than half the antennal scale; stylocerite narrow, sharp pointed, attached laterally to 1st article which exceeds it by one-fifth, other articles subequal.

Antenna (d): scale concave at outer edge; disto-lateral spine small about even with blade; peduncle about half scale.

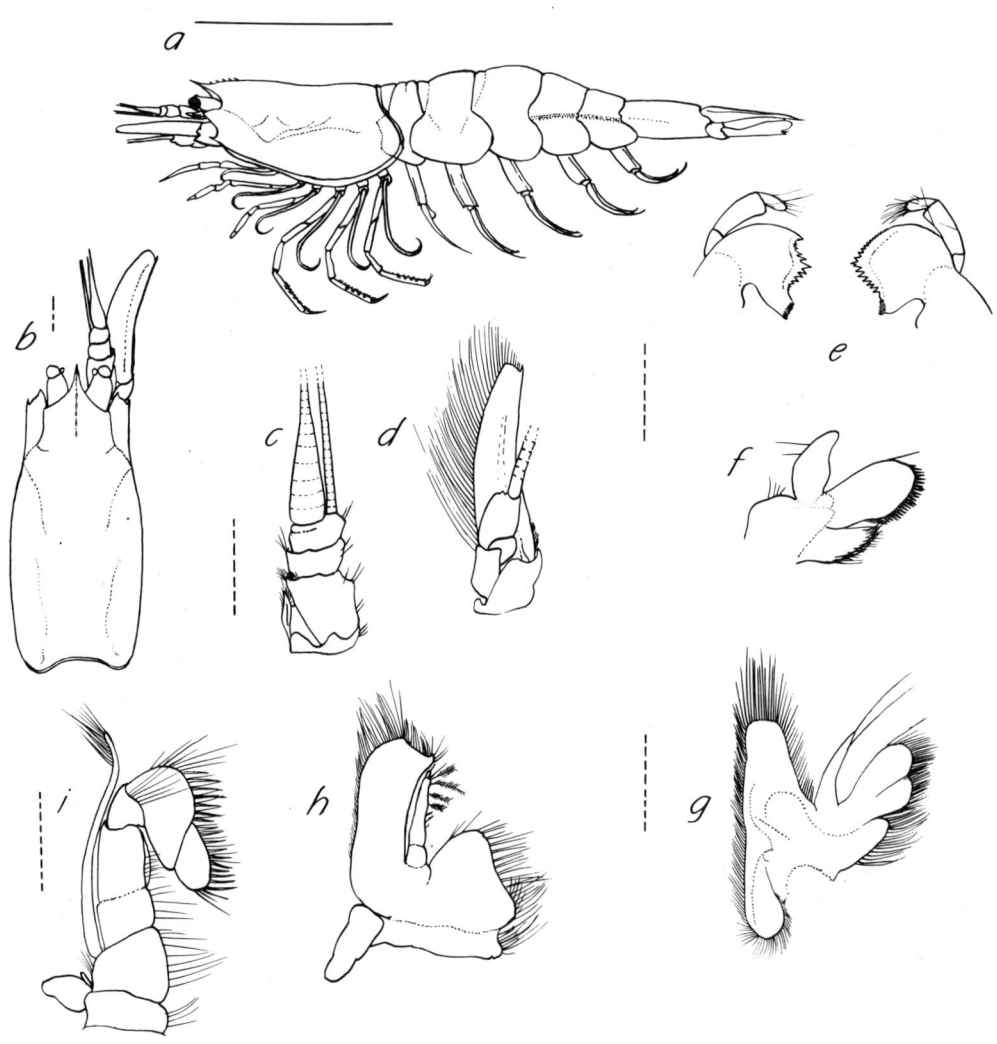

Fig. 34. *Hymenodora glacialis*: *a*, whole shrimp from left side; *b*, carapace in dorsal aspect; *c*, antennule; *d*, antenna; *e*, mandible; *f*, maxillule; *g*, maxilla; *h*, first maxilliped; *i*, second maxilliped; solid line = 10 mm, broken line = 1 mm.

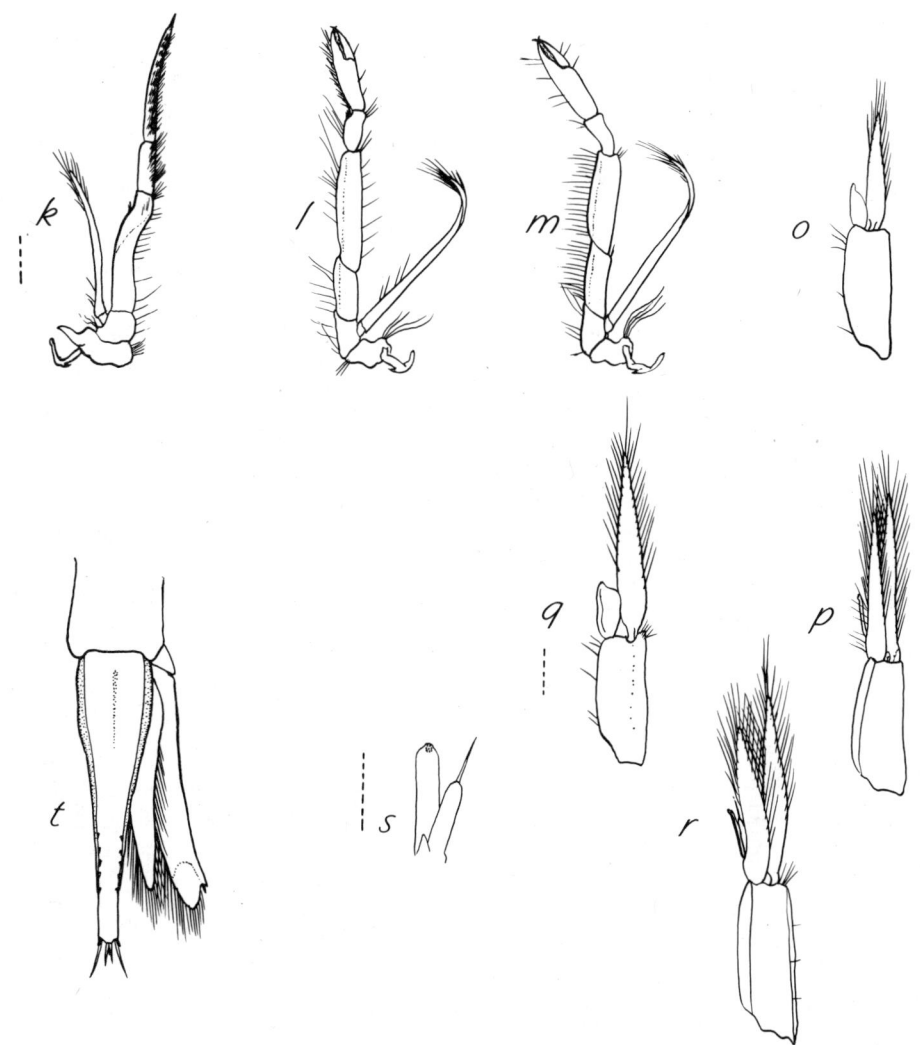

Fig. 35. *Hymenodora glacialis*: *k*, third maxilliped; *l*, first pereopod; *m*, second pereopod; *o*, F first pleopod; *p*, F second pleopod; *q*, M first pleopod; *r*, M second pleopod; *s*, appendix masculina; *t*, telson; broken line = 1 mm.

Mandible (e): right and left dissimilar: left incisor pointed, with 12 teeth to largest central tooth above and 6 below, right with only 6 above (one at corner large) and 5 below. Molar small, oval at crown, with serrate edge. Palp 3-segmented.

Maxillule (f): distal endite subovate with edge rows of short strong spines and setae; proximal endite almost as wide but pointed and with curved setae; non-branching endopodal palp with few sub-apical setae.

Maxilla (g): distal endite subequally bilobed, proximal reduced to an oblique oval with long curved setae; endopod long and tapering with apical seta; scaphognathite with posterior lobe rounded shorter than anterior, with short setae.

First maxilliped (h): Endite sub-triangular lower lobe small, central lobe or palp with faint proximal suture and very small distal segment, slightly shorter than exopod, the latter widening distally and obliquely truncate with corner a probable incipient lash. Epipod short, thick, rounded distally.

Second maxilliped (i): pediform, compressed, distal segment longer than wide; exopod long and slender; epipod short and wide, without podobranch but short small cylindrical lobe outside.

Third maxilliped (k): leg-like, five segments, a single distal spine; exopod about as long as longest segment; epipod a short rounded projection with an attached curved arm with grooming hook at elbow.

First and second pereopods (l,m): chelate; ischium and merus compressed with thinner lower edge; exopods long and slender; epipods armlike, distally extending into the branchial chamber and with hook at elbow.

Third to fifth pereopods: moderately longer than first two, all with exopods; dactyl long and sharp in III and IV and short and rounded with few apical setae and a curved spine in V, propodus with lateral setae and distal spine.

Pleopods: female I (o) endopod is short with a terminal projection without hooks; female II (p) endopod subequal to exopod, appendix interna slender with subapical patch of hooks. Male I (q) endopod is short and somewhat expanded distally with inner corner projecting slightly with patch of hooks; male II (r) endopod subequal to exopod, and appendix interna longer than appendix masculina, the latter (s) with only one apical seta.

RANGE OF DISTRIBUTION

Arctic region in both eastern and western North Atlantic southward to at least 30° N latitude, and South Atlantic off Argentina and Tristan da Cunha; also from Arabian Sea, southern Indian Ocean and eastern Pacific off Oregon and Panama; mesopelagic and bathypelagic (Chace 1986). Butler (1980) reports it from off British Columbia. Depths from superficial water layers to 3 900 m (Sivertsen and Holthuis 1956).

Distribution records in the area of reference are shown in Fig. 36.

BIOLOGY

Size to 12 mm cl (Sivertsen and Holthuis 1956). Specimens examined were: male 10 mm and female 8 mm cl, the latter nonovigerous in March.

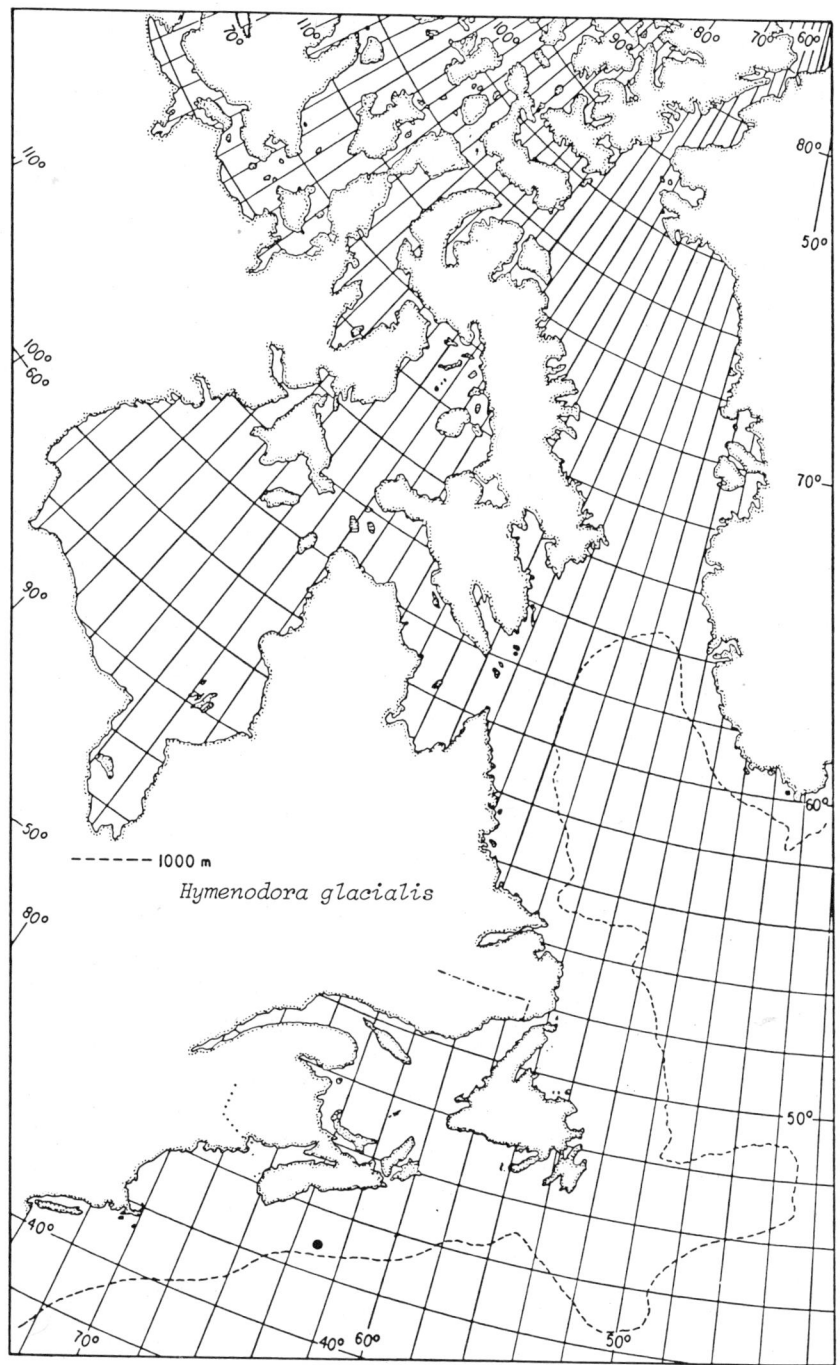

Fig. 36. *Hymenodora glacialis*, distribution records in the area of reference.

Genus *Notostomus* A. Milne-Edwards, 1881
Chace 1986

Carapace deep with high dorsal serrate ridge, toothed along entire length and with longitudinal carinae along lateral surface; also with an oblique carina joining lateral carinae just behind hepatic area. Rostrum with more teeth above than below, larger on rostrum and on arch above it than on rest of mid-dorsal carina. Small antennal and large branchiostegal spine, the carina from the latter continuous with the sub-hepatic carina. Abdomen carinate middorsally on all somites and with carina produced as a spine on 3rd to 6th somites; palp of first maxilliped with three divisions; 2nd maxilla with submarginal lamella forming rounded and pointed lobes outside proximal endite.

Notostomus elegans A. Milne-Edwards, 1881
Chace 1986: 56, fig. 30; Crosnier et Forest 1973: 56, figs. 15 and 16a, b.
(Figures 37 and 38)

DISTINGUISHING CHARACTERISTICS

Carapace with high mid-dorsal carina with numerous small teeth almost reaching posterior edge, also continued but slightly larger on the rostrum above and below. Rostrum long with two lateral carinae at base, upper short, lower not confluent with orbit but above it and ending just above anterior end of gastro-orbital carina which reaches the posterior edge. Just below it at the posterior edge the infra-gastro-orbital carina is longer than in *N. robustus*; an infra-subhepatic carina is between the subhepatic and submarginal carinae (lacking in *N. robustus*). Merus of pereopod II with 4 strong spines disto-laterally on flexor edge; appendix masculina rounded distally and with only a few apical spines.

DESCRIPTION

Integument thin, smooth and shiny and flexible but firm and with some rugosity (vermicular) above lateral carinae on carapace. Colour reddish.

Rostrum long with many teeth below and above continuous with a series of numerous smaller teeth along the high middorsal carina almost reaching posterior edge of carapace; rostrum with two lateral carinae, the lower intersecting with the orbital rim but passing it and ending just above the anterior end of the long gastro-orbital carina, the upper ending just above the orbit and turning slightly toward the median carina. The wide base of the rostrum forms a roof over the orbits.

Carapace deep, dorsal serrate carina arching above the orbit and almost reaching the posterior edge; antennal spine small branchiostegal large; gastro-orbital longitudinal carina begins just below the lower rostral and extends in a slightly sinuous arc to the posterior edge, it is joined about one-third the way along by the oblique post-hepatic which reaches the sub-hepatic below, the latter begins at the branchiostegal and extends to the posterior edge; below it are two other carinae, the infra-subhepatic almost as long as the submarginal which runs almost all the way along the lower infolding edge of the carapace. A shorter carina, the infra-gastro-orbital extends from the posterior edge two-thirds toward the post-hepatic and lies close to the gastro-orbital.

Abdominal somites 1 and 2 carinate, but posteriorly with a median deep notch; other somites sharply carinate each carina produced as a dorso-median spine; pleura rounded, without ventral spines, first and second with slight concavity; low lateral carina on pleura of somites 4 and 5.

Telson (t) tapering to subtruncate tip with central spine and one lateral at each corner, proximally a deep median sulcus becoming shallow distally and forming dorso-lateral ridges with four pairs of tiny spines on distal half. Inner branch of uropod about equal to telson and slightly shorter than outer branch.

Eye moderate, cornea at least as wide as stalk, a small inner tubercle distally on stalk; a small oscellus near superior lateral margin of cornea.

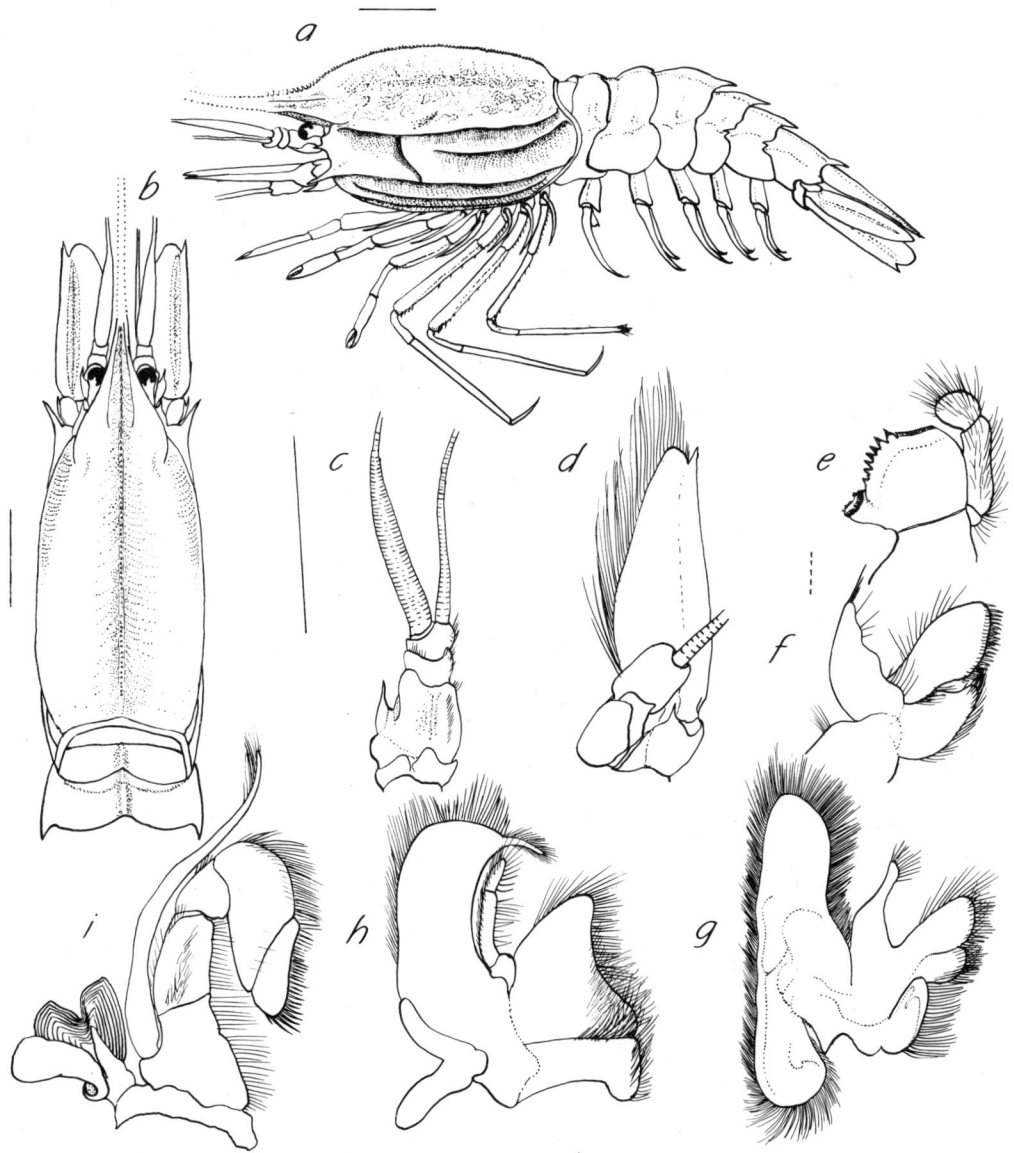

Fig. 37. *Notostomus elegans*: *a*, whole shrimp from left side; *b*, carapace in dorsal aspect; *c*, antennule; *d*, antenna; *e*, mandible; *f*, maxillule; *g*, maxilla; *h*, first maxilliped; *i*, second maxilliped; solid line = 10 mm, broken line = 1 mm.

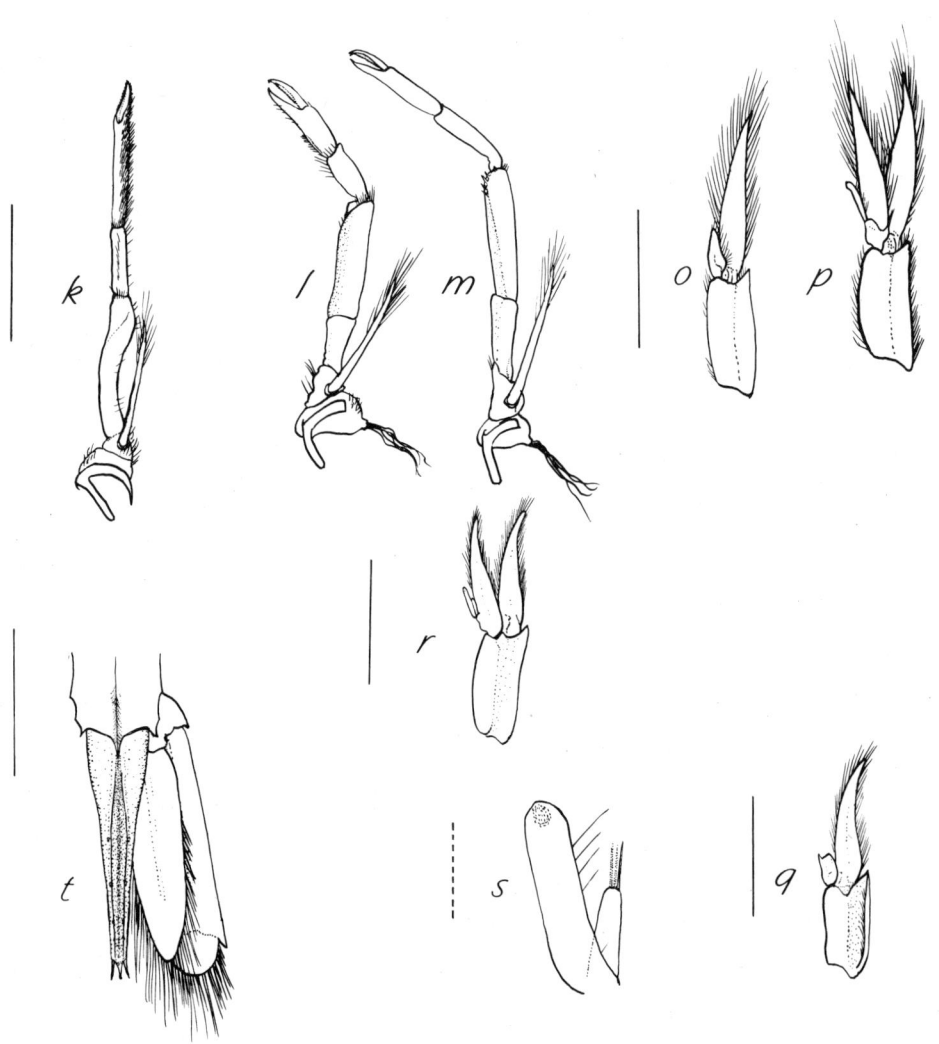

Fig. 38. *Notostomus elegans*: *k*, third maxilliped; *l*, first pereopod; *m*, second pereopod; *o*, F first pleopod; *p*, F second pleopod; *q*, M first pleopod; *r*, M second pleopod; *s*, appendix masculina; *t*, telson; solid line = 10 mm, broken line = 1 mm.

Antennular peduncle (c) with distal articles about equal, all about equal to half antennal scale; stylocerite with sharp attenuate point about even with distal edge of 1st article.

Antennal scale (d) with outer edge slightly convex, deep longitudinal sulcus dorsally, spine about even with blade, length about 3.8 times width (much shorter than carapace); peduncle short; base with strong ventral spine.

Mandibles (e) slightly dissimilar: left incisor with low serrate anterior edge about equal to edge with teeth, right similar but with a tooth at anterior corner of serrate edge. Molars small, surface subtriangular with cutting edges, the right with more teeth than left.

Maxillule (f) with subequal endites, proximal pointed, endopod large uniramous.

Maxilla (g): distal endite about equally bilobed; proximal reduced, oval, outer marginal lamina with small upper lobe or papilla; endopodal palp curved; scaphognathite with larger anterior lobe inner edge slanted, posterior lobe rounded with curved setae.

First maxilliped (h): distal endite subtriangular edge concave, proximal short with thickened edge; endopodal palp or central lobe with three segments. Exopod wide, lamellate, with extension of distal inner corner as short lash. Epipod short, bilobed.

Second maxilliped (i): compressed, five segments, distal subtriangular joined diagonally to next segment, exopod slender long; epipod short with podobranch.

Third maxilliped (k): four apparent segments; distal with shovel-like, hollow and pointed apex with lateral fringe of strong short spines, segment triangular in cross-section; exopod long slender, epipod L-shaped with hook at angle, vertical fleshy endpiece and proximally a long sharp spine.

Pereopod I (l): chelate, stout; merus and ischium compressed laminate at one side; carpus distally with setose sulcus crosswise; dactyl with bifid tip over-reaching single spine of finger of propodus. Exopod long slender; epipod L-shaped with hook at angle; setobranch.

Pereopod II (m): chelate, more slender than I; merus and ischium laminate at one side; propodus longer than carpus, fingers short, dactyl bifid. Exopod long, slender; epipod L-shaped with hook at angle; setobranch.

Pereopods III & IV (a): moderately longer than II; merus and ischium with lateral row of spines, propodus very long, dactyl slender, sharp. Exopod long, slender; epipod L-shaped with hook at angle in III; rudimentary in IV; setobranchs.

Pereopod V (a): not quite as long as III and IV and slightly more slender; propodus long, dactyl only a short curved spine surrounded by a number of brush and long setae. Exopod; no podobranch or setobranch.

Pleopods: female I (o) with short pointed endopod fringed with long fine setae and with an apical patch of very few hooks; female II (p) endopod and exopod subequal, appendix interna with subapical patch of hooks. Male I (q) short sub-rectilinear with corner tooth distally which has patch of hooks. Male II (r) with appendix interna rounded and with oval patch of hooks distally, and longer than slightly tapering appendix masculina (s) with few moderate setae apically.

RANGE OF DISTRIBUTION

Known with certainty from the North Atlantic (Crosnier et Forest 1973) and the Phillipines (Chace 1986); in the west Atlantic from Gulf of Mexico, Bermuda and the Bahamas to 43° N latitude (on the Nova Scotian Banks); in the east from coast of Spain and Portugal to possibly South Africa (Crosnier et Forest 1973). Also possibly from the Pacific: the Phillipines and Indonesia and off Ecuador (Chace 1986). Depths 450–5 380 m (Crosnier et Forest 1973).

Distribution records in the area of reference are shown in Fig. 39.

BIOLOGY

Lengths of specimens examined: Male — 32 mm cl, possibly immature; Female — 41 mm cl, nonovigerous in February.

Size to 45 mm cl (Chace 1986).

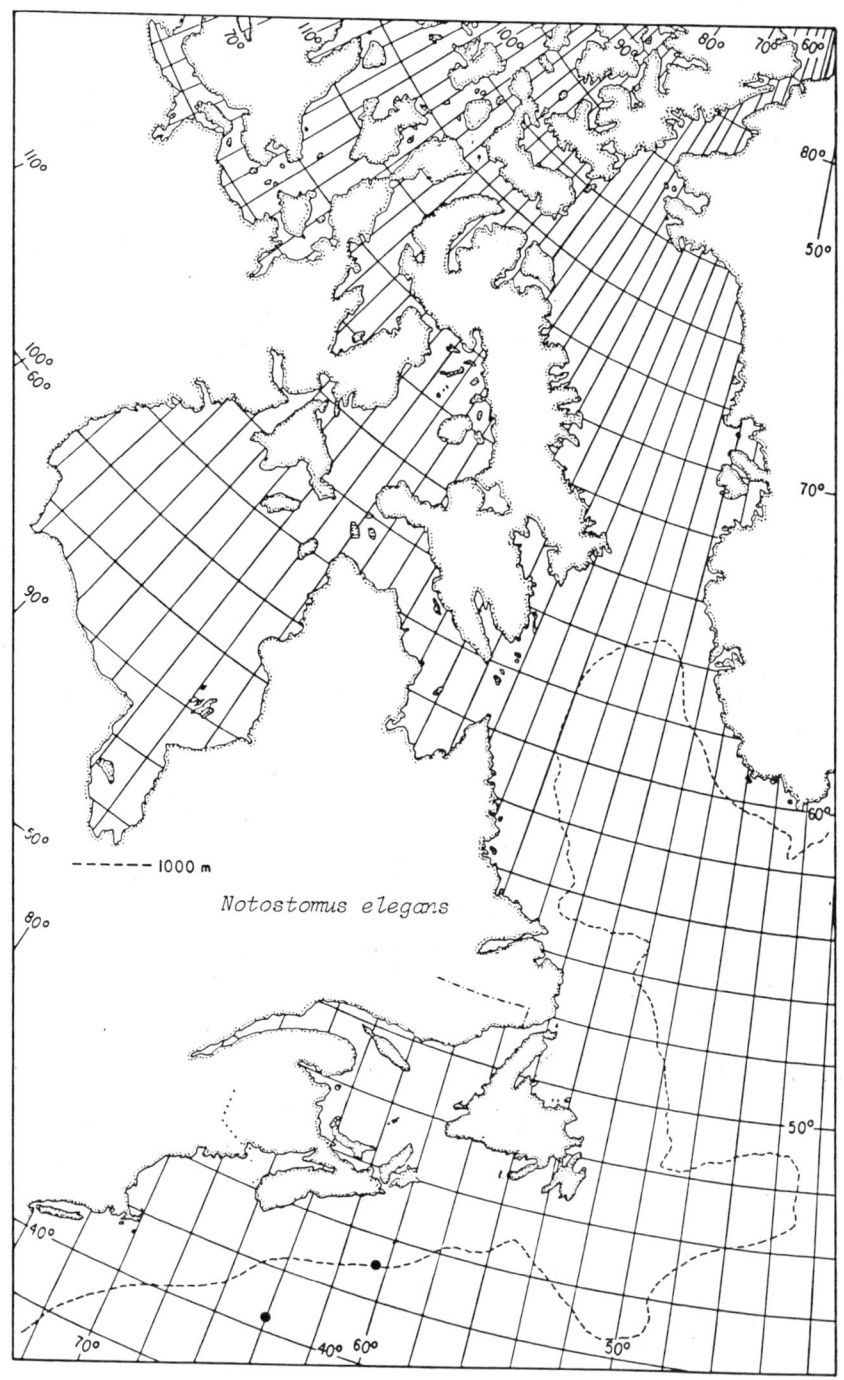

Fig. 39. *Notostomus elegans*, distribution records in the area of reference.

Notostomus robustus Smith, 1884
Chace 1940: 168–170, fig. 41, 1986: 53,
figs. 28 and 29;
(Figures 40 and 41)

DISTINGUISHING CHARACTERISTICS

Carapace with high dorsal ridge denticulate for its whole length; rostrum short and pointed with teeth above and below (about 15/4) with teeth on rostrum and arch above orbit larger than on rest of mid-dorsal carina; posteriorly carina (infra-gastro-orbital) below top one (gastro-orbital) very short; no carina between subhepatic and submarginal; appendix masculina tapering slightly to apex with about 5 long fine spines.

DESCRIPTION

Integument soft and showing some wrinkles. Colour scarlet (Chace 1940).

Rostrum short, pointed, with 2–4 ventral and about 15 smaller dorsal teeth continuous with still smaller teeth on median carina of carapace arching high above orbit and running level to descend to posterior edge; lateral carina on rostrum confluent with orbit but passing it and ending just above lateral gastro-orbital carina which descends from behind orbit to meet post-hepatic oblique carina and extends horizontally to meet posterior edge: just below this point is the short infra-gastro-orbital carina; connecting with the oblique post-hepatic carina is the horizontal sub-hepatic extending from the branchiostegal spine to the posterior edge, and parallel to it below is the weaker submarginal carina. Anterior spines are the small antennal and strong branchiostegal.

Abdomen strongly compressed with sharp middorsal carina on all somites ending in posterior spinous process on 3rd to 6th somites; all pleura rounded except for a small postero-lateral tooth on 5th and an oblique carina and sulcus on 4th and 5th. Mid-ventral tubercle on each somite and pre-anally.

Telson (t) with two dorso-lateral ridges joined proximally and a median sulcus between them; two pairs of tiny spinules on distal half; tapering to narrow tip with central spine and two pairs of superimposed lateral spines; inner branch of uropod shorter and narrower than outer which is slightly longer than telson.

Eyes moderate, cornea globular, tubercle on distal inner edge of stalk.

Antennule (c) 1st article about equal to 2nd and 3rd combined; stylocerite spine about even with distal 1st article.

Antenna (d) disto-lateral spine slightly exceeds blade; scale length about 3 times width, about half cl; peduncle less than half scale.

Mandibles (e) slightly dissimilar, left with fewer teeth on incisor and molar; palp with 3 segments.

Maxillule (f) distal endite ovate rounded at edge; proximal slightly smaller pointed; endopod large uniramous.

Maxilla (g) distal endite with proximal lobe slightly smaller, pointed; proximal endite reduced, rounded, submarginal lamella with small pointed lobe and larger rounded.

Maxilliped I (h) distal endite large, edge almost straight, proximal short with thickened edge; endopodal palp or central lobe with three segments, not as long as exopod, latter with short lash at inner corner; epipod sock-like.

Maxilliped II (i) compressed, distal endite inserted diagonally on next, wider than long; exopod long and slender; epipod with podobranch.

Maxilliped III (k) five segments, distal long with shovel-like tip fringed with strong spines; longest segment with distal tooth outside; exopod about equal to longest segment; epipod straplike with sharp spine on proximal end and fleshy part at right angles, grooming hook at "elbow".

Pereopods: I (l) chelate, moderate, shorter than 2nd, dactyl bifid; exopod short slender; epipod L-shaped, fleshy terminal part shorter than straplike part, hook at angle. II (m) more slender than I, chela smaller, dactyl bifid; exopod. III–V (a) long and

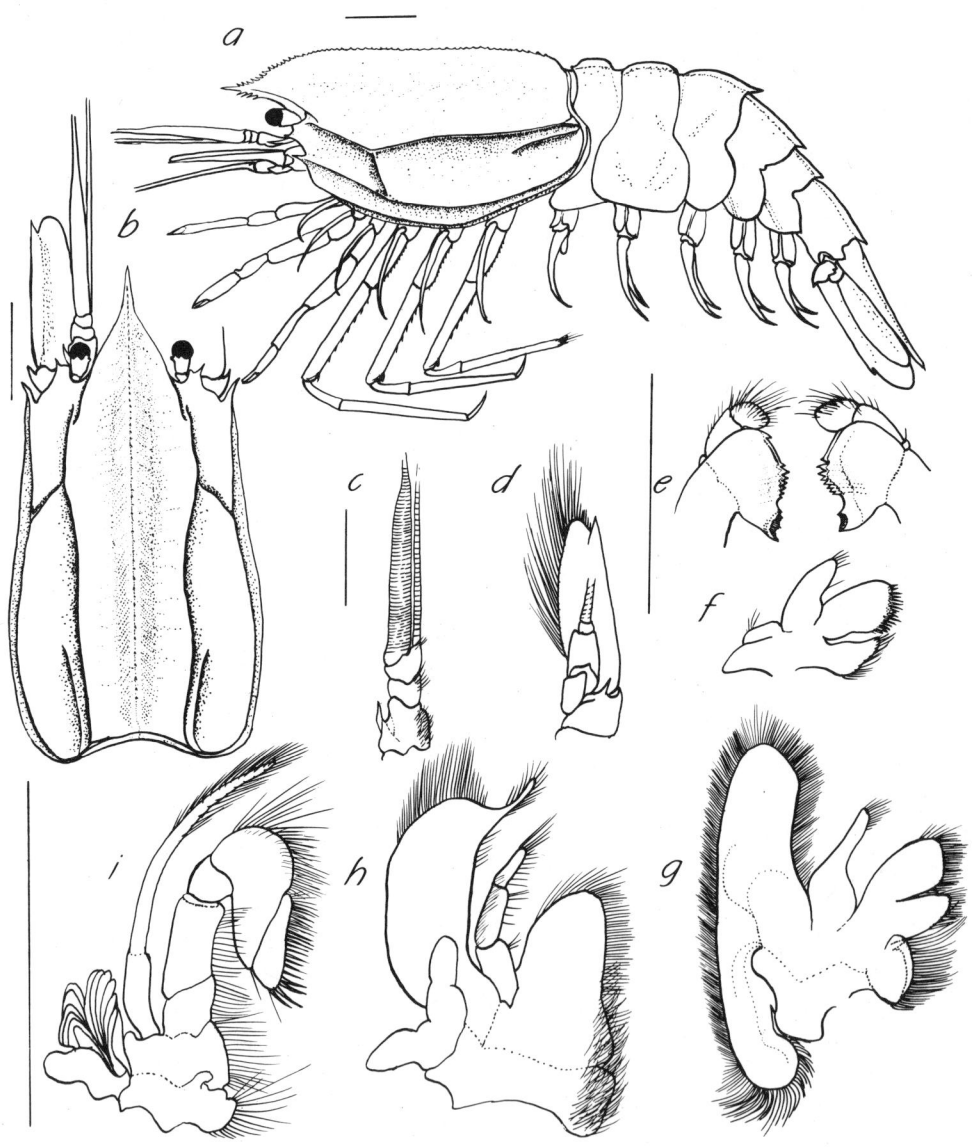

Fig. 40. *Notostomus robustus*: *a*, whole shrimp from left side; *b*, carapace in dorsal aspect; *c*, antennule; *d*, antenna; *e*, mandible; *f*, maxillule; *g*, maxilla; *h*, first maxilliped; *i*, second maxilliped; solid line = 10 mm.

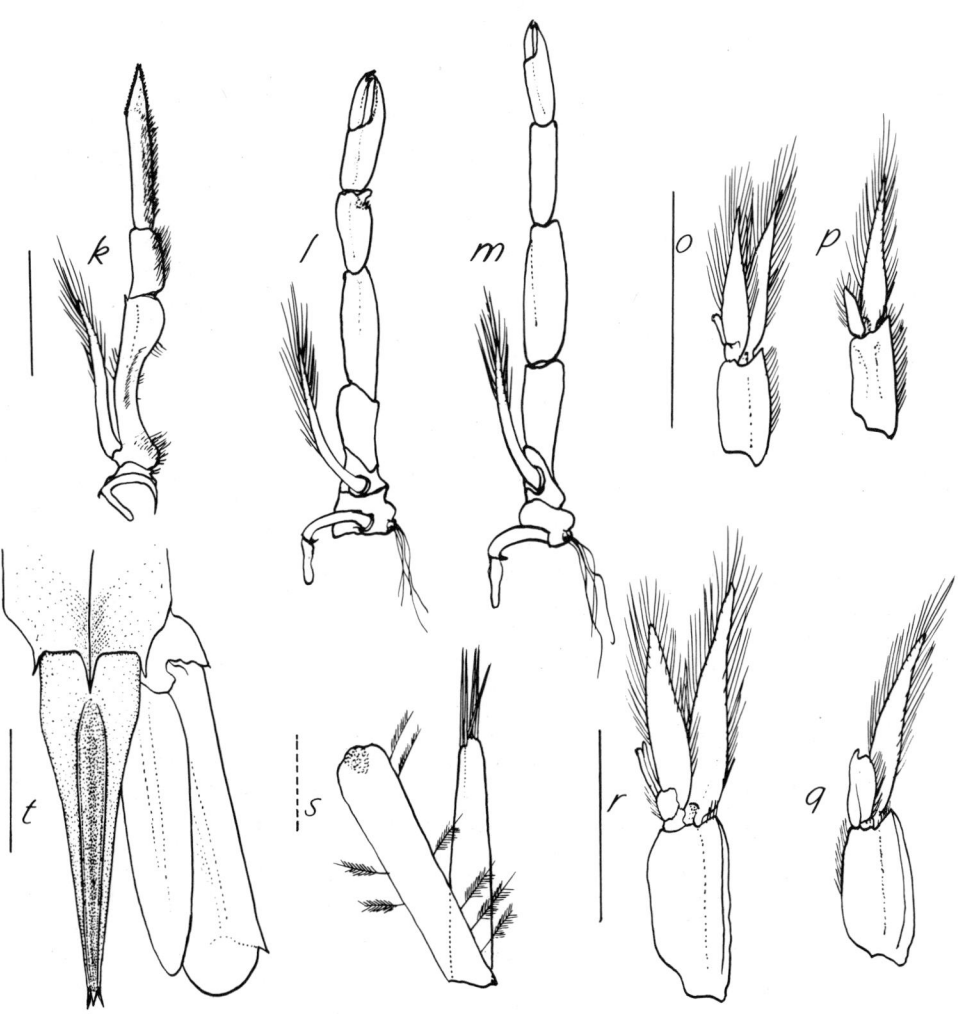

Fig. 41. *Notostomus robustus*: *k*, third maxilliped; *l*, first pereopods; *m*, second pereopod; *o*, F first pleopod; *p*, F second pleopod; *q*, M first pleopod; *r*, M second pleopod; *s*, appendix masculina; *t*, telson; solid line = 10 mm, broken line = 1 mm.

moderately stout; all with exopods; epipod on III and IV; spinules on ischium and merus, strong spine distal on merus; dactyl slender on III and IV, short and with hook and brush setae on V.

Pleopods: female I (o) endopod short about 1/5 exopod; female II (p) endopod and exopod subequal, appendix interna short, with subapical patch of hooks. Male I (q) endopod short sub-rectilinear, distal projection and notch, projection with hooks. Male II (r) appendix interna stouter but about as long as tapered appendix masculina (s) with about 5 moderate apical spines; a.i. with apical patch of hooks.

RANGE OF DISTRIBUTION

Western North Atlantic only, betwen Bahamas, Bermuda and Nova Scotian Banks (43° N lat., 60° W long.). Depths 850–3 000 m (Sivertsen and Holthuis 1956).

Distribution records in the area of reference are shown in Fig. 42.

BIOLOGY

Size to 51 mm (Sivertsen and Holthuis 1956). Specimens examined: male 37 mm cl; female 16 and 19 mm.

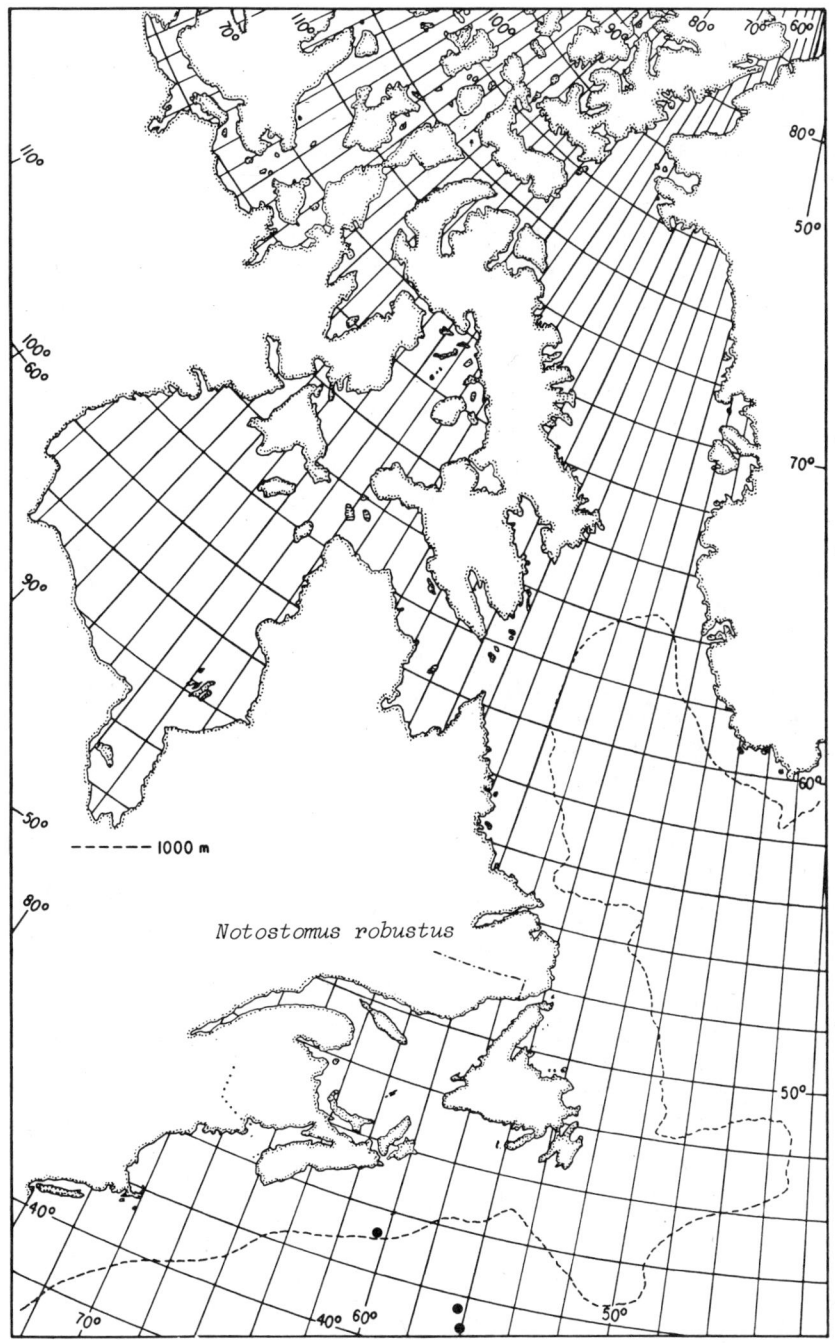

Fig. 42. *Notostomus robustus*, distribution records in the area of reference.

Genus *Oplophorus* H. Milne-Edwards, 1837
Chace 1940: 184; 1986: 57, figs. 32, 33.

Body laterally compressed; integument smooth, shiny; rostrum long, with as many as or more teeth dorsally as ventrally. Carapace not denticulate dorsally; abdomen with median carinae posteriorly produced as long spines on 3rd to 5th somites; 6th somite shorter than 5th; telson acute posteriorly; eye large, cornea at least as wide as stalk; antennal scale with lateral teeth proximal to distolateral spine; mandibles somewhat similar, incisor toothed along cutting edge; 1st maxilliped with central lobe or palp divided into 3 segments, distal small; 3rd maxilliped and 1st pereopod with wide and rigid exopods. Appendix masculina with few distal setae; eggs large and few (less than 50). (Modified from Chace 1986).

Oplophorus spinosus (Brullé, 1839)
Palaemon spinosus Brullé, 1839;
Oplophorus spinosus, Sivertsen and Holthuis 1956: 19, fig. 15, pl. III, figs, 1, 2.
(Figures 43 and 44)

DISTINGUISHING CHARACTERISTICS

As for Genus *Oplophorus* given above.

DESCRIPTION

Integument smooth, shiny. Colour whitish translucent background with reddish spots.

Carapace with strong suborbital lobe, moderate antennal and branchiostegal spines, the latter supported by a carina which continues to the posterior edge; below this carina the ventral margin lowers for posterior half and in between is a short scalloped carina with sharp points and two oval loops posteriorly; rostrum about equal to cl, slender, tapering to sharp point, with spines 5–10/5–8; at base of rostrum two lateral carinae begin, the lower continuing past the edge of the orbit and the upper extending backward and slightly converging towards the mid-dorsal for about one-third cl, forming a shallow sulcus on each side of the median carina which extends posteriorly about three-quarters cl; from the orbit a depression extends obliquely to hepatic area forming a sulcus which becomes less distinct as it rises above the branchial area. Posterior edge of carapace with prominent projection at level of anterior projection on pleuron of 1st abdominal somite.

Abdomen (a): tergum of first somite rounded, pleuron with antero-lateral projection and ventral gap; second with two dorsal carinae; 3rd to 5th somites each with sharp mid-dorsal carina produced posteriorly as long sharp spine; 6th somite shorter than 5th and without dorsal carina. Telson (t) tapering to long sharp apex and with three pairs of dorsolateral spines on posterior half; branches of uropod about equal and 0.8 telson length.

Eye large, wide cornea with ocellus, an inner and outer tubercle on eyestalk distally.

Antennule (c) with short stylocerite, about half 1st article; 2nd and 3rd articles subequal.

Antenna (d) scale rigid, stout, outer edge slightly concave and with about 14 teeth, tip sharp with inner barb on distal sixth; base with long, stout fixed ventral spine; length about 1.1 cl and 1.0 rostrum.

Mandibles (e) slightly dissimilar, curved edge of left incisor with teeth decreasing in size from centre, of right with larger tooth at upper corner; molars squarish, with hollow crown and sharp edges, more spinous in right.

Maxillule (f): endites wide, short; endopodal palp obscurely bifurcate, lower "branch" only with seta.

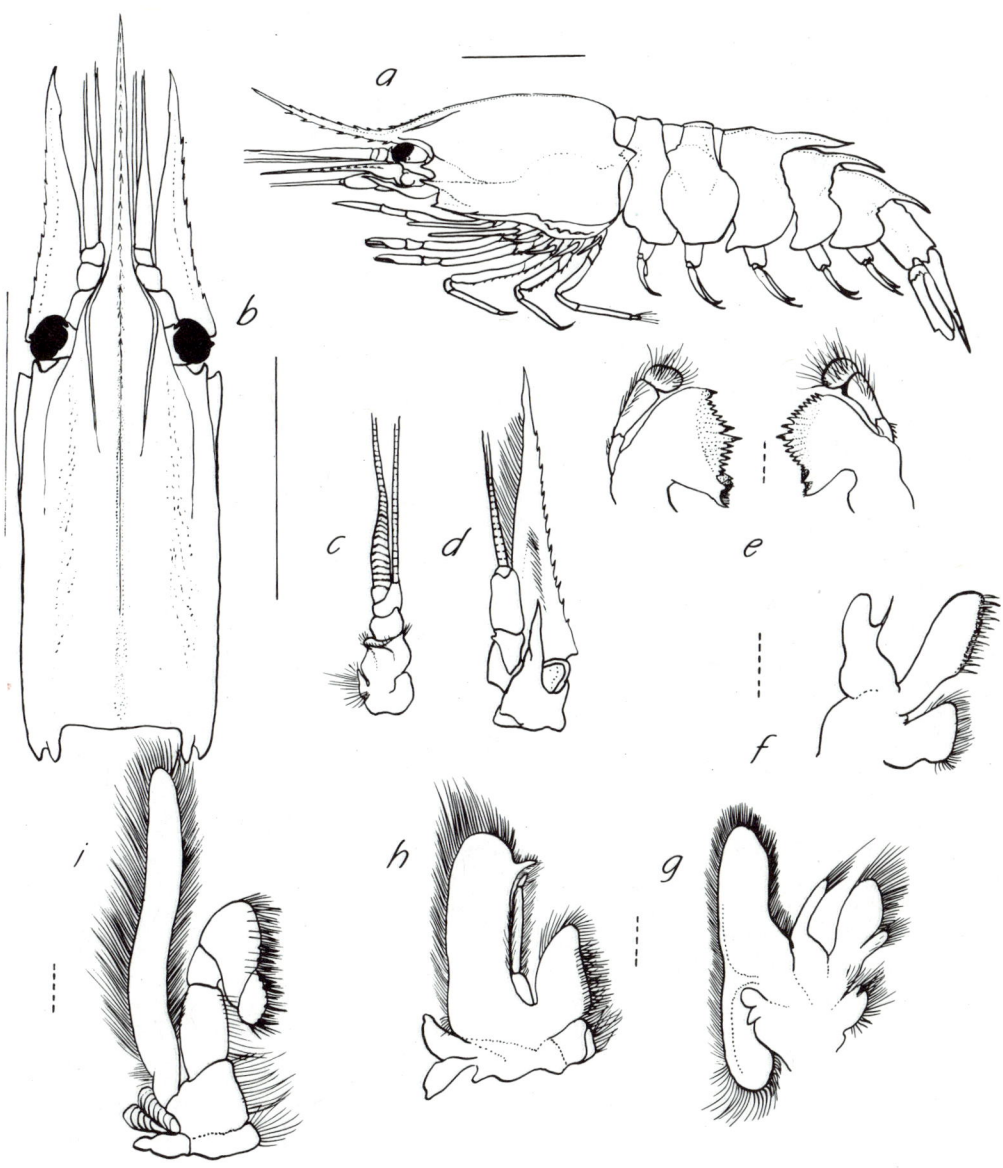

Fig. 43. *Oplophorus spinosus*: *a*, whole shrimp from left side; *b*, carapace in dorsal aspect; *c*, antennule; *d*, antenna; *e*, mandible; *f*, maxillule; *g*, maxilla; *h*, first maxilliped; *i*, second maxilliped; solid line = 10 mm, broken line = 1 mm.

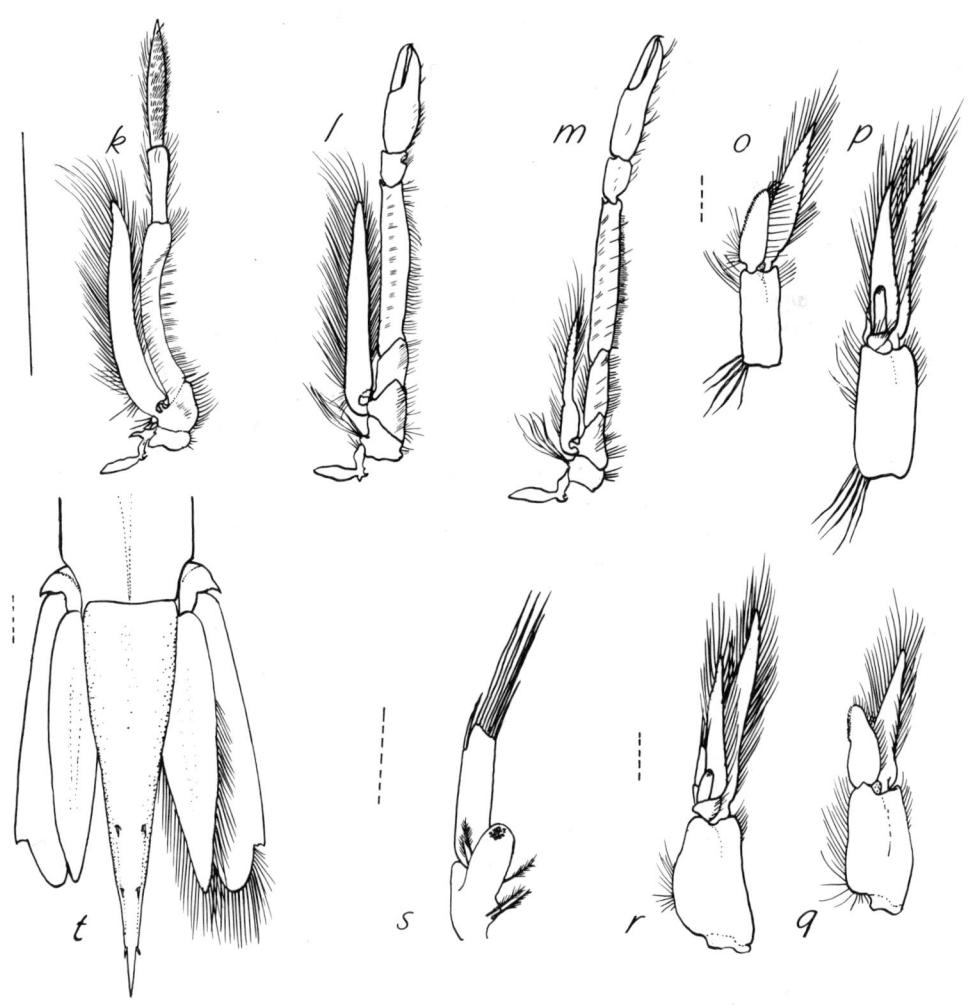

Fig. 44. *Oplophorus spinosus*: *k*, third maxilliped; *l*, first pereopod; *m*, second pereopod; *o*, F first pleopod; *p*, F second pleopod; *q*, M first pleopod; *r*, M second pleopod; *s*, appendix masculina; *t*, telson; solid line = 10 mm, broken line = 1 mm.

Maxilla (g): distal endite unequally bilobed, proximal reduced, ovate, with a small upper lobe or papilla but no apparent marginal lamella; posterior lobe of scaphognathite the smaller, rounded and with curved setae.

First maxilliped (h): central lobe with three segments, the distal small; exopod rectilinear but distally rounded and with inner corner projection an incipient lash; epipod bilobed.

Second maxilliped (i): endopod compressed; distal segment longer than wide; exopod rigid, flat, long, fringed with setae; epipod with podobranch.

Third maxilliped (k): leglike but with 5 segments, distal compressed and with about 15 transverse rows of setae; exopod rigid, flat, longer than longest segment of endopod; epipod with short rigid part with hook at end and a thick fleshy endpiece extending into branchial chamber, also proximally a strong, curved fixed spine.

First pereopod (l): chelate, moderately robust; chevrons of setae on underside of palm; exopod rigid, flat, more than half length of endopod; epipod with short rigid part with hook and thick fleshy part at right angles; setobranch.

Second pereopod (m): chelate, more slender than First; exopod not rigid, less than half endopod; epipod as in First; setobranch.

Third and Fourth pereopods (a): non-chelate, slightly longer than Second; spines on ischium and merus; dactyl short, curved and sharp; exopods short, flexible; epipod reduced on Fourth; setobranchs.

Fifth pereopod (a): non-chelate, more slender than Fourth; dactyl short, rounded and setose; exopod short, flexible; no epipod or setobranch.

Pleopods: female I (o) with endopod somewhat shorter than exopod, inner edge with hooks over more than half its length; female II (p) endopod with appendix interna compressed, about 0.3 endopod and with oval patch of hooks distally. Male I (q) with endopod about 0.6 exopod, inner edge with hooks over less than half its length. Male II (r) endopod with appendix interna short, compressed, with few plumose setae; and longer appendix masculina (s) with 3–4 long apical and one subapical setae.

RANGE OF DISTRIBUTION

In the west North Atlantic at about 46° N lat. near the Gulf Stream (55° W long.) (specimens examined), and western and eastern subtropical North Atlantic, central South Atlantic, Indian Ocean, southern Japan, off Hawaii, seamounts west of North America, and northeast of Easter Island (Chace 1986). Depths 0–1 800 m (Sivertsen and Holthuis 1956).

Distribution records for the area of reference are shown in Fig. 45.

BIOLOGY

Size to 17 mm cl in males and 14 mm in females (Sivertsen and Holthuis 1956, p. 19).

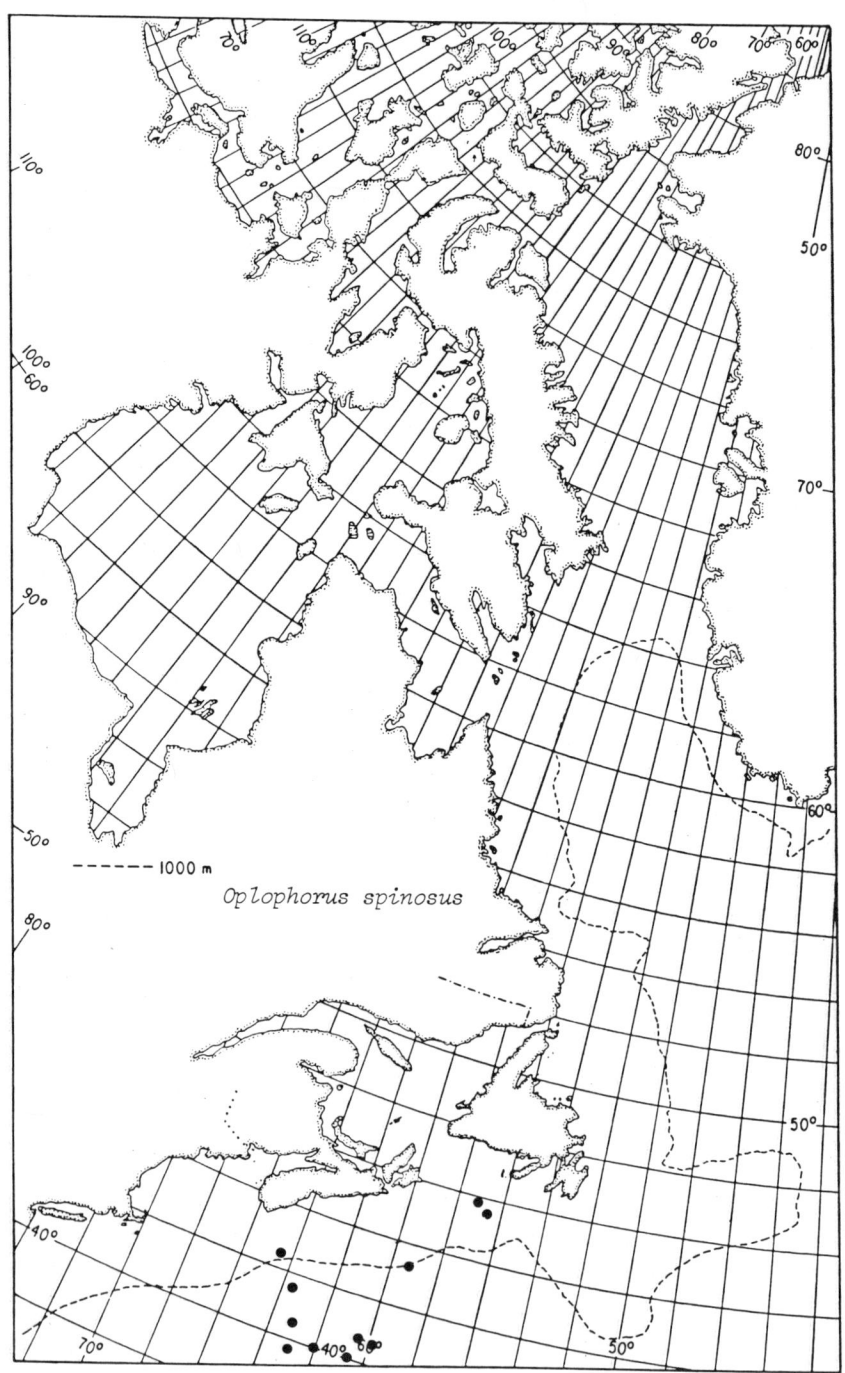

Fig. 45. *Oplophorus spinosus*, distribution records in the area of reference.

Genus *Systellaspis* Bate, 1888
Chace 1986: 61, figs 34–36.

Shell firm; rostrum with more dorsal teeth than ventral, rostral series forms a crest over the orbit; eye with cornea at least as wide as eyestalk; incisor of mandible toothed along entire cutting edge; 2nd maxilla with papilla or small lobe and submarginal lamina on proximal endite; abdomen with 6th somite longer than 5th; telson posteriorly with a sharply pointed endpiece with lateral spines. Eggs large and few (less than 50).

Systellaspis debilis (A. Milne-Edwards, 1881)
Acanthephyra debilis A. Milne-Edwards, 1881.
Systellaspis debilis, Chace 1940: 181-184; 1986: 65.
(Figures 46 and 47)

DISTINGUISHING CHARACTERISTICS

Rostrum long and narrow with strong teeth above and below and forming a crest of four spines over the orbit; all legs with exopods; abdomen compressed, sharply carinate on 3rd somite dorsally, carina produced as a sharp spine over 4th; posterior mid-dorsal spines also on 4th to 6th somites and on 4th and 5th a series of denticles on each side of the spine at posterior edge; telson long and with a long sharp tail-piece with small lateral spines.

DESCRIPTION

Integument with numerous small pits, firm, shiny. Colour scarlet red with appendages tinged with salmon-orange, light organs bluish purple (Chace 1940).

Rostrum slender and usually longer than carapace (1.8 cl in specimens examined), curving upward from middle to sharp tip and with crest of 3 or 4 spines above orbit, rostral spines $9-15+4/7-8$. Carapace with suborbital lobe, antennal and branchiostegal spines, the latter with a short carina. In freshly preserved specimens purplish spots or luminiscent organs appear largest in hepatic area and also as a thick broken line along ventral edge of carapace, also a thick vertical organ extends from ventral thoracic area behind legs. A number of other small spots are scattered over the carapace.

Abdomen slightly compressed; first somite with rounded tergum and pointed lateral extension anteriorly; second somite rounded also and with lateral light organ near articulation; all other somites have a similarly placed light organ; the 4th and 5th have no dorsal carina but the third has a posterior extension of its carina as a long spine, and the 4th and 5th a shorter spine; tergum of 4th with 8 posterior denticles and of 5th with 4 posterior denticles between articulation and median spine; 6th somite has no carina but posteriorly 2 dorso-lateral spines; pleura of 5th has a posterior notch with spine. Ventrally a low spine at centre is on all somites as well as a strong anal spine.

Telson (t) is narrow and about 1.4 times length of 6th somite, tapering to a sharp endpiece which has three lateral spinules and two pairs of flanking spines one above the other; dorso-laterally are also 5 pairs of spines on distal two-thirds and mid-dorsally an oval light organ on proximal third. Outer branch of uropod longer than inner, shorter than telson.

Eyes globular with ocellus dorsally and inner and outer tubercles distally on eyestalk, also a light organ extends as a line the length of the eyestalk ventrally.

Antennule (c): peduncle reaching about half length of antennal scale; stylocerite slightly curved and carinate almost as long as 1st article; 2nd and 3rd articles subequal.

Antenna (d): scale about 0.8 times cl, length about 4.5 times width; spine exceeds scale; base with strong ventral spine; peduncle less than 0.5 length of scale.

Mandibles (e): slightly dissimilar: left incisor with about 10 large and 10 small teeth, the centre largest, joined to small molar with curved subtriangular edges and hollow crown; right somewhat similar except for larger tooth at upper corner of row of small ones.

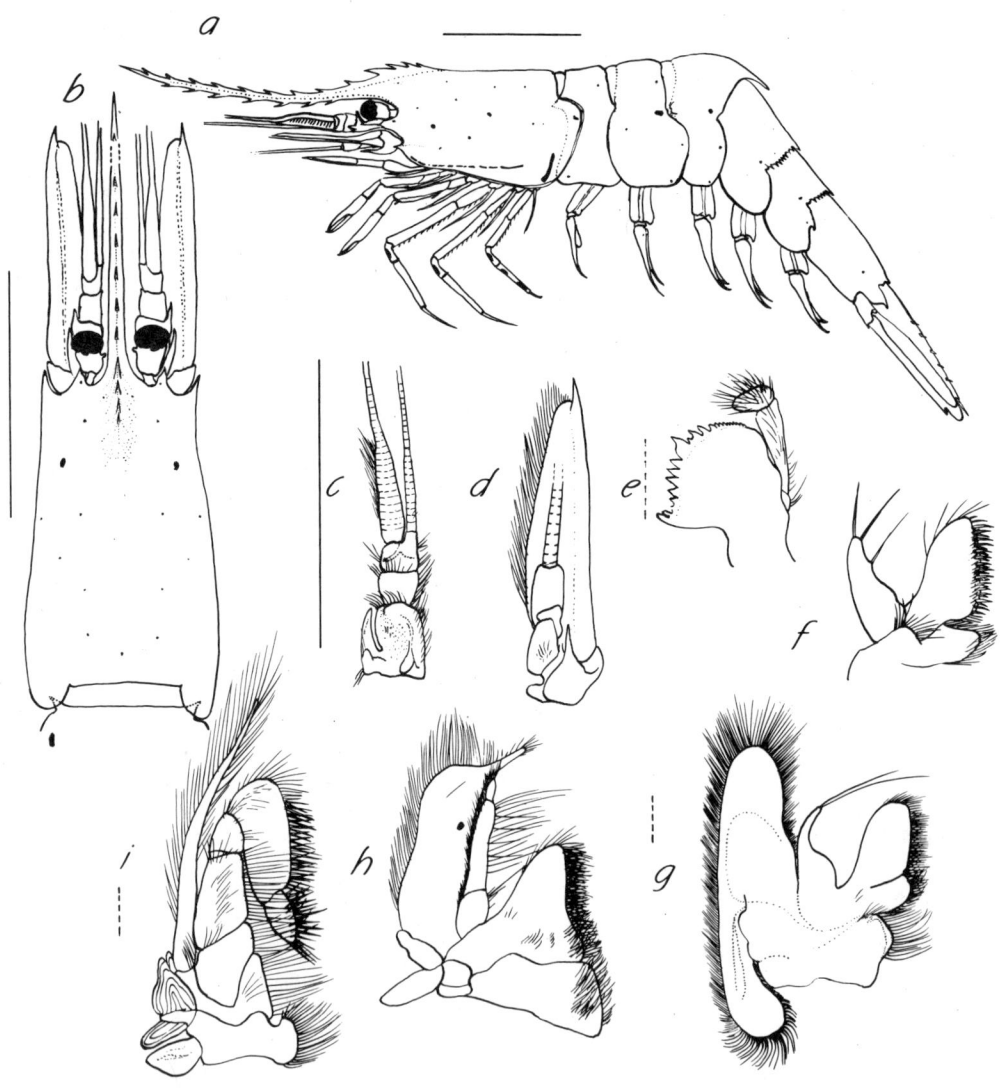

Fig. 46. *Systellaspis debilis*: *a*, whole shrimp from left side; *b*, carapace in dorsal aspect; *c*, antennule; *d*, antenna; *e*, mandible; *f*, maxillule; *g*, maxilla; *h*, first maxilliped; *i*, second maxilliped; solid line = 10 mm, broken line = 1 mm.

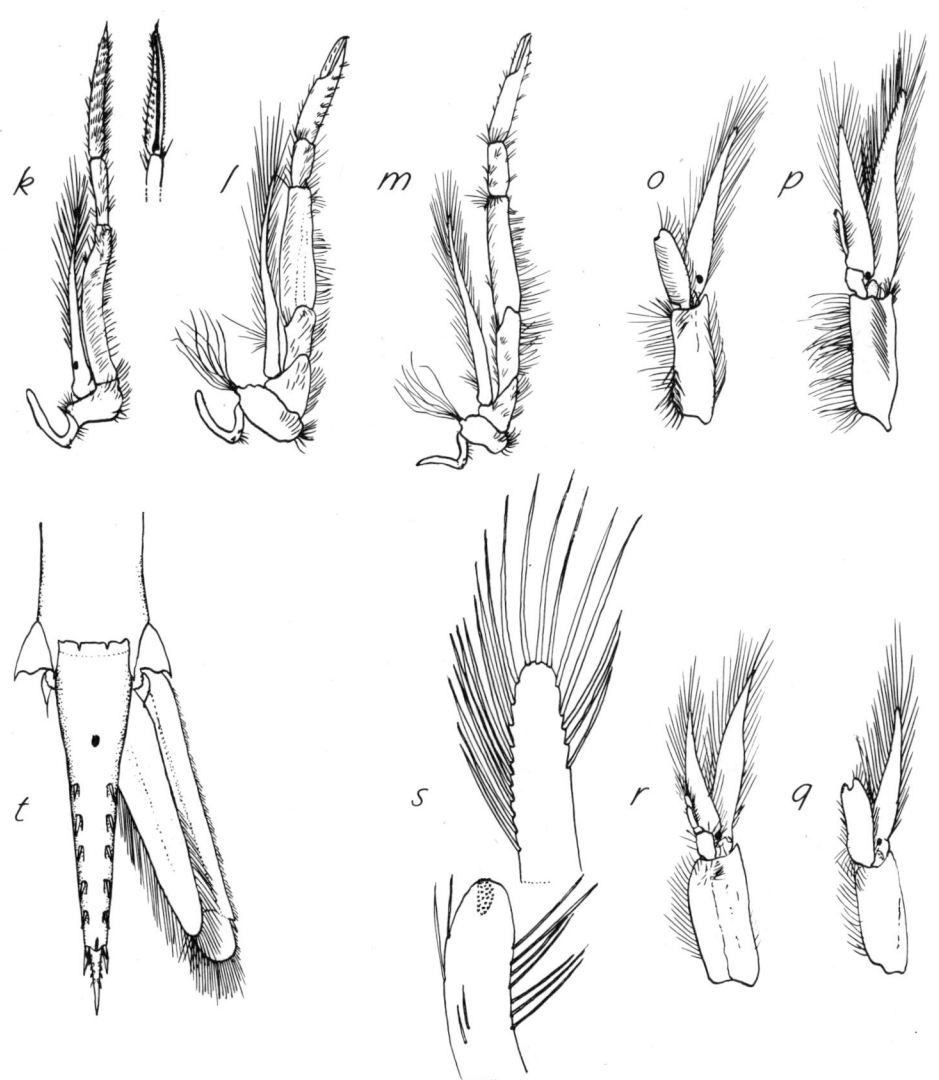

Fig. 47. *Systellaspis debilis*: *k*, third maxilliped; *l*, first pereopod; *m*, second pereopod; *o*, F first pleopod; *p*, F second pleopod; *q*, M first pleopod; *r*, M second pleopod; *s*, appendix masculina; *t*, telson. All to same scale except *s*. Actual length of *k* = 12.0 mm; length of *s* = 1.1 mm.

Maxillule (f): distal endite very wide, about twice proximal; endopodal palp incipiently biramous with two apical and one subapical setae.

Maxilla (g): distal endite unequally bilobed; one twice width of other; proximal reduced, ovate, with outer marginal lamella and a small upper lobe or papilla; posterior lobe of scaphognathite rounded and with curved setae.

First maxilliped (h): endites subtriangular; central lobe with three segments the distal smallest; exopod with inner corner projection and light organ on distal third laterally; epipod small, bilobed.

Second maxilliped (i): compressed, distal segment longer than wide; exopod with lash; small epipod with podobranch.

Third maxilliped (k): leglike but with five segments; distal segment long with spine at tip and ventrally a light organ — a thick line as long as segment; exopod with a small circular light organ proximally; epipod with strong proximal spine, a strap-like part with terminal hook and fleshy part at right angles.

Pereopods: I (l) chelate, moderately stout; exopod reaching carpus; epipod and setobranch present. II (m) more slender than I, merus with distal spine. III and IV (a) longer than first two, ischium and merus with flexor row of spines, light organ on carpus, dactyl about 0.6 propodus, slender and sharp; epipod reduced on IV; V shorter than IV, dactyl shorter also, tapering and with distal curved spine and setae.

Pleopods: Female I (o) endopod short, about half exopod, sub-rectilinear pointed at outer side, inner "shoulder" with patch of hooks distally; II (p) appendix interna slender with lateral setae and distal patch of hooks.

Male I (q) endopod about half exopod, notched distally, inner projection with patch of hooks. A light organ proximally on exopods; II (r) appendices interna short with a few lateral setae and masculina (s) long with about 12 long setae on one side and 7 on the other including one apical. A light organ proximally on endopods.

RANGE OF DISTRIBUTION

Western Atlantic from south of Greenland to Gulf of Mexico and Bahamas, eastern Atlantic from Faroe Islands to Angola, also South Africa, Indian Ocean, Philipines, Indonesia and Hawaii; mesopelagic between 650 and 800 m during the day and 150 m at night approximately (Chace 1986). Recorded depths between 25 and 3 000 m (Sivertsen and Holthuis 1956).

Distribution records for the area of reference are shown in Fig. 48.

BIOLOGY

Largest males and females about 14 mm cl, smallest ovigerous females 11 mm. Eggs about 4 × 2 mm. Monthly sampling from a population of this species from near Bermuda showed that more than 50% of the females were ovigerous in all samples while about 6% of these had eyed embryos (Chace 1940).

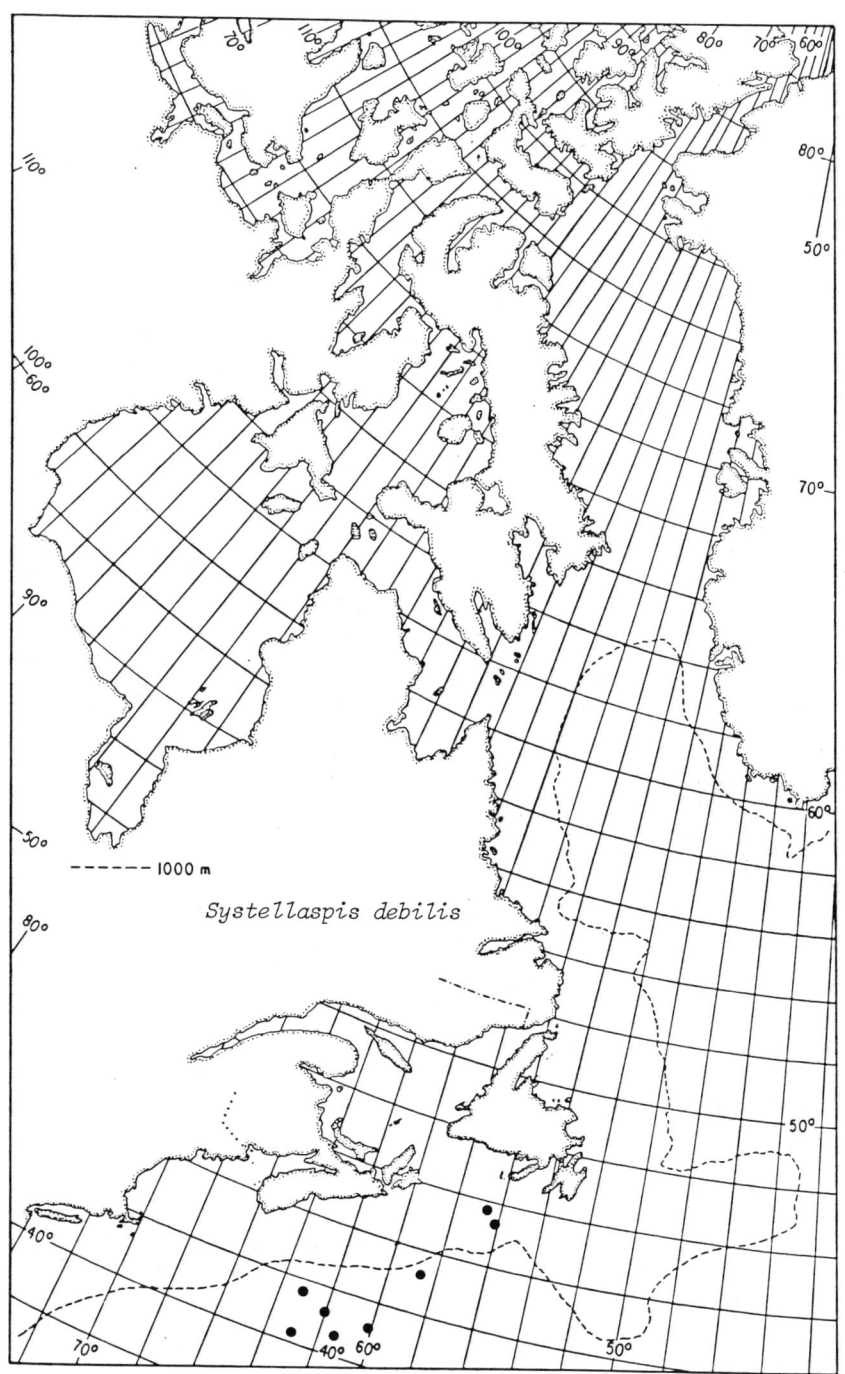

Fig. 48. *Systellaspis debilis*, distribution records in the area of reference.

Family NEMATOCARCINIDAE Smith, 1884
Crosnier et Forest 1973: 96; Holthuis 1955: 11;
de Man 1920: 72.

Pereopods III–V very long and slender, carpus several times longer than propodus; ischio-meral joint larger than others; first four pairs of pereopods with exopods and epipods; mandible with a palp; first pair of legs usually more slender and shorter than second. Rostrum with small dorsal spines close together and on anterior carapace, few ventrally.

This family has only one genus: two species are reported from the area of reference.

Genus *Nematocarcinus* A. Milne-Edwards, 1881
Crosnier et Forest 1973: 96–98;
Holthuis 1955: 18, fig. 5.

The genus has the same characteristics as the family: the last three pairs of legs are very long and slender and likely to be broken away through handling, but usually the parts remaining will have the joint between the merus and ischium which is expanded and different from any others. The rostrum has many small spines above continued on to the carapace for a short distance, but very few or no teeth ventrally. All but fifth legs with epipods and exopods. Inner branch of uropod shorter than telson, outer branch longer than telson.

Nematocarcinus cursor A. Milne-Edwards, 1881
Crosnier et Forest 1973: 105–116, figs 29b, 30j–m, 31c–d.
(Figures 49 and 50)

DISTINGUISHING CHARACTERISTICS

Rostrum horizontal, dorsally serrate with less than 20 small sharp spines, (plus 6 or so on carapace) and ventrally with 0–2 teeth; pleura of 5th abdominal somite forming an angle of about 90 degrees, armed at the tip with a well-developed tooth; joint of leg between ischium and merus typcially swollen. Seven pairs of tiny spines dorso-laterally on telson.

DESCRIPTION

Integument smooth and shiny but minutely punctate. Colour not available.

Rostrum straight, horizontal, with about 15 small sharp spines dorsally, six of which are on the carapace, one or no tooth ventrally, lateral carina confluent with orbit; low rostral crest continued on the carapace as a carina for about half its length; antennal and pterygostomian spines present; a faint cervical groove.

Abdomen with rounded terga and pleura except 5th which has a posterior ventro-lateral spine.

Telson (t) slender with narrow subtruncate tip and seven pairs of tiny dorso-lateral spines on distal two-thirds. Terminally two pairs of spines, the inner longer. Outer branch (exopod) of uropod with concave edge and axillary spine, longer than inner and than the telson, the inner shorter than the telson.

Antennule (c) 1st article of peduncle about equal to other two combined, 2nd and 3rd subequal; stylocerite about two-thirds the length of 1st article, pointed distally, foliaceous, expanded with a blunt point laterally; thick part of dorsolateral flagellum very short.

Antenna (d) scale length about 4 times width, outer edge slightly concave, tapered, disto-lateral spine exceeded by blade; peduncle short, only a little longer than width of scale.

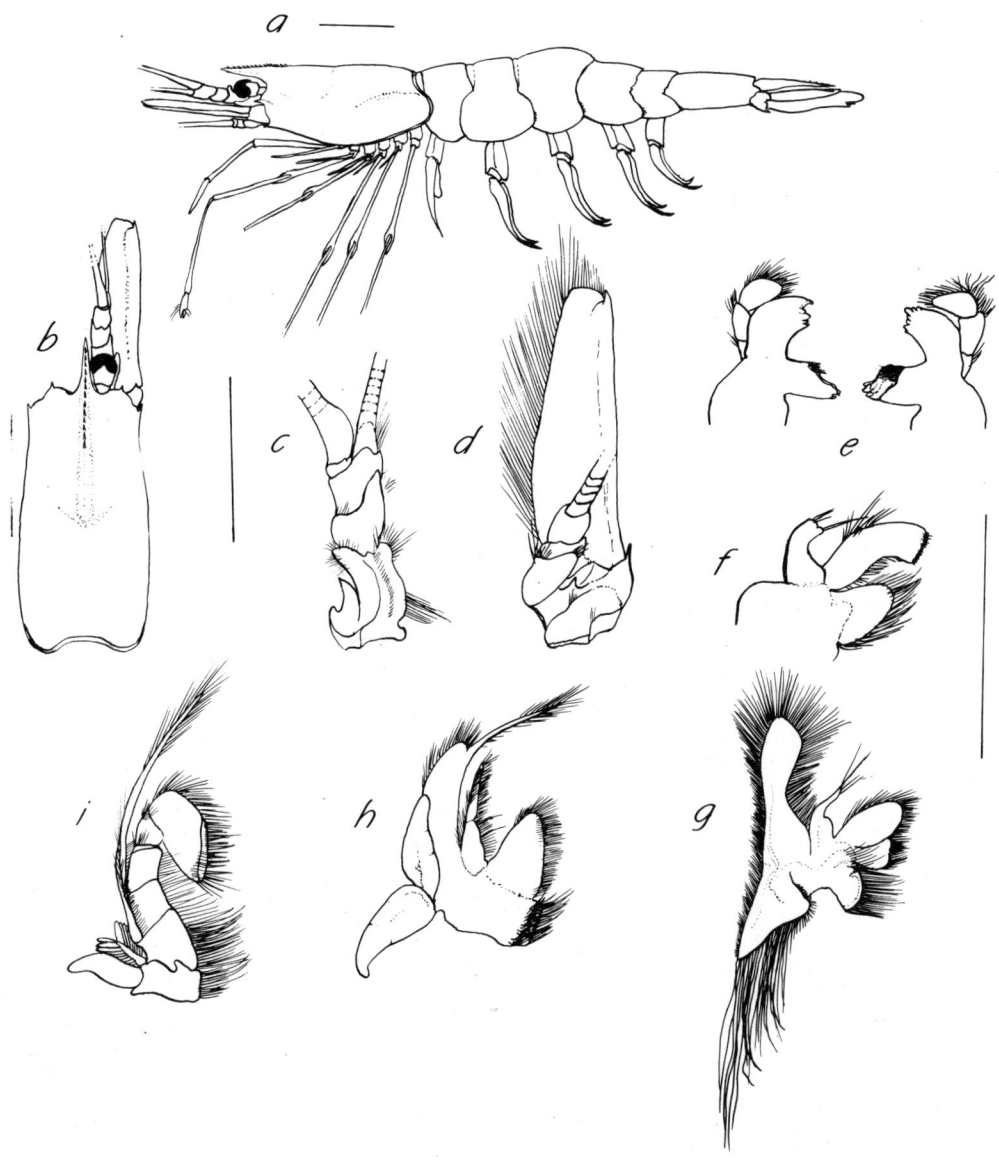

Fig. 49. *Nematocarcinus cursor*: *a*, whole shrimp from left side; *b*, carapace in dorsal aspect; *c*, antennule; *d*, antenna; *e*, mandible; *f*, maxillule; *g*, maxilla; *h*, first maxilliped; *i*, second maxilliped; solid line = 10 mm.

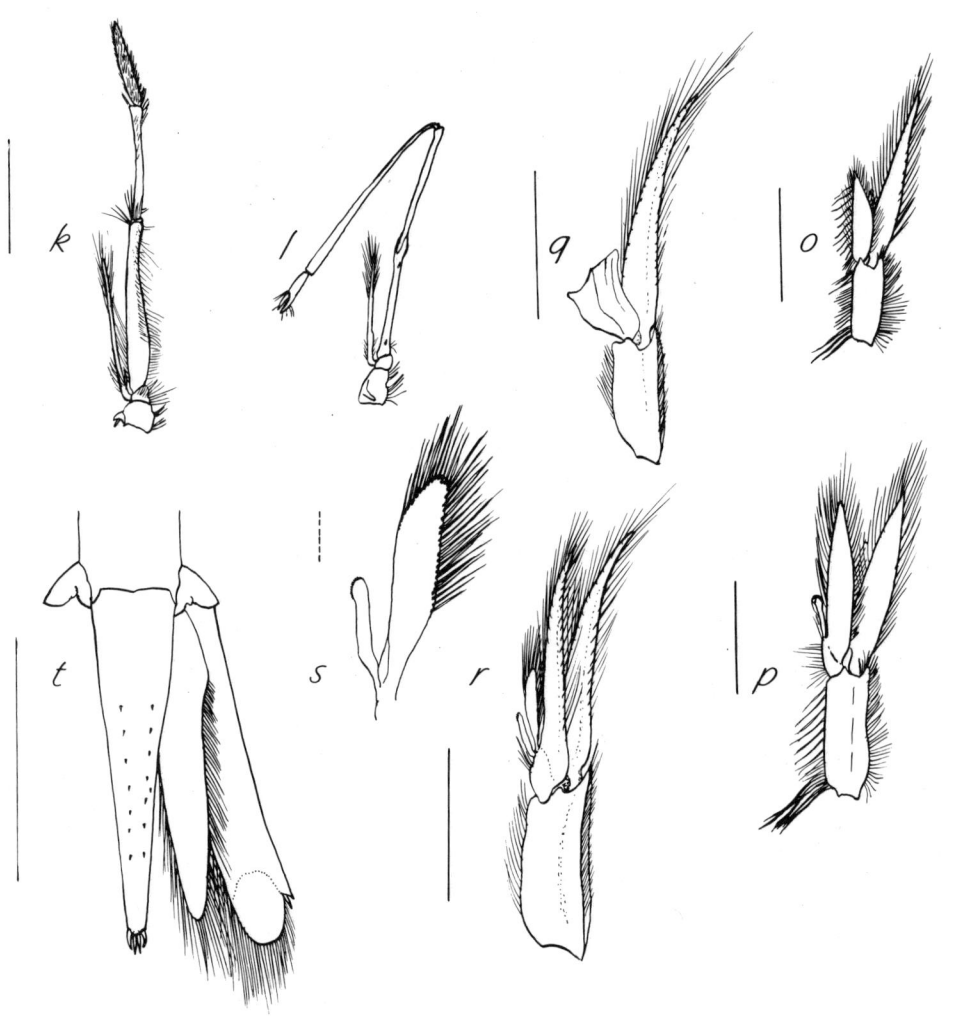

Fig. 50. *Nematocarcinus cursor*: k, third maxilliped; l, first pereopod; m, second pereopod; o, F first pleopod; p, F second pleopod; q, M first pleopod; r, M second pleopod; s, appendix masculina; t, telson; solid line = 10 mm, broken line = 1 mm.

Mandibles (e) somewhat dissimilar: left incisor with smaller teeth on each side of central tooth than right, left molar with rounded crown, right with hollow crown. Palp with three segments.

Maxillule (f) distal endite narrower but longer than proximal; endopod curved, obscurely bifid.

Maxilla (g) distal endite unequally bilobed; proximal reduced to rounded single lobe with long curved setae, also with an outer marginal lamella part of which is papillate; endopod small with neck curved inward; scaphognathite with long narrow irregular anterior lobe and short triangular posterior lobe which has very long terminal setae.

Maxilliped I (h) distal endite large subtriangular, proximal endite very much smaller; endopod with long distal part and shorter and wider proximal segment, shorter than lobe or lamellate part of exopod or its long corner flagellum; epipod large, subequally bilobed.

Maxilliped II (i) leglike, compressed, distal segment much wider than long; exopod slender, longer than endopod; epipod with podobranch.

Maxilliped III (k) leglike, proximal segments stout, but two distal ones slender; exopod slender, about three-quarters the length of longest segment; epipod small, falcate, but with a straplike part attached which lacks grooming claws but has marginal spinous setae.

Pereopods: I (l) slender, chelate, carpus longer than other segments, ischium joined to merus diagonally with expansion around joint; exopod shorter than ischium; epipod small, straplike, with no terminal grooming hooks. II slender, longer than I, chelate, with expanded ischio-meral joint; exopod; epipod straplike with no hook. III–V very long (broken away in specimens examined) with expanded ischio-meral joint (remaining); III and IV with exopod and epipod.

Pleopods: female I (o) with wide tapered endopod shorter than narrow exopod; female II (p) endopod subequal to exopod, with appendix interna expanded toward tip and apical patch of hooks; male I (q) wide subtriangular endopod shorter than narrow exopod; male II (r) endopod and exopod subequal, appendix interna shorter than appendix masculina (s) the latter with two rows of moderate spines apically continuous with numerous spines along one side in several rows.

RANGE OF DISTRIBUTION

West Indies and east coast of Florida to about 38° N latitude; depths about 330–1 240 m (Crosnier et Forest 1973).

Records of distribution in the area of reference are in Fig. 51.

BIOLOGY

Lengths 21 mm cl in male and 25 mm cl in female examined.

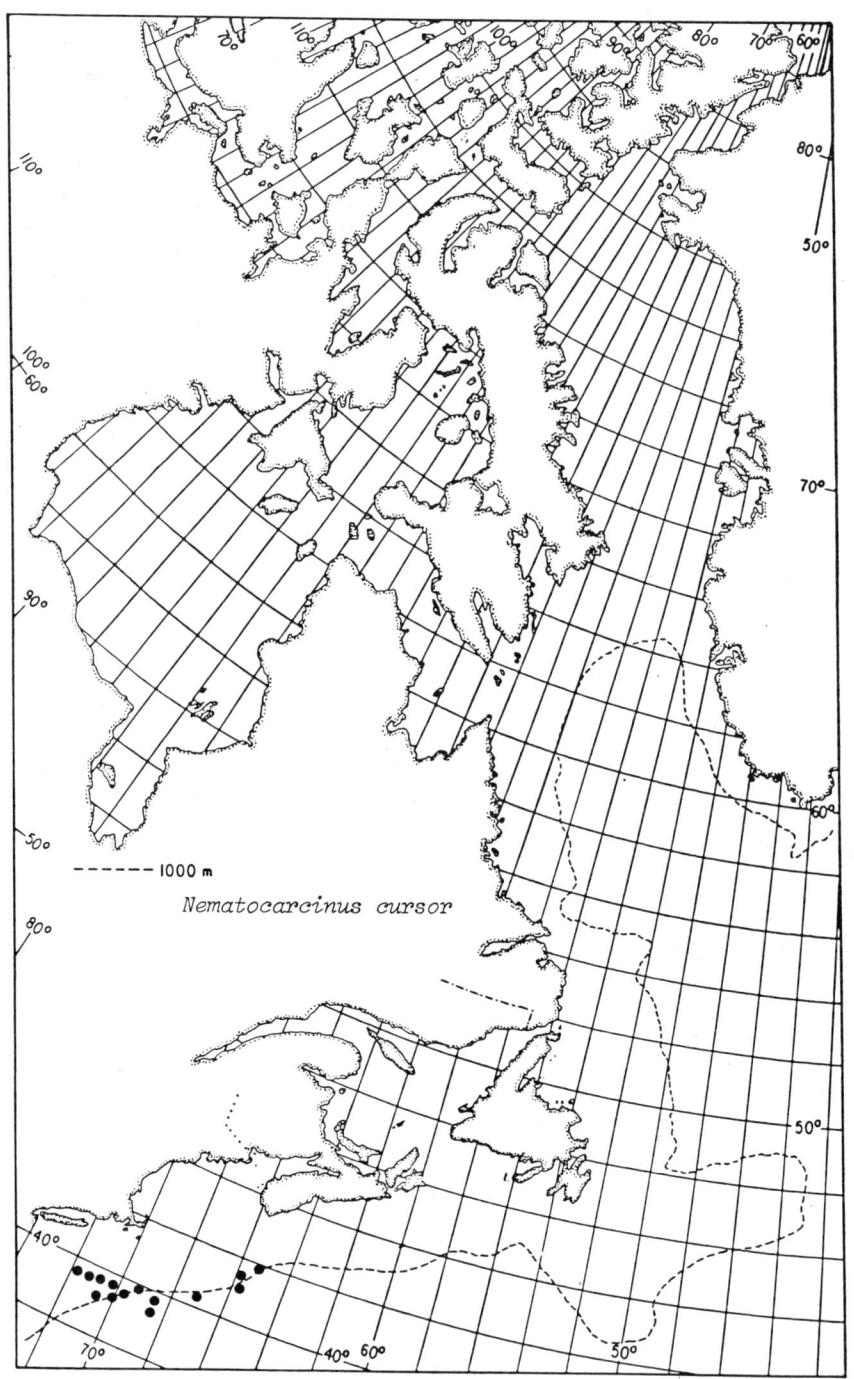

Fig. 51. *Nematocarcinus cursor*, distribution records in the area of reference.

Nematocarcinus rotundus Crosnier et Forest, 1973
Crosnier et Forest 1973: 103, figs 29c, 30f–i, 31e–f.
(Figures 52 and 53)

DISTINGUISHING CHARACTERISTICS

Rostrum slightly ascending, with spines reaching almost to tip dorsally, spines close together except near tip, and extending on to carapace on a low carina; one or two ventral teeth near tip; legs slender and long, the last three pairs very long and likely to be broken off, but the ischio-meral or third joint may remain and it is expanded more than others; the 5th abdominal somite has rounded pleura with no or a very small ventro-lateral tooth. Telson has four pairs of tiny dorsolateral spines.

DESCRIPTION

Integument apparently smooth but punctate to somewhat scaly miscropically and may feel rough to a probe. Colour not available.

Rostrum slightly ascending, moveable spines small and close together except near tip, a slight lateral ridge is confluent with orbit, about four of rostral series of spines on carapace behind orbit.

Carapace with suborbital lobe and antennal and pterygostomian spines; also a faint cervical depression and groove.

Abdomen with rounded terga and pleura; 5th somite with slightly pointed pleura but no ventro-lateral spine.

Telson (t) slender, tapering to subtruncate narrow tip with two pairs of spines, the outer longer; four pairs of tiny dorsolateral spines on distal half, proximal pair double in some. Outer branch of uropod is slightly concave on outer edge, with axillary spine, slightly longer than telson and inner branch.

Eye large with globular cornea.

Antennule (c) 1st article about equal to other two combined; 3rd slightly longer than 2nd; stylocerite with pointed foliaceous tip (about 0.8 length of 1st article) and thickened at base with lateral blunt projection.

Antenna (d) scale about 0.7 cl, length 3.5 times width; disto-lateral spine about even with blade; basal article with distal spine ventrally; peduncle only about one-third scale.

Mandibles (e) somewhat dissimilar: left incisor pointed and with 6 teeth, the right with 7 teeth and straightish; molars slightly different. Palp of three segments, the distal narrow.

Maxillule (f) distal endite narrow, curved; proximal about as long but larger, ovate; endopod curved, tip truncate, not clearly bifid, lower edge with strong seta.

Maxilla (g) distal endite unequally bilobed, proximal lobe with concave edge; proximal endite reduced, rounded, unilobate, with outer marginal lamella part of which is papillate; endopod short with short neck; anterior lobe of scaphognathite narrow from wide base, longer than triangular posterior lobe, the latter with long terminal setae.

Maxilliped I (h) distal endite subtriangular, rounded, about twice as wide as proximal endite, the latter with thick edge and rows of setae; endopod almost as long as lobe of exopod, the lash at inner corner longer than either; epipod bilobed, each about equal, subtriangular, narrow.

Maxilliped II (i) leglike, compressed, distal endite much wider than long; exopod slender, about as long as endopod; epipod with podobranch.

Maxilliped III (k) leglike, five segments, proximal stout and distal two slender; exopod slender, almost as long as longest segment; epipod falcate with attached straplike part without terminal hook.

Pereopods: I (l) slender, chelate, carpus longer than other segments; ischio-meral joint diagonal, expanded; exopod shorter than ischium; epipod straplike, without terminal

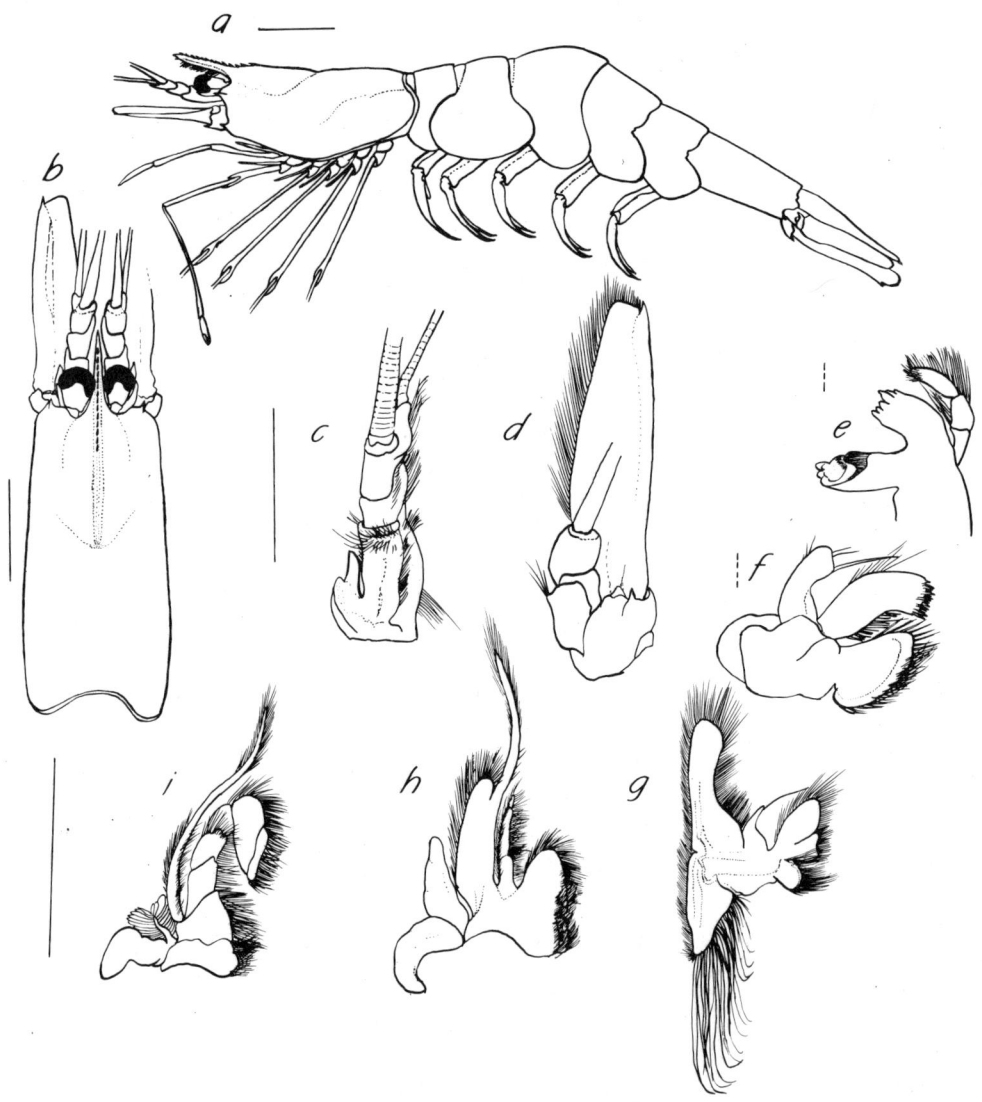

Fig. 52. *Nematocarcinus rotundus*: *a*, whole shrimp from left side; *b*, carapace in dorsal aspect; *c*, antennule; *d*, antenna; *e*, mandible; *f*, maxillule; *g*, maxilla; *h*, first maxilliped; *i*, second maxilliped; solid line = 10 mm.

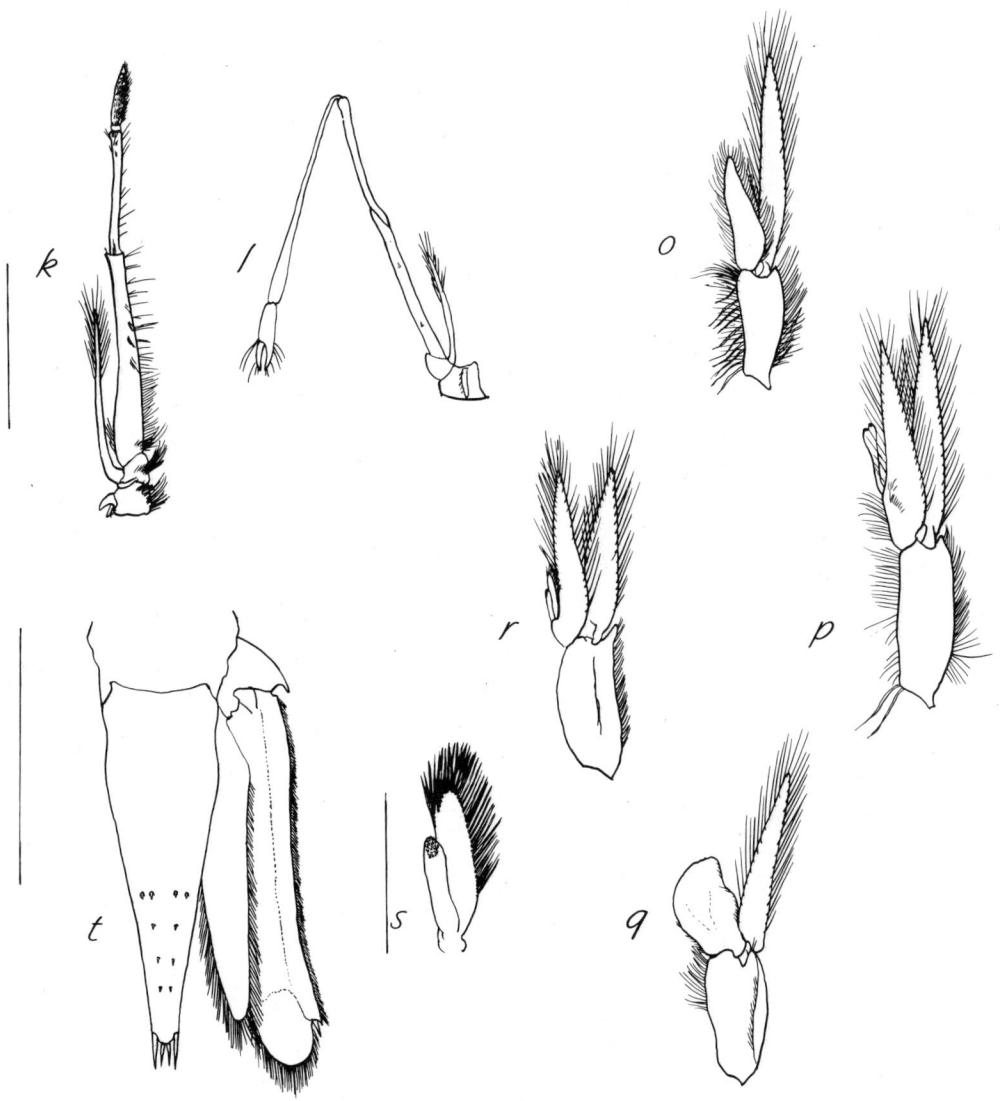

Fig. 53. *Nematocarcinus rotundus*: *k*, third maxilliped; *l*, first pereopod; *m*, second pereopod; *o*, F first pleopod; *p*, F second pleopod; *q*, M first pleopod; *r*, M second pleopod; *s*, appendix masculina; *t*, telson; solid line = 10 mm.

hook. II slender, chelate, longer than I, exopod short, epipod straplike; III–V very long, non-chelate, all with expanded ischio-meral joint, all except V with exopod and epipod.

Pleopods: female I (o) endopod much shorter than exopod and with sharp point; female II (p) endopod and exopod subequal, appendix interna expanded and bifid distally, each tip with patch of hooks. Male I (q) large squarish endopod shorter than narrow exopod; male II (r) endopod about equal to exopod, appendix interna with apical patch of hooks and shorter than masculina (s) the latter rounded apically with series of marginal long fine setae, most laterally (on one side).

RANGE OF DISTRIBUTION

West Indies, Gulf of Mexico and east coast of the USA as far north as 40° N latitude; depths 700–1 570 m (Crosnier et Forest 1973).

Records of distribution in the area of reference are in Fig. 54.

BIOLOGY

Lengths to 19 mm cl in males and 27 mm cl in females (Crosnier et Forest 1973).

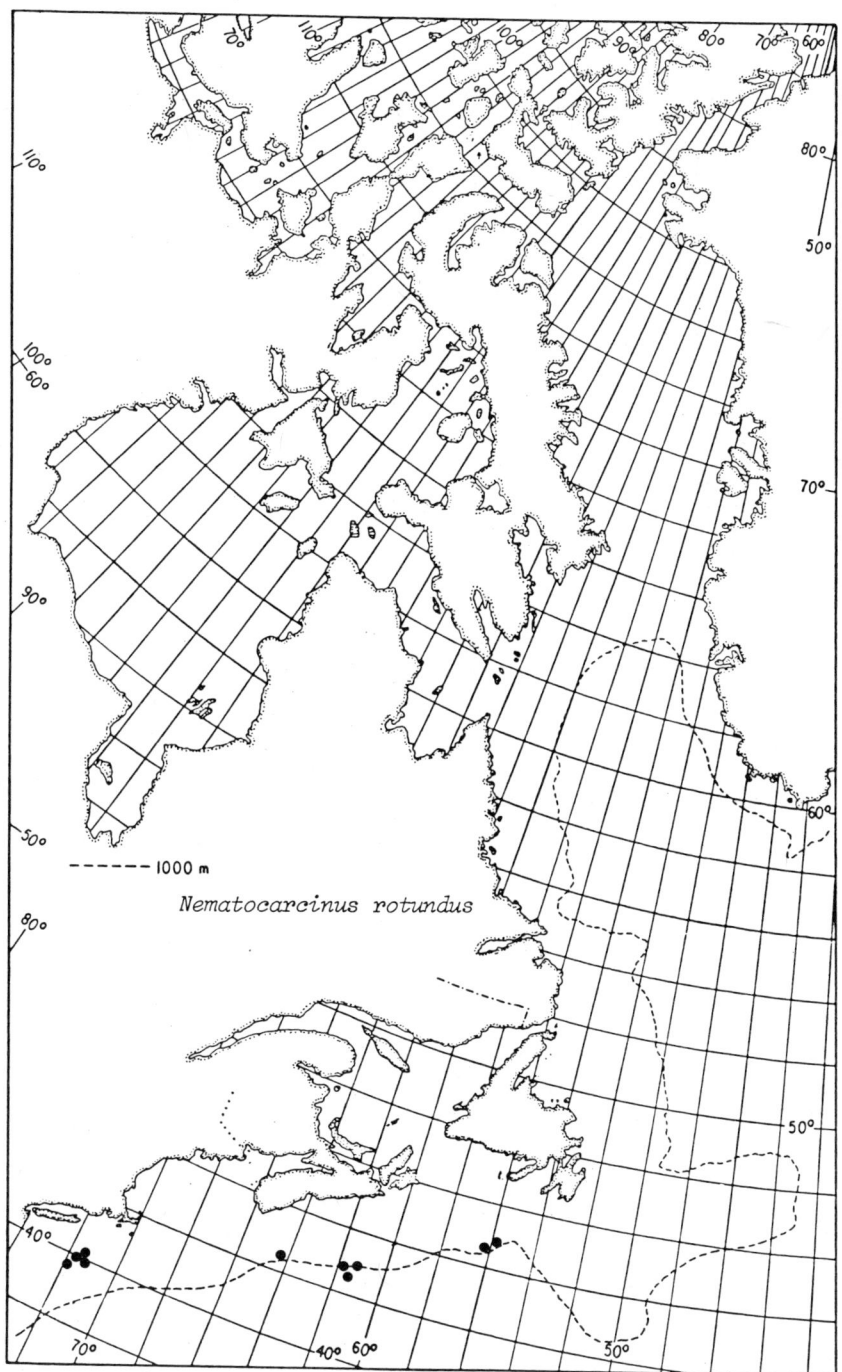

Fig. 54. *Nematocarcinus rotundus*, distribution records in the area of reference.

Family PASIPHAEIDAE Dana, 1852
Butler 1980: 52; Williams 1984: 57.

Rostrum short; mandible with large toothed incisor; third maxillipeds with exopod; all pereopods with exopods; first two pairs of pereopods larger than others, with long chelae and fingers that are pectinate on cutting edge.

Two genera and three species are represented in the area of reference.

Key to genera and species of the Family Pasiphaeidae
(After Holthuis 1955)

1 Rostrum a short post-frontal spine; mandible without a palp; carapace with branchiostegal spine .... *Pasiphaea*
— Heavy reddish shrimps; 1–5 spines on distal lower edge of basis of 2nd legs .... *Pasiphaea tarda*
— Slender almost translucent shrimps; 7–12 spines on distal lower edge of basis of 2nd legs .... *Pasiphaea multidentata*

Rostrum a short spinous projection of the front of the carapace; mandible with a palp; no branchiostegal spine .... *Parapasiphaea sulcatifrons*

Genus *Parapasiphaea* Smith, 1884

Butler 1980: 57; Crosnier et Forest 1973.

Rostrum a projection of the front of the carapace; first two legs long and stout, larger than the rest, with long chelae, the long fingers pectinate on cutting edge; carapace mid-dorsally carinate forming a hump on front half, the carina anteriorly sulcate; mandible with small slender two-segmented palp; no branchiostegal spine.

The eight known species of the genus are pelagic in deep water; only one species is present in the area of reference.

Parapasiphaea sulcatifrons Smith, 1884

Butler 1980: 58, fig.; Chace 1940: 126, fig. 6;
Crosnier et Forest 1973: 142, fig 41.
(Figures 55 and 56)

DISTINGUISHING CHARACTERISTICS

Rostrum a short projection of the front of the carapace; middorsal carina rising high on the front half of the carapace with a groove in it from the tip of the rostrum to just ahead of the highest point; first two legs longer and stouter than the others, both with long chelae and fingers which are pectinate on cutting edges; no anterior spines on carapace but a ventro-lateral notch; abdomen with short postero-median spine on 4th somite; telson subtruncate at narrow tip with 4 pairs of small spines.

DESCRIPTION

Integument smooth and shiny. Colour scarlet, the carapace and chelae light; eyes amber; antennae salmon-orange (Chace 1940).

Rostrum a short pointed projection of the front of the carapace, the latter with middorsal carina with a narrow sulcus (about one-third its length) continued on the carapace, rising to a hump on anterior half and levelling out to reach the posterior edge. No spines on carapace but a lateral horizontal groove from anterior edge to near the posterior edge, joined near its centre by a short oblique inferior groove and by other less distinct superior grooves. There is a narrow anterior ventral notch.

Abdomen with pleura and terga rounded, except on 4th somite which has a short dorsal carina with a posterior median extension over the 5th; 6th has a short median posterior spine without carina.

Telson (t) with high sides but rounded lateral ridges and median sulcus, tapering to a subtruncate tip with 4 pairs of small spines; inner branch of uropods tapering to pointed tip, shorter than outer branch but longer than telson.

Eyes with globular cornea slightly wider than stalk, the latter with inner distal pointed tubercle.

Antennule (c) 1st article about equal in length to other two; 3rd longer than 2nd; 1st with inner ventro-lateral spine near middle; stylocerite thick with sharp pointed tip reaching distal 1st article.

Antenna (d) scale with outer edge convex, length about three times width; disto-lateral spine exceeding blade; peduncle greater than half scale.

Mandible (e) no molar; incisor wide, compressed, slightly curved, with two major and about 6 minor teeth plus a small corner tooth. Palp slender with two short segments.

Maxillule (f) endites subequal, distal rounded with single row of strong setae, proximal slightly pointed; endopod longer than both with subapical projection (possibly bifid).

Maxilla (g) endites reduced to slight rounded projection; endopod slender; narrow anterior lobe of scaphognathite longer than rounded posterior lobe.

Maxilliped I (h) endites reduced, endopod short; exopod with narrow lobe with subapical projection, lash oval, foliaceous; epipod with slender unequal lobes.

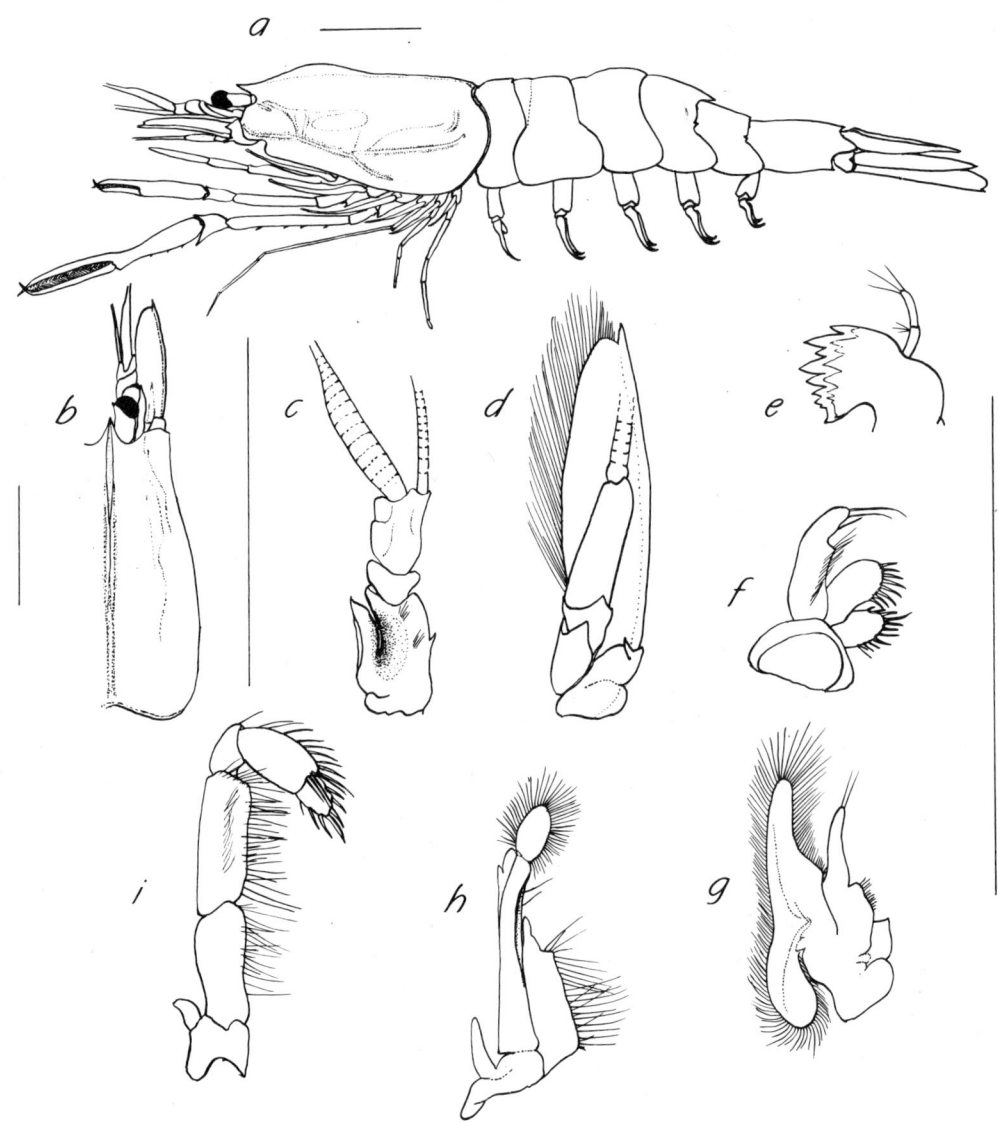

Fig. 55. *Parapasiphaea sulcatifrons*: *a*, whole shrimp from left side; *b*, carapace in dorsal aspect; *c*, antennule; *d*, antenna; *e*, mandible; *f*, maxillule; *g*, maxilla; *h*, first maxilliped; *i*, second maxilliped; solid line = 10 mm.

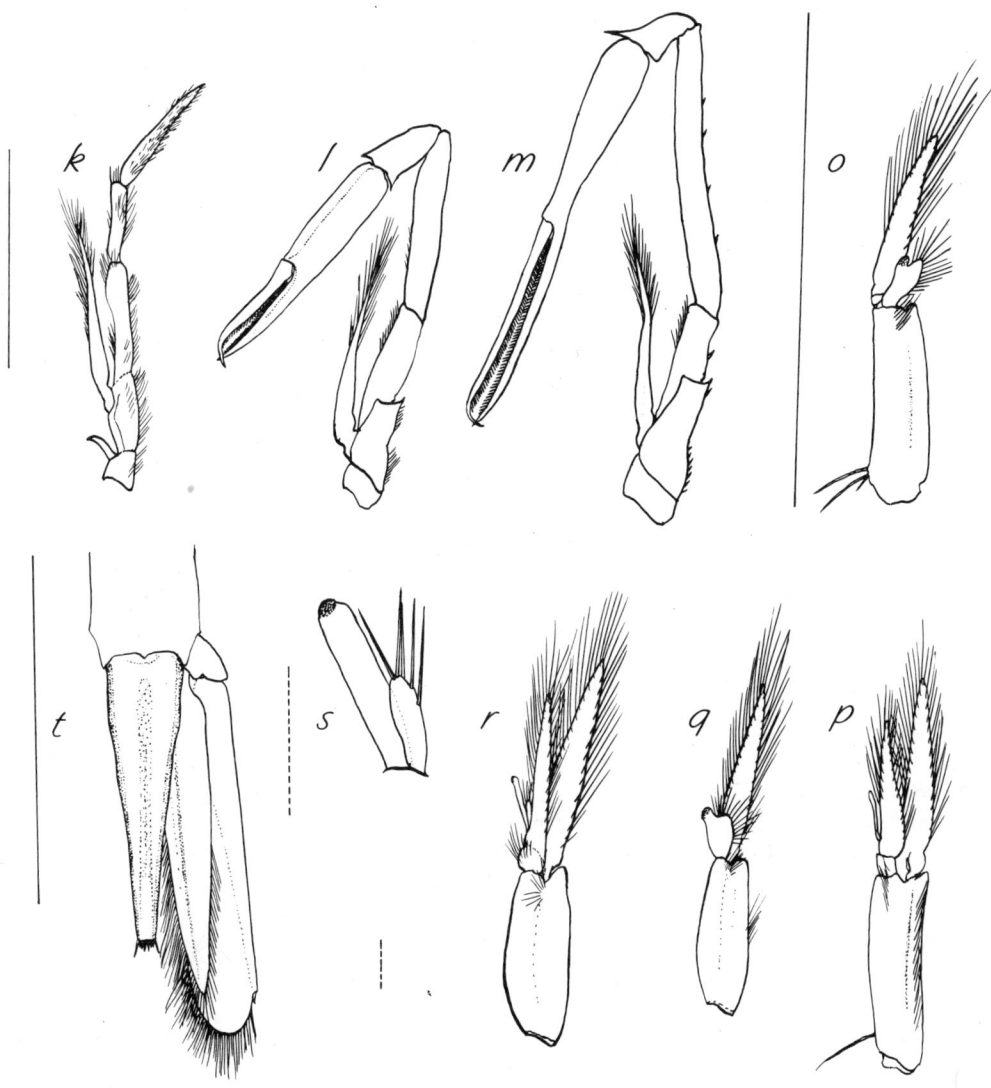

Fig. 56. *Parapasiphaea sulcatifrons*: k, third maxilliped; l, first pereopod; m, second pereopod; o, F first pleopod; p, F second pleopod; q, M first pleopod; r, second pleopod; s, appendix masculina; t, telson; solid line = 10 mm, broken line = 1 mm.

Maxilliped II (i) leglike, slightly compressed, distal segment longer than wide; epipod small, rudimentary.

Maxilliped III (k) leglike, five segments, distal tapered with sharp apical spine; exopod longer than any one segment, tapered; epipod short, pointed.

Pereopods: I (l) shorter and slightly smaller than II, chelate, chela smaller than in II; propodal finger sharp tapered, dactyl strongly curved, cutting edges of both pectinate. Exopod about as long as merus, with long slender point; distal edge of basis with strong tooth. II (m) longer than I, large chela, palm inflated proximally, shorter than fingers: propodal slender finely tapered to sharp point, dactyl strongly curved at tip, both with cutting edges pectinate; merus elongate with 4 mesial spines, ischium with 1 spine, basis with distal edge a tooth and with 1 distal and several proximal spines, distal edge of carpus a sharp spine. III long and slender, sharp dactyl; exopod. IV shorter than others, dactyl rounded; exopod. V moderate, slender, shorter than III, dactyl rounded; exopod.

Pleopods: female I (o) endopod very short, expanded toward tip with side projection and patch of hooks giving biramous appearance; protopod longer than exopod; female II (p) endopod shorter than exopod, appendix interna long and sinuous with apical patch of hooks. Male I (q) endopod short, wide, with inner projection with grappling hooks; exopod much longer, slender; male II (r) endopod very slender but proximally widened, shorter than exopod, appendix interna with round patch of hooks and about twice as long as appendix masculina (s), the latter with two apical long and three distolateral fine spinous setae.

RANGE OF DISTRIBUTION

Reported from Greenland and Iceland to Bermuda and the Gulf of Mexico in the western Atlantic, and in the eastern Atlantic from Ireland to off the coast of Angola (Crosnier et Forest 1973). In the Pacific off the west coast of the United States (Pearcy and Forss 1966) and off the coast of British Columbia to Queen Charlotte Sound (Butler 1980). Depths 500–5 340 m (Smith 1884; Sivertsen and Holthuis 1956).

Records of distribution in the area of reference are in Fig. 57.

BIOLOGY

Lengths to 23 mm cl in males and 26 mm cl in females. Smallest females ovigerous were 22 mm cl, and ovigerous females were taken in April and from June to September. The smallest number of juveniles were taken in April to June, and the largest number in July and August, occurring mostly in depths of 900–1 800 m (Chace 1940). Eggs were 3.1 × 4.6 mm in size and about 15–20 carried per female (Crosnier et Forest 1973).

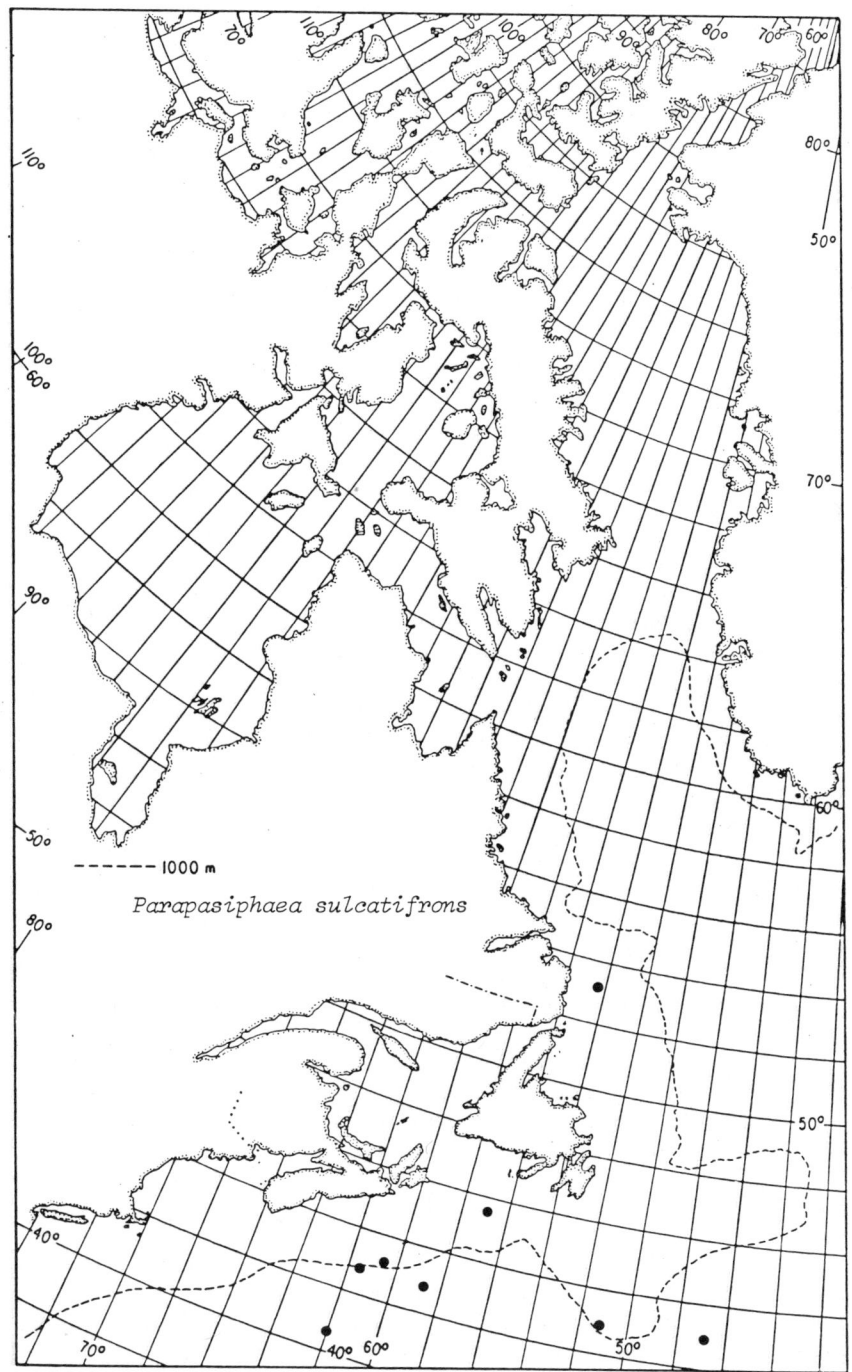

Fig. 57. *Parapasiphaea sulcatifrons*, distribution records in the area of reference.

Genus *Pasiphaea* Savigny, 1816
Butler 1980: 52; Chace 1940: 122.

Carapace and abdomen compressed considerably; rostrum represented by a post-frontal or epigastric spine; carapace with branchiostegal spine; mandible without palp; third maxilliped without epipod.

Of the approximately 30 species known, only two are reported from the area of reference.

Pasiphaea multidentata Esmark, 1866
Sivertsen and Holthuis 1956: 27, figs. 19–21;
Rathbun 1929: 5, fig. 1; Williams 1984: 60, fig. 40.
"Pink glass shrimp" — "Sivade rose"
(Figures 58 and 59)

DISTINGUISHING CHARACTERISTICS

Rostrum short, a short thin post-frontal spine rises above it; carapace and abdomen compressed with sharp dorsal carina; first two legs much more robust and longer than others, both with strong chelae the long fingers of which are pectinate on cutting edge; on the lower edge of the basis of the 2nd leg are 7–12 sharp spines; fourth leg shortest; all legs with exopods; telson narrowly and deeply forked.

DESCRIPTION

Integument smooth, shiny. Colour translucent whitish background with many dots (chromatophores) reddish and other colours, concentrated in some areas.

Rostrum short and rounded anterior edge of carapace, much less conspicuous than the post-frontal thin fixed spine projecting forward above it, this spine is the anterior projection of a sharp median carina which extends to near the posterior edge of the carapace where it turns down abruptly; carapace tapers anteriorly to narrow frontal edge with small sharp branchiostegal spine and slightly hollowed orbit.

Abdomen very compressed, all somites with sharp median carina beginning at half the 1st and ending at about posterior four-fifths of the 6th. Pleura rounded but pointed at 5th and with a ventro-lateral keel and small posterior spine on 6th.

Telson (t) dorsally sulcate and with lateral ridges without spines, tapered to a narrow fork each tip of which has a strong spine and six inner spines decreasing in size toward the fork; both branches of the uropod are longer than the telson, the outer longer and with a strong disto-lateral spine (diaeresis at about two-thirds length distally, away from spine).

Eye large, cornea globular angled outward on stalk.

Antennule (c) 1st article about equal to other two combined, 3rd longer than 2nd; stylocerite slender appearing twisted and with sharp acuminate tip reaching distal edge of 1st article.

Antenna (d) scale length about three times width, blade curving and narrowed distally, much exceeded by spine; outer edge slightly convex; basal article with strong disto-ventral spine; peduncle slightly less than half scale.

Mandibles (e) similar; wide incisor with about 9 major teeth, two larger ones with subsidiary teeth; lower corner with gap and corner tooth; no palp or molar.

Maxillule (f) proximal endite short and rounded; distal wide, expanded from narrow base; endopod straight, tapered at single apex.

Maxilla (g) endites greatly reduced; endopod short from wide base, separated from narrow anterior lobe which is only slightly longer than wider, rounded posterior lobe.

Maxilliped I (h) endites greatly reduced, very short endopod possibly fused for most of its length with the exopod, the latter with long tapered lobe and short ovate, pointed lash; epipod small.

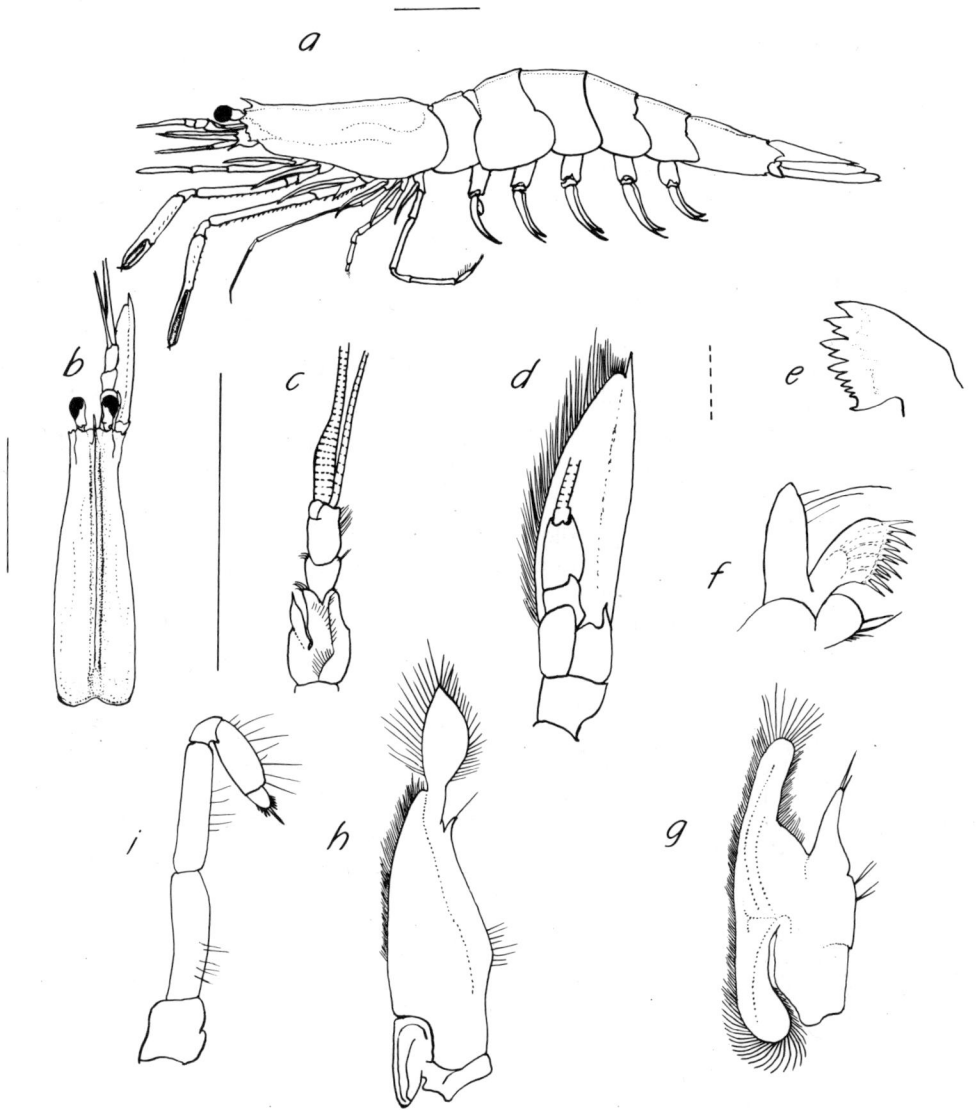

Fig. 58. *Pasiphaea multidentata*: *a*, whole shrimp from left side; *b*, carapace in dorsal aspect; *c*, antennule; *d*, antenna; *e*, mandible; *f*, maxillule; *g*, maxilla; *h*, first maxilliped; *i*, second maxilliped; solid line = 10 mm, broken line = 1 mm.

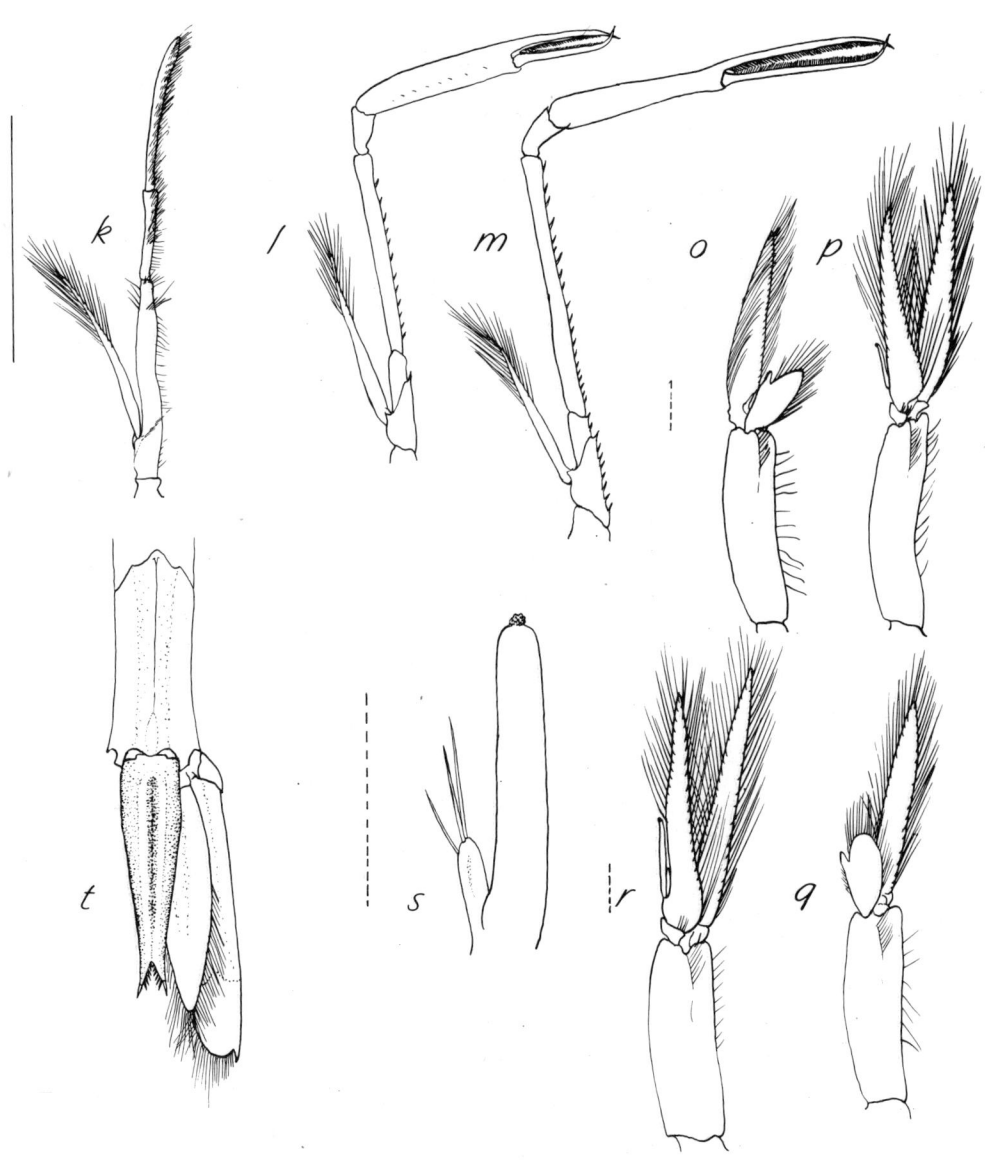

Fig. 59. *Pasiphaea multidentata*: *k*, third maxilliped; *l*, first pereopod; *m*, second pereopod; *o*, F first pleopod; *p*, F second pleopod; *q*, M first pleopod; *r*, M second pleopod; *s*, appendix masculina; *t*, telson; solid line = 10 mm, broken line = 1 mm.

Maxilliped II (i) leg-like, with 6 segments, the distal longer than wide; no epipod or exopod.

Maxilliped III (k) five segments, the distal longest, setose but without spines; exopod tapered, about as long as longest segment.

Pereopods: I (l) long, robust, chelate, fingers shorter than palm, cutting edges pectinate; merus with flexor row of about 12 sutured spines, basis with none. II (m) longer than I, robust, chelate, fingers about as long as palm, cuttting edges pectinate; merus with about 20 sutured spines, basis with 7–12, ischium with 1. III slender, about as long as I, dactyl sharp; IV shortest, not as slender as III, dactyl rounded at tip; V about as long as III but stouter, dactyl spatulate, with strong brush setae. An exopod on all pereopods.

Pleopods: I (o) female, endopod short, mitt-shaped, inner appendix with disto-lateral patch of hooks; exopod narrow, long; II (p) female endopod and exopod subequal, slender appendix interna near base with disto-lateral patch of hooks. I (q) male endopod short, mitt-shaped, with disto-lateral patch of hooks, exopod narrow; II (r) male endopod and exopod subequal, narrow; appendix interna with apical patch of hooks, more than twice as long as short appendix masculina (s), the latter with only three long apical spines.

RANGE OF DISTRIBUTION

In the western Atlantic from southeast of Greenland to Massachusetts including the Gulf of St. Lawrence and the Gulf of Maine; in the eastern Atlantic from Iceland and Norway to the British Isles, Bay of Biscay and the Mediterranean to the Adriatic. Depths 10–2 000 m (Sivertsen and Holthuis 1956; Zariquiey Alvarez 1968; Apollonio 1969).

Records of distribution in the area of reference are in Fig. 60.

BIOLOGY

Lengths to 30 mm cl in males and females (Squires 1965a).

Most females 25 mm cl or more were ovigerous in December in Hermitage Bay at temperatures of 4.5–5.5° C and probably spawned annually. Egg diameter from 2.2 to 2.4 mm (Squires 1965a). In the Gulf of Maine at temperatures of 4.1–7.3° C, Apollonio (1969) concluded that this species produced two clutches of eggs during the year, and that the second was slightly less in number than the first clutch (about 60–120 eggs in the first and 40–90 eggs in the second).

Stomach contents were mainly crustacean fragments including copepods and euphausiids (Apollonio 1969; Squires 1965a).

Parasitization of the anterior edge of the carapace by protistans of the family Ellobiopsidae may result in distortion of the shape of the rostral area (Sivertsen and Holthuis 1956).

The pelagic shrimps *Pasiphaea tarda* and *Sergestes arcticus* were taken in the same hauls with this species in Hermitage Bay, and also semi-demersal species *Pandalus borealis* and *Spirontocaris lilljeborgi* (Squires 1965a).

FISHERY

Of minor importance only, the species is found mixed with other deep-sea shrimps at the fish markets of NE Spain when landed by Spanish trawlers (Zariquiey Alvarez 1968). Also reported in the Genoa fish markets when landed by Italian fishermen (Brian 1941; Holthuis 1980).

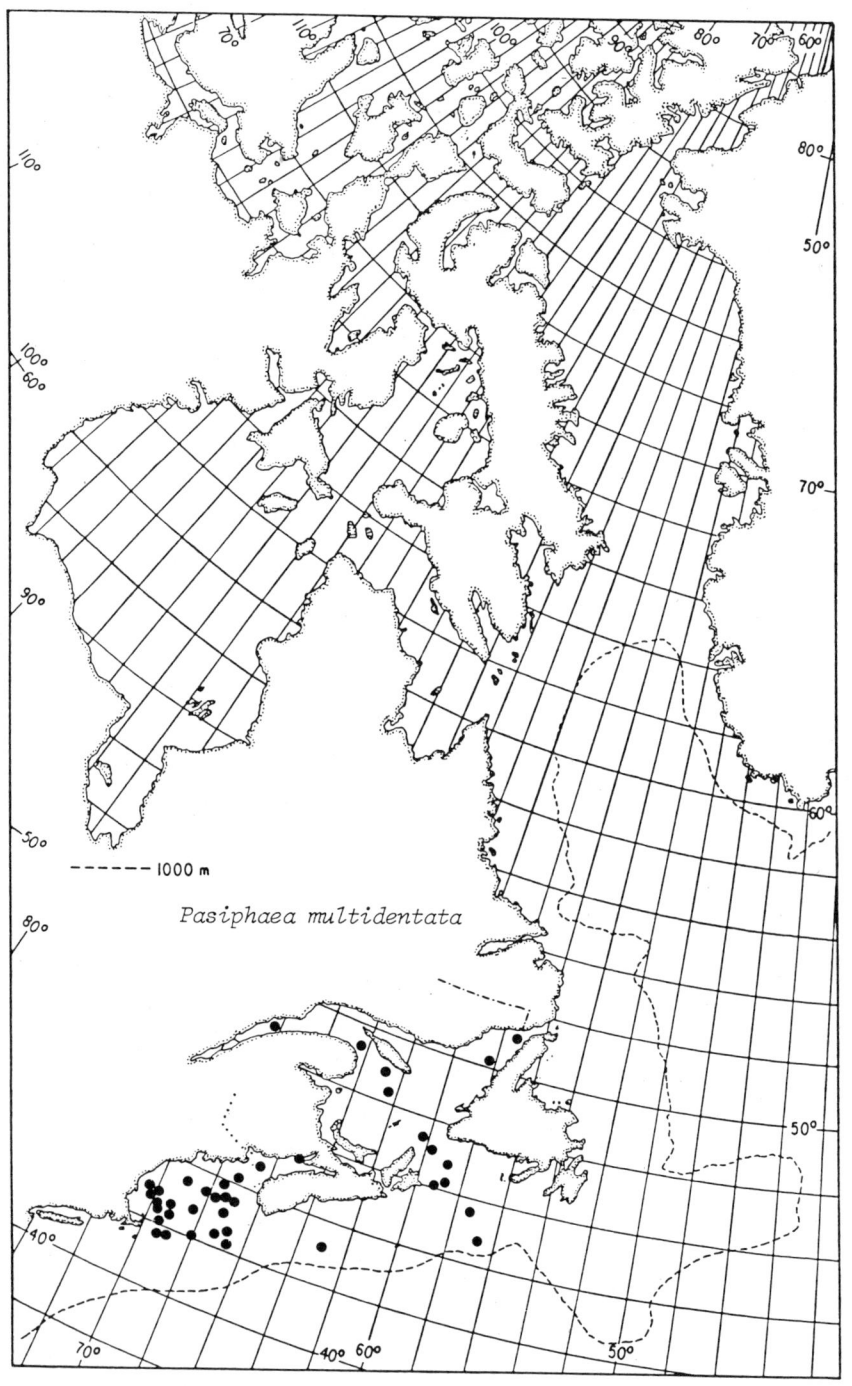

Fig. 60. *Pasiphaea multidentata*, distribution records in the area of reference.

Pasiphaea tarda Krøyer, 1845
Butler 1980: 56, 57 fig.; Sivertsen and
Holthuis 1956: 23–27, figs. 17, 18.
(Figures 61 and 62, Plate 1b)

DISTINGUISHING CHARACTERISTICS

Rostrum short, a thin post-frontal spine lies close above and ahead of it, turning down at the tip; first two pairs of legs more robust and longer than others, chelate, with long slender fingers pectinate on cutting edges; few (0–5) spines on the lower edge of the basis of the second leg; body compressed with sharp dorsal carinae, crimson in colour and heavier than *Pasiphaea multidentata*; all legs with exopods; mandible with incisor only.

DESCRIPTION

Integument smooth and shiny. Colour crimson with some appendages translucent.

Rostrum short, at median edge of carapace, exceeded by a sharply pointed, blade-like, post-frontal fixed spine curving downward from slight hump on mid-dorsal anterior carapace.

Carapace compressed, widening and deepening posteriorly with median carina sharp in front, extending to near the posterior edge. A low lateral carina extends from hepatic area over part of branchial area; above it is a shallow depression and a less distinct carina curving upward towards the posterior edge of the carapace. Small sharp branchiostegal spine with short low carina, antennal spine rounded, exceeded by postorbital lobe. Ventral notch at anterior edge.

Abdomen strongly compressed with sharp mid-dorsal carina on all somites (on posterior half of 1st somite), on the 6th it branches into two at posterior four-fifths; 6th also has a lateral arched carina.

Telson (t) about equal in length to 6th somite; with dorsal sulcus, tapering slightly to wide fork, each tip with a strong spine and series of about 15 spinules to mid-fork; outer branch of uropod with disto-lateral spine (separated from diaeresis), longer than inner branch, both longer than telson.

Eye moderate, cornea globular, wider than short stalk.

Antennule (c) 1st article about equal to other two sub-equal articles; stylocerite with keel proximally and short tapered point distally, reaching end of 1st article.

Antenna (d) scale length slightly more than 3 times width, outer edge convex, spine exceeding blade; peduncle reaches less than half scale; basal article with sharp disto-ventral spine and lateral curved projection near junction with blade.

Mandible (e) incisor only with 3 major teeth and decreasing series of 6 plus sharp corner.

Maxillule (f) proximal endite ovate, as long as distal, the latter expanding from narow base to straight spinous edge; endopod straight, thick, with subapical setae.

Maxilla (g) endites greatly reduced; endopod straight, shorter than long anterior lobe of scaphognathite, the latter narrower than short rounded posterior lobe.

Maxilliped I (h) endites reduced; endopod moderate, partly fused with exopod, the latter with reduced lobe and distal spatulate lash. Epipod small.

Maxilliped II (i) leglike, only slightly compressed, distal segment longer than wide with a strong apical spine.

Maxilliped III (k) distal segment sharp-pointed, not as long as longest segment; exopod slightly longer than the latter.

Pereopods: I (l) robust, chelate, chela long, slender fingers with sharp points crossing over each other, almost as long as palm, cutting edges pectinate; merus with 7 strong spines on mesial edge; exopod. II (m) more robust and longer than I, chelate, fingers slender, with sharp curved points, cutting edges pectinate (teeth larger than in I), as long

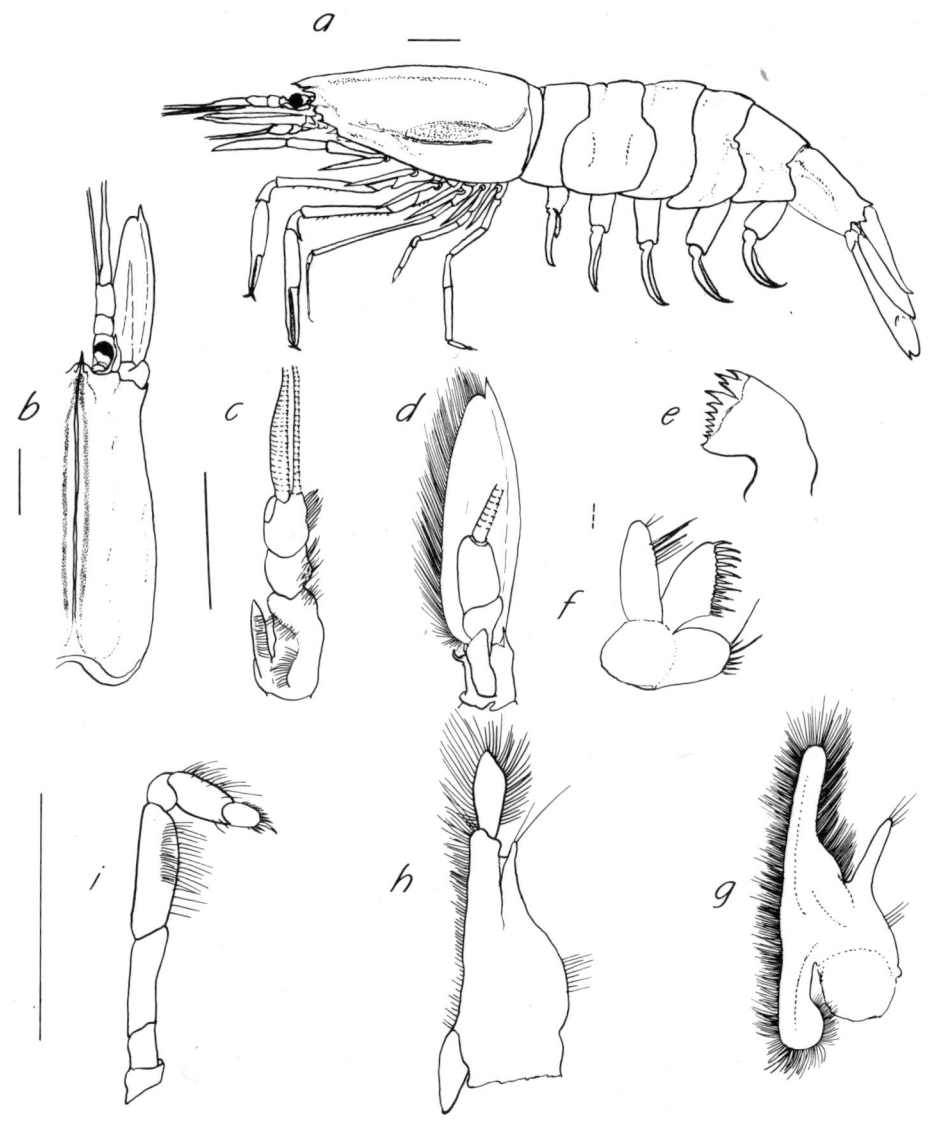

Fig. 61. *Pasiphaea tarda*: *a*, whole shrimp from left side; *b*, carapace in dorsal aspect; *c*, antennule; *d*, antenna; *e*, mandibles; *f*, maxillule; *g*, maxilla; *h*, first maxilliped; *i*, second maxilliped; solid line = 10 mm, broken line = 1 mm.

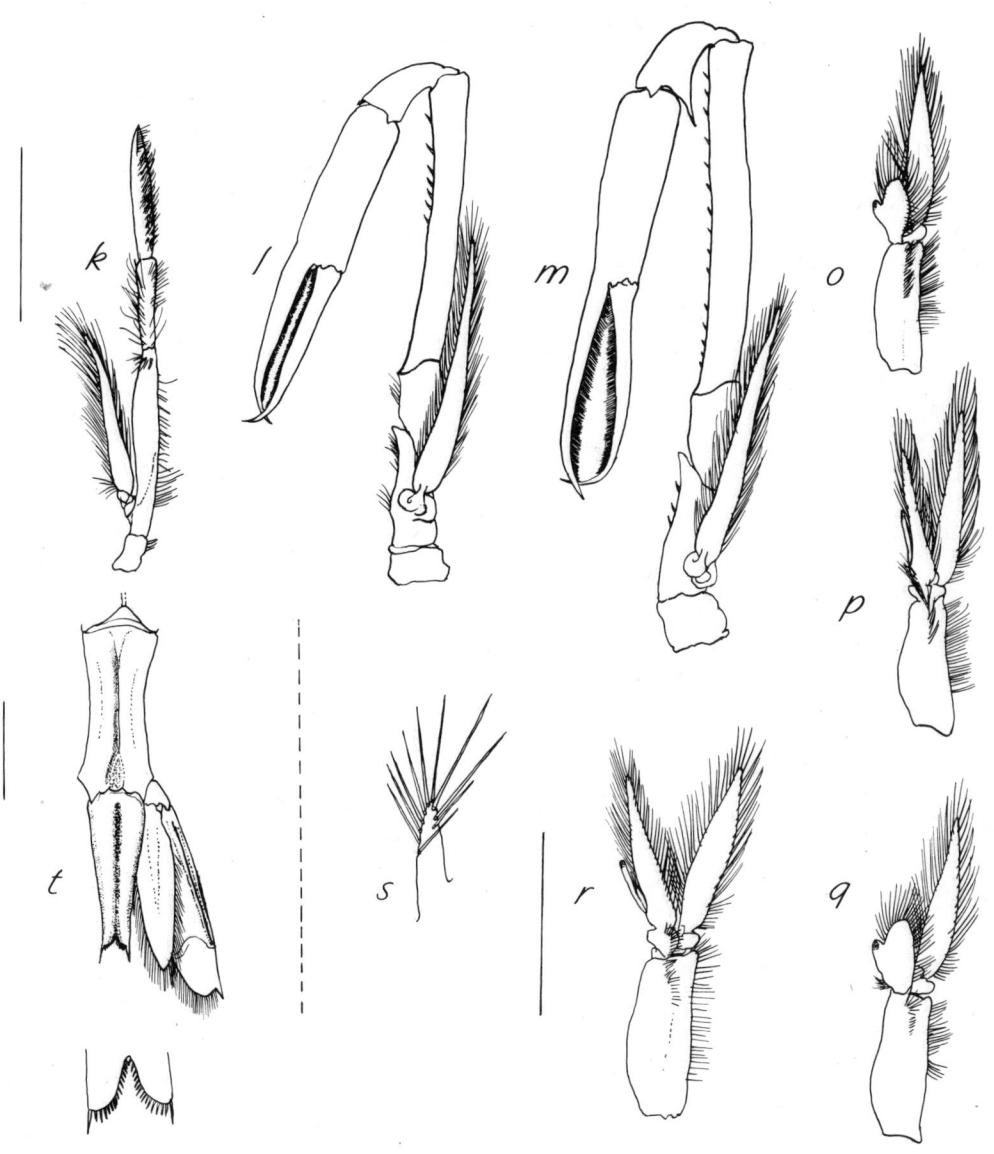

Fig. 62. *Pasiphaea tarda*: *k*, third maxilliped; *l*, first pereopod; *m*, second pereopod; *o*, F first pleopod; *p*, F second pleopod; *q*, M first pleopod; *r*, M second pleopod; *s*, appendix masculina; *t*, telson; solid line = 10 mm, broken line = 1 mm.

as palm; basis with 3 (1–5) sharp spines on lower edge, merus with 14, carpus with strong distal spine; exopod. III very slender, about as long as I, tapering to sharp dactyl; exopod. IV short, about 0.4 length of III, slender; dactyl ovate, compressed; exopod. V about as long as III, moderately slender; dactyl compressed, subovate, with a few long setae; exopod.

Pleopods: female I (o) endopod short, mitt-shaped, lateral projection with patch of hooks; female II (p) endopod and exopod subequal, narrow, appendix interna long with apical patch of hooks; male I (q) endopod short, mitt-shaped, lateral projection with patch of hooks; male II (r) endopod subequal to exopod, appendix interna with apical patch of hooks, longer than appendix masculina (s), the latter with about 10 apical and subapical long spines.

RANGE OF DISTRIBUTION

In the western Atlantic from Greenland and Hudson Strait to off South Carolina, and in the eastern Atlantic from Jan Mayan to the British Isles and as far south as 9°27′ S latitude off the coast of Angola (Butler 1980; Crosnier et Forest 1973). In the Pacific from Unalaska to Oregon and off Ecuador, South America (Butler 1980).

Records of distribution in the area of reference are in Fig. 63.

BIOLOGY

Lengths to 55 mm cl in males and 75 mm cl in females (Smith 1884c).

Females were first mature at 38 mm cl. Egg diameter 3.5 mm. About 50% of mature females collected off Newfoundland in May were ovigerous, one with advanced embryos. Farther north 40% of the mature females collected were ready to lay eggs in autumn. Two of the ovigerous females with advanced embryos in eggs also had large ova in the ovaries indicating that they would extrude eggs after hatching (Squires 1965a).

Stomach contents were euphausiids, chaetognaths, shrimp fragments and beak of a small squid (Squires 1965a).

Off Labrador 14% of specimens collected were infested with a species of the Ellobiopsidae; 10% on the Grand Banks of Newfoundland. Distortion of the rostral area was caused by this parasite (Squires 1965a).

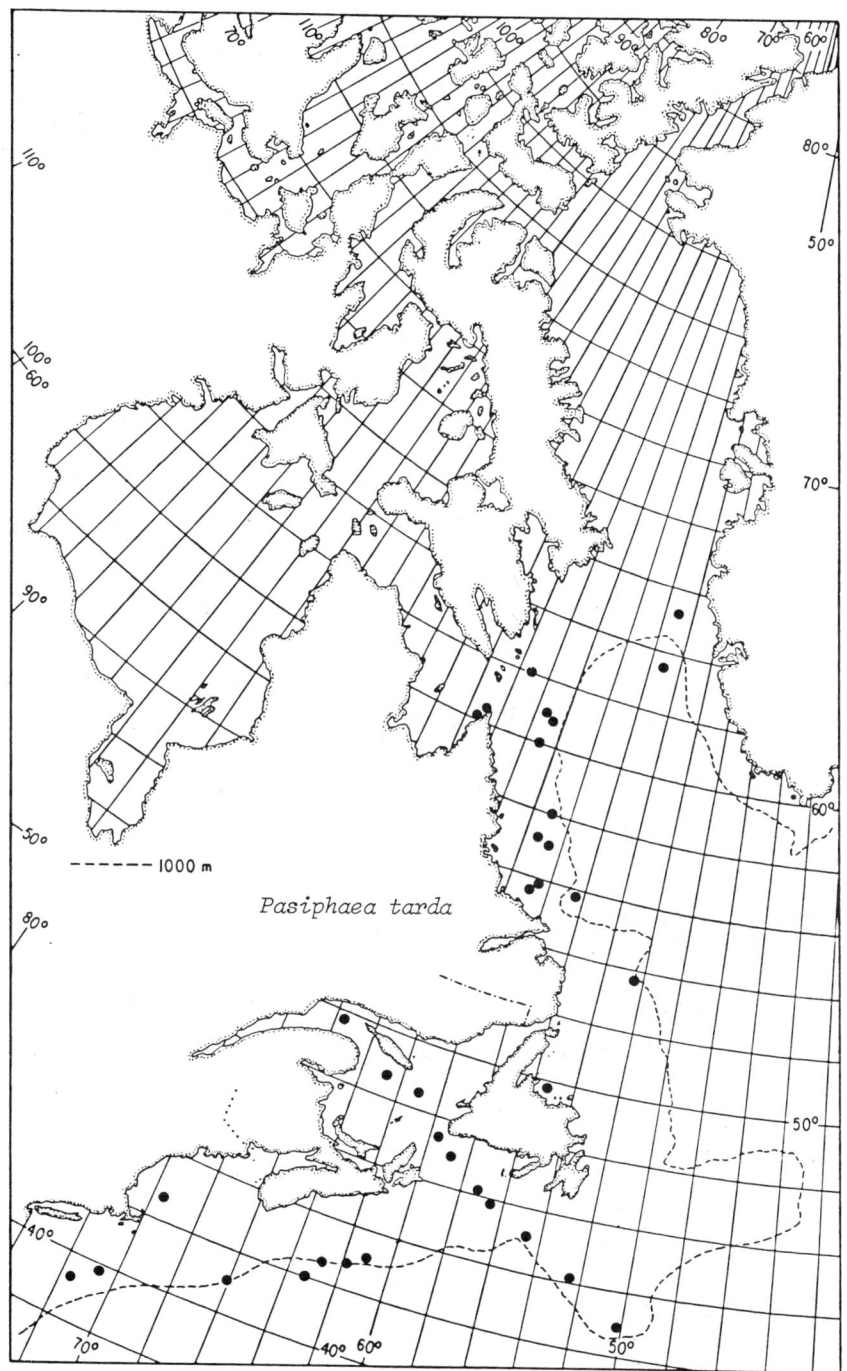

Fig. 63. *Pasiphaea tarda*, distribution records in the area of reference.

Family PALAEMONIDAE Samouelle, 1819
Holthuis 1952; Rathbun 1929: 19; Williams 1984: 63.

Second pair of chelate legs larger than first; carpus of second pair not subdivided; dorso-lateral flagellum of antennule divided into two; rostrum usually with teeth and not moveable; mandible usually with incisor process.

Key to species in area of reference
(After Williams 1984)

1 Hepatic spine present, branchiostegal absent; second legs enlarged (missing) .... *Macrobrachium* sp. (?)

   Hepatic spine absent, branchiostegal present; second legs not greatly enlarged 2

2 Carapace without branchiostegal groove ventral to antennal spine; endopod of first pleopod of male with appendix interna; mandible with palp .... *Leander tenuicornis*

   Carapace with branchiostegal groove, endopod of first pleopod of male without appendix; mandible without palp 3

3 Rostrum with 2 teeth of rostral series behind orbit, teeth reaching to tip, 3–5 ventral teeth; dactyl of second leg with 2 teeth, fixed finger with 1 tooth on cutting edge .... *Palaemonetes (P.) vulgaris*

   Rostrum with only 1 tooth behind orbit, teeth not reaching to tip; fingers of second leg without teeth on cutting edge .... *Palaemonetes (P.) pugio*

Genus *Leander* Desmarest, 1849
Holthuis 1952: 167.

Rostrum with strong dorsal and ventral teeth; carapace with antennal and branchiostegal spines, the latter behind anterior margin; no branchiostegal groove; telson with two pairs of dorsolateral and terminal spines, the latter with plumose setae between them; mandible with a two-segmented palp; maxillipeds with exopods; first pleopod of male with an appendix interna.

Leander tenuicornis (Say, 1818)
Holthuis 1952: 159, Pl. 41, figs. a–g, Pl. 42, figs. a–f; Williams 1984: 65, fig. 45.
(Figures 64 and 65)

DISTINGUISHING CHARACTERISTICS

Second pair of legs much larger than first, fingers longer than palm, carpus not segmented; rostrum large, wider blade in female than in male, dorsal and ventral teeth strong, 2 behind orbit, ventral partially hidden by rows of setae; mandible with palp; antennule with 3 flagella (one branched near base).

DESCRIPTION

Integument smooth, shiny. Colour reddish brown with lighter brown mottling or streaks, and whitish bands at front and posterior carapace and end of abdomen (Sivertsen and Holthuis, 1956: Pl. I, fig. 3).

Rostrum deep and arched in female (ascending and narrower in male) with 9–11 + 2/6–7 spines, smaller distally, intercalated by fine setae, two rows of setae attached above spines ventrally; rostrum about equal to cl in females, longer in males.

Carapace with short median carina with 2 spines behind orbit; suborbital lobe, strong antennal and moderate branchiostegal spines, the latter with short carina.

Abdomen with rounded terga and pleura on all somites except 5th with a ventroposterior spine.

Telson (t) short and wide, slightly tapering to subtruncate end with short fixed central spine flanked by pair of very long moveable spines and a short pair at the corners, also a lateral seta at each side of central spine and a pair of plumose setae parallel to but not as long as long spines; two pairs of dorsolateral spines. Inner branch of uropod longer than telson but shorter than outer branch, the latter with conspicuous row of short setae along outer edge and axillary spine.

Eye large with globular cornea and ocellus.

Antennule (c) 1st article with lateral expansion and distal spine, longer than other two equal articles; stylocerite very sharp pointed almost as long as basic part of 1st article; fused area of branching dorso-lateral flagellum much shorter than free part.

Antenna (d) scale length 3.2 times width; disto-lateral spine about even with blade; peduncle reaches less than half scale; basal article with ventro-lateral spine.

Mandible (e) molar with strong sharp cusps; incisor wide with three sharp teeth; palp with two segments.

Maxillule (f) endites about equal, curved toward each other; endopod clearly biramous.

Maxilla (g) proximal endite reduced; distal endite bilobed narrow; endopod moderate, thin, curved; anterior lobe of scaphognathite narrow from wide base, longer than wider axe-shaped posterior lobe.

Maxilliped I (h) distal endite wide, subtriangular, proximal short, thick; endopod very short, not as long as short lobe of exopod but lash very long; epipod large bilobed, anterior lobe the larger.

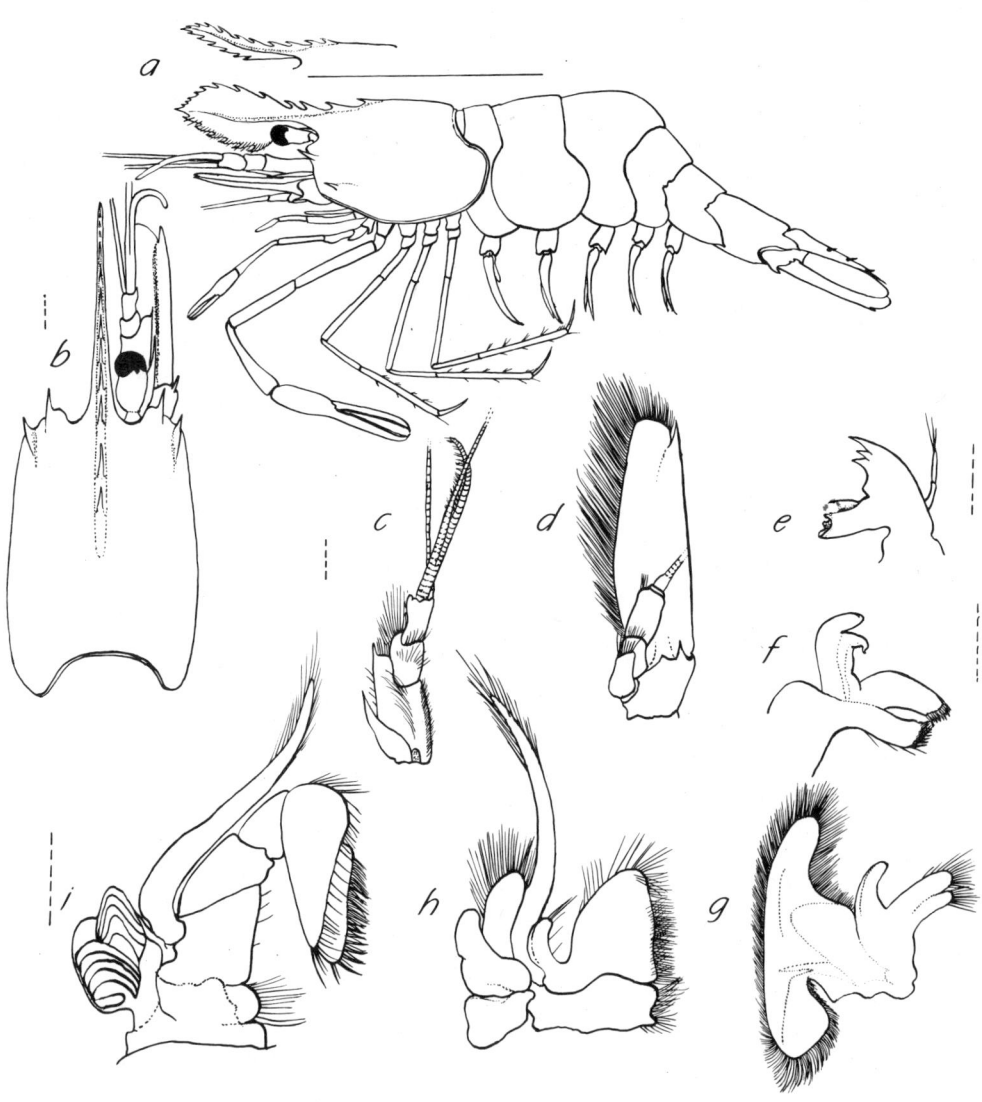

Fig. 64. *Leander tenuicornis*: *a*, whole shrimp from left side; *b*, carapace in dorsal aspect; *c*, antennule; *d*, antenna; *e*, mandible; *f*, maxillule; *g*, maxilla; *h*, first maxilliped; *i*, second maxilliped; solid line = 10 mm, broken line = 1 mm.

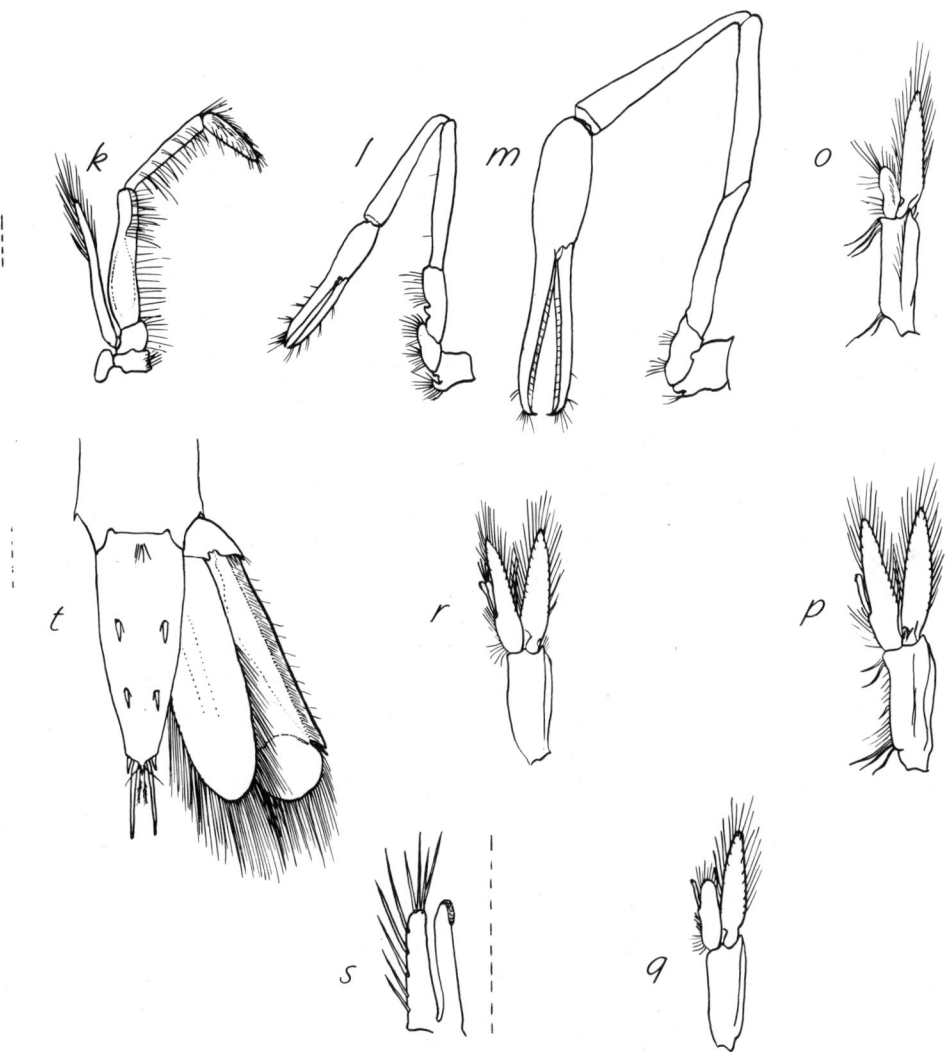

Fig. 65. *Leander tenuicornis*: *k*, third maxilliped; *l*, first pereopod; *m*, second pereopod; *o*, F first pleopod; *p*, F second pleopod; *q*, M first pleopod; *r*, M second pleopod; *s*, appendix masculina; *t*, telson; broken line = 1 mm.

Maxilliped II (i) compressed; distal segment much wider than long; exopod with long lash; epipod rounded with large podobranch.

Maxilliped III (k) with 5 segments, the longest with disto-lateral expansion; exopod slender, long; short thin epipod.

Pereopods: I (l) chelate, shorter and more slender than II; fingers longer than palm; propodus longer than carpus, about equal to merus. II (m) chelate, much larger than I; propodus longer than carpus or merus; fingers slender longer than inflated palm, cutting edges with fine close setae. III–V about equal, moderately long and slender; dactyls long smooth, sharp; propodus with row of moveable spines on flexor surface.

Pleopods: female I (o) endopod much shorter than exopod, slightly curved; female II (p) endopod and exopod equal, appendix interna slender with disto-lateral patch of hooks. Male I (q) endopod short compresed, with appendix interna with disto-lateral patch of hooks; male II (r) endopod about equal to exopod, appendix interna with disto-lateral patch of hooks, slightly longer than appendix masculina (s), the latter with 4 apical long fine spines and a lateral series of about 5 long spines.

RANGE OF DISTRIBUTION

Tropical and subtropical waters of the world except west coast of the Americas; Grand Banks and mouth of Bay of Fundy (Wigley 1970; Williams and Wigley 1977) to Falkland Islands in west Atlantic (Holthuis 1952; Bruce 1974). Found among floating sargassum in the open sea or on grass or *Porites* flats in shallow water (Chace 1972).

Records of distribution in the area of reference are in Fig. 66.

BIOLOGY

Lengths to 47 mm body length. Ovigerous females of 23 mm body length have been reported (Holthuis 1952).

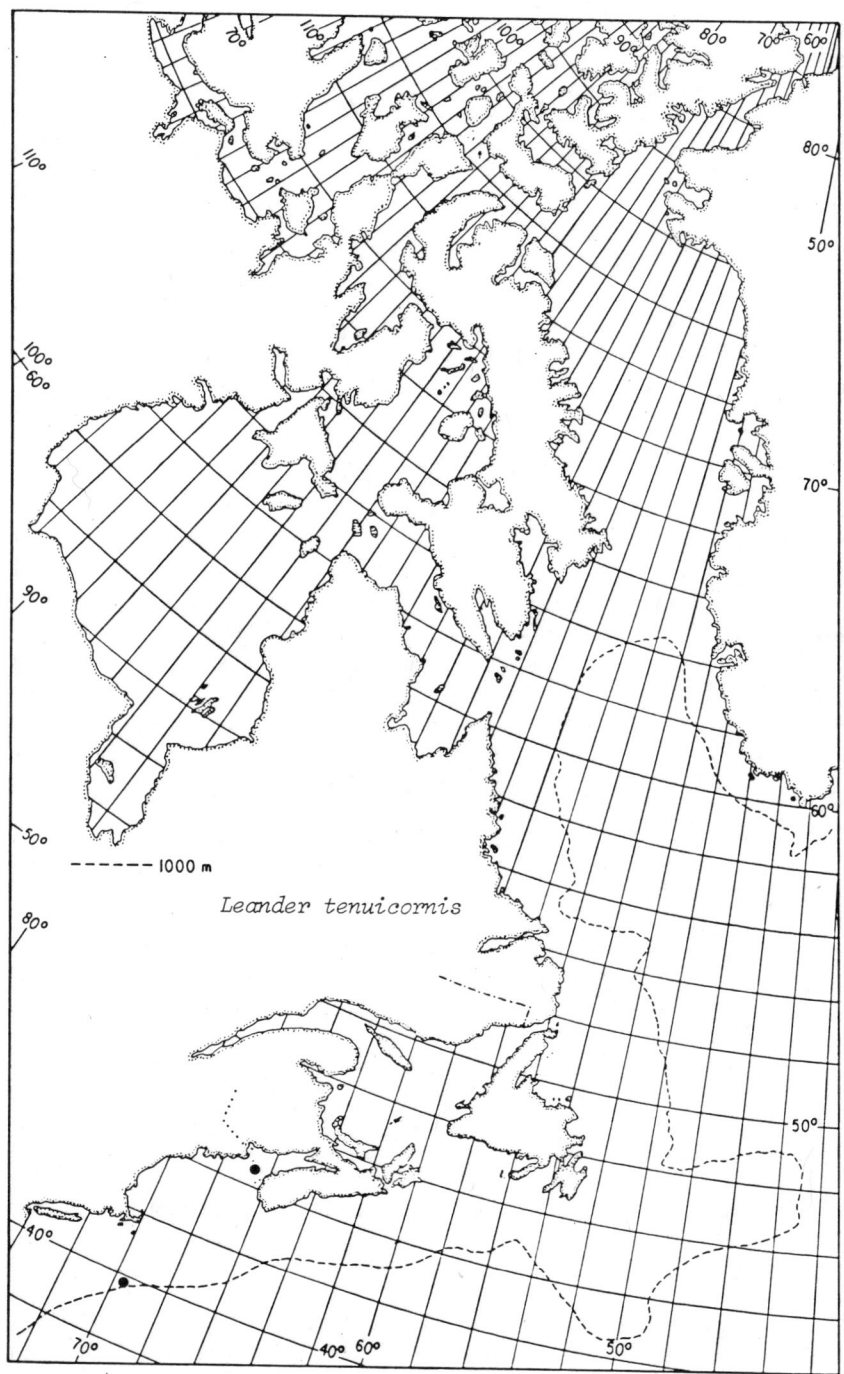

Fig. 66. *Leander tenuicornis*, distribution records in the area of reference.

Genus *Macrobrachium* Bate, 1868
Holthuis 1952: 10; Chace and Hobbs 1969: 89.

Carapace with strong antennal and hepatic spines in line, with branchiostegal groove just below and behind them; last three legs with dactyl simple (without accessory spine); rostrum strong with dorsal and ventral teeth, some behind orbit; mandible with three-segmented palp; exopods on maxillipeds; first pleopod of male without appendix interna.

Twenty-six species of this genus are known from America but all inhabit fresh water, although not restricted to fresh water in their development (Holthuis 1952).

(Collection of the following species (three specimens including 2 females and 1 male) in the Fundian Channel south of Nova Scotia is well-documented. The 2nd pereopods were missing in all specimens, a plausible accident when such legs are extra large. Since identification of species in this genus is frequently dependant on description of the 2nd pereopods, I have not advanced a name for it at this time. The specimens are on deposit at the Atlantic Reference Centre, St. Andrews, N. B., catalog No. ARC 8958232).

Macrobrachium new species (?)
(Figures 67 and 68)

DISTINGUISHING CHARACTERISTICS

Large rostrum with strong fixed spines (12–14) dorsally, 4 of which are behind orbit, and 2–4 ventrally; the only other spines on carapace are a strong antennal at anterior edge and a hepatic spine a short distance behind it; behind the spines and slightly below them is a branchiostegal groove; the dorso-lateral flagellum of the antennule is divided into two near its beginning; antennal scale is longer than the rostrum; telson is short, stubby, and has two pairs of dorso-lateral spines, terminally also two pairs of spines and a bunch of long setae between them at the centre.

DESCRIPTION

Integument, thin, smooth, shiny. Colour when living not available.

Rostrum thin at edges but stout, almost straight but slightly descending towards tip, lateral supporting ridge confluent with orbit; large fixed spines intercalated with setae, slightly larger distally, continued on to short median carina on carapace, about 4 behind orbit, 9–10 + 4 dorsally and 2–4 ventrally; rostrum reaches about distal end of antennular peduncle.

Carapace with sharp antennal spine just behind anterior margin, with a short carina, followed by the hepatic spine and slightly below and behind it a short branchiostegal groove. No other spines. A small ventral expansion at posterior edge.

Abdomen with all terga and pleura rounded except at 5th somite with a short pointed tip ventro-laterally; 6th somite only slightly longer than 5th.

Telson (t) only slightly longer than 6th somite, rounded with two pairs of spinules sunk into the integument dorsally on posterior half, bluntly pointed terminally with two pairs of spines, the inner longer, lateral to several long setae. Outer branch of uropod wide, with axillary spine, about equal in length to inner branch, longer than telson.

Eye with large pigmented and faceted cornea on short tapered stalk, with small ocellus.

Antennule (c) 1st article with outer edge lamellate extending distally with a sharp spine, inner edge thickened with small ventral spine; 2nd article with disto-ventral extension reaching almost distal end of 3rd; stylocerite tapering to sharp point, short, reaching only about 0.3 1st article; dorso-lateral flagellum divided, branches fused a short distance at base.

Antenna (d) scale length about 3 times width, disto-lateral spine exceeded by blade; ventro-distal spine on basal article; peduncle reaching about half scale.

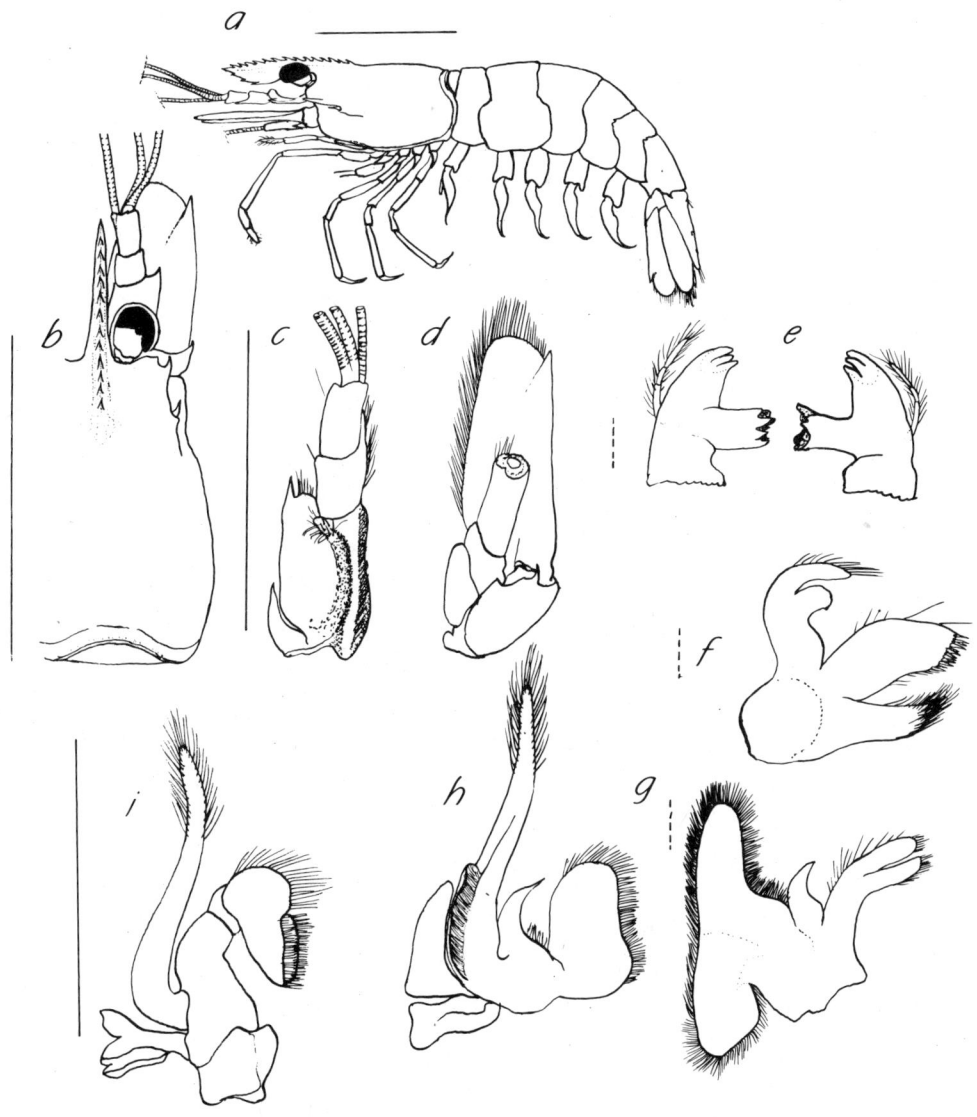

Fig. 67. *Macrobrachium* new species (?): *a*, whole shrimp from left side; *b*, carapace in dorsal aspect; *c*, antennule; *d*, antenna; *e*, mandible; *f*, maxillule; *g*, maxilla; *h*, first maxilliped; *i*, second maxilliped; solid line = 10 mm, broken line = 1 mm.

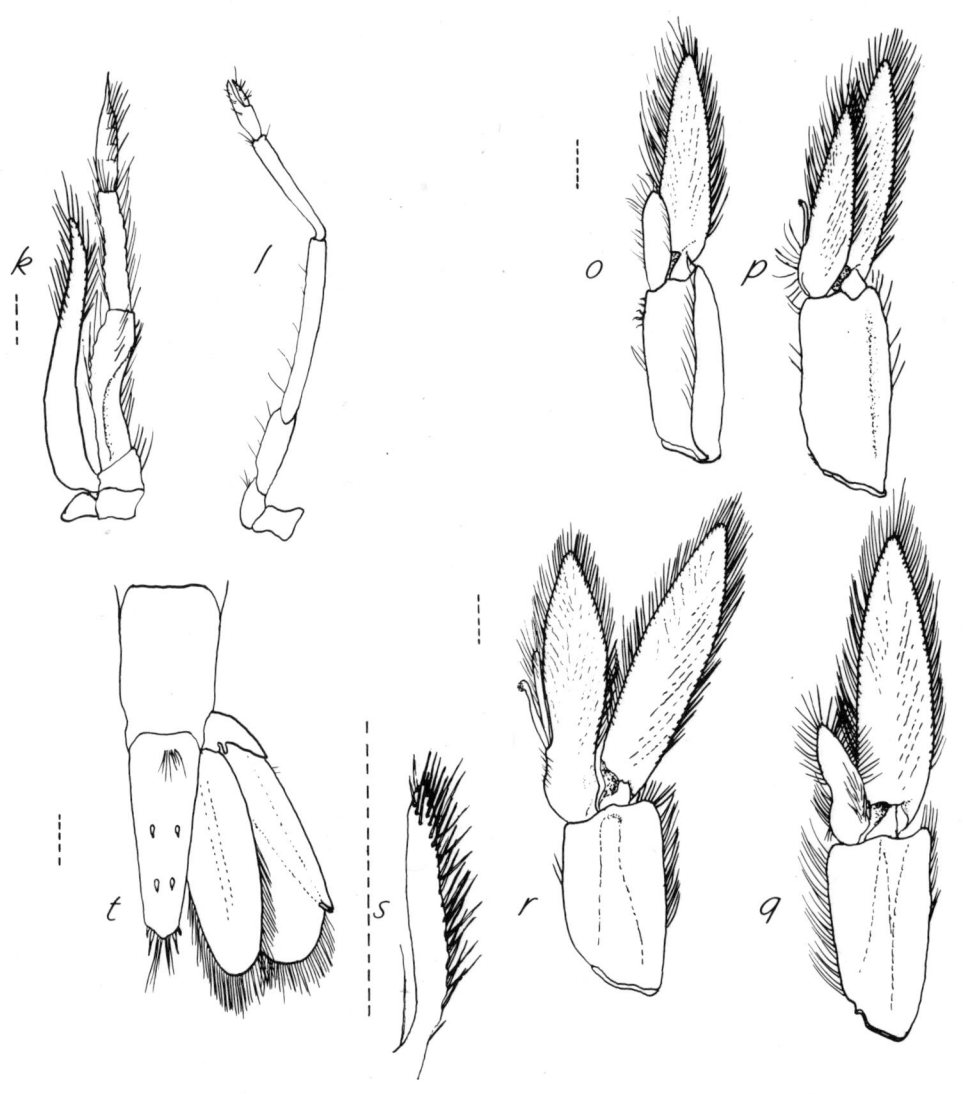

Fig. 68. *Macrobrachium* new species (?): *k*, third maxilliped; *l*, first pereopod; *m* (missing); *o*, F first pleopod; *p*, second pleopod; *q*, M first pleopod; *r*, M second pleopod; *s*, appendix masculina; *t*, telson; broken line = 1 mm.

Mandible (e) incisors similar from wide base with three teeth; molars slightly different in numbers of sharp surface cusps, short. Palp slender, with three segments.

Maxillule (f) distal endite wider and longer than proximal, the latter tapered and slightly curved; endopod clearly biramous, distal ramus narrower and pointed.

Maxilla (g) distal endite bilobed, equal, narrow; proximal endite completely reduced; endopod short, sharp pointed; anterior lobe of scaphognathite narrow from wide base, slightly longer than posterior axe-shaped lobe.

Maxilliped I (h) distal endite sub-rectilinear, wide, proximal reduced; endopod short, sub-triangular in shape and cross-section, with knobby edge distally and sharp point, about as long as lobe of long exopod; epipod with two different lobes, anterior subtriangular and almost as long as exopod lobe.

Maxilliped II (i) compressed, distal segment much wider than long; exopod long; epipod with small podobranch.

Maxilliped III (k) robust, heavier than 1st leg; distal segment with long acuminate tip; long stout exopod; small rudimentary epipod.

Pereopods: I (l) slender, chelate, ischium short and expanded; II with short expanded ischium as in I. Other segments broken away and lost. III–V moderately long and slender, dactyl smooth and sharp, propodus with series of spines on flexor edge; propodus of V with also horizontal rows of setae.

Pleopods: female I (o) endopod short, sub-rectilinear, exopod long wide; female II (p) endopod appreciably shorter than exopod, appendix interna slender with apical hooks. Male I (q) endopod short, pointed, slightly curved, exopod much longer; male II (r) endopod slightly shorter than exopod, appendix interna with subapical patch of hooks, shorter than appendix masculina (s), the latter with two rows of apical spines and numerous lateral spines.

RANGE OF DISTRIBUTION

Taken in the Fundian Channel, off Nova Scotia, Lat. 43°28′ N, 67°32′ W long., depth 218 m, temperature 7.5° C at bottom, 14.3° C at surface, on July 11, 1974, by the *A. T. Cameron* on Trip 225 (Station 18); collector Dr. J. G. Scott (Fig. 69).

The specimens are on deposit at the Atlantic Reference Centre, Huntsman Marine Science Centre, St. Andrews, N. B.

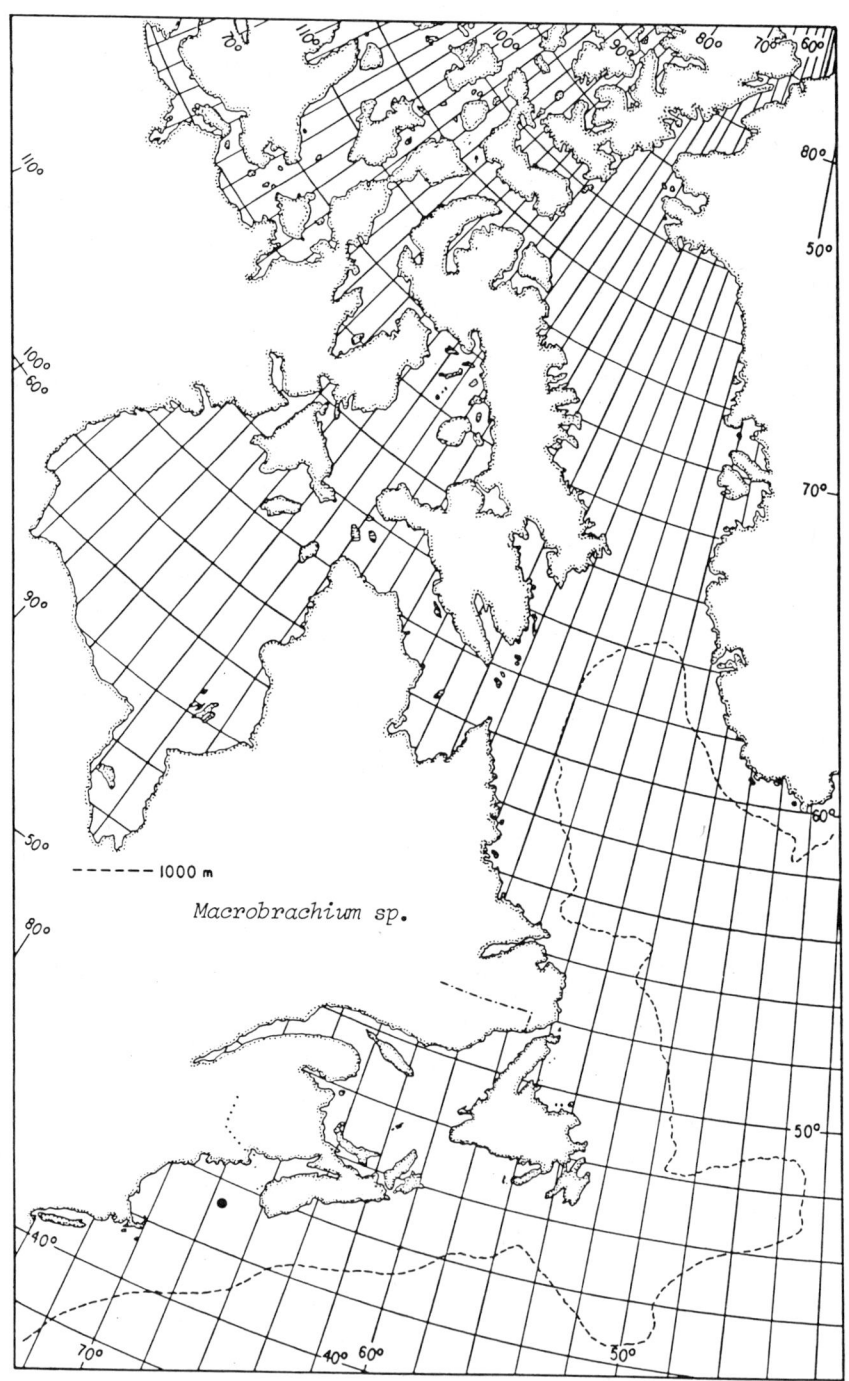

Fig. 69. *Macrobrachium* new species (?), distribution record in the area of reference.

Genus *Palaemonetes (Palaemonetes)* Heller, 1869
Holthuis 1952: 199; Rathbun 1929: 19.

Carapace with antennal and branchiostegal spines but no hepatic spine; branchiostegal groove; rostrum large with dorsal and ventral teeth; mandible without palp; upper antennular flagellum with 2 branches, fused near base; second pair of legs larger than first.

Of the ten species of the subgenus *Palaemonetes* in American waters only two are present in the area of reference.

Palaemonetes (Palaemonetes) pugio Holthuis, 1949
Holthuis 1949: 95, figs 2m–o; 1952: 244, pl. 55,
figs g–l; Williams 1984: fig. 51;
Fleming 1969: 445, ff, figs 7, 9.
(Figures 70 and 71)

Distinguishing Characteristics

Dagger-shaped tip of rostrum (teeth set back from tip) reaching end of scale, slightly turned up; only one tooth of rostral series behind orbit, ventral teeth 2–4 mostly 3; no tooth on cutting edge of chela of second legs; upper flagellum of antennule with fused part (of 9-14 segments) of both branches slightly shorter than free part of shorter branch; carpus of 2nd leg longer than palm.

Description

Integument smooth, shiny. Colour translucent.

Rostrum moderately high and straight but midrib curved slightly upward at tip (confluent with orbit), bare and sharp near tip but with strong teeth above and below 8–9/2–4.

Carapace with short mid-dorsal carina and only one spine of rostral series behind orbit; branchiostegal spine, larger than antennal, is at anterior margin just below a curved branchiostegal groove; low carinae extend back from the rostro-orbital confluence and from the antennal spine.

Abdomen with all terga and pleura rounded except on 5th somite which has a small ventro-posterior spine; 6th somite about 0.8 telson.

Telson (t) moderately tapered to subtruncate tip with short thick central spine and two pairs of lateral spines, the inner much longer than the outer; two pairs of dorso-lateral spines on posterior half; a faint median sulcus.

Eye large, cornea globular; ocellus.

Antennule (c) 1st article longer than other two combined, expanded on outer edge with distal short spine; ventral extension of 2nd reaches half 3rd article; fusion of dorso-lateral flagellum (about 9 annulations) one-third length of short branch in males and one-half length in females; stylocerite tapered to sharp point, only about one-third length of 1st article.

Antenna (d) scale length about 3 times width, blade exceeding disto-lateral spine; peduncle reaching about one-third scale; basal article with strong disto-ventral spine; flagellum about 1.2 times body length, 7 times cl.

Mandible (e) incisor moderate with four apical teeth, the corner ones sharper and slightly longer than the others; molar with ridged centre and 3 sharp fringing cusps. No palp.

Maxillule (f) proximal endite narrow tapering, smaller than wide slightly pointed distal endite; endopod bifurcate, the distal branch small, proximal wider, curved at tip.

Maxilla (g) proximal endite reduced to rounded areas; distal endite bilobed, both narrow and well-defined; endopod short, tapered; anterior lobe of scaphognathite moderately wide, from wider base, sloped slightly at tip; slightly longer than rounded posterior lobe.

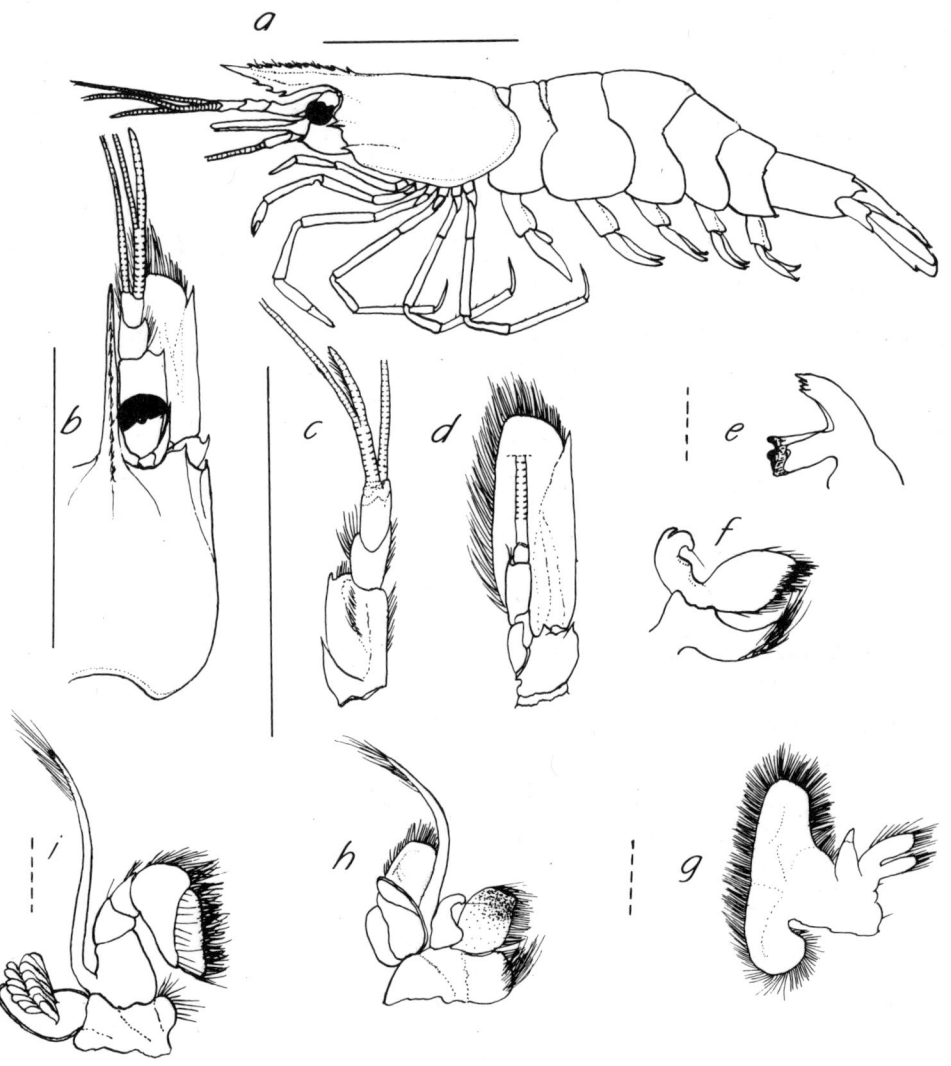

Fig. 70. *Palaemonetes pugio*: *a*, whole shrimp from left side; *b*, carapace in dorsal aspect; *c*, antennule; *d*, antenna; *e*, mandible; *f*, maxillule; *g*, maxilla; *h*, first maxilliped; *i*, second maxilliped; solid line = 10 mm, broken line = 1 mm.

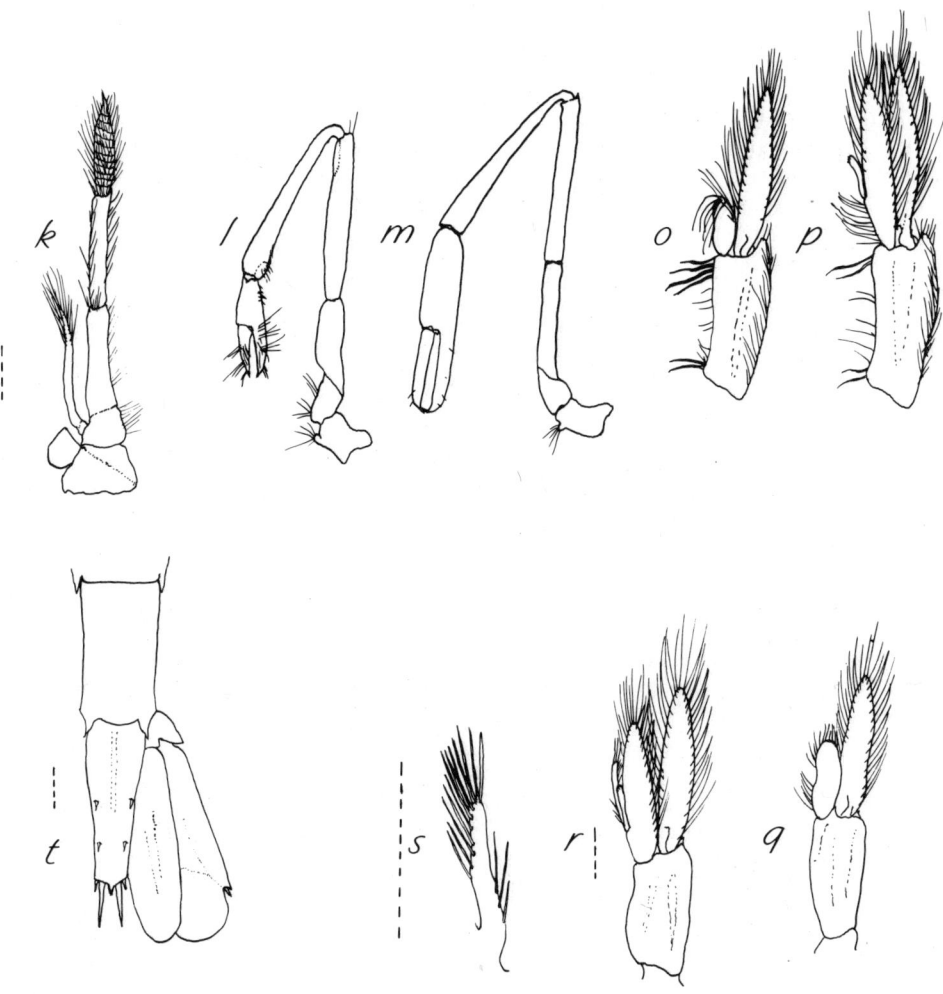

Fig. 71. *Palaemonetes pugio*: *k*, third maxilliped; *l*, first pereopod; *m*, second pereopod; *o*, F first pleopod; *p*, F second pleopod; *q*, M first pleopod; *r*, M second pleopod; *s*, appendix masculina; *t*, telson; broken line = 1 mm.

Maxilliped I (h) distal and proximal endites almost equal, proximal thickened at edge; endopod very short from wide base, curled at tip, only about half as long as wide lobe of exopod, lash slender, long; epipod foliaceous, bilobed.

Maxilliped II (i) compressed, distal segment long but wider than long; exopod slender, long; epipod saucer-like, with podobranch.

Maxilliped III (k) stout, shorter than 1st leg; distal segment with long acuminate tip and spiral rows of setae; exopod about as long as longest segment; epipod short, rudimentary.

Pereopods: I (l) moderately slender, chelate, fingers slightly longer than palm, carpus much longer than propodus. II (m) longer than I and chela much larger, palm longer than fingers; propodus longer than carpus; no tooth on cutting edges of fingers; III–V long and slender, non-chelate, dactyls long and sharp.

Pleopods: female I (o) endopod much shorter than narrow exopod; female II (p) endopod and exopod sub-equal, appendix interna with subapical patch of hooks. Male I (q) endopod rectilinear-ovate slightly bent, about half as long as exopod, no appendix interna. Male II (r) endopod shorter than exopod, appendix interna shorter than appendix masculina (s) the latter slightly crooked with 4 long spines apically continuous with 7 on one side and a separate group of 3 on the other proximally.

Range of Distribution

From Verte R., 3 mi. W. St. Modiste, Quebec (47°51′ N lat.) through Nova Scotia to Corpus Christi, Texas. In estuaries especially among submerged vegetation (Williams 1984).

Records of distribution in the area of reference are in Fig. 72.

Biology

Lengths of body to 33 mm in males and 50 mm in females. Males slightly different from females in morphology.

Ovigerous females have been reported at all times in the year in the southern part of its range (Rouse 1970). Experiments have shown that this species can be induced to breed at any time in the year at suitable temperatures: they spawned in 5 weeks, the eggs hatched in 15–19 days and larvae metamorphosed in 25–47 days (Little 1968). In nature at optimum conditions it is possible that they could produce a brood every two months.

Stomach contents have been observed to be detritus with associated bacteria and pennate diatoms (Williams 1984).

Parasitization with bopyrids in collections from Nova Scotia have been reported (Williams 1984).

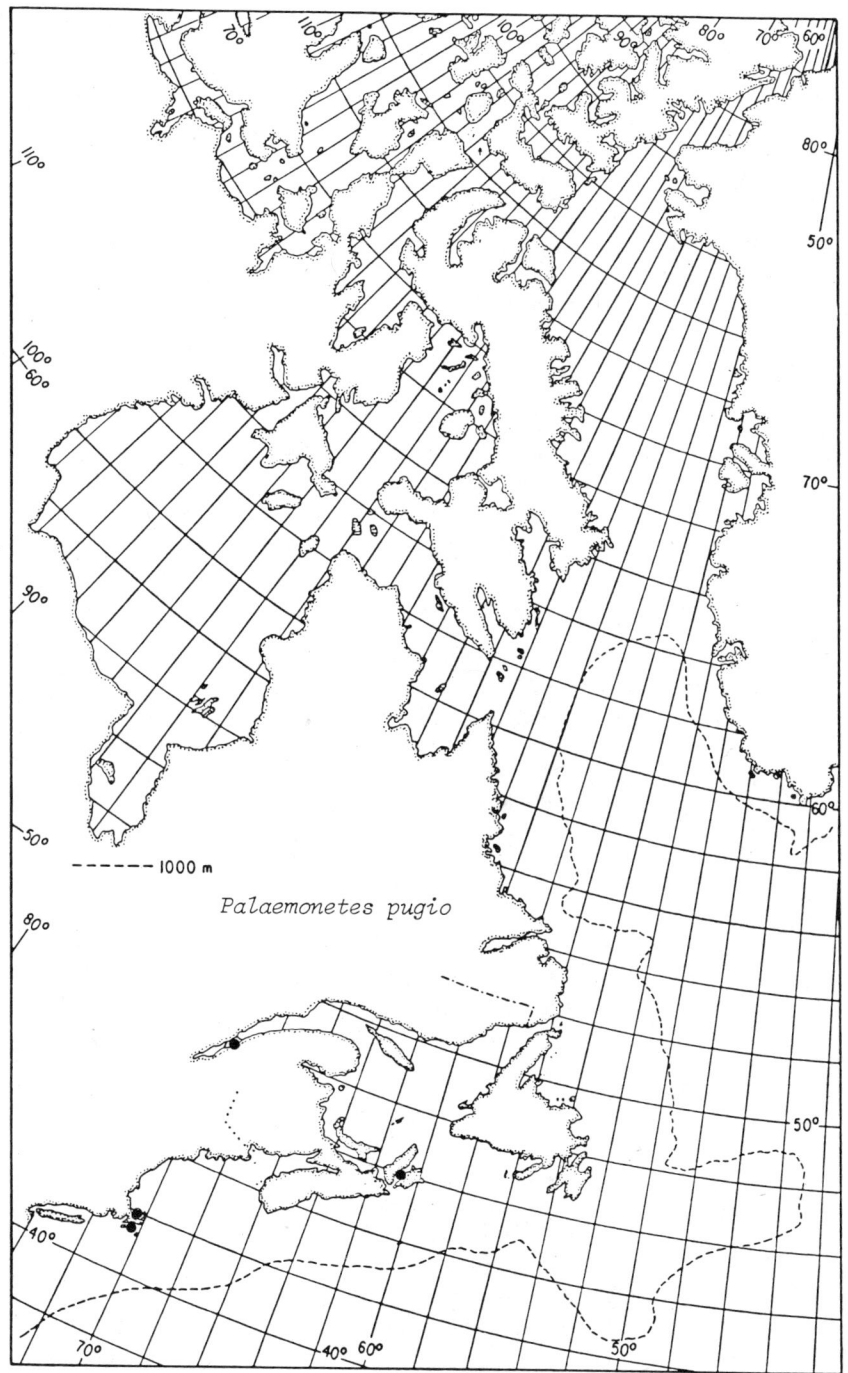

Fig. 72. *Palaemonetes pugio*, distribution records in the area of reference.

Palaemonetes (Palaemonetes) vulgaris Say, 1818
Holthuis 1952: 231, pl. 54, figs. f–l;
Rathbun 1929: 19, fig. 23.
"Marsh shrimp" — "Bouquet des marais"
(Figures 73 and 74)

DISTINGUISHING CHARACTERISTICS

Rostrum about as long as carapace, compressed and deep, with moderately large teeth reaching almost to tip dorsally and with two behind orbit, number 7 + 2/4–5; dactyl of second leg with 2 teeth and propodus with one tooth on cutting edge; antennal spine strong, branchiostegal moderate and moderate branchiostegal groove; 2nd leg larger than 1st; antennule with 3 flagella, upper one divided into two branches — joined over about 7–9 annulations.

DESCRIPTION

Integument smooth, shiny. Colour translucent.

Rostrum about as long as carapace, teeth reaching to tip and with 2 behind orbit, slightly larger distally, with intercalated setae; lateral carina thickening proximally and confluent with orbit; depth of rostrum about 0.25 length, reaching to or beyond scale.

Carapace with short mid-dorsal carina, prominent antennal spine and moderate branchiostegal spine and groove, small suborbital lobe.

Abdomen with rounded terga and pleura; 6th somite with ventro-posterior spine and short supporting carina.

Telson (t) rounded and slightly tapering to subtruncate but rounded tip with strong central spine flanked by two pairs of spines — the inner very much the larger — and a pair of strong plumose setae, longer than spines. Inner branch of uropod slightly shorter than outer but 1.3 times longer than telson, outer with a ventral submarginal row of setae and axillary spine shorter than outer spine.

Eye large, cornea globular wider than stalk, with dorsal ocellus.

Antennule (c) 1st article with outer lateral wing extended distally with short apical spine; stylocerite sharp, reaching about half 1st article; 2nd and 3rd articles subequal; fused area of dorso-lateral flagellum about 7–9 annulations, shorter free branch about 1.5 times fused part.

Antenna (d) scale about 0.8 time cl, length about 3 times width, with distal narrow area from strong convexity proximally, tip of blade exceeding disto-lateral spine; basal article with ventro-lateral spine distally.

Mandible (e) incisor with four teeth, molar with sharp cusps at edge of cutting surface. No palp.

Maxillule (f) narrow curved proximal endite about same length as ovate pointed distal endite; endopod bifurcate with very short distal branch and subapical branch with short spine.

Maxilla (g) distal endite subequally bilobed, narrow and separate; proximal endite reduced; endopod short; anterior lobe of scaphognathite rounded, longer than rounded posterior lobe.

Maxilliped I (h) distal endite wide, squarish, with slightly concave edge; proximal endite about equal to distal; endopod robust, rounded at tip, not as long as lateral lobe of exopod, or long lash; epipod unequally bilobed, both sub-rectilinear.

Maxilliped II (i) compressed, distal segment much wider than long; exopod slender, longer than endopod; epipod large, with podobranch.

Maxilliped III (k) leglike, distal segment with apical spine; exopod slender, about as long as longest segment; epipod moderate, squarish.

Pereopods: I (l) slender, chelate, chela short, fingers shorter than palm, carpus longer than merus; II (m) much stronger than I, chelate, carpus shorter than palm (in female:

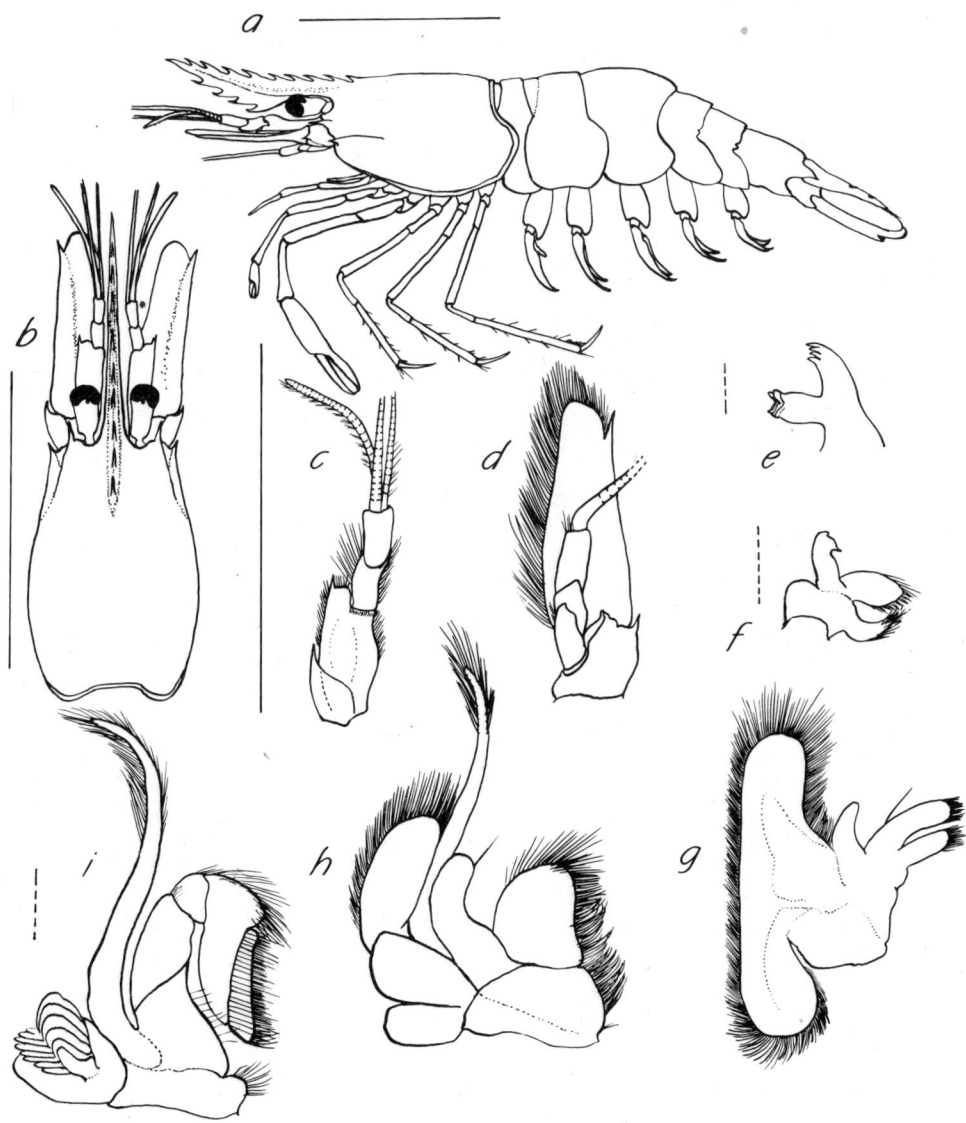

Fig. 73. *Palaemonetes vulgaris*: *a*, whole shrimp from left side; *b*, carapace in dorsal aspect; *c*, antennule; *d*, antenna; *e*, mandible; *f*, maxillule; *g*, maxilla; *h*, first maxilliped; *i*, telson; solid line = 10 mm, broken line = 1 mm.

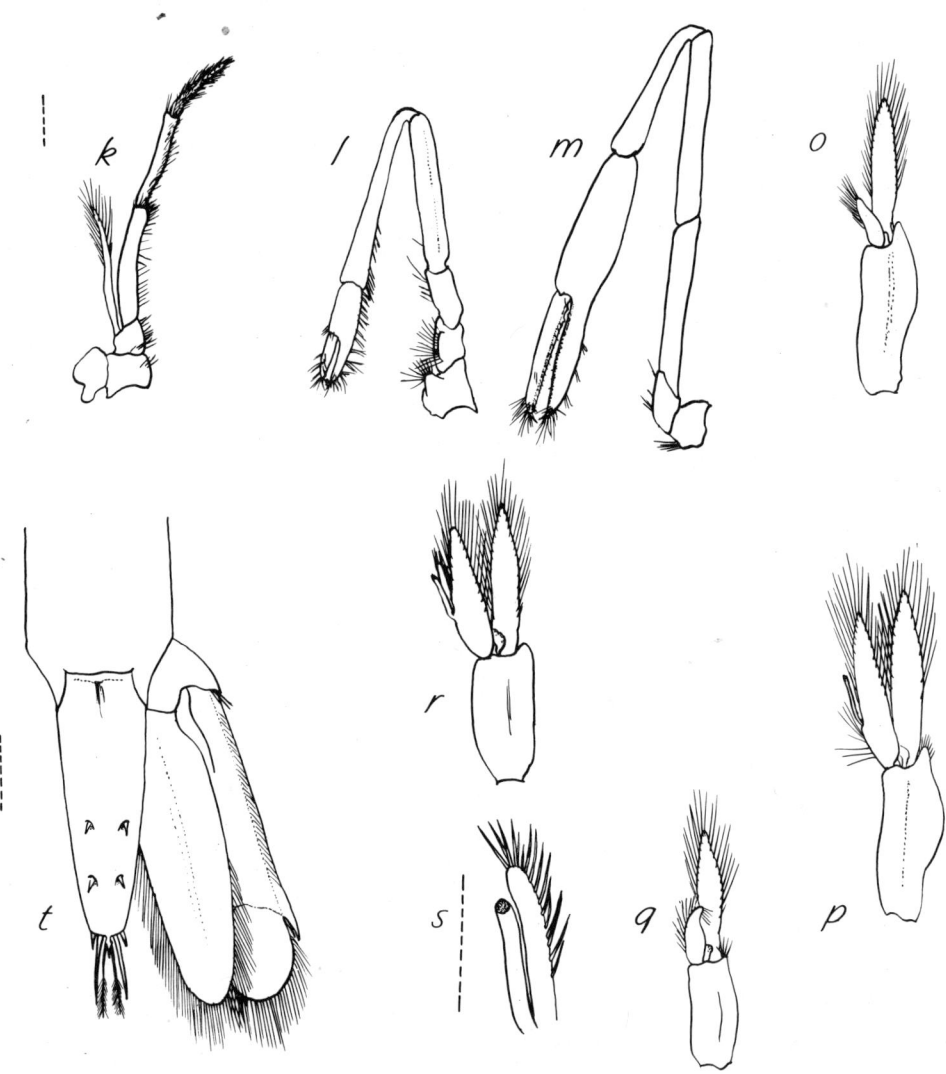

Fig. 74. *Palaemonetes vulgaris*: *k*, third maxilliped; *l*, first pereopod; *m*, second pereopod; *o*, F first pleopod; *p*, F second pleopod; *q*, M first pleopod; *r*, M second pleopod; *s*, appendix masculina; *t*, telson; broken line = 1 mm.

longer in male), 2 teeth proximally on dactyl, and 1 tooth on propodus on cutting edges; III–V long and slender, dactyl smooth and sharp, propodus with series of spines on both sides.

Pleopods: female I (o) endopod about one-third length of exopod; female II (p) endopod and exopod subequal, appendix interna long and slender, with apical hooks; male I (q) endopod almost half exopod, curved at tip; male II (r) endopod and exopod subequal, appendix interna with apical hooks, shorter than appendix masculina (s) the latter with about 4 apical spines and a series of about 11 spines on one side.

RANGE OF DISTRIBUTION

From southern Gulf of St. Lawrence (Bousfield and Laubitz 1972) to Mexico. Estuarine in beds of submerged vegetation (Williams 1984).

Records of distribution in the area of reference are in Fig. 75.

BIOLOGY

Length of body 30 mm in males and 42 mm in females. Females are first mature at about 5 mm cl or 22 mm body length.

In southern North Carolina females breed continuously during the summer (from April to September), releasing eggs a couple of days after the brood being carried is hatched. Limiting conditions for breeding seems to be only temperature (and salinity) and food sufficiency (Williams 1984). Williams (1984) reviews experimental work done on this species.

FISHERY

This shrimp may not necessarily be used as food, because of its small size, however it has been used for bait in recreational fisheries in New York and New Jersey (Holthuis 1980).

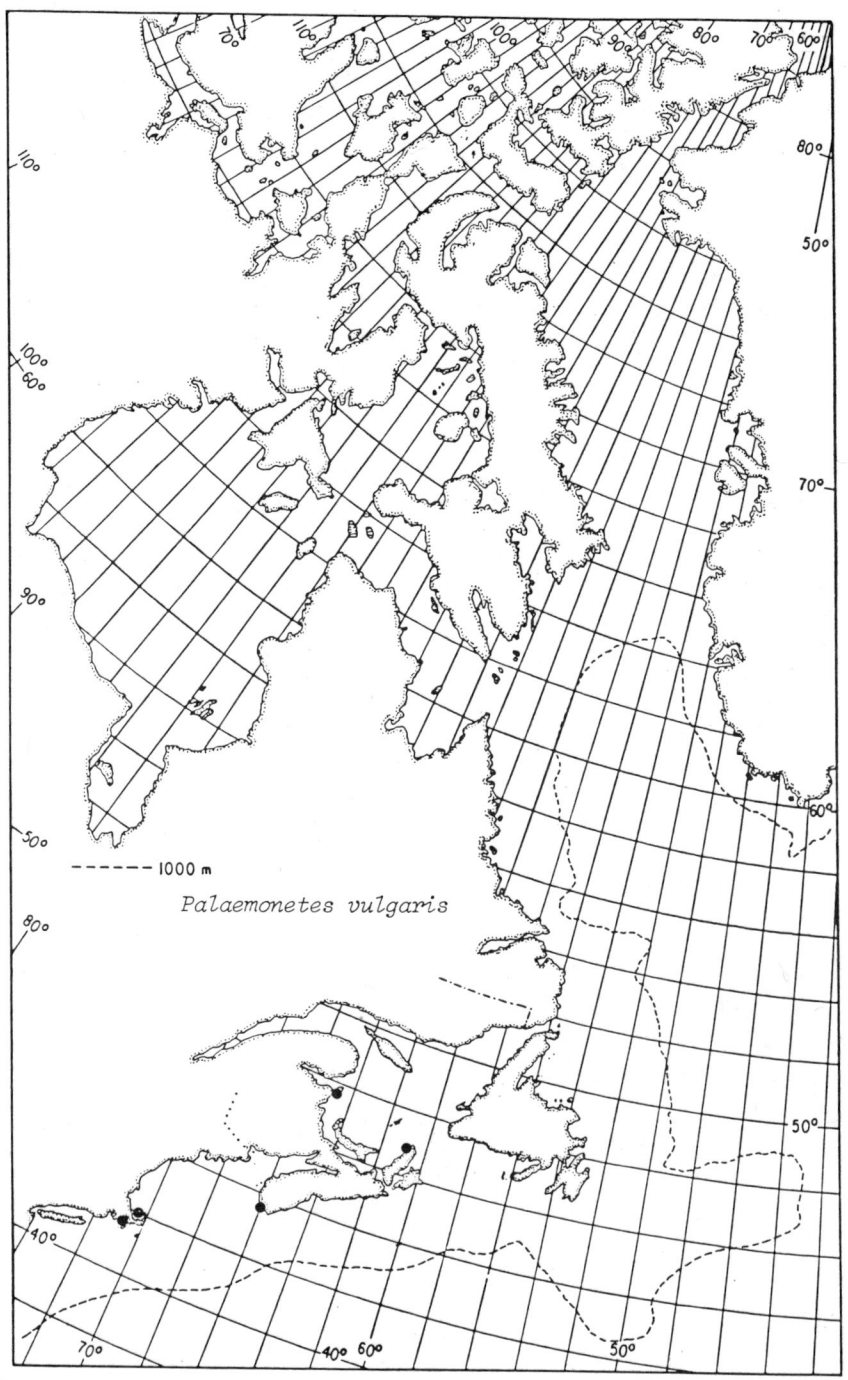

Fig. 75. *Palaemonetes vulgaris*, distribution records in the area of reference.

Family HIPPOLYTIDAE Bate, 1888
Holthuis 1947: 3; Smaldon 1979: 50.

First pair of legs chelate, shorter and heavier than second pair; second pair chelate with carpus subdivided into two or more segments; no exopods on pereopods; eyes free, moderately large.

Key to Genera and Species of the Hippolytidae

1	Carpus of second legs with less than seven segments	2
	Carpus of second legs with seven or more segments	4
2	Carpus of second legs with two segments	*Caridion gordoni*
	Carpus of second legs with three segments	3
3	Rostrum with deep blade truncate at tip; lower anterior edge of carapace with several small teeth	*Latreutes fucorum*
	Rostrum shallow with long pointed tip; lower anterior edge of carapace smooth, without teeth	*Hippolyte coerulescens*
4	Carpus of second legs with nine segments	*Bythocaris*
	Pleura of 2nd–5th abdominal somites ventrolaterally spiniform	*B. spinipleura*
	Pleura of 2nd–5th somites not spiniform	5
	Carpus of second legs with seven segments	6
5	Peduncle of antenna less than one-half scale	*B. payeri*
	Peduncle of antenna more than one-half scale	*B. gracilis*
6	Carapace with one or no supraorbital spines at each side	9
	Carapace with two supraorbital spines at each side	*Spirontocaris* 7
7	Median dorsal teeth not reaching posterior half of carapace and similar in size; one supraorbital spine smaller than other	*Spirontocaris phippsi*
	Median dorsal teeth reaching at least posterior two-thirds of carapace and of markedly different sizes; supraorbital spines equal	8
8	Median dorsal teeth almost reaching posterior margin of carapace; tergum of third abdominal somite produced posteriorly as a long spine over fourth; apex of rostrum with two spinous tips	*Spirontocaris spinus*
	Median dorsal teeth reaching about two-thirds length of carapace; apex of rostrum a single long point	*Spirontocaris lilljeborgi*
9	Carapace with no supraorbital spines	*Eualus* 10
	Carapace with one supraorbital spine at each side	*Lebbeus* 13
10	Rostrum with no dorsal spines	*Eualus fabricii*
	Rostrum with dorsal spines	11

11 Rostrum longer than carapace .... *Eualus gaimardi*
    Lobe on third abdominal somite in most and occasionally with weak hook; colour uniform .... *E. gaimardi gaimardi*
    Lobe on third abdominal somite with strong hook; abdomen striped ...
    .... *E. gaimardi belcheri*

    Rostrum shorter than carapace 12

12 Rostrum thin and translucent with crest of many small teeth dorsally, and rounded ventrally with one or two teeth .... *Eualus macilentus*

    Rostrum thick at base, small, with few dorsal teeth only .... *Eualus pusiolus*

13 Rostrum short, about even with 1st article of antennule ..... *Lebbeus microceros*

    Rostrum longer than peduncle of antennule 14

14 Carapace and abdomen rough, high median spines on carapace, pleura produced ventrally forming strong spines .... *Lebbeus groenlandicus*

    Carapace and abdomen smooth, pleura rounded .... *Lebbeus polaris*

Genus *Bythocaris* G. O. Sars, 1870

Carpus of second leg with nine segments; rostrum very short, slightly more than anterior edge of carapace; supraorbital spine on each side of base of rostrum; faint median carina about half length of carapace; mandible without palp or incisor process.

Sivertsen and Holthuis (1956) discuss the difficulty of separating *B. gracilis* and *B. payeri* from different areas of the North Atlantic where taken by the *Michael Sars* North Atlantic Deep-sea Expedition of 1910.

These are small deep-sea shrimps, three species of which have appeared in our collections.

Bythocaris gracilis Smith, 1885
Bythocaris gracilis, Sivertsen and Holthuis
1956: 32, figs. 22.
(Figures 76 and 77)

DISTINGUISHING CHARACTERISTICS

From specimens examined this species can be separated from *B. payeri* as follows: antennal scale is narrower (0.3 length); peduncle of antenna is equal to or more than one-half blade; outer branch of uropod is narrower (0.36 length); legs are proportionately longer. It differs from *B. spinipleura* in rounded pleura on all abdominal somites.

DESCRIPTION

Integument smooth, thin, flexible. Colour unavailable.

Carapace with supraorbital spine as lateral wing to base of rostrum and rising above it on each side, each with a short but prominent carina; moderate antennal and strong branchiostegal spines, the latter set back from the anterior edge and with a short carina.

Abdomen: pleura of 4th and 5th somites with a ventro-lateral spine (in females only); 6th almost twice length of 5th, telson about 2.6 times 5th. No ventral or anal spines in males or females.

Telson (t) moderately wide, tapering over distal two-thirds to somewhat truncate tip with a central and one lateral spine on each side. Dorso-laterally on posterior third are two pairs of minute spines. Outer branch of uropods longer than telson, inner branch slightly shorter.

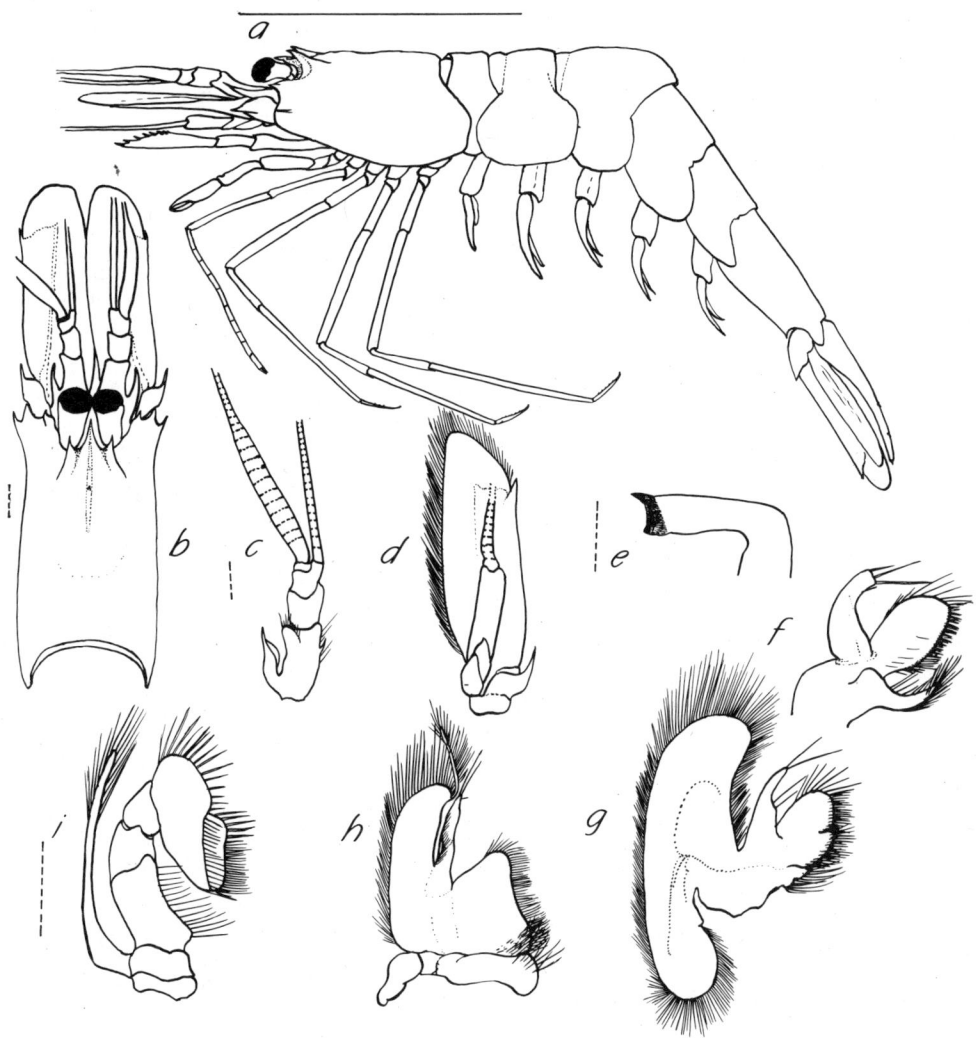

Fig. 76. *Bythocaris gracilis*: *a*, whole shrimp from left side; *b*, carapace in dorsal aspect; *c*, antennule; *d*, antenna; *e*, mandible; *f*, maxillule; *g*, maxilla; *h*, first maxilliped; *i*, second maxilliped; solid line = 10 mm, broken line = 1 mm.

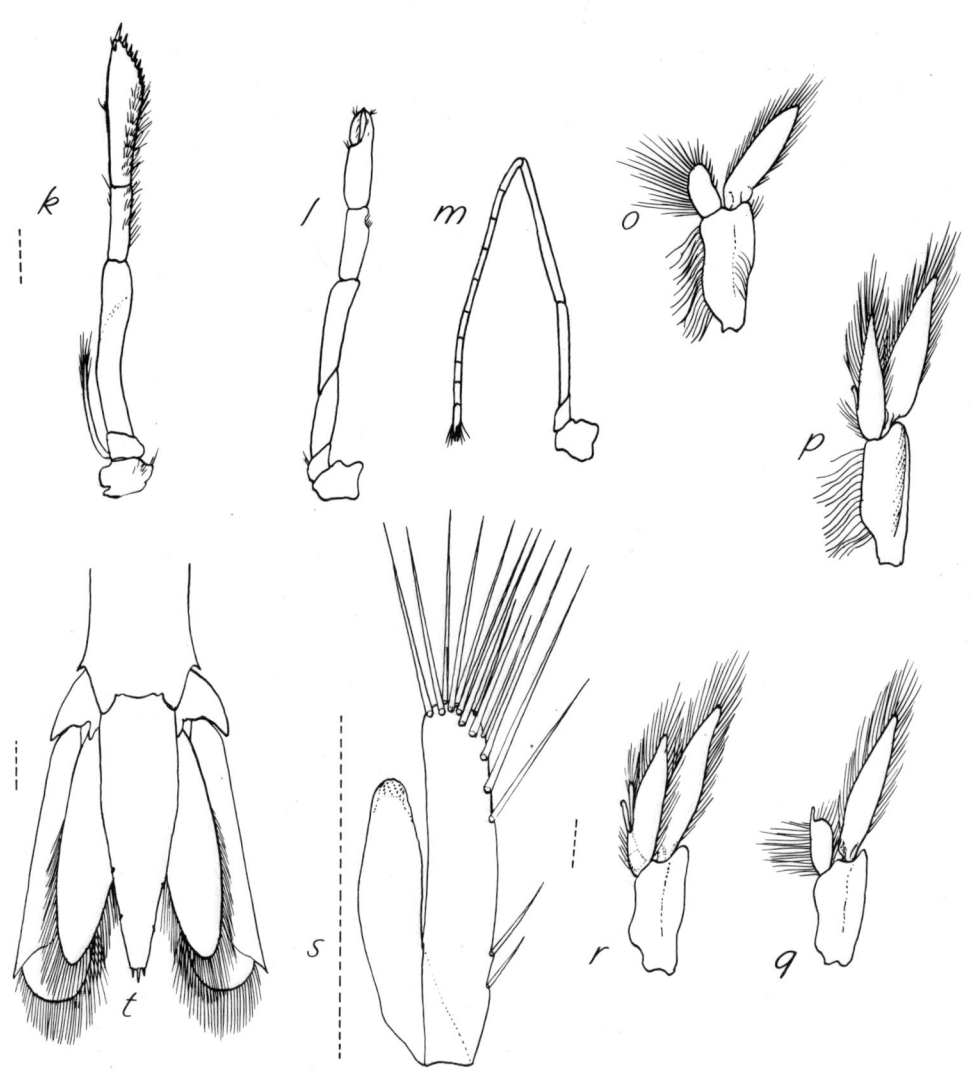

Fig. 77. *Bythocaris gracilis*: *k*, third maxilliped; *l*, first pereopod; *m*, second pereopod; *o*, F first pleopod; *p*, F second pleopod; *q*, M first pleopod; *r*, M second pleopod; *s*, appendix masculina; *t*, telson; broken line = 1 mm.

Eye moderate, cornea globular.

Antennule (c): 1st article about twice eye length, with inner spine on distal third; stylocerite almost as long as first article; 3rd article about half length of 2nd, both together shorter than first.

Antenna (d): scale longer than carapace, width about 0.3 length; disto-lateral spine much exceeded by scale; peduncle equal to or more than one-half scale.

Mandible (e): without palp or incisor; molar with central ridge and sharp tooth at upper corner, with short brush setae along inner edge of crown and along outer edge.

Maxillule (f): distal endite ovate with strong spines along leading edge; proximal endite pointed, narrow, curved, with long apical spines; endopod with 3 apical setae.

Maxilla (g): Proximal endite reduced; distal sub-equally bilobed; endopod tapering from wide base; scaphognathite with longer slightly curved anterior lobe, posterior lobe narrower, rounded with moderate setal fringe.

Maxilliped I (h): Endite subtriangular, lower lobe short; endopodal palp not clearly segmented, reaching tip of exopod; exopod with sharply narrowed apical lash about half length of exopod; epipod short sock-like (larger in *B. payeri*).

Maxilliped II (i): Flattened, leglike, with about 7 segments; dactyl inserted along edge of propodus (wider than long); exopod slender, almost as long as endopod.

Maxilliped III (k): large, leglike, distal segment expanded like a shovel, with strong apical spines and inner marginal row of about 6 spines. Exopod short, less than or equal to half longest segment; epipod rudimentary.

Pereopods: I (l) chelate, stout, shorter than third maxilliped or other legs, tips of chelae dark. II (m) chelate slender, carpus with 9 divisions, chelae small with strong setae distally. III–V (a) long; dactyls slender, sharp, with few short spinous setae on flexor edge; V with longest propodus.

Pleopods: Female I (o) endopod short, rounded at tip with long lateral setae, exopod about twice as long; II (p) endopod and exopod sub-equal, appendix interna short, with sub-apical patch of hooks. Male I (q) endopod short, rounded at tip and with thin outer finger-like apical projection. II (r) appendix interna with hooks, shorter than appendix masculina.

Appendix masculina (s) with about 14 apical and sub-apical spines and two shorter spines proximally.

RANGE OF DISTRIBUTION

West Greenland to east coast of the United States at 35° N latitude; eastern Atlantic from SW of Ireland; depths 550–1 900 m (Sivertsen and Holthuis 1956).

Distribution records in the area of reference are in Fig. 78.

BIOLOGY

Size to 8 mm cl (Female 7 mm cl, Male 5 mm cl in specimens examined).

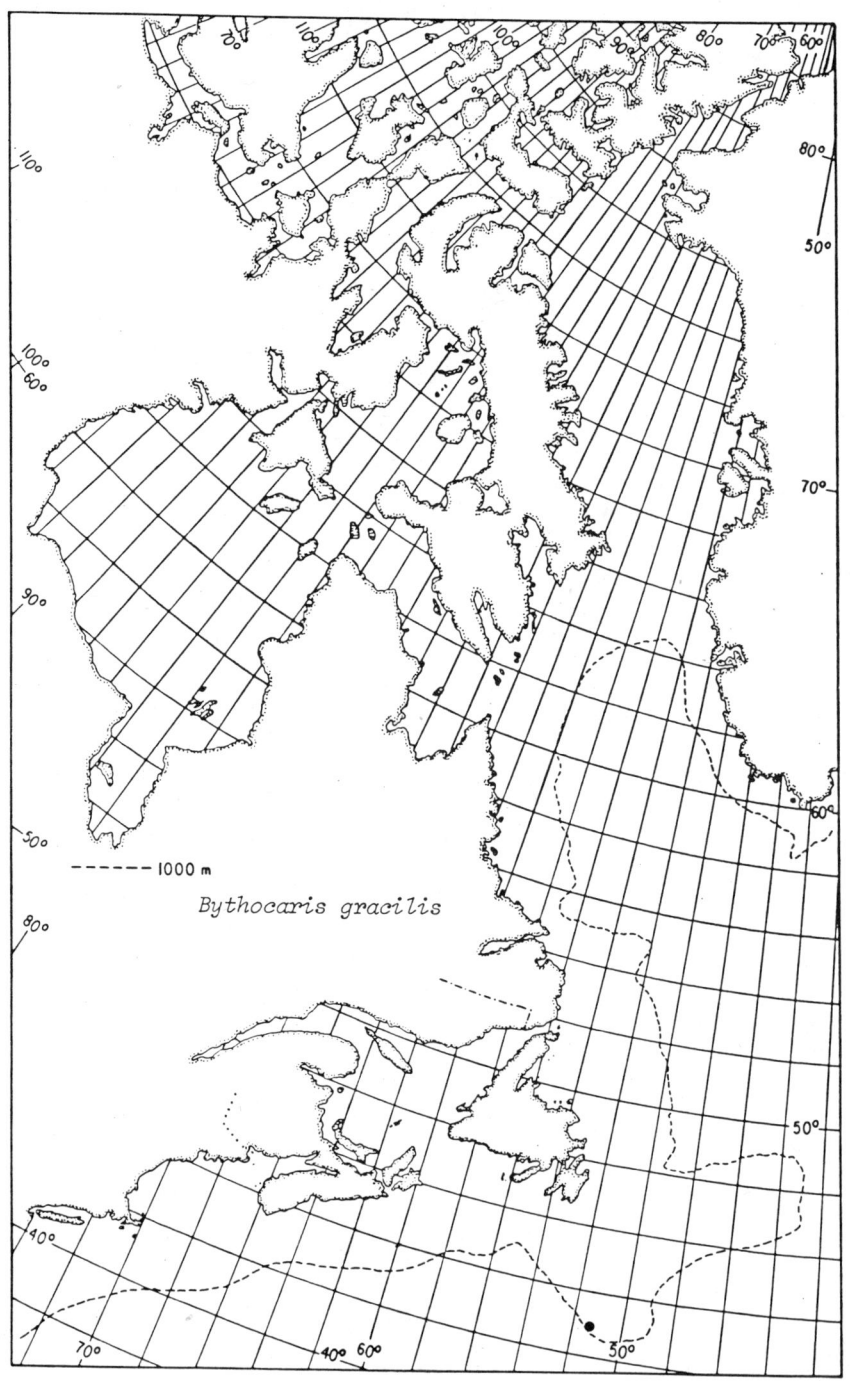

Fig. 78. *Bythocaris gracilis*, distribution records in the area of reference.

Bythocaris payeri (Heller, 1875)
Bythocaris payeri, Sivertsen and Holthuis 1956:
34, figs. 23, 24, Bowman and Manning 1972: key.
(Figures 79 and 80)

DISTINGUISHING CHARACTERISTICS

From specimens examined this species can be separated from *B. gracilis* as follows: antennal scale is wider (0.4 length); peduncle of antenna is less than one-half scale; outer branch of uropod is wider (0.44 length); legs are proportionately shorter. It differs from *B. spinipleura* in rounded pleura of second to fifth abdominal somites.

DESCRIPTION

Integument smooth, shiny. Colour pale red.

Carapace smooth with faint median ridge extending from between the lateral spines (supraorbital spines and short carinae) at base of rostrum to about two-thirds posteriorly (b). Rostrum slightly depressed with short median ridge and hollow behind it; postorbital lobe and sharp antennal spine; strong branchiostegal spine set back from anterior edge.

Abdomen with all pleura and terga rounded but 6th with small ventral spine and slight flare.

Telson (t) rounded dorsally, posterior two-thirds tapering to narrow tip with central fixed spine and two pairs of spines lateral to it, the inner stouter; outer branch of uropod about as long as telson and very wide (0.44 times length).

Eyes moderate, cornea globular.

Antennule (c): first article of peduncle about equal to other two together, the 3rd about 0.6 the 2nd; a small spine on inner side of first article; stylocerite almost reaches distal end of first article.

Antenna (d): Scale very wide, 0.4 times length; disto-lateral spine much exceeded by blade; peduncle less than half scale.

Mandible (e): without incisor process or palp; left and right somewhat similar, right with more and sharper cusps; crown hollow; outer edge with close setal fringe, inner with straight cutting edge with two cusps below and one above.

Maxillule (f): Distal endite ovate, leading edge with short spines followed by long spines; proximal endite narrow, curved with strong distal and fine lateral setae; endopod uniramous.

Maxilla (g): anterior lobe of scaphognathite longer than posterior, slightly curved inward at tip, posterior lobe rounded with short setal fringe; endopod tapering from narrow base, two apical setae; distal endite about equally bilobed, proximal much reduced.

Maxilliped I (h): distal endite subtriangular, proximal almost equal to distal; endopodal palp not reaching end of exopod, not distinctly subdivided; exopod almost rectilinear with sharply differentiated lash at distal inner corner; epipod fleshy, subtriangular in outline.

Maxilliped II (i): leg-like but compressed; distal segment wider than long, possibly seven segments; exopod slender almost as long as endopod.

Maxilliped III (k): leg-like, stout, five segments, distal expanded, shovel-like with row of about 8 strong spines along inner edge of expansion; very short exopod, rudimentary epipod.

Pereopods: I (l): stout, shorter than third maxilliped or other legs, chelate, fingers black-tipped. II (m) chelate, carpus with 9 divisions, chela small, fingers slender, distally setose. III–V (a) moderately stout and longer than others, dactyls long and slender with a few sharp spines on flexor edge.

Pleopods: Female I (o) with short sub-circular endopod, fringed with long setae; II (p) endopod about equal to exopod, appendix interna short with row of hooks from apex to two-thirds length. Male I (q) short ovate endopod with short sub-apical appendix with

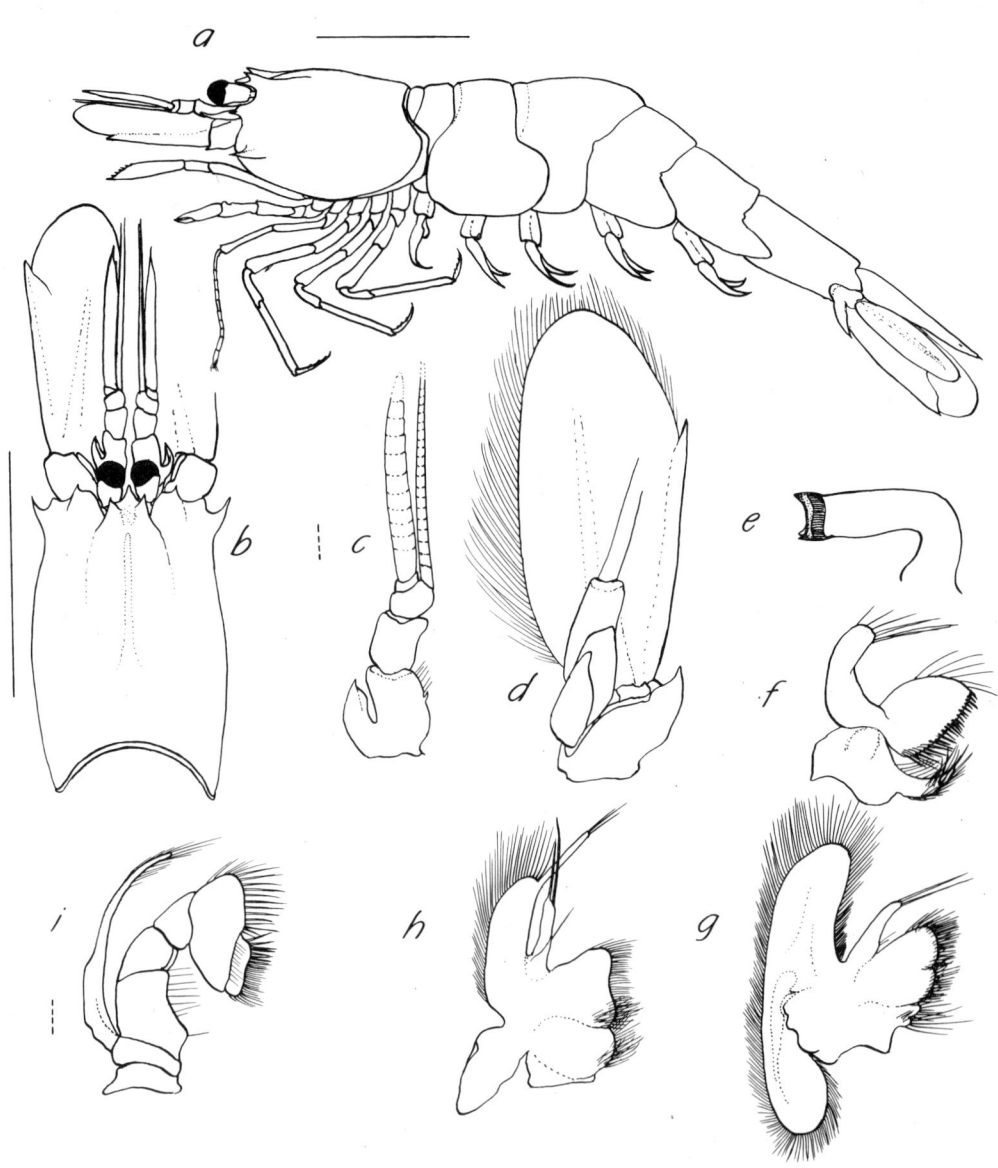

Fig. 79. *Bythocaris payeri*: *a*, whole shrimp from left side; *b*, carapace in dorsal aspect; *c*, antennule; *d*, antenna; *e*, mandible; *f*, maxillule; *g*, maxilla; *h*, first maxilliped; *i*, second maxilliped; solid line = 10 mm, broken line = 1 mm.

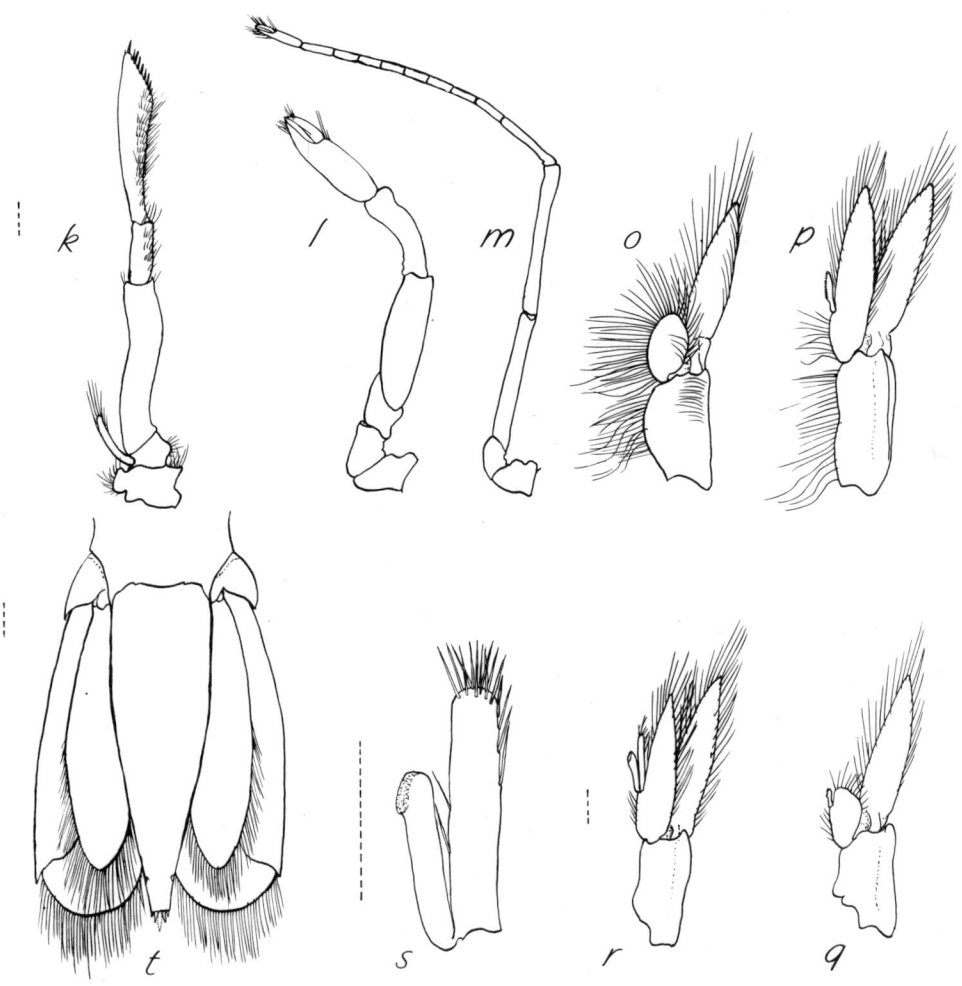

Fig. 80. *Bythocaris payeri*: *k*, third maxilliped; *l*, first pereopod; *m*, second pereopod; *o*, F first pleopod; *p*, F second pleopod; *q*, M first pleopod; *r*, M second pleopod; *s*, appendix masculina; *t*, telson; broken line = 1 mm.

distal patch of hooks; II (r) endopod slightly shorter than exopod and with appendix masculina longer and stouter than appendix interna, the latter with subapical patch of hooks.

Appendix masculina (s) with crown of about 9 spines and several spines down one side to about distal one-third.

RANGE OF DISTRIBUTION

Reported from the Arctic Ocean and from Greenland east to the Kara Sea and south to Iceland, the Faroes and the Shetland Islands (Sivertsen and Holthuis 1965). Also off Cape Dyer, Baffin Island and south to Newfoundland (Squires 1965a). Depths 180–1 000 m. Temperatures −1.1 to 1.2° C.

Distribution records in the area of reference are in Fig. 81.

BIOLOGY

Range of lengths in males 5–7 mm, average 6 mm cl; and in females 7–11 mm, average 9 mm cl.

All females taken in August were ovigerous, and in one large ova were also present. Eggs 2.2 mm diameter, almost all with eyed embryos.

Crustacean remains and sponge spicules were found in the few stomachs examined (Squires 1965a).

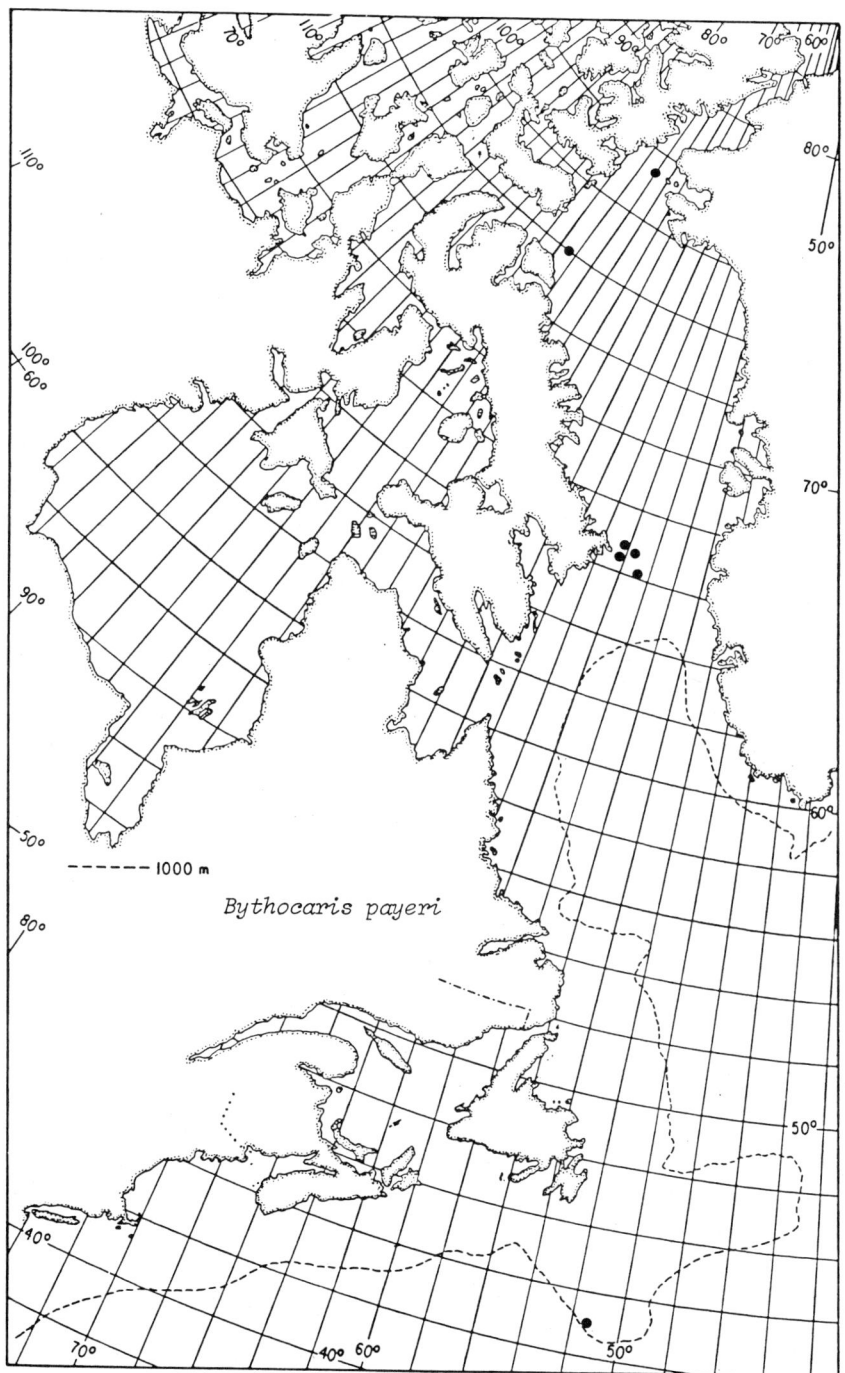

Fig. 81. *Bythocaris payeri*, distribution records in the area of reference.

Bythocaris spinipleura, new species
(Figures 82 and 83)

DISTINGUISHING CHARACTERISTICS

Rostrum moderately long, sharp, reaching about half the first article of antennular peduncle and as far as cornea of eye; carapace with short anterior median ridge with two small teeth; pleura of all somites including first with spiniform process; telson with three (or four) dorsolateral spines, terminally truncate with a median notch and two long slender spines each with a short spine on each side.

This species differs from *B. payeri* and *B. gracilis* as described in this work and from *B. simplicirostris* (Sars, 1912) and *B. leucopis* (Sars, 1885) in having all abdominal pleura with a spiniform process. It differs also from *B. nana* (Smith, 1885) since in addition to the spiniform pleura it has a longer rostral process than its supraorbital spines (shorter in *B. nana*).

The holotype is a male 4.5 mm cl, and paratype also a male 5.0 mm cl both taken from cod stomachs off Bonavista Bay, Newfoundland (Lat. 48°49′ N, Long. 51°30′ W, depth 309 m, and Lat. 49°24′ N, Long 52°28′ W, 340 m) (Fig. 84), the first on April 4, 1981, and the second on November 29, 1986, and are deposited in the United States National Museum of Natural History, Smithsonian Institution, Washington, USA, Nos. USNM 221919, and USNM 221920, respectively.

DESCRIPTION

Integument smooth, thin but firm; colour not available.

Carapace with moderately long sharp rostrum, the tip reaching about half the first article of antennular peduncle; a median low ridge behind rostrum reaches about half carapace and has two small teeth at about half its length; supraorbital spine large, rising above rostrum; antennal and branchiostegal spines only, the latter set back from anterior edge.

Abdomen smooth, terga rounded, all pleura including first with ventrolateral spiniform process, most pronounced on fifth, with flaring ridge on sixth.

Telson (t) wide, slightly tapered, with four pairs of dorsolateral spines, and terminally truncate with a small median notch and two long slender spines on each side, each with a small spine on each side of base. Inner branch of uropod shorter than outer, both shorter than telson; no axillary spine at diaeresis.

Antennule (c) 1st article with inner spine on distal two-thirds, longer than 2nd and 3rd subequal articles; dorsolateral flagellum long, dilated over most of its length (about three times longer than peduncle); ventromesial flagellum slender, only slightly longer than dorsolateral; stylocerite flared outward, sharp attenuate, not quite as long as 1st article.

Antenna (d) scale wide, length 2.8 times width, blade much exceeding distolateral spine; outer spine of basal article long, sharp; peduncle does not reach half blade.

Mandible (e) with molar only, cutting edge almost straight with minor sharp cusps at corners.

Maxillule (f) distal endite short and wide, ovate, proximal narrow, curved; endopod moderate, obscurely bifid.

Maxilla (g) proximal endite reduced, distal subequally bilobed; endopod slender, shorter than subrectilinear anterior lobe of scaphognathite, latter sloped distolaterally at outer edge; posterior lobe narrower and shorter, subtriangular.

First maxilliped (h) endites wide, subequal; endopod slender, with distal narrow, small segment; exopod lobe large, almost as long as endopod, with lash; epipod sock-shaped.

Second maxilliped (i) leglike but compressed, wide; distal segment wider than long; exopod long, slender; epipod not in evidence.

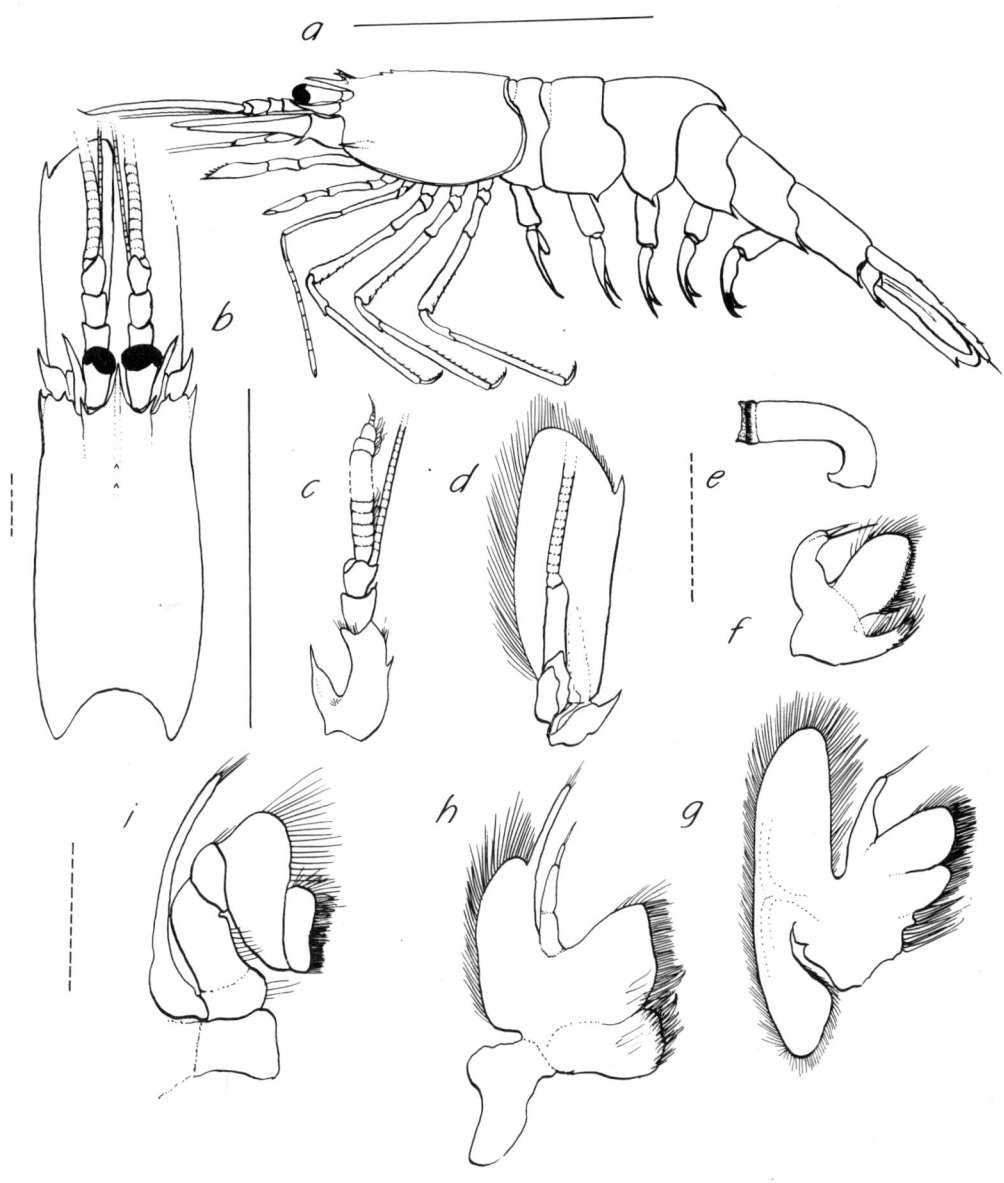

Fig. 82. *Bythocaris spinipleura*: *a*, whole shrimp from left side; *b*, carapace in dorsal aspect; *c*, antennule; *d*, antenna; *e*, mandible; *f*, maxillule; *g*, maxilla; *h*, first maxilliped; *i*, second maxilliped; solid line = 10 mm, broken line = 1 mm.

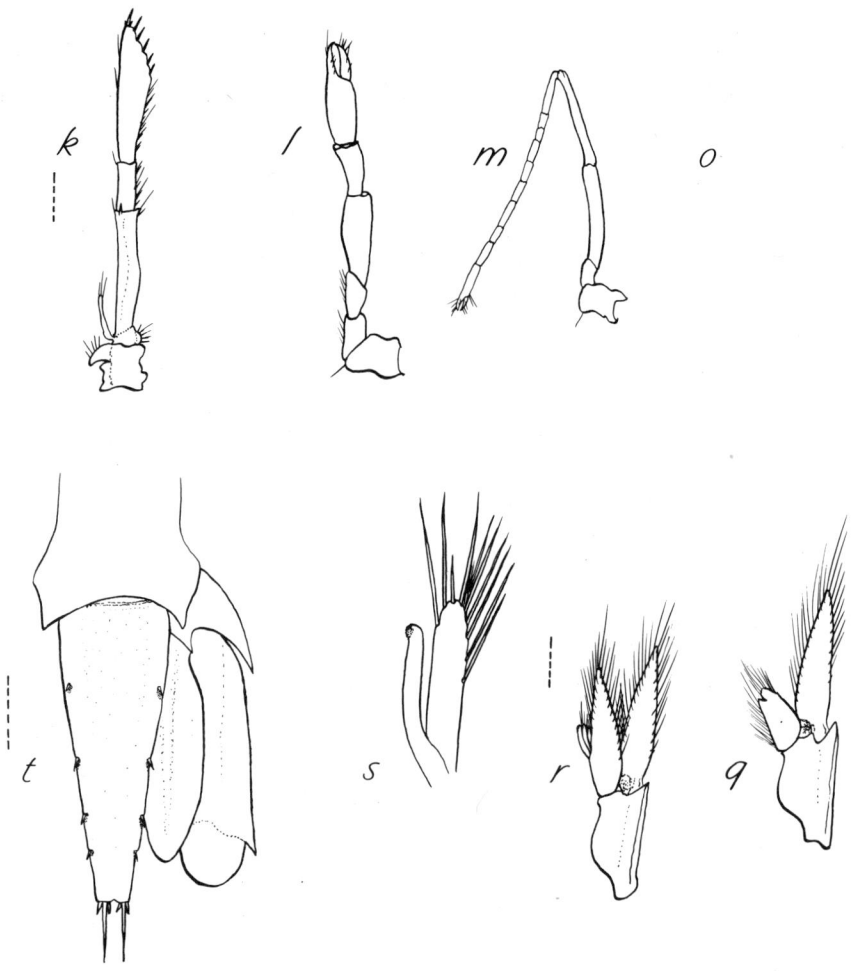

Fig. 83. *Bythocaris spinipleura*: *k*, third maxilliped; *l*, first pereopod; *m*, second pereopod; *o*, F first pleopod; *p*, F second pleopod; *q*. M first pleopod; *r*, M second pleopod; *s*, appendix masculina; *t*, telson; broken line = 1 mm.

Third maxilliped (k) stout, leglike, distally shovel-shaped with 6 strong spines along edge; distal spine on ischio-merus. Exopod short, slender; epipod rudimentary, beak-like.

Pereopods: I (l) stout, slightly shorter than third maxilliped, tips of fingers sharp, dark, simple. II (m) slender, elongate carpus with 9 divisions; about twice length of I. III–V (a) similar, longer and stouter than II, merus with series of spines on ventral edge; dactyl short, slender, with series of about 5 dark spines, not biunguiculate.

Pleopods: Female, I (o), II (p); no females collected. Male I (q) endopod short with short apical segment, exopod three times longer; protopod dilated; II (r) endopod slightly shorter than exopod, appendix interna slightly curved, with apical patch of hooks; appendix masculina (s) slightly longer than a.i., with one short and two long fine spines apically, plus one long seta outside and two irregular rows of about 10 long spines on inner edge.

DISTRIBUTION

Records from the area of reference include only two specimens from stomachs of cod (*Gadus morhua*) taken off Bonavista Bay, Newfoundland (Fig. 84).

BIOLOGY

The specimens examined were adult males 4.5 and 5.0 mm cl.

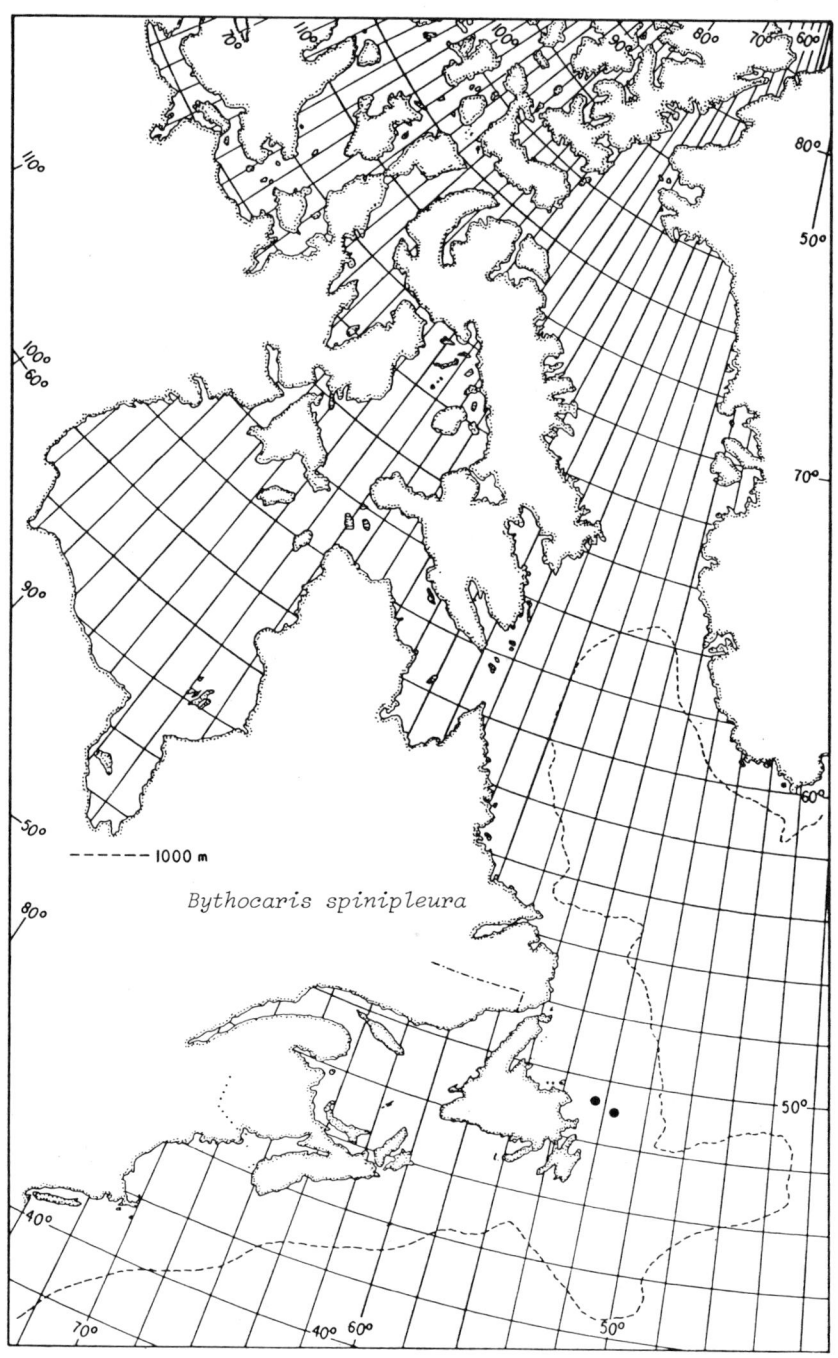

Fig. 84. *Bythocaris spinipleura*, distribution records in the area of reference.

Genus *Caridion* Goes, 1863
Allen 1967: 11, 49; Smaldon 1979: 53.

Carpus of second leg with two segments; supraorbital spine absent; mandible with incisor process, mandibular palp with three segments.

Only one species of this genus in the area of reference.

Caridion gordoni (Bate, 1958)
Allen 1967: 49 (key); Smaldon 1979: 53, fig. 18.
(Figures 85 and 86)

DISTINGUISHING CHARACTERISTICS

Carpus of second legs with only two segments; no supraorbital spine; rostrum about three-quarters the length of the carapace, slightly longer than the antennular peduncle and with 6–10 dorsal teeth and 1 or 2 ventral teeth; antennal spine only on anterior margin of carapace.

DESCRIPTION

Integument firm, smooth and shiny. Colour reddish translucent.

Carapace without supraorbital spines and with large antennal but no pterygostomian spine. Crest of rostrum with one or two spines behind orbit and a median dorsal carina about half length of carapace. Orbit has a double margin, the outer being confluent with the lateral carina on the rostrum.

Rostrum exceeds antennular peduncle, formula 6–10/1–2, slightly descending; with lateral carina confluent with upper margin of the orbit.

Abdomen with rounded pleura and terga, except ventro-lateral spine on 5th somite. Telson (t) slightly longer than the 6th somite has two pairs of dorso-lateral spines; terminally a small fixed centre spine is flanked by three pairs of spines the second pair of which is longer than the others. The outer branch of the uropod has an axillary spine; it is longer than the telson.

Eyes large, cornea globular.

Antennule (c): first article exceeds length of other two together, 3rd shorter than 2nd; stylocerite slightly longer than the 1st article; flagella almost equal in length.

Antenna (d): scale wide about 0.38 length; peduncle about 0.5 length of scale; scale exceeding disto-lateral spine.

Mandibles (e): only slightly different: incisor of left with 5 teeth, of right with 4; palp three-segmented; molar with hollow crown, a sharp cutting inner edge with a pointed cusp at each end and an uneven edge with fringe of bristles.

Maxillule (f): endites widely separated, the distal ovate and the proximal narrow, tapering, slightly curved at tip; endopod obscurely bifid.

Maxilla (g): Distal endite large, unequally bilobed, proximal reduced to two small rounded lobes; endopod slender, tapering; scaphognathite with larger anterior lobe distally uneven, posterior lobe rounded with long even setae.

Maxilliped I (h): distal endite subtriangular with concave inner edge, proximal much shorter; endopod very slender, untwisted; exopod large with sharply differentiated lash at inner corner distally; epipod large, bilobed, both subtriangular in outline.

Maxilliped II (k): somewhat leg-like but compressed, appearing to have seven segments, merus and ischium conspicuously hollowed; exopod long, slender; epipod small, ovate.

Maxilliped III (l): slender, distal segments with parallel horizontal rows of setae; exopod very short and slender; epipod small, triangular, with strap-shaped part with grooming hooks.

Pereopods: I (l) stout, chelate, fingers slightly shorter than palm; strap-shaped epipod with grooming hooks. II (m) chelate, slender, shorter than I, fingers longer than palm;

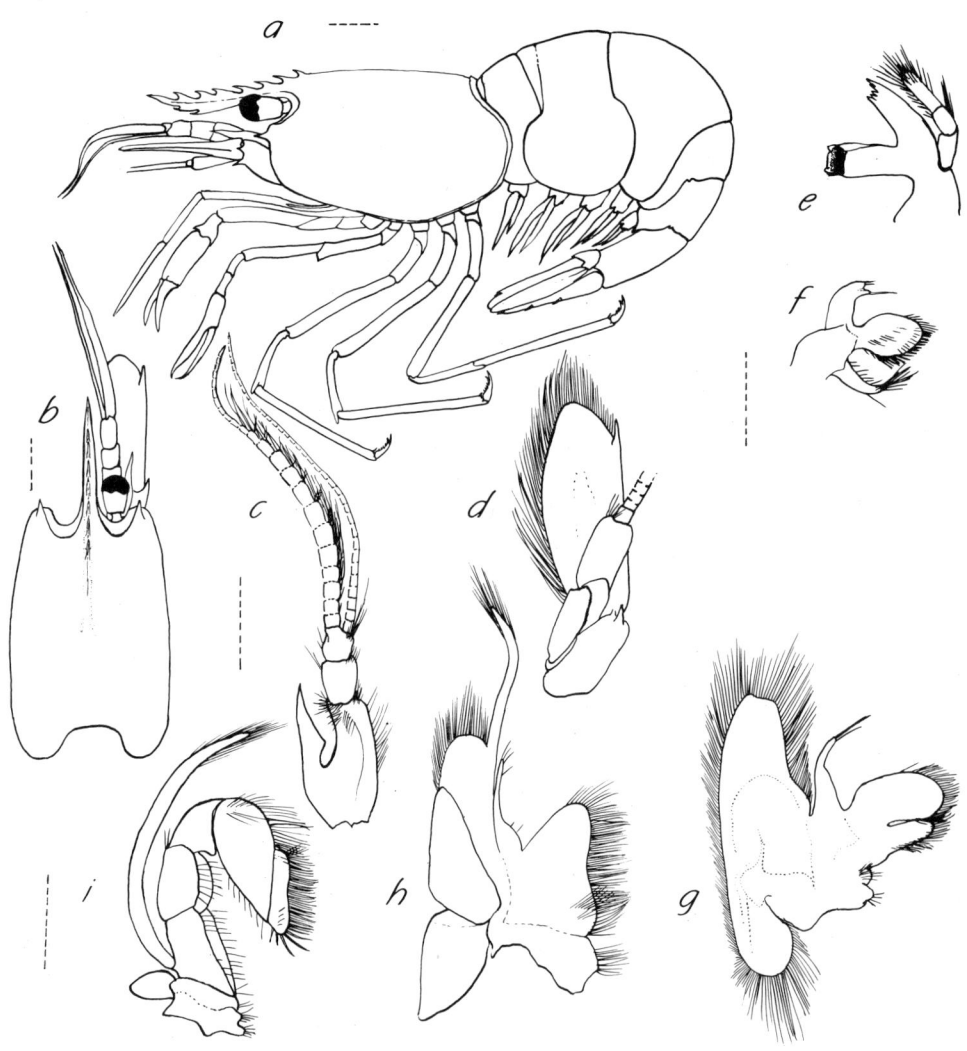

Fig. 85. *Caridion gordoni*: *a*, whole shrimp from left side; *b*, carapace in dorsal aspect; *c*, antennule; *d*, antenna; *e*, mandible; *f*, maxillule; *g*, maxilla; *h*, first maxilliped; *i*, second maxilliped; broken line = 1 mm.

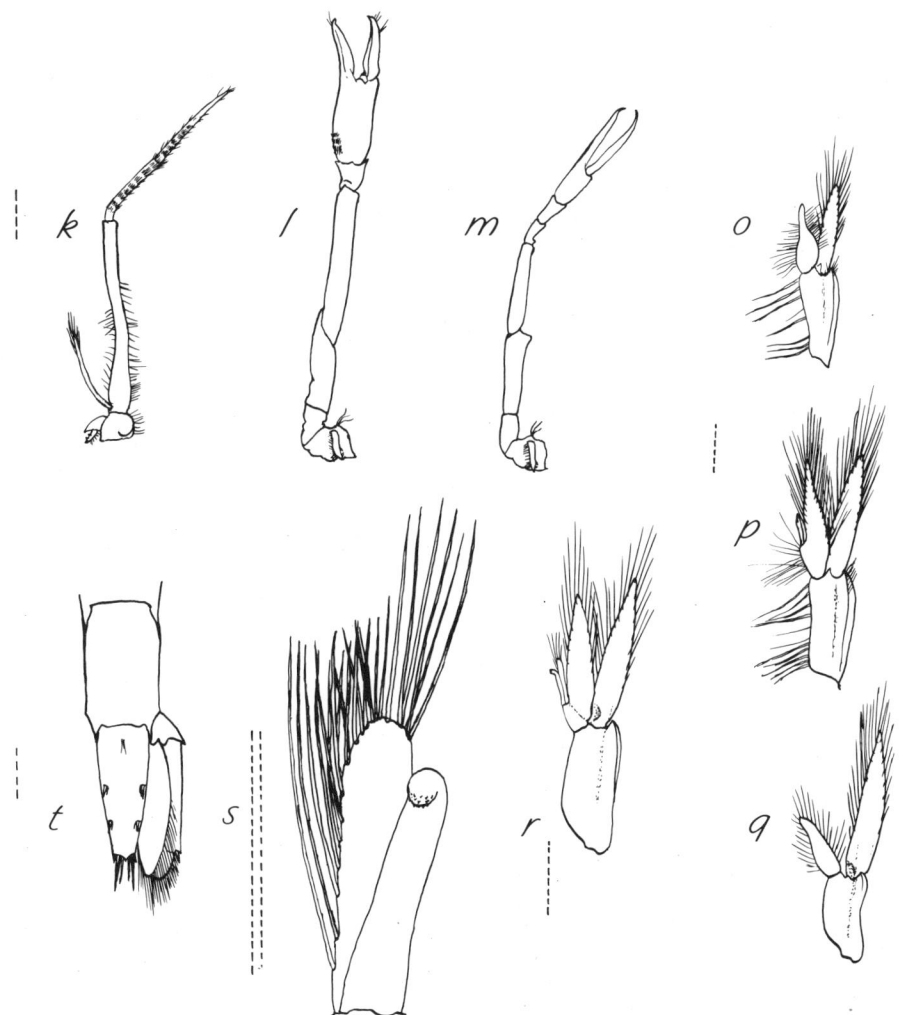

Fig. 86. *Caridion gordoni*: k, third maxilliped; l, first pereopod; m, second pereopod; o, F first pleopod; p, F second pleopod; q, M first pleopod; r, M second pleopod; s, appendix masculina; t, telson; broken line = 1 mm.

ischium with slight distal expansion; epipod strap-shaped with grooming hooks. III–V slender and longer than others, III–IV with strap-shaped epipods; dactyls stout, biunguiculate and with a few mesial spines.

Pleopods: Female I (o) endopod somewhat shorter than exopod and flask-shaped with no setae or hooks at apex; II (p) endopod about equal to exopod, appendix interna with disto-lateral patch of hooks. Male I (q) endopod only half as long as exopod, with setae but no hooks apically; II (r) endopod shorter than exopod, appendix interna shorter than appendix masculina, the latter (s) with about five long spines and a bunch of shorter spines apically and a series of long lateral spines reaching same distance toward tip.

RANGE OF DISTRIBUTION

Southwestern Newfoundland to Chesapeake Bay; northern Europe; (Frost 1936, Rathbun 1929); fairly common in Clyde Sea area, in muddy sand and gravel (Allen 1967); depths 10–300 m (Smaldon 1979).

Distribution records in area of reference are in Fig. 87.

BIOLOGY

Total lengths from 19 to 27 mm (Rathbun 1929). Specimen examined 4 mm cl (female).

Allen (1967) refers to observations of ovigerous females in March–May and October with post-larval stages in July in the Clyde Sea area.

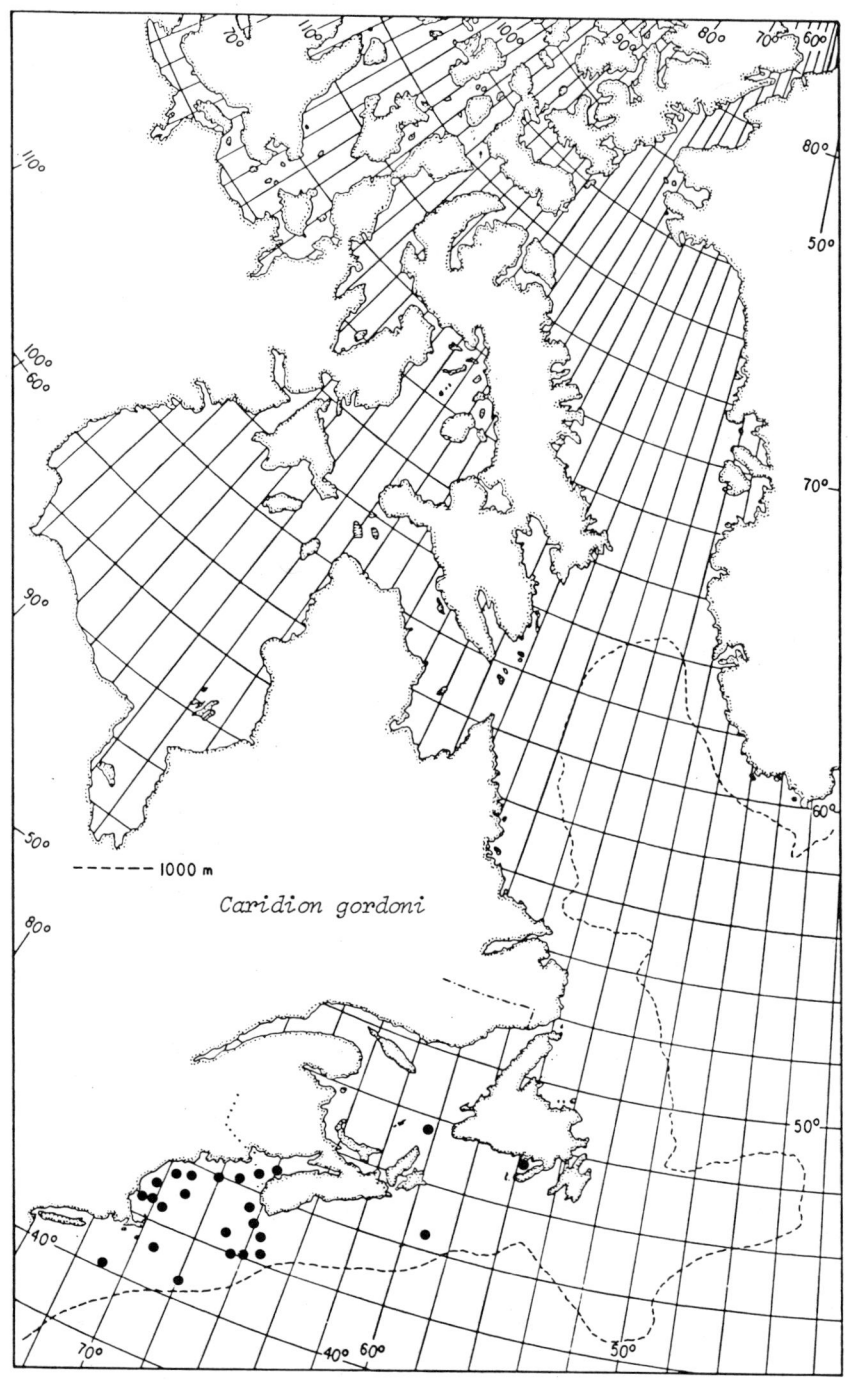

Fig. 87. *Caridion gordoni*, distribution records in the area of reference.

Genus *Eualus* Thallwitz, 1892
Holthuis 1947: 10, 43; 1950: 46; 1955: 100.

Supraorbital spine absent; third maxilliped with exopod; carpus of second leg with 7 divisions; mandible with incisor and palp of two segments; only antennal and pterygostomian spines present on anterior edge of carapace.

Four species and one subspecies of the 19 known species of this genus are present in the area of reference.

Eualus fabricii (Kroyer, 1841)
Spirontocaris fabricii, Rathbun 1929: 15, fig 15.
Eualus fabricii, Holthuis 1947: 10, 43
(Figures 88 and 89)

DISTINGUISHING CHARACTERISTICS

Rostrum longer than carapace, with no teeth dorsally except on crest above orbit at base of rostrum, about four fixed spines with two behind the orbit; first leg only with epipod, lacking in some specimens.

DESCRIPTION

Integument smooth and shiny. Colour red spots over whitish translucent background.
Eyes moderately large with short tapering stalk and globular cornea.
Carapace usually has 4 fixed median spines (1–7) at base of rostrum with 2 or 3 behind orbit, but no dorsal teeth on the rostrum: about 3 ventrally (1–3). Blade of rostrum is usually deep but may also be shallow in some specimens, it is sharply pointed, about 1.4 times cl, slightly lower than level of dorsal carapace but has slight rise distally, lateral ridge is confluent with orbit. Median dorsal carina reaches about half carapace. A strong antennal and small pterygostomian spine only at anterior edge. Ventrally a pair of long curved sternal spines are between pereopods III–V and a small one between pereopods II.

Abdomen with all terga and pleura rounded except 4th and 5th with a postero-ventral spine each, and the 6th with a postero-lateral spine and flared ridge. Ventrally in males a pair of spines on each of somites one and two and one spine each on other somites; anal spine present in males and females.

Telson (t) with four pairs of dorso-lateral spines and three pairs of terminal spines, the second longest. Outer branch of uropod about even with telson and inner branch, outer with axillary spine at diaeresis.

Antennule (c): 1st article longer than other two combined, 2nd about twice length of 3rd; distal spine on 2nd and 3rd articles; short thick flagellum about equal length of peduncle, ventro-mesial flagellum about 3 times length of other; stylocerite very sharp pointed, exceeding 1st article.

Antenna (d): scale exceeding disto-lateral spine, slightly longer than carapace, width about 0.3 times length; peduncle more than half length of scale, flagellum about 5 times length of carapace; basal article with strong spine at outer edge.

Mandibles (e): slightly dissimilar on molar surface; also incisor of right with 5 teeth distally and of left with 4 teeth; distal segment of palp about twice proximal segment.

Maxillule (f): distal endite ovate with wide base, proximal narrow and tapering to curved tip; endopod bifid with single seta at tip of each short ramus.

Maxilla (g): distal endite about equally bilobed, proximal with a narrow lobe without setae and a wide rounded lobe with long curved setae; endopod short; scaphognathite with long rectangular anterior lobe and short rounded posterior lobe with even setal fringe.

Maxilliped I (h): proximal endite almost as large as distal, with diverging rows of setae at leading edge; distal endite broadly subtriangular in outline; endopod not clearly

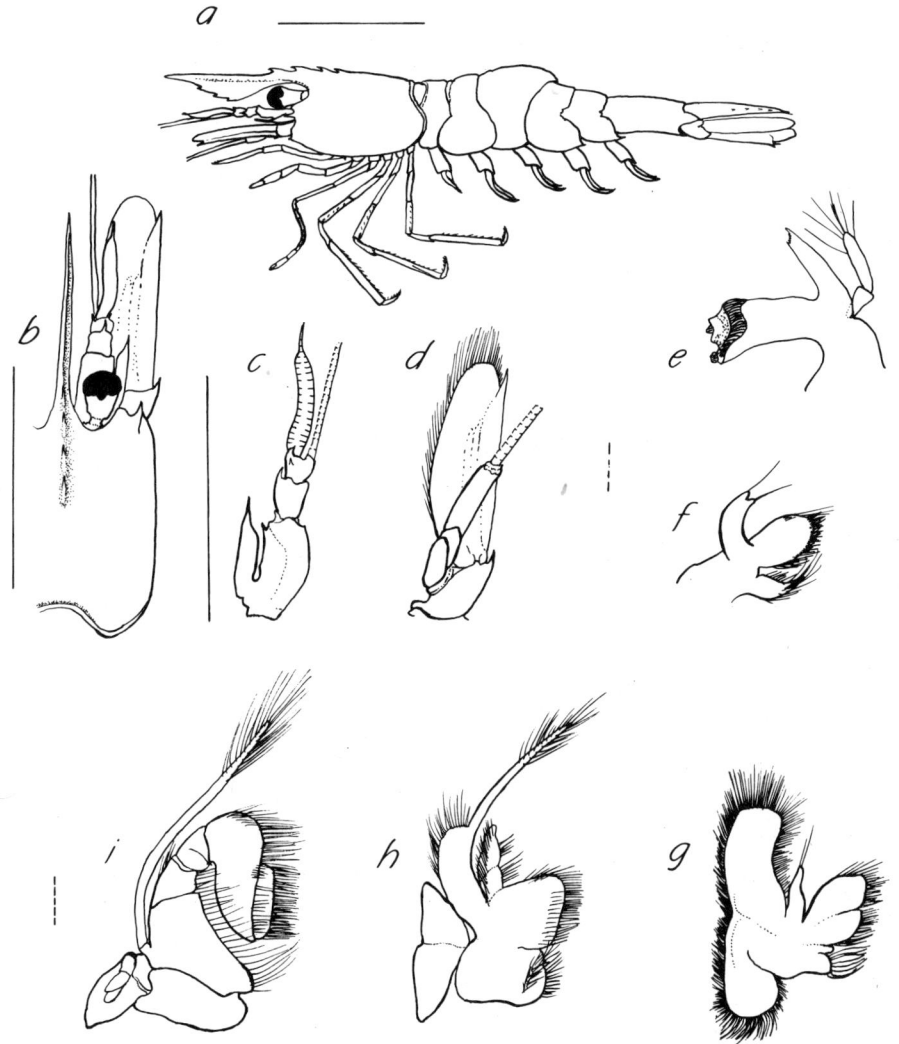

Fig. 88. *Eualus fabricii*: *a*, whole shrimp from left side; *b*, carapace in dorsal aspect; *c*, antennule; *d*, antenna; *e*, mandible; *f*, maxillule; *g*, maxilla; *h*, first maxilliped; *i*, second maxilliped; solid line = 10 mm, broken line = 1 mm.

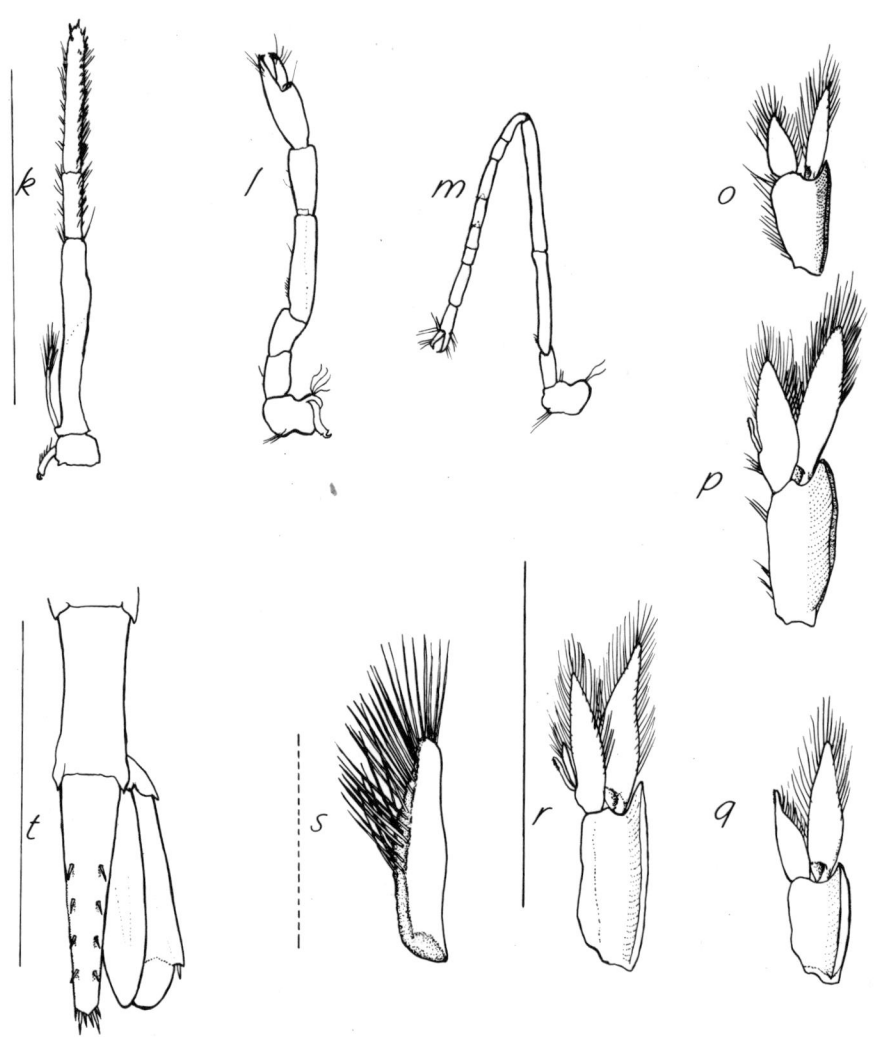

Fig. 89. *Eualus fabricii*: *k*, third maxilliped; *l*, first pereopod; *m*, second pereopod; *o*, F first pleopod; *p*, F second pleopod; *q*, M first pleopod; *r*, second pleopod; *s*, appendix masculina; *t*, telson; solid line = 10 mm, broken line = 1 mm.

divided, reaching to or slightly beyond end of exopod which has long lash at inner corner; epipod bilobed each lobe subtriangular.

Maxilliped II (i): leglike, compressed, distal segment wider than long; exopod long and slender; epipod moderate, with small podobranch.

Maxilliped III (k): leglike, stout, longer than 1st leg, with seven apical dark spines in an oval on distal segment; appears to have four segments only; exopod very short and slender; epipod straplike with grooming hooks distally.

Pereopods: I stout, chelate, dactyl with dark spines distally forming bifid tip into which spine of propodal finger fits; epipod straplike with distal grooming hooks. II slender, chelate, carpus with seven divisions; dactyl bifid as in I; no epipod. III–V moderately long and slender; dactyls biunguiculate followed by five strong spines on flexor edge; merus with row of outer lateral spines the distal one longest; propodus also with row of spinules along flexor edge.

Pleopods: Female I (o) endopod short, sub-triangular, with fringing setae; exopod narrow; protopodite with wing-like expansion. Female II (p) endopod almost as long as exopod, moderately wide; appendix interna with patch of hooks near apex. Male I (q) endopod has apical finger-like extension with patch of hooks. Male II (r) Appendix interna with apical patch of hooks, shorter than appendix masculina (s) which has distally about 5 long setae and laterally a single series of long setae becoming double proximally as appendix thickens.

RANGE OF DISTRIBUTION

Northwest Atlantic from Hudson Bay, Foxe Basin and west Greenland to Cape Cod; Chukchi Sea, Bering Sea to British Columbia; Okhotsk Sea to Sea of Japan (Butler 1980; Williams 1984). Depths 4–275 m and temperatures -1.5 to 4.5° C (Williams 1984).

Records of distribution in the area of reference are in Fig. 90.

BIOLOGY

Carapace lengths 12 mm in males and 14 mm in females, averages 7 and 9, respectively.

Almost all females examined from Newfoundland and Labrador were ovigerous in spring and autumn; in Foxe Basin about 64% were potentially ovigerous in autumn, and in Hudson Bay about 54%. In Foxe Basin females were first mature at 5 mm cl and in Hudson Bay at 8 mm cl. In view of the possibility of females in these populations spawning three times in a lifetime, annual spawning could take place with the numbers observed potentially ovigerous (Squires 1965a, 1968a).

Feeding was on phytobenthos with occasionally crustacean and polychaete fragments.

Predation was common by cod, seals and beluga whales (Squires 1965a, 1967a).

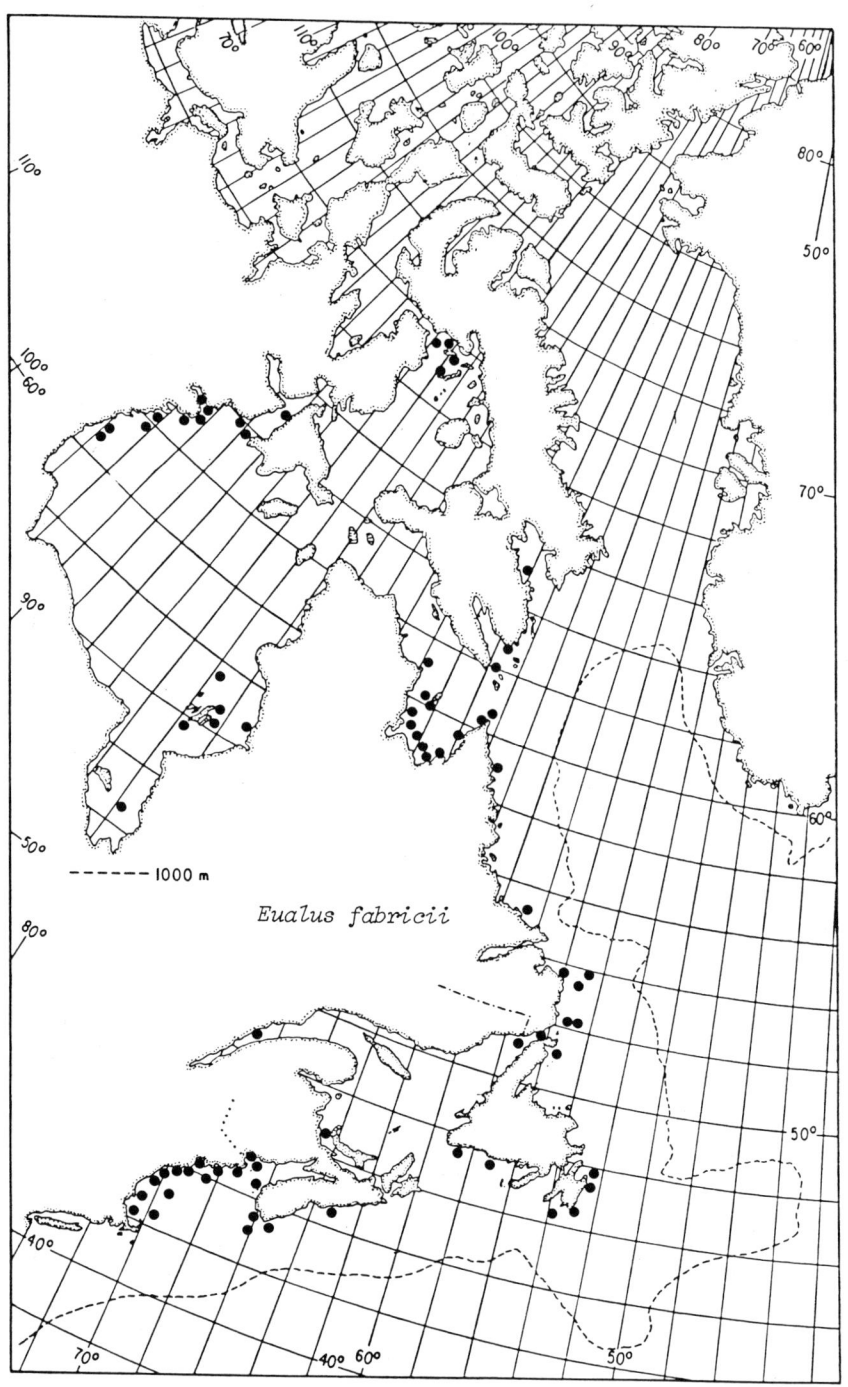

Fig. 90. *Eualus fabricii*, distribution records in the area of reference.

Eualus gaimardi belcheri Bell, 1855
Spirontocaris gaimardi belcheri Rathbun, 1929: 16, fig. 17;
Eualus belcheri, Couture et Trudel 1968: 875, fig. 14a.
(Figures 91 and 92, Plate 2)

DISTINGUISHING CHARACTERISTICS

Lobe with very strong black-tipped hook on 3rd abdominal somite; no supraorbital spine; teeth dorsally as well as ventrally on rostrum; rostrum about 1.3 times cl; distinct red striping around abdomen (persisting in preservative for few years); epipod on first legs only (sometimes on 2nd).

DESCRIPTION

Integument smooth, shiny. Colour reddish with well-defined red banding on dull whitish background on abdomen.

Rostrum slightly longer than carapace, with dorsal and ventral teeth; strong antennal spine with short carina and weak pterygostomian spine on anterior edge of carapace; median carina reaching about half carapace with crest of 4 fixed spines above orbit, usually 3 behind orbit; paired strong sternal spines ventrally between legs III–V in males only, single spine between 2nd legs also in females.

Abdomen with pronounced dorsal lobe and hooked spine (black-tipped) on 3rd somite; all other terga and pleura rounded except postero-ventral spine on 4th–6th somites, 6th with flared supporting carina; in males paired ventral spines on 1st and 2nd somites, single ventral spines on 3rd to 5th somites, also anal spine in males and females.

Eyes large, short tapering eyestalk, globular cornea.

Antennule (c): 1st article longer than other two combined, 3rd about half 2nd; 2nd and 3rd with distal spine; stylocerite with sharp point reaching distal 2nd article.

Antenna (d): Scale exceeding disto-lateral spine; peduncle about half length of scale; scale almost as long as carapace; basal article with strong outer spine.

Mandible (e): molars slightly different, right with more surface irregularities; incisor with finger-like tooth at distal corner and 4 smaller teeth; distal segment of palp slightly longer than proximal.

Maxillule (f): distal endite sub-ovate, large, proximal narrow slightly curved; endopod bifid.

Maxilla (g): distal endite bilobed, proximal lobe slightly larger; proximal endite reduced, bilobed, distal lobe small pointed and with fine setae, proximal lobe rounded with long curved setae; endopod short, with distal neck; scaphognathite with anterior lobe longer, rectilinear but slightly curved, posterior somewhat axe-shaped, rounded and with moderate setal fringe.

Maxilliped I (h): endites almost equal from broad base, posterior with double leading edge and long setal fringes; endopod two-segmented reaching distal end of lobe of exopod; exopod with lash longer than lamellate part; epipod bilobed, anterior lobe triangular.

Maxilliped II (i): leg-like but compressed; endopod with six segments, distal much wider than long; exopod long, slender; epipod single lobe with podobranch.

Maxilliped III (k): moderately stout, longer than first leg; four apparent segments, distal with 9 strong black spines apically; very short slender exopod; proximal lateral expansion with strap-like epipod (with terminal grooming hooks).

Pereopods: I (l) stout, shorter than others, chelate, fingers black-tipped, dactyl bifid; strap-like epipod. II slender, carpus with seven divisions; dactyl bifid; epipod absent. III–V long, about equal in length; dactyls biunguiculate followed by series of strong black spines; outer series of spines on merus, series of spinules on flexor edge of propodus.

Pleopods: Female I (o) endopod short, sub-triangular, tip with setae only; female II (p) appendix interna moderate with apical patch of hooks. Male I (q) endopod short with apical finger-like extension with terminal patch of hooks; male II (r) appendix interna

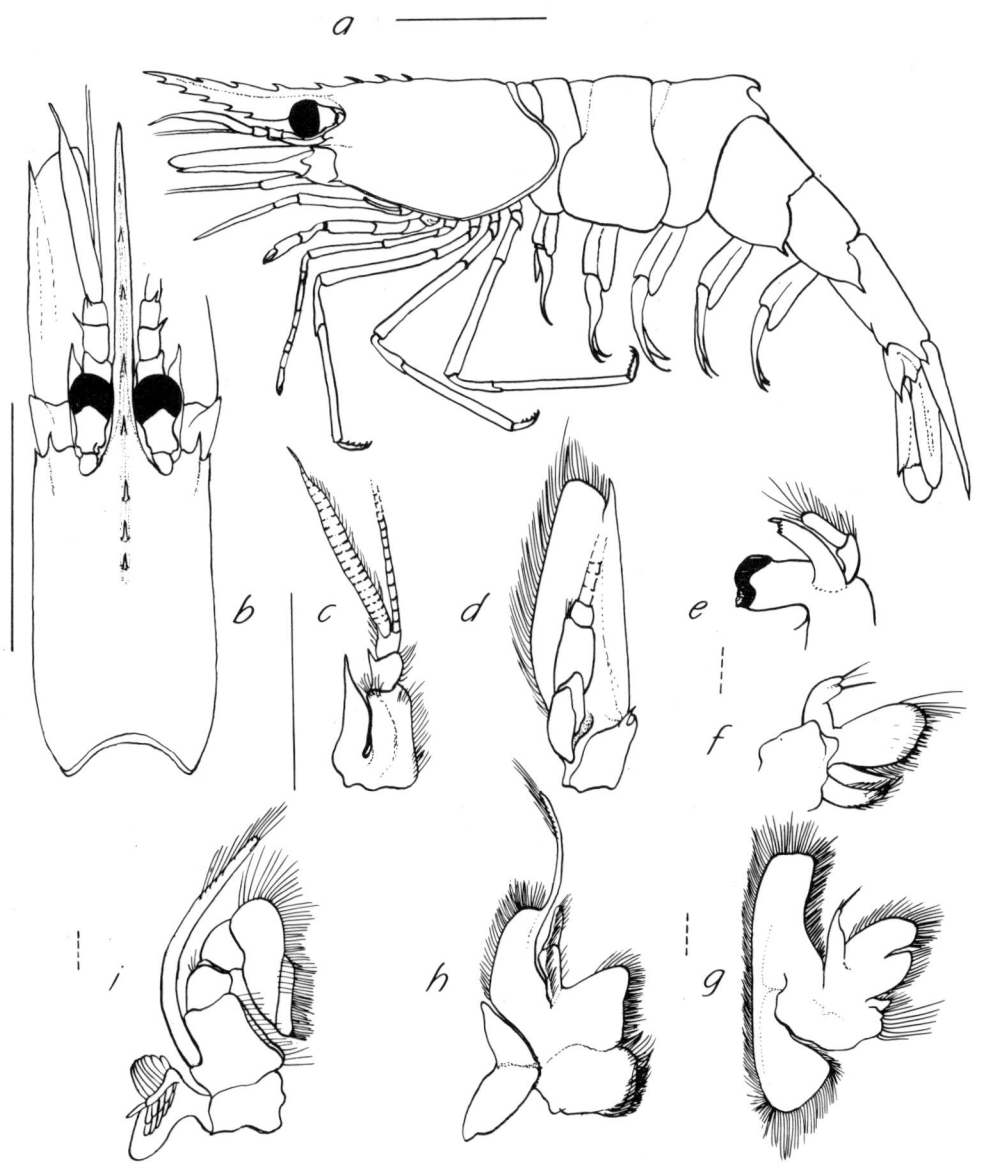

Fig. 91. *Eualus gaimardi belcheri*: *a* whole shrimp from left sides; *b*, carapace in dorsal aspect; *c*, antennule; *d*, antenna; *e*, mandible; *f*, maxillule; *g*, maxilla; *h*, first maxilliped; *i* second maxilliped; solid line = 10 mm, broken line = 1 mm.

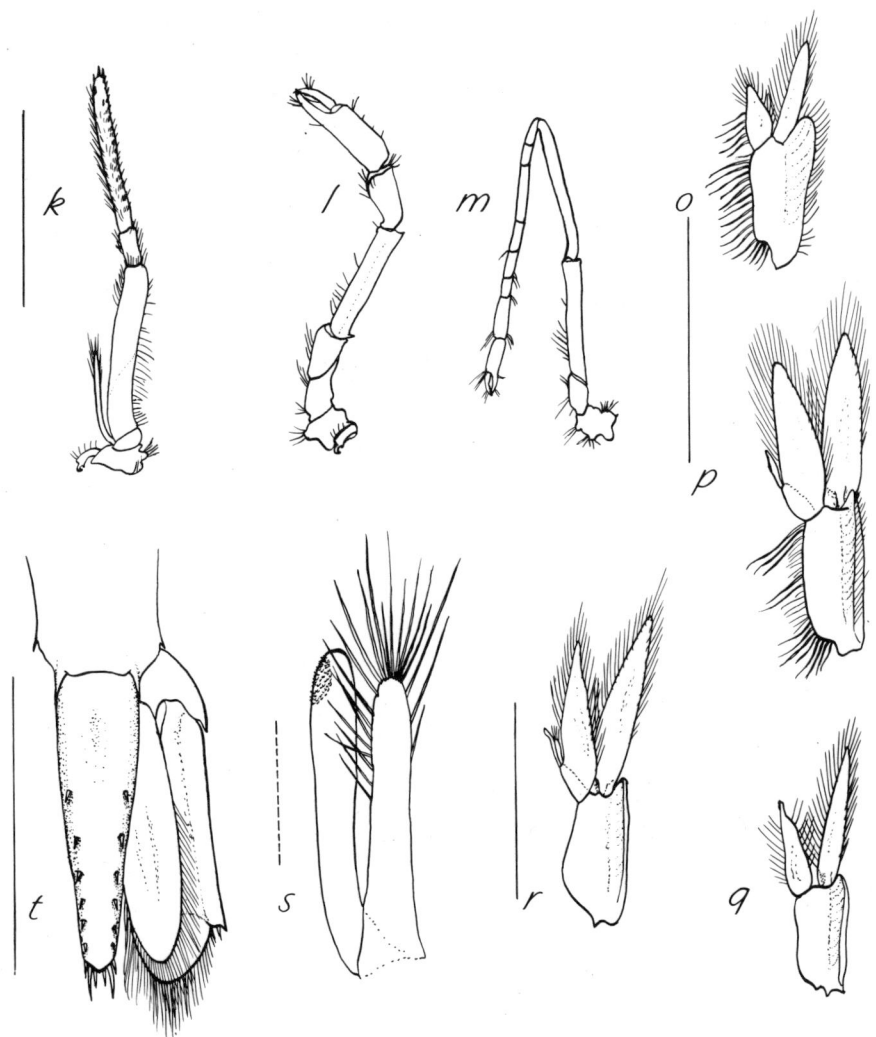

Fig. 92. *Eualus gaimardi belcheri*: *k*, third maxilliped; *l*, first pereopod; *m*, second pereopod; *o*, F first pleopod; *p*, F second pleopod; *q*, M first pleopod; *r*, M second pleopod; *s*, appendix masculina; *t*, telson: solid line = 10 mm, broken line = 1 mm.

longer than masculina, the latter (s) with about 6 long apical spines and two series of about 15 long spines on one side.

RANGE OF DISTRIBUTION

In world distribution not separated from *E. g. gaimardi* which is in the eastern Atlantic from Spitzbergen to the North Sea, and western Atlantic from Greenland and Baffin Island to Cape Cod; also in the Arctic Ocean from Point Barrow and north of Siberia, and in the North Pacific as far south as Sitka (Holthuis 1947). In the area of reference specifically from Ungava Bay and Trinity Bay, depths 135–290 m and temperatures of -1.1 to 0.5 C.

Distribution records in the area of reference are in Fig. 93.

BIOLOGY

Carapace lengths in males to 16 mm and females to 22 mm, averages 14 and 16 mm, respectively (Squires 1965a).

Females were first mature at 12 mm cl and males at 6 mm. In Trinity Bay at a temperature of about 1° C, 70% of the females were ovigerous in autumn; in Foxe Basin at lower temperatures only 50% were potentially ovigerous, suggesting that in the colder situation spawning might occur only every second year (Squires 1965a; 1967a).

Feeding appeared to be on crustaceans, including ostracods and euphausiids; also present occasionally were polychaetes and pelecypods, and detritus and foraminiferans (Squires 1965a).

Associated species were chiefly *Pandalus montagui*, *Eualus macilentus*, *Lebbeus polaris*, *Pandalus borealis*, *Sabinea septemcarinata* and *Argis dentata* (Squires 1965a).

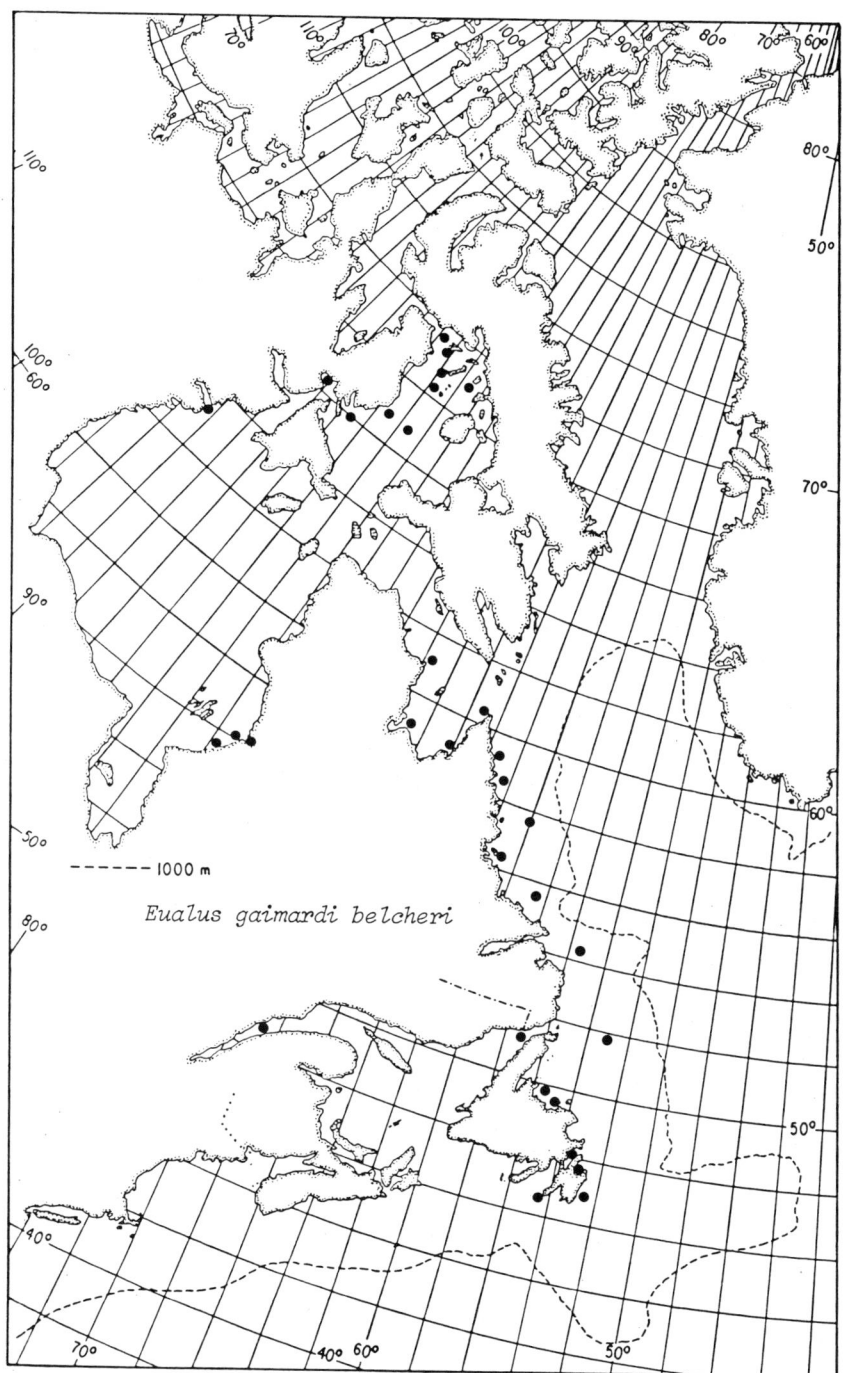

Fig. 93. *Eualus gaimardi belcheri*, distribution records in the area of reference.

Eualus gaimardi gaimardi (H. Milne-Edwards, 1837)
Spirontocaris gaimardi, Rathbun 1929:16, fig.16.
Eualus gaimardi, Holthuis 1947:10, 43.
(Figures 94 and 95)

DISTINGUISHING CHARACTERISTICS

Rostrum is slightly longer than carapace and has teeth above as well as below; a lobe is present on the 3rd abdominal somite (may be reduced or lacking in some) and this is strongly hooked in adults as in *E. g. belcheri* but with colourless and not black tippped spines; epipods on the first two pereopods, except in some larger ones with epipod on 1st only; this subspecies appears to have many intermediates as far as the structure of the lobe on the third abdominal somite is concerned: in *E. g. belcheri*, on the other hand, all appear to have the strong hook on the lobe and there are strong red stripes on the abdomen (Plate 2), persisting in preserved specimens for some years. Samples in considerable numbers from deep water in Trinity Bay, Newfoundland, and Ungava Bay, Quebec, have all been of the subspecies *belcheri* without intermediates: samples in smaller numbers from shallow water in coastal areas of the Arctic and Newfoundland have contained intermediates and were characteristically the subspecies *gaimardi*.

DESCRIPTION

Integument smooth, shiny. Colour whitish translucent background with brownish red or greenish markings, eggs green.

Rostrum 1.4 times longer than carapace, with long pointed tip and dorsal and ventral teeth, formula 7–8/3–4, with 3 on the crest above the orbit and a short low carina on the carapace; rostrum exceeds the antennal scale and short (dorso-lateral) flagellum of the antennule. Antennal and pterygostomian spines only present on carapace, the antennal has a short carina. Strong sternal spines in pairs between 3rd to 5th legs.

Abdomen with a short carina and lobe with hooked spine (especially in adults) on third somite, hook and lobe reduced in some specimens. Pleura all rounded except ventro-lateral spines on 4th–6th somites; 6th somite about twice length of 5th; mid-ventral spines on each of somites 1–5, the one on 5th directed posteriorly (1st and 2nd only have pairs of spines).

Telson (t) is 1.3 times length of 6th somite, tapering to somewhat truncate tip which has three pairs of terminal spines; five pairs of dorso-lateral spines.

Eye with short tapering stalk and large globular cornea. Antennule (c): 1st article longer than 2nd and 3rd combined; 2nd and 3rd sub-equal, each with sharp distal spine, also 1st with ventral spine on distal third; dorso-lateral flagellum 1.3 times length of peduncle, ventro-mesial about 2.8 times. Stylocerite with sharp attenuate point almost reaching distal 2nd article.

Antenna (d): Scale length 3.3 times width, blade exceeding spine by about 2.2 mm; basal article with strong ventro-mesial spine distally; flagellum 4.6 times length of carapace; peduncle about half length of scale.

Mandibles (e): molars with slightly different grinding surfaces, left with fewer cusps than right; incisors with finger-like tooth at the corner followed by three small teeth; palp with distal segment slightly longer than proximal.

Maxillule (f): proximal endite narrow, curved; distal endite subovate with broad base; endopod curved, bifid.

Maxilla (g): distal endite almost equally bilobed, large; proximal endite small, unequally bilobed, the distal lobe pointed; endopod short with slender distal portion; scaphognathite with slightly curved anterior lobe — distally slightly concave — and shorter somewhat axe-shaped posterior lobe with moderate setal fringe.

Maxilliped I (h): endites broad, almost equal; endopodal palp reaching almost as far as expanded part of exopod, somewhat twisted, not clearly segmented; exopod with long lash (longer than lobe); epipod large, bilobed, both pointed.

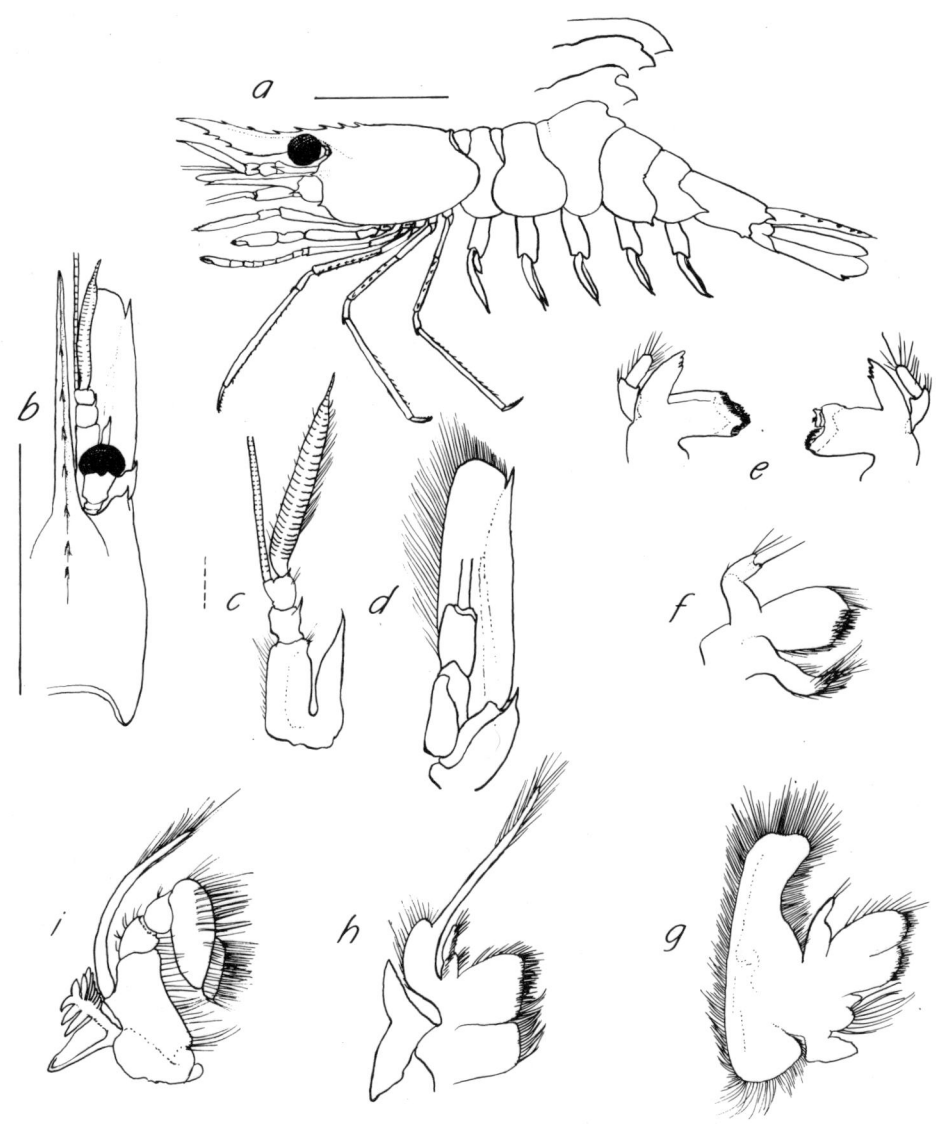

Fig. 94. *Eualus gaimardi gaimardi*: *a*, whole shrimp from left side; *b*, carapace in dorsal aspect; *c*, antennule; *d*, antenna; *e*, mandible; *f*, maxillule; *g*, maxilla; *h*, first maxilliped; *i*, second maxilliped; solid line = 10 mm, broken line = 1 mm.

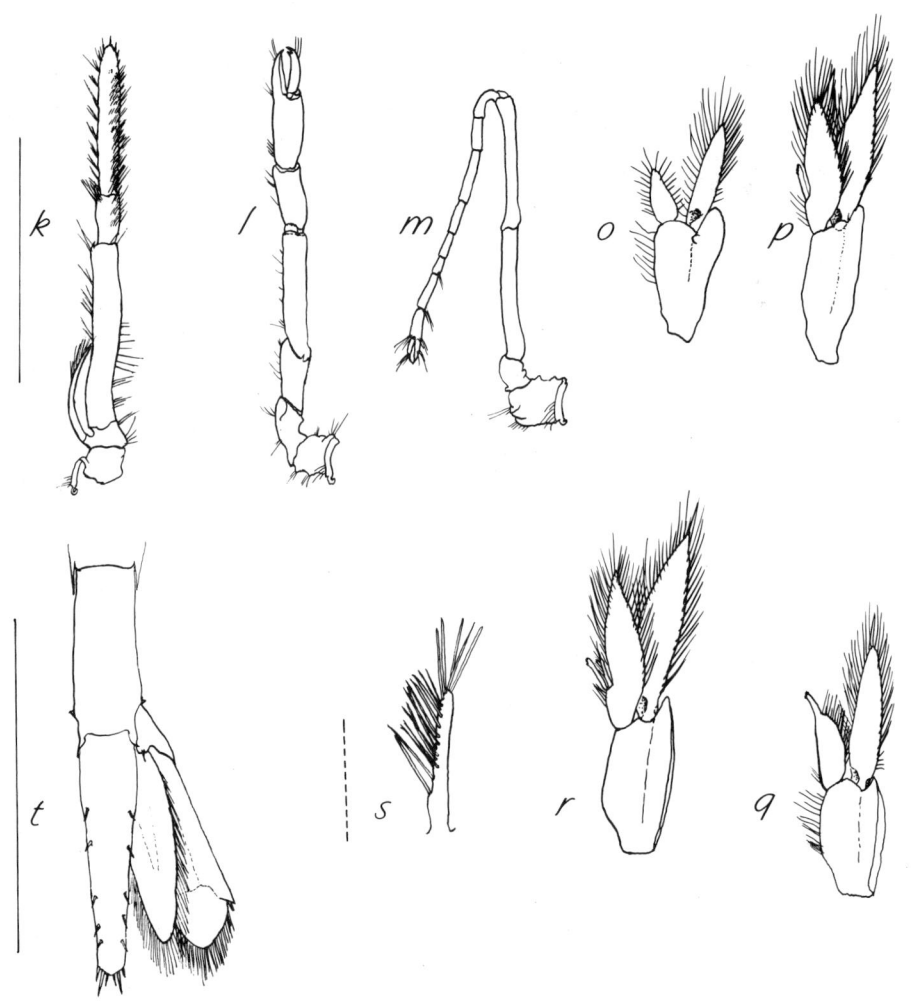

Fig. 95. *Eualus gaimardi gaimardi*: *k*, third maxilliped; *l*, first pereopod; *m*, second pereopod; *o*, F first pleopod; *p*, F second pleopod; *q*, M first pleopod; *r*, M second pleopod; *s*, appendix masculina; *t*, telson; solid line = 10 mm, broken line = 1 mm.

Maxilliped II (i): leg-like but compressed; of six segments, distal wider than long; exopod slender, about as long as endopod; epipod saucer-like, with podobranch.

Maxilliped III (k): stout, slightly longer than first leg; about five segments, distal with apex edged with eight strong black spines; exopod very short; epipod strap-like with terminal grooming hooks.

Pereopods: I (l) stout, shorter than others; chelate, fingers shorter than palm, dactyl bifid with two spines that the spine of the propodal finger fits between; epipod strap-like with terminal grooming hooks. II (m) slender, longer than I but shorter than others, carpus with seven divisions; epipod present in some specimens only. III–V about equal in length, few outer spines on merus and series of spines on flexor edge of propodus; dactyl biunguiculate with a few shorter spines on flexor surface, spines black.

Pleopods: female I (o) endopod short subtriangular, exopod narrow; protopodite with wing-like lateral expansion. Female II (p) endopod slightly shorter than exopod, appendix interna with sub-apical patch of hooks. Male I (q) endopod pointed and with apical finger-like extension with terminal patch of hooks. Male II (r) endopod with appendices interna and slightly shorter masculina, the latter (s) with about 5 apical long spines and lateral series of about 14 shorter spines.

RANGE OF DISTRIBUTION

Circumarctic southward to the North Sea in Europe and to Cape Cod in America; in the Arctic Ocean it has been reported from Point Barrow and off Siberia and in the north Pacific as far south as Sitka, Alaska. Depths from 10 to 900 m; temperatures −1.0 to 3.8° C (Williams and Wigley 1977).

Records of occurrences in the area of reference are in Fig. 96.

BIOLOGY

Lengths to 9 mm cl in males and 14 mm in females, averages 6 and 9 mm, respectively.

In Foxe Basin about 33% of females collected were potentially ovigerous in August and September, and in Hudson Bay about 41%, indicating cold stressed populations that would probably only spawn every second year. Most specimens were taken in negative temperatures. First maturity of females was at 9 mm cl.

Food in stomachs was largely phytobenthos, ostracods, foraminiferans, crustacean and polychaete fragments; occasionally gammarid amphipods, bivalves, small gastropods, hydroids and red algae.

Predators were cod on the Grand Banks, ringed and bearded seals from Ungava Bay and beluga whales from Hudson Bay (Squires 1957, 1965a, 1967a).

Most frequently associated decapod species were *Pandalus montagui*, *Lebbeus polaris*, *Spirontocaris spinus*, *Sabinea septemcarinata*, *Argis dentata* and *Sclerocrangon boreas* (Squires 1965a).

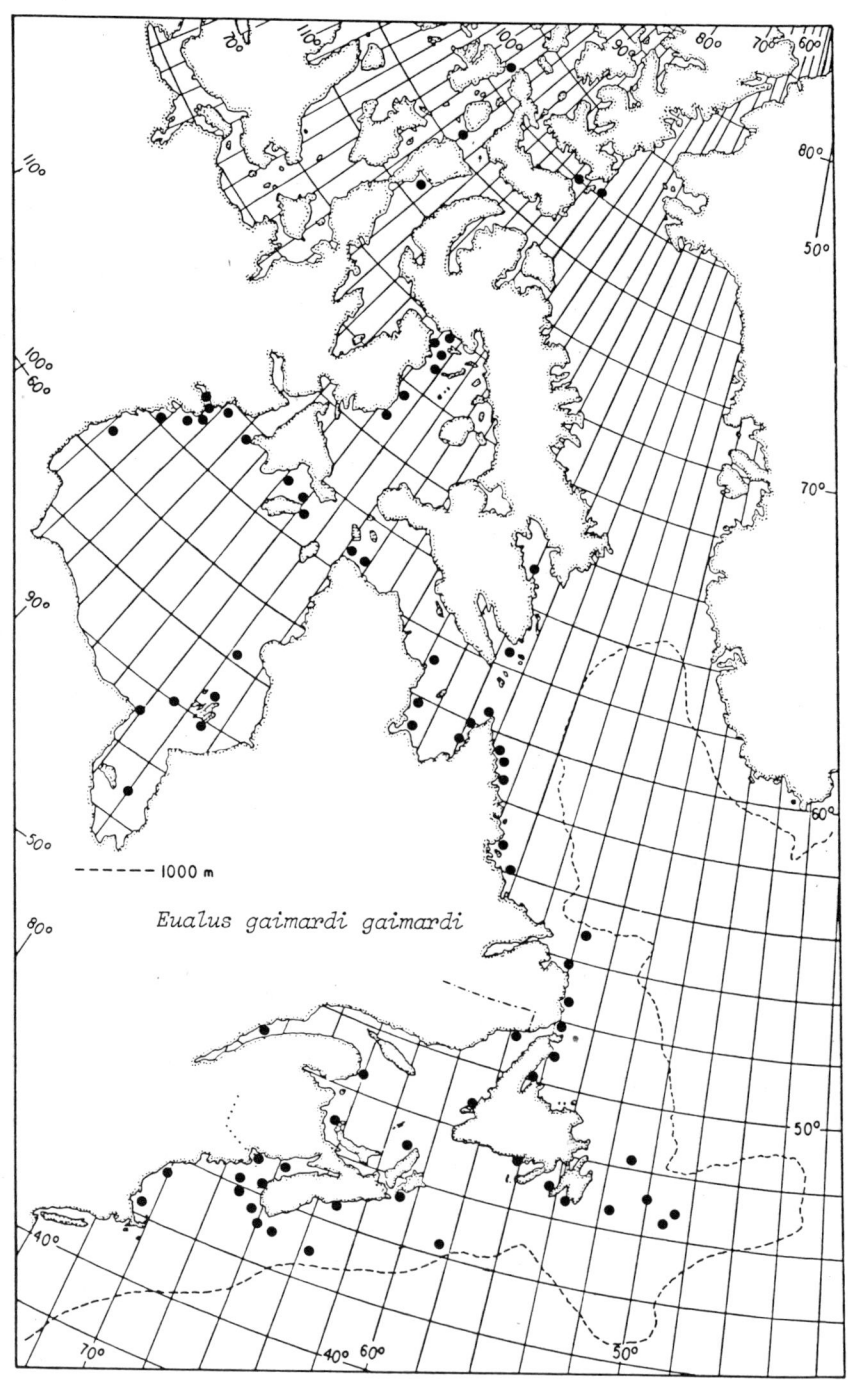

Fig. 96. *Eualus gaimardi gaimardi*, distribution records in the area of reference.

Eualus macilentus (Krøyer, 1842)
Spirontocaris macilenta, Rathbun 1929: 16, fig. 18.
Spirontocaris stoneyi Rathbun, 1902; Rathbun 1929: 17, fig. 20.

(Figures 97 and 98)

DISTINGUISHING CHARACTERISTICS

No supraorbital spines; rostrum shorter than carapace, slightly convex dorsally and ventrally, several small close teeth above and 0–3 below; rostrum thin and transparent with lateral ridge confluent with orbit; epipods on 1st to 3rd legs; telson with three pairs of dorso-lateral spines.

DESCRIPTION

Integument smooth, shiny. Colour whitish translucent with reddish spots.

Rostrum 0.55 times carapace length; translucent and with a straight horizontal rib confluent with orbit, reaches about distal edge of 2nd antennular peduncle; small teeth in close series above, rounded, a few (1–3) on carapace, teeth below near tip giving a truncated appearance, formula 7–16/0–4, 0–3 on carapace; rostrum is more pointed and narrow in young immature specimens (*E. stoneyi* forms); antennal scale as long as carapace, peduncle reaching about half scale.

Abdomen with all pleura and terga rounded except 4th–6th with posterior ventral sharp spine; 6th somite 1.5 times length of 5th and about equal to telson.

Telson (t) long and tapering with three pairs of dorso-lateral spines; terminally three pairs of spines the 2nd laterals longest.

Eye large, globular cornea, short tapering eyestalk.

Antennule (c): 1st article three times length of 3rd, with ventral spine on distal third; stylocerite thin and flat about as long as 1st article; short distal spine on 2nd and 3rd articles; ventro-mesial flagellum about as long as carapace.

Antenna (d): scale exceeds spine; scale narrows perceptibly at about distal third; peduncle half or less than half scale in length.

Mandible (e): incisor with 4 teeth, distal one longest; molars slightly different, right with more surface irregularities than left; palp of two segments, distal slightly shorter than proximal.

Maxillule (f): distal endite ovate base thick; proximal endite narrow, curved at tip; endopod bifid.

Maxilla (g): proximal endite reduced, rounded, with long curved setae; distal about equally bilobed; endopod short; anterior lobe of scaphognathite longer than posterior, squarish at tip, posterior axe-shaped rounded, with moderate even setal fringe.

Maxilliped I (h): proximal endite squarish with double edge and long curved setae; distal endite subtriangular, edge slightly concave; endopod irregularly fusiform with surface row of setae, about as long as lobe of exopod; slender lash about twice length of endopod; epipod large, bilobed.

Maxilliped II (i): pediform, compressed, distal segment much wider than long; exopod long slender; epipod with podobranch.

Maxilliped III (k): endopod with five segments, distal armed at the tip with six strong black spines; exopod small, slender; epipod strap-like with terminal hook.

Pereopods: I (l): stout, chelate, dactyl bifid; epipod strap-like with terminal hook. II (m): slender, chelate; dactyl bifid; carpus with seven divisions; epipod strap-like with terminal hook. III–V (a): long, slender; III only with strap-like epipod as in I and II; dactyl slender, without spines except at tip; spines on merus and spinules on propodus on flexor edge.

Pleopods: female I (o) endopod moderately wide, pointed with setal fringe and no hooks; exopod narrow less than twice endopod. II (p) endopod about as long as exopod, appendix interna with sub-apical patch of hooks. Male I (q) endopod fusiform, with

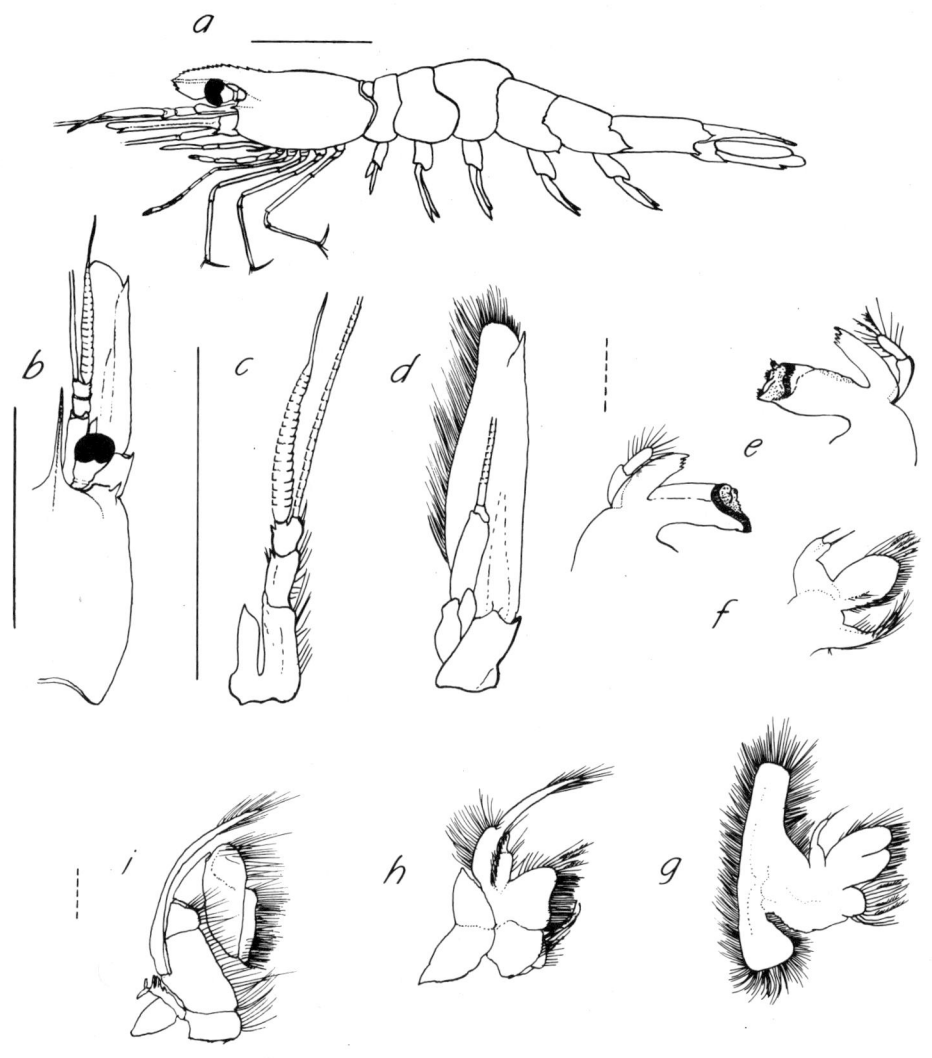

Fig. 97. *Eualus macilentus*: *a*, whole shrimp from left side; *b*, carapace in dorsal aspect; *c*, antennule; *d*, antenna; *e*, mandible; *f*, maxillule; *g*, maxilla; *h*, first maxilliped; *i*, second maxilliped; solid line = 10 mm, broken line = 1 mm.

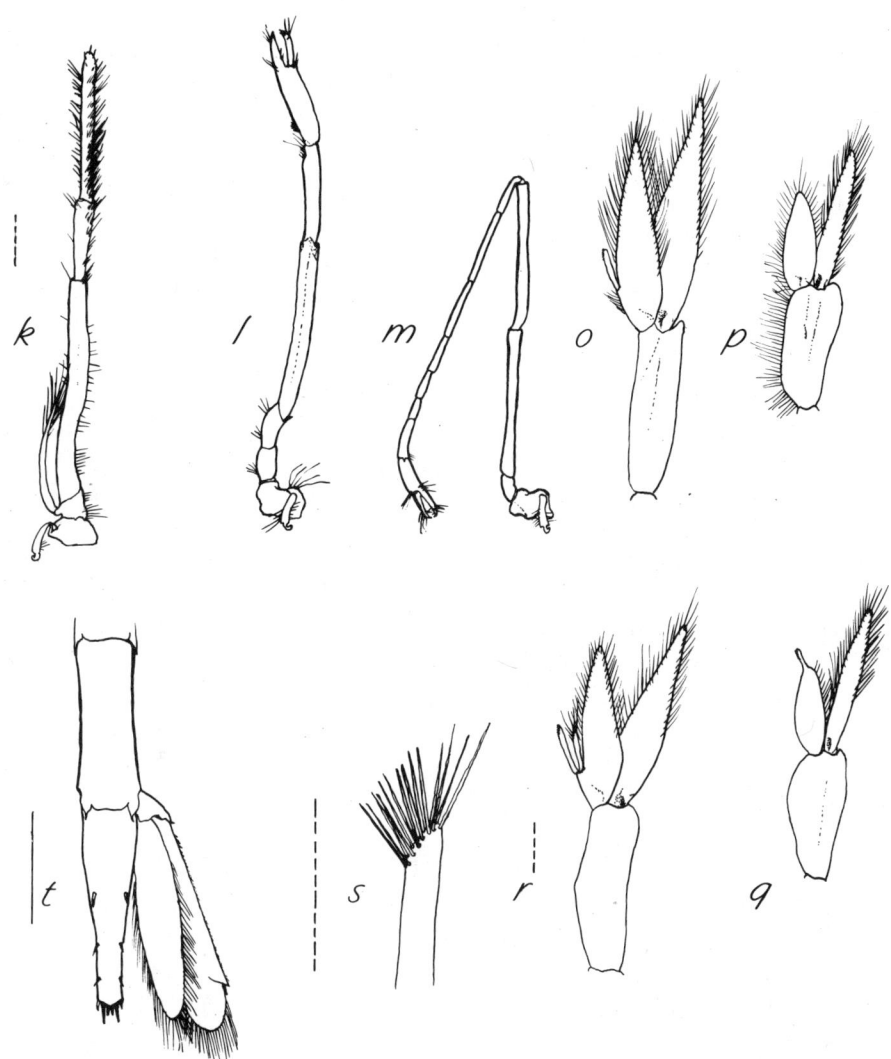

Fig. 98. *Eualus macilentus*: *k*, third maxilliped; *l*, first pereopod; *m*, second pereopod; *o*, F first pleopod; *p*, F second pleopod; *q*, M first pleopod; *r*, M second pleopod; *s*, appendix masculina; *t*, telson; solid line = 10 mm, broken line = 1 mm.

terminal projection with distal hooks; II (r) endopod slightly shorter than exopod, appendix masculina shorter than interna; appendix masculina (s) with four apical long spines and a double series of five spines each on one side.

Range of Distribution

West Atlantic from Hudson Bay, Foxe Channel and Greenland to Nova Scotia; in the northern Pacific from the Okhotsk and Bering Seas, and in the Arctic Ocean from off Siberia. Depths 55–540 m (Holthuis 1947; Squires 1967a).

Rathbun (1913) reviewed the Labrador records of this species from Fish Island to Square Island in 28–137 m. In later surveys it was taken in southern Foxe Channel and Ungava Bay and in Hudson Bay (mainly in Richmond Gulf: 95–100 m). It was found also in deep Newfoundland bays and Labrador fjords . Generally it occurred in small numbers from shallow water in the north but in much greater numbers from 200 to 300 m in the south at temperatures lower than 0° C (Squires 1957; 1965a; 1967a).

Records of distribution in the present area of reference are shown in Fig. 99.

Biology

Lengths in males and females were 14 and 16 mm cl, averages 9 and 11 mm cl, respectively.

In the Newfoundland–Labrador area 100% of mature females spawned annually and in Hudson Bay 93% were potentially ovigerous, indicating full adaptation to temperatures below 0° C. Females were first mature at 8 mm cl. In Foxe Channel, however, only 62% were potentially ovigerous and may have been under stress in this shallow water environment (Squires 1967a).

Stomach contents were mainly phytobenthos, crustaceans including ostracods and euphausiids and foraminiferans; also in fewer instances were hydroids, polychaetes and sponge spicules (Squires 1967a).

Predators were cod from the Grand Banks (134–141 m), from Fortune Bay (183 m) and from off Bonavsita (285–322 m) (Squires 1965a).

Although *E. macilentus* occurred in fairly large numbers in samples from Newfoundland, *Pandalus montagui* which was invariably present dominated the samples. *P. borealis* was also occasionally present in these catches as well as *Argis dentata* and *Sabinea septemcarinata* (Squires 1965a).

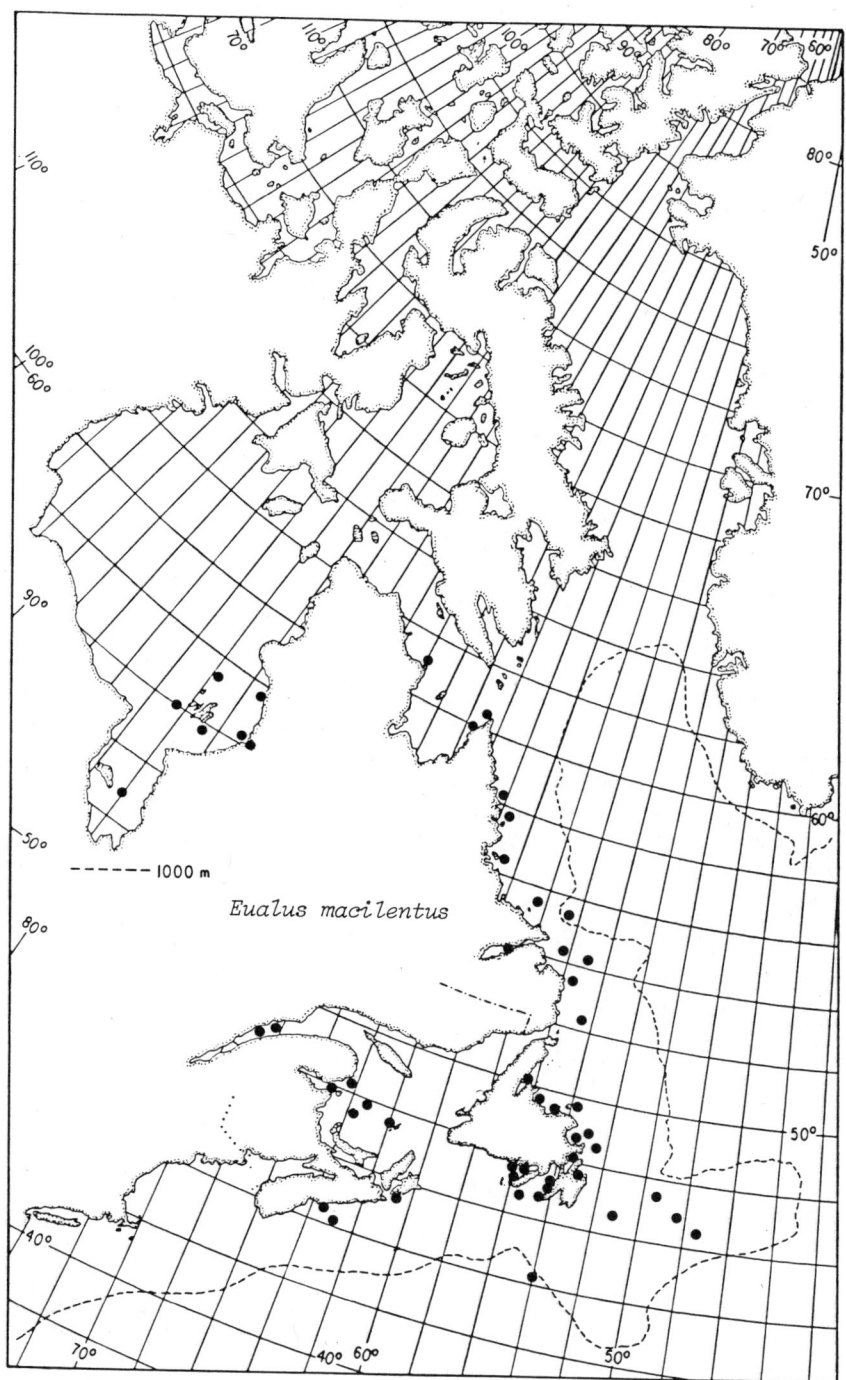

Fig. 99. *Eualus macilentus*, distribution records in the area of reference.

Eualus pusiolus (Kroyer, 1841)
Spirontocaris pusiola, Rathbun 1929: 17, fig. 19.
Eualus pusiolus, Holthuis 1947: 11; Williams 1984: 113, fig 78.
(Figures 100 and 101)

DISTINGUISHING CHARACTERISTICS

No supraorbital spines; rostrum short, pointed, slightly descending, with a few dorsal teeth only; epipods on first three legs; a small shrimp.

DESCRIPTION

Integument smooth, shiny. Colour a few red to orange spots over whitish translucent background (Leim 1921); dark brownish green like colour of Laminaria among which it occurs, and with stripes of red brown (Greve 1963).

Rostrum sharply pointed, rarely bidentate, short, does not reach farther than eye, only very slightly descending, lateral carina extends on to the carapace past the orbit, dorsal spines two to five; rostral carina short on carapace with one tooth behind the orbit. Antennal spine with short carina, pterygostomian spine small.

Abdomen with rounded terga and pleura except 5th and 6th somites with posterior ventral spine; 6th somite 1.7 times 5th. Telson (t) only slightly tapered with four pairs of dorso-lateral spines and 3 pairs of terminal spines. Branches of uropod slightly longer than telson.

Eye large with globular cornea and short tapered stalk,

Antennule (c): 1st article exceeding other two combined, 2nd longer than 3rd, all with strong distal spine, also on 1st a small inner spine on distal nine-tenths. Stylocerite longer than 1st article.

Antenna (d): scale length 2.5 times width, peduncle more than half scale; distolateral spine about even with lamella; flagellum about as long as the body.

Mandible (e): incisor narrow, longer than palp; with three distal teeth on left and four on right incisor; distal segment of palp slightly longer than proximal. Molar stout.

Maxillule (f): proximal endite sharply tapered, curved at tip; distal endite wide at base, ovate; endopod curved and obscurely bifurcate.

Maxilla (g): proximal endite reduced, distal lobe small; distal endite almost equally bilobed; endopod short; scaphognathite with long anterior lobe squarish at tip, posterior lobe rounded and with fringe of setae only slightly longer terminally.

Maxilliped I (h): distal endite subtriangular, somewhat wider than proximal; proximal with frontal edge double and triangular in outline. Endopod with two segments, exceeding short lobe of exopod, lash about three times as long as lobe; epipod large, bilobed, each subtriangular.

Maxilliped II (i): endopod compressed; distal segment much wider than long; exopod long slender; epipod with small podobranch.

Maxilliped III (k): distal segment of endopod long with eight strong black spines in oval apically; third segment with stout spine distally; exopod short, slender; epipod strap-like with terminal hook.

Pereopods: I (l) chelate, stout, shorter than third maxilliped and legs; dactyl bifid; strap-like epipod. II (m) chelate, slender, carpus seven-segmented, dactyl bifid, epipod strap-like. III–V moderately stout, III only with epipod, merus with strong distal spine, propodus with many spinules and distal bunch of bristles, dactyl biunguiculate and with strong black spines on flexor surface.

Pleopods: female I (o) endopod wide, pointed, as long as narrow exopod, with setal fringe and no hooks; female II (p) appendix interna with subapical patch of hooks. Male I (q) endopod with distal extension and apical patch of hooks; male II (r) appendix interna about twice length of masculina, the latter (s) with apically a star of about 10 setae.

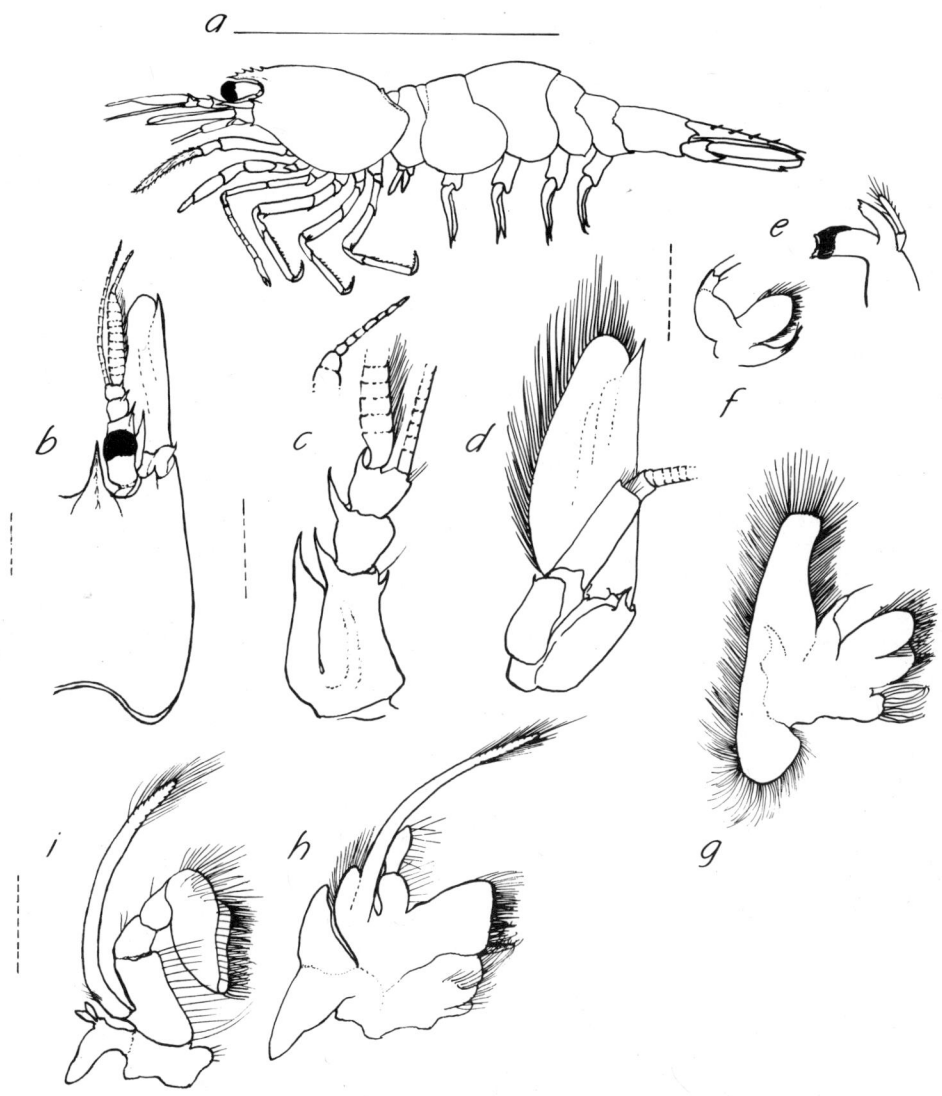

Fig. 100. *Eualus pusiolus*: *a*, whole shrimp from left side; *b*, carapace in dorsal aspect; *c*, antennule; *d*, antenna; *e*, mandible; *f*, maxillule; *g*, maxilla; *h*, first maxilliped; *i*, second maxilliped; solid line = 10 mm, broken line = 1 mm.

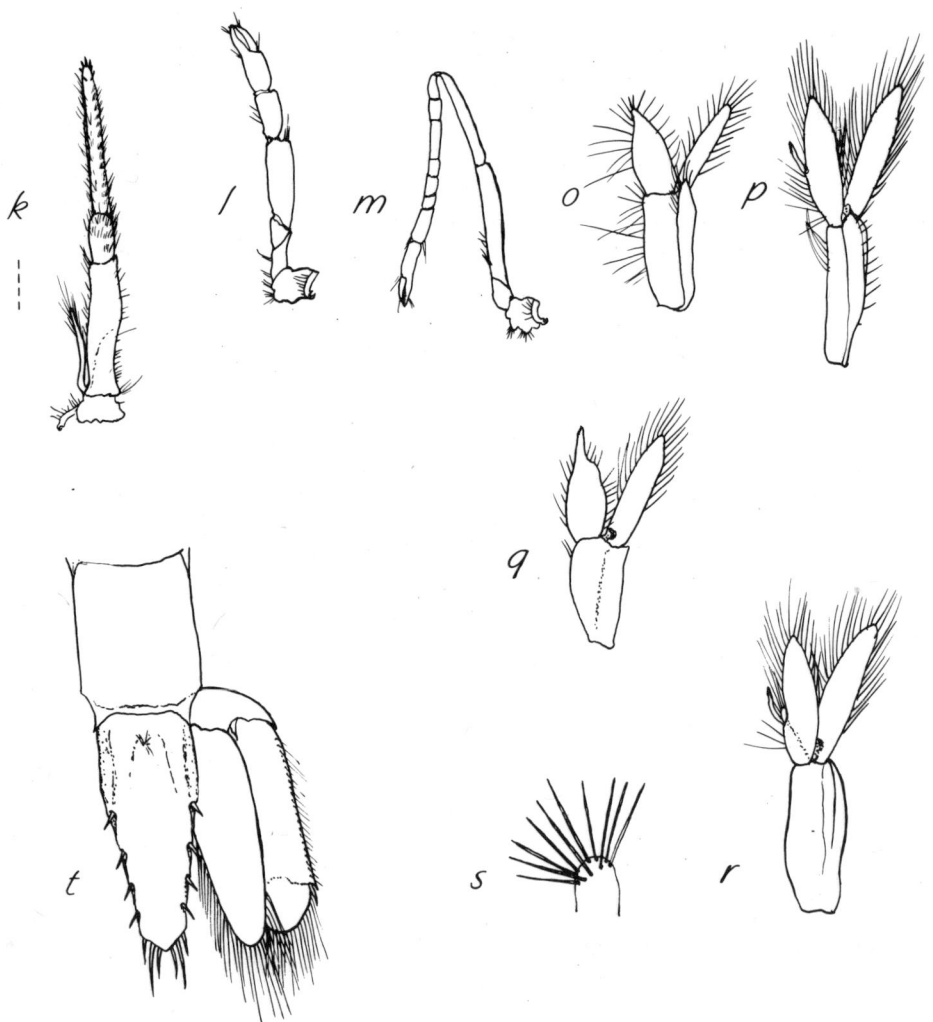

Fig. 101. *Eualus pusiolus*: *k*, third maxilliped; *l*, first pereopod; *m*, second pereopod; *o*, F first pleopod; *p*, F second pleopod; *q*, M first pleopod; *r*, M second pleopod; *s*, appendix masculina; *t*, telson; broken line = 1 mm.

Range of Distribution

North Atlantic from the Murman Sea to Channel Islands and Catalonian coast of Spain (Mediteranean) in Europe, and Gulf of St. Lawrence to Virginia in America (Zariquiey Alvarez 1968; Williams 1984). Arctic Ocean and North Pacific from the Chukchi and Bering Seas to British Columbia and Washington; Okhotsk Sea and Sea of Japan (Butler 1980; Williams 1984). Depths 0–500 m (Holthuis 1947).

Records in the area of reference are shown in Fig. 102.

Biology

Length of body to 30 mm (Couture and Trudel 1968); 4 mm cl in females in our collection. Egg diameter 0.6 mm.

In the Clyde Sea area of Scotland ovigerous females were seen in early months of the year, larvae hatched in April and eggs were laid in April and May (Allen 1967). Two broods were produced each 8–10 weeks, a female carrying 100–300 eggs (Allen 1966; Smaldon 1979).

Cod was a predator on the southwest edge of the Grand Banks (Squires 1965a).

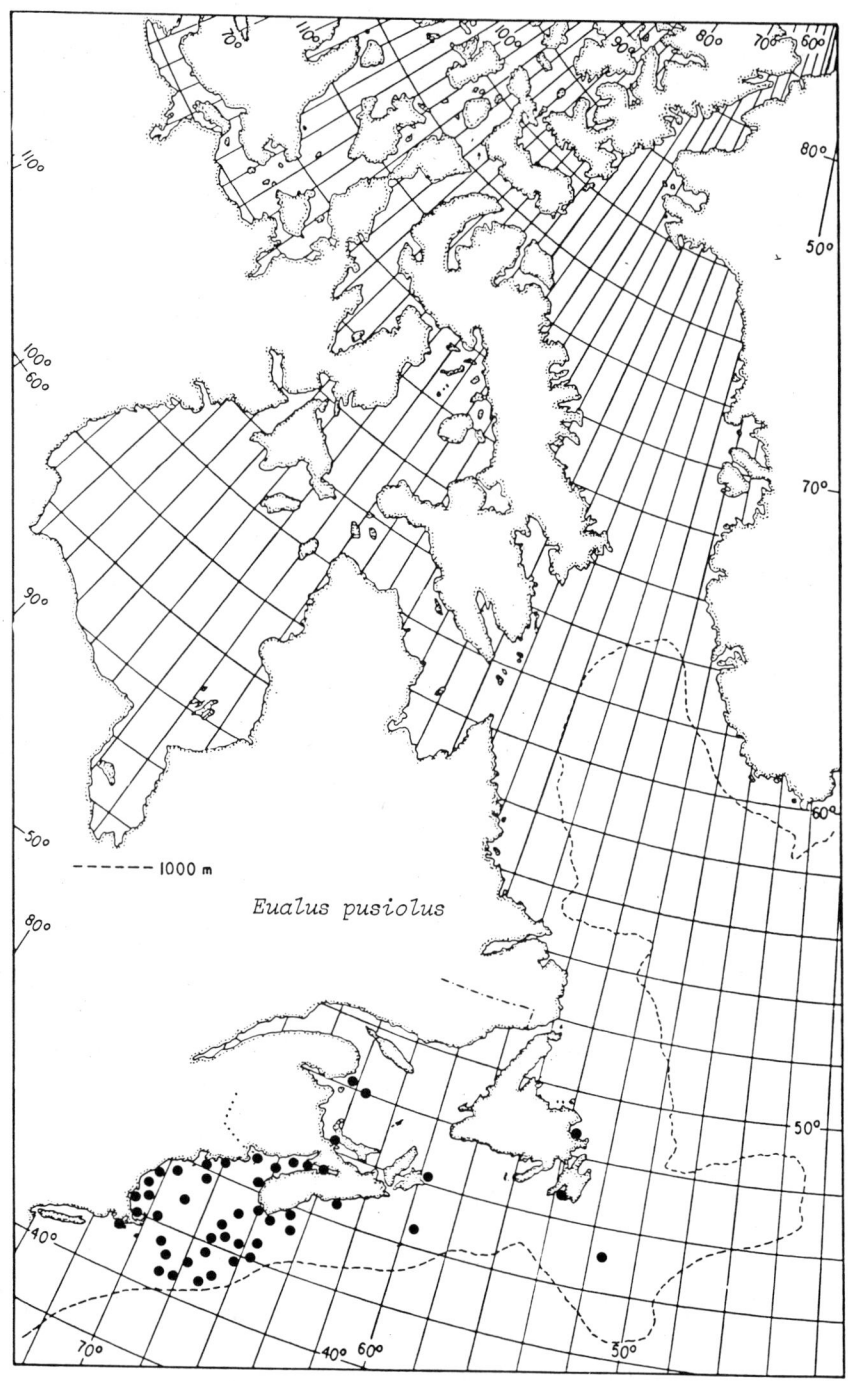

Fig. 102. *Eualus pusiolus*, distribution records in the area of reference.

Genus *Hippolyte* Leach (1814)
Holthuis 1947: 53; Williams 1984: 115;
Zariquey Alvarez 1968: 117.

Carapace with supraorbital, antennal and branchiostegal spines; carpus of second legs with 3 divisions; mandible with incisor and molar processes but no palp; third maxilliped with exopod but no epipod; telson truncate, with 2 pairs of dorso-lateral spinules; carpus of legs III–V subprehensile.

Hippolyte coerulescens (Fabricius, 1775)
Hippolyte coerulescens, Holthuis 1947: 15, 53;
Chace 1972: 111, figs. 42-43; Williams 1984: 116, fig. 80.
(Figures 103 and 104)

Distinguishing Characteristics

Large supraorbital spine; carpus of 2nd leg with 3 segments; rostrum sharp pointed, thick at base, descending; strong dorsolateral spine on fifth abdominal somite; telson truncate with two pairs of dorsolateral spines on posterior third.

Description

Integument smooth, shiny. Colour brownish yellow banding, pattern imitating vesicles of *Sargassum* (Gurney 1936).

Carapace with large supraorbital, small antennal and branchiostegal spines, the latter over-reaching the anterior edge. Rostrum stout at base, tapering to sharp point, descending, a small tooth dorsally and ventrally; rostrum about 0.8 times cl in female, shorter in male; a small tubercle on median ridge just behind supraorbital spine.

Abdomen with rounded terga and pleura except large spinous process dorso-laterally on 5th and 6th somites; pleura of second somite large.

Telson only slightly tapering to truncate tip with four pairs of terminal spines, and two pairs of tiny dorso-lateral spines on posterior third; outer branch of uropod longer than telson, with disto-lateral spine quite near tip.

Eyes large, moderately long eyestalk, globular cornea.

Antennule (c): 1st article longer than other two combined, 2nd and 3rd subequal, peduncle about half antennal scale; stylocerite very short; dorsolateral flagellum shorter than peduncle.

Antenna (d): scale length about three times width and almost as long as cl, spine slightly exceeding blade; peduncle less than half scale; basal article with small outer spine.

Mandible (e): without palp; incisor with four apical teeth.

Maxilule (f): proximal endite narrow, curved; distal ovate from narrow base; endopod short, uniramous, curved.

Maxilla (g): proximal endite reduced, rounded single lobe; distal unequally bilobed, proximal lobe small; endopod short, curved; scaphognathite with very wide anterior lobe and very narrow posterior lobe.

Maxilliped I (h): distal endite wide; proximal narrow; endopod not segmented, as long as lobe of exopod; lash of exopod about as long as lamellate part; epipod bilobed.

Maxilliped II (i): slightly compressed, leg-like, six segments; distal longer than wide; exopod long and slender; epipod unequally bilobed.

Maxilliped III (k): leg-like with five segments; longer than first pereopod; distal with pointed tip and single spine; exopod short; no epipod.

Pereopods: I (l) chelate, stout, palm inflated; no epipod. II (m) chelate, slender; carpus with three divisions. III–V longer than others, moderately stout and similar; propodus with series of spines on flexor edge; dactyl short, stout, with few spines. propodus of male thickened in middle.

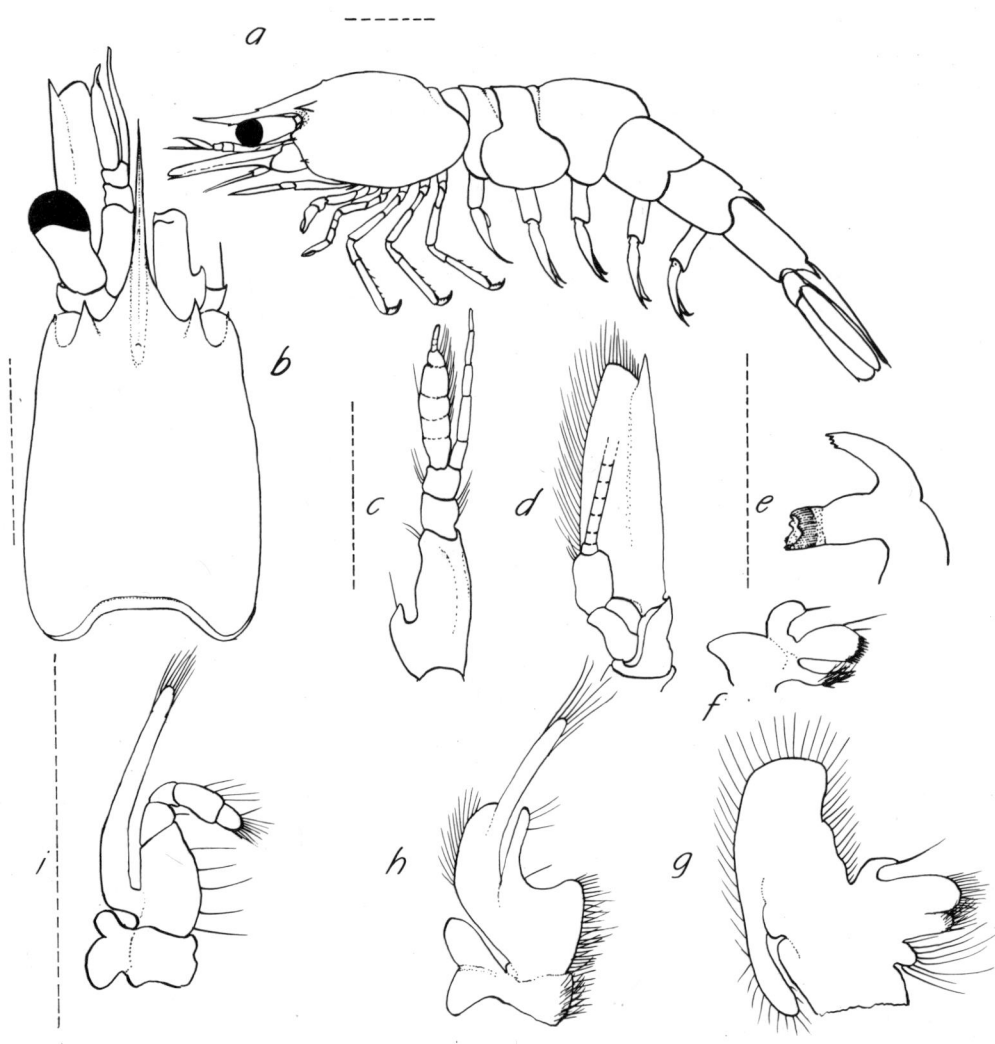

Fig. 103. *Hippolyte coerulescens*: *a*, whole shrimp from left side; *b*, carapace in dorsal aspect; *c*, antennule; *d*, antenna; *e*, mandible; *f*, maxillule; *g*, maxilla; *h*, first maxilliped; *i*, second maxilliped; broken line = 1 mm.

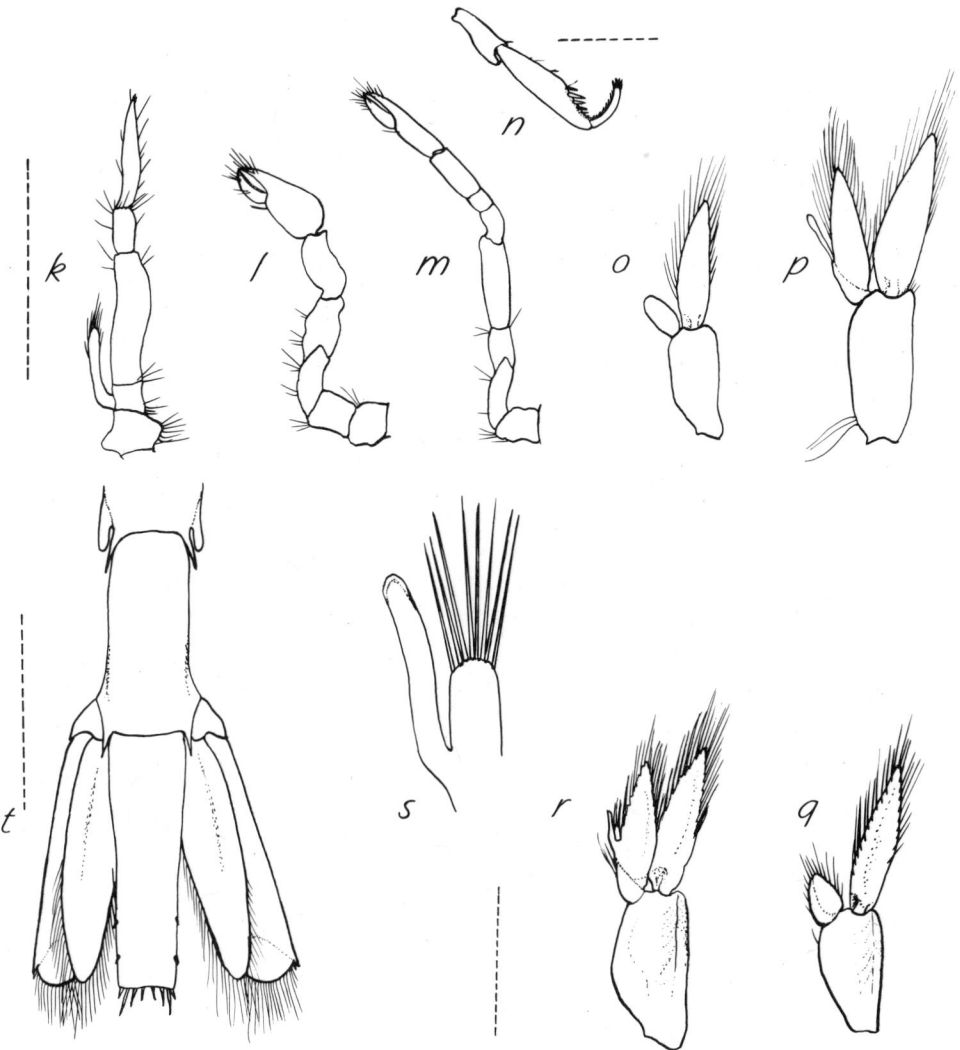

Fig. 104. *Hippolyte coerulescens*: *k*, third maxilliped; *l*, first pereopod; *m*, second pereopod; *n*, third pereopod; *o*, F first pleopod; *p*, F second pleopod; *q*, M first pleopod; *r*, M second pleopod; *s*, appendix masculina; *t*, telson; broken line = 1 mm.

Pleopods: Female I (o) endopod very short, oval, with no setae; Female II (p) endopod almost as long as exopod, appendix interna long, slightly expanded at apex, with patch of hooks. Male I (q) endopod subtriangular about one-third or less length of exopod; male II (r) endopod and exopod subequal, appendix interna about twice length of appendix masculina (s), the latter with about 8 long fine spinous setae apically.

RANGE OF DISTRIBUTION

In tropical and subtropical Atlantic Ocean associated with floating gulfweed (*Sargassum* sp.) (Chace 1972; Sivertsen and Holthuis 1956) including just south of the Grand Banks in the Gulf Stream and throughout the Sargasso Sea.

Records of distribution in the area of reference are shown in Fig. 105.

BIOLOGY

Lengths in females to 9 mm cl (Sivertsen and Holthuis 1956); egg largest diameter 0.6 mm, number carried about 190 in specimen 16 mm body length (Crosnier et Forest 1973).

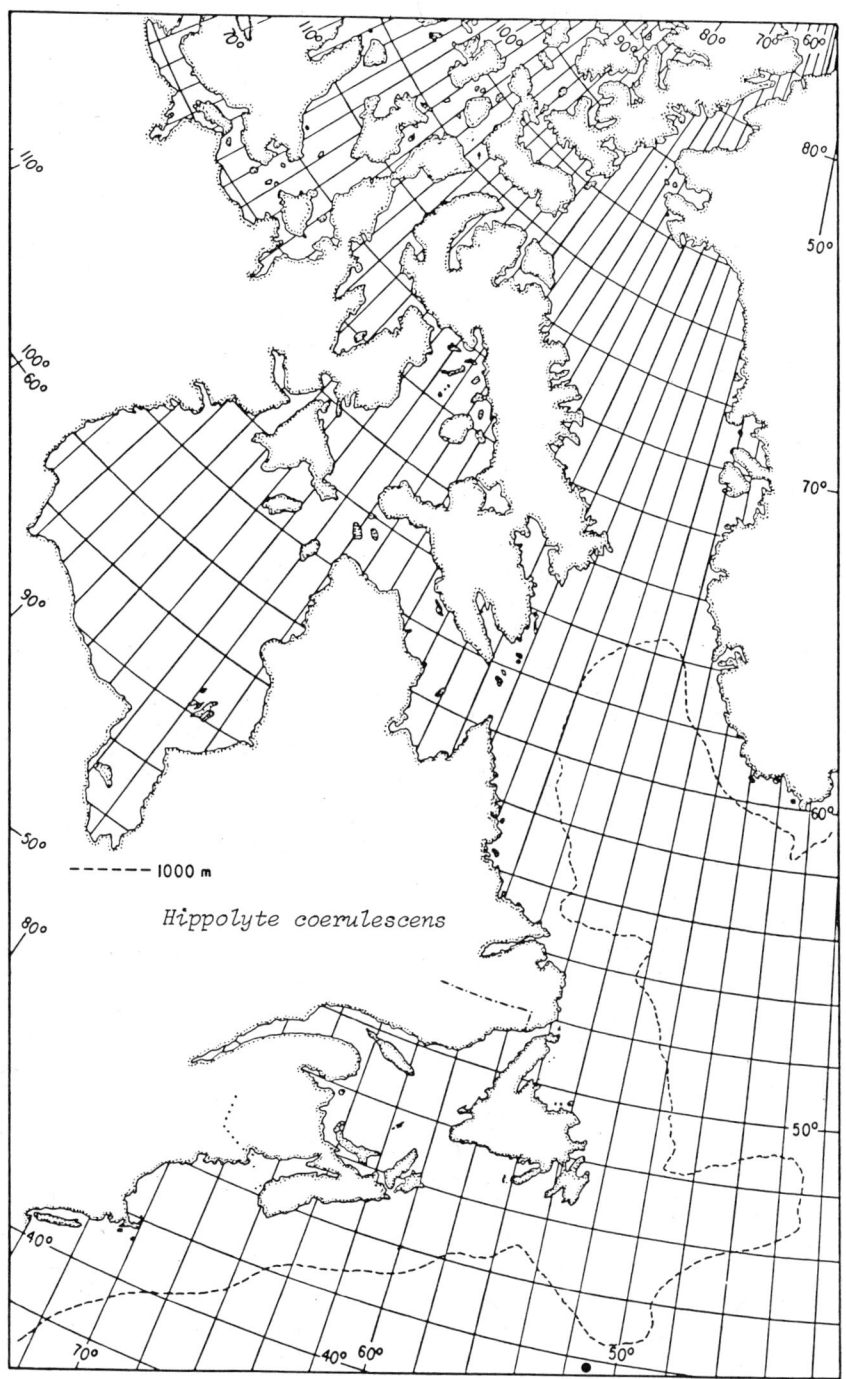

Fig. 105. *Hippolyte coerulescens*, distribution records in the area of reference.

Genus *Latreutes* Stimpson, 1860
Holthuis 1947: 59; Rathbun 1929: 18.

Carpus of second leg with three divisions; mandibles without incisor process or palp; rostrum deep; pterygostomian angle of carapace with series of small teeth; no supraorbital spine.

Latreutes fucorum (Fabricius, 1798)
Latreutes fucorum Rathbun 1929: 18, fig. 22;
Williams 1984: 119, fig. 84.
(Figures 106 and 107)

DISTINGUISHING CHARACTERISTICS

Rostrum deep, almost as long as carapace, anteriorly somewhat truncate with five small teeth but no dorsal or ventral teeth; a small spine middorsally just behind orbit; carpus of second legs with three divisions; no supraorbital spine; 5–8 small teeth at antero-lateral (pterygostomian) angle of the carapace.

DESCRIPTION

Integument smooth, shiny. Colour often nearly transparent, sometimes with body pale yellow, yellowish green, greenish brown, brown, red, black with white spots and bars; bright blue patches on dorsal and lateral surfaces; often mottled, striped or barred and corresponding in pattern to irregularly coloured bits of weed (Williams 1984).

Rostrum as long as the carapace, deep with posterior ventral keelson, subtruncate with 5–7 spinules at tip but no dorsal or ventral spines, slightly upturned and with lateral carina confluent with orbit; carapace with middorsal spine just behind orbit, a small spine just below suborbital lobe set back from the anterior margin and a series of 5–8 small teeth at antero-lateral angle.

Abdomen with pleura and terga rounded; telson (t) tapering to subtruncate tip with a central spine and two pairs of flanking spines, also two pairs of dorso-lateral spines on distal half; inner branch of uropod longer than telson and outer branch, the latter with axillary spine at disto-lateral spine.

Eyes large, cornea subglobular about equal in diameter to stalk which widens proximally; small tubercle disto-mesially near edge of cornea.

Antennule (c): 1st article with disto-mesial tooth, concave laterally, longer than other two combined; 3rd with distal spine, shorter than 2nd; stylocerite short, rounded.

Antenna (d): tapering to narrow tip and long spine, outer concavity, reaching about 3/4 length of rostrum; basal article with distal tooth; peduncle less than half scale.

Mandible (e): short stout molar process; no incisor or palp.

Maxillule (f): proximal endite narrow tapering to point; distal endite subovate, pointed; endopod without setae, uniramous.

Maxilla (g): distal endite with short equal lobes; proximal endite reduced to single rounded lobe with few curved setae; scaphognathite with wide anterior lobe tapering and rounded distally, posterior lobe narrower rounded with short setal fringe; endopod moderate.

Maxilliped I (h): distal endite subtriangular, proximal about equal but with thick double leading edge; endopod longer than lobe of exopod, lash also longer; epipod about equally bilobed.

Maxilliped II (i): leg-like compressed; distal segment longer than wide; exopod inserted at about middle of longest segment (basis-ischium); epipod large, without podobranch.

Maxilliped III (k): leg-like stout, distal segment with apical and subapical series of strong spines; exopod short; rudimentary epipod, rounded.

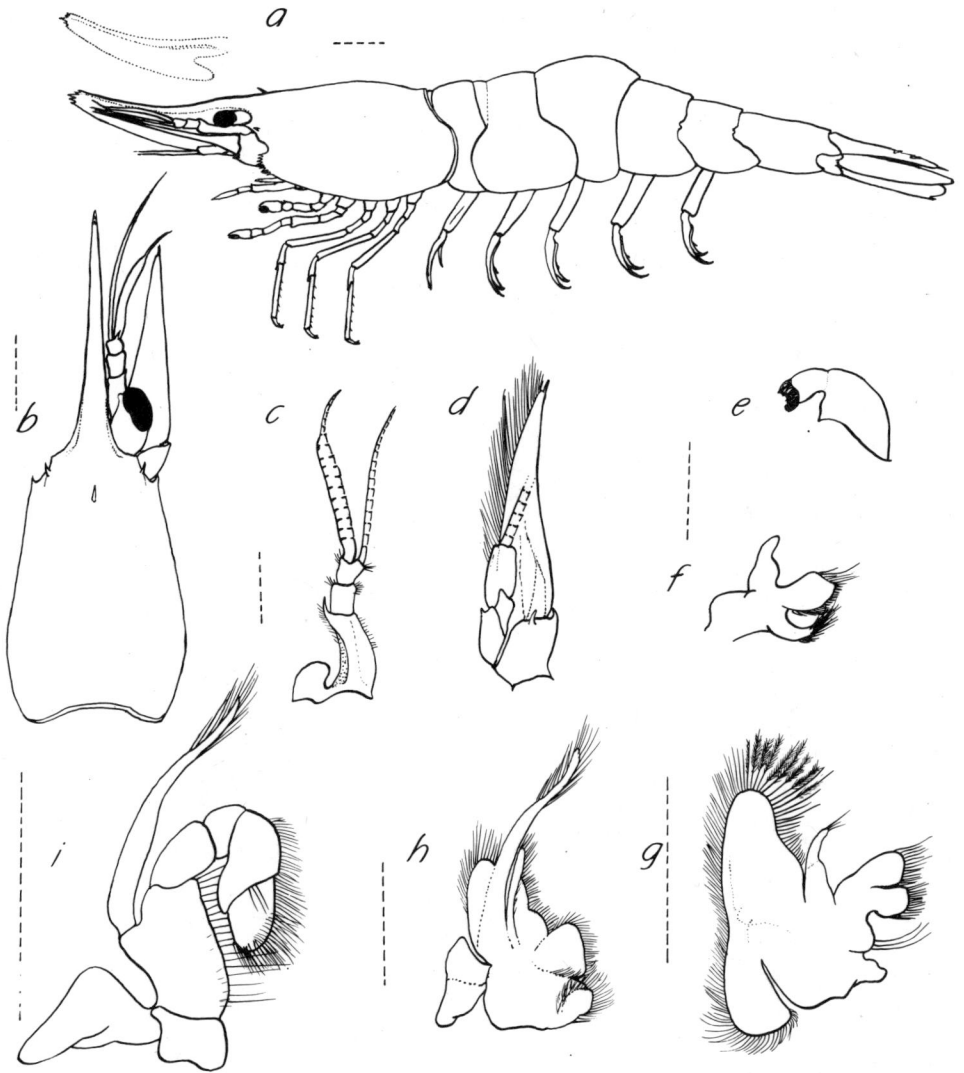

Fig. 106. *Latreutes fucorum*: *a*, whole shrimp from left side; *b*, carapace in dorsal aspect; *c*, antennule; *d*, antenna; *e*, mandible; *f*, maxillule; *g*, maxilla; *h*, first maxilliped; *i*, second maxilliped; broken line = 1 mm.

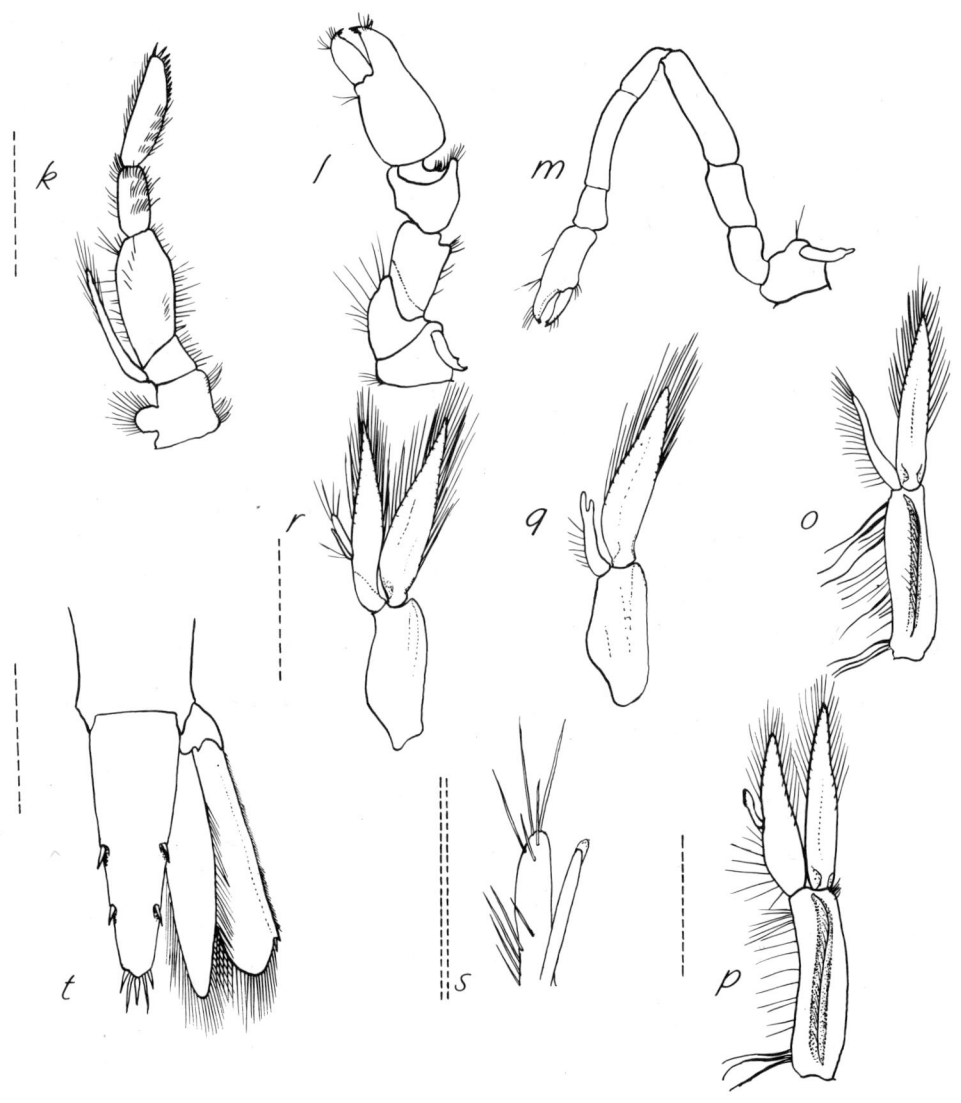

Fig. 107. *Latreutes fucorum*: *k*, third maxilliped; *l*, first pereopod; *m*, second pereopod; *o*, F first pleopods; *p*, F second pleopod; *q*, M first pleopod; *r*, M second pleopod; *s*, appendix masculina; *t*, telson; broken line = 1 mm.

Pereopods: I (l) chelate, subequal, short but very stout; dactyl with two black distal teeth, finger of propodus with three, palm inflated; carpus cup-shaped; epipod straplike with terminal hook. II (m) more slender and longer than I; chelate, fingers have distally two denticles each; carpus with three divisions the middle longest; epipod straplike as in I. III–V moderately long and slender, dactyl unarmed and tapering to a sharp point; propodus with a few sharp spines along flexor edge; III only with straplike epipod.

Pleopods: female I (o) endopod narrow tapering and curved, much more slender than exopod; female II (p) appendix interna inserted at about half endopod, expanded distally and with patch of hooks. Male I (q) endopod much smaller than exopod narrow and distally biramous; II with appendix interna much more slender and shorter than masculina (s); the latter with 3 long and two short spines apically, separated from lateral series of about 5 moderate spines.

RANGE OF DISTRIBUTION

In masses of gulf weed (*Sargassum* sp.) or on grass flats in western North Atlantic between 10 and 50° N; also Azores and Cape Verde Islands (Williams 1984).

Records of distribution in area of reference are in Fig. 108.

BIOLOGY

Lengths to 9 mm cl in females (ovigerous at 4–9 mm cl) (Sivertsen and Holthuis 1956).

A branchial parasite, *Probopyrinella latreuticola* has been reported for this species (Markham 1977).

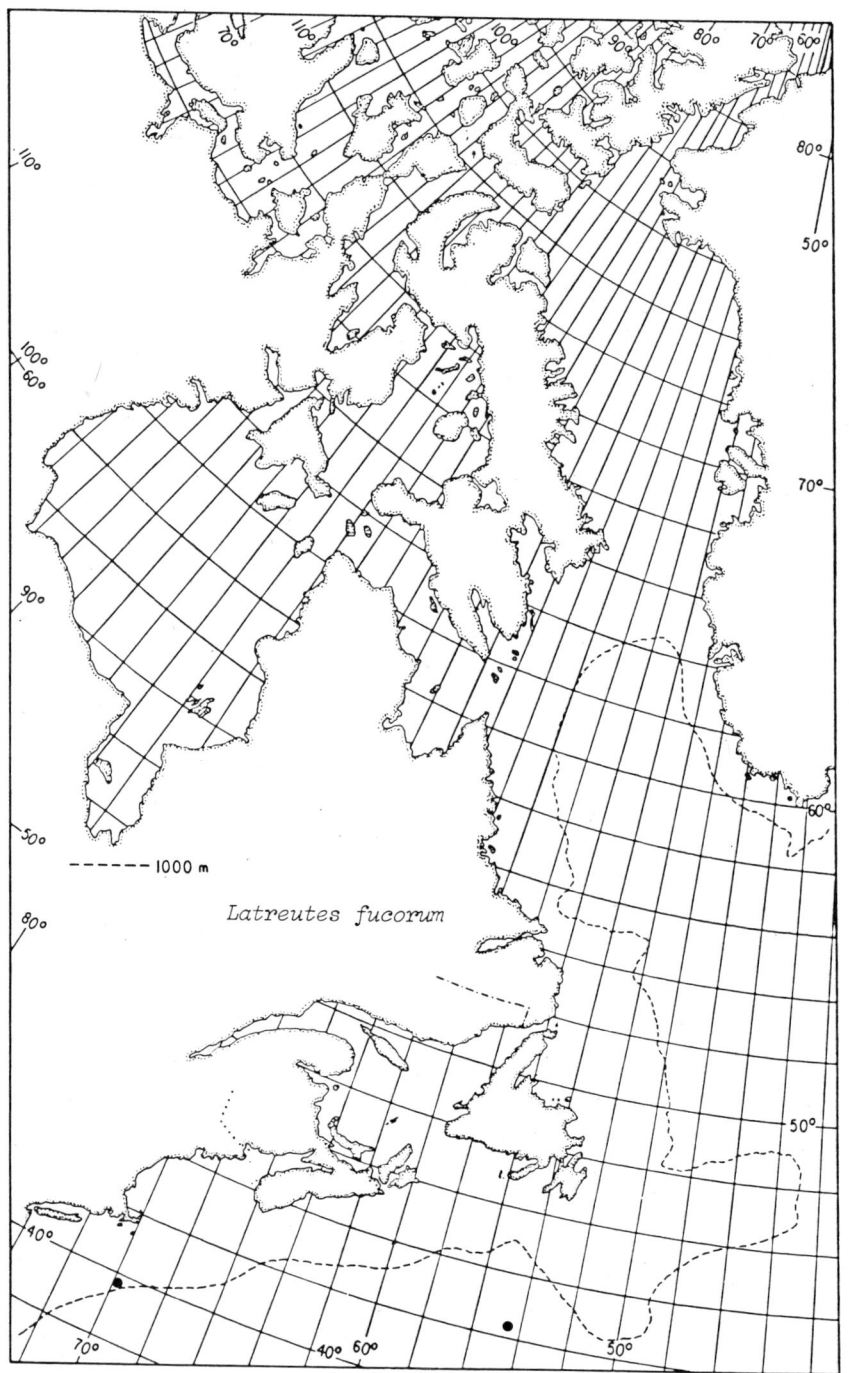

Fig. 108. *Latreutes fucorum*, distribution records in the area of reference.

Genus *Lebbeus* White, 1847
Holthuis 1947: 38.

One supraorbital spine present; third maxilliped without exopod but with epipod; carpus of 2nd leg with seven divisions; rostrum almost always with spines.

Of the 18 known species only 3 are present in the area of reference.

Lebbeus groenlandicus (Fabricius, 1775)
Spirontocaris groenlandica, Rathbun 1929: 11, fig. 8;
Lebbeus groenlandicus, Williams 1984: 122, fig. 86
(Figures 109 and 110, Plate 3a)

DISTINGUISHING CHARACTERISTICS

Carapace rough with four strong middorsal teeth directed forward, one strong supraorbital spine; abdominal pleura with strong ventral spines; rostrum moderate with strong dorsal and smaller ventral teeth; carpus of slender 2nd leg with seven divisions.

DESCRIPTION

Integument thick, rigid, rough, covered with scattered short setae. Colour brownish red or brownish green on whitish translucent background.

Carapace with one large supraorbital spine; rostrum moderate, about 0.5 cl, with strong dorsal teeth and smaller ventral teeth, usually 3/2–4; strong middorsal carina with four prominent teeth with black spinous tip directed forward; antennal spine large with prominent carina; pterygostomian spine moderate, with short submarginal carina.

Abdomen stout, pleura with ventral spines: respectively, 3, 1, 2, 2, 2, 1 from 1st to 6th somites: lateral carinae near points of articulation of 3rd to 5th somites.

Telson (t) with wide sulcus dorsally, proximally a tubercle with setae, and lateral ridges with continuous row of about 9 pairs of strong black-tipped spines, distally rounded with two pairs of short spines on each side of a row of several long setae; outer branch of uropod about as long as inner and both shorter than telson.

Eye moderate, cornea globular with small ocellus.

Antennule (c): stylocerite slender, sharp, almost reaching the distal end of the peduncle; 1st article about as long as other two together, 2nd and 3rd subequal, all three with very strong distal spine.

Antenna (d): antennal scale length about 4 times width, disto-lateral black-tipped spine slightly exceeds the blade; peduncle about 0.7 times length of scale; basal article with two strong outer distal spines.

Mandible (e): molar heavy, edged with fine setae and cutting cusps on surface; incisor laminate, tapering, with four distal teeth; palp with two segments the distal slightly longer than proximal.

Maxillule (f): distal endite oval, wide, proximal narrow tapering to curved tip; endopod short, bifid.

Maxilla (g): distal endite large, bilobed, proximal lobe larger; proximal endite bilobed, proximal lobe very small; endopod small; scaphognathite with longer anterior lobe narrowing distally, posterior wider axe-shaped but rounded.

Maxilliped I (h): distal endite subtriangular, laterally concave; proximal endite slightly smaller but with thicker edge and double rows of setae; endopod two-segmented, twisted, not quite as long as lamellar part of exopod, lash about as long, slender and tapering; epipod large, bilobed.

Maxilliped II (i): compressed, six-segmented; distal segment wider than long; exopod slender, longer than endopod; epipod with podobranch.

Maxilliped III (k): stout, about as long as first leg; five segments, distal with apical row of 8 strong black spines; no exopod; epipod strap-like with terminal hook.

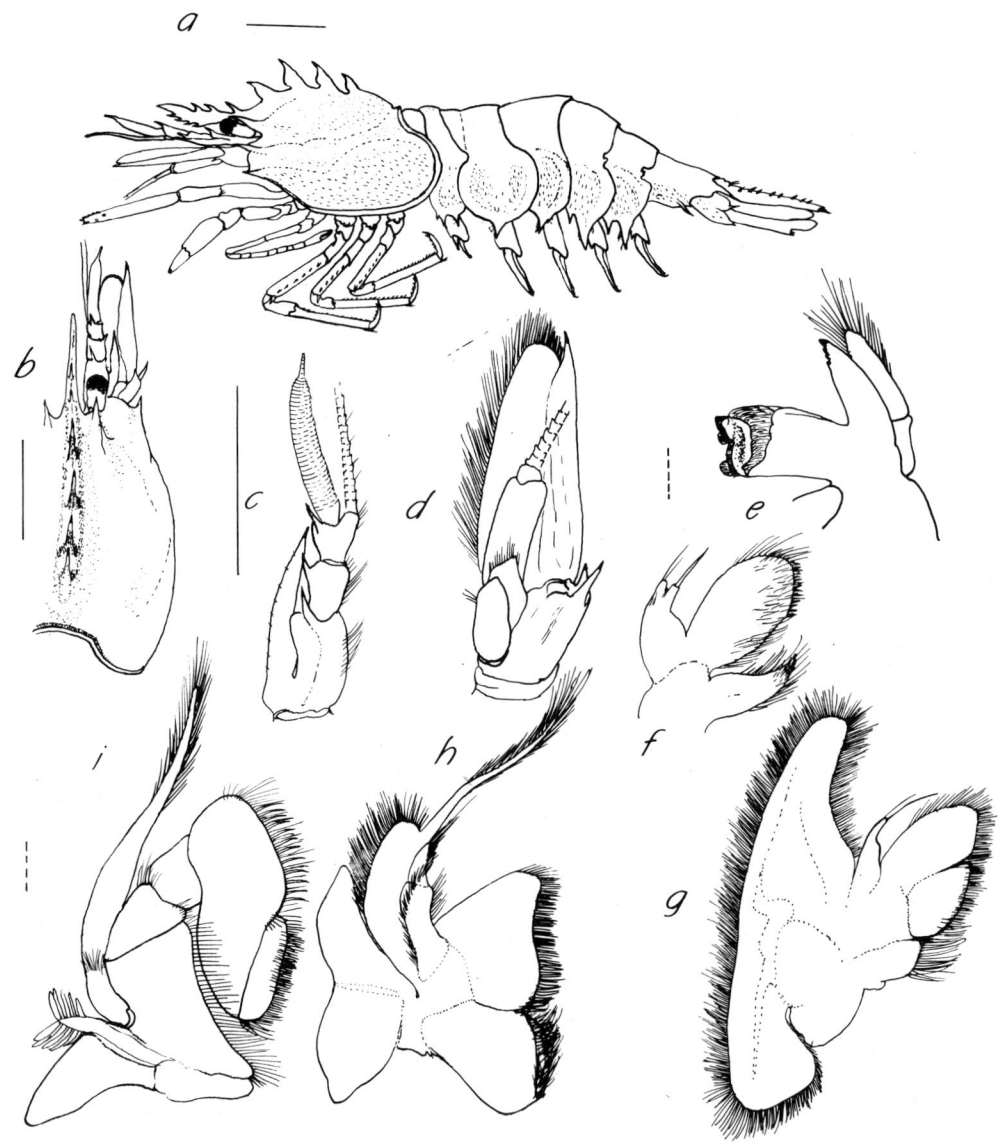

Fig. 109. *Lebbeus groenlandicus*: *a*, whole shrimp from left side; *b*, carapace in dorsal aspect; *c*, antennule; *d*, antenna; *e*, mandible; *f*, maxillule; *g*, maxilla; *h*, first maxilliped; *i*, second maxilliped; solid line = 10 mm, broken line = 1 mm.

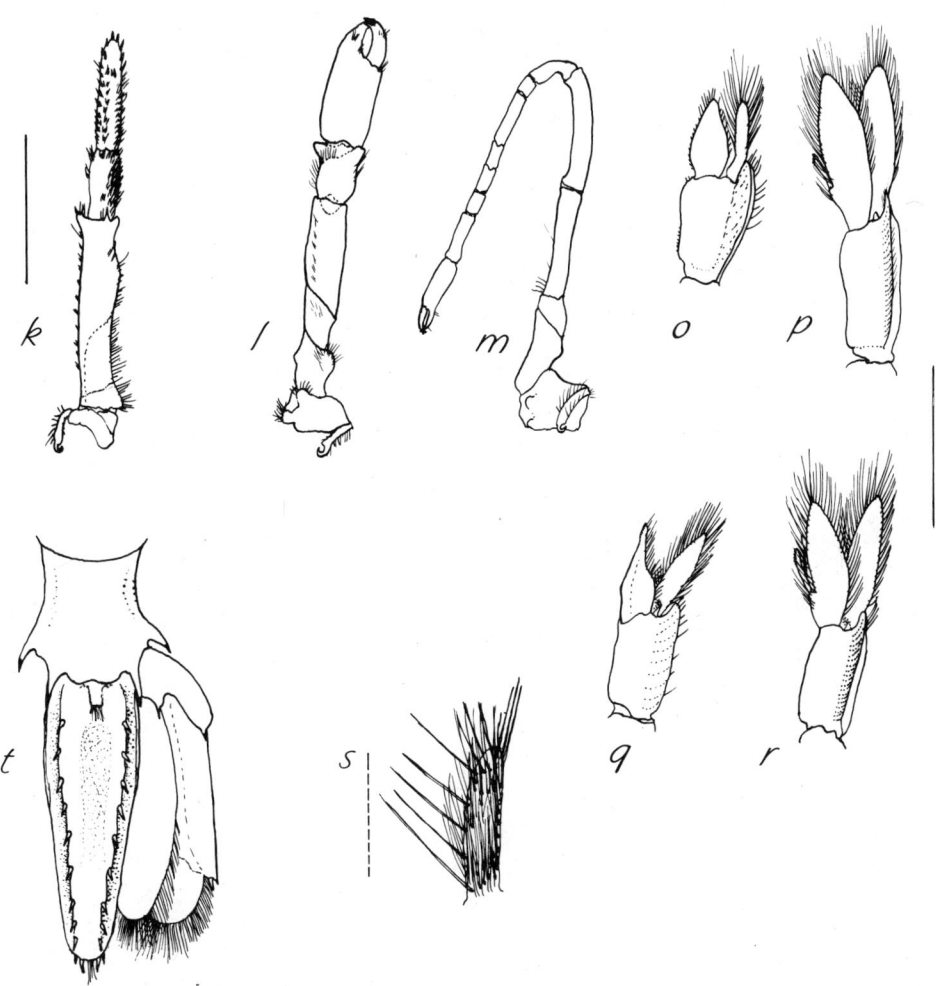

Fig. 110. *Lebbeus groenlandicus*: *k*, third maxilliped; *l*, first pereopod; *m*, second pereopod; *o*, F first pleopod; *p*, F second pleopod; *q*, M first pleopod; *r*, M second pleopod; *s*, appendix masculina; *t*, telson; solid line = 10 mm, broken line = 1 mm.

Pereopods: I (l) chelate, very stout; fingers black-tipped, dactyl bifid; epipod strap-like with hook. II (m) chelate, slender, fingers faintly black-tipped, dactyl bifid; carpus with seven divisions; strap-like epipod with hook. III–V non-chelate, longer than others; III only with epipod; dactyl stout, biunguiculate, with about five other sharp spines; outer row of dark-tipped spines on merus, longest distally; series of spinules along flexor edge of propodus.

Pleopods: female I (o) endopod wide, tapering, as long as narrow exopod, protopodite with large wing-like expansion; II (p) endopod and exopod about equal, appendix interna with sub-apical patch of hooks. Male I (q) endopod wide at base and tapering to long point with row of hooks, exopod more slender but equal in length to endopod. Male II (r) endopod and exopod subequal, appendix interna slightly longer than masculina, latter (s) with about 12 apical and lateral long spines and numerous fine setae.

RANGE OF DISTRIBUTION

Hudson Bay to Greenland and south to Rhode Island; arctic Canada and Alaska; southern Chukchi Sea through Bering Sea to Puget Sound, and Sea of Okhotsk to Vladivostok (Williams 1984). Depths 2–314 m and temperatures −1.4 to 9.4° C (Williams and Wigley 1977).

Records of distribution in the area of reference are in Fig. 111.

BIOLOGY

Lengths to 27 mm cl in males and 28 mm cl in females.

Of the 46 females taken in August and September in Foxe Basin 68% were potentially ovigerous, possibly indicating annual spawning in that area. In Hudson Bay females first matured at 15 cl (20 specimens only) but in Foxe Basin at 10 mm cl (46 specimens examined) (Squires 1967a).

Usually only a small number of this species was captured with larger numbers of *Lebbeus polaris*, *Pandalus montagui* and *Spirontocaris spinus*.

Stomach contents included phytobenthos, crustacean fragments, polychaetes and hydroids, also occasionally euphausiids, foraminiferans, ostracods, ophiurans, small bivalves, gastropods and rhodophyte fragments (Squires 1965a).

Predators on this species were cod in the Newfoundland area and bearded seals in Ungava Bay (Squires 1957, 1965a).

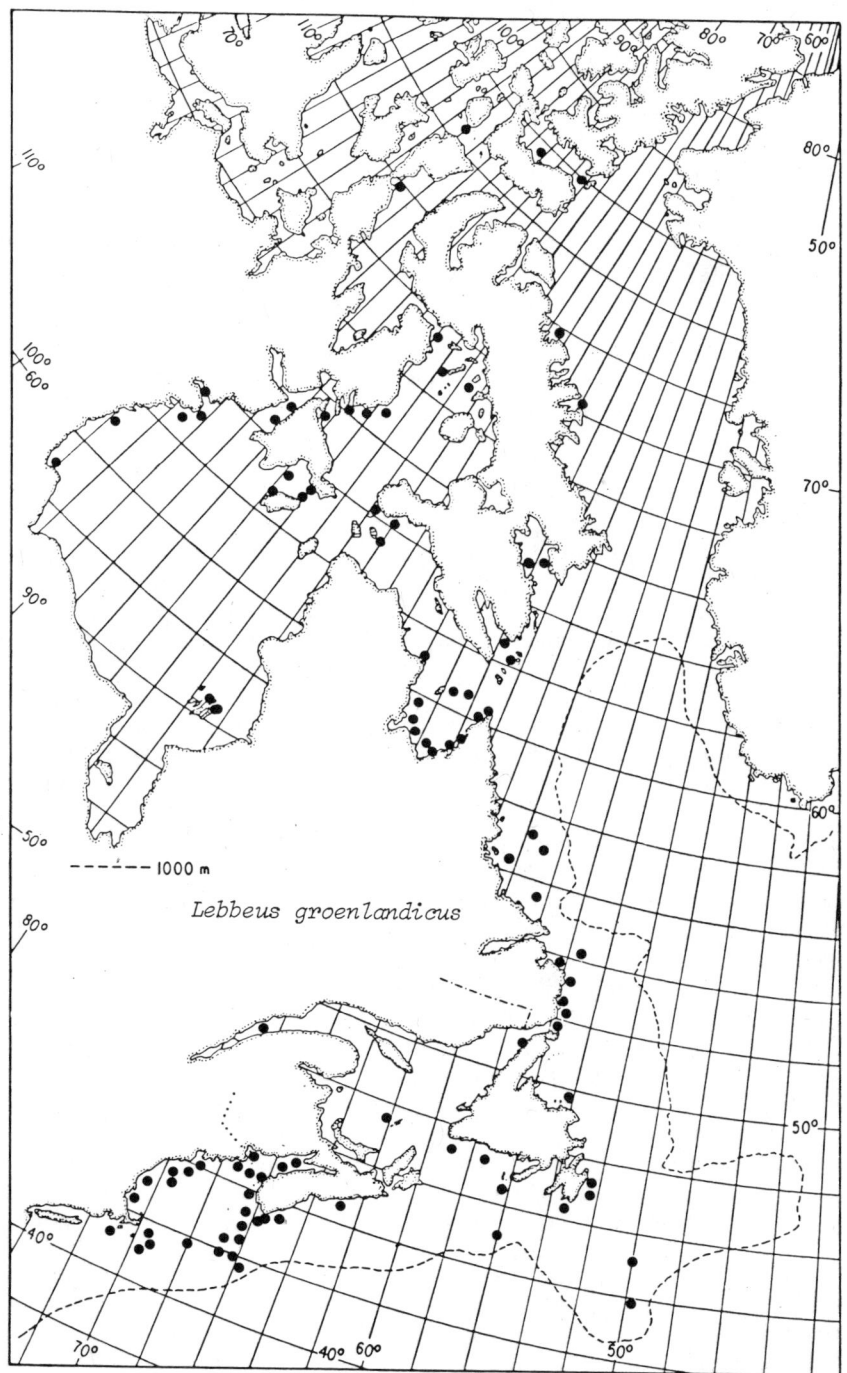

Fig. 111. *Lebbeus groenlandicus*, distribution records in the area of reference.

Lebbeus microceros (Krøyer, 1841)
Spirontocaris microceros, Rathbun 1929:12, fig. 10;
S. zebra, 1929: 13, fig. 11.
Spirontocaris zebra Leim 1921: 133, pls. 2–3.
Lebbeus microceros, Holthuis 1947; 9, 40.
(Figures 112 and 113)

DISTINGUISHING CHARACTERISTICS

Rostrum short with dorsal spines close together; one supraorbital spine; carpus of 2nd leg with seven divisions; no exopod on third maxilliped but a strap-like epipod; epipod on first three legs.

DESCRIPTION

Integument smooth, shiny. Coloured with bright brownish red to orange stripes dorso-ventrally, oblique on abdomen, intermediate areas bluish on carapace and whitish on abdomen; also white, orange and bluish bandings on appendages (Leim 1921; Rathbun 1929). (Considerable variation in colouring between specimens, colours not lasting long in preservative).

Rostrum short, just reaching distal edge of 1st article of antennule but exceeding eyes, dorsal teeth close set usually 4–5/0–1 with 2 behind orbit, slightly descending.

Carapace with low mid-dorsal carina including two teeth at front and continuing to about half carapace; strong supraorbital spine, postorbital lobe, small antennal spine and moderate pterygostomian spine; paired spines on sterna of 2nd to 4th legs.

Abdomen with smooth rounded terga and pleura except ventro-posterior spine on 4th to 6th somites; weak spines ventrally, paired on 1st to 3rd somites, single on 4th and 5th; anal spine.

Telson (t) slightly tapering and subtruncate with 4–5 pairs of dorso-lateral spines and 3 pairs of terminal spines; one tiny ventral spine near tip. Branches of uropod subequal, slightly longer than telson, outer branch with axillary spine and row of short setae on outer edge.

Eye moderate, cornea globular on short tapered stalk; small ocellus.

Antennule (c): 1st article longer than others combined, with 2 strong spines distally and one laterally, also one on distal edge of 2nd and 3rd articles; peduncle reaching about two-thirds of antennal scale; stylocerite with long attenuate sharp point reaching distal 2nd article.

Antenna (d): scale length three times width; disto-lateral spine exceeded by blade; peduncle more than half scale; basal article with small ventral spine.

Mandible (e): molars similar but right with flatter surface; incisor with 3 teeth; distal segment of palp slightly longer than proximal.

Maxillule (f): distal endite wide, ovate, pointed; proximal narrow, curved, tapering to point; endopod bifid.

Maxilla (g): distal endite unequally bilobed, proximal wider; proximal endite reduced, one lobe very small other rounded with long curved setae; endopod short; anterior lobe of scaphognathite long, sub-rectilinear but slightly tapering; posterior lobe very short, rounded with short setal fringe.

Maxilliped I (h): distal endite subtriangular, slightly larger than proximal, latter with double leading edge; endopod two-segmented, about as long as lobe of exopod, lash about twice as long; epipod bilobed, large.

Maxilliped II (i): leg-like, compressed; distal segment much wider than long; exopod longer than endopod; epipod with podobranch.

Maxilliped III (k): stout, about as long as first leg; distal segment with 9 strong terminal black spines; longest segment with strong curved spine distally outside; epipod strap-like with terminal hook.

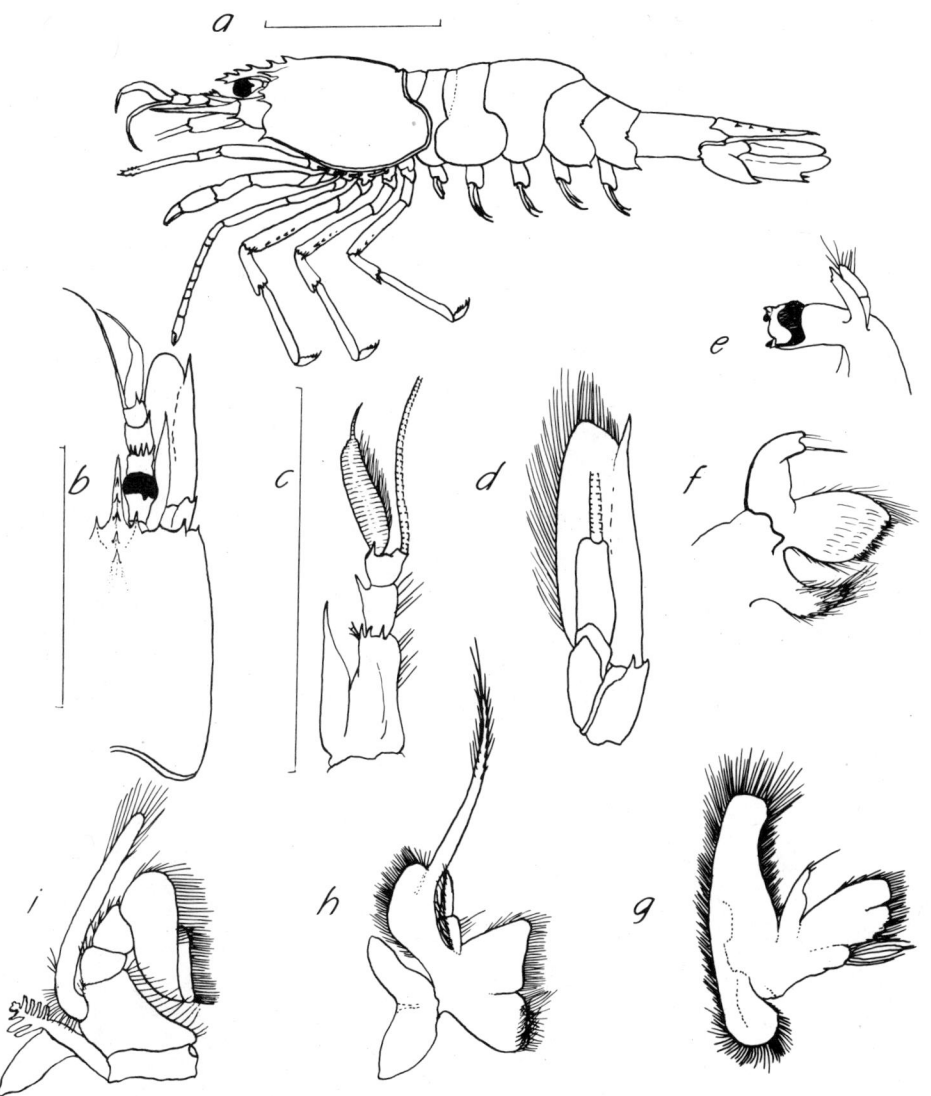

Fig. 112. *Lebbeus microceros*: *a*, whole shrimp from left side; *b*, carapace in dorsal aspect; *c*, antennule; *d*, antenna; *e*, mandible; *f*, maxillule; *g*, maxilla; *h*, first maxilliped; *i*, second maxilliped; solid line = 10 mm.

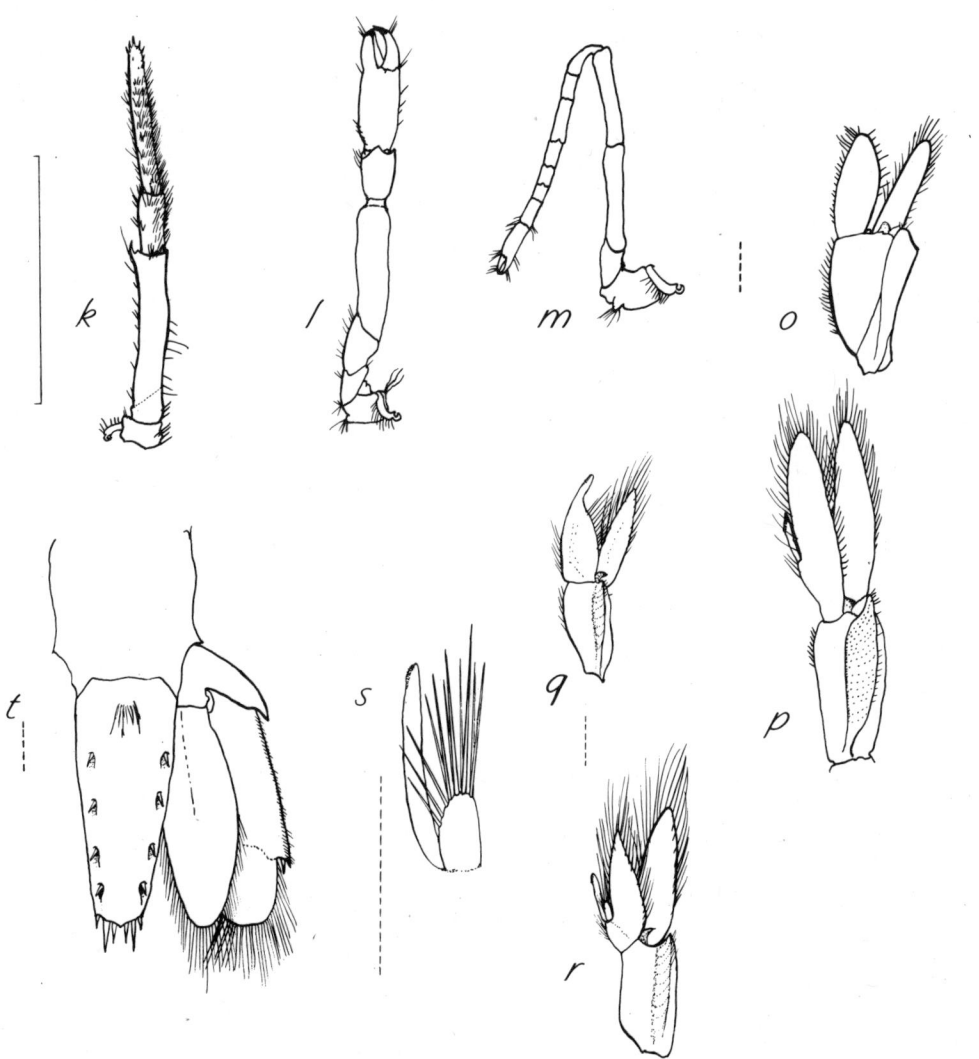

Fig. 113. *Lebbeus microceros*: k, third maxilliped; l, first pereopod; m, second pereopod; o, F first pleopod; p, F second pleopod; q, M first pleopod; r, M second pleopod; s, appendix masculina; t, telson; solid line = 10 mm, broken line = 1 mm.

Pereopods: I (l) chelate, stout, shorter than others; dactyl bifid, curved unequal black spines between which terminal spine of propodus fits; epipod straplike with terminal hook. II (m) chelate, slender, carpus with 7 divisions; dactyl bifid; epipod as in I. III–V (a) longer than others, III stoutest; dactyl biunguiculate, stout, with series of 4 spines on flexor edge; merus has series of spines outside, the sub-teminal longest; also numerous spinules along posterior edge of propodus; III only with epipod.

Pleopods: female I (o) endopod ovate, as long as narrower exopod, protopodite with wing-like expansion; female II (p) endopod and exopod subequal, appendix interna with subapical patch of hooks. Male I (q) endopod slightly longer and wider than exopod and with distal finger-like projection and subapical patch of hooks; male II (r) endopod slightly shorter than exopod, appendix interna with subapical patch of hooks and much longer than appendix masculina (s), the latter with two rows of about 4 + 4 long fine spines and an outer row of 3 similar distolateral spines.

RANGE OF DISTRIBUTION

Foxe Basin and Southern Greenland to Newfoundland, Nova Scotia and New Brunswick (Squires 1965a); also possibly from Bering Sea to off Kamchatka. Depths 8–80 m (Holthuis 1947).

Records of distribution in the area of reference are in Fig. 114.

BIOLOGY

Lengths of males to 9 mm cl and females to 14 mm cl; egg diameter 2.4 mm (Squires 1965a).

Stomach contents were phytobenthos.

Predation by cod occurred at Ramea, Newfoundland. (Squires 1965a).

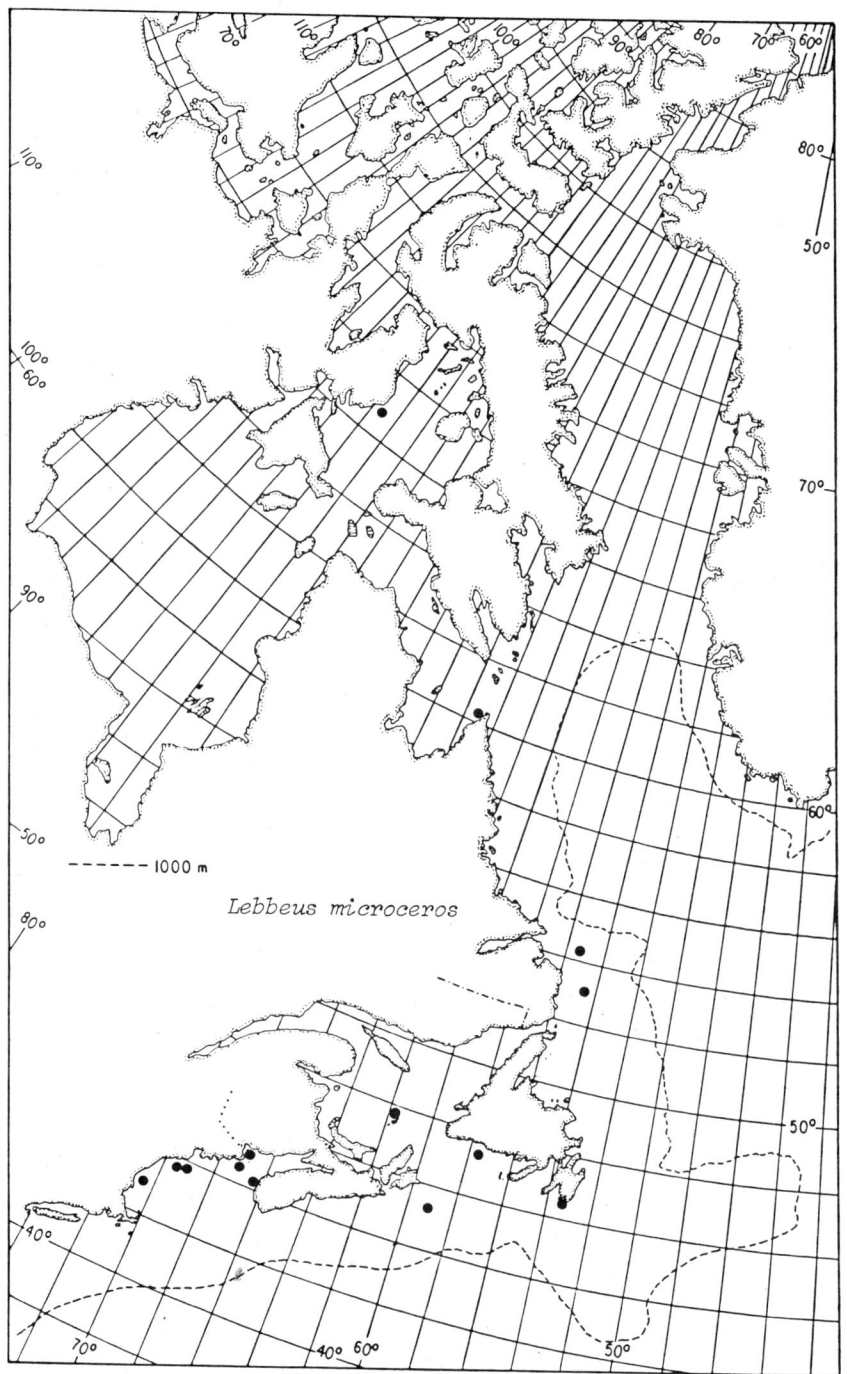

Fig. 114. *Lebbeus microceros*, distribution records in the area of reference.

Lebbeus polaris (Sabine, 1821)
Spirontocaris polaris, Rathbun 1929: 12, fig. 9;
Lebbeus polaris, Holthuis 1947: 9, 39.
(Figures 115 and 116, Plate 3b)

DISTINGUISHING CHARACTERISTICS

One supraorbital spine; relatively long rostrum with teeth above and below (some with smooth rostrum); carpus of 2nd leg with 7 divisions; no exopod on third maxilliped, epipod on first two legs; abdomen with rounded terga and pleura.

DESCRIPTION

Integument smooth and shiny. Colour: bright red and orange chromatophores over whitish translucent background forming spots or stripes over carapace and abdomen (Plate 3b).

Carapace with one supraorbital spine with short carina, a moderate antennal and small pterygostomian spines; a low mid-dorsal carina extends backward to more than half carapace; rostrum long (0.8 times cl) with dorsal and ventral teeth (0–8/0–5) or sometimes without teeth.

Abdomen with rounded terga and pleura except 5th and 6th somites with ventroposterior spines. Ventrally are small spines on cross carinae of somites, a pair on first and single spine on 2nd to 5th, no anal spine.

Telson (t) with 6–7 sharp dorso-lateral spines, and four pairs of terminal spines at subtruncate tip; outer branch of uropod slightly longer than inner and about equal to telson length; axillary spine at distolateral spine of outer branch.

Eye large, cornea globular, much wider than tapered stalk.

Antennule (c): 1st article longer than other two combined, 2nd longer than 3rd, all three with strong distal spine; stylocerite with long attenuate point about equal first two articles of peduncle, the latter about half length of scale.

Antenna (d): scale length 2.7 times width, 0.8 times cl; distolateral spine exceeded by blade; peduncle more than one-half scale; basal article with strong ventro-lateral spine.

Mandible (e): molars similar but left with more rounded surface; incisor with 5 teeth; both palp segments about equal.

Maxillule (f): distal endite wide almost circular; proximal endite narrow but thick at base and tapering to curved tip; endopod bent, bifid.

Maxilla (g): distal endite equally bilobed; proximal reduced, one lobe small pointed, other rounded larger; endopod short; scaphognathite with rectilinear anterior lobe longer than posterior rounded lobe; setal fringe short.

Maxilliped I (h): distal endite large, subtriangular; proximal small, thick, double edged; endopod with two segments, reaching distal lamellar part of exopod, lash about twice as long; unequally bilobed large epipod.

Maxilliped II (i): leg-like, compressed; distal segment much wider than long; exopod longer than endopod; epipod with small podobranch.

Maxilliped III (k): longer than first leg; distal segment with strong apical spines; straplike epipod with terminal grooming hooks.

Pereopods: I (l) stout, chelate, shorter than others; tips of fingers black spinous, propodal sharp fitting between two unequal distal spines of the dactyl; epipod straplike with hook. II (m) slender, chelate, carpus with seven divisions; epipod straplike with terminal hook. III–V (a) slightly longer than II, non-chelate; dactyl stout, biunguiculate and with row of 4 strong black spines.

Pleopods: female I (o) endopod wide pointed, exopod narrow slightly longer; protopod with strong wing-like expansion. Female II (p) endopod and exopod subequal, appendix interna expanded distally with subapical patch of hooks. Male I (q) endopod with long apical point, exopod longer. Male II (r) appendix interna expanded at tip with

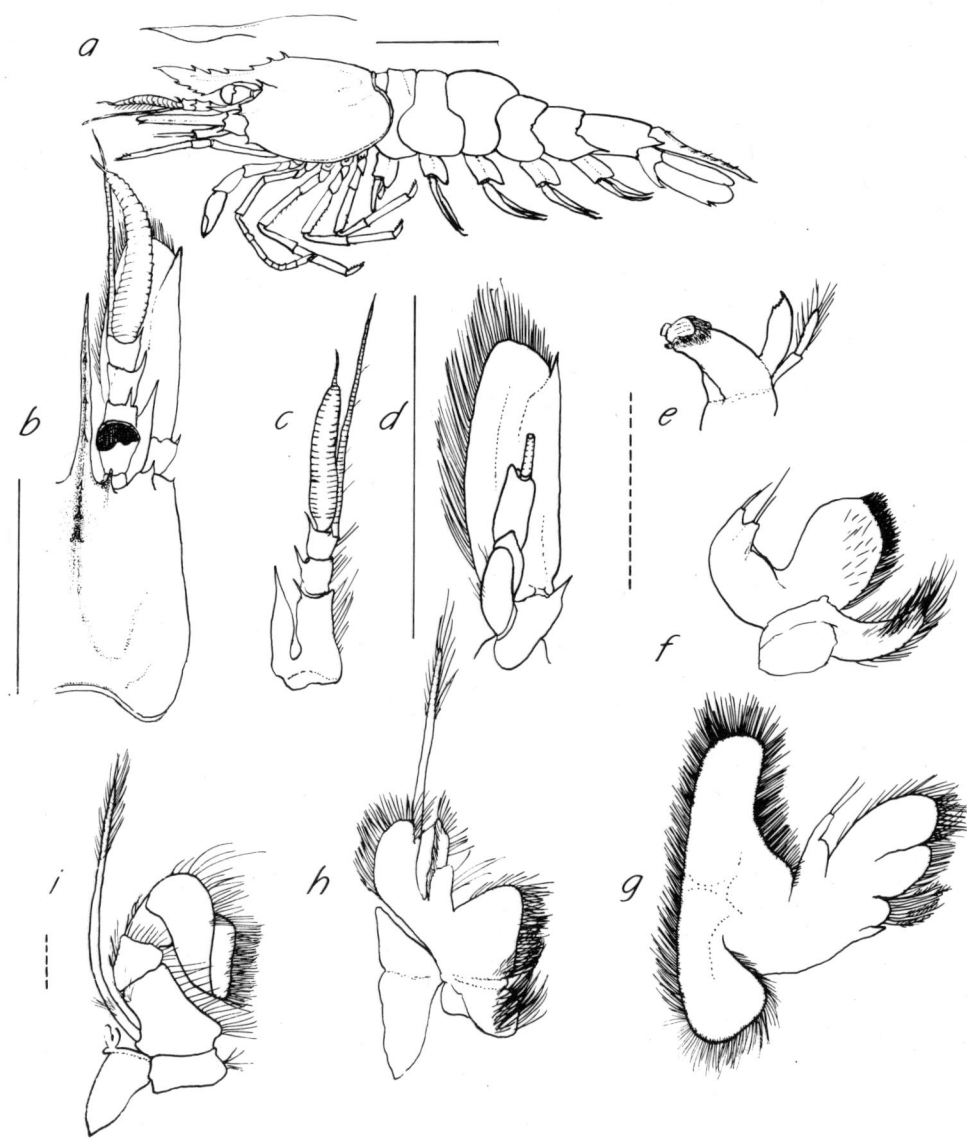

Fig. 115. *Lebbeus polaris*: *a*, whole shrimp from left side; *b*, carapace in dorsal aspect; *c*, antennule; *d*, antenna; *e*, mandible; *f*, maxillule; *g* , maxilla; *h*, first maxilliped; *i*, second maxilliped; solid line = 10 mm, broken line = 1 mm.

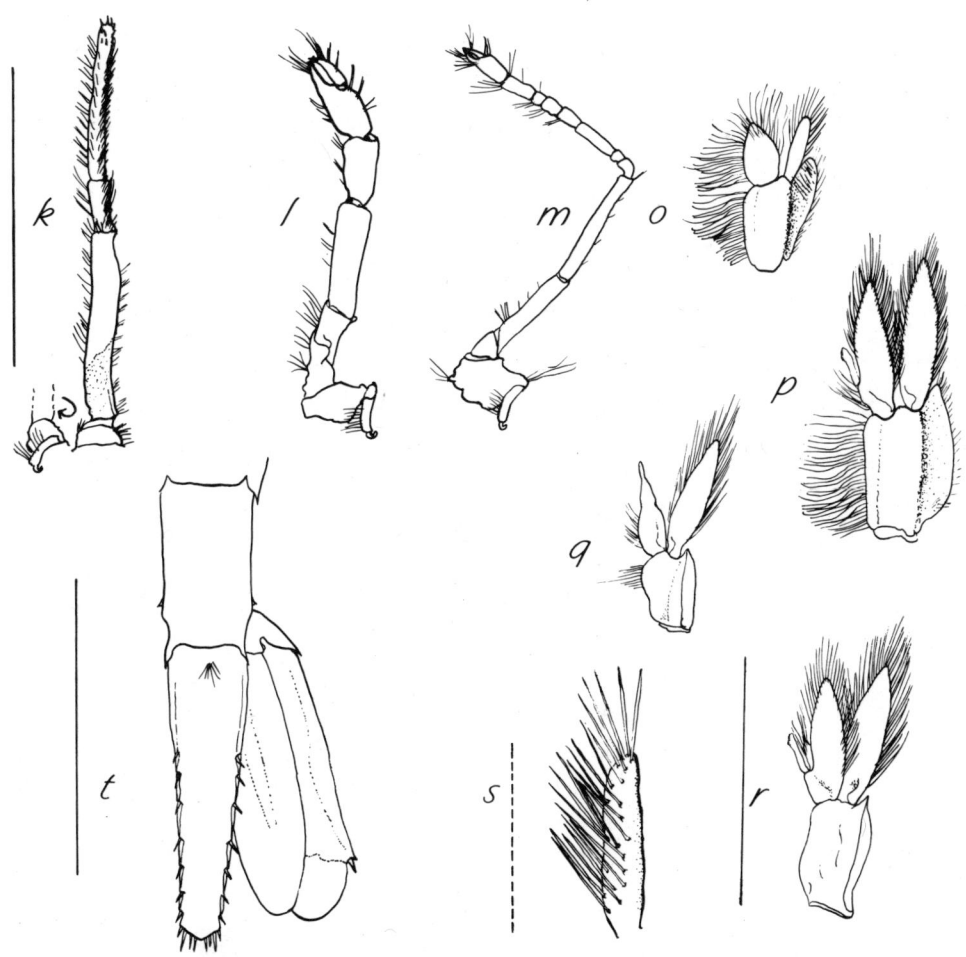

Fig. 116. *Lebbeus polaris*: *k*, third maxilliped; *l*, first pereopod; *m*, second pereopod; *o*, F first pleopod; *p*, F second pleopod; *q*, M first pleopod; *r*, M second pleopod; *s*, appendix masculina; *t*, telson; solid line = 10 mm, broken line = 1 mm.

subapical patch of hooks, longer than appendix masculina (s) which has three apical long spines continuous with two lateral rows of about 20 long spines.

Range of Distribution

Circumarctic to the Hebrides in Europe and Chesapeake Bay in America; through Bering Sea to British Columbia and Okhotsk Sea (Williams 1984). Depths 0–930 m (Holthuis 1947).

Records of distribution in the area of reference are in Fig. 117.

Biology

Lengths to 18 mm cl in males and 20 mm cl in females.

Percentage potentially ovigerous was high (78% in 274 mature females) indicating annual spawning; three major year groups appeared to be present (Squires 1965a).

Phytobenthos, crustacean fragments, ostracods and gammarid amphipods occurred in most stomach contents; polychaetes, forameniferans, hydroids and rhodophytes occurred less frequently, and pteropods, small gastropods, bivalves, mysids and sea urchin remains only occasionally (Squires 1965a).

Predators were cod, seals and murres (Tuck and Squires 1955; Squires 1965a).

Species also present in catches were mostly *Spirontocaris spinus*, *Eualus gaimardi gaimardi*, *Pandalus montagui*, *Sabinea septemcarinata* and *Pagurus pubescens* (Squires 1965a).

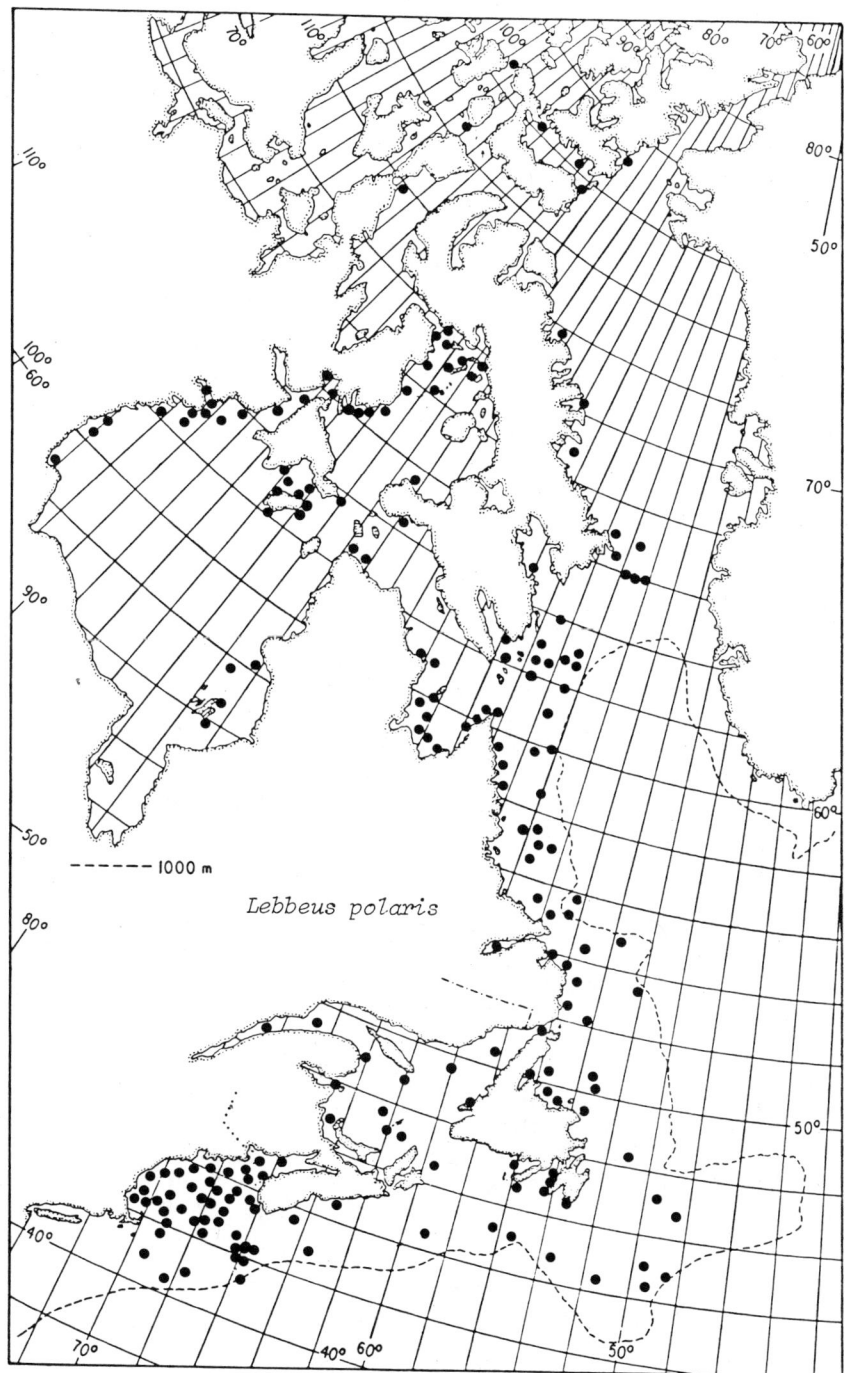

Fig. 117. *Lebbeus polaris*, distribution records in the area of reference.

Genus *Spirontocaris* Bate, 1888
Holthuis 1947: 36; Hayashi 1977: 155.

Carapace with large spines on dorsal carina and 2 or more supra-orbital spines; rostrum prominent with dorsal and usually ventral spines; eyes large with an inner tubercle on eyestalk; carpus of 2nd leg with seven divisions; exopod on third maxilliped; telson with more than three pairs of dorsolateral spines.

Only three of the 19 known specis of this genus are present in the area of reference.

Spirontocaris lilljeborgi (Danielssen, 1895)
Spirontocaris lilljeborgi, Rathbun 1929: 14, fig. 13.
(Figures 118 and 119, Plate 4a)

Distinguishing Characteristics

Carapace with 2 supraorbital spines at each side, and a high mid-dorsal carina reaching about two-thirds the length of the carapace with 4 or 5 large sparsely dentate teeth directed forward; rostrum ends in a long tapering point anteriorly; carpus of 2nd leg with 7 divisions; third maxilliped has an exopod; tergum of 3rd abdominal somite not produced as a spine over 4th.

Description

Integument smooth, shiny. Colour mottled bright red with yellow and white spots (Plate 4a).

Rostrum with distal long spine, slightly ascending including lateral carina but rounded above and below with serrate teeth: only a few accessory teeth or small teeth on shoulder of larger spines, unlike *S. spinus* which has many accessory teeth; rostral formula 9–12 + 4–6/3–5. Rostrum length about 0.7 times cl. Carapace with two supraorbital spines, the upper one larger, also antennal and pterygostomian spines and pointed suborbital angle; high median carina with continuous series of rostral-carapacial spines (5 or 6 on carapace) reaching about posterior two-thirds. Ventrally are long sternal spines between 2nd to 5th legs.

Abdomen with rounded terga and pleura except small ventro-lateral spine on 4th to 6th somites; third somite tergum not produced dorsally as a spine over 4th. Ventrally are small spines on cross carinae of 1st to 5th somites, paired on 1st and 2nd. Telson (t) slightly tapering to subtruncate tip, with four pairs of dorso-lateral spines and terminally a moderate central spine flanked by three pairs of spines, the second pair largest.

Eye large with globular cornea and inner tubercle on eyestalk.

Antennule (c): 1st article greater than other two, the 2nd about twice 3rd, distal spine on all three; stylocerite large, sharp point reaching distal 2nd article; thick flagellum almost as long as slender ventro-mesial.

Antenna (d): scale reaching beyond tip of rostrum; disto-lateral spine exceeds blade; peduncle more than half scale; basal article with ventro-lateral spine distally.

Mandible (e): minor differences in surfaces of molars, right more expanded; incisor with 3 teeth; distal segment of palp about twice proximal.

Maxillule (f): distal endite ovate, pointed; proximal endite narrow, tapering to twisted tip; endopod bent, bifid.

Maxilla (g): distal endite subequally bilobed; proximal reduced to short rounded lobes; endopod narrow, short; scaphognathite with anterior lobe sub-rectilinear distally squarish, longer than rounded posterior lobe.

Maxilliped I (h): distal endite wider than proximal, latter triangular in cross-section with leading triangular face fringed with setae; endopod longer than lobe of exopod, lash twice as long; bilobed epipod, each lobe subtriangular.

Maxilliped II (i): leglike, compressed; distal segment much wider than long; exopod slender, long; epipod with podobranch.

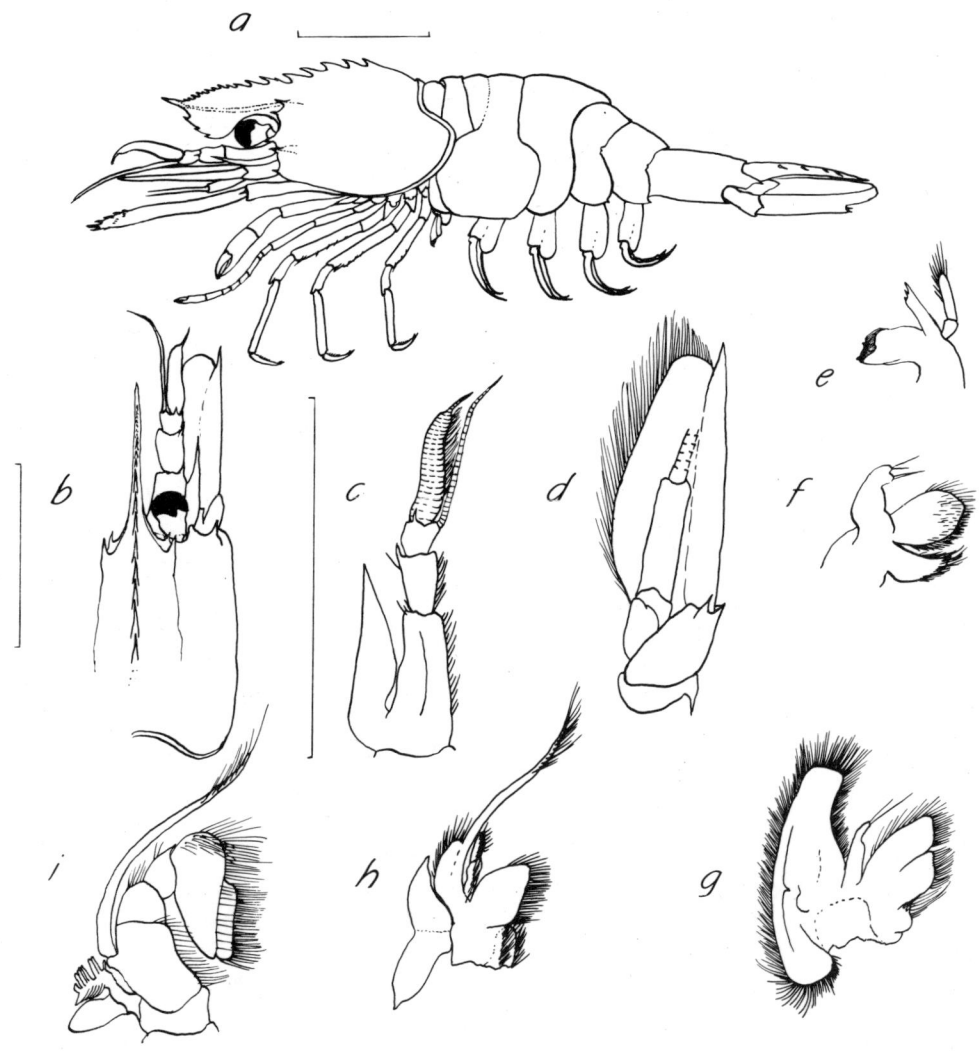

Fig. 118. *Spirontocaris lilljeborgi*: *a*, whole shrimp from left side; *b*, carapace in dorsal aspect; *c*, antennule; *d*, antenna; *e*, mandible; *f*, maxillule; *g*, maxilla; *h*, first maxilliped; *i*, second maxilliped; solid line = 10 mm.

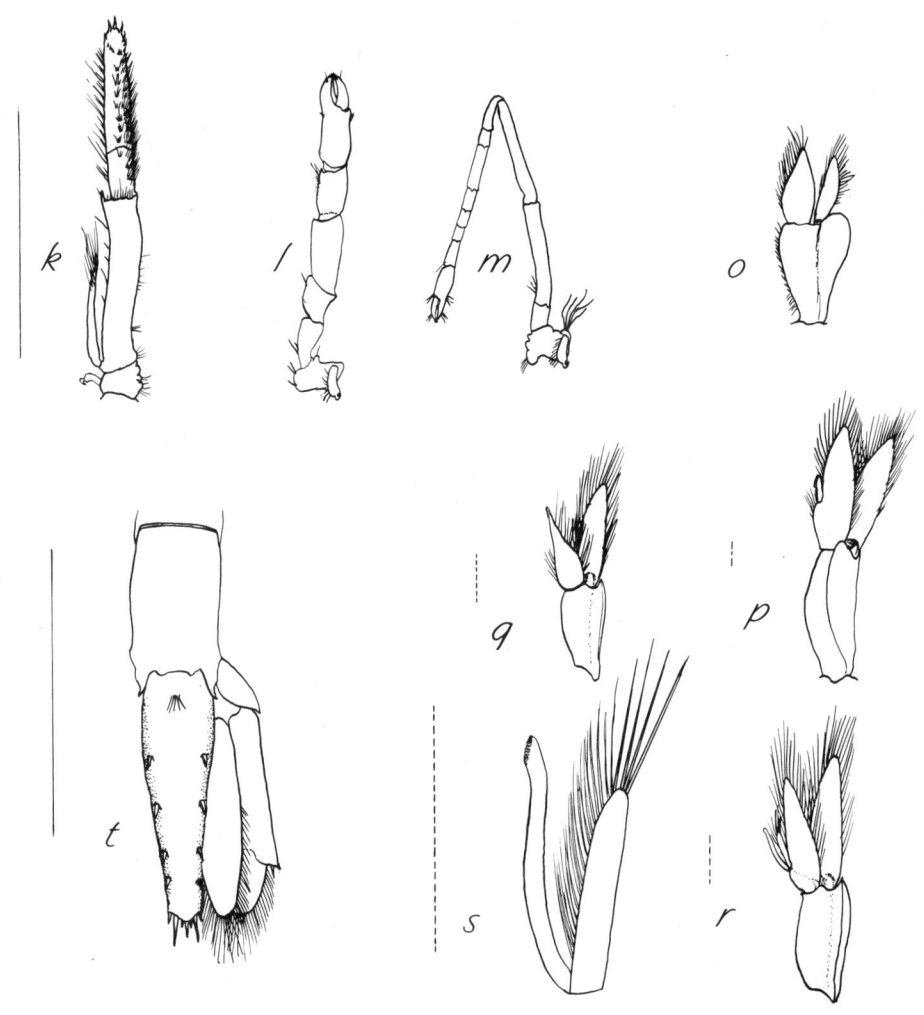

Fig. 119. *Spirontocaris lilljeborgi*: *k*, third maxilliped; *l*, first pereopod; *m*, second pereopod; *o*, F first pleopod; *p*, F second pleopod; *q*, M first pleopod; *r*, M second pleopod; *s*, appendix masculina; *t*, telson; solid line = 10 mm, broken line = 1 mm.

Maxilliped III (k): stout, longer than first leg; distal segment with apical row of 6 strong black spines; longest segment has a strong spine outside distally; exopod slender, longer than half longest segment; epipod straplike with terminal hook.

Pereopods: I (l) chelate, stout, shorter than others; dactyl bifid; epipod straplike with terminal hook. II (m) slender, chelate, dactyl bifid; carpus with seven divisions; epipod straplike with terminal hook. III–V (a) longer than others, dactyl slender, biunguiculate, with row of spines; merus with outer row of spines the distal one longest; propodus with series of spinules on posterior edge.

Pleopods: female I (o) endopod wide tapering to a point, exopod narrow shorter than endopod; protopod with winglike expansion. Female II (p) endopod and exopod subequal, appendix interna short with subapical patch of hooks. Male I (q) endopod with elongate tip and patch of hooks distolaterally, exopod slightly longer; male II (r) endopod slightly shorter than exopod; appendix interna compressed, slightly curved, with distolateral patch of hooks, longer than appendix masculina (s), the latter stout and with 4 apical spinous setae stouter than the many inner lateral setae.

RANGE OF DISTRIBUTION

North Atlantic from Murman Sea, Spitzbergen, Iceland to south coast of England and west and southwestern Ireland in Europe, and from Foxe Channel and Davis Strait (Greenland) to off Delaware Bay in America; also arctic Alaska but not in the Pacific (Holthuis 1947; Smaldon 1979; Williams 1984). Depths 20–1 200 m but mostly 35–90 m (Holthuis 1947).

Records of distribution in the area of reference are in Fig. 120.

BIOLOGY

Lengths to 7 mm cl in males and 15 mm cl in females.

In late autumn in the Newfoundland area 75% of the mature females were ovigerous probably indicating annual spawning. Allen (1962) found that a few females taken off Northumberland survived to lay a second batch of eggs in the following year. Pike (1954) showed that only a small proportion of females breed in their first year, but most in the second year and a few in their third year in the Firth of Clyde, Scotland. Temperatures were at about 4–14° C during most of the year.

The parasite, *Hemiarthrus abdominalis* is found on this species in Europe.

This species was taken with larger numbers of *Sergestes arcticus* and *Pasiphaea multidentata* as well as in the *Pandalus borealis* communities and with *Pontophilus norvegicus* (Squires 1965a).

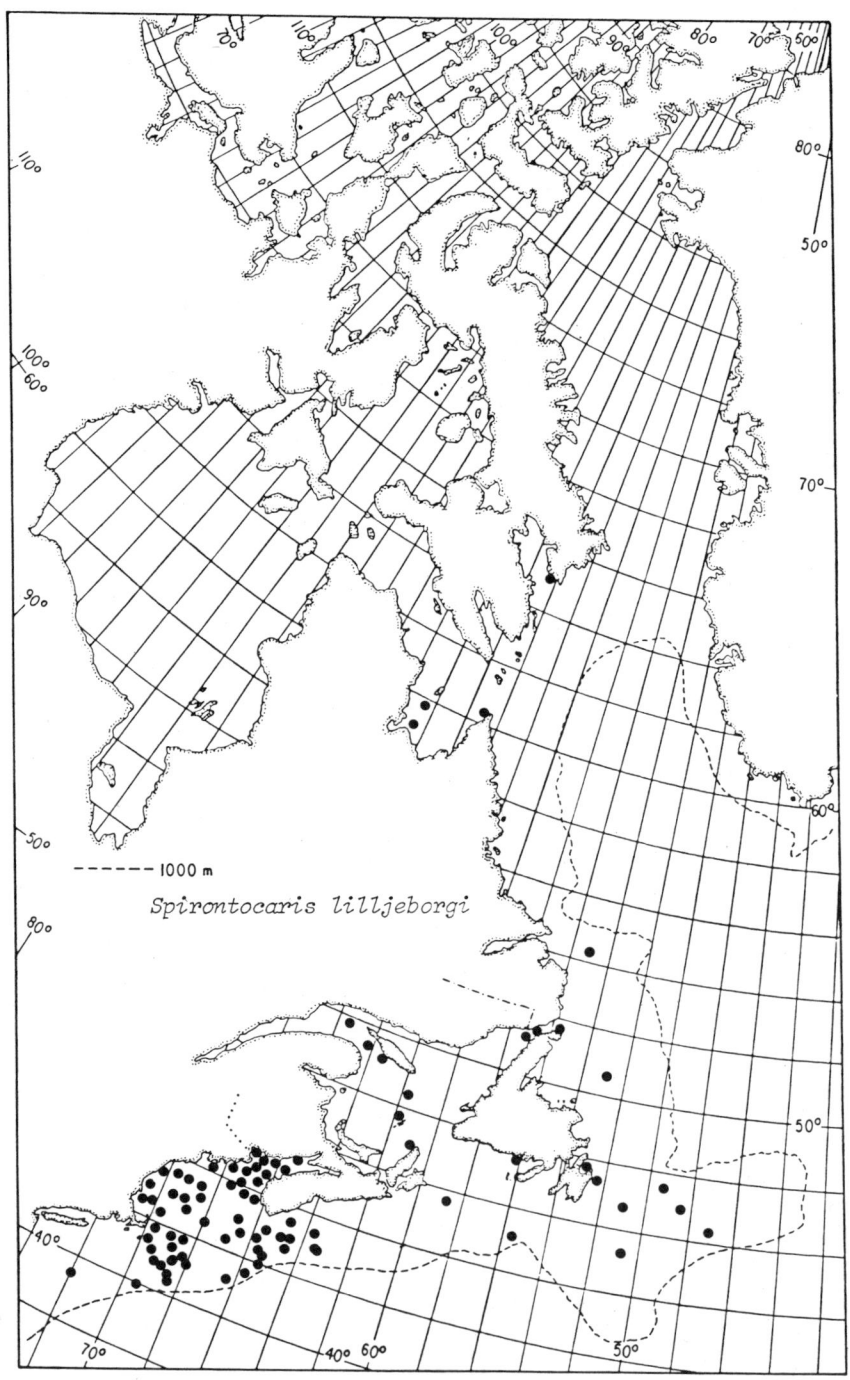

Fig. 120. *Spirontocaris lilljeborgi*, distribution records in the area of reference.

Spirontocaris phippsi (Krøyer, 1841)
Spirontocaris phippsi, Holthuis 1947: 8, 37.
Spirontocaris phippsi, Rathbun 1929: 13, fig. 12.
(Figures 121 and 122)

DISTINGUISHING CHARACTERISTICS

Small but robust with a large triangular rostrum with several even teeth above and on the forward edge below, teeth continued on carina on carapace to about half carapace length; two supraorbital spines; exopod on third maxilliped; epipods on first three legs; carpus of 2nd leg with seven divisions.

DESCRIPTION

Integument smooth, shiny. Colour mottled reddish brown on lighter background.

Rostrum straight, sharp-edged and with sharp teeth above and below on forward edge, formula 5–7 + 3–4/4–6; lateral carina confluent with orbit. Carapace with two unequal supraorbital spines and moderate antennal and small pterygostomian spines, postorbital lobe pointed; median carina rises above level of rostrum and reaches about half carapace and has 4–5 fixed spines of slightly different sizes.

Abdomen with terga and pleura rounded except very small ventro-lateral spine on 4th and larger one on 5th and 6th somites, tergum of 3rd somite extended over 4th but rounded not spinous.

Telson (t) tapered evenly with four pairs of sharp dorso-lateral spines; subtruncate tip with central bunch of setae flanked by two pairs of spines.

Eye large, cornea globular, with ocellus, stalk with inner tubercle.

Antennule (c): 1st article longer than other two combined, 3rd shorter than 2nd, all with strong distal spine; stylocerite with long sharp point reaching beyond second article; a ventro-mesial spine on distal third of 1st article; flagella short, almost equal in length.

Antenna (d): scale length 2.4 times width, disto-lateral spine exceeded by blade; peduncle reaches about half blade; flagellum slightly longer than body; basal article with strong distal spine ventrally.

Mandible (e): molars slightly dissimilar, surface of left more hollow and with fewer cusps; incisors similar with 3 teeth; palp with two equal parts.

Maxillule (f): distal endite wide slightly pointed, proximal narrow tapering to twisted apex; endopod bent slightly, bifid.

Maxilla (g): distal endite large, equally bilobed; proximal reduced with one lobe very small; endopod small; anterior lobe of scaphognathite squarish distally, much longer than rounded posterior lobe.

Maxilliped I (h): distal endite only slightly wider than proximal, the latter with long setae on each side of frontal facet; endopod two-segmented, as long as lobe of exopod, lash longer; epipod about equally bilobed.

Maxilliped II (i): leglike, compressed, distal segment much wider than long; exopod slender, longer than endopod; epipod with podobranch.

Maxilliped III (k): stout, longer than first leg; distal segment with apical ring of curved black spines; longest segment with two strong distal spines; slender exopod more than half longest segment; epipod straplike with terminal grooming hook.

Pereopods: I (l) chelate, stout, dactyl bifid, epipod straplike; shorter than others; II (m) chelate, slender, carpus with seven divisions, epipod straplike; III–V with series of outer spines on merus, the distal longest, also many spinules on posterior edge of propodus, dactyl slender, biunguiculate and with flexor spines; epipod on III only.

Pleopods: female I (o) endopod wide, fusiform, longer than narrow exopod; protopod with lateral winglike expansion; female II (p) endopod and exopod subequal, appendix interna with apical patch of hooks. Male I (q) endopod with attenuate tip and patch of hooks, exopod slightly longer, narrow; male II (r) endopod with appendix interna slightly

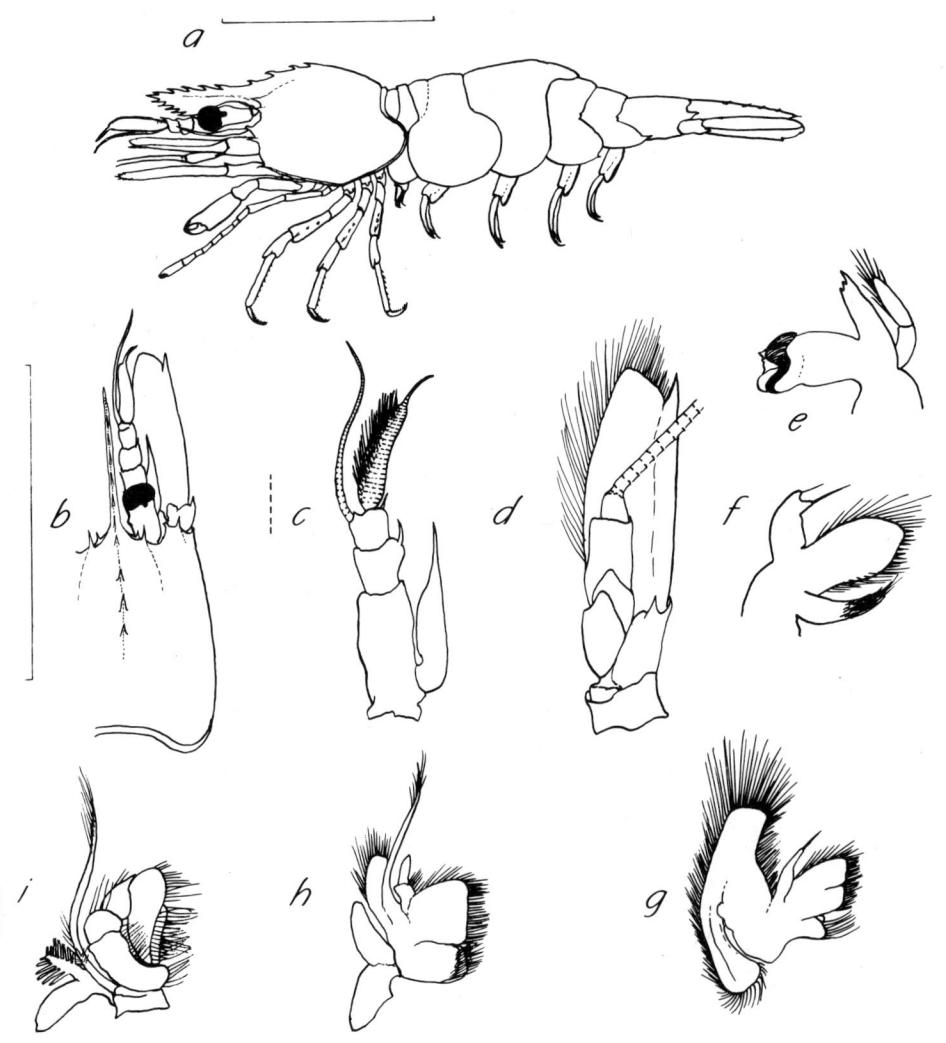

Fig. 121. *Spirontocaris phippsi*: *a*, whole shrimp from left side; *b*, carapace in dorsal aspect; *c*, antennule; *d*, antenna; *e*, mandible; *f*, maxillule; *g*, maxilla; *h*, first maxilliped; *i*, second maxilliped; solid line = 10 mm, broken line = 1 mm.

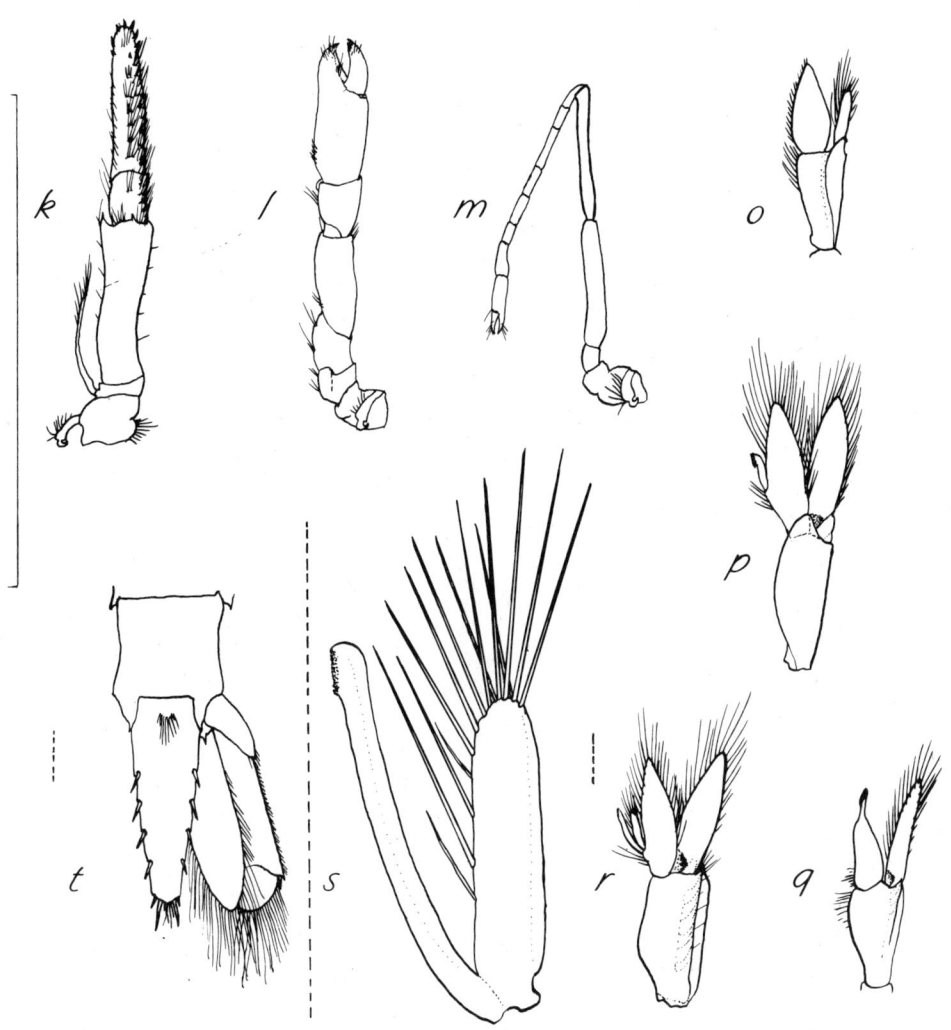

Fig. 122. *Spirontocaris phippsi*: *k*, third maxilliped; *l*, first pereopod; *m*, second pereopod; *o*, F first pleopod; *p*, F second pleopod; *q*, M first pleopod; *r*, M second pleopod; *s*, appendix masculina; *t*, telson; solid line = 10 mm, broken line = 1 mm.

curved and with distolateral patch of hooks, longer than appendix masculina (s), the latter stout and with two rows of about 5 apical long fine setae and about eight similar inner lateral setae in continuous series.

RANGE OF DISTRIBUTION

Circumpolar, North Atlantic from Cornwallis Island and Hudson Bay to Martha's Vineyard in America, and Spitzbergen to southern Norway and Britain in Europe; Arctic Ocean north of Siberia, Beaufort Sea and Bering Sea to Siberian east coast in the North Pacific (Allen 1967; Holthuis 1947; Williams 1984). Depths 10–270 m (Holthuis 1947).

Records of distribution in the area of reference are in Fig. 123.

BIOLOGY

Lengths 7 mm cl in males and 10 mm cl in females.

Females were first mature at 6 mm cl. In Foxe Basin about 70% were potentially ovigerous in autumn indicating possible annual spawning in this area (Squires 1965a).

Stomach contents consisted largely of foraminiferans and phytobenthos as well as ostracods, mysids and copepods. A few had hydroids, small bivalves and rhodophytes (Squires 1965a; 1967a).

Other species present in catches with this species were most frequently *Spirontocaris spinus* and *Eualus fabricii* but also *Lebbeus groenlandicus* and *Pandalus montagui* (Squires 1965a).

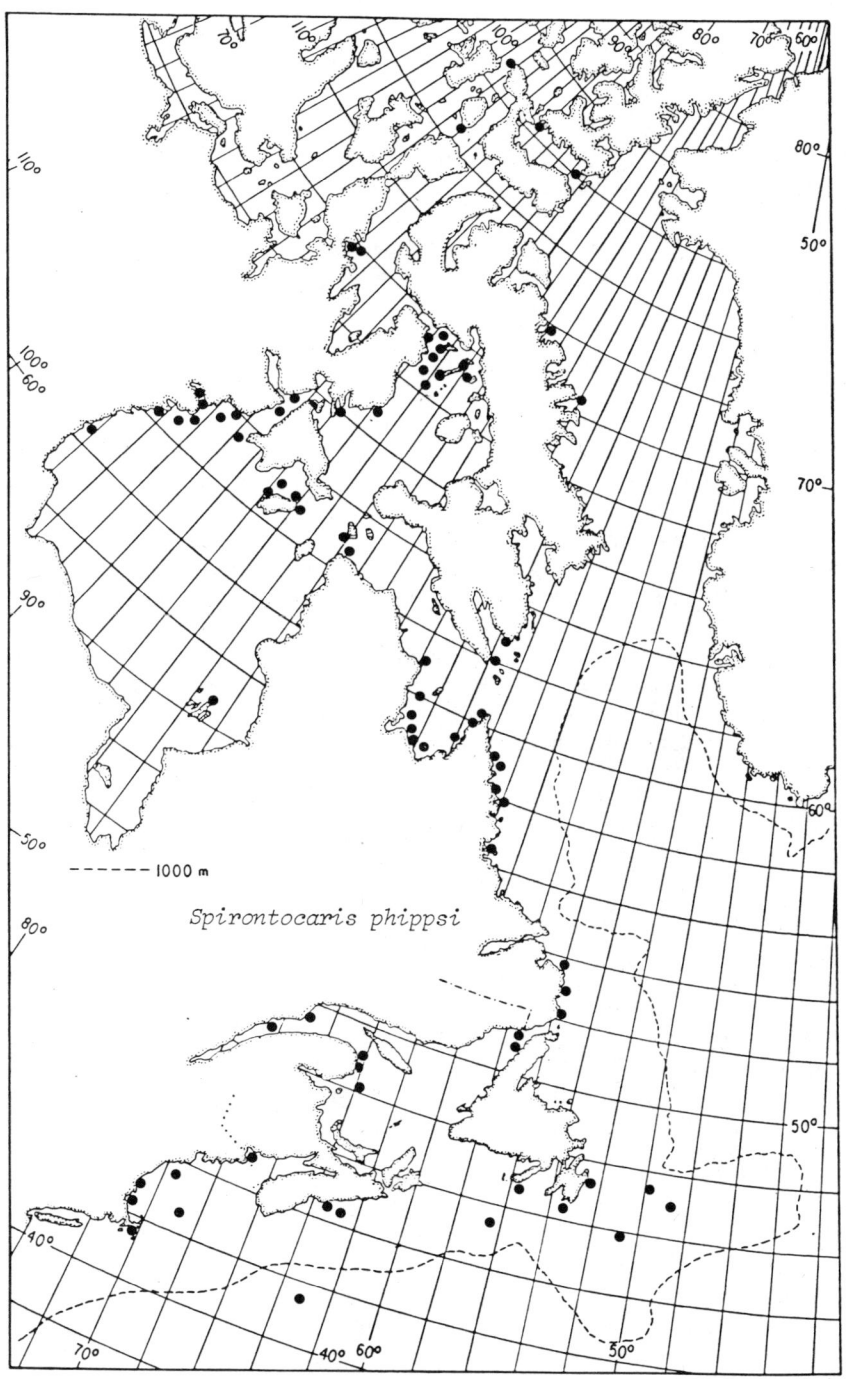

Fig. 123. *Spirontocaris phippsi*, distribution records in the area of reference.

Spirontocaris spinus (Sowerby, 1805)
Spirontocaris spinus, Holthuis 1947: 8; 37.
Rathbun 1929: 14, fig 14.
(Figures 124 and 125, Plate 4b)

DISTINGUISHING CHARACTERISTICS

Carapace with high dorsal carina almost reaching posterior margin and with large serrate teeth with small secondary teeth on them; rostrum with numerous small dorsal teeth and distally subtruncate with major and minor points; two unequal supraorbital spines; third abdominal somite has posterodorsal margin produced to form a spine over 4th, stronger in male; carpus of 2nd leg with seven divisions.

DESCRIPTION

Integument smooth, shiny. Colour dull red with brown, bright red and green mottling on translucent background, sometimes white markings on the legs (Plate 4b).

Rostrum subtruncate with often more than one spinous tip, with a deep blade with few teeth below but many small sagittate teeth above; length about 0.7 times carapace length; lateral carina horizontal but turning up at tip, not confluent with orbit, ending at larger supraorbital spine.

Carapace with two unequal supraorbital spines, the larger with strong carina, strong antennal spine also with carina but pterygostomian spine small; middorsal carina high with about five large teeth, often with secondary small teeth or sagittations, reaching the posterior edge of the carapace.

Abdomen with third somite tergum dorsally carinate and produced as a hollow spine over the 4th; pleura rounded except 4th–6th with ventro-lateral spine. Telson (t) with four dorso-lateral spines and three pairs of terminal spines.

Eye with globular cornea larger than tapered stalk which has a proximal protuberance inside.

Antennule (c): articles decreasing in size distally, all with strong distal spine; stylocerite with attenuate sharp point reaching farther than distal 3rd article; dorsolateral flagellum shorter than peduncle; ventral spine on 3rd article.

Antenna (d): scale reaches farther than tip of rostrum, from orbit 0.8 times cl; distolateral spine exceeds blade; peduncle reaches about half scale; basal article with two ventro-lateral spines outside.

Mandible (e): molars heavy, similar except for surface differences; incisor with 4 teeth; distal segment of palp longer than proximal.

Maxillule (f): distal endite ovate, slightly pointed; proximal tapered, short; endopod bent at middle, obscurely bifid.

Maxilla (g): distal endite large, bilobed, proximal lobe larger; proximal endite reduced, bilobed, distal lobe small; scaphognathite with axe-shaped posterior lobe smaller than squarish tipped anterior lobe. Endopod small, tapered.

Maxilliped I (h): distal endite large, slightly concave; proximal smaller with thick frontal edge; endopod with two segments, longer than lobe of exopod, lash twice as long; epipod large, bilobed.

Maxilliped II (i): endopod leglike, compressed, distal segment much wider than long; exopod slender, about as long as endopod; epipod with podobranch.

Maxilliped III (k): stout, about as long as first leg; distal segment with apical ring of 7 strong black spines; exopod slender, about half as long as longest segment; epipod straplike with terminal grooming hook.

Pereopods: I (l): chelate, stout, shorter than others, dactyl bifid; epipod straplike with terminal hook. II (m) chelate, slender, carpus with seven divisions, dactyl bifid; epipod straplike with terminal hook. III–V (a) slightly shorter than second leg; dactyl biunguiculate.

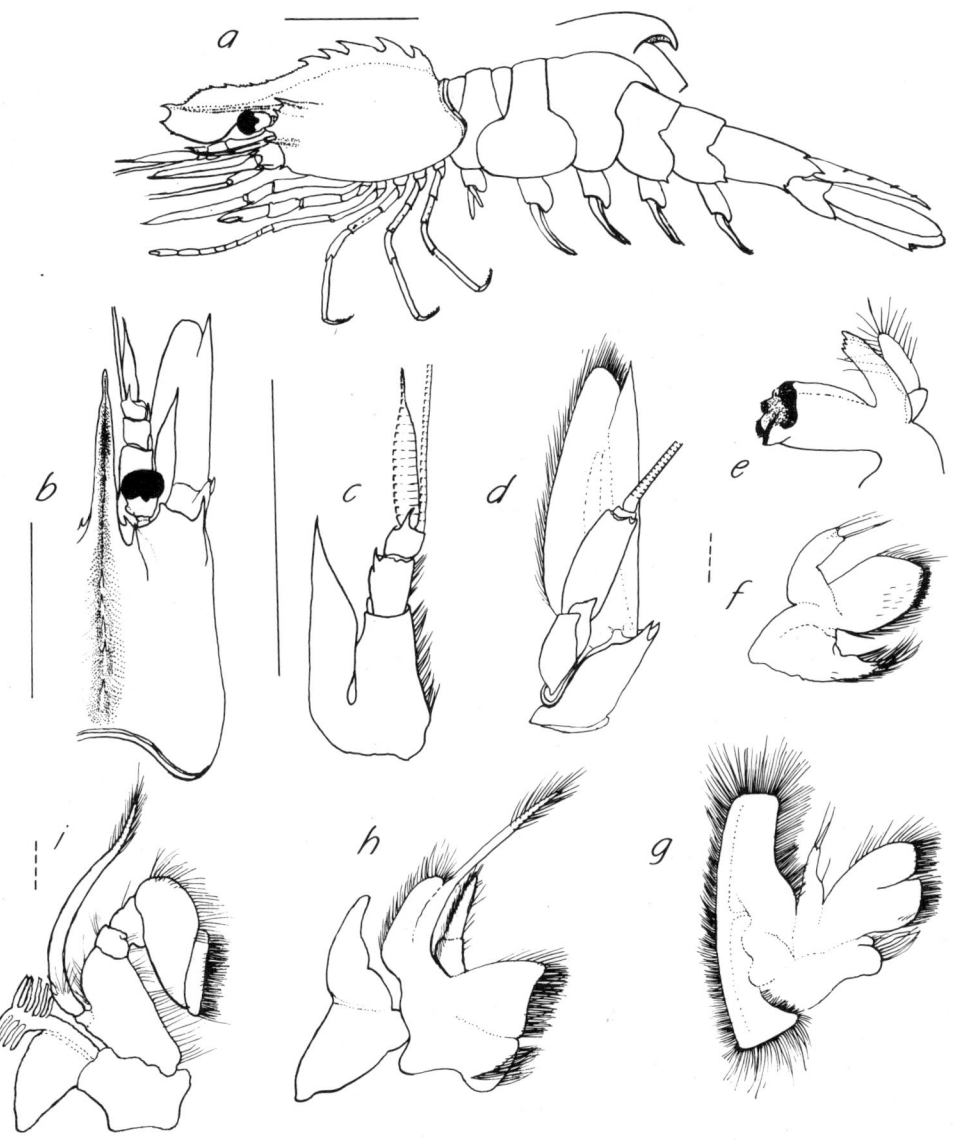

Fig. 124. *Spirontocaris spinus*: *a*, whole shrimp from left side; *b*, carapace in dorsal aspect; *c*, antennule; *d*, antenna; *e*, mandible; *f*, maxillule; *g*, maxilla; *h*, first maxilliped; *i*, second maxilliped; solid line = 10 mm, broken line = 1 mm.

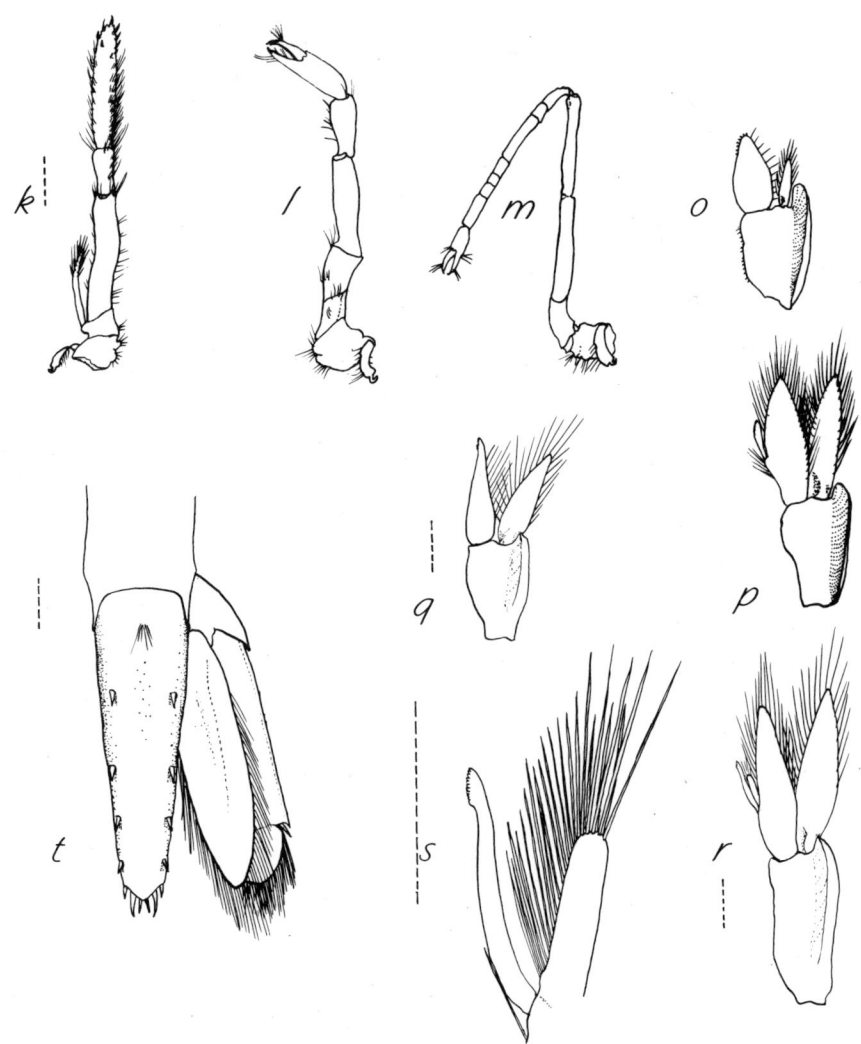

Fig. 125. *Spirontocaris spinus*: *k*, third maxilliped; *l*, first pereopod; *m*, second pereopod; *o*, F first pleopod; *p*, F second pleopod; *q*, M first pleopod; *r*, M second pleopod; *s*, appendix masculina; *t*, telson; broken line = 1 mm.

Pleopods: female I (o) endopod wide tapered, longer than narrow exopod; large winglike expansion of protopod; female II (p) endopod and exopod about equal; appendix interna expanded distally with subapical patch of hooks. Male I (q) endopod tapered to narrow tip with hooks, about equal in size to exopod. Male II (r) appendix interna sinuous, longer than masculina, expanded tip with hooks; masculina (s) with about 7 apical long spines continuous with lateral row of about 15 long spines.

Range of Distribution

Circumarctic from Hudson Bay, Foxe Basin and Greenland to Massachusetts Bay in America, and from Spitzbergen to northern North Sea and Irish Sea in Europe; Alaska to Puget Sound in the North Pacific and Bering Sea to Sea of Japan (Williams 1984). Depths 5–465 m (Heegaard 1941).

Records of distribution in the area of reference are in Fig. 126.

Biology

Lengths in males to 12 mm cl and in females to 17 mm cl. Females were first mature at 8 mm cl.

About 70% of the females from Foxe Basin (101 specimens) were potentially ovigerous in autumn suggesting annual spawning (Squires 1965a). Off Northumberland, U.K., Allen (1962; 1966) found that females had one batch of 350–850 eggs per year and carried them 12–13 weeks. He estimated a life span of 2½years for the species in that area.

Stomach contents were phytobenthos, foraminiferans, ostracods, crustacean fragments, small bivalves, hydroids and sponge spicules, and rarely polychaetes and small gastropods (Squires 1965a).

The isopod parasite, *Hemiarthrus abdominalis*, infested 18% of the females and 3% of the males from Foxe Basin, Hudson Strait and northern Labrador. It was also seen in specimens from the Newfoundland coast (Squires 1965a) and off Northumberland, England (Allen 1962).

Predators observed were cod and beluga whales (Squires 1965a; 1967a).

Species also present in catches were *Spirontocaris phippsi*, *Lebbeus polaris*, *Pandalus montagui* and *Sabinea septemcarinata* and occasionally several others (Squires 1965a).

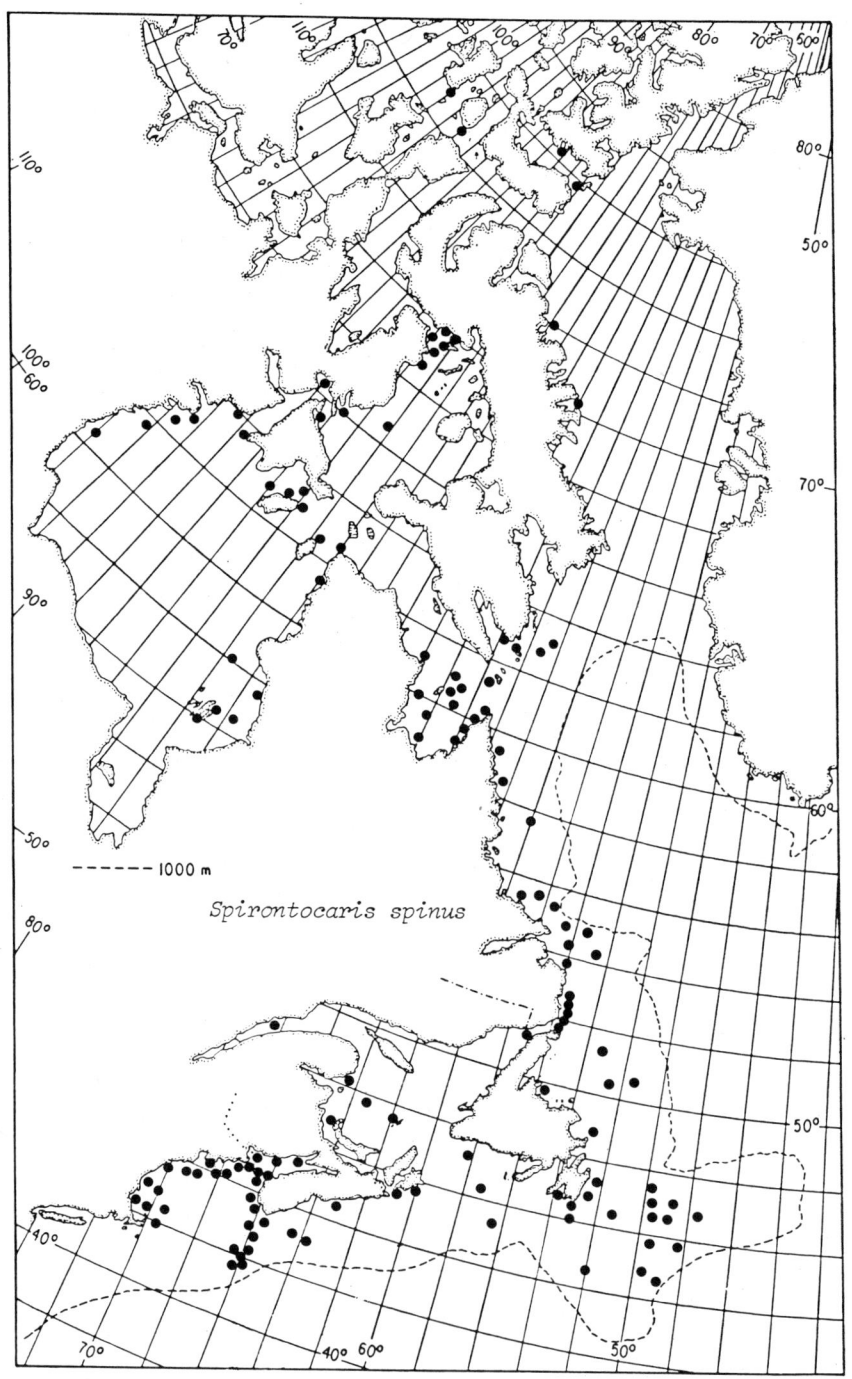

Fig. 126. *Spirontocaris spinus*, distribution records in the area of reference.

Family PANDALIDAE Haworth, 1825
Butler 1980: 122; Chace, 1986: 9; Williams 1984: 149; Zariquiey Alvarez 1968: 96.

Rostrum long and slender, compressed, armed with spines and teeth; first pair of legs with no or a very small chela; 2nd legs chelate, unequal, carpus with few or many divisions; mandible with palp and incisor and molar processes widely separate; 1st maxilliped with lash on exopod; exopod of uropod with moveable axillary spine at outer edge.

Key to species of Pandalidae in area of reference
(Adapted from Allen, 1967)

1	Pereopods with epipods	2
	Pereopods without epipods	5
2	Third maxilliped with an exopod .... *Dichelopandalus leptocerus*	
	Third maxilliped without exopod	3
3	Carpus of 2nd pereopod on right side with 4 or 5 divisions .... *Pandalus propinquus*	
	Carpus of 2nd pereopod on right side with at least 20 divisions	4
4	Prominent dorsal lobe on 3rd abdominal somite .... *Pandalus borealis*	
	No dorsal lobe on 3rd abdominal somite .... *Pandalus montagui*	
5	Small moveable spine at mid-dorsal posterior edge of 3rd abdominal somite. Very long rostrum .... *Stylopandalus richardi*	

Genus *Dichelopandalus* Caullery, 1896
Williams 1984: 150; Zariquiey Alvarez 1968: 112.

Third maxilliped with an exopod; carapace and abdomen covered with short, irregular, punctate ridges each with short setae; rostrum turning up slightly near tip which is bifid, dorsal spines all moveable, ventrally strong teeth; 2nd pair of pereopods unequal, the right shorter and its carpus with 5 divisions.

Dichelopandalus leptocerus (Smith, 1881)
Rathbun 1929: 10, fig. 7; Williams 1984: 150, fig. 106;
Couture and Trudel 1968: 865, fig. 6.
(Figures 127 and 128)

DISTINGUISHING CHARACTERISTICS

Body roughish, covered with many small irregular ridges with short, bristle-like setae; second pair of legs unequal, the right shorter, its carpus with five divisions; exopod on third maxilliped; rostrum almost straight, with small tooth near tip which appears bifid, other dorsal spines moveable.

DESCRIPTION

Integument roughish, covered with low, short, irregular ridges with many short setae. Colour brick red.

Carapace with strong antennal and small pterygostomian spines; middorsal carina about half length of carapace and with 2 post-orbital spines continued from the dorsal rostral series; the rostrum has 8–9 + 2/6 moveable spines dorsally and curved teeth ventrally plus an extra tooth above at the tip giving bifid appearance, all are intercalated with short setae except near the tip; rostrum almost straight.

Abdomen with rounded terga and pleura except the 5th and 6th somites with a ventrolateral spine on pleuron, the latter also with supporting carina.

Telson (t) narrow, tapered to near subtruncate tip, with about 6 pairs of dorso-lateral spines, and three pairs of terminal spines, the 2nd pair much longer than the others; the ventral surface of the telson is covered with short setae. Uropod slightly longer than telson, outer branch slightly longer than inner branch.

Eye large, cornea globular, with short stalk and small ocellus.

Antennule (c) 1st article longer than other two combined, the 3rd a little longer than the 2nd; dorsolateral flagellum thickened over an area reaching almost to the tip of the rostrum; stylocerite truncate, foliaceous, less than half length of 1st article.

Antenna (d) scale very narrow, more than five times as long as wide; distolateral spine exceeding the blade; basal article with disto-ventral tooth; peduncle less than half scale.

Mandible (e) incisor thin, curved, with 5 terminal teeth; molar longer with three edge cusps rounded and chitinous; palp slightly longer than incisor, with 3 segments.

Maxillule (f) distal endite almost circular but slightly pointed; proximal endite narrow tapering to point; endopod bifurcate, each short branch with fine spine and a few setae.

Maxilla (g) distal endite large, bilobed, the distal lobe almost circular; proximal endite reduced, bilobed, proximal lobe squared off, the distal small and pointed; endopod short with narrow neck and subapical fine spine; anterior lobe of scaphognathite rounded at tip, inner edge concave, about equal in length to posterior triangular lobe, the latter with very long setae.

Maxilliped I (h) distal endite large, rounded, proximal shorter, with thickened edge; endopod with first segment distally twisted and with a short narrow apical segment, not as long as lobe of exopod, the latter with a lash about as long; epipod large, bilobed, anterior lobe larger, both subtriangular, inflated.

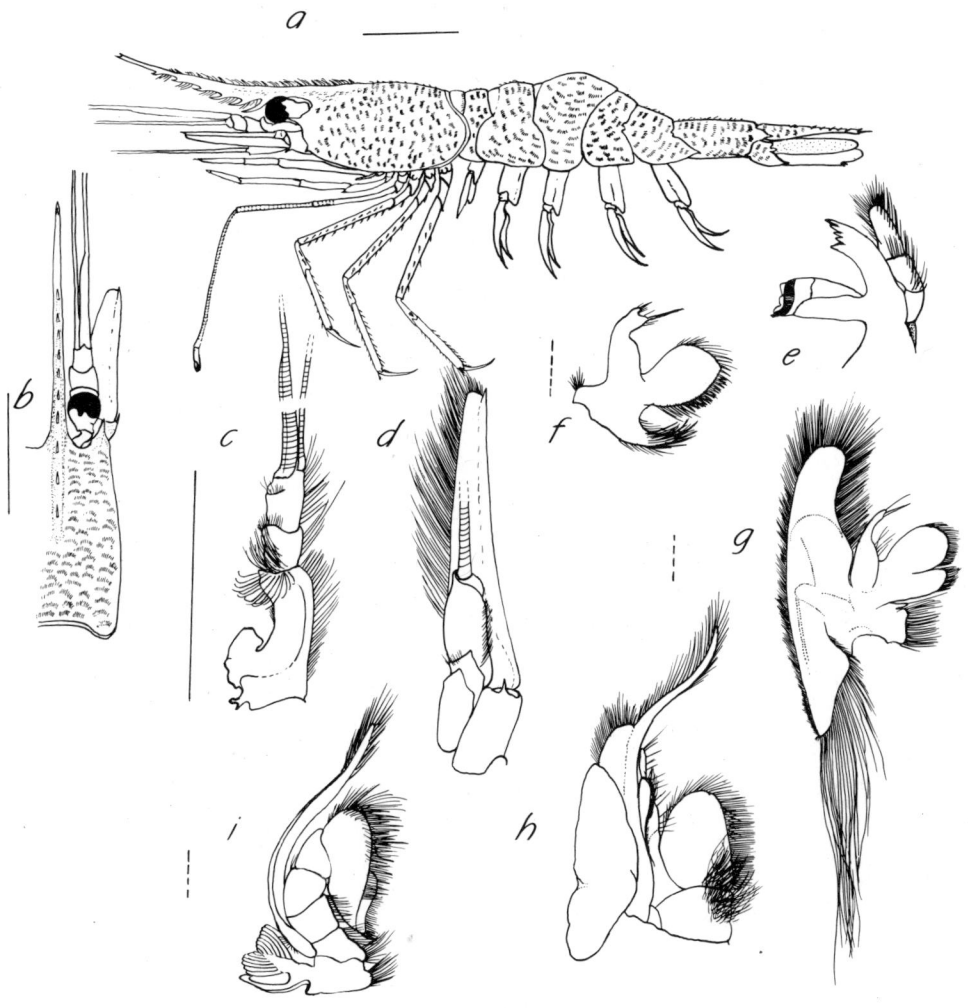

Fig. 127. *Dichelopandalus leptocerus*: *a*, whole shrimp from left side; *b*, carapace in dorsal aspect; *c*, antennule; *d*, antenna; *e*, mandible; *f*, maxillule; *g*, maxilla; *h*, first maxilliped; *i*, second maxilliped; solid line = 10 mm, broken line = 1 mm.

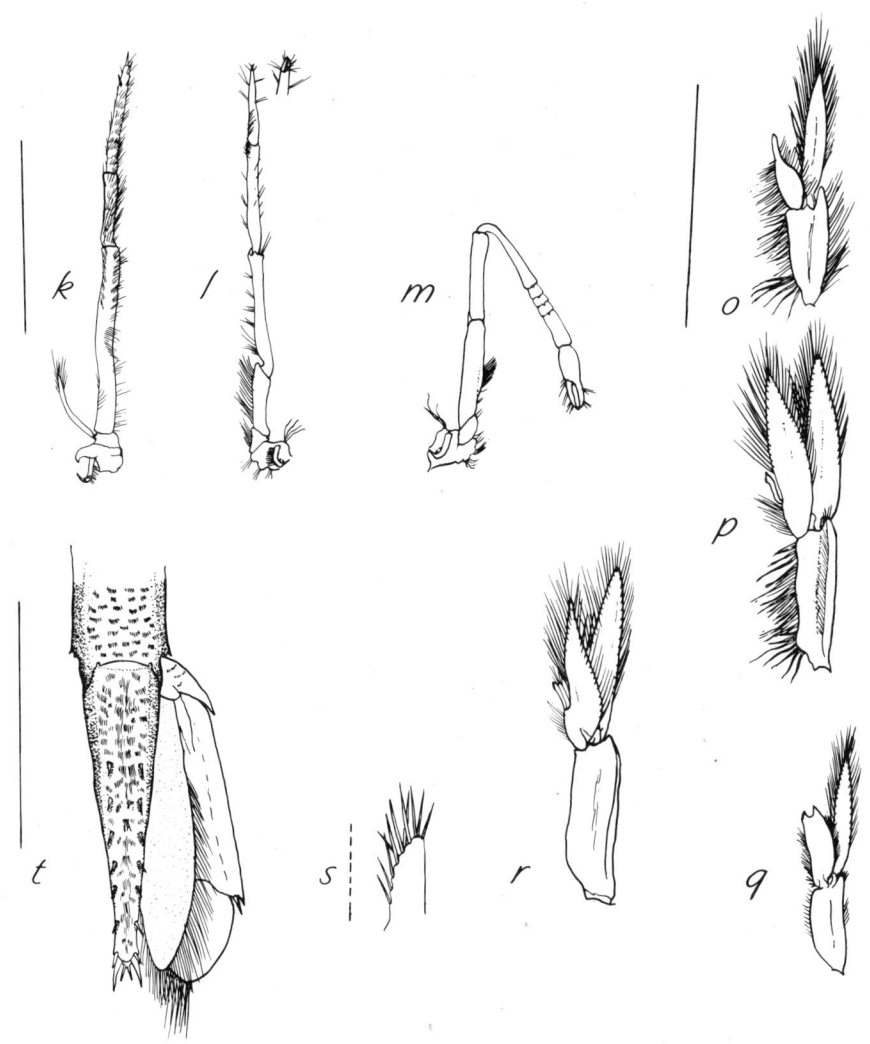

Fig. 128. *Dichelopandalus leptocerus*: k, third maxilliped; l, first pereopod; m, second pereopod; o, F first pleopod; p, F second pleopod; q, M first pleopod; r, M second pleopod; s, appendix masculina; t, telson; solid line = 10 mm, broken line = 1 mm.

Maxilliped II (i) leglike, compressed, distal segment much wider than long; exopod long, slender; epipod with podobranch.

Maxilliped III (k) about as long as first leg, distal segment with tip a spiral of four strong spines; exopod very short and slender; epipod with two parts, basal one falcate with a straplike attachment curved and with apical point and subapical grooming hooks.

Pereopods: I (l) microscopically chelate, merus with strong curve proximally, carpus longest segment, propodus long, tapered to small chela, epipod straplike, curved with sharp point and subapical hooks. II (m) right shorter than left, carpus with 5 divisions (3 short ones together); left very long annulated carpus (about 50 divisions), also merus and part of ischium annulated, very slender. III–V non-chelate, moderately long and stout, with 2 series of spines on merus and spinules on propodus; dactyl long and sharp.

Pleopods: female I (o) exopod narrow longer than jar-shaped endopod with distal neck; female II (p) endopod and exopod about equal, appendix interna moderate with subapical patch of hooks; male I (q) narrow exopod longer than wider endopod, the latter with distal notch and inner part with hooks subapically; male II (r) endopod slightly shorter than exopod with appendix interna with terminal hooks and appendix masculina (s) with 3 apical and about seven lateral fine spines.

RANGE OF DISTRIBUTION

Gulf of St. Lawrence and St. Mary's Bay, Newfoundland, to off Oregon Inlet, North Carolina (Squires 1966; Williams 1984); also Shumagin Bank, Alaska (Rathbun 1929). Depths 0–790 m (Williams and Wigley 1977).

Records of distribution in the area of reference are in Fig. 129.

BIOLOGY

Lengths to 17 mm cl in males and 20 mm cl in females (Squires 1965a); body length 90 mm in males and 98 mm in females (Williams 1984).

Females appeared to reach first maturity at 16 mm cl, and in those taken in autumn 61% (of 42 specimens) were potentially ovigerous, possibly suggesting annual spawning. In spring captures some females were ovigerous and with eyed embryos in eggs in June, and may hatch their eggs late.

Stomach contents included crustacean remains and some whole euphausiids, mysids and copepods (Squires 1965a).

A predator was the harp seal (*Phoca groenlandica*) in the Gulf of St. Lawrence (Squires 1965a).

A chytrid-like parasite from the gills of this species was obtained from samples from off Nova Scotia to off Long Island with incidences of 52%, 54% and 95% in June, January and October, respectively, occurring at 79 of 126 sampling stations (Uzmann and Haynes 1968).

In catches with this species were frequently *Pandalus propinquus*, and occasionally it was found among much larger numbers of *Pandalus borealis* (Squires 1965a).

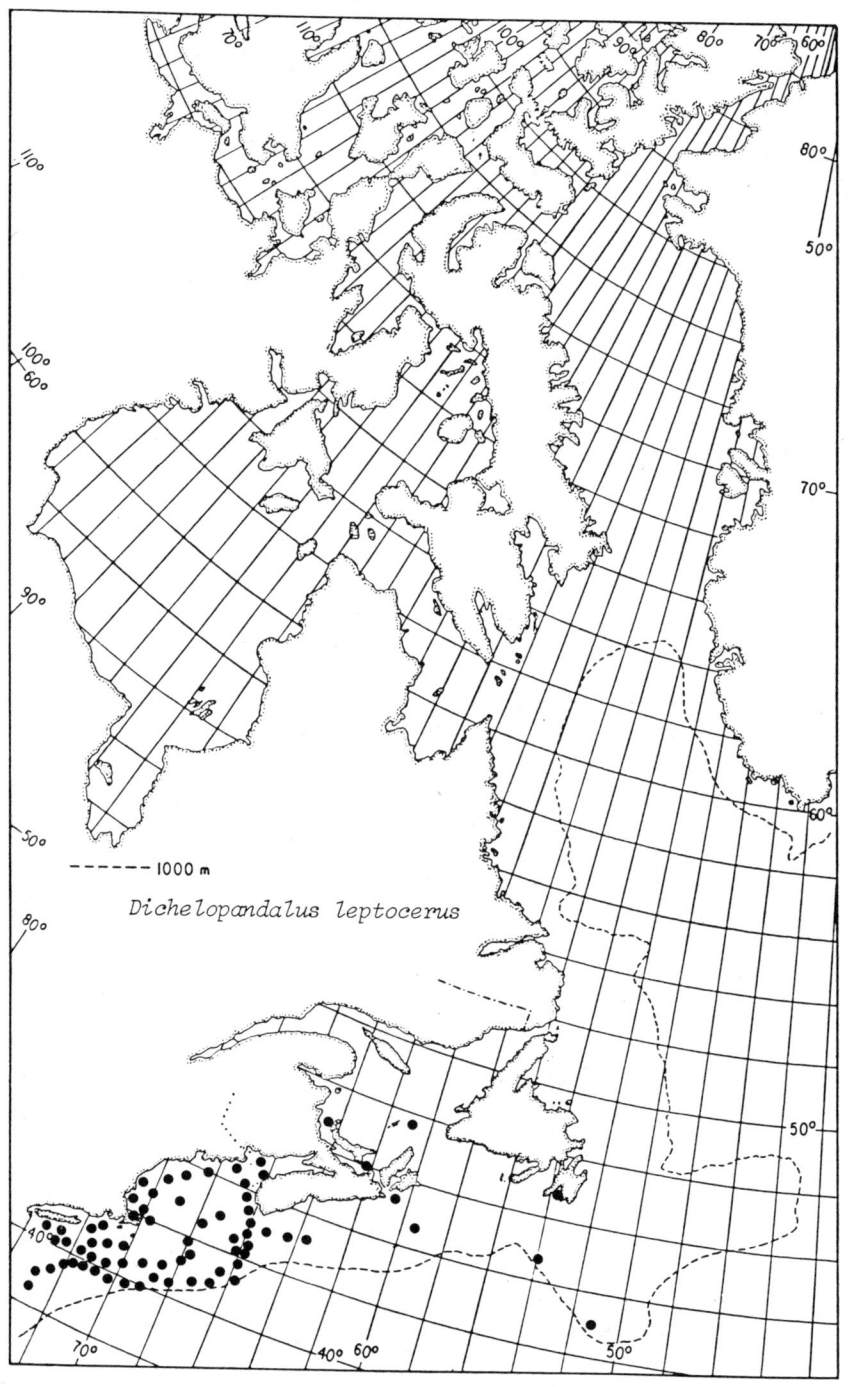

Fig. 129. *Dichelopandalus leptocerus*, distribution records in the area of reference.

Genus *Pandalus* Leach, 1814
Butler 1980: 126; Rathbun 1929: 7;
Williams 1984: 151.

Rostrum long, prominent, with mostly moveable spines dorsally and fixed rigid teeth ventrally; rostral series continued on median dorsal carina to about half carapace; third maxilliped without an exopod; mandible with three-segmented palp; 2nd pair of legs unequal, carpus with many annulations, fewer on shorter right leg of the pair; epipod on all legs but V.

About 17 species of the genus are known, only three of which are in the area of reference.

Pandalus borealis Krøyer, 1838
Rathbun 1929: 8, fig. 4;
Williams 1984: 151, fig. 107.
"Pink shrimp" — "Northern shrimp" — "Crevette nordique"
(Figures 130 and 131)

DISTINGUISHING CHARACTERISTICS

Rostrum long and narrow, slightly arched above the eyes, terminal half slightly ascending but straightening out at the tip, with continuous row of similar moveable spines intercalated with short setae, about 12–16 spines dorsally and 6–9 fixed short teeth ventrally; dorsally on third somite of the abdomen is a short carina forming a spine or spine-tipped lobe in front of the posterior margin; small median spine on posterior edge of fourth somite; second pair of legs unequal, the right shorter, carpus annulated: on the left about 58 and on the right about 25 annulations.

DESCRIPTION

Integument shiny but covered with minute rugosities and some microscopic pits. Colour bright pink.

Rostrum very long and narrow, slightly arched over the eyes, curving upward slightly on distal half but straightening out at single sharp tip, also with a lateral carina which makes it very rigid, dorsally are strong moveable spines (except distal 1–2) interspersed with fine plumose setae and ventrally moderate fixed teeth alternating (not paired) with dorsal spines (12–16/6–8) with about 4 behind orbit on median carina which reaches about distal third of carapace.

Carapace with strong antennal and small pterygostomian spines.

Abdomen with a prominent short carina and spine on middorsal third somite, also a short spine on median posterior edge of 4th somite; pleuron of 2nd somite large almost reaching edge of carapace; pleura of 4th and 5th with ventrolateral spine, 6th with ventroposterior spine.

Telson (t) slightly longer than 6th somite; tapered to subtruncate tip (bluntly pointed) with three pairs of terminal spines; about eight (7–10) pairs of dorsolateral spines; outer branch of uropod about equal in length to telson, the inner shorter.

Eye large, cornea globular with ocellus, much larger than short stalk.

Antennule (c) 1st article longer than other two combined, with forward extension under 2nd; 3rd shorter than 2nd; stylocerite short, rounded, with long fringing setae; dorsolateral flagellum thickened for most of its length, with attenuate slender tip reaching almost to tip of rostrum.

Antenna (d) scale length about 3 times width; blade slightly exceeds spine; peduncle does not quite reach middle of scale; basal article with ventral spine.

Mandible (e) left incisor with 6 teeth, right with 4; left molar with hollow crown and 3 major cusps, right with central ridge and edge of about 5 cusps; palp with 3 segments the distal longest.

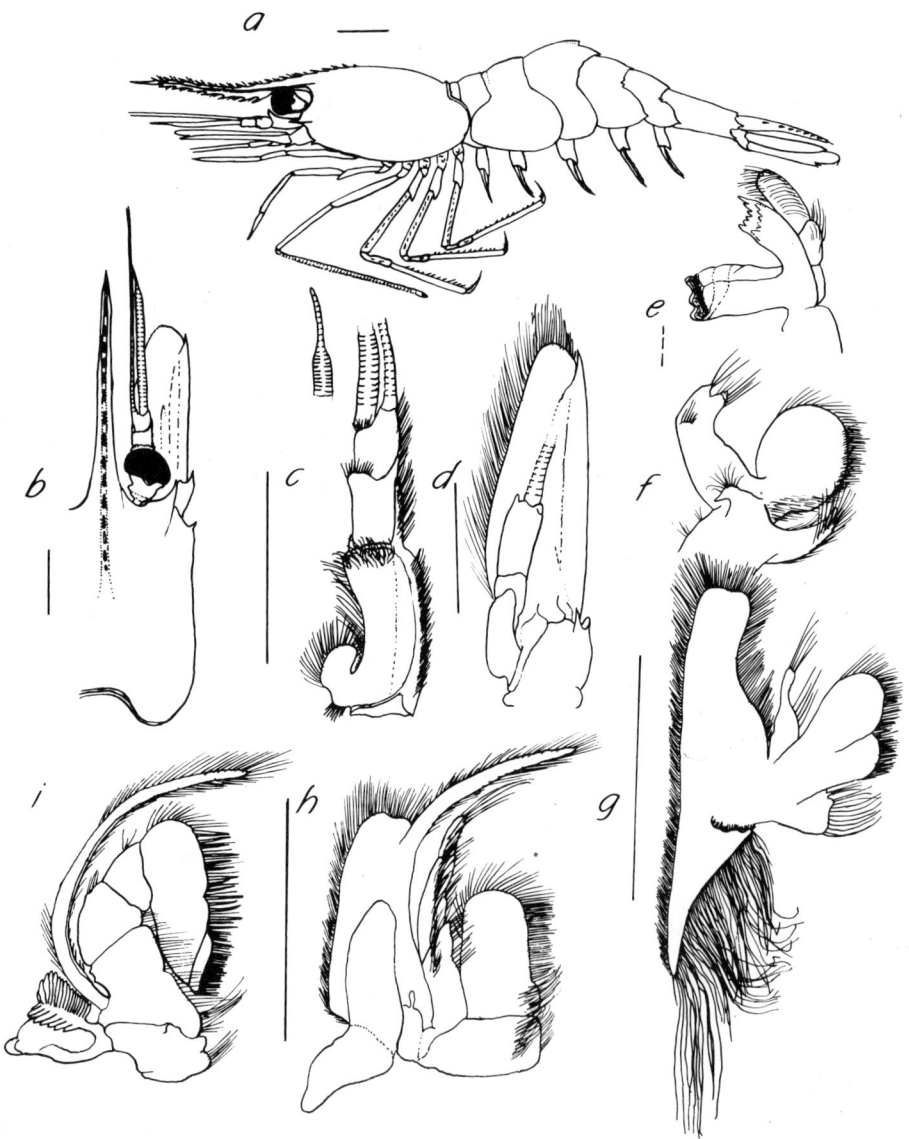

Fig. 130. *Pandalus borealis*: *a*, whole shrimp from left side; *b*, carapace in dorsal aspect; *c*, antennule; *d*, antenna; *e*, mandible; *f*, maxillule; *g*, maxilla; *h*, first maxilliped; *i*, second maxilliped; solid line = 10 mm, broken line = 1 mm.

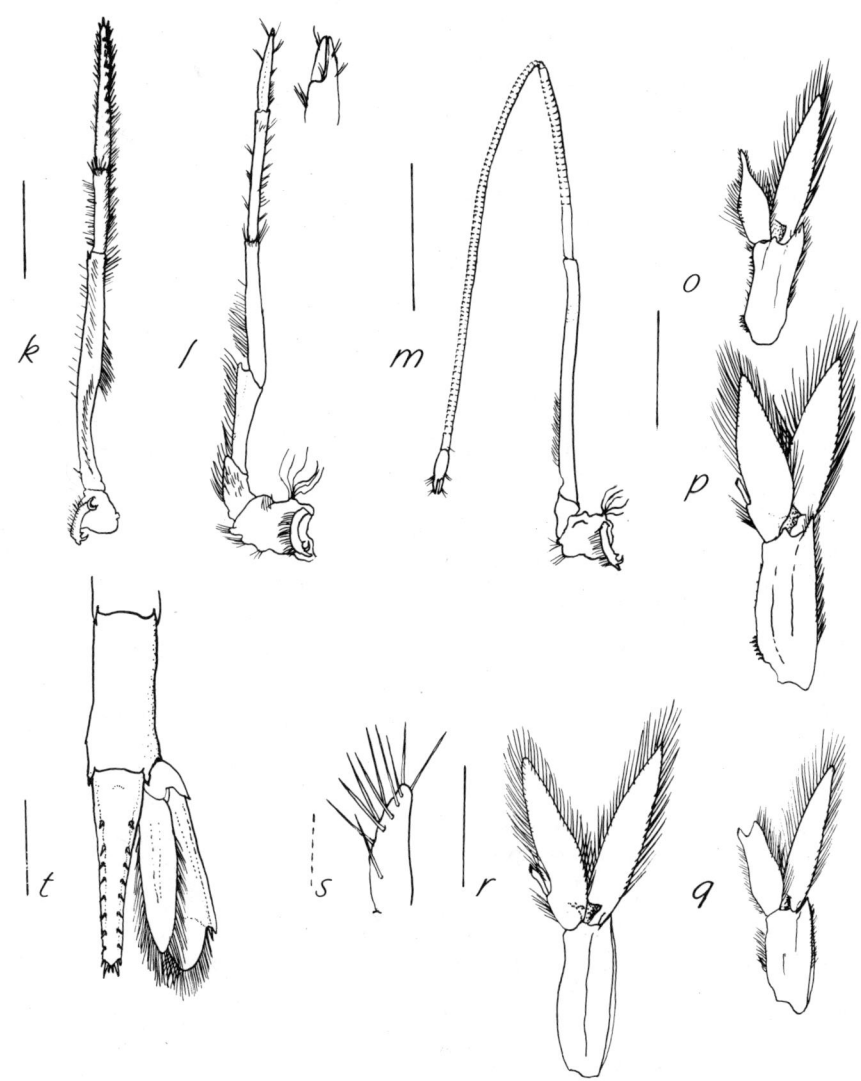

Fig. 131. *Pandalus borealis*: *k*, third maxilliped; *l*, first pereopod; *m*, second pereopod; *o*, F first pleopod; *p*, F second pleopod; *q*, M first pleopod; *r*, M second pleopod; *s*, appendix masculina; *t*, telson; solid line = 10 mm; broken line = 1 mm.

Maxillule (f) distal endite almost circular in outline, proximal narrow and tapering to a point; endopod curved distally with 2 apices giving bifurcate appearance; inner pointed exopod.

Maxilla (g) distal endite with large slightly unequal lobes, proximal somewhat reduced with one large straight-edged lobe and a tiny pointed lobe; endopod short somewhat expanded at tip; anterior lobe of scaphognathite long, slightly depressed at tip, inner edge concave, base wide, about as long as strongly pointed triangular-shaped posterior lobe with inner and apical fringe of very long setae.

Maxilliped I (h) distal endite large, pillar-shaped with thickened lower edge, proximal shorter with thickened edge; endopod slightly longer than lobe of exopod, the latter slightly depressed apically and with long lash or flagellum; epipod bilobed, anterior lobe longer.

Maxilliped II (i) leglike, compressed, distal segment much wider than long; exopod long, slender; epipod large, with podobranch.

Maxilliped III (k) leglike, slender, about as long as first leg; distal segment with a few strong apical spines; distal spine on longest segment; straplike epipod with subapical grooming hooks.

Pereopods: I (l) slender, chelate (propodus long tapering apically and with a short dactyl); distolateral projection on ischium, ischio-meral joint expanded; epipod straplike with pointed apex and subapical hook or claw; setobranch. II (m) slender, long, chelate, carpus with many annulations, merus also annulated in part, right shorter than left; epipod straplike with subapical claw; setobranch; III–V non-chelate, moderately slender and long; merus, carpus and propodus with series of spinules, dactyl long and sharp, with 6–9 proximal spines, only III and IV with epipod and setobranch.

Pleopods: female I (o) exopod narrow longer than fusiform long-necked endopod; female II (p) endopod and exopod about equal, appendix interna short, with distolateral hooks. Male I (q) narrow exopod longer than endopod, the latter with a distolateral projection with rows of hooks apically; male II (r) endopod and exopod about equal, appendix interna about as long as appendix masculina (s) the latter with two apical fine long spines continuous with two short rows of about eight fine lateral spines. In transitional forms pleopod I retains a small lateral appendix near the tip and the appendix masculina is lost.

Range of Distribution

From Greenland southward to Martha's Vineyard in the western Atlantic, and Iceland, Novaya Zemlya, Franz Josef Land and Spitzbergen to the British Isles in the eastern Atlantic. (In the Pacific possibly a different species, *P. eous* Makarov 1935, from the Bering Sea (and part of the Chukchi Sea) southward to British Columbia and stragglers as far as San Diego in California, and also in the western North Pacific as far as Hokkaido and Honshu in Japan and South Korea to 35°30′ N). Depths 10–500 m (Parsons and Fréchette 1989).

Distribution in the area of reference is shown in Fig. 132.

Biology

Lengths to 24 mm cl in males and 35 mm cl in females.

This species is a protandrous hermaphrodite, it functions as a male for part of its life and then changes sex to function as a female. There are some variations in its life history depending principally upon environmental temperatures, i.e. primary females may occur in areas where temperatures are usually high (averaging 6–14° C) throughout the year, and some may remain as males all their lives in areas where temperatures are continuously below 0° C (Squires 1968b). Also environmental temperatures may affect rates of growth, age at which sex change occurs and years spent as males or females (Rasmussen 1953; 1967; Horsted and Smidt 1956; Allen 1959; Squires 1965a; 1968b).

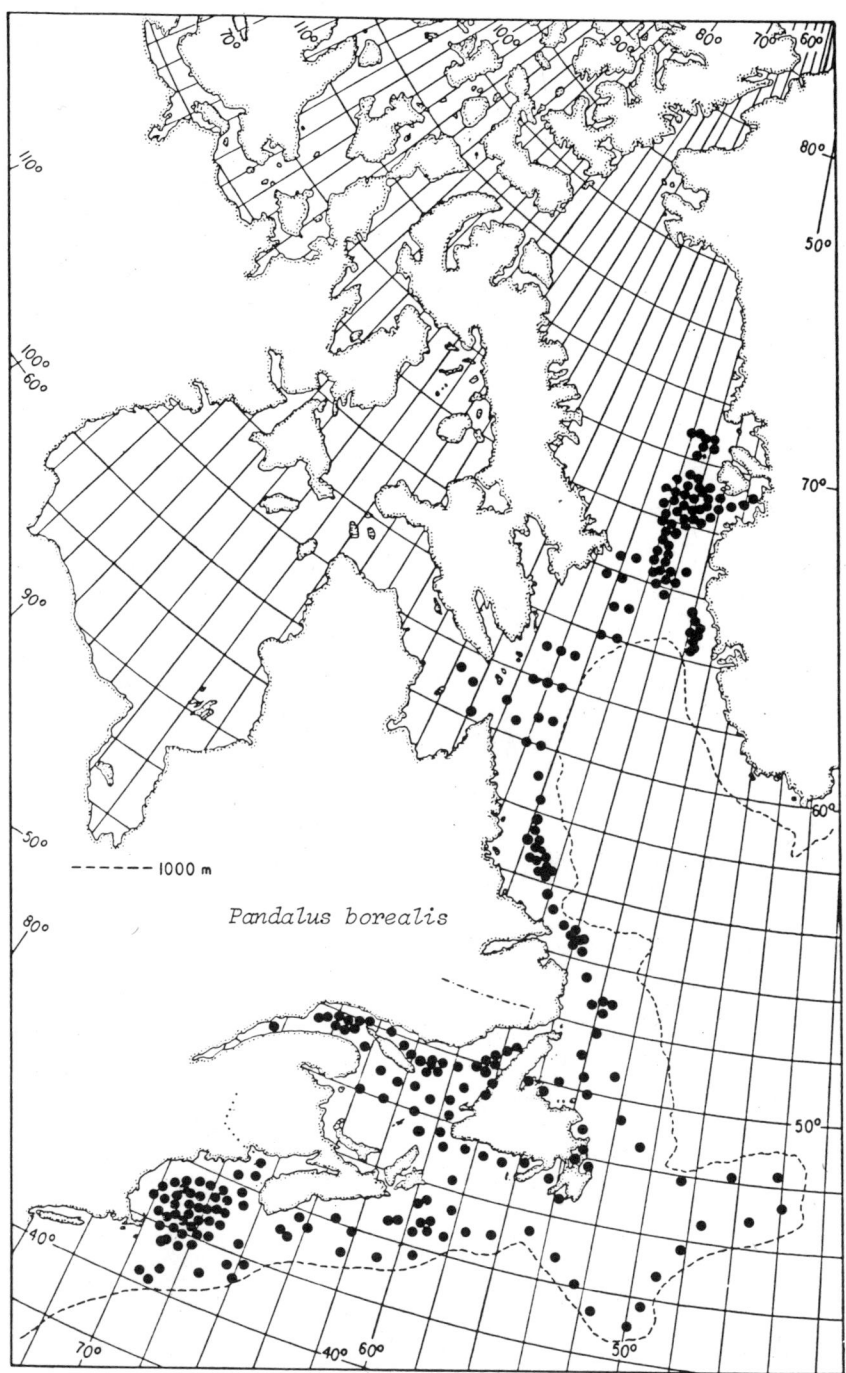

Fig. 132. *Pandalus borealis*, distribution records in the area of reference.

Fecundity has been shown to vary greatly according to individual size and area, averaging 2 400 eggs (800–3 400) in the Gulf of Maine (Haynes and Wigley 1969), 800–4 300 in the present area of reference (Parsons and Tucker 1986), 1 300 eggs (180–3 800) in Greenland (Horsted and Smidt 1956) and 300–1 500 off Northumberland, England (Allen 1966). At reasonably optimum temperatures (3–5° C) females of this species are found to be almost 100% potentially ovigerous in autumn and spawn at least annually (Squires 1965a).

Stomach contents include fragments of crustaceans such as euphausiids, amphipods and copepods, and also benthic organisms such as polychaetes and foraminiferans (Squires 1965a).

Predators on this species in the Northwest Atlantic are mostly cod (*Gadus morhua*) and Greenland halibut (*Reinhardtius hippoglossoides*).

Parasites reported for the species in this area are bopyrid isopods, but their incidence is low. However, protozoans which infect muscle tissues (microsporidia) may cause culling of the catch in some northern areas, while those that cause mortality of eggs may affect recruitment in some years in the Gulf of Maine (Parsons and Khan 1986; Parsons and Fréchette 1989).

This species occurred most frequently in relatively large numbers and with no other species in the catches (Squires 1965a).

FISHERY

Commercial fisheries for this species have developed in several areas where concentrations are sufficient for harvesting. In the Northwest Atlantic these include the Gulf of Maine, the Gulf of St. Lawrence, off the coast of Labrador and in Davis Strait off the coasts of Greenland and Baffin Island. Parsons and Fréchette (1989) give an account of these fisheries. The largest at present is the offshore Davis Strait fishery yielding an average of 41 000 t annually during 1980–88. Yields from other fisheries at present are 11 000 t off Labrador, 14 000 t in the Gulf of St. Lawrence and about 3 300 t in the Gulf of Maine.

In the Pacific, commercial fisheries for several pandalids during the 1970s totalled about 80 000 t for the eastern North Pacific, including about 20 000 t for Alaska; also about 35 000 t for Japanese and Soviet catches in the western North Pacific (Williams 1984).

Pandalus montagui Leach, 1814
Rathbun 1929: 8, fig. 5; Simpson et al 1970:
1230, fig. 1; Williams 1984: 154, fig. 108.
"Striped pink shrimp" — "Crevette esope"
(Figures 133 and 134, Plate 5a)

DISTINGUISHING CHARACTERISTICS

Rostrum with no dorsal spines anteriorly and turned upward slightly, deeper than in *P. borealis* and with larger ventral teeth, tip bifid; second pereopods unequal, the right shorter with about 20 annulations on the carpus; third abdominal somite with no dorsal carina, lobe or spine and no posterior spine on 4th somite dorsally; reddish stripes obliquely across carapace and abdomen (eventually lost in preservative), and a short whitish oblique bar surrounded by darker colour at the hepatic area of the carapace (seen through the integument; D. G. Parsons, pers. comm.).

DESCRIPTION

Integument smooth and shiny but with minute pits and rugosities. Colour pale translucent with reddish stripes obliquely across the carapace and abdomen, and a short whitish oblique bar at the hepatic area surrounded by darker colour. (The latter not seen in females with fully developed ova in the ovaries: D. G. Parsons, personal communication) (Plate 5a).

Rostrum long, 1.3 times cl, moderately deep, slightly ascending distally, with no spines above on distal half but series extending on to median carina of carapace, ventral teeth strong decreasing in size distally (about 6–8/5–6 with 4–5 on carapace); short setae intercalated between spines or teeth on rostrum; lateral carina strong almost confluent with orbit.

Carapace with strong antennal and weak pterygostomian spines; median carina reaches about half carapace.

Abdomen with rounded terga and pleura except fourth and fifth somites with ventroposterior spine, longer and turned down on fifth.

Telson (t) gradually tapering to subtruncate, rounded tip, the latter with three pairs of terminal spines; 4–5 pairs of dorsolateral spines. Inner branch of uropod slightly longer than telson, outer branch slightly longer than inner.

Eye large, cornea globular with ocellus, eyestalk short with small inner tubercle.

Antennule (c) 1st article longer than other two combined, 3rd shorter than 2nd; stylocerite short and rounded and about one-third length of 1st article; distolateral flagellum with terminal narrow part as long as thickened proximal part.

Antenna (d) scale length about 3.5 times width; spine about even with blade; strong distal spine ventrally on basal article. Peduncle less than half scale.

Mandible (e) with 4 teeth on left and 5 on right incisor distally; left molar with 3 rounded cusps and even cutting edge, right with 4 and irregular cutting edge; palp with three segments, distal as long as other two, longer than incisor.

Maxillule (f) proximal endite narrow, slightly curved at tip; distal endite large, rounded; endopod slightly curved and bifurcate.

Maxilla (g) proximal endite reduced, proximal lobe squarish, wide, distal lobe very tiny, pointed; distal endite unequally bilobed, large; endopod short, with short neck; anterior lobe of scaphognathite subrectilinear with concave inner edge; longer than posterior triangular lobe, the latter with very long point apically and long setae along slightly concave side.

Maxilliped I (h) distal endite long, rounded apically, proximal short with thickened edge; endopod longer than lobe of exopod, the latter with long lash; epipod equally bilobed.

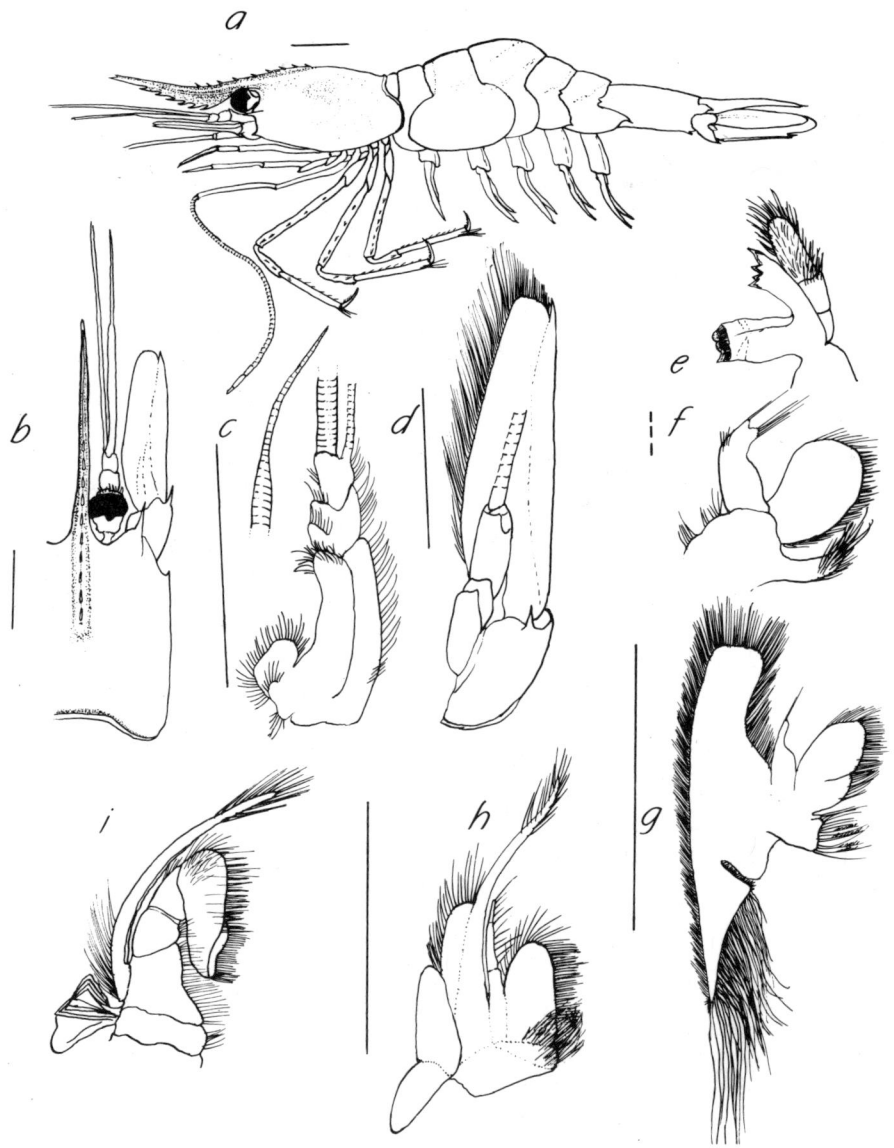

Fig. 133. *Pandalus montagui*: *a*, whole shrimp from left side; *b*, carapace in dorsal aspect; *c*, antennule; *d*, antenna; *e*, mandible; *f*, maxillule; *g*, maxilla; *h*, first maxilliped; *i*, second maxilliped; solid line = 10 mm, broken line = 1 mm.

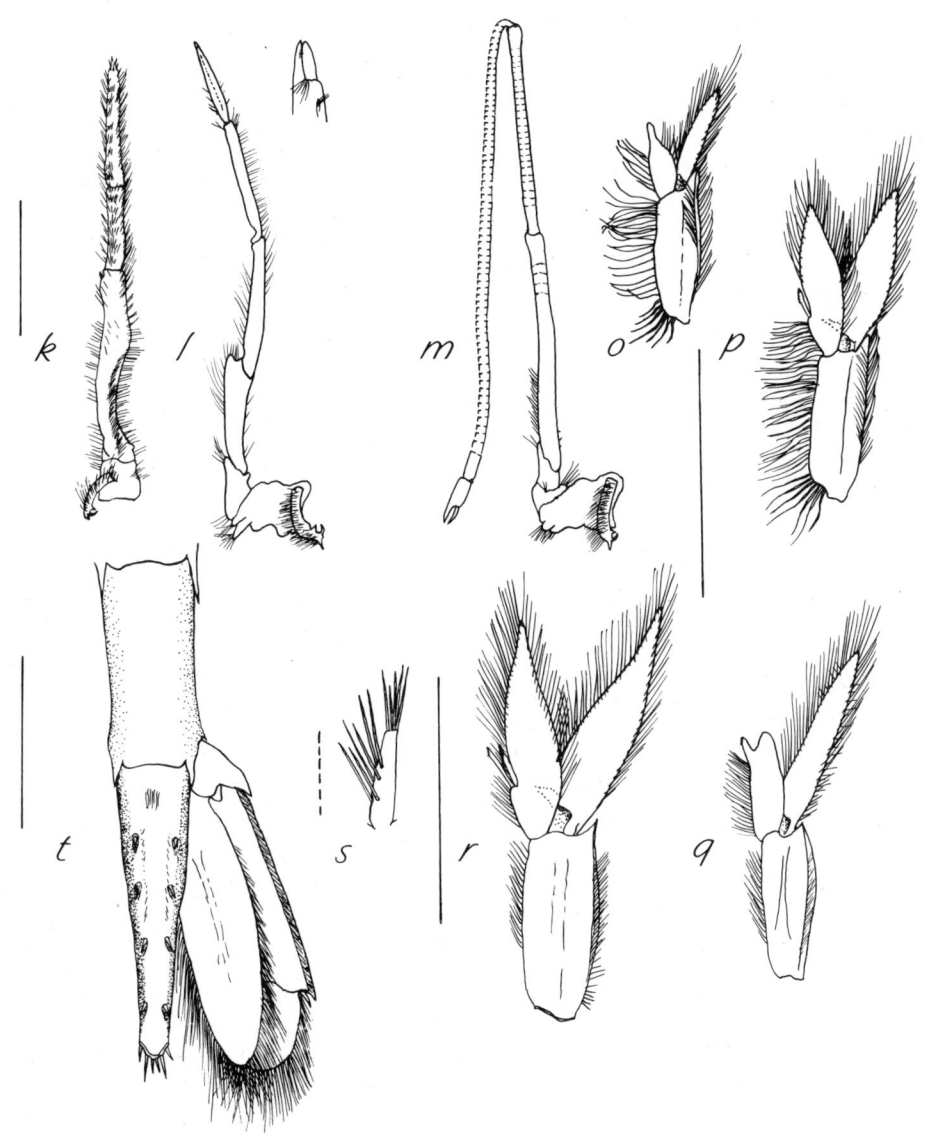

Fig. 134. *Pandalus montagui*: *k*, third maxilliped; *l*, first pereopod; *m*, second pereopod; *o*, F first pleopod; *p*, F second pleopod; *q*, M first pleopod; *r*, M second pleopod; *s*, appendix masculina; *t*, telson; solid line = 10 mm, broken line = 1 mm.

Maxilliped II (i) leglike, compressed, distal segment sinuous in shape, very short but very wide. Exopod long slender. Epipod with podobranch.

Maxilliped III (k) somewhat stouter but shorter than first pereopod; strong spines at tip of distal segment, spine distolaterally on longest segment; epipod straplike, pointed terminally and with grooming claw.

Pereopods: I (l) minutely chelate; propodus long, tapered; ischium expanded distally with distolateral projection; epipod straplike, pointed teminally and with claw. II (m) chelate very much longer than I, right shorter than left, right carpus with about 25 annulations, merus also annulated; epipod with terminal point and claw; III–V stouter than others, long, merus and carpus with strong outer series of spines (one only on V) and propodus with a lateral row of fine spines and setae; dactyl of V shorter than in III and IV, but long and sharp with proximal row of spines. No epipod on V only.

Pleopods: female I (o) narrow exopod longer than fusiform endopod with short neck; female II (p) exopod and endopod subequal, appendix interna short with subapical patch of hooks; male I (q) endopod short with distolateral projection with apical patch of hooks; male II (r) exopod slightly longer than endopod, appendix interna with hooks, slightly longer than appendix masculina (s) with subtruncate apex with five fine spines, separated from two rows of six lateral spines.

RANGE OF DISTRIBUTION

Greenland and Hudson Bay to Rhode Island in the western Atlantic; Iceland, the White Sea and Norway to the western Baltic, the North Sea and British Isles in the eastern Atlantic.

Records of distribution in the area of reference are in Fig. 135.

BIOLOGY

Lengths to 22 mm cl in males and 29 mm cl in females.

As in *P. borealis*, this species is a protandrous hermaphrodite, spending the early part of its life as a functional male and changing to a female following a short transitional phase. In the area of reference populations of this species are found in shallow inshore areas where temperatures are variable but may be sub-zero (°C) for the greater part of the year, and in fishing areas where it was taken temperatures were invariably close to 0° C or where arctic water was in evidence. From inspection of length groups it appeared that in its life history males spent about 1 year as immatures, (7 mm cl) 2 years as functional males (9–12 mm cl) , and passed through an intersex phase (average 12 mm cl) to become females at 3–4 years of age (14–17 mm cl), afterwards producing eggs at 5 (19–21 mm cl) and possibly 6 years of age (22–26 mm cl). Percentages potentially ovigerous would indicate annual spawning (Squires 1965a).

At Gaspé, Québec, where temperatures of the water where *P. montagui* was taken were mostly above 0° C (−0.6 to 14° C), primary females were found, and maturity appeared to take place faster than in areas of lower temperature. Males were first mature at 13 mm cl in their second year and transformed females at an average of 17 mm cl the following year, growth was slower during the spawning years and a maximum size reached was 20 mm cl (Couture et Trudel 1969a; 1969b).

In the eastern Atlantic, however, the species is found in warmer water (4–13° C), especially in the British Isles where there is a small fishery for it (of about 400 t annually) in coastal areas near the Wash and the Thames estuary (Mistakidis 1957). Also in the British Isles the proportion of primary females (30–35%) is high (Allen 1963) and egg production may occur at least twice. Numbers of eggs produced varies according to size of the individual averaging 2 500 for some of the larger shrimp (range 136–3 796). In males maturity may also occur twice before transforming (Allen 1963; 1966). (See also Parsons et al 1983.)

In the area of reference this species was taken in catches with *Eualus macilentus*, *Eualus pusiolus*, *Lebbeus polaris*, *Spirontocaris spinus*, *Pandalus borealis* and *Pagurus pubescens* (Squires 1965a).

FISHERY

A fishery for this species has begun recently (1985) in Hudson Strait, near Resolution Island, and in Ungava Bay, nearby. The yearly amount taken was about 1 000 t in 1987. (Parsons, D. G., personal communication).

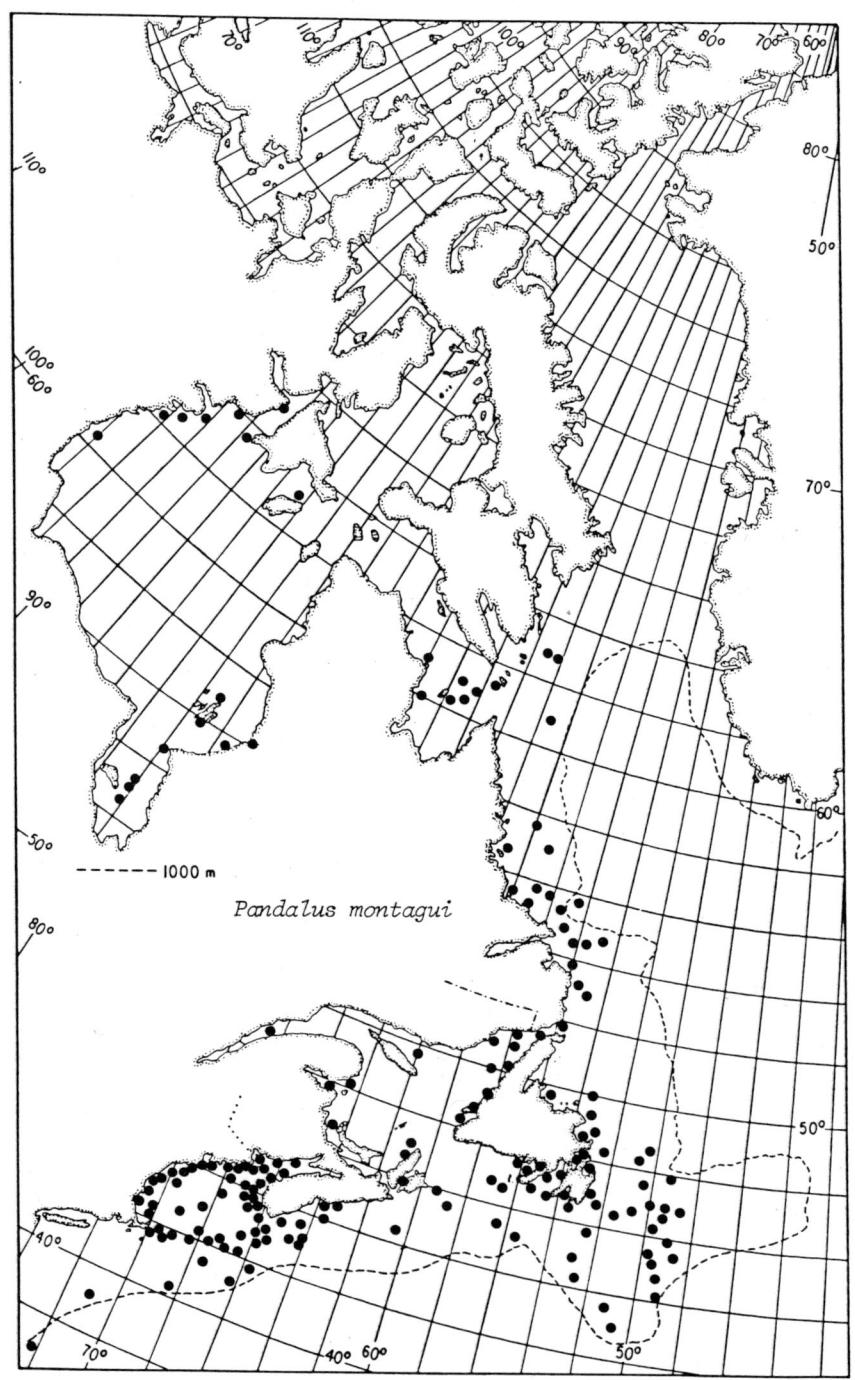

Fig. 135. *Pandalus montagui*, distribution records in the area of reference.

Pandalus propinquus G. O. Sars, 1869
Rathbun 1929: 9, fig.6; Williams 1984: 156, fig.
109; Squires 1965a: 65.
(Figures 136 and 137)

DISTINGUISHING CHARACTERISTICS

Rostrum strongly upturned, tip bifid, no dorsal spines on anterior two-thirds, moveable spines above, very strong teeth below; carpus of 2nd right pereopod with only five divisions, of left with about 25 divisions; scale of antenna strongly tapered and narrow at tip; third to fifth legs longer than second, dactyl moderately stout, spinous.

DESCRIPTION

Integument smooth, shiny. Colour uniform reddish on semi-translucent background, rostrum yellowish towards tip.

Rostrum long and moderately deep, strongly ascending on distal half (at about 45 degrees), no spines above on distal two-thirds, dorsally about 6 moveable spines plus 3 on carapacial carina, ventrally about 7 strong curved teeth proximally longest; lateral carina confluent with orbit.

Carapace with median carina reaching about half way to posterior margin; strong antennal and weak pterygostomian spines.

Abdomen with terga and pleura rounded except ventro-posterior spines on 4th to 6th somites.

Telson (t) slightly tapered, with four pairs of dorsolateral spines and three pairs of terminal spines, second pair of the latter much longer than others. Inner branch of uropod longer than telson and outer branch.

Eye large, cornea globular and with ocellus, much larger than short stalk.

Antennule (c) 1st article longer than other two equal articles, 2nd with distal tooth; stylocerite broad, truncate, less than half 1st article.

Antenna (d) scale length about 4 times width, outer edge concave, strongly narrowed toward truncate tip with spine exceeding blade; basal article with ventral spine distally; peduncle not reaching half scale.

Mandible (e) incisor expanded slightly towards tip, with six teeth on left one; molar long, with 3 rounded cusps at the edge; palp with distal segment squarish at tip.

Maxillule (f) proximal endite tapered and slightly curved, narrow; distal endite subcircular, large; endopod slightly constricted and curved toward bifurcate tip.

Maxilla (g) distal endite large with almost equal lobes; proximal reduced, one lobe a small point, other wide but short and with straight edge; endopod short with curved neck; anterior lobe of scaphognathite subrectilinear distally from wide base, about equal in length to triangular posterior lobe with blunt tip and long setae.

Maxilliped I (h) distal endite squarish but rounded distally, larger than proximal, the latter with thick edge; endopod not as long as lobe of exopod and with narrow apical segment, exopod with long lash; anterior lobe of epipod longer than posterior lobe, large, inflated.

Maxilliped II (i) leglike, compressed; distal segment wider than long, both it and next segment with strong spines; exopod long and slender; epipod spatulate, with podobranch.

Maxilliped III (k) long and slender, about as long as first leg, with a strong distal spine and strong lateral spines on distal segment. Epipod in two parts the terminal straplike with sharp point and subapical grooming claw.

Pereopods: I (l) minutely chelate, slender; ischium expanded distaly at joint with merus; epipod straplike with sharp distal point and subapical claw; setobranch. II (m) (right) chelate, shorter than left, carpus with five divisions (left with about 25 divisions); epipod straplike, with sharp point apically and subapical claw; setobranch. III–V mod-

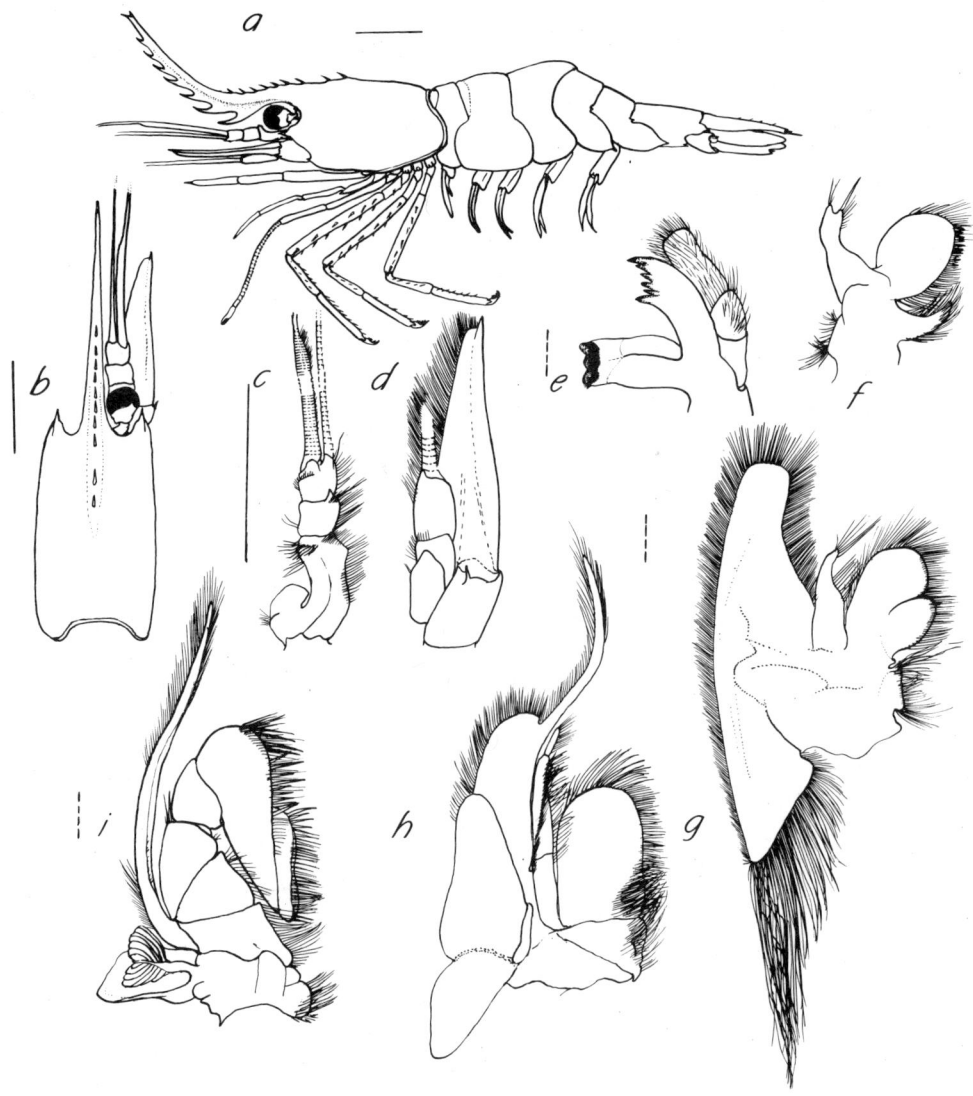

Fig. 136. *Pandalus propinquus*: *a*, whole shrimp from left side; *b*, carapace in dorsal aspect; *c*, antennule; *d*, antenna; *e*, mandible; *f*, maxillule; *g*, maxilla; *h*, first maxilliped; *i*, second maxilliped; solid line = 10 mm, broken line = 1 mm.

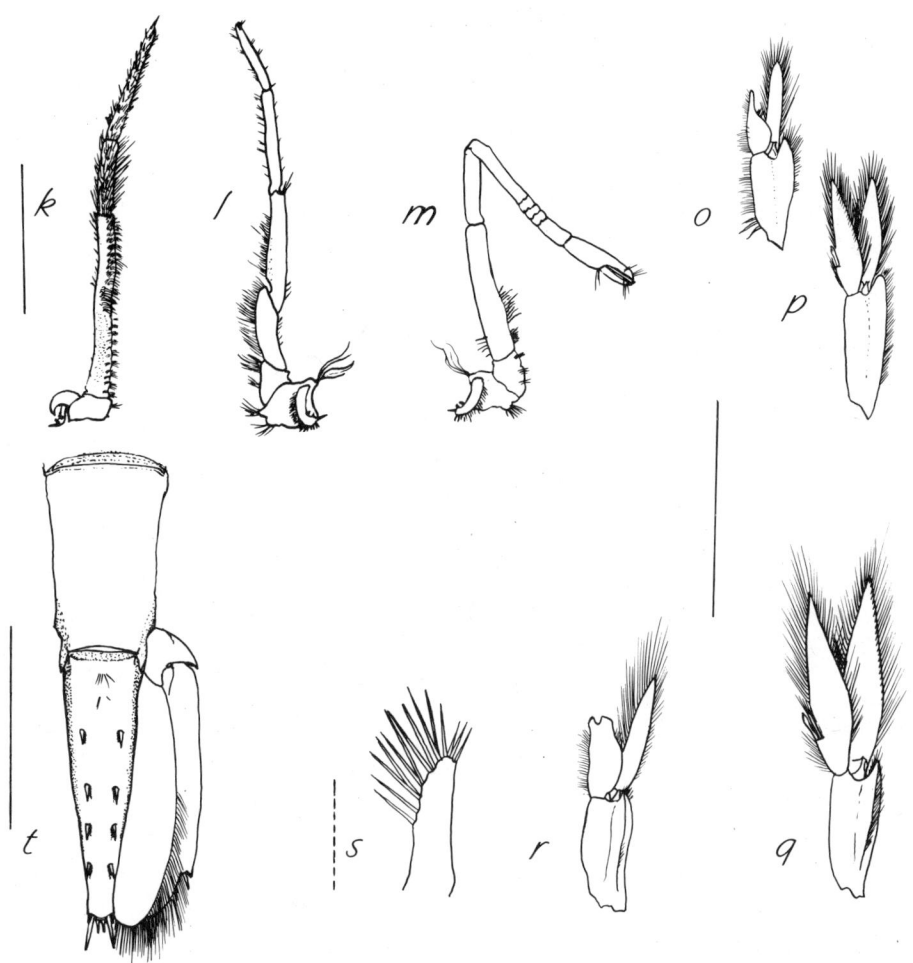

Fig. 137. *Pandalus propinquus*: *k*, third maxilliped; *l*, first pereopod; *m*, second pereopod; *o*, F first pleopod; *p*, F second pleopod; *q*, M first pleopod; *r*, M second pleopod; *s*, appendix masculina; *t*, telson; solid line = 10 mm, broken line = 1 mm.

erately stout, longer than II; two series of spines on merus and carpus and one series on propodus; dactyl robust, short, with strong spines. No epipod on V.

Pleopods: female I (o) narrow exopod longer than endopod, the latter with wide rounded base and neck-like apex. II (p) endopod about equal to exopod; appendix interna short, with distolateral patch of hooks. Male I (q) exopod narrow, longer than wide endopod, the latter with truncate divided apex and inner projection with distolateral patch of hooks; male II (r) appendix interna with distolateral hooks, slightly longer than appendix masculina (s) which has four apical spines continuous with a lateral series of about 10 fine spines.

RANGE OF DISTRIBUTION

From the North Atlantic only; from Davis Strait to Delaware Bay including the Gulf of St. Lawrence and Gulf of Maine in America; and from east Greenland, Iceland and Norway (to 69° N) to the British Isles and Bay of Biscay in Europe (Heegaard 1941; Squires 1965a). Depths 20–2 180 m and temperatures 2.6 to 9.9° C (Williams and Wigley 1977).

Records of occurrence in the area are in Fig. 138.

BIOLOGY

Lengths to 19 mm cl in males and 20 mm cl in females.

Because males and females were both about 9–20 mm in carapace length (80 males and 65 females) in catches throughout the area of reference, it is presumed that this species is not a protandrous hermaphrodite as in *Pandalus borealis* and *P. montagui*.

At all lengths taken some males (49%) were mature (with large vas deferens), and 82% of the females were potentially ovigerous indicating annual spawning (Squires 1965a).

Stomach contents were phytobenthos and crustacean fragments including euphausiids, copepods, amphipods and isopods, as well as polychaetes and small bivalves (Squires 1965a).

Taken in catches with this species were *Dichelopandalus leptocerus* and *Pandalus borealis* in most instances (Squires 1965a).

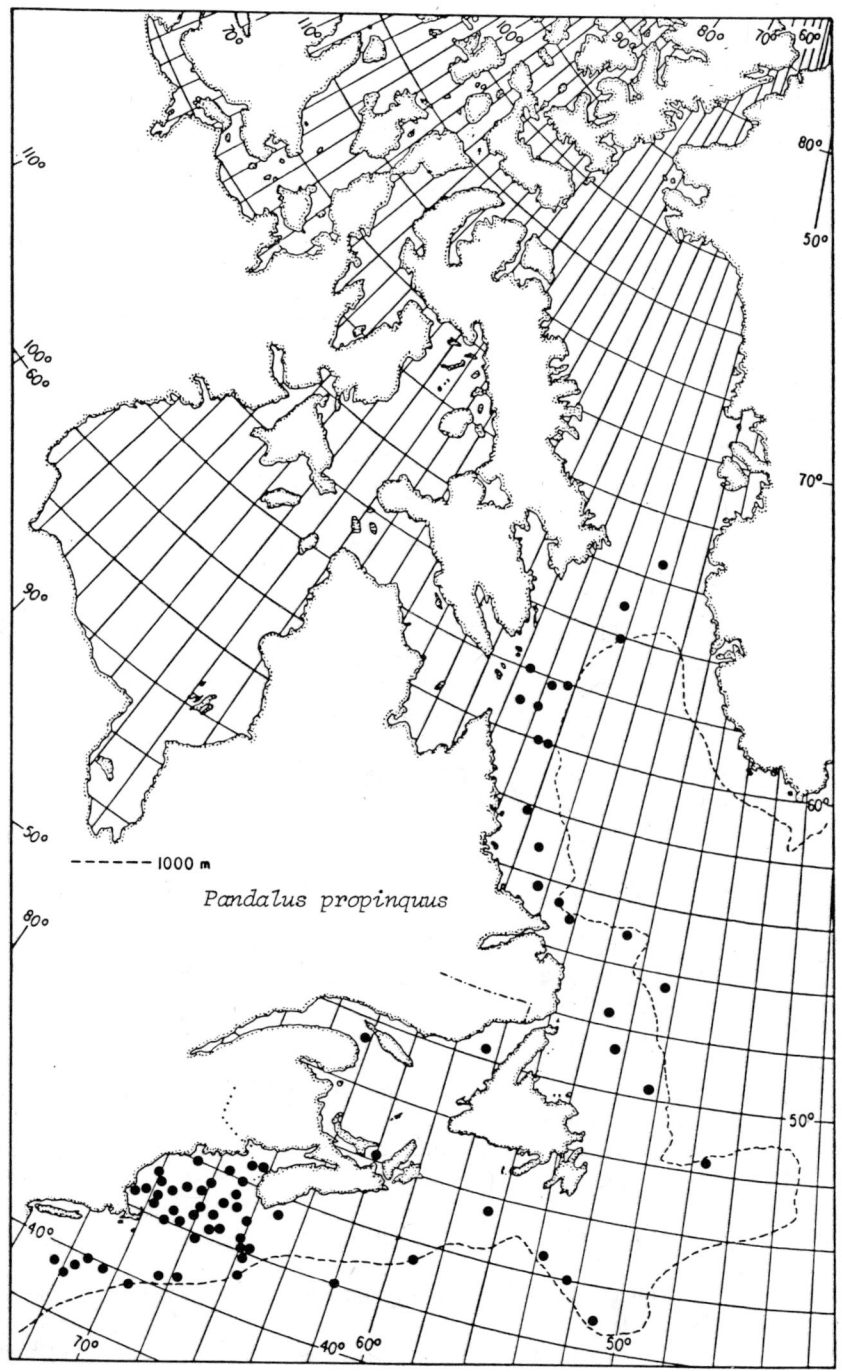

Fig. 138. *Pandalus propinquus*, distribution records in the area of reference.

Genus *Stylopandalus* Coutière, 1905
Chace 1985: 135.

Rostrum very long, attached immovably to carapace, with mostly fixed spines dorsally and ventrally; carapace dorsally carinate on anterior half; 3rd abdominal somite bearing small moveable posterior middorsal spine or stylet; cornea wider than eyestalk; 2nd maxilliped with terminal segment longer than wide; 3rd maxilliped with exopod; pereopods without epipods, 2nd pair subequal, carpus with 11 divisions.

Stylopandalus richardi Coutière, 1905
Chace 1985: 135, fig. 62 a-p.
Parapandalus richardi, Crosnier et Forest 1973:
224: fig. 69 b; Chace 1940: 192, fig. 58–61.
(Figures 139 and 140)

DISTINGUISHING CHARACTERISTICS

Very small shrimp with rostrum about three times longer than carapace; rostrum slender, tapering to sharp point, with many fixed spines above and below; a median posterior spine or stylet on third abdominal somite; 2nd pair of legs subequal, with about 11 divisions of the carpus; exopod on 3rd maxilliped.

DESCRIPTION

Integument evenly pitted or punctate, some of the pits with a short seta; surface scales. Colour bright scarlet red, paler on whitish translucent abdomen.

Rostrum very long, about 3 times as long as carapace, fixed teeth dorsally 15–21, the first 2 separate from the others: a small sutured spine followed by a large fixed spine, proximal spines closer together ventrally following a clear space (16–27); rostrum only slightly compressed, almost round in cross-section.

Carapace with small antennal and pterygostomian spines; as seen from above carapace joins rostrum almost at right angles; median carina reaches almost one-half posteriorly; one carinal spine behind orbit.

Abdomen with rounded terga and pleura; moveable median spine or stylet at posterior edge of 3rd somite; 6th about 3 times as long as 5th, with ventro-posterior spine.

Telson (t) long and narrow, almost as long as 6th somite, tapering to subtruncate tip with a central and three pairs of lateral spines; four pairs of dorsolateral spines of different sizes, the third pair (from apex) largest; inner branch of uropod shorter than telson, the outer longer.

Eye large, cornea globular, larger than eyestalk.

Antennule (c) 1st article longer than other two together, 2nd and 3rd subequal; stylocerite acuminate tip reaching distal 1st article.

Antenna (d) scale length 7 times width, tapering to narrow tip exceeded by strong spine; peduncle only one-quarter length of scale; ventro-distal spine on basal article.

Mandible (e) incisor pointed, tipped with seven teeth partly lateral, the corner ones largest; molar with even edge and corner brush; palp slightly longer than incisor, distal segment longest.

Maxillule (f) distal endite ovate, slightly larger than proximal, the latter slightly curved; endopod bifurcate.

Maxilla (g) distal endite large with two rounded lobes; proximal endite reduced, unilobate, truncate; endopod short, thick at base; anterior lobe of scaphognathite subrectilinear, rounded apically, much longer than axe-shaped posterior lobe with moderate setae.

Maxilliped I (h) distal endite large, proximal short; endopod longer than lobe of exopod, lash long; epipod bilobed anterior one longer, inflated.

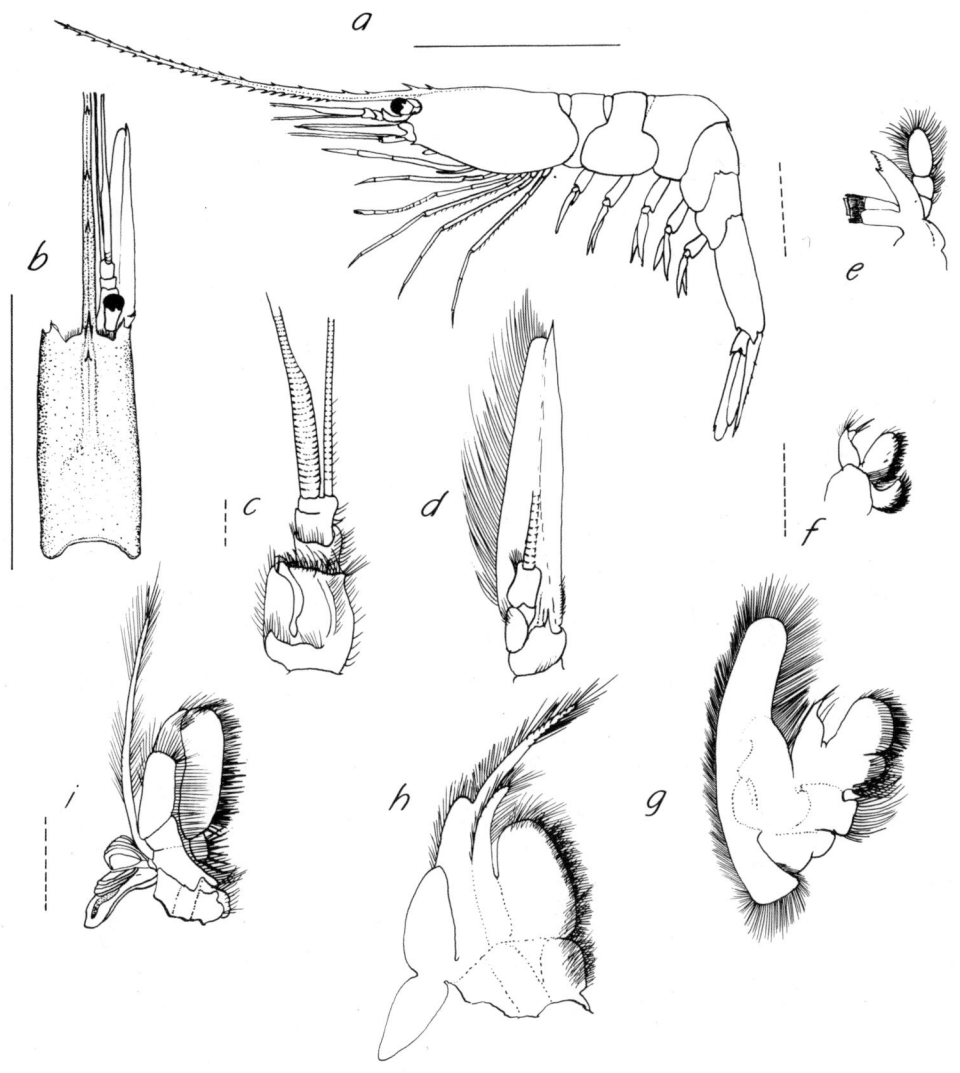

Fig. 139. *Stylopandalus richardi*: *a*, whole shrimp from left side; *b*, carapace in dorsal aspect; *c*, antennule; *d*, antenna; *e*, mandible; *f*, maxillule; *g*, maxilla; *h*, first maxilliped; *i*, second maxilliped; solid line = 10 mm, broken line = 1 mm.

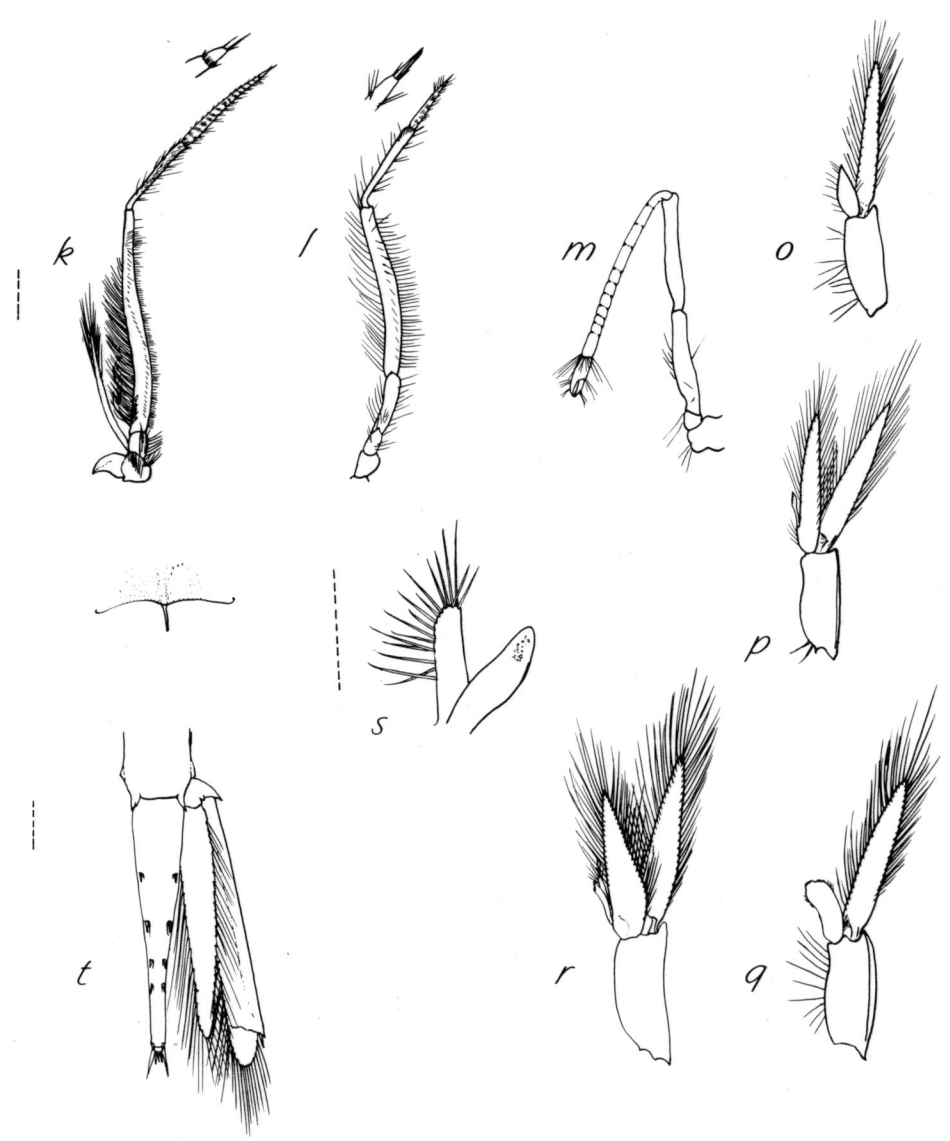

Fig. 140. *Stylopandalus richardi*: *k*, third maxilliped; *l*, first pereopod; *m*, second pereopod; *o*, F first pleopod; *p*, F second pleopod; *q*, M first pleopod; *r*, M second pleopod; *s*, appendix masculina; *t*, telson; broken line = 1 mm.

Maxilliped II (i) leglike, slightly compressed, distal segment longer than wide with strong lateral and apical spines; exopod long and slender; epipod with podobranch.

Maxilliped III (k) leglike, slender, distal segment tapered with sharp spine and seta apically; exopod short, slender; epipod falcate; longer than 1st pereopod.

Pereopods: I (l) slender, without chela but apex of tapered propodus with two strong straight spines and a few setae; merus very long. II (m) chelate, right chela slightly larger, carpus of left with 10 and right with 11 divisions. III–V slender, tapering, merus with series of prominent spinules, dactyl without spines but sharp terminally.

Pleopods: female I (o) narrow exopod much longer than ovate sharp-pointed endopod; female II (p) narrow endopod and exopod subequal, appendix interna slightly expanded at tip with distolateral hooks; male I (q) endopod slightly expanded toward tip, with terminal hooks, much shorter than exopod; male II (r) endopod and exopod subequal, appendix interna with distolateral hooks, slightly shorter than appendix masculina (s) the latter with apex of 6 fine spines continuous with lateral series of about 10 spines.

Range of Distribution

In the western Atlantic from off Newfoundland (48° N) to Bermuda and the Gulf of Mexico; in the eastern Atlantic from the Bay of Cadiz, the Azores, the Canary Islands and Madeira to the west African coast to Angola. Also in the Indo-West Pacific — the Seychelles, Arabian Sea, Gulf of Bengal, Malayasia to west of Australia (Crosnier et Forest 1973). Chace (1985) believes it to be in all tropical and temperate seas at depths of 0–3 600 m.

Records from the area of reference are shown in Fig. 141.

Biology

Lengths to 7 mm cl in males and 9 mm cl in females.

Diurnal migrations have been observed from considerable depths during the day to near the surface at night. Ovigerous females were taken at various stations across the Atlantic from May to July (Sivertsen and Holthuis 1956), and may be ovigerous at 6 mm cl (Crosnier et Forest 1973). Chace (1940) found that 60–100% of adult females were ovigerous, and 15–30% were carrying eyed eggs, in every month in which towing was done (April–September).

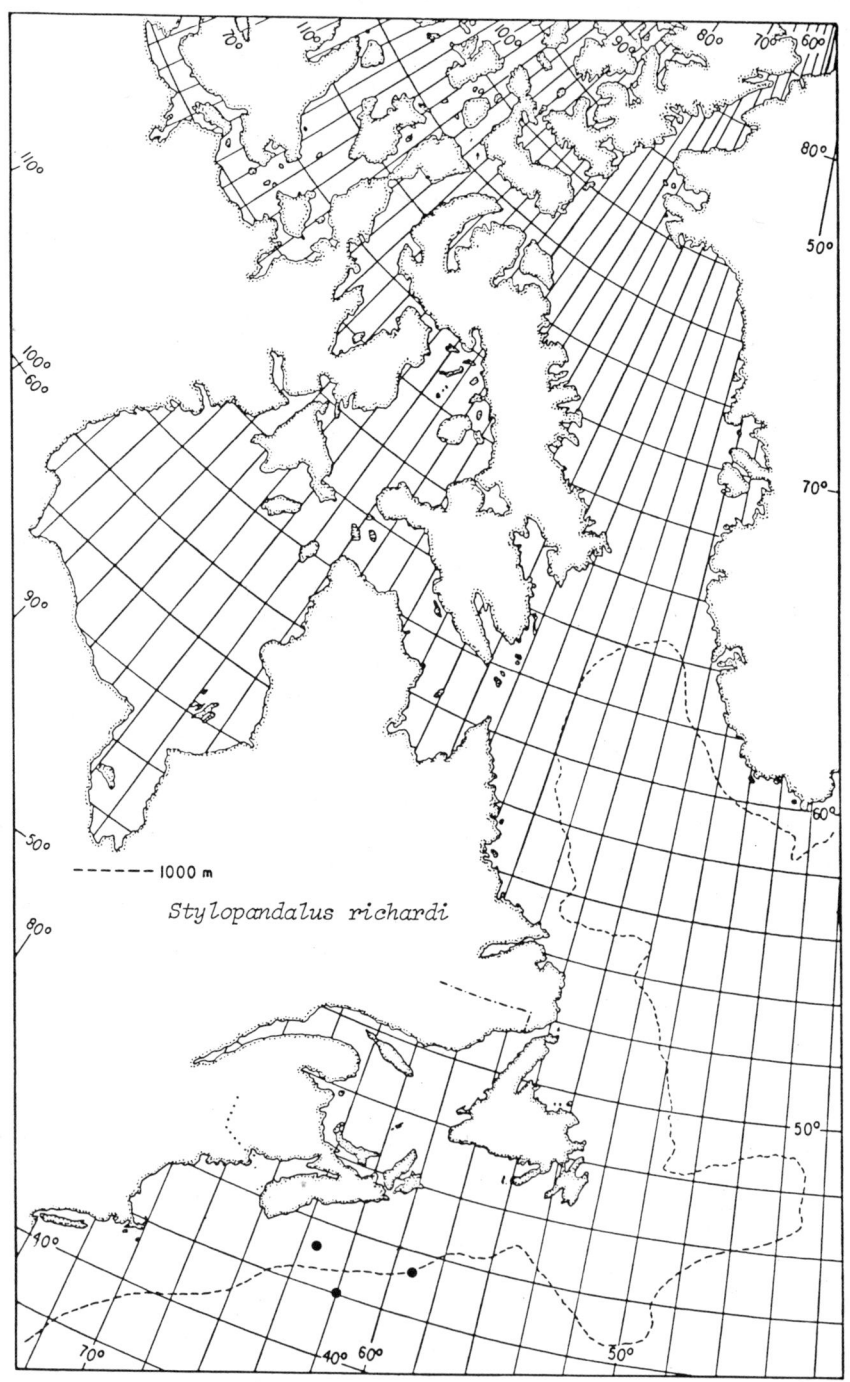

Fig. 141. *Stylopandalus richardi*, distribution records in the area of reference.

Family CRANGONIDAE White, 1847

First pair of legs subchelate, stouter than second. Second legs slender, equal, small chela or none. Mandible without incisor process or palp.

Key to species of the family Crangonidae
(Adapted from Allen, 1967)

1	Second pereopods chelate	4
	Second pereopods not chelate	2
2	Rostrum obtuse	*Sabinea septemcarinata*
	Rostrum acute	3
3	Rostrum short, without spines	*Sabinea sarsi*
	Rostrum with strong spines	*Sabinea hystrix*
4	Dactyl of 4th and 5th pereopods flat and wide; no rostrum, both eyes in a single socket	*Argis dentata*
	Dactyl of 4th and 5th pereopods not flat or wide; rostrum present	5
5	First pereopods with an exopod; two lateral carinae, the first with 2 spines and the second with 1	6
	First pereopod without an exopod	7
*6	Rostrum short, not reaching edge of cornea of eye	*Pontophilus brevirostris*
	Rostrum reaching beyond edge of cornea of eye	*Pontophilus norvegicus*
7	Carapace without strong sculpture	*Crangon septemspinosa*
	Carapace with strong sculpture	8
8	No dorsal carina on abdominal somites 3–5	*Metacrangon jacqueti agassizi*
	Dorsal carina on each of 3rd–5th abdominal somites	9
9	Rostrum short and flattened above, with an acuminate tip and rounded keel	*Sclerocrangon boreas*
	Rostrum prominent with a strong ascending spine, the keel also pointed below	*Sclerocrangon ferox*

Genus *Argis* Kroyer, 1842
Butler 1980: 77, fig. 13
Holthuis 1955: 134.

Rostrum reduced to a spine and both eyes contained in a single socket formed by fusion of the rostral, post orbital and antennal plates or carinae; first leg stout, sub-chelate, second slender, chelate; dactyls of 4th and 5th legs wide, flattened, shovel-like, for burrowing in sand. Sternal plates of 3rd to 5th legs with a long sharp spine triangular in cross-section.

Of the 12 known species only one is recorded from the area of reference.

Argis dentata (Rathbun, 1902)
Rathbun 1929: 21, fig. 27, *a-d*.
Squires 1964b: 461-466, fig. 1B.
(Figures 142 and 143, Plates 5b and 6a)

DISTINGUISHING CHARACTERISTICS

Both eyes included in a single socket which almost covers them like a hood, rostrum reduced to central spine of socket; body stout, depressed, carina on carapace with 2 spines, carinae on 2nd to 5th abdominal somites, two on 6th, the latter ending in strong spines posteriorly; 1st legs sub-chelate, stout; 2nd legs chelate, slender; dactyls of 4th and 5th legs wide and flat, shovel-like.

DESCRIPTION

Integument thick, rigid, appearing smooth but punctate and in some parts of carapace covered with minute spines pointing forward. Colour brownish bands and mottling on greyish background (Plates 5b and 6a).

Front of carapace slightly rising to form single eye socket with blunt rostrum exceeded by fused postorbital lobe and antennal carina with spine; strong branchiostegal and moderate hepatic spines with carinae; also mid-dorsal carina with 2 strong fixed spines, and lateral carina from antennal plate to posterior margin; hollows in front of hepatic spine and above lateral carina. Pterygostomian spine present.

Abdomen with single median carina on 1st to 5th somites and two parallel carinae on 6th; these carinae have no spines but are pointed posteriorly on 5th and 6th; pleura are rounded except on 4th and 5th with ventro-lateral spine posteriorly; ventrally are central spines on 1st to 4th somites and an anal spine.

Telson (t) with prominent lateral ridges and deep sulcus between them, two pairs of dorso-lateral spines on distal third and a terminal triangular-shaped central spine with a pair of lateral spines at base. Outer branch of uropod with slightly convex outer edge, no axillary spine, shorter than inner branch, both shorter than telson.

Eyes moderate, eyestalk slightly wider than cornea, small tubercle.

Antennule (c) 1st article exceeds others combined, 3rd shorter than 2nd; stylocerite wider than article, with blunt point; dorso-lateral flagellum shorter and narrower than ventro-mesial.

Antenna (d) scale length is only about twice width, spine slightly exceeding blade, outer edge slightly convex; peduncle reaches as far forward as distal blade; spine ventrally on basal article.

Mandible (e) molars similar, three long pointed cusps, the longest with a subsidiary tooth. No incisor or palp.

Maxillule (f) proximal endite rounded, distal curved, slightly tapering; endopod truncate at tip, with lateral expansion.

Maxilla (g) endites greatly reduced to slightly rounded areas; endopod as long as anterior lobe of scaphognathite, about equal to long posterior lobe, the latter with long distal setae.

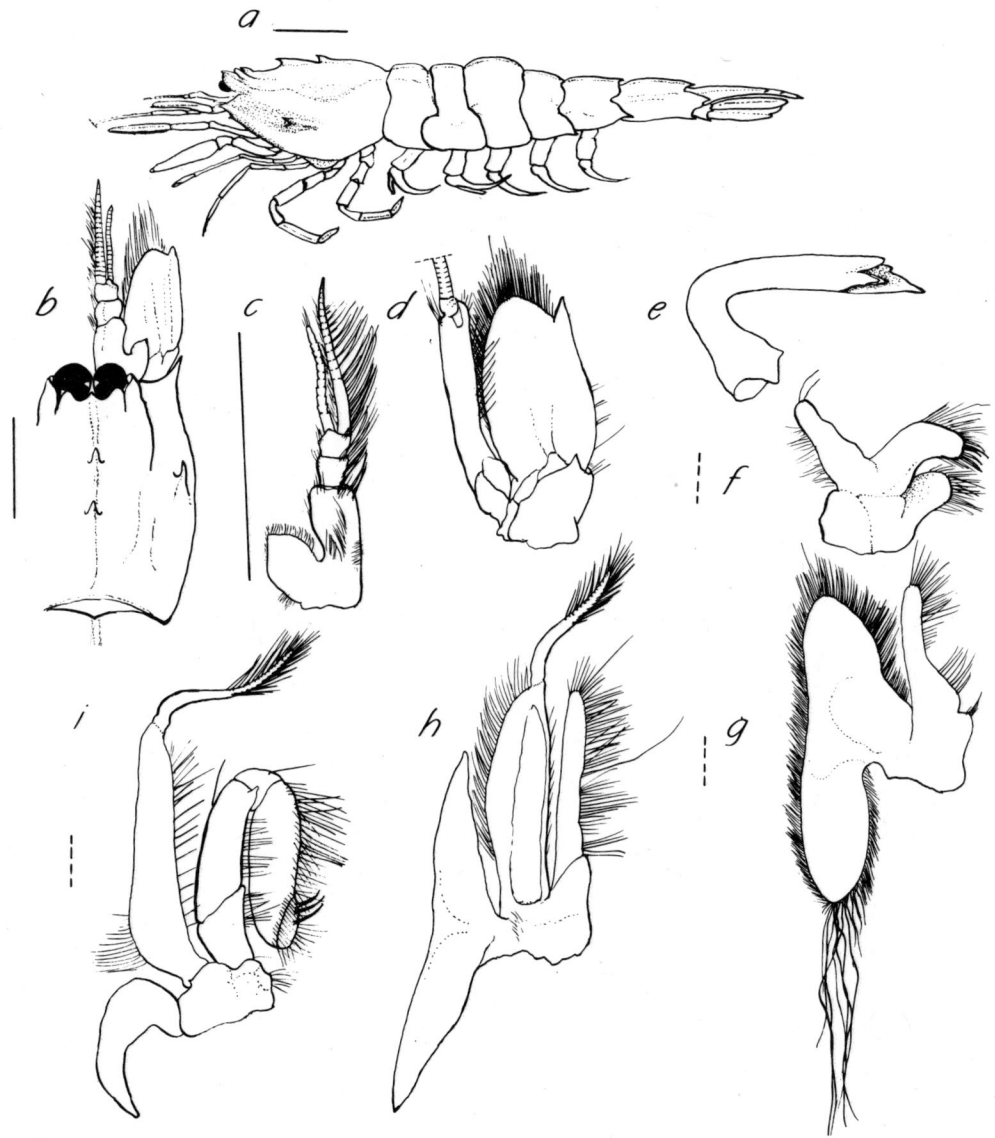

Fig. 142. *Argis dentata*: *a*, whole shrimp from left side; *b*, carapace in dorsal aspect; *c*, antennule; *d*, maxilla; *e*, mandible; *f*, maxillule; *g*, maxilla; *h*, first maxilliped; *i*, second maxilliped; solid line = 10 mm, broken line = 1 mm.

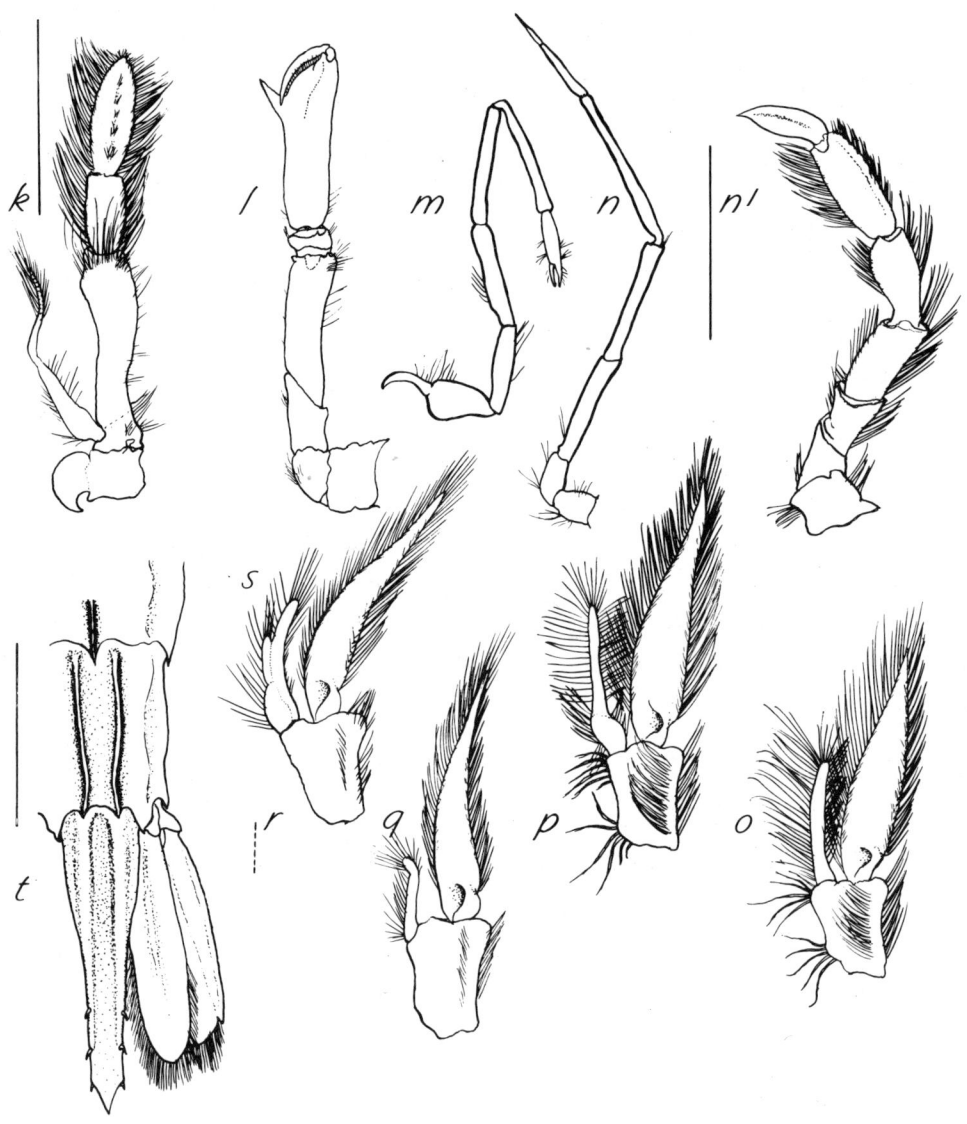

Fig. 143. *Argis dentata*: *k*, third maxilliped; *l*, first pereopod; *m*, second pereopod; *n*, third pereopod; *n1*, fifth pereopod; *o*, F first pleopod; *p*, F second pleopod; *q*, M first pleopod; *r*, M second pleopod; *s*, appendix masculina; *t*, telson; solid line = 10 mm, broken line = 1 mm.

Maxilliped I (h) endites completely reduced; endopod very long, reaching tip of lamellate part of exopod, latter with moderate lash; epipod with long narrow, pointed lobes.

Maxilliped II (i) leglike, slightly compressed, distal segment small, attached diagonally to long next segment and with three prominent curved spines; exopod long, with lash; epipod moderate, falcate.

Maxilliped III (k) stout, distal segment wide, shovel-like; exopod about as long as longest segment; epipod stubby, falcate.

Pereopods: I (l) subchelate, stout. II (m) chelate, slender, shorter than I; epipod narrow, distally falcate. III (n) very slender, dactyl simple, longer than I and II. IV and V (n') longer and stouter than II and III, with compressed and widened propodus and dactyl, the latter shovel-like.

Pleopods: female I (o) endopod much shorter and narrower than exopod; female II (p) endopod much narrower and shorter than exopod, expanded and rounded proximally apparently of two segments (no appendix interna on any pleopod). Male I (q) endopod narrow, short, slightly curved distally; male II (r) endopod with thicker proximal segment, the distal with partly fused appendix masculina(s) at inner edge, both with long slender spinous setae apically and laterally.

RANGE OF DISTRIBUTION

Hudson Bay, Canadian Arctic islands and northwest Greenland to Nova Scotia in the North Atlantic; Arctic Ocean from Cambridge Bay through Beaufort Sea, and Bering Sea to the Sea of Japan and San Juan Islands (Washington) in the North Pacific (Butler 1980; Squires 1965 a, b; 1967a; 1968a; 1969). Depths intertidal to 2 000 m (Vinogradov 1950).

Records of occurrence in area of reference are in Fig. 144.

BIOLOGY

Lengths of mature males 10–17 mm cl and mature females 10–27 mm (Squires 1965a). These populations from temperatures mostly of less than 0° C (-1.4 to $1.8°$ C) had low numbers of females potentially ovigerous in autumn (about 35% in 49 specimens examined).

Fréchette et al (1970) demonstrated that in some populations (temperatures -0.2 to $3.2°$ C) this species is protandrous with a change of sex from male to female of some individuals, while others are primary females, and it was suggested that males lived only 2 years while females reached 5 years of age.

Stomach contents were phytobenthos, crustacean fragments, foraminiferans, small bivalves, gastropods, ostracods and polychaetes (Squires 1965a).

Predators on this species were cod, Brunnich's Murre at Akpatok Island, Beluga whales in Hudson Bay and ringed, bearded and harbour seals in Ungava Bay (Squires 1957; 1965a; 1967a. Tuck and Squires 1955).

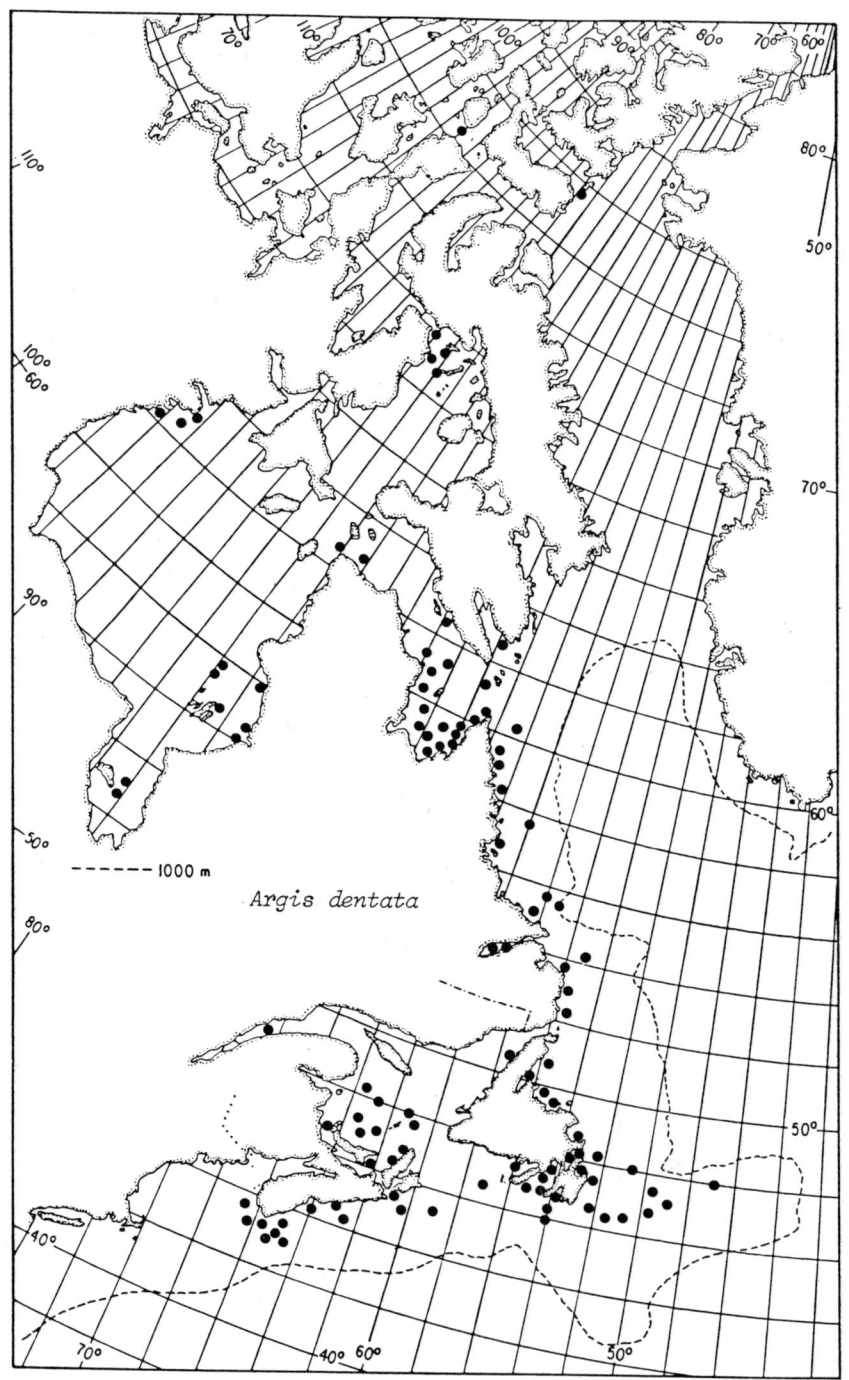

Fig. 144. *Argis dentata*, distribution records in the area of reference.

Genus *Crangon* Fabricius, 1798
Butler 1980: 95, fig. 14; Holthuis 1955: 134.

Mottled greyish shrimp with relatively short carapace, depressed, first legs stoutish subchelate; rostrum short, tip rounded, hollowed dorsally; one spine on median carapace a short distance behind rostrum; one hepatic spine lateral to median spine; eyes large in separate orbits; 4th and 5th legs longer and stouter than 2nd and 3rd, dactyls slightly flattened; 3rd maxilliped with arthrobranch.

Crangon septemspinosa Say, 1818
Crago septemspinosus, Rathbun 1929: 20, fig. 24
Crangon septemspinosa, Haefner 1979: 2; Squires 1965a: 78; Squires and Figueira 1974, pl. 1, fig. 1.
(Figures 145 and 146)

DISTINGUISHING CHARACTERISTICS

A greyish shrimp with body depressed, first legs stout subchelate, three conspicuous spines on the carapace: one median dorsal behind the short rostrum and the other two at each side (in the hepatic region); eyes large in separate orbits; antennal scale large; 4th and 5th legs longer and stouter than 2nd and 3rd; abdomen as wide as carapace, smooth, tapering to slender 6th somite and telson.

DESCRIPTION

Integument finely punctate covered with many short setae. Colour sandy grey mottling but variable in lightness, dark spots on each side of 4th abdominal somite and a dark spot dorsally on 5th somite.

Rostrum short, about 0.2 cl, as long as eyestalk, hollowed dorsally and rounded at the tip with many long setae below; sharp lateral edge is confluent with orbit but forms a low carina sweeping back from the orbital fissure to about half carapace.

Carapace with low median carina extending from base of rostrum to moderate median spine on anterior fifth; hepatic spine with short carina; moderate antennal, strong branchiostegal and weak pterygostomian spines; ventral edge of carapace forming slight lobe at second pereopod; sternum of 3rd pereopods with strong spine.

Abdomen with smooth rounded terga and pleura; 6th somite about 0.8 times telson. Midventral spine on 1st to 4th somites; anal spine.

Telson (t) tapering to subtruncate tip with central fixed spine flanked by 3 pairs of moveable spines; proximally a faint sulcus and two pairs of small dorsolateral spines on posterior third; outer branch of uropod with slightly convex edge slightly shorter than inner branch and about equal to telson.

Eyes large, cornea globular, dorsal tubercle.

Antennule (c) 1st article slightly longer than other two combined, 3rd longer than 2nd; stylocerite wide at base subtriangular, point reaching distal 1st article; dorsolateral flagellum about as long as peduncle, ventromesial longer.

Antenna (d) scale length 2.8 times width; distolateral spine exceeds blade; peduncle about half scale; flagellum equal to total body length.

Mandible (e) molar with two rounded cusps equal in width to longer pointed cusp with subsidiary tooth.

Maxillule (f) proximal endite small rounded, distal large curved; endopod only slightly curved near tip with subapical spine.

Maxilla (g) endites reduced to slight expansion at base of endopod which is moderate in length and curved slightly; scaphognathite with subtriangular anterior lobe wide at base, posterior lobe more narrow, rounded and with setae only a little longer than mesial fringe.

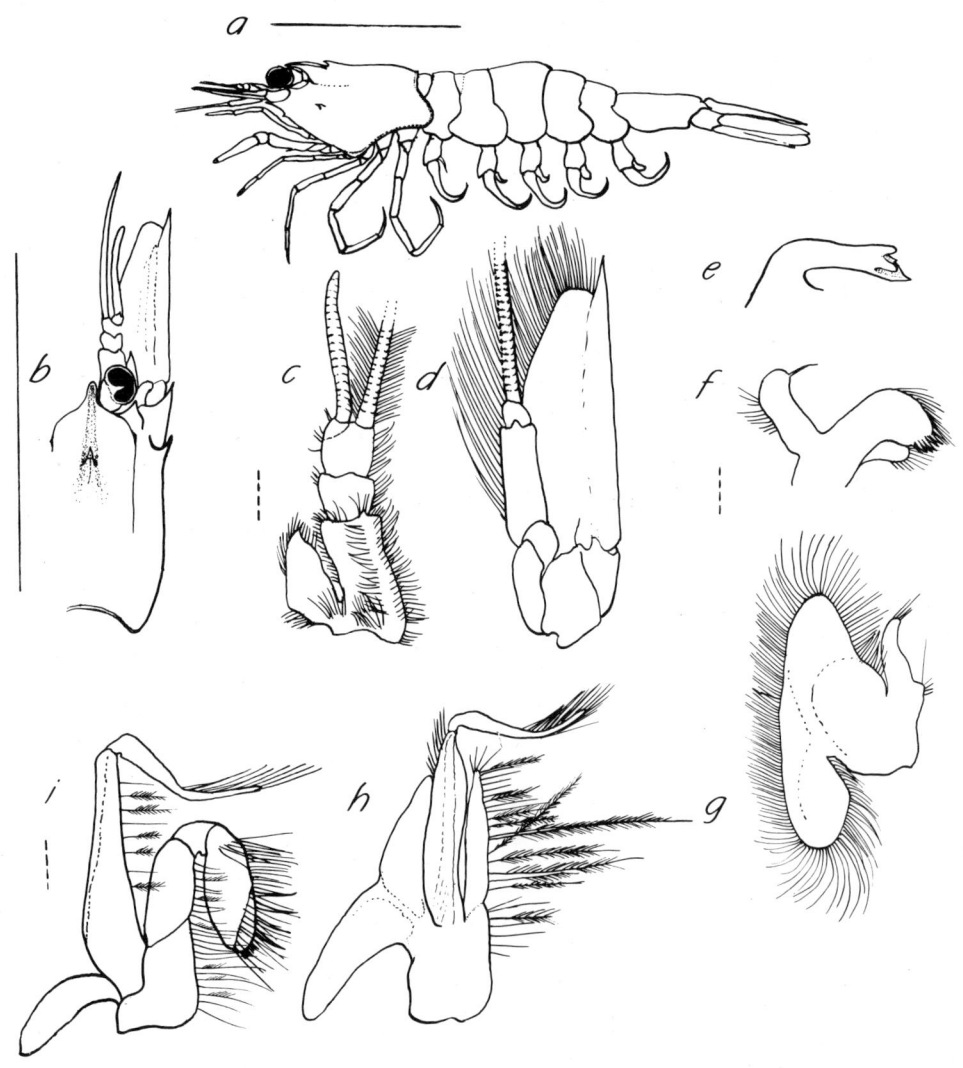

Fig. 145. *Crangon septemspinosa*: *a*, whole shrimp from left side; *b*, carapace in dorsal aspect; *c*, antennule; *d*, antenna; *e*, mandible; *f*, maxillule; *g*, maxilla; *h*, first maxilliped; *i*, second maxilliped; solid line = 10 mm, broken line = 1 mm.

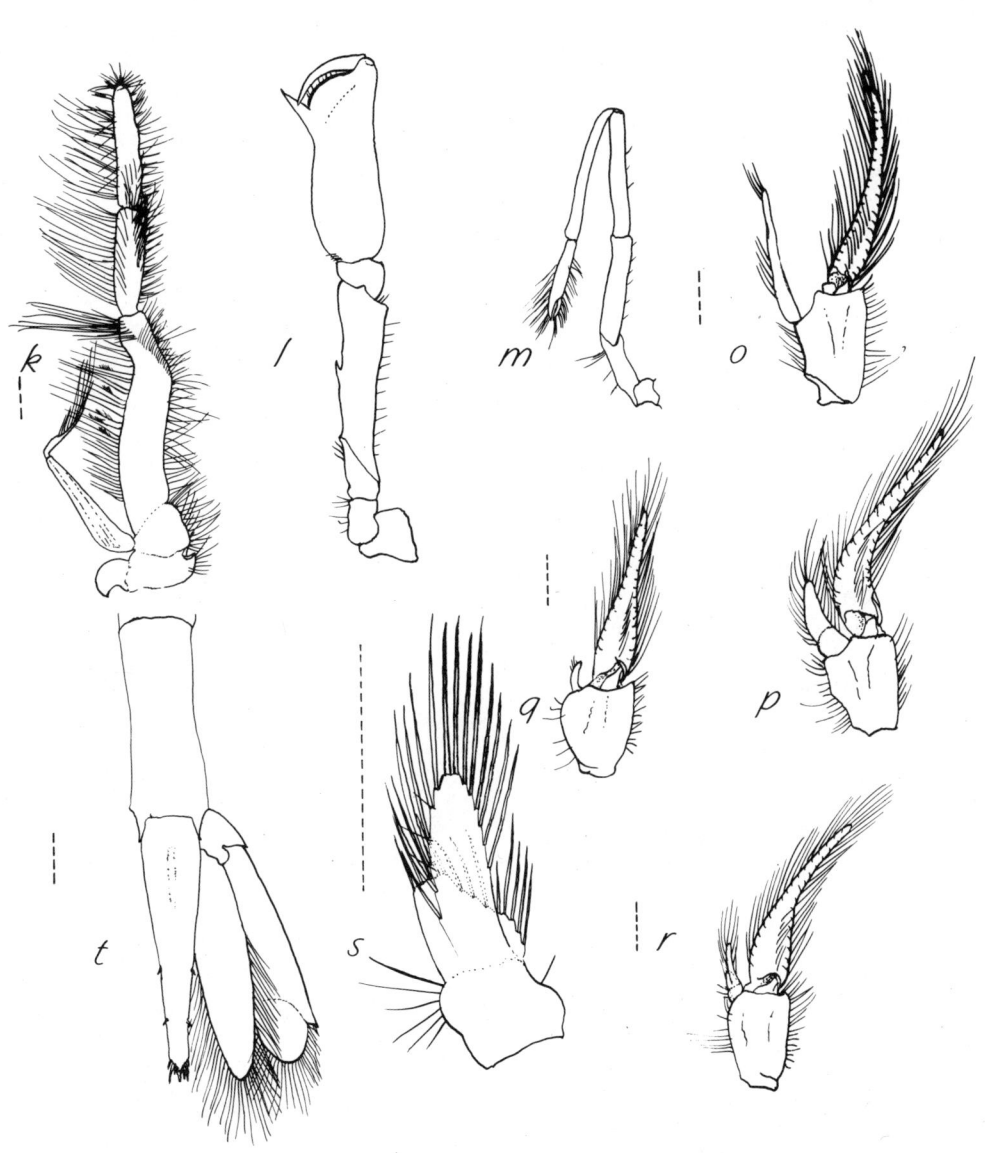

Fig. 146. *Crangon septemspinosa*: *k*, third maxilliped; *l*, first pereopod; *m*, second pereopod; *o*, F first pleopod; *p*, F second pleopod; *q*, M first pleopod; *r*, M second pleopod; *s*, appendix masculina; *t*, telson; broken line = 1 mm.

Maxilliped I (h) endites completely reduced, endopod long with long lateral plumose setae; exopod longer than endopod and with moderate lash; long narrow bilobed epipod each lobe subtriangular.

Maxilliped II (i) endopod compressed, five segments, the distal attached diagonally to next and with four strong spines; exopod long with long lash; epipod a single lobe.

Maxilliped III (k) stout, as long as first leg, with five segments, the distal with apical crown of setae; exopod moderate, with apical curved lash; epipod rudimentary, falcate.

Pereopods: I (l) stout, massive palm, subchelate; middle fixed spine on merus. II (m) slender, chelate, about as long as I; III (a) slender, slightly longer than II, finely tapered distally; IV and V more robust than II and longer, propodus and dactyl slightly compressed.

Pleopods: female I (o) endopod slender, about one-third length of exopod; female II (p) endopod subtriangular about one-quarter length of exopod. Male I (q) endopod very short and narrow, exopod long, broad at base but tapered to slender tip; male II (r) endopod appearing two-segmented, much shorter than exopod, tapered and truncate, shorter than appendix masculina (s), the latter slightly tapered with two long apical spines and about six long lateral spines on each side.

RANGE OF DISTRIBUTION

From northern Gulf of St. Lawrence to east Florida, in North Atlantic only. Depths 0–90 m, rarely to 450 m.

Records of distribution in the area of reference are in Fig. 147.

BIOLOGY

Lengths to 12 mm cl in males and females; body lengths males to 47 mm, females to 70 mm (Price 1962).

Females were first mature at 4 mm cl and produced two batches of eggs during the summer at Port au Port Bay, Newfoundland, in the eel grass (*Zostera* sp.) community where water temperatures were 8–25° C from May to September (Squires 1965a).

Other decapod crustacean species present occcasionally were *Pagurus acadianus* and *Cancer irroratus*.

Stomach contents were mainly mysids but also amphipods, small gastropods, and small bivalves (Squires 1965a).

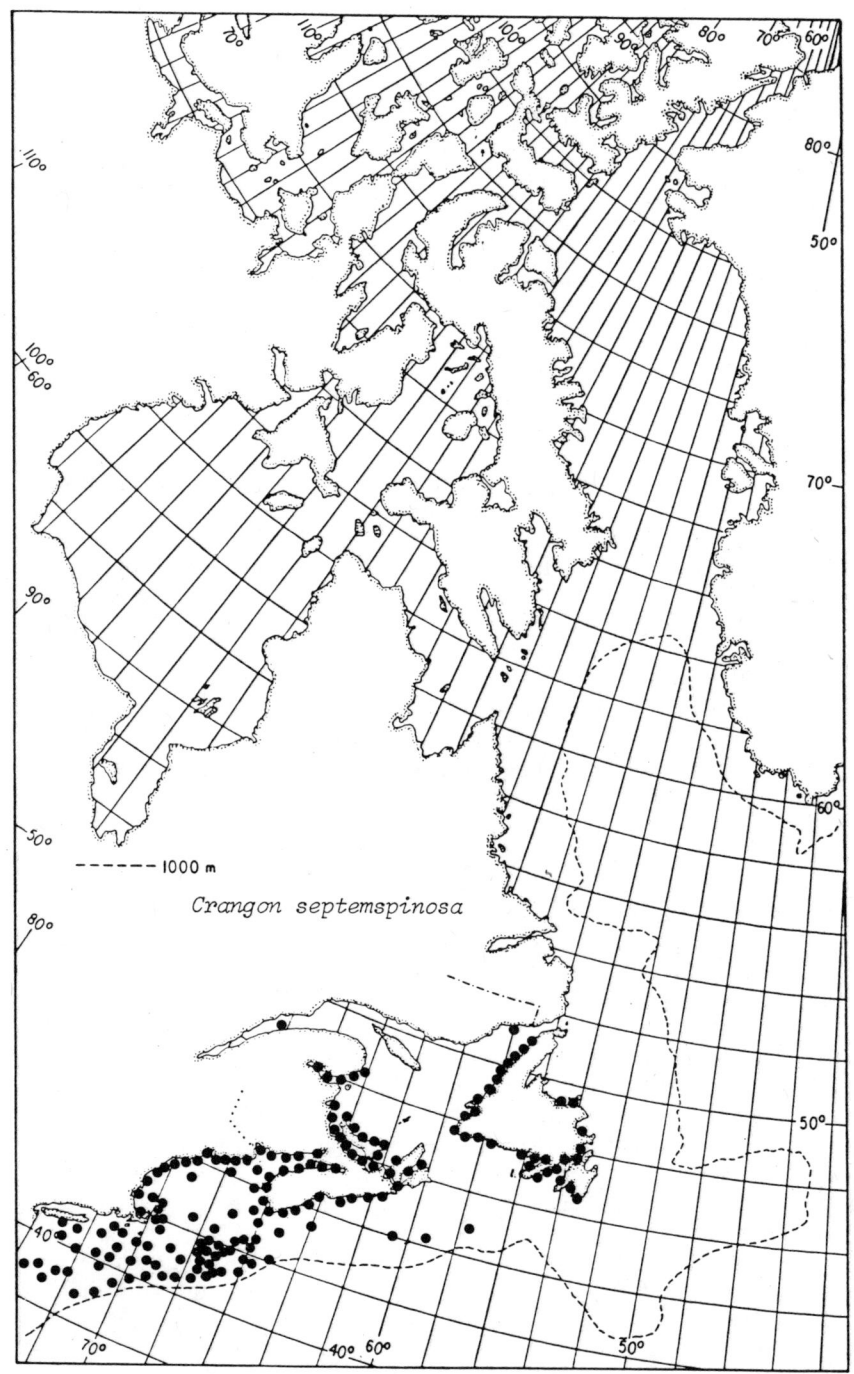

Fig. 147. *Crangon septemspinosa*, distribution records in the area of reference.

Genus *Metacrangon* Zarenkov, 1965
Butler 1980: 113, fig. 16; Zarenkov 1965:
1764, Fig. 4; Crosnier et Forest 1973: 233.

Carapace strongly carinate and with strong spines, abdomen also carinate and with pleural spines ventrally; rostrum short and horizontal with a very strong spine behind it which reaches over and ahead of it; strong branchiostegal and hepatic spines with prominent carina; a submedian spine with short carina; postero-ventral spines of 6th abdominal somite supported by flared ridges.

Metacrangon jacqueti agassizi (Smith, 1882)
Crosnier et Forest 1973: 233-239, fig. 76c.
Sclerocrangon jacqueti (A. Milne-Edwards),
Sivertsen and Holthuis 1956: 40.
(Figures 148 and 149)

DISTINGUISHING CHARACTERISTICS

Carapace and abdomen sculptured, carinate and with strong median frontal spine rising above the short rostrum; two other median spines on the carapace, the middle one small, also a submedian spine at each side; branchiostegal and hepatic spines strong and with supporting carinae; 6th abdominal somite with two dorsal carinae not quite reaching posterior edge; pleura of somites 1–3 with ventral spine, 5 and 6 with ventroposterior spine and 6 with posterolateral spine.

DESCRIPTION

Integument rough in appearance, punctate and with short curled setae, especially in hollows. Colour whitish brown with yellowish markings.

Carapace strongly sculptured; three mid-dorsal spines, the large anterior rising above and exceeding the short horizontal rostrum, the large posterior near the posterior edge of the carapace, and between the two a small spine: these spines are very thin in cross-section; at the anterior edge are strong branchiostegal and antennal spines each with a prominent carina; also submedian spine with short carina and hepatic spine with long carina. Anteriorly a hollow on each side of anterior median spine and posteriorly a hollow and cross ridge on each side of posterior median spine. Ventrally a central rounded lobe at base of 3rd pereopod.

Abdominal somites each with a mid-dorsal carina — the 6th with two — and lateral carina at level of articulation; pleura of 1st to 3rd with a single ventral spine, 4th rounded, 5th ventro-laterally pointed sharply and 6th with posterior vental spine and flared supporting carina. Also 6th has sharp curved spines above lateral to base of telson and two dorsal carinae which do not reach the rounded posterior end.

Telson (t) flat, tapering to small lateral spines and sharply tapered tail with two plumose setae at each side; also two pairs of tiny dorso-lateral spines on distal half. Inner branch of uropod shorter than telson but longer than outer branch

Eye large, cornea globular .

Antennule (c) 1st article longer than others combined, 2nd longer than 3rd; dorso-lateral flagellum longer than ventromesial; stylocerite curved, blunt point not reaching distal 1st article.

Antenna (d) scale length 2.4 times width, disto-lateral spine exceeding blade; peduncle with 3rd segment long, almost reaching end of scale; basal article with two strong ventro-lateral spines.

Mandible (e) molar with two sharp fangs, each with low subsidiary tooth.

Maxillule (f) proximal endite small, tapering to rounded point; distal bent at middle, moderate; endopod straight with subapical point with setae.

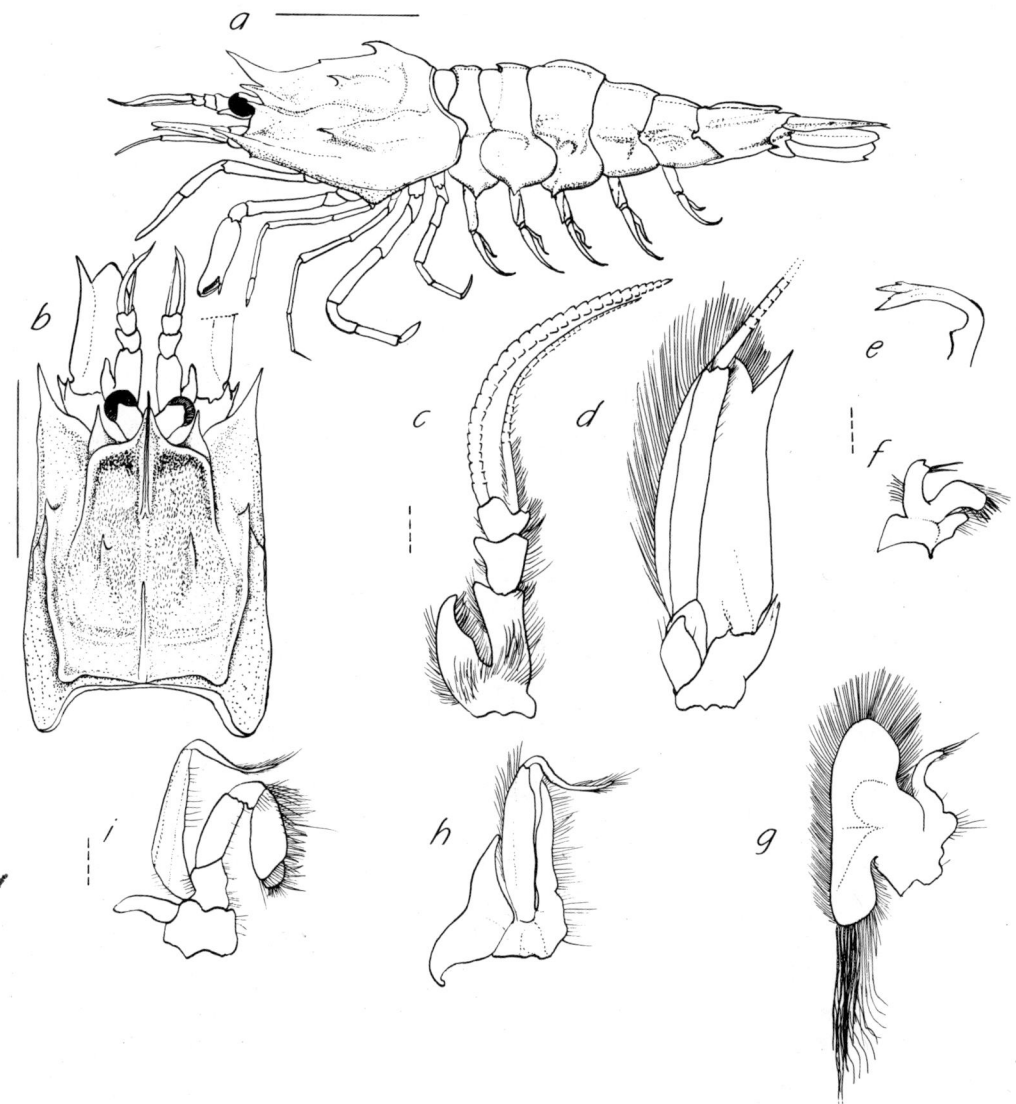

Fig. 148. *Metacrangon jacqueti agassizi*: *a*, whole shrimp from left side; *b*, carapace in dorsal aspect; *c*, antennule; *d*, antenna; *e*, mandible; *f*, maxillule; *g*, maxilla; *h*, first maxilliped; *i*, second maxilliped; solid line = 10 mm, broken line = 1 mm.

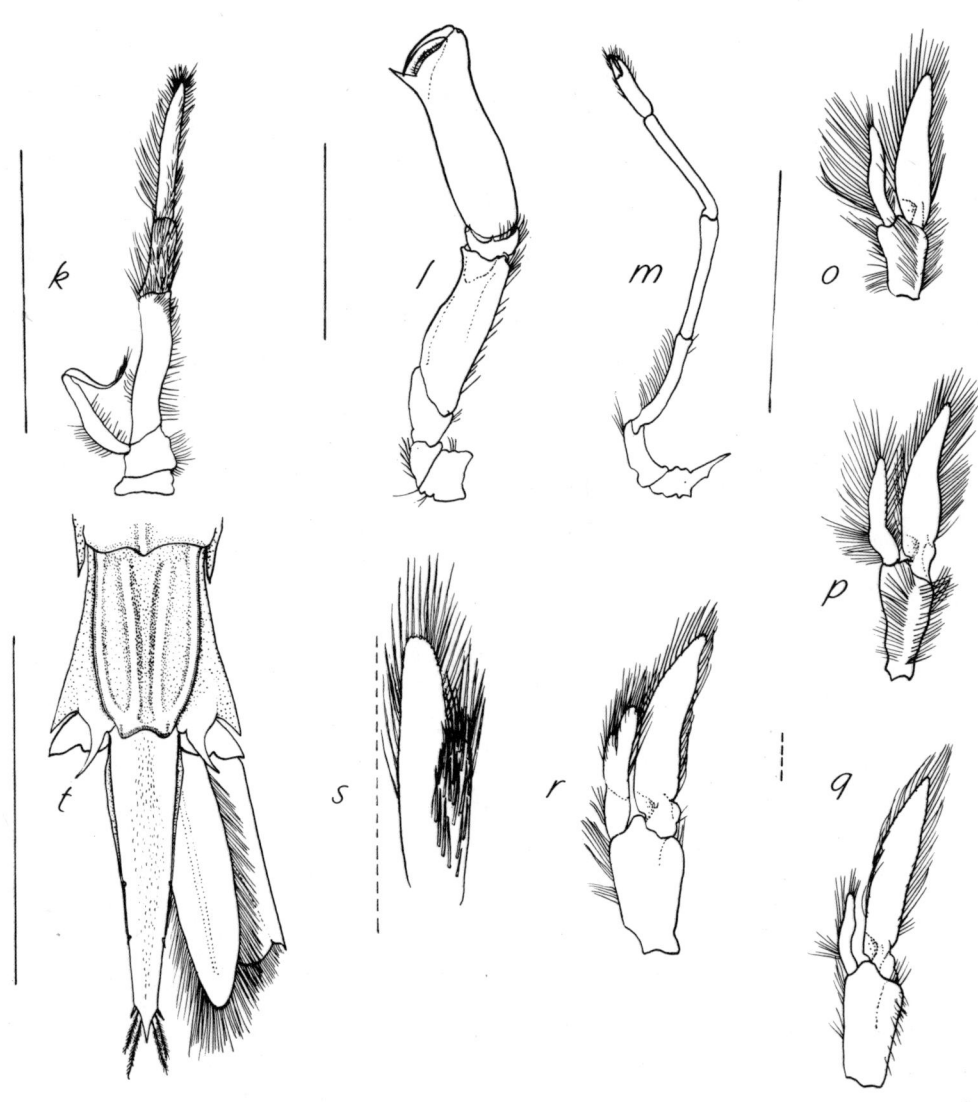

Fig. 149. *Metacrangon jacqueti agassizi*: k, third maxilliped; l, first pereopod; m, second pereopod; o, F first pleopod; p, F second pleopod; q, M first pleopod; r, M second pleopod; s, appendix masculina; t, telson; solid line = 10 mm, broken line = 1 mm.

Maxilla (g) distal endite reduced to rounded projection with few setae; proximal almost completely reduced; endopod short, curved; anterior lobe of scaphognathite wide, slanted inside; posterior lobe shorter and narrower, distally with very long annulated setae.

Maxilliped I (h) endites completely reduced; endopod long, almost as long as exopod; exopod with lash; epipod large, bilobed, each lobe subtriangular.

Maxilliped II (i) endopod little compressed, with 6 segments, distal inserted diagonally in next, with 3 strong spines; exopod long, with lash; epipod a single lobe.

Maxilliped III (k) leg-like, stout, with 5 segments, distal with setae only; exopod shorter than longest segment, with lash; no epipod.

Pereopods: I (l) stout, with massive propodus, subchelate; II (m) chelate, slender, short pointed rudimentary epipod. III (a) more slender than others, tapering to fine point; IV–V unequal, IV longer and stouter with wide flattened dactyl; V with sharp slender dactyl.

Pleopods: female I (o) endopod narrow sinuous, shorter than exopod; female II (p) endopod two-segmented narrower and shorter than exopod; male I (q) endopod smaller than in female, sinuous; Male II (r) endopod two-segmented, wide appendix masculina (s) with numerous short setae.

Range of Distribution

Western Atlantic only, between 31°57′ N and 42°59′ N Depths 480–1754 m (Crosnier et Forest 1973).

Records of distribution in the area of reference are in Fig. 150.

Biology

Lengths of specimens examined 12 mm cl in male and 13 mm cl in female; greatest egg diameter 2.3 mm.

A specimen (9 mm cl) taken SE of Newfoundland (42°59′ N, 51°15′ W), depth 1 100 m, had a bopyrid isopod in the left branchial chamber (Sivertsen and Holthuis 1956).

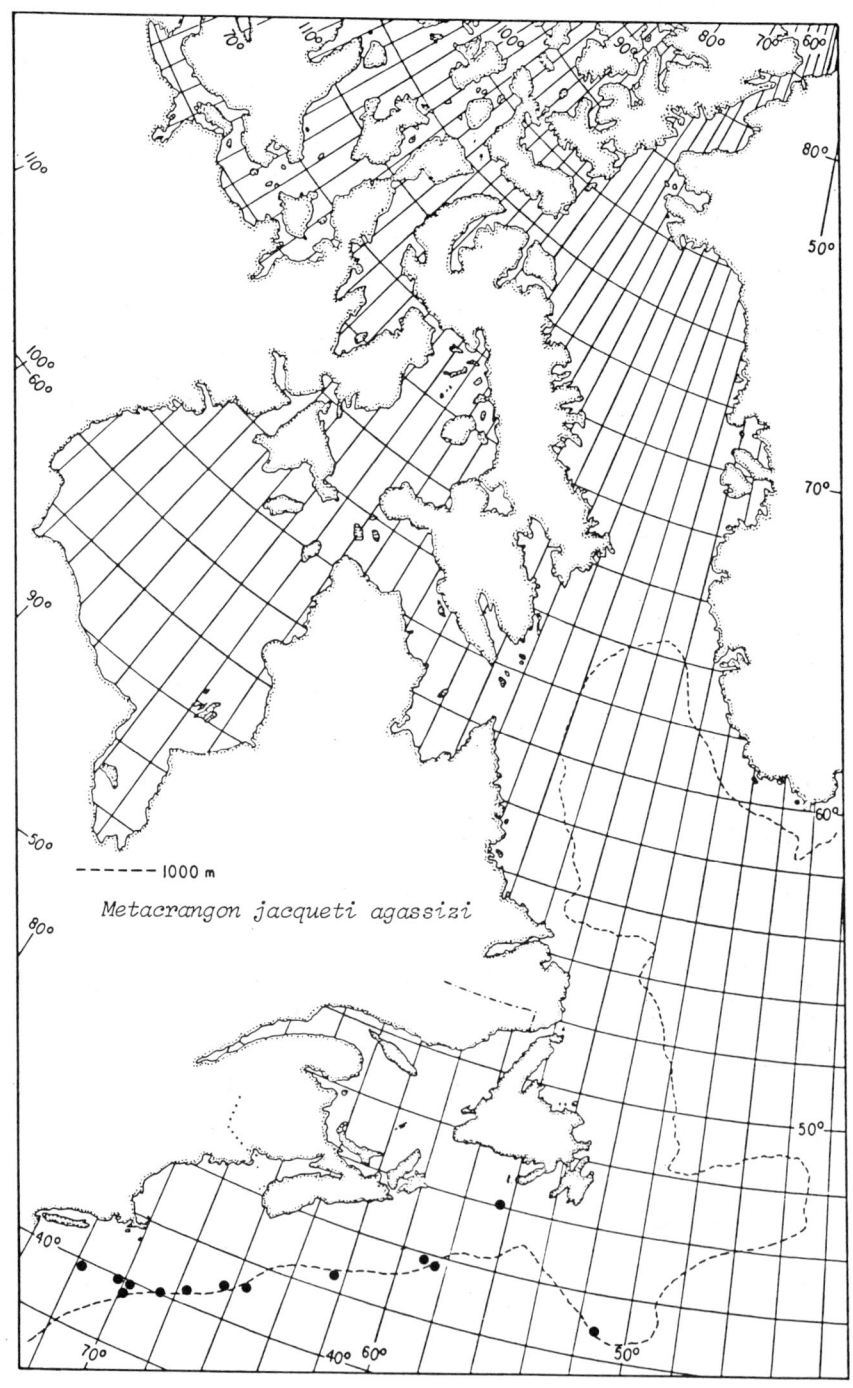

Fig. 150. *Metacrangon jacqueti agassizi*, distribution records in the area of reference.

Genus *Pontophilus* Leach, 1817
Holthuis 1955: 136; Rathbun 1929: 21;
Williams 1984: 160.

First legs subchelate, second chelate and very short, not reaching to end of merus of first; carapace with median and two lateral carinae with spines; rostrum short; abdominal somites not dorsally carinate except 6th with 2 carinae; pleopods with appendices internae.

Pontophilus brevirostris Smith, 1881
Williams 1984: 161, fig. 113.
(Figures 151 and 152)

DISTINGUISHING CHARACTERISTICS

Carapace with mid-dorsal carina with 3–4 spines, a long first lateral carina with 2–3 spines and a short second lateral carina with only one spine; rostrum very short with a tooth at each side of base; abdomen with first four somites rounded, fifth with low diverging carinae and 6th with two parallel carinae; telson tapering to narrow tip with short central spine and 2 pairs of superimposed lateral spines.

DESCRIPTION

Integument smooth, shiny, with some pubescence. Colour brownish.

Rostrum short, slightly hollow above, with a spine at each side of base rising above rostrum; suborbital spine small and sharp, antennal strong and acute reaching farther forward than rostrum, pterygostomian small. Carapace with three prominent spines on median carina and a small spine or tubercle in front of them and a hollow just behind rostrum; also two or three spines on a long lateral carina and below it a short lateral carina with one spine.

Abdomen wider and thicker at 3rd somite, tapering quickly to long narrow 6th somite; pleura and terga rounded except ventro-lateral spine on 5th and 6th, and faint dorsolateral parallel carinae on 6th somite.

Telson (t) longer than 6th, narrow and tapering to fixed central terminal spine flanked by a small pair above and a long pair of spines below at lateral edge of narow tip; dorsally a faint sulcus and two pairs of tiny dorsolateral spines. Inner branch of uropod longer than telson and outer branch.

Eyes large with globular carina.

Antennule (c) 1st article longer than other two combined, 3rd half length of 2nd; 1st with strong ventral spine at middle. Stylocerite broad with sharp acuminate tip almost reaching distal edge of 1st article. Dorsolateral flagellum shorter than peduncle and ventromesial flagellum.

Antenna (d) scale length 2.7 times width; spine about even with tip of blade; 3rd article of peduncle very long, almost reaching tip of blade.

Mandible (e) molar with two fangs, lower with proximal ridge. No incisor or palp.

Maxillule (f) proximal endite greatly reduced, small rounded at tip; distal endite expanded and curved distally with apex of long spines; endopod moderate, obscurely bifid with subapical spine.

Maxilla (g) endites almost completely reduced, endopod short; wide anterior lobe of scaphognathite slightly tapered, narrower and shorter posterior lobe rounded and with some long fine setae distally.

Maxilliped I (h) endites completely reduced; slender endopod two-segmented, shorter than wide exopod which has curved lash; epipod unequally bilobed.

Maxilliped II (i) six segments, distal very small with brush setae; exopod long, with lash; epipod with podobranch.

Maxilliped III (k) leg-like, five segments, distal with setae only; exopod short, with lash; epipod rudimentary, falcate.

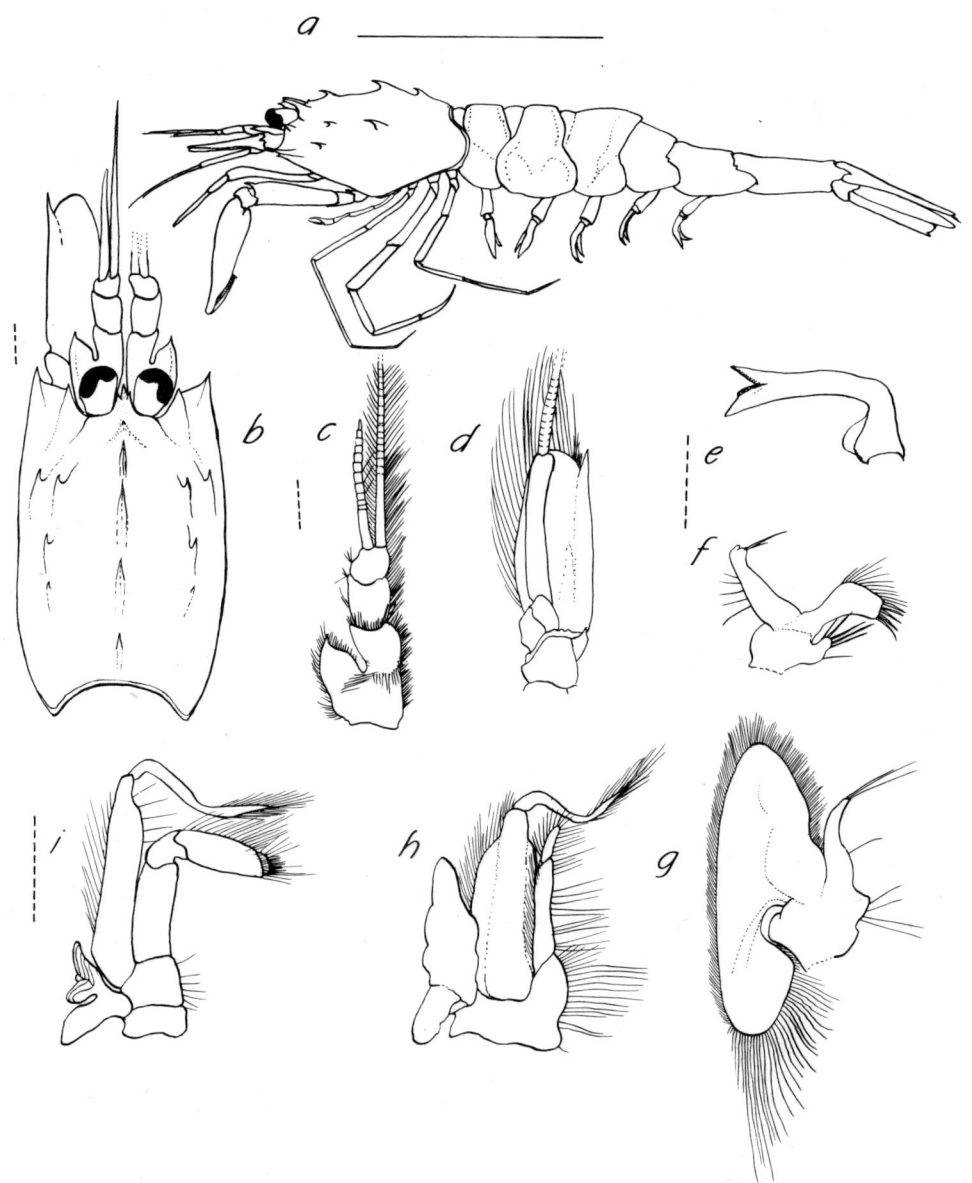

Fig. 151. *Pontophilus brevirostris*: *a*, whole shrimp from left side; *b*, carapace in dorsal aspect; *c*, antennule; *d*, antenna; *e*, mandible; *f*, maxillule; *g*, maxilla; *h*, first maxilliped; *i*, second maxilliped; solid line = 10 mm, broken line = 1 mm.

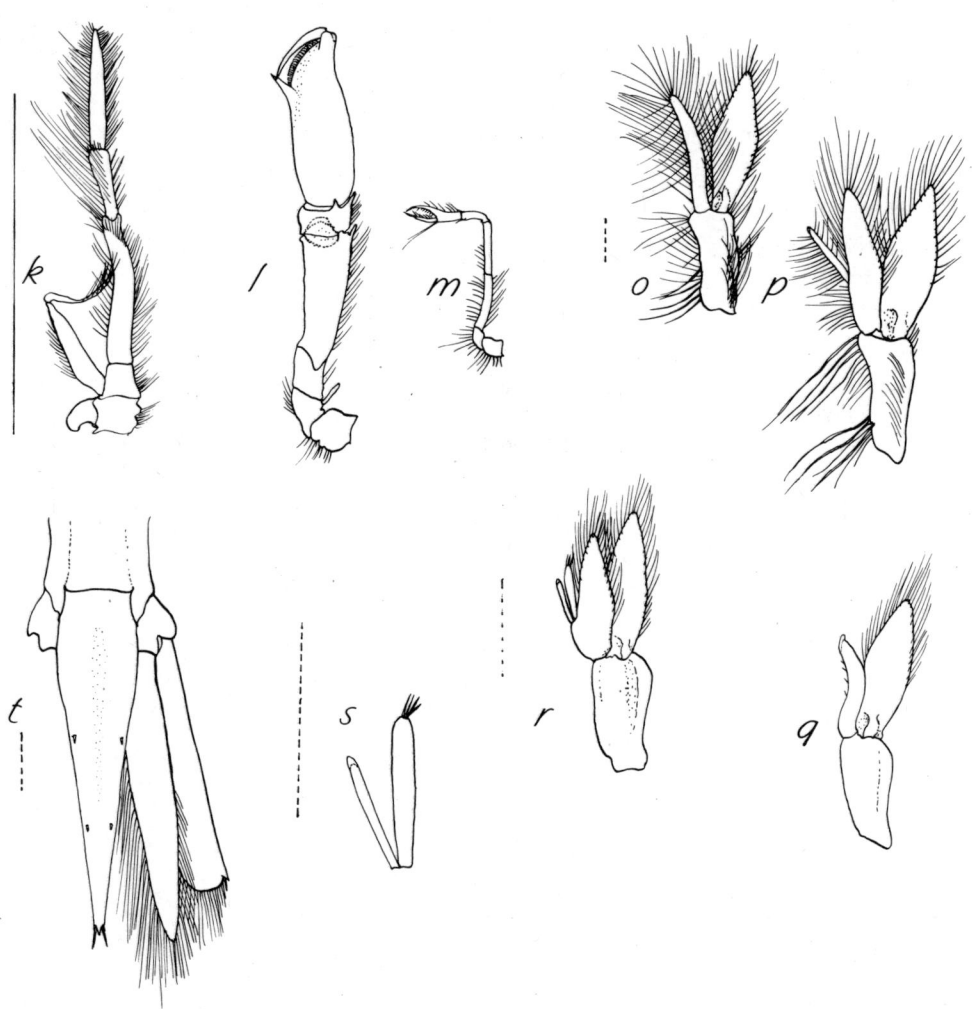

Fig. 152. *Pontophilus brevirostris*: k, third maxilliped; l, first pereopod; m, second pereopod; o, F first pleopod; p, F second pleopod; q, M first pleopod; r, second pleopod; s, appendix masculina; t, telson; solid line = 10 mm, broken line = 1 mm.

Pereopods: I (l) subchelate, stout; two distal spines on carpus, one distally on merus; slender exopod, short, not reaching distal ischium. II (m) chelate, slender and short, fingers curved sharp-pointed. III non-chelate, long and slender tapering to slender straight dactyl. IV moderately stout long, dactyl compressed and curved, about as long as propodus. V about as long as IV but more slender, tapering to slender dactyl.

Pleopods: female I (o) endopod narrow almost as long as exopod, fringed with long setae, exopod ovate; female II (p) endopod only slightly smaller than exopod, appendix interna long inserted near middle of endopod, terminal patch of hooks. Male I (q) endopod narrow, sinuous, tip modified with few hooks; male II (r) endopod almost same size as exopod, appendix interna shorter than appendix masculina (s) latter with a few short apical setae.

RANGE OF DISTRIBUTION

Gulf of Maine to Gulf of Mexico off Dry Tortugas and Cuba (Williams and Wigley 1977; Pequegnat 1970; Williams 1984).

Records of distribution in area of reference are in Fig. 153.

BIOLOGY

Length of body to 37 mm in males and 36 mm in females (Williams 1984); 7 mm cl in specimens examined.

Females were ovigerous in April and from June to September (Williams 1984).

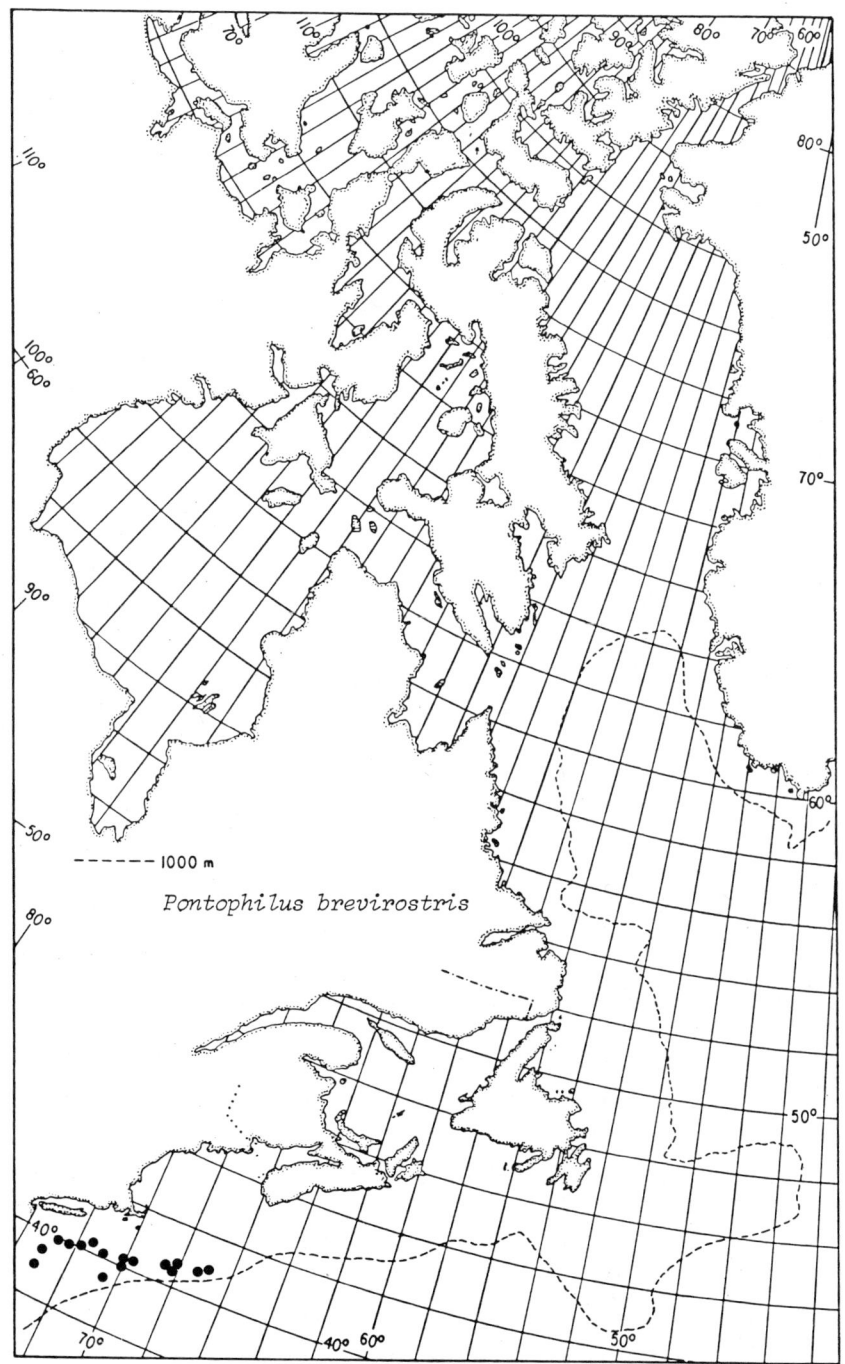

Fig. 153. *Pontophilus brevirostris*, distribution records in the area of reference.

Pontophilus norvegicus (M. Sars, 1861)
Rathbun 1929: 21, fig. 26; Williams 1984: 162, fig 115.
(Figures 154 and 155)

DISTINGUISHING CHARACTERISTICS

Rostrum sharp, not quite exceeding eyes, pointing slightly upwards, spines at its base not as high; prominent middorsal carina with three strong teeth, a small tubercle and a small spine in front of them; first lateral carina with two teeth and second lateral carina with one tooth; first leg subchelate, stout, second chelate short, slender; abdomen thickest at 3rd somite, tapering quickly to narrow 6th which has two dorsal carinae.

DESCRIPTION

Integument thin, smooth, with scattered short fine setae dorsal to first lateral carina.

Colour of carapace and abdomen pale dull reddish brown, often mottled and darkened on last three somites, and with whitish oblique bands on carapace, also whitish traces on first two abdominal somites (Kemp 1910).

Rostrum slightly ascending, not quite exceeding the eyes, two basal fixed spines joining rostrum at middle, a bunch of ventral setae near apex and a few lateral setae.

Carapace with prominent dorsal carina with four teeth, the most anterior small with a tubercle behind it; also two lateral carinae, the more dorsal almost reaching the posterior edge with two teeth and the other shorter with only one tooth; anterior edge with short sharp post-orbital, strong antennal and small pterygostomian spines.

Abdomen with 2 low parallel carinae on 6th somite; pleura and terga smooth, rounded, 5th only with ventro-lateral spine posteriorly.

Telson (t) slender and 1.3 times length of 6th somite, tapered to a point at subtruncate end with two pairs of moderate lateral spines; two pairs of dorsolateral spines; inner branch of uropod about even with telson, longer than outer branch.

Eyes large, cornea globular.

Antennule (c) 1st article longer than 2nd and 3rd combined, 2nd longer than 3rd, 1st with ventral spine at middle; stylocerite wide, acuminate tip not reaching end of 1st article; thick flagellum longer than peduncle, slightly shorter than thin flagellum.

Antenna (d) scale length about 2.3 times width, spine slightly exceeded by blade; peduncle not as long as scale; basal article with distal outer spine.

Mandible (e) molar with two fangs, one with subsidiary tooth, the other with a proximal pectinate shelf.

Maxillule (f) proximal endite small, distal subovate, curved; endopod long, straight, with subapical spine.

Maxilla (g) endites reduced to minor protuberances; endopod short; anterior lobe of scaphognathite tapered, subtriangular, slightly longer than rounded posterior lobe which has a few long distal setae.

Maxilliped I (h) endites completely reduced; endopod slender, two-segmented, as long as lamellate part of exopod, lash of exopod slightly longer; anterior lobe of epipod longer than endopod, subtriangular, posterior lobe much shorter.

Maxilliped II (i) six segments, the distal inserted diagonally, with two strong spines; exopod long with curved lash; epipod with podobranch.

Maxilliped III (k) five segments, distal with crown of setae distally; exopod plus lash about as long as longest segment; epipod rudimentary.

Pereopods: I (l) subchelate, stout, small distal spine on carpus; exopod very tiny. II (n) chelate, slender, short; curved fingers slender sharp; III (m) long, slender, tapered to slender sharp dactyl; IV about as long but stouter than III or V, dactyl slightly compressed, curved; V slightly shorter than IV.

Pleopods: female I (o) endopod much narrower and shorter than exopod, sinuous, sparingly setose and with hooked spine on proximal third outside (not present in some);

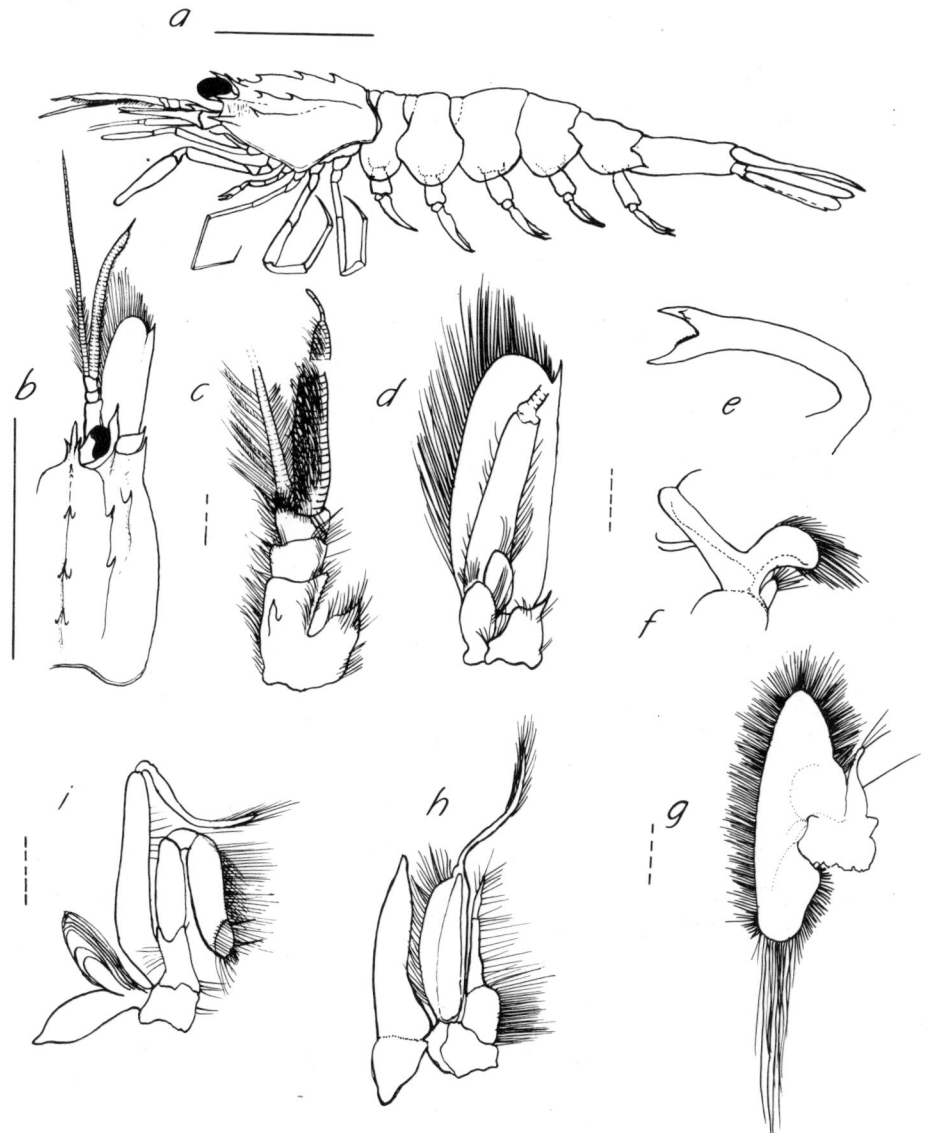

Fig. 154. *Pontophilus norvegicus*: *a*, whole shrimp from left side; *b*, carapace in dorsal aspect; *c*, antennule; *d*, antenna; *e*, mandible; *f*, maxillule; *g*, maxilla; *h*, first maxilliped; *i*, second maxilliped; solid line = 10 mm, broken line = 1 mm.

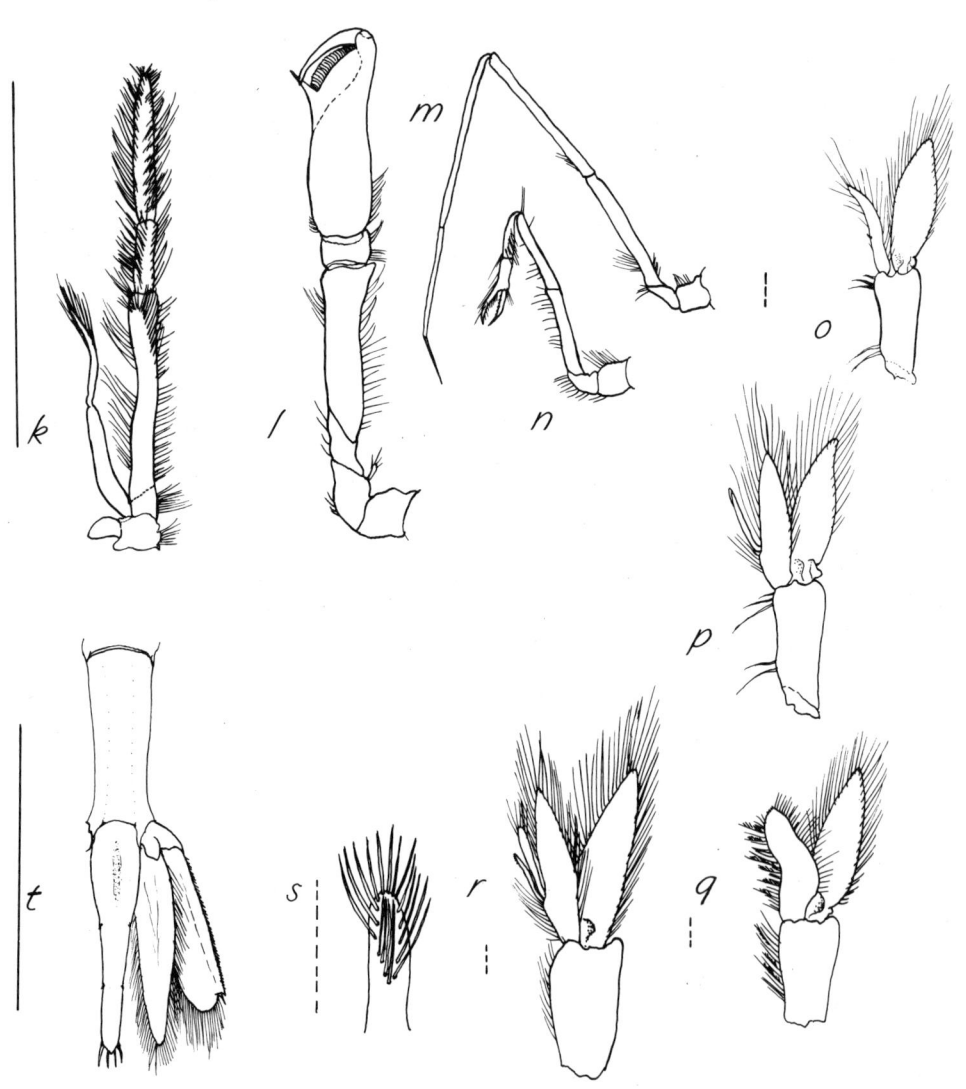

Fig. 155. *Pontophilus norvegicus*: k, third maxilliped; l, first pereopod; m, second pereopod; n, third pereopod; o, F first pleopod; p, F second pleopod; q, M first pleopod; r, M second pleopod; s, appendix masculina; t, telson; solid line = 10 mm, broken line = 1 mm.

female II (p) endopod almost as wide and as long as exopod, appendix interna long with apical hooks. Male I (q) endopod wide, sinuous, shorter than exopod, with apical patch of hooks; Male II (r) appendix interna more slender and shorter than appendix masculina (s) which has apical and subapical moderate spines (about 17).

RANGE OF DISTRIBUTION

Greenland to Maryland and points southeast (Williams and Wigley 1977); Iceland; Spitzbergen; Murman coast and northwestern Europe including British Isles to Bay of Biscay; Balearic Islands (Forest 1965; Williams 1984). Depths 50–1450 m (Sivertsen and Holthuis 1956).

Records of distribution in area of reference are in Fig. 156.

BIOLOGY

Lengths to 12 mm cl in males and 19 mm cl in females (Squires 1965a); body lengths in males to 55 mm and in females to 76 mm (Wollebaek 1908).

Females first reached maturity at 10 mm cl, males at 9 mm cl. About 90% of the females (78 examined) were potentially ovigerous in autumn indicating possible annual spawning in these populations. Egg greatest diameter was 1.3 mm (one 1.8 mm in November) (Squires 1965a).

Stomach contents included phytobenthos, crustacean fragments, polychaetes and foraminiferans (Squires 1965a).

Other decapod crustacean species in catches with this species were *Sergestes arcticus*, *Pasiphaea tarda*, *Pandalus borealis*, *Pandalus montagui*, *Sclerocrangon ferox*, *Munida tenuimana* and *Munidopsis curvirostra*.

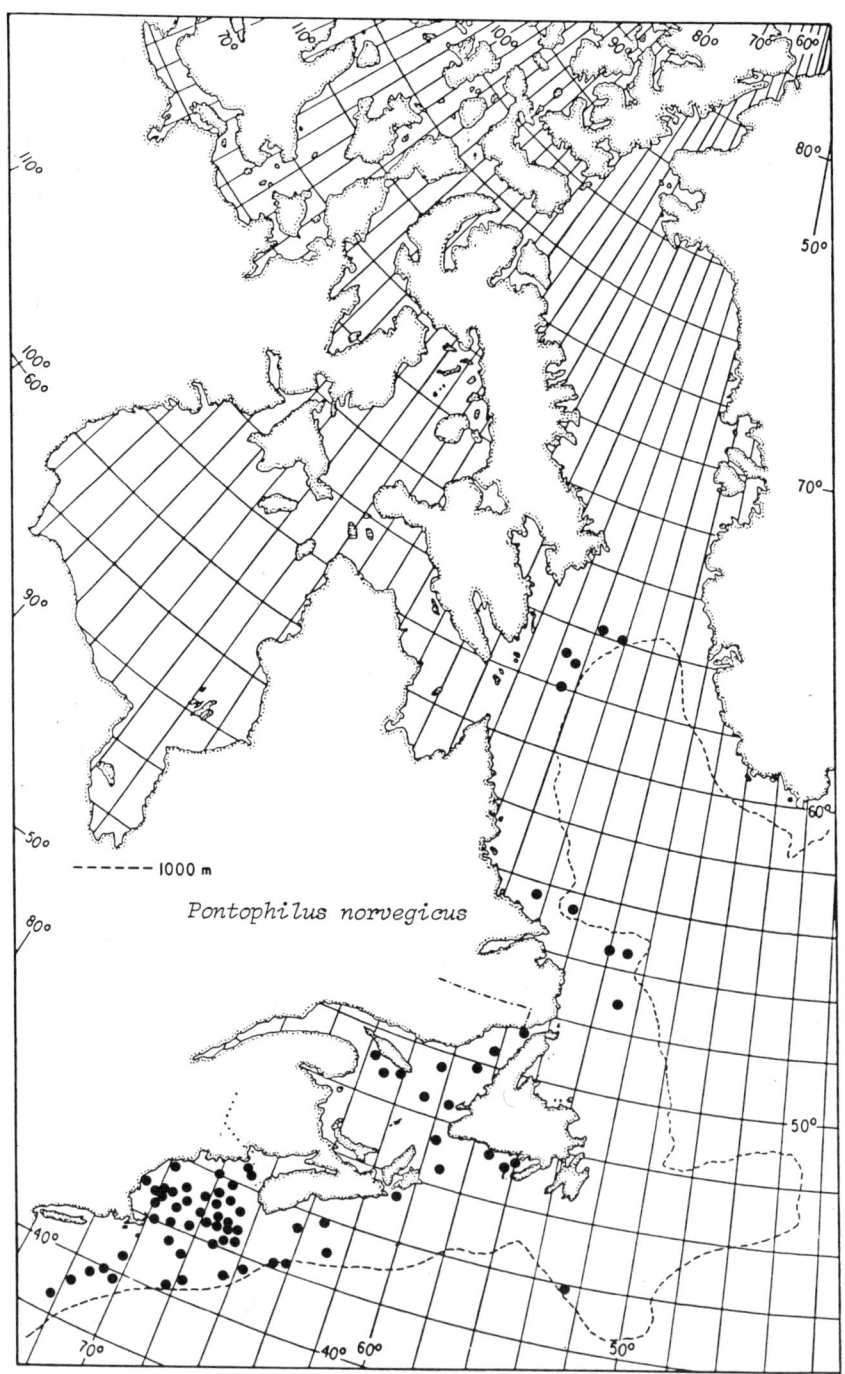

Fig. 156. *Pontophilus norvegicus*, distribution records in the area of reference.

Genus *Sabinea* Ross, 1835
Holthuis 1955: 133; Rathbun 1929: 22.

Second legs simple, not chelate; carapace with a median carina and 3 lateral carina at each side, all seven with teeth; first legs large, subchelate; rough-looking, abdominal somites strongly carinate. Long sharp spine triangular in crosssection on sternum of 3rd legs.

Sabinea hystrix (A. Milne-Edwards, 1881)
Crosnier et Forest 1973: 232, fig. 73 c–d;
Sivertsen and Holthuis 1956: 40.
(Figures 157 and 158)

DISTINGUISHING CHARACTERISTICS

Very rough-looking and spiny; rostrum ascending, strong and sharp with one ventral spine on distal third and a pair laterally on proximal third; median carina with about 8 prominent strong spines, other six carina with many strong spines; very heavy antennal spine at anterior lower corner; abdominal somites all with spiny carinae, the 5th with two diverging and the 6th with two parallel carinae, heavily spinose.

DESCRIPTION

Integument rigid, heavily armoured, covered with short setae (pilose).

Rostrum spike-like, ascending, almost as long as carapace; a spine on each side above near base with strong oblique supporting carina confluent with orbit, and a ventral spine on distal third; triangular in cross-section with base of triangle upwards.

Carapace with one middorsal carina and three lateral carinae on each side all extending to posterior edge of carapace; about 8 large high teeth on middorsal and on others laterally 7, 6 and 9 smaller teeth (8, 6, 8; 8, 6, 10), respectively; anterior edge with sharp postorbital, extremely heavy sharp antennal and small pterygostomian spines; ventro-lateral margin of carapace is turned under anteriorly with a ventral lobe at about middle of margin; ventrally large spines on sterna of 2nd and 3rd legs and small spines with carina on sterna of 4th and 5th; a large spine behind coxa in males, a knob in females.

Abdomen with a double dorsal and 2 lateral carinae each with spine on 1st somite; divided carina dorsally on 2nd; long mid-dorsal with 2–5 small spines on 3rd; long mid-dorsal with 3 large spines on 4th; divergent carinae with 2 spines on each on 5th; and 2 parallel each with 6–7 spines on 6th (also ventro-lateral carina with posterior spine); pleura of 1st and 2nd somites with 1 ventral spine, and 3rd–5th with 2.

Telson (t) with dorso-lateral carina without spines; tapering to strong terminal spine. Branches of uropod shorter than telson, the outer shortest.

Eye very large, cornea globular, stalk short.

Antennule (c) 1st article longer than others combined; 2nd and 3rd subequal; stylocerite very wide, the attenuate point not reaching distal end of 1st article; dorso-lateral flagellum slightly compressed almost as long as carapace, ventromesial slender slightly longer.

Antenna (d) scale reaches about middle of rostrum, length 3.3 times width, spine exceeds blade, peduncle almost as long as scale; a very strong sharp spine ventrally on basal article.

Mandible (e) molar with two fangs, each with small subsidiary tooth.

Maxillule (f) proximal endite reduced, short with rounded tip; distal with slight curve; endopod rounded distally with subapical lobe probably representing incipient ramus.

Maxilla (g) endites reduced to slight rounded protuberances; endopod short with separation from anterior lobe of scaphognathite, lobe slightly tapered, wider and longer than rounded posterior lobe, latter with long distal setae.

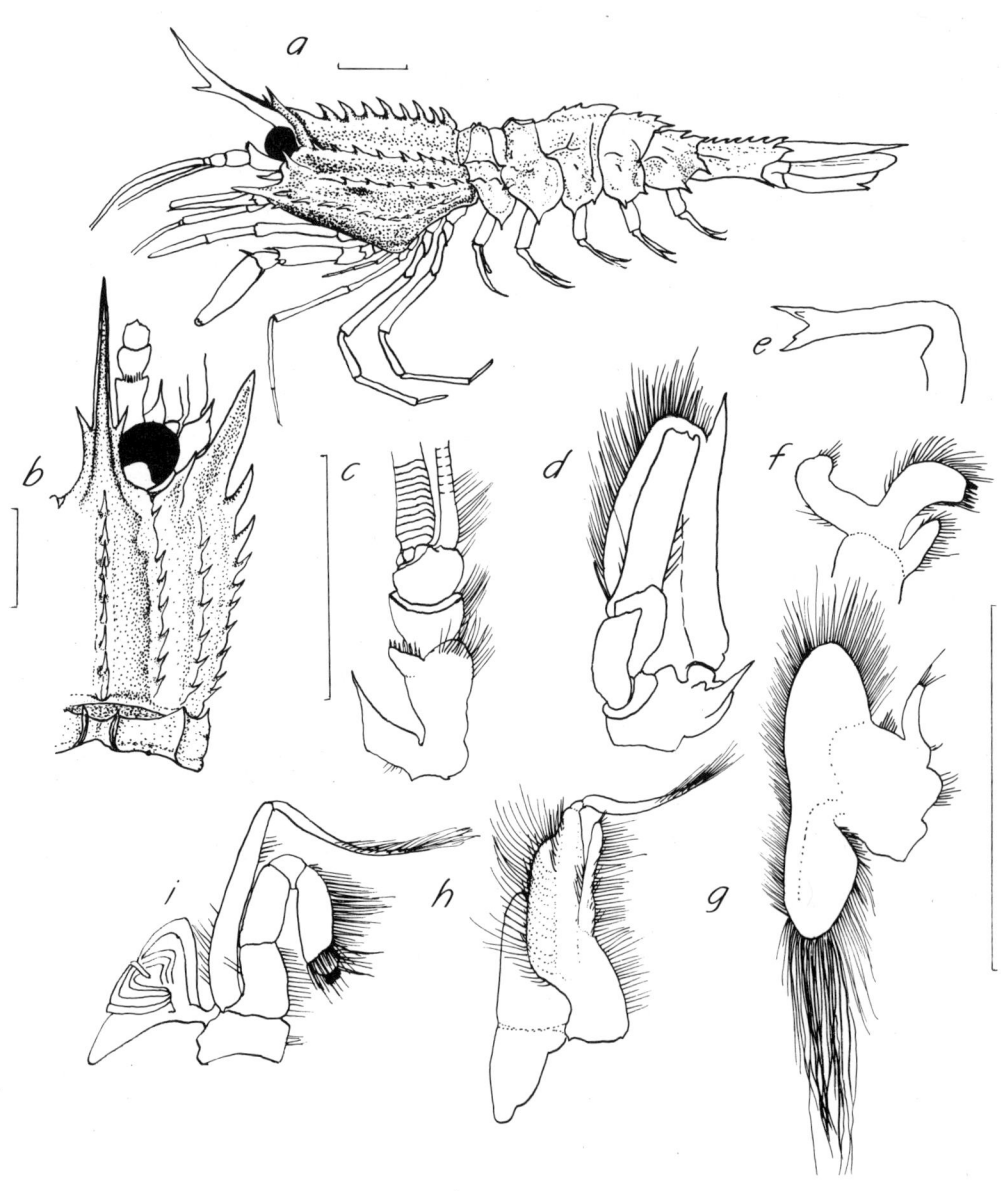

Fig. 157. *Sabinea hystrix*: *a*, whole shrimp from left side; *b*, carapace in dorsal aspect; *c*, antennule; *d*, antenna; *e*, mandible; *f*, maxillule; *g*, maxilla; *h*, first maxilliped; *i*, second maxilliped; solid line = 10 mm.

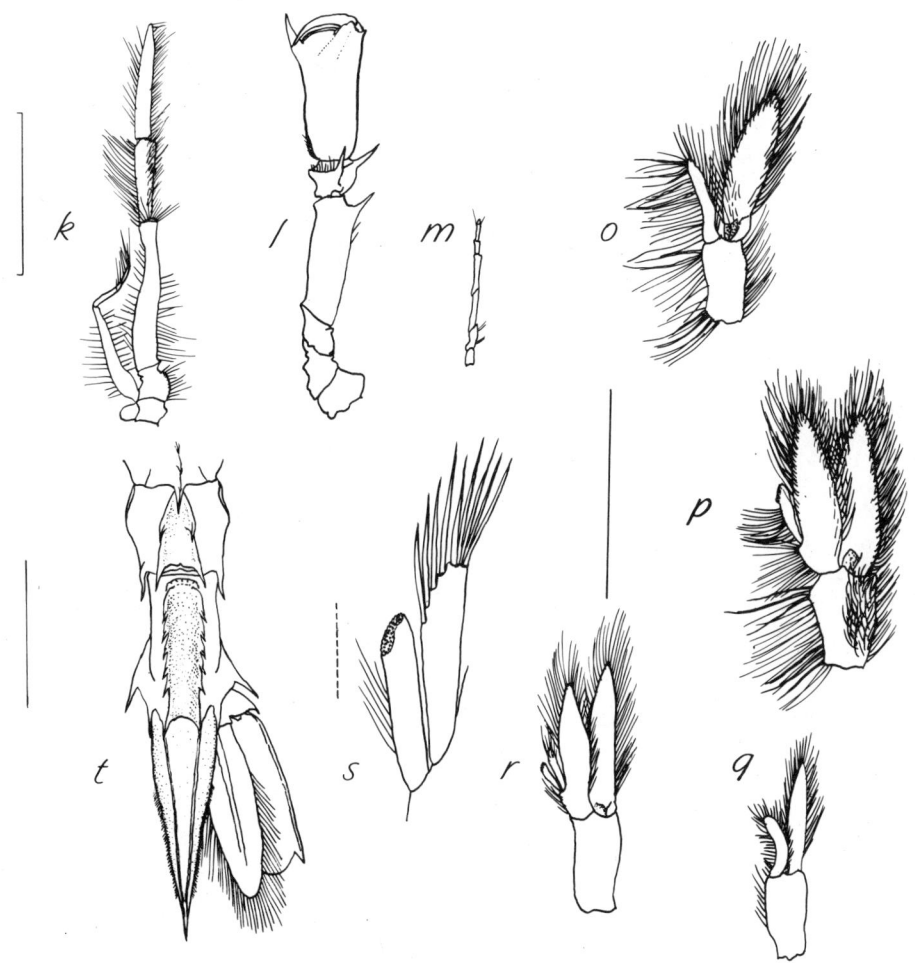

Fig. 158. *Sabinea hystrix*: k, third maxilliped; l, first pereopod; m, second pereopod; o, F first pleopod; p, F second pleopod; q, M first pleopod; r, M second pleopod; s, appendix masculina; t, telson; solid line = 10 m, broken line = 1 mm.

Maxilliped I (h) endites completely reduced; endopod slender almost as long as lamellate part of exopod, lash about as long; epipod almost equally bilobed, anterior lobe reaching two-thirds exopod.

Maxilliped II (i) endopod only slightly compressed, six segments, distal almost as long as wide, with bunch of apical brush setae; exopod slender, long, with lash; epipod with podobranch.

Maxilliped III (k) moderate, about as long as 1st leg; unarmed; short exopod; rudimentary rounded epipod.

Pereopods: I (l) subchelate, stout; strong sharp spines on carpus (2 distally) and merus (1 distally); II (m) not chelate, very short and slender; III long and slender, tapering to fine sharp dactyl; IV and V about as long as III but stouter; dactyl not tapered but sharp at tip.

Pleopods: female I (o) endopod more slender and shorter than exopod; female II (p) endopod slightly longer than exopod, appendix interna expanded distally with patch of hooks; male I (q) endopod narrow, curved, about half as long as exopod; male II (r) appendix interna expanded distally with patch of hooks, slightly shorter than appendix masculina (s) with about 5 apical and 4 subapical spines.

RANGE OF DISTRIBUTION

Western Atlantic from Davis Strait and Iceland to the West Indies (Guadaloupe). Depths 550–3 600 m (Sivertsen and Holthuis 1956).

Records of distribution in the area of reference are in Fig. 159.

BIOLOGY

Lengths 19 mm cl in males and 24 mm cl in females.

A female taken SE of Newfoundland was ovigerous in June; egg diameter 2.2 mm (Sivertsen and Holthuis 1956).

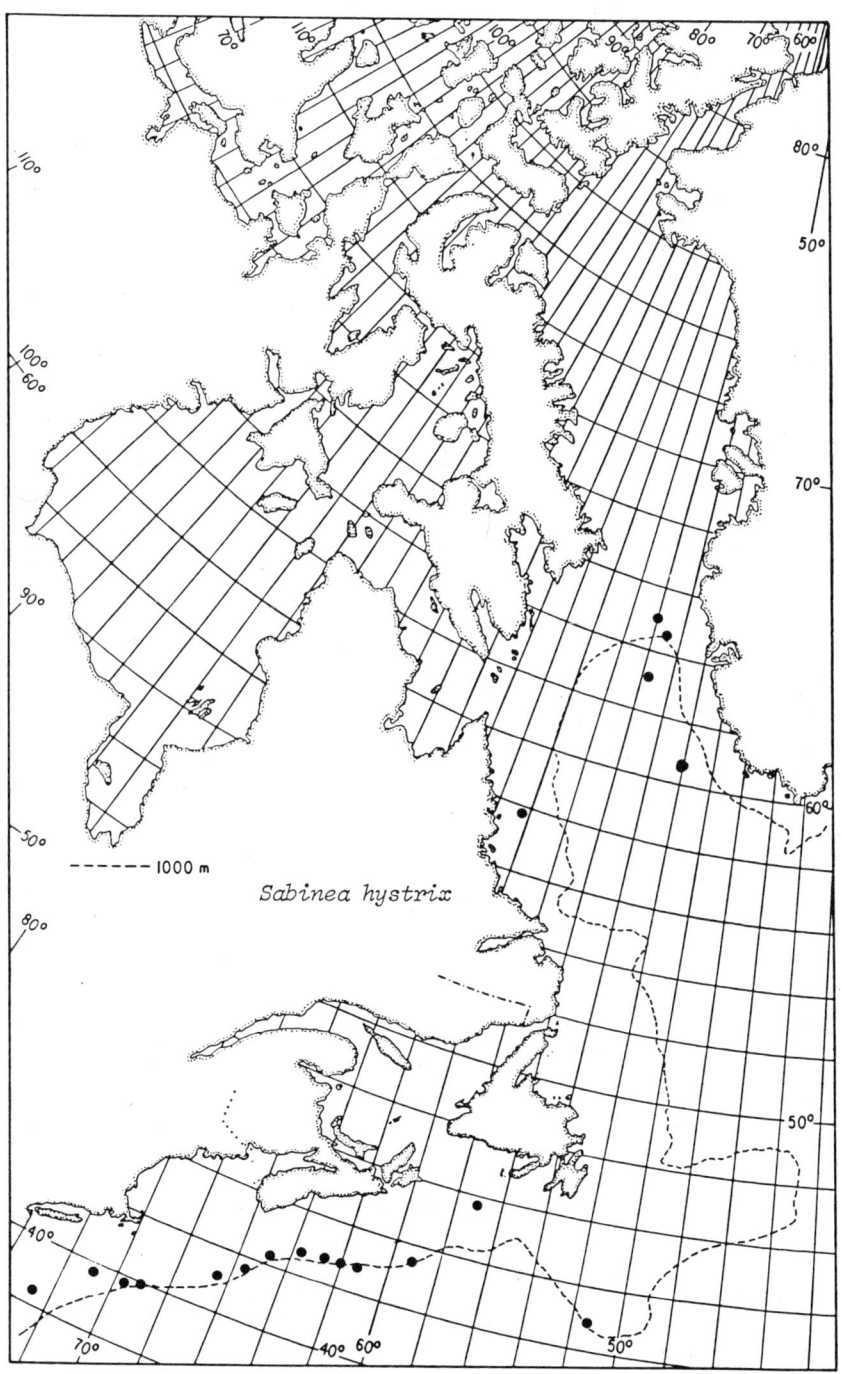

Fig. 159. *Sabinea hystrix*, distribution records in the area of reference.

Sabinea sarsi Smith, 1879
Rathbun 1929: 22, fig. 29; Williams 1984:
163, fig. 116; Squires 1965a: 69.
(Figures 160 and 161)

DISTINGUISHING CHARACTERISTICS

Seven spiny carinae on the carapace as in *Sabinea septemcarinata* but with a longer rostrum compressed vertically at the tip and reaching beyond the eyes; abdomen carinate on all somites, with double carinae on the 1st, 5th and 6th; 5th with divergent spiny carinae, not spiny in *S. septemcarinata*.

DESCRIPTION

Integument rigid, thick, covered with short setae. Colour brownish mottling on greyish background.

Rostrum exceeding eyes, hollow dorsally but with median carina rising distally to form compressed rounded vertical tip, lateral carina confluent with orbit.

Median carina of carapace reaching posterior margin, with 5 teeth, the anterior one small and below the others; first lateral beginning at orbital fissure, with 10 major teeth; second beginning at postorbital spine, with 11 major teeth; third beginning at antennal spine, with 10 major teeth, all carinae reaching posterior edge; very small pterygostomian spine. Anterior ventral edge of carapace turned under, lobe at about middle of ventral margin.

First somite of abdomen with two short dorsal carina with anterior spine, a lateral short vertical carina and two horizontal lower carina the lowest with anterior spine; 2nd–4th somites with median carina, the 4th with also an oblique lateral carina; 5th with diverging dorsal carinae each with 4 spines plus posterior spine, and a lower lateral carina with 2 spines; 6th with 2 dorsal parallel carinae each with about 12 teeth. Small spine on ventro-lateral edge of 4th and two spines on ventro-lateral edge of 5th somite.

Telson (t) tapered to subtruncate tip with central spinous projection flanked by a single spine at each corner; two pairs of dorsolateral spines.

Eyes large, cornea globular, on very short stalk.

Antennule (c) 1st article longer than other two combined; 2nd and 3rd subequal. Stylocerite wide with attenuate point not reaching distal 1st article. Thick flagellum about as long as peduncle, thin one slightly longer.

Antenna (d) scale length about 2.8 times width; spine exceeding blade; peduncle about half blade; small ventral spine on basal article.

Mandible (e) molar with two unequal fangs, both with small subsidiary tooth.

Maxillule (f) distal endite expanded distally curved; proximal moderate sub-rectilinear, rounded at tip; endopod with subapical projection with setae, probably incipient second ramus.

Maxilla (g) endites reduced to rounded projection with setae; endopod short; anterior lobe of scaphognathite sub-rectilinear, rounded, longer than posterior lobe which is sub-triangular with many distal long setae.

Maxilliped I (h) endites completely reduced; endopod slender with 2 segments, as long as exopod, lash somewhat shorter; anterior lobe of epipod as long as endopod, posterior lobe only about half as long, both subtriangular in outline.

Maxilliped II (i) with six segments, distal inserted diagonally on next segment, with two strong curved spines; exopod and lash about as long as endopod; epipod with podobranch.

Maxilliped III (k) with 5 segments, moderate, not as long as first leg, unarmed except with setae; exopod short, including lash about as long as longest segment; epipod rudimentary.

Pereopods: I (l) subchelate, stout; 2 strong spines distally on merus; II (m) non-chelate, slender, very short reaching only about half merus of I; III (n) very slender,

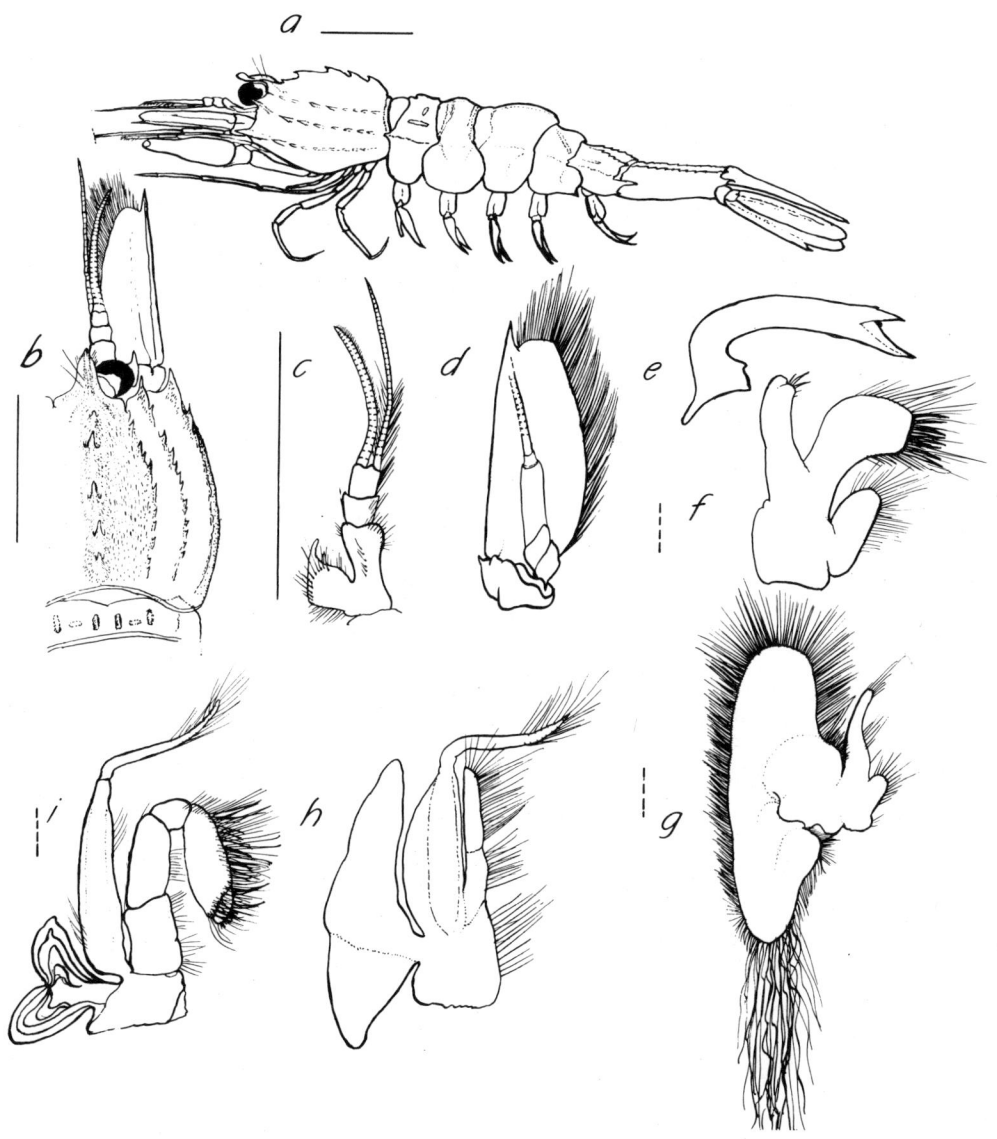

Fig. 160. *Sabinea sarsi*: *a*, whole shrimp from left side; *b*, carapace in dorsal aspect; *c*, antennule; *d*, antenna; *e*, mandible; *f*, maxillule; *g*, maxilla; *h*, first maxilliped; *i*, second maxilliped; solid line = 10 mm, broken line = 1 mm.

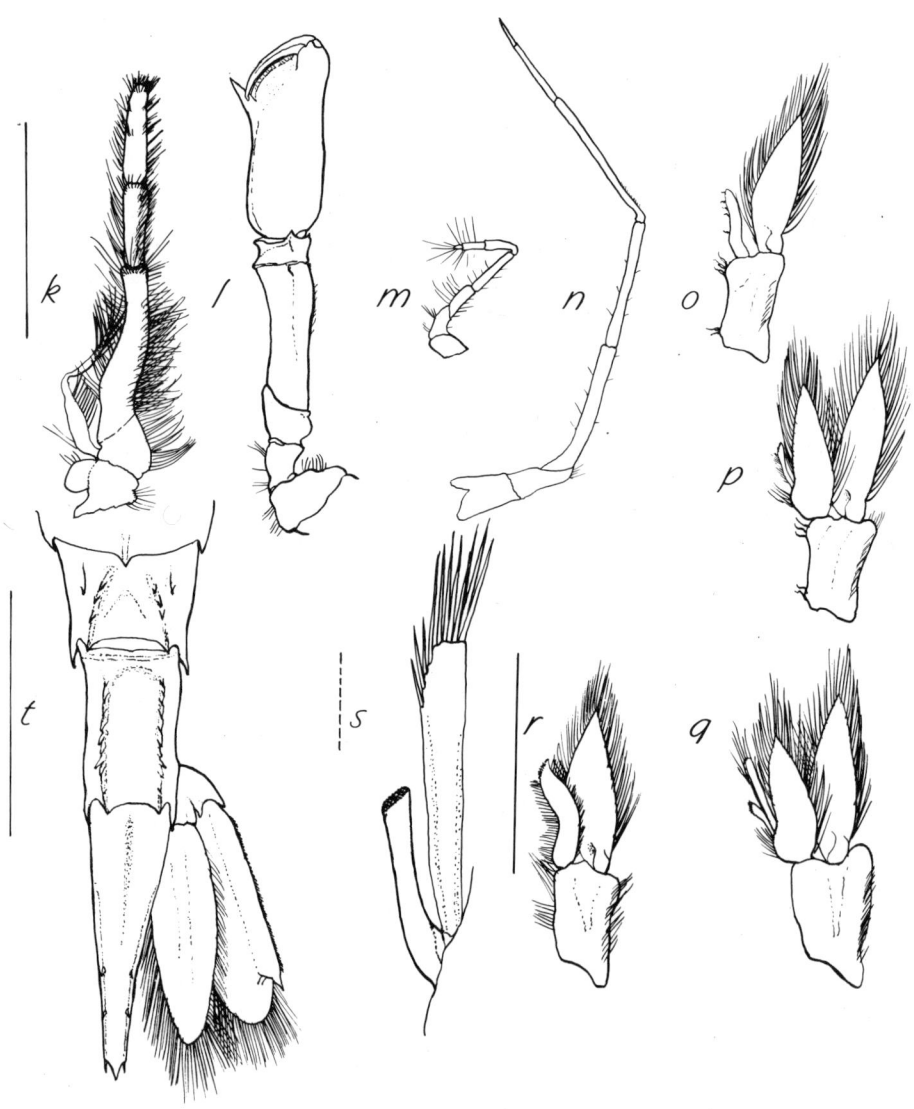

Fig. 161. *Sabinea sarsi*: *k*, third maxilliped; *l*, first pereopod; *m*, second pereopod; *n*, third pereopod; *o*, F first pleopod; *p*, F second pleopod; *q*, M first pleopod; *r*, M second pleopod; *s*, appendix masculina; *t*, telson; solid line = 10 mm, broken line = 1 mm.

long, tapering to sharp straight dactyl. IV and V not quite as long as III, somewhat stouter, dactyl somewhat compressed, curved, very sharp.

Pleopods: female I (o) short and narrow sinuous endopod, exopod twice as long and wide; female II (p) endopod slightly shorter than exopod, appendix interna expanded toward tip with subapical patch of hooks; male I (r) endopod about two-thirds as long as exopod, somewhat narrow, sinuous, with subapical patch of hooks; male II (q) endopod shorter than exopod, appendix interna only about half as long as appendix masculina (s) which has apically about 6 long spines and 4 laterally.

RANGE OF DISTRIBUTION

North Atlantic only: Davis Strait to 40° N off the east coast of the USA; Iceland; northern Europe. Depths 48–710 m (Williams and Wigley 1977).

Records of distribution in the area of reference are in Fig. 162.

BIOLOGY

Lengths to 14 mm cl in males and 20 mm cl in females (Squires 1965a); body lengths 62 mm in males and 72 mm in females (Williams 1984).

Females were 88% potentially ovigerous in autumn at temperatures of -0.6 to $4.1°$ C, indicating possible annual spawning.

It was taken in populations of *Pandalus borealis* and occasionally alone or with *Pandalus montagui*, *Sabinea septemcarinata*, *Sergestes arcticus*, *Pasiphaea tarda* and *Lithodes maja* (Squires 1965a).

Stomach contents were phytobenthos, amphipods, copepods, crustacean remains and polychaetes (Squires 1965).

Fontaine (1977) reports a relationship between this shrimp and the prosobranch gastropod *Lora cancellata cancellata* in the northwest Atlantic with ootheca attached to the abdominal sterna of the shrimp.

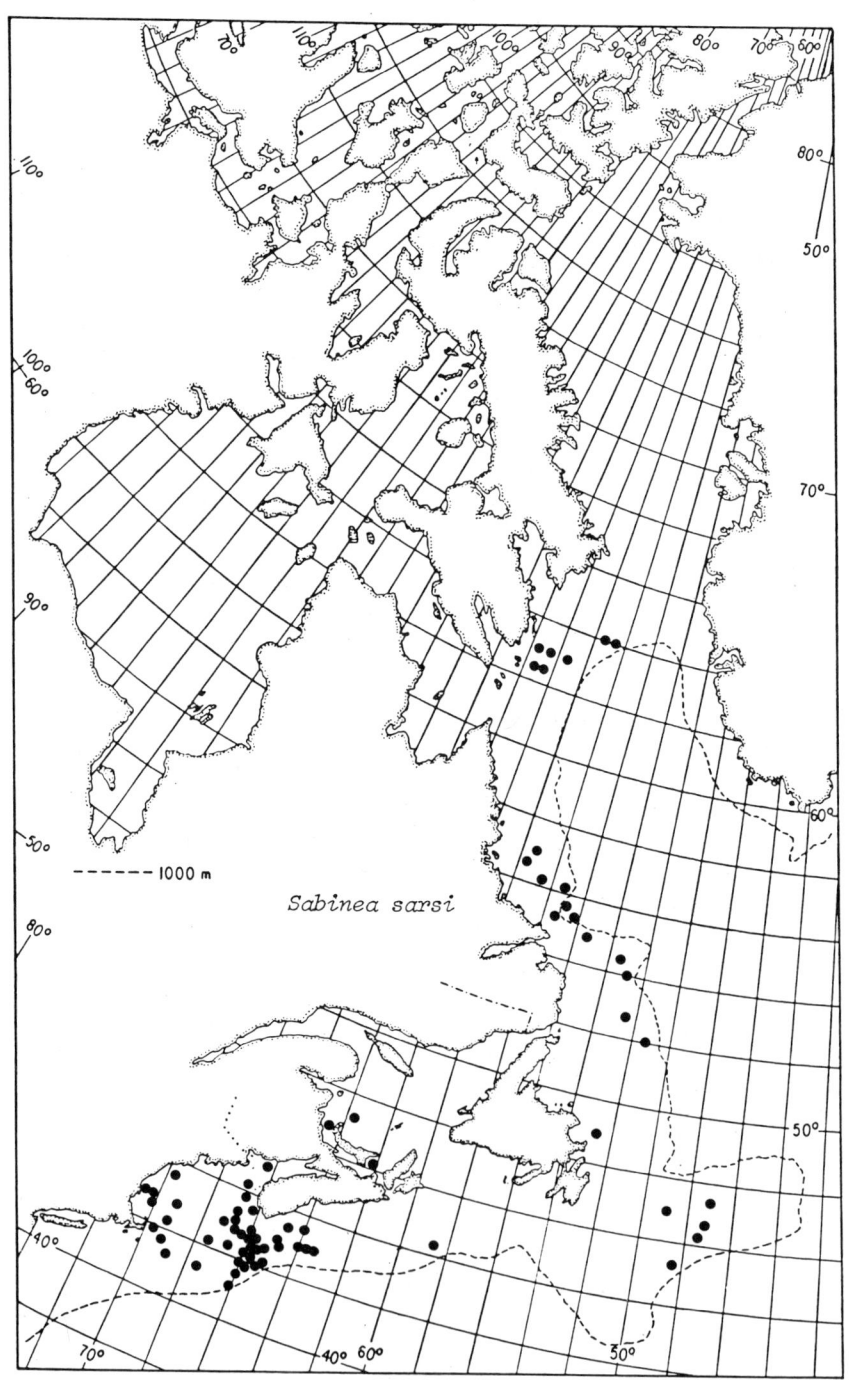

Fig. 162. *Sabinea sarsi*, distribution records in the area of reference.

Sabinea septemcarinata (Sabine, 1824)
Rathbun 1929: 22, fig. 28a–c; Williams 1974:
13, figs. 34A–B (key); 1984: 164, fig. 117.
(Figures 163 and 164)

DISTINGUISHING CHARACTERISTICS

Rostrum short, rounded, not exceeding eyes; carapace with seven longitudinal carinae, not as clearly toothed as in *S. sarsi*; abdominal somites all with dorsal carinae, divergent on 5th but without teeth; telson subtruncate with central spine and 4 or 5 pairs of lateral spines, also 4 pairs of dorso-lateral spines.

DESCRIPTION

Integument covered with very short reddish setae above first lateral carina, and small rounded tubercles with occasional short red setae below first lateral carina, especially anteriorly — even on eyestalks. Colour brownish red mottling on greyish background, slightly darker than *S. sarsi*.

Carinae on carapace with teeth as follows: mid-dorsal 4 plus a tubercle anterior to them; 1st lateral 11–12; 2nd lateral 11–13; 3rd lateral 12–13 plus the strong antennal; posteriorly the teeth become smaller. Dorsal carina begins on rostrum with slight hollow on each side, edge of rostrum confluent with orbit. Moderate post-orbital, strong antennal and small pterygostomian spines. Anterior edge of carapace turned under ventrally and with ventral lobe at about middle.

Lateral spine on 1st abdominal somite anteriorly is even with 2nd lateral on carapace; 1st somite has also two dorsal and a short lateral carina; the 2nd–4th with one dorsal carina, the 4th with an oblique lateral carina also; the 5th with two divergent carinae without teeth, and the 6th with two parallel dorsal carinae with small teeth. Pleura of 5th with a strong posterior lateral spine, similar to 6th.

Telson (t) with dorso-lateral carinae with 4–5 pairs of spinules on distal half, and tip rounded with a central and 4–5 pairs of lateral spines.

Antennule (c) 1st article longer than other two, 3rd shorter than 2nd; ventral spine on proximal third of 1st article; stylocerite compressed, wide, reaching about distal 3/4 of 1st article; dorso-lateral flagellum thick, shorter than ventromesial.

Antenna (d) scale length about 2.4 times width; distolateral spine exceeding blade; peduncle more than half scale; flagellum four times length of scale, 2.7 times length of carapace.

Mandible (e) molar with two unequal fangs, each with a subsidiary small tooth.

Maxillule (f) distal endite expanded from base, ovate, with long spines on short leading edge; proximal ovate, slightly tapered; endopod obscurely bifid with subapical incipient ramus with sharp spine.

Maxilla (g) endites reduced to slight protuberances with setae; endopod moderate; anterior lobe of scaphognathite with wide base rounded at tip, longer than posterior lobe, tongue-shaped with distal long setae.

Maxilliped I (h) endites almost completely reduced; endopod with two segments, as long as expanded part of exopod, lash a bit shorter; epipod large equally bilobed, each sub-triangular, about two-thirds the length of exopod.

Maxilliped II (i) six segments, the distal diagonally attached to next, with two strong spines and distal brush setae; exopod plus lash about as long as endopod; epipod with podobranch.

Maxilliped III (k) slightly compressed, apically setose, with a few lateral spines on distal segment; exopod short, plus lash equal to longest segment; rudimentary epipod.

Pereopods: I (l) subchelate, stout; merus compressed, laterally carinate; carpus short, with mesial spine. II (m) non-chelate, very short, reaching about middle of merus. III (n) non-chelate, very long tapering to sharp dactyl. IV and V robust, about as long as III, dactyl slightly compressed, curved, sharp.

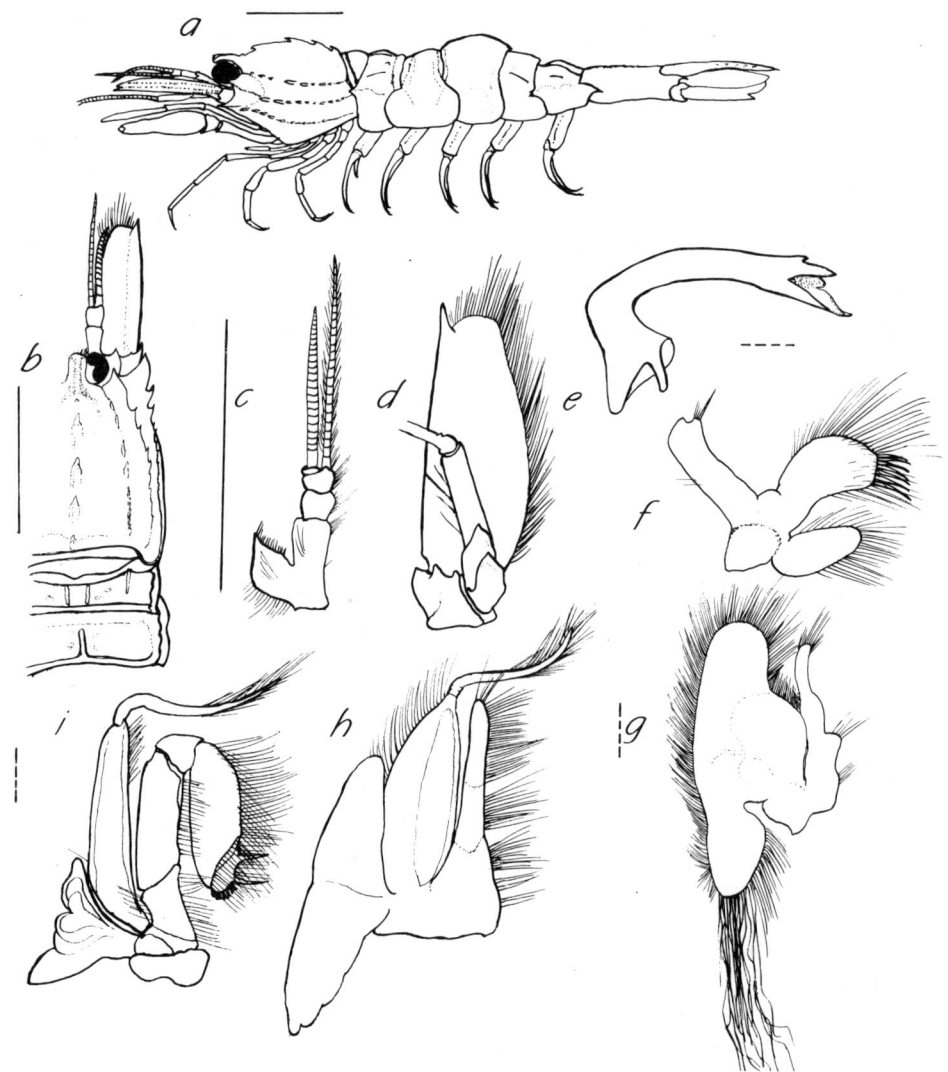

Fig. 163. *Sabinea septemcarinata*: *a*, whole shrimp from left side; *b*, carapace in dorsal aspect; *c*, antennule; *d*, antenna; *e*, mandible; *f*, maxillule; *g*, maxilla; *h*, first maxilliped; *i*, second maxilliped; solid line = 10 mm, broken line = 1 mm.

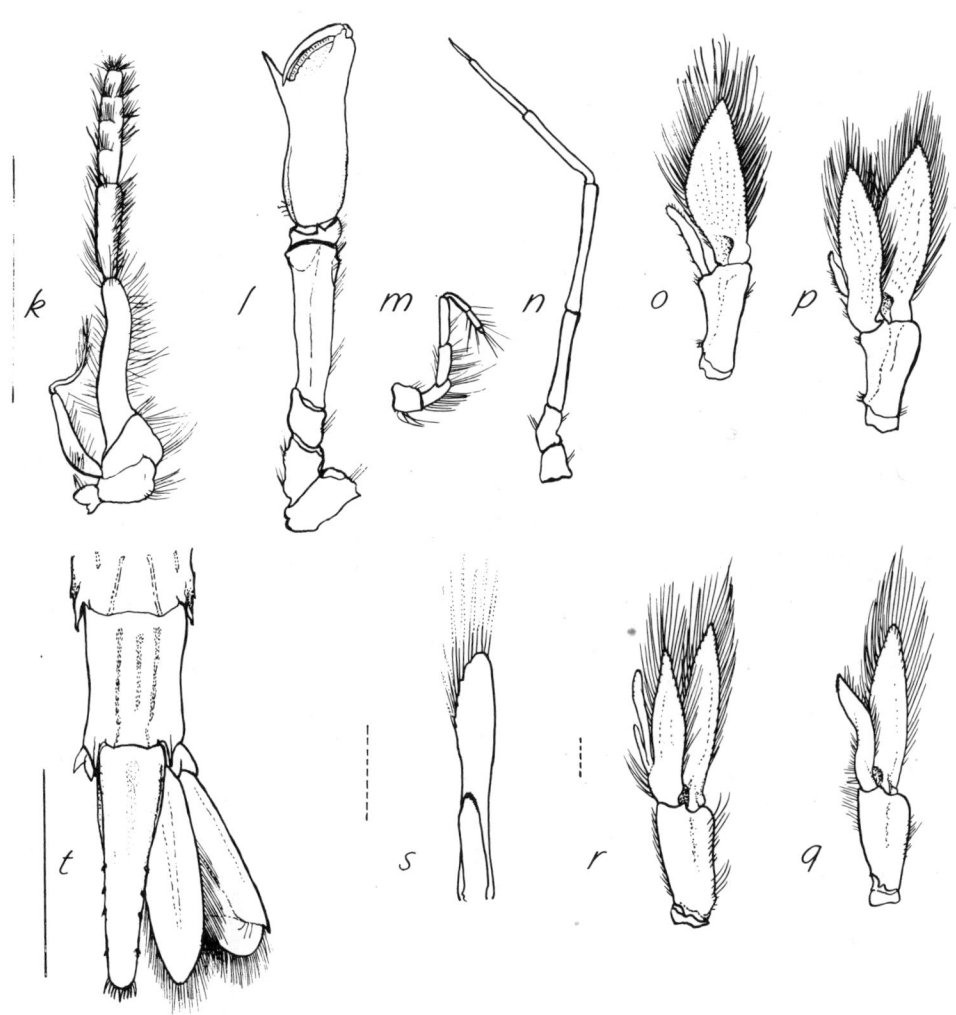

Fig. 164. *Sabinea septemcarinata*: k, third maxilliped; l, first pereopod; m, second pereopod; n, third pereopod; o, F first pleopod; p, F second pleopod; q, M first pleopod; r, M second pleopod; s, appendix masculina; t, telson; solid line = 10 mm, broken line = 1 mm.

Pleopods: female I (o) endopod narrow and much shorter than exopod, a few flexible spines laterally outside and a few setae but no hooks; female II (p) appendix interna narrowing distally and with subapical pad of hooks. Male (q) endopod sinuous, about two-thirds as long as exopod, with distal patch of hooks; male II (r) endopod slightly shorter than exopod, with appendix interna less than half as long as appendix masculina (s), the latter expanded slightly distally and with about nine long fine spines.

RANGE OF DISTRIBUTION

Arctic Canada and Alaska to Point Barrow and Chukchi Sea (Makarov 1941); Hudson Bay and Greenland to Massachusetts Bay; Iceland; Kara, White and Barents Seas to British Isles and Faroes in Europe (Williams 1984; Allen 1967). Depths 0–10 to 406 m (Heegaard 1941).

Records of distribution in the area of reference are in Fig. 165.

BIOLOGY

Lengths to 18 mm cl in males and 21 mm cl in females.

First maturity of females was 15 mm cl (26 specimens) in Hudson Bay and 10 mm (163 specimens) from Foxe Basin, Newfoundland and Labrador. In Hudson Bay 46% were potentially ovigerous in autumn and in the other areas about 52%. In Hudson Bay temperatures were -1.4 to $0.7°$ C and in other areas 1.4 to $5.2°$ C, although closer to $0°$ C where taken in most instances.

Other cold water species of decapods taken in the same hauls were *Lebbeus polaris*, *Pandalus montagui* and *Argis dentata*.

Stomach contents were phytobenthos, crustacean fragments (mostly gammarid amphipods, ostracods and cumaceans) and occasionally polychaetes (Squires 1965a).

Predators were cod, beluga whales and bearded seals (Squires 1957; 1965a; 1968a).

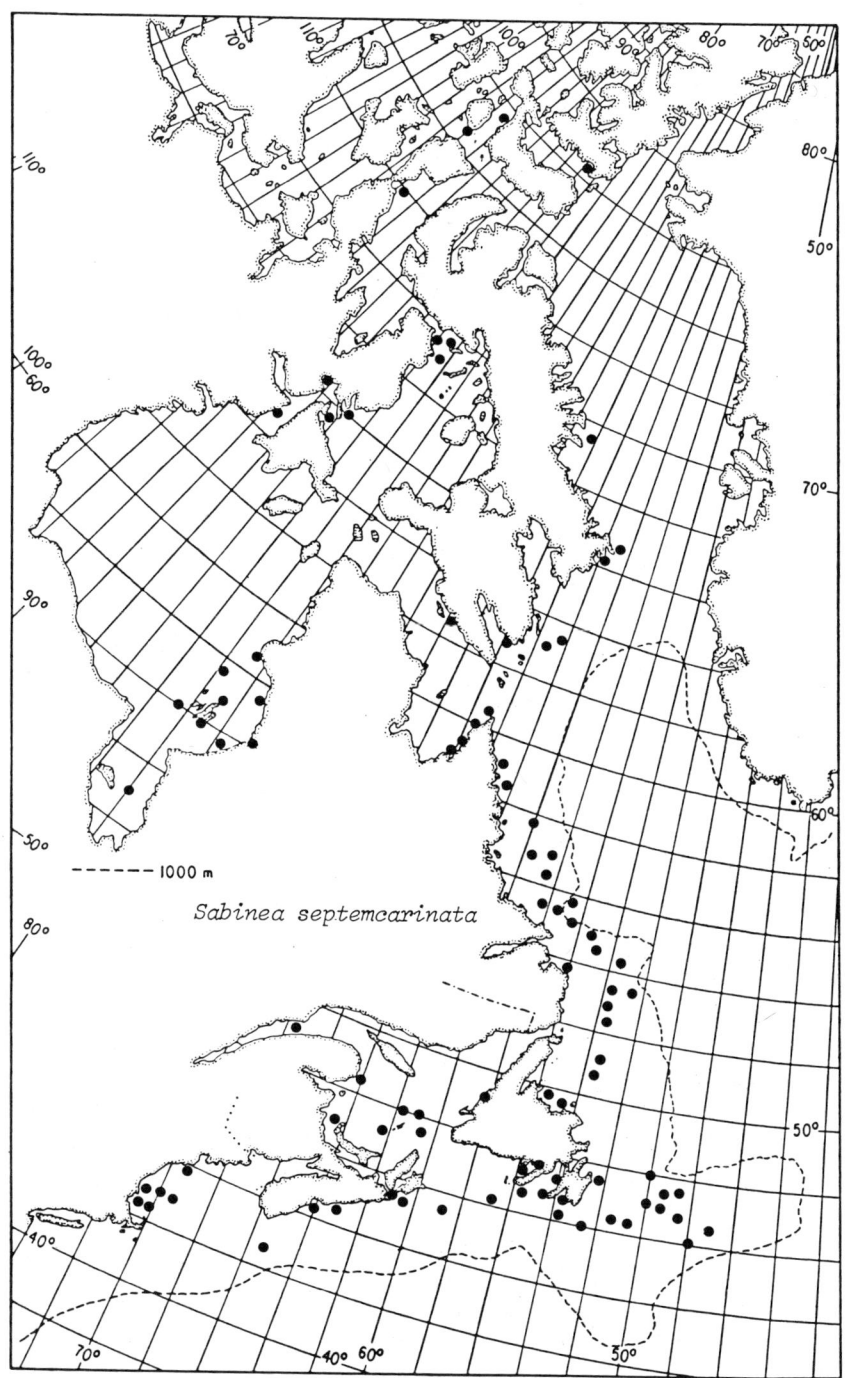

Fig. 165. *Sabinea septemcarinata*, distribution records in the area of reference.

Genus *Sclerocrangon* G. O. Sars, 1883
Butler 1980: 90; Holthuis 1955: 136;
Rathbun 1929: 20; Zarenkov 1965: 1764, fig. 6.

First pair of legs subchelate; second pair of legs chelate; dactyls of 4th and 5th legs not broadened; 2nd and 3rd legs subequal in length to other legs but much more slender; broad and heavy arctic shrimps; body surface sculptured, pleura of abdomen with ventral spines, dorsal carinae of 6th abdominal somite pronounced and pointed posteriorly, carapace with mostly three teeth on strong median carina. Endopod of 2nd pleopod reduced, smaller than large appendix masculina; arthrobranch absent on 3rd maxilliped; strong sharp spine on sternal plate of 3rd legs.

Of the six species suggested by Zarenkov (1965) two occur in the area of reference.

Sclerocrangon boreas (Phipps, 1774)
Butler 1980: 90–92, fig.; Rathbun 1929:
20, fig. 25; Williams 1984: fig. 118.
(Figures 166 and 167, Plate 6b)

DISTINGUISHING CHARACTERISTICS

Rostrum short, hollowed dorsally, with an acuminate tip, broadening rapidly with edge arching and confluent with the orbit, ventrally a narrow rounded keel. Carapace heavily sculptured, strong median carina with three broad and bluntish teeth the middle one sometimes divided into two. Second leg as long as the first but very slender and chelate, first leg stout subchelate.

DESCRIPTION

Integument rough, rigid, with many tubercles, interspersed with many short recurved setae. Colour reddish brown (Plate 6b).

Carapace with a rough median dorsal carina with three rough bluntish teeth, the middle one double, appearing sharper and more pronounced in specimens from the Arctic (Squires 1957); irregular lateral tuberculate carinae from moderate post-orbital and strong antennal spines extending posteriorly and including hepatic spine on anterior third, and intersecting with oblique tuberculate ridges toward median carina. From small pterygostomian spine anterior ventral edge of carapace is turned inward as far as middle ventral lobe.

Rostrum with acuminate point and widening tuberculate edges curving over orbits, short, but exceeding smallish eyes and strong antennal spines at the anterior corners of carapace, its short ventral keel is rounded and narrow.

Abdomen has mid-dorsal and lateral carinae on each somite, single on 1st and 2nd but double on 3rd to 6th dorsally; lateral carinae are less distinct but formed by rows of tubercles. Pleura are ventrally straight with a small posterior spine on 2nd to 6th somites, the 6th with flared supporting ridges.

Telson (t) with lateral tuberculate ridges with three pairs of spines on posterior third, last pair at base of triangular terminus with spine at tip. Uropod branches wide, the inner one shorter than telson but longer than outer.

Eyes smallish, the cornea smaller than stalk, with a small dorsal tubercle.

Antennule (c) 1st article longer than other two, 2nd and 3rd subequal; stylocerite short, about as wide as long, wing-like, tip reaches about half 1st article; flagella subequal, shorter than peduncle.

Antenna (d) scale length about 2 times width, disto-lateral spine sligthly exceeded by blade; peduncle is almost as long as scale; basal article with outer spine.

Mandible (e) molar with two unequal fangs, each with small subsidiary tooth.

Maxillule (f) distal endite large, slightly curved, with five long robust spines apically; proximal endite moderate, only slightly tapered to rounded tip; endopod straight with expanded round tip.

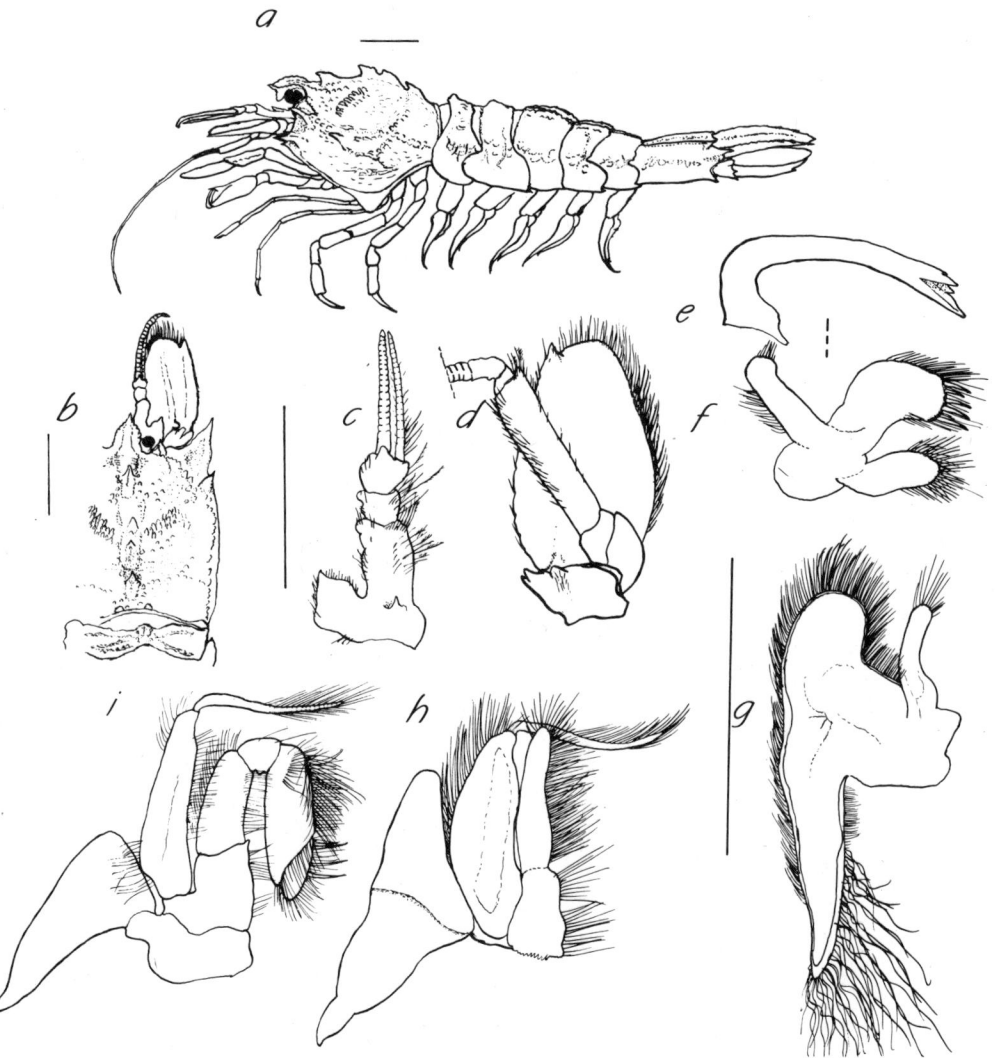

Fig. 166. *Sclerocrangon boreas*: *a*, whole shrimp from left side; *b*, carapace in dorsal aspect; *c*, antennule; *d*, antenna; *e*, mandible; *f*, maxillule; *g*, maxilla; *h*, first maxilliped; *i*, second maxilliped; solid line = 10 mm, broken line = 1 mm.

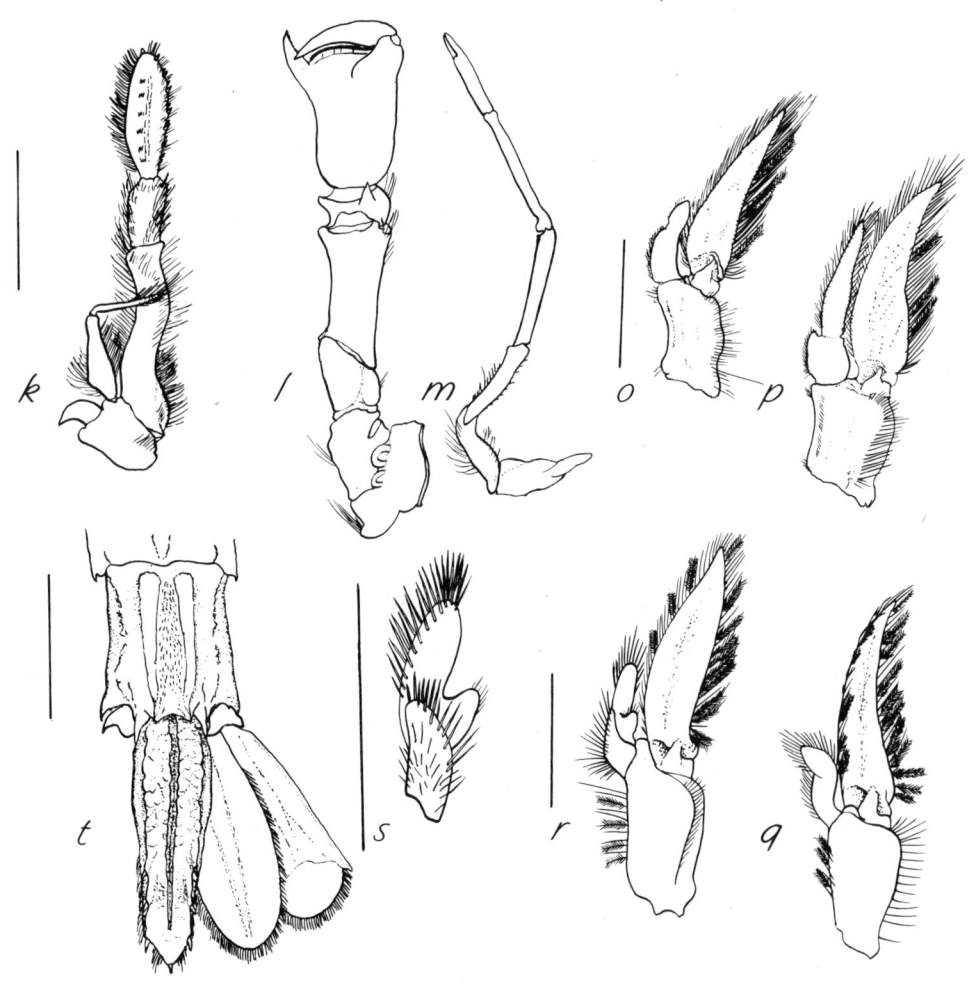

Fig. 167. *Sclerocrangon boreas*: *k*, third maxilliped; *l*, first pereopod; *m*, second pereopod; *o*, F first pleopod; *p*, F second pleopod; *q*, M first pleopod; *r*, M second pleopod; *s*, appendix masculina; *t*, telson; solid line = 10 mm.

Maxilla (g) endites greatly reduced; endopod reaches tip of anterior broad rounded lobe of scaphognathite, the latter much shorter than posterior narrow tapering lobe which has numerous apical and lateral long setae.

Maxilliped I (h) endites reduced; endopod reaches slightly farther forward than tip of exopod, lash not as long. Epipod large, bilobed, posterior lobe longer.

Maxilliped II (i) leglike compressed, distal segment attached to next diagonally and with three strong spines; exopod and lash about as long as endopod; epipod large, triangular in outline.

Maxilliped III (k) stout, compressed, distal segment ovate fringed apically and laterally with short strong spines; exopod short, with lash; epipod rudimentary, falcate.

Pereopods: I (l) massive, subchelate, strong distal spine on carpus; palm distally only slightly oblique with very strong corner spine. II (m) slender, chelate, palm much longer than fingers, about as long as I; epipod rudimentary. III slender, non-chelate, tapering to slender sharp dactyl, about as long as II. IV and V stoutish about as long as others, dactyl slender tapering, sharp.

Pleopods: female I (o) endopod less than half as long as exopod, narrow, extended distal part with setae but no hooks. Female II (p) endopod two segmented, first segment wider but shorter than second, much narrower than exopod. Male I (q) endopod much shorter and narrower than exopod, sinuous, without distal hooks. Male II (r) endopod wide at base, tapering rounded, shorter than appendix masculina (s), mitt-shaped with about 7 apical moderate spines continuous with about 16 lateral spines, lateral projection with fine setae only.

RANGE OF DISTRIBUTION

Possibly circumpolar; Hudson Bay to west and east Greenland and south to Cape Cod in America; Kara and White Seas to Iceland and British Isles and Faroes in Europe; arctic Canada and north coast of Alaska and Siberia (near Bering Strait) in the Arctic Ocean; Chukchi and Bering seas to British Columbia in the north Pacific (various authors). Depths 0–400 m (Heegaard 1941).

Records of distribution in the area of reference are in Fig. 168.

BIOLOGY

Lengths to 25 mm cl in males and 35 mm cl in females.

Females were first mature at 23 mm cl. Those potentially ovigerous in autumn were about 65% in 33 specimens, and 85% in specimens possibly indicating annual spawning. Eggs were 2.9 mm in greater diameter.

Fecundity is up to 488 eggs per female (Williams 1984). Development of larvae is brief and they remain with the female rather than becoming planktonic. In one female from Cornwallis Island 158 larvae were counted, but some could have been lost in handling (most larvae were Stage III with uropods enclosed, about 2.5 mm cl, but a few were Stage IV with uropods free, 3 mm cl. Specimens lent through courtesy of B. W. Fallis, Freshwater Institute, Winnipeg, Manitoba).

Species of decapod crustaceans present in catches with this species was primarily *Sabinea septemcarinata*, but also occuring were *Spirontocaris spinus*, *Eualus fabricii* and *Argis dentata*. (Squires 1965a).

Stomach contents were phytobenthos, polychaetes, crustacean fragments (gammarids, ostracods, isopods and copepods), small bivalves, ophiuroids, sponge spicules and foraminiferans (Squires 1965a, 1968a, 1969).

Predators were cod, longhorn sculpin, ringed, bearded and harbour seals and beluga whales (Squires 1957, 1965a, 1967a, 1968a).

It is possible that this species is a protandric hemaphrodite (Bernier and Poirier 1981).

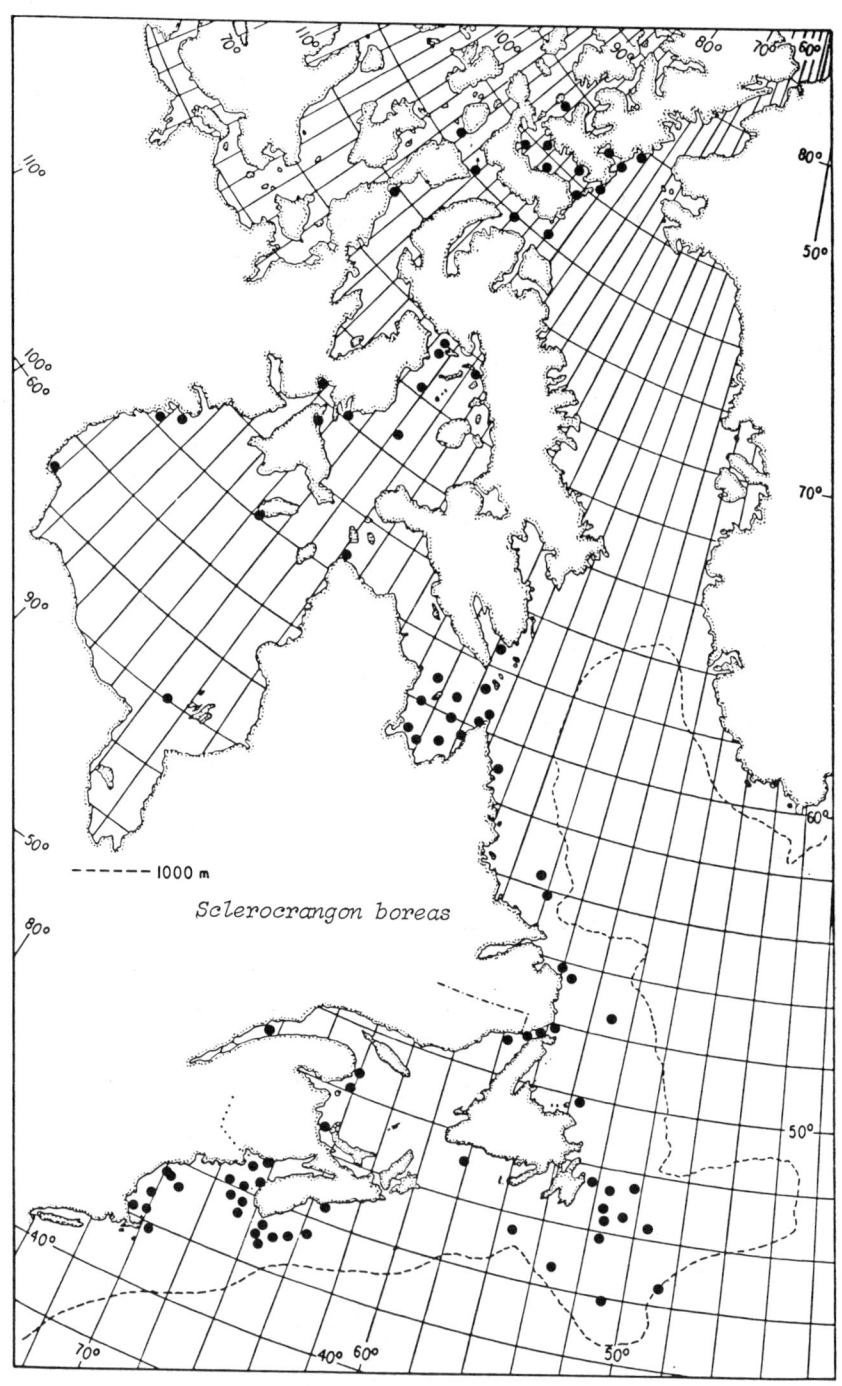

Fig. 168. *Sclerocrangon boreas*, distribution records in the area of reference.

Sclerocrangon ferox G. O. Sars, 1877
Blacker 1957: 14; de Man 1920; Squires 1965a.
(Figures 169 and 170)

DISTINGUISHING CHARACTERISTICS

Somewhat similar in appearance to *S. boreas* but spines are longer and more numerous; also the rostrum is longer, pointing upward and spike-like and the ventral keel is deeper and pointed; the three teeth on median carina are sharp and with accessory teeth; two hepatic spines with strong carina; ventral spines on abdominal pleura are sharp and include two on somites 2nd–4th and four on 5th.

DESCRIPTION

Integument rigid, thick, covered with short pointed setae. Colour brownish.

Carapace strongly but plainly sculptured (less tubercles than in *S. boreas*) and with a high median carina and 3 sharp teeth with accessory teeth (3, 1 and 1 posteriorly). Rostrum a blade exceeding eyes, about 0.2 cl, rising high as a sharp tooth and extending below eyes as a ventral keel compressed and with many lateral setae, edge confluent with orbit and extending back as a ridge to the hepatic region; postorbital spine moderate, antennal very strong and sharp, pterygostomian small, two hepatic spines strong and with long carina; band along ventral edge of carapace forming a ventral lobe at position of 2nd legs.

Abdominal somites with median carinae, first with 1 or 2 spines (also a strong lateral spine overlapping posterior edge of carapace), second with 1 median spine, fifth with carina produced posteriorly as a spine and sixth with two parallel carinae and distally on each two sharp spines in line. Ventrally the pleura have 1, 2, 2, 2, 4 and 1 sharp spines on somites 1–6 respectively; 5th somite also has a sharp spine posterioly on tergum above articulation point.

Telson (t) with strong mid-dorsal sulcus proximally and dorso-lateral ridges with one pair of spines at about the middle, and distally two pairs of spines at base of triangular-shaped tail spine.

Eyes moderate, cornea globular, smaller than diameter of short bulbous stalk.

Antennule (c) 1st article longer than other two combined, 3rd about half 2nd; subequal flagella shorter than peduncle; stylocerite rounded, with sharp acuminate tip reaching about half 1st article.

Antenna (d) scale length 2.8 times width, 0.5 times cl, scale exceeds small distolateral spine; peduncle about 4/5 length of scale; strong outer spine on basal article.

Mandible (e) molar divided into two unequal fangs, the lower longer and sharper, each with a proximal squared-off tooth.

Maxillule (f) proximal endite sub-rectilinear rounded at tip; distal endite expanded at middle, curved, with 4 long spines at distal edge; endopod moderate with subapical tuft of setae.

Maxilla (g) endites reduced to faint rounded areas; endopod sinuous, moderate, not as long as very wide anterior lobe of scaphognathite; posterior lobe very narrow and pointed distally with long apical and lateral setae.

Maxilliped I (h) endites reduced to slight expansion of base; endopod short, joined near middle of exopod and about half length of exopod, lash longer; epipod large with anterior lobe larger than posterior.

Maxilliped II (i) compressed, leglike, six segments, the distal attached diagonally, short but wide, with four spines at outer corner; exopod and lash about as long as endopod; epipod unilobed, sack-like.

Maxilliped III (k) five segments, distal pointed with one apical spine only and setal tufts; exopod small and very short, with lash not as long as longest segment; slightly longer than 1st leg. Epipod rudimentary, falcate.

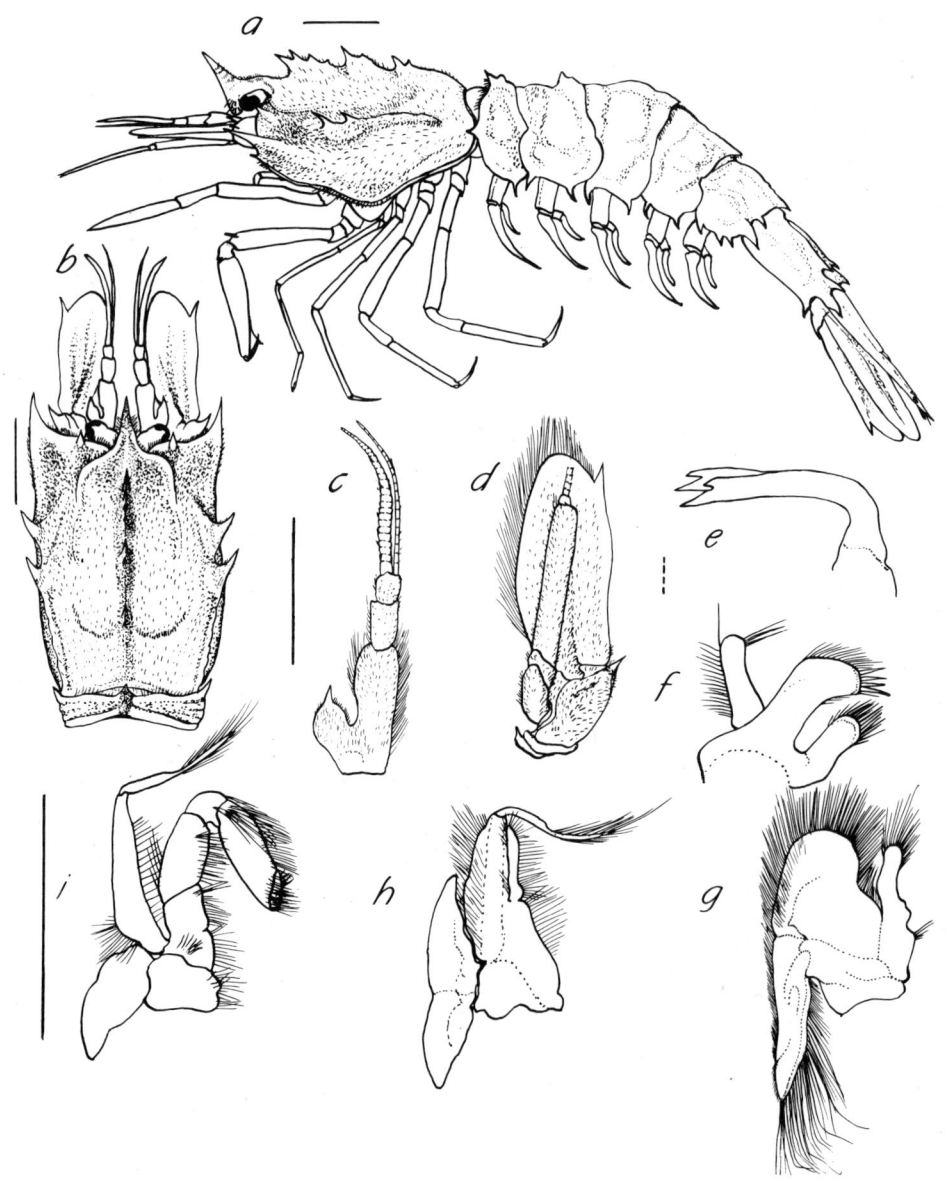

Fig. 169. *Sclerocrangon ferox*: *a*, whole shrimp from left side; *b*, carapace in dorsal aspect; *c*, antennule; *d*, antenna; *e*, mandible; *f*, maxillule; *g*, maxilla; *h*, first maxilliped; *i*, second maxilliped; solid line = 10 mm, broken line = 1 mm.

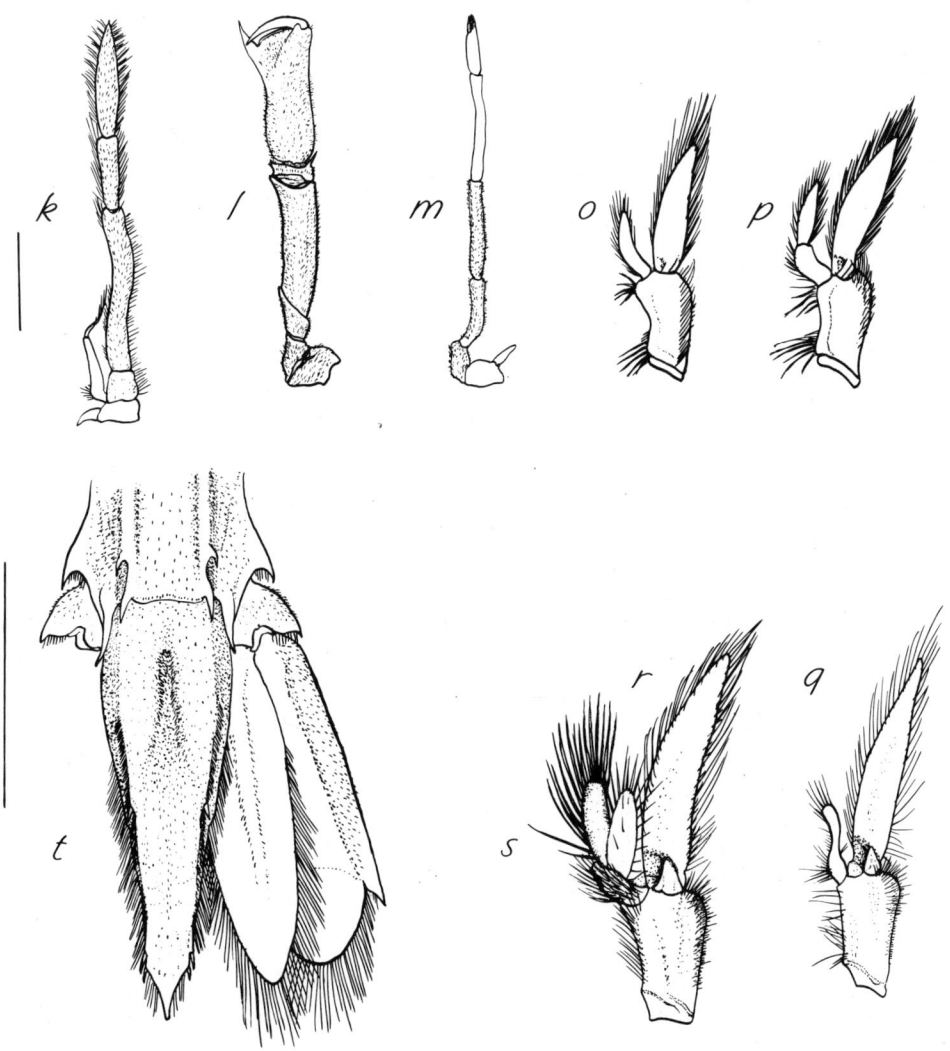

Fig. 170. *Sclerocrangon ferox*: k, third maxilliped; l, first pereopod; m, second pereopod; o, F first pleopod; p, F second pleopod; q, M first pleopod; r, M second pleopod; s, appendix masculina; t, telson; solid line = 10 mm.

Pereopods: I (l) subchelate, very stout, pilose, carpus with sharp spine distally; distal edge of palm only slightly oblique; II chelate, with fingers much shorter than palm, carpus and propodus without setae; epipod rudimentary, tiny. III slender, as long as others, tapering to slender sharp dactyl; IV and V stoutish, about as long as III, dactyl round, curved, 0.6 propodus.

Pleopods: female I (o) endopod narrow, curved, about half as long as exopod; female II (p) endopod two-segmented, narrower and shorter than exopod. Male I (q) endopod much shorter than exopod, sinuous with apical setae but no hooks; male II (r) endopod much shorter and narrower than exopod, also slightly shorter than somewhat curved and truncate appendix masculina (s), the latter with many terminal and outer lateral long spines.

RANGE OF DISTRIBUTION

Possibly circumpolar; in the Arctic Ocean — East Siberian Sea — and in the eastern North Atlantic from the Kara Sea to the Shetlands; in the western North Atlantic from east Greenland and Baffin Bay to the eastern slope of the Grand Banks, including Ungava Bay to the north. Depths 90–1 000 m (de Man 1920; Squires 1965a).

Records of distribution in the area of reference are in Fig. 171.

BIOLOGY

Lengths to 24 mm cl in males and 31 mm cl in females.

Males were first mature at 15 mm and females at 21 mm. In 57 females collected off Baffin Island and northern Labrador 80% were potentially ovigerous in August and September, indicating possible annual spawning (Squires 1965a). Number of eggs in a clutch is about 133; greater egg diameter 3 mm (Zarenkov 1965).

Species in the hauls with *S. ferox* were *Bythocaris payeri*, *Sabinea sarsi*, *Pandalus borealis*, *Pandalus montagui*, *Spirontocaris spinus* and *Lebbeus polaris*.

Stomach contents were phytobenthos, crustacean fragments (including gammarid amphipods), polychaetes, ophiuroids, gastropods, bivalves and sponge spicules (Squires 1965a).

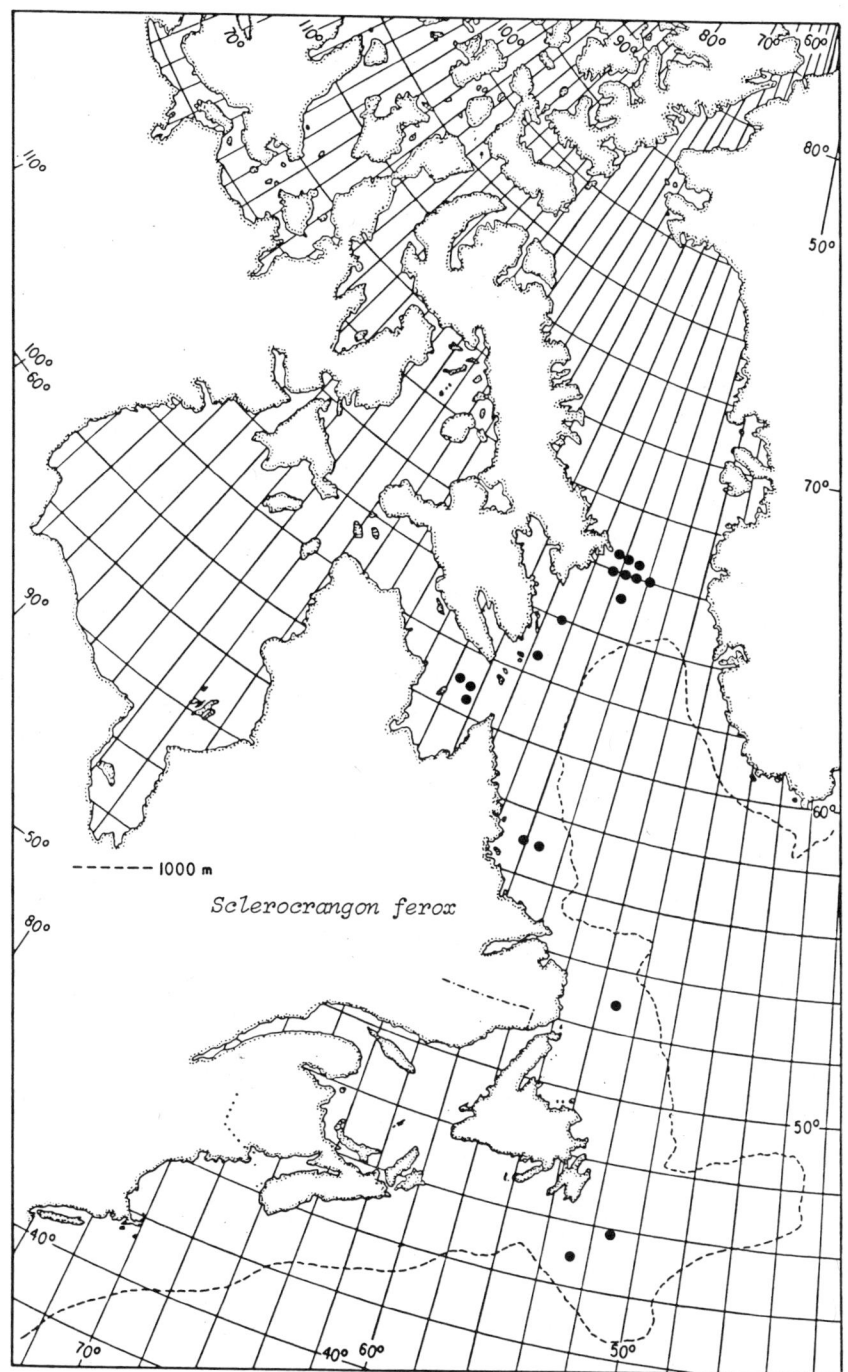

Fig. 171. *Sclerocrangon ferox*, distribution records in the area of reference.

PLATES

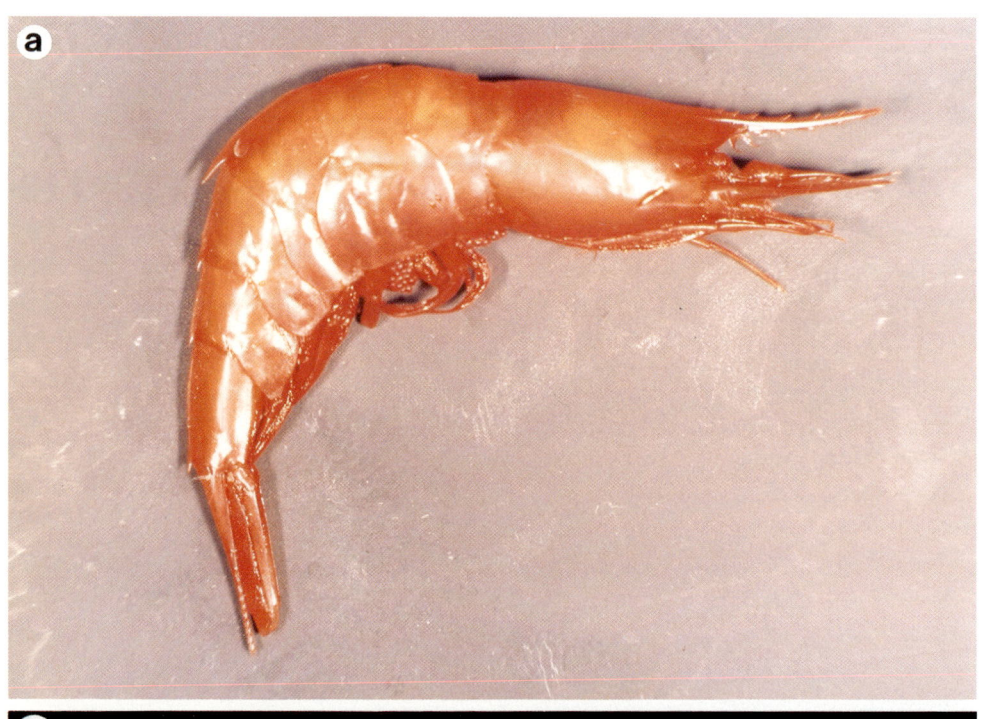
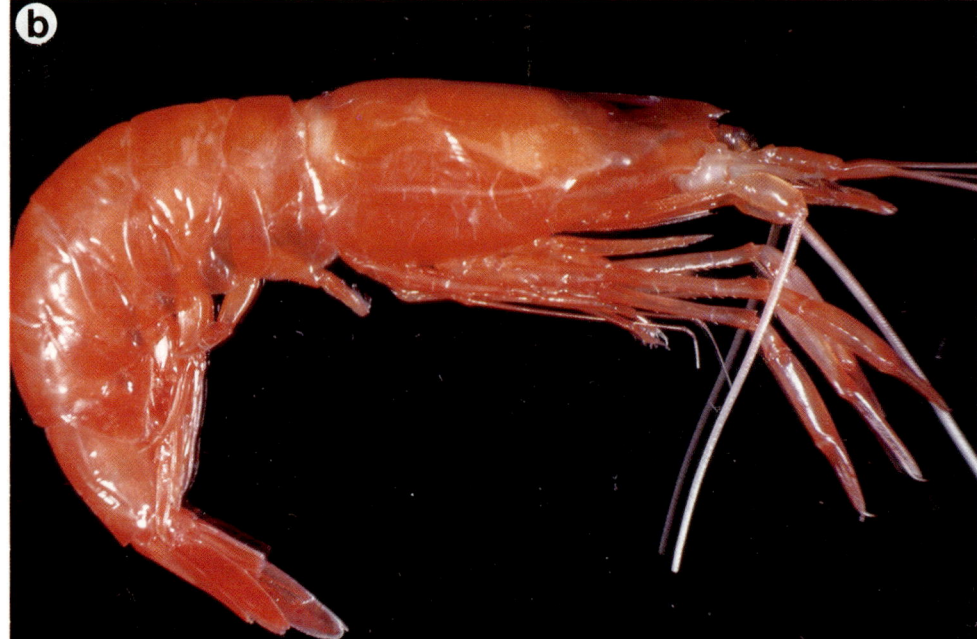

PLATE 1. a, *Acanthephyra pelagica*; b, *Pasiphaea tarda*.

PLATE 2. *Eualus gaimardi belcheri*.

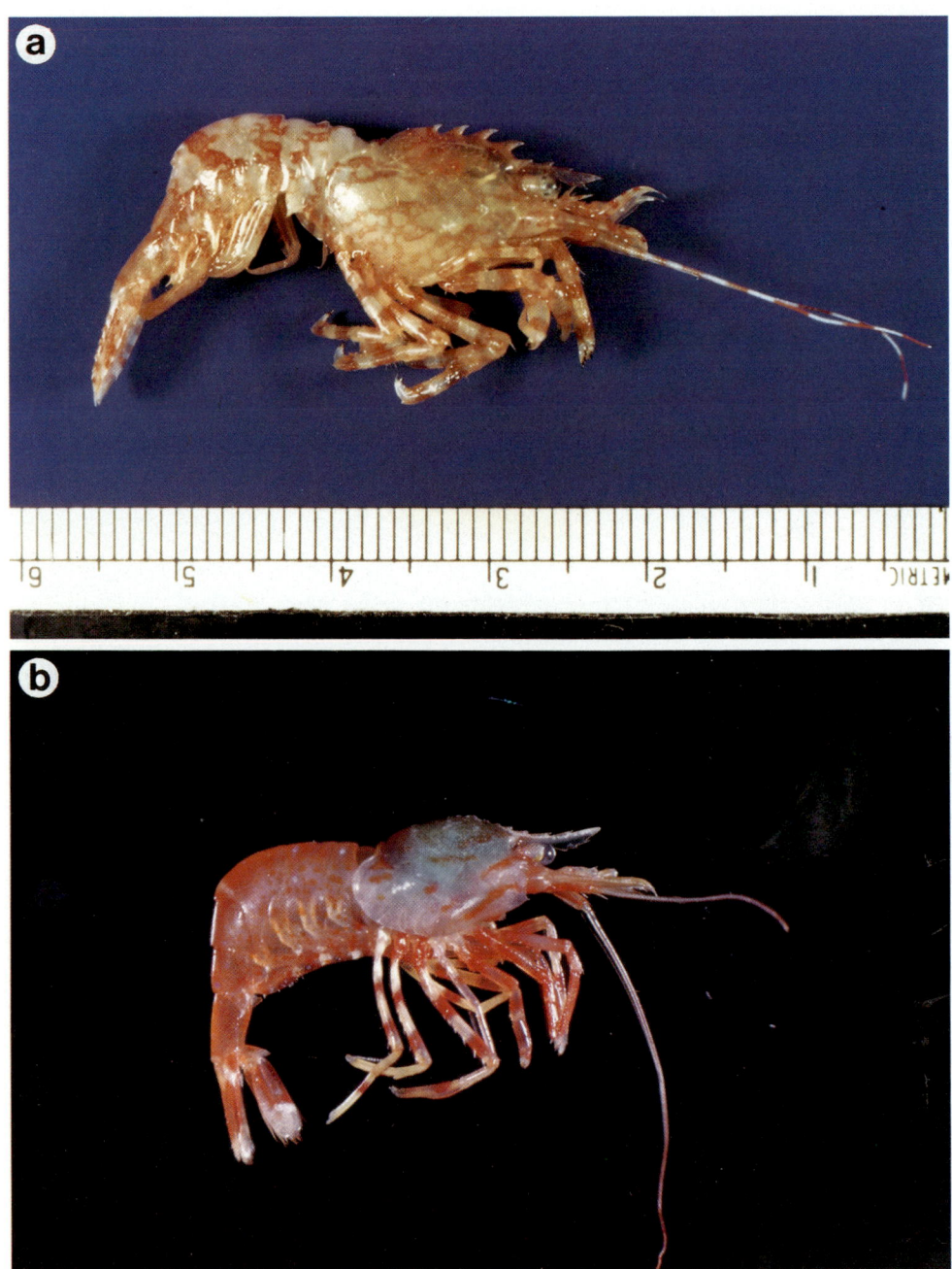

Plate 3. a, *Lebbeus groenlandicus*; b, *Lebbeus polaris*.

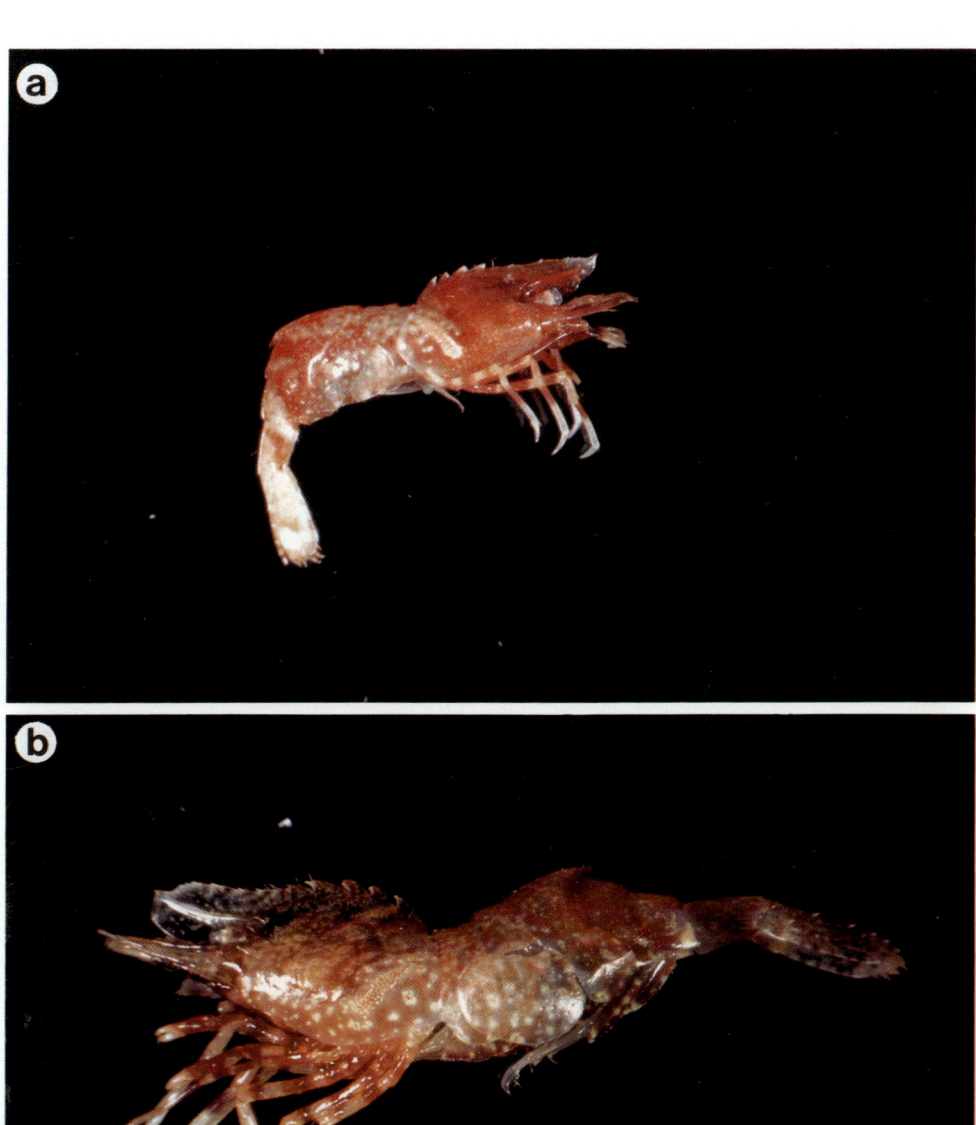

PLATE 4. a, *Spirontocaris lilljeborgi*; b, *Spirontocaris spinus*.

Plate 5. a, *Pandalus montagui*; b, *Argis dentata*, side view.

PLATE 6. a, *Argis dentata*, dorsal view; b, *Sclerocrangon boreas*.

PLATE 7. *Axius serratus*.

PLATE 8. *Lithodes maja*, male: a, dorsal view; b, ventral view.

PLATE 9. *Lithodes maja*, female: a, dorsal view; b, ventral view.

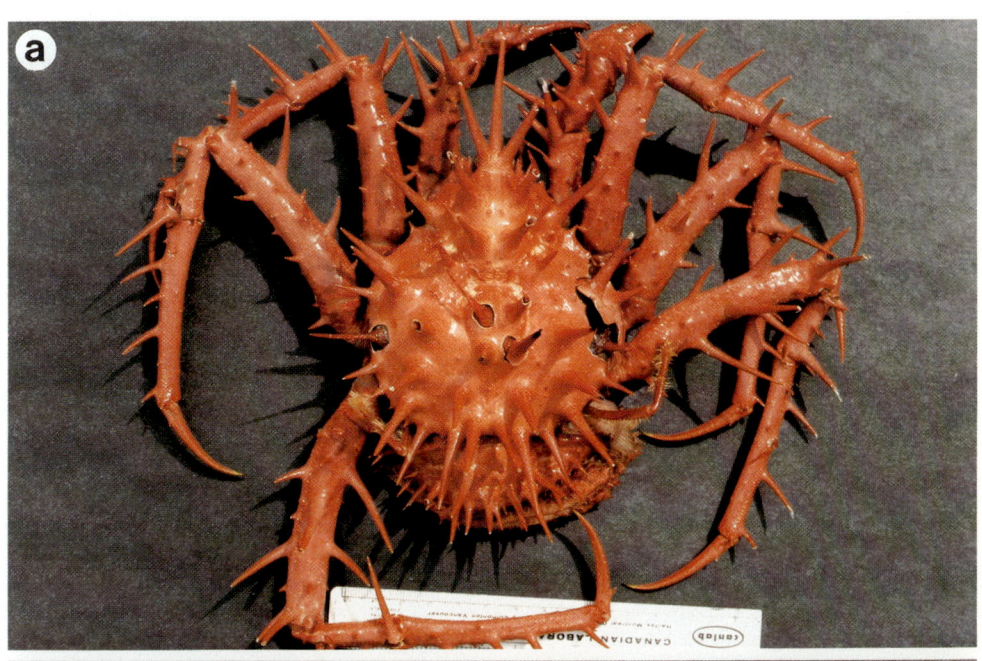

PLATE 10. *Neolithodes grimaldii*, female: a, dorsal view; b, ventral view.

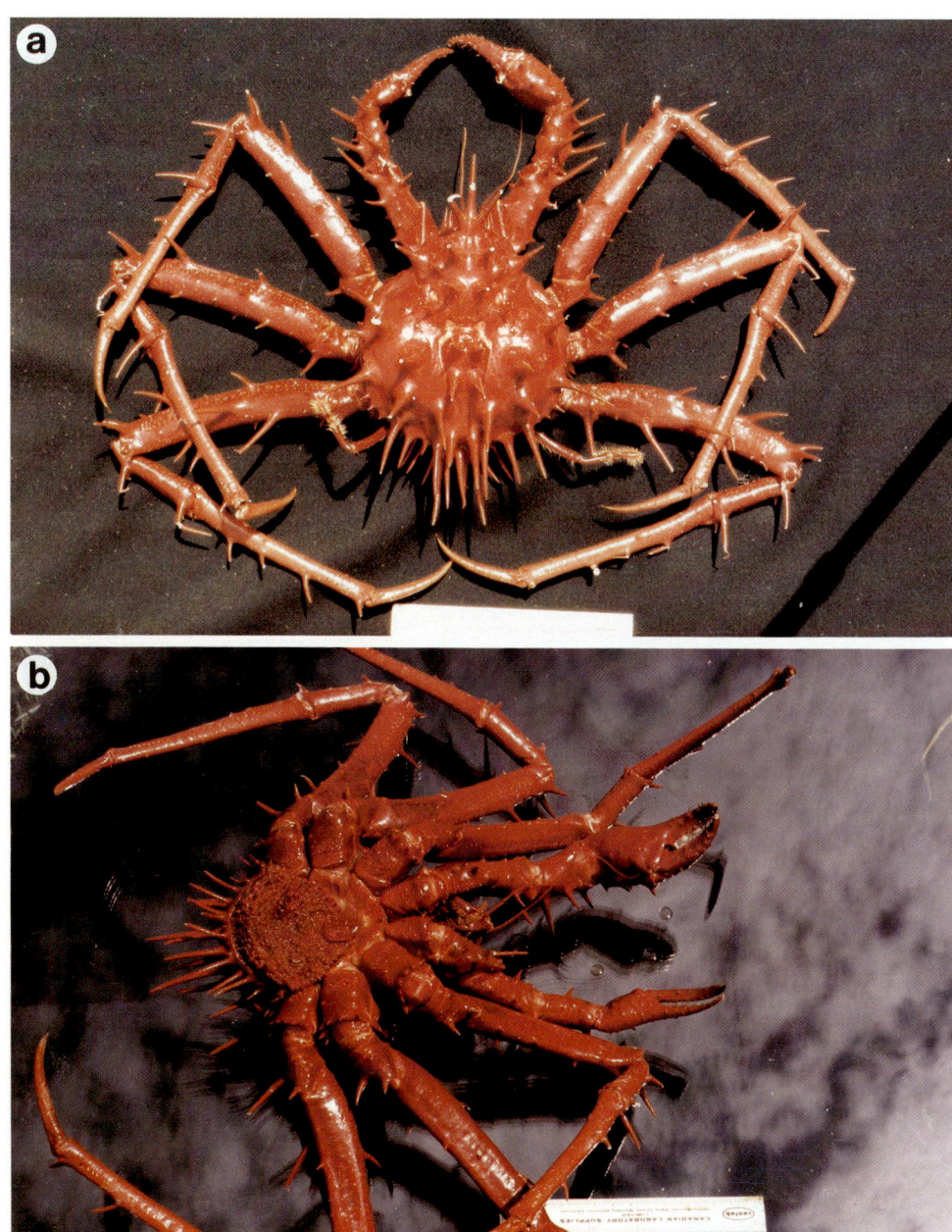

PLATE 11. *Neolithodes grimaldii*, male: a, dorsal view; b, ventral view.

Suborder	REPTANTIA
Infraorders	ASTACIDEA
	THALASSINIDEA
	PALINURA
	ANOMALA
	BRACHYURA

Family NEPHROPIDAE Dana, 1852
Holthuis 1974: 732-738; Williams 1984: 167.

Rostrum with lateral and sometimes ventral teeth; eyes on moveable stalks; carapace with postorbital but no branchiostegal spine; postcervical groove the most conspicuous (several other grooves described by Holthuis 1974), it crosses the dorsomedian groove which extends from the point of the rostrum to the posterior margin; abdominal pleura large usually ending in a ventral point; telson with posterolateral or lateral spines; outer branch of uropod usually with suture; all maxillipeds and pereopods (except 5th) have a large epipod; first pereopods large with strong chela, 2nd and 3rd slender and chelate, 4th and 5th non-chelate.

Only one genus and species in the area of reference.

Genus *Homarus* Weber, 1795
Holthuis 1974: 815; Williams 1984: 167.

Surface smooth; rostrum prominent with lateral and sometimes ventral teeth; carapace with supraorbital, postorbital and antennal spine; dorsomedian groove from tip of rostrum to posterior margin; postcervical groove most distinct dorsally and urogastric groove just behind it but fainter; abdomen smooth, pleura pointed; telson narrowing posteriorly, with posterolateral spine at each side, tip rounded; outer uropod with suture; first legs with large cutting and larger crushing claw.

Homarus americanus H. Milne-Edwards, 1837
Corrivault et Tremblay 1948: 16, figs. 1–5;
Holthuis 1974: 818; Williams 1984: 168, fig. 119.
"American lobster" — "Le homard americain"
(Figures 172 and 173)

DISTINGUISHING CHARACTERISTICS

The predominantly dark greenish colour of the American lobster with its claws and legs reddish underneath is distinct from the predominantly dark bluish colour of the European lobster with its claws and legs orangish underneath. The enormous claws of the first legs and the smooth exterior are unmistakeable features; the body is subcylindrical and robust, the rostrum short and with lateral upturned teeth, and the abdomen with large sharp-pointed pleura; the carapace has only a few small anterior spines: the supraorbital, postorbital and antennal; a distinct cervical (postcervical precisely) groove crosses the dorsomedian groove at about half its rostral-carapacial length; the 2nd and 3rd maxillipeds and legs I–IV have a large epipod and podobranch; legs I–III are chelate.

DESCRIPTION

Integument heavily calcified, smooth and shiny, but with numerous small pits and short setae. Colour is dark mottled green with reddish trim and undersides.

Rostrum with strong upturned apical spine and three pairs of upward curving lateral spines, and a ventral spine near the tip.

Carapace subcylindrical with pronounced postcervical groove and less distinct urogastric behind it; cervical groove short, placed well ahead of postcervical though connected to it by faint branch; hepatic and gastro-orbital grooves faint. A median dorsal groove extends the length of the carapace and out on the rostrum. A strong marginal groove curves downward at about halfway along the ventral edge and widens out as it turns upward at posterior edge.

Abdomen with smooth rounded terga; pleura of 2nd to 6th somites sharp-pointed ventrally; 2nd to 5th somites with spines on ventral ridges in males and immature females.

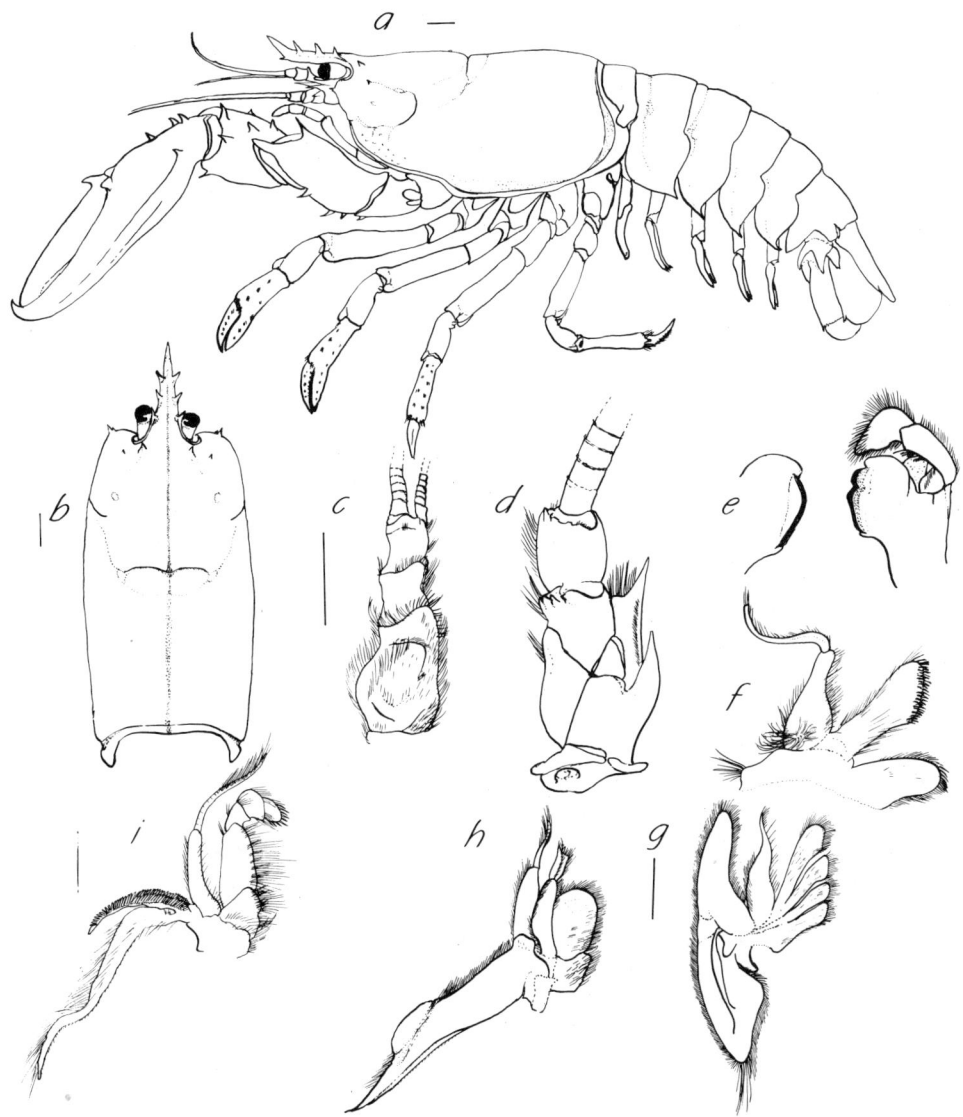

Fig. 172. *Homarus americanus*: *a*, whole lobster from left side; *b*, carapace in dorsal aspect; *c*, antennule; *d*, antenna; *e*, mandible; *f*, maxillule; *g*, maxilla; *h*, first maxilliped; *i*, second maxilliped; solid line = 10 mm.

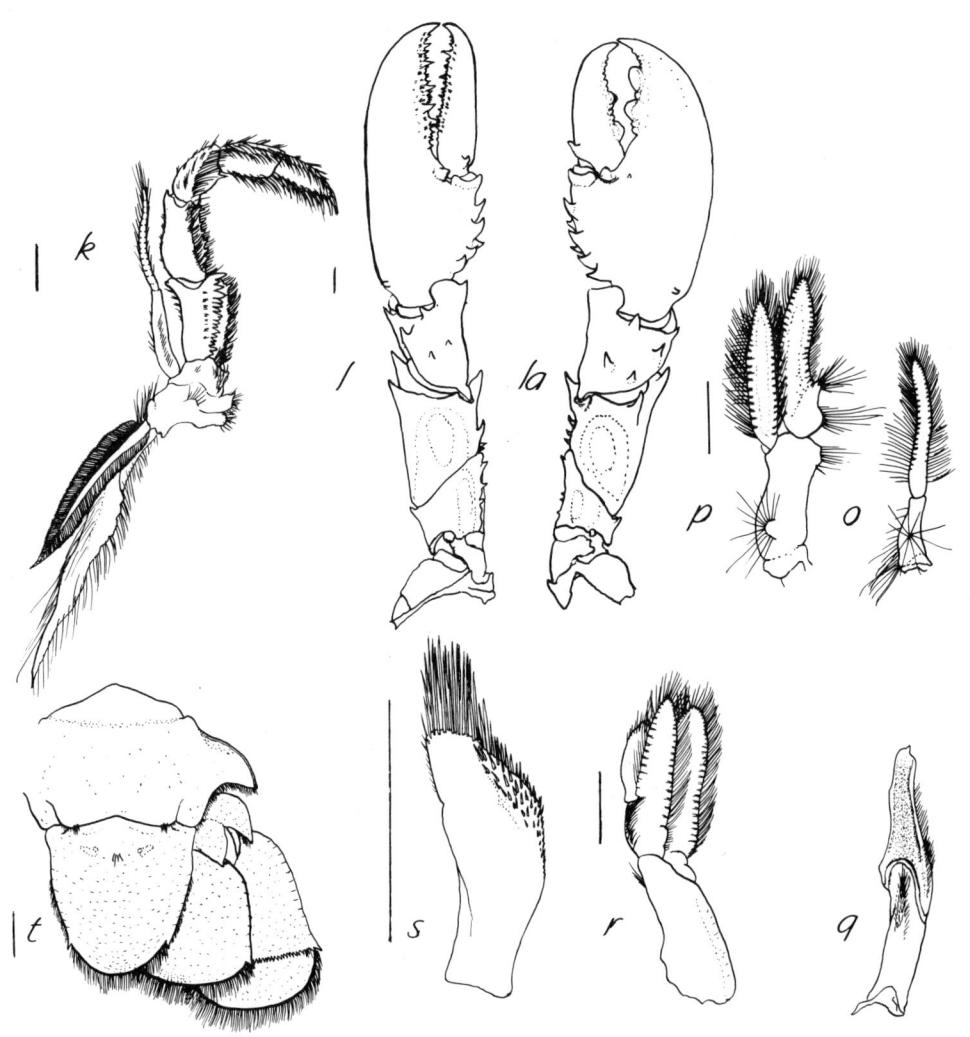

Fig. 173. *Homarus americanus*: *k*, third maxilliped; *l*, first pereopod right; *la*, first pereopod left; *o*, F first pleopod; *p*, F second pleopod; *q*, M first pleopod; *r*, M second pleopod; *s*, appendix masculina; *t*, telson; solid line = 10 mm.

Telson (t) wide, tapering as it gradually narrows to a pair of distolateral spines at posterior four-fifths and becomes well-rounded terminally with a short setal fringe. Branches of uropod longer than telson, the outer longer, the inner with spine at distal corner, the outer with transverse suture with numerous small sagittate teeth and outer spine.

Eye moderate, cornea globular with diameter about equal to diameter of eyestalk.

Antennule (c) stout, 1st article longer than other two combined, with ventrodistal tooth; 2nd and 3rd subequal; flagella slender, outer twice as thick as inner.

Antenna (d) basal article with strong distolateral spine; scale short and wide with distolateral spine about half as long as blade; peduncle with two short outer distolateral spines on second article and one of the same on 3rd; flagellum stout tapering, slightly longer than body.

Mandible (e) heavily calcified but incisor with chitinous edge, left with two rounded teeth and straight edge, right with curved untoothed edge; molar shelf lies behind incisor and is slightly longer; palp three-segmented.

Maxillule (f) distal endite expanded slightly toward spinous edge, proximal tapered to rounded tip, about equal in length; endopod short from wide base, with long curved lash.

Maxilla (g) endites almost equally bilobed, long and slender; endopod sinuous, slender, about even with tip of narrow anterior lobe of scaphognathite, the latter slightly longer than triangular posterior lobe (with a few long setae at its apex)

Maxilliped I (h) endites rounded; endopod or central lobe with distal segment triangular in cross-section and shape; exopod with short lash; epipod very long with distolateral lobe.

Maxilliped II (i) leglike, compressed, exopod with multiarticulate lash; epipod long and ribbon-like but tapered and with many spinules along one side and setae on the other; podobranch.

Maxilliped III (k) leglike, merus and ischium triangular in cross-section, ischium with crista dentata plus row of moderate teeth on inner edge, merus with row of teeth on inner edge, carpus with inner distolateral spine; exopod short with multiarticualte lash; epipod very long with expanded edge and tapering distally and with many setae; also with podobranch.

Pereopods: I unequal, chelate, enormous, left (l) the crusher (sometimes right) wide and heavy with few rounded teeth, fingers about equal to palm in length; the other (cutter (m)) has many small sharp teeth, and a couple of large ones outside the cutting edge, fingers longer than palm, the latter with a few very strong teeth at inner edge; dactyl of both with proximal spine. II and III slender, chelate; IV and V non-chelate, dactyl stout but sharp; II–IV with long epipod and podobranch.

Pleopods: female I (o) uniramous, slender sinuous endopod; female II (p) protopod stout with proximal setose tubercle, endopod and exopod subequal; male I (q) rigid, the endopod overlapping and locked to protopod, tapered, with small acuminate point distally, and forming a hollow along inner surface; male II (r) endopod and exopod equal, the former with appendix masculina (s) slightly curved, with apical fine spines and on convex side many short spines.

RANGE OF DISTRIBUTION

From the Straits of Belle Isle, Newfoundland, to Cape Hatteras, North Carolina (occasionally as far south as Rich Inlet, ENE Wilmington, NC). Depths mostly from 4 to 50 m but on the edge of the continental shelf may reach 480 m (Holthuis 1974; Williams and Williams 1981).

Records of distribution in the area are in Fig. 174.

BIOLOGY

Total lengths (including rostrum) to 643 mm in males and 610 mm in females (Wolff 1978).

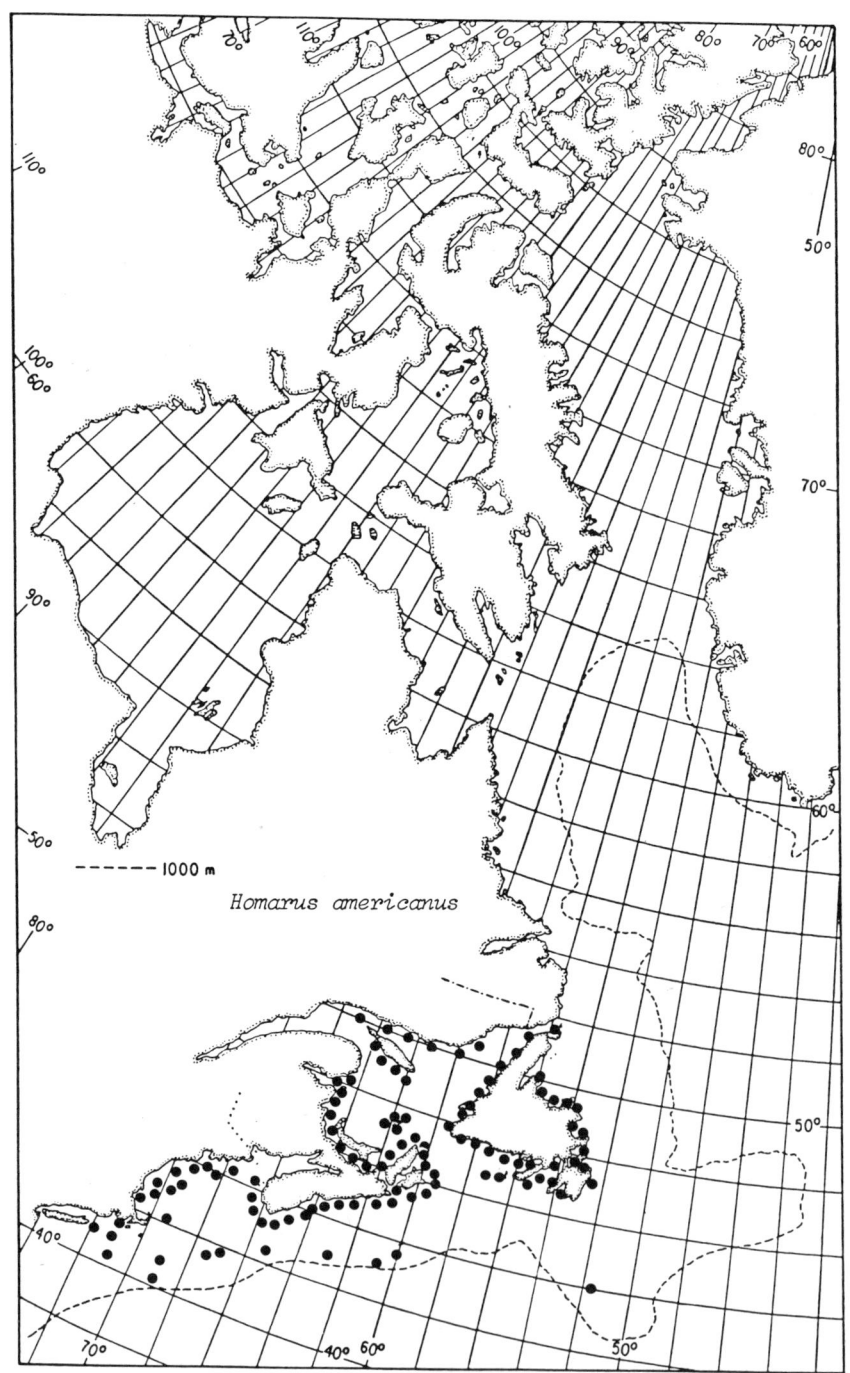

Fig. 174. *Homarus americanus*, distribution records in the area of reference.

Williams (1984) refers to the various extensive studies made on this species throughout its range.

Reproduction and growth are influenced by prevailing ambient temperatures, but since it is a cool temperate species there are limits to this influence. Studies have indicated that eggs are carried for as long as eleven months before hatching, and at least in the northern part of its range, spawning is biennial: the percentage potentially ovigerous in samples is close to 50%, and ova in ovaries of those carrying eggs are almost invariably small (Squires 1970; Squires et al 1971, 1973). First maturity has been shown to occur at smaller sizes where annual warm conditions last longer such as in the southern Gulf of St. Lawrence (Templeman 1935, 1936a and 1944), and growth per moult, although varying considerably throughout its range, is greater under the more optimum prevailing conditions of temperature, i.e., in the more southerly or stable parts of its range (studies reviewed by Ennis 1986b).

Larval transport is thought to be through surface drift so that lobster populations would be replenished from expatriate sources or from local sources if caught up in local surface gyres (Templeman 1937, 1939; Templeman and Tibbo 1945; Scarratt 1964; Dadswell 1979). On the other hand the larvae may not be entirely passive drifters but able to maintain their position near parental grounds through behavioural responses to environmental stimuli which result in vertical migrations, for example, to take advantage of counter currents (Squires 1965, 1970; Squires et al 1971; Caddy 1979). Ennis (1983) gives a review of the behavioural responses of various decapod crustaceans including lobster to environmental conditions such as surface drift. Studies of swimming reactions to currents indicate positive rheotaxis for all larval stages of the lobster (Ennis 1986a).

Ecological studies of a local lobster population in Newfoundland indicate that nocturnal activity is the rule and that lobsters remain in shelters during the day. Also that during periods of very low temperatures (below 0° C) in the winter, activity of lobsters is greatly reduced (Ennis 1984a; 1984b).

Social orientation in lobsters with order of dominance between individuals has been shown in behaviour studies (Douglis 1946; Squires 1967b).

Stomach contents in specimens examined from Newfoundland were rock crab (*Cancer irroratus*), hermit crab, other crustacean remains, opercula of gastropods and gastropod remains, small bivalve shells including *Mytilus edulis*, polychaete remains, brittle stars, small sea stars, sea cucumber, fish bones, and pieces of *Chondrus crispus* and kelp (Squires 1970).

FISHERY

Lobsters are fished over almost the full extent of their range through mainly an inshore trap fishery, although there are landings from a deepwater continental shelf trawl fishery off the United States and Canada.

Ennis (1986b) reviews the history of commercial landings of lobster in Canada and the United States which has varied considerably over the years but at present approximates 45 000 metric tonnes annually. The fishery is well-regulated through a minimum size limit (in some areas a maximum limit as well), limited entry into the fishery (with controlled numbers of traps per fisherman as well) and release of ovigerous females.

Because of the minimum size limit, rates of fishing can be estimated from the percentage of first year recruits to the fishery appearing in the catch in any area, when the approximate growth per moult at these sizes is known (Squires 1970; Anthony 1980; Ennis et al 1982; Ennis 1983; Campbell and Mohn 1983).

Causes of annual fluctuations in catch are difficult to determine. Variability in recruitment as a result of temperature variations has been suggested (Dow 1969, 1977, 1978); also changes in freshwater runoff from the St. Lawrence River (Sutcliffe 1973) and fishery-induced variability in egg production (Pringle and Duggan 1984). Failure of larval recuitment has also been cited as causing low production in specific areas especially where man-made obstruction to larval movement has occurred (Dadswell 1979).

Family AXIIDAE Huxley, 1879
Williams 1984: 185.

Carapace with rostrum and cervical groove, lacking linea thalassinica. Antennular flagella well developed; both moveable and fixed antennal thorns present though sometimes minute. First legs with large chelae, second legs with small chelae (Williams 1984).

Key to species of Axiidae in the area of reference
(Modified from Williams 1984)

1   Eyes with pigment; outer branch of uropod without transverse suture; antennal thorns large .... *Axius serratus*

    Eyes without pigment; outer branch of uropod with transverse suture; antennal thorns small .... *Calocaris templemani*

Genus *Axius* Leach, 1815
de Man 1925: 8; Williams 1984: 187;
Zariquiey Alvarez 1968: 223.

Eyes with pigment; rostrum triangular with spines at each side anterolaterally; cervical groove prominent; no transverse suture on outer branch of uropod; antennal thorns large; first pereopods with very large chela, second pair with moderate chela.

Axius serratus Stimpson, 1852
Rathbun 1929: 25, fig. 32; Williams 1984: 187, fig. 130.
(Figures 175 and 176, Plate 7)

DISTINGUISHING CHARACTERISTICS

With its large chelae of first legs (unequal) it looks like a small lobster but is pale brownish in colour; cervical groove well-marked; rostrum elongate triangular with 5–8 forward directed spines along edge and the distal one turned up; eyes coloured black; no suture on outer branch of uropod; abdomen somewhat depressed, wider in the middle than carapace.

DESCRIPTION

Integument firm, smooth but with occasional pits and setae. Colour pale translucent brown as in Plate 7.

Rostrum prominent with 5–8 forward directed teeth along edges and an upturned distal tooth slightly larger than others, its shape is elongate triangular and hollow above but has a median ridge extending on carapace.

Carapace with edges of rostrum continued as acute wavy carinae on gastric area plus two short mesial carinae, and median carina almost as far as deep cervical groove; near posterior margin is a middorsal mound with central pit, and just in front of it a faint oblique groove toward branchial area; posterior margin forms a deep notch, the projecting side of which fits into lappet on 1st abdominal somite.

Thelycum (u): sterna of 3rd and 4th pereopods form a deeply recessed receptacle, the two sides of which project anteriorly to sharp points and are rounded posteriorly with a central join reaching anterior recess.

Abdomen wider centrally than at the ends; first somite with dorsal and dorso-lateral lappets in tergum and ventrally pointed pleura; all other pleura rounded, the second overlapping first and third, except the 6th which has a marginal tooth at its proximal expansion.

Telson (t) somewhat squarish in outline but with slight lateral expansion and small lateral tooth proximally; a pair of small spines dorsally near centre, a lateral spine at each distal corner and a prominent central spine terminally; branches of uropod are very wide and about even with end of telson; inner branch has row of spines on central rib, an outer distolateral spine and a proximal small spine; outer branch a series of spines at outer distal corner.

Eye with dark pigment, cornea globular, slightly larger than reverse tapered eyestalk.

Antennule (c) 1st article longer than subequal 2nd and 3rd combined; 1st slightly hollowed on one side with a small fixed spine on distal two-thirds; no stylocerite.

Antenna (d) 1st article of peduncle with strong inner spine distally, 2nd article longest of three; basal article with long distal outer spine, and thorn longer than 2nd article of peduncle.

Mandible (e) heavy calcareous incisor with corneous edge of five irregular teeth; molar narrow ovate, lying behind incisor with both ends visible on each side; palp with 3 segments.

Maxillule (f) distal endite narrow slightly curved and expanded distally; proximal endite foot-shaped moderately wide; endopod slender, with curved lash.

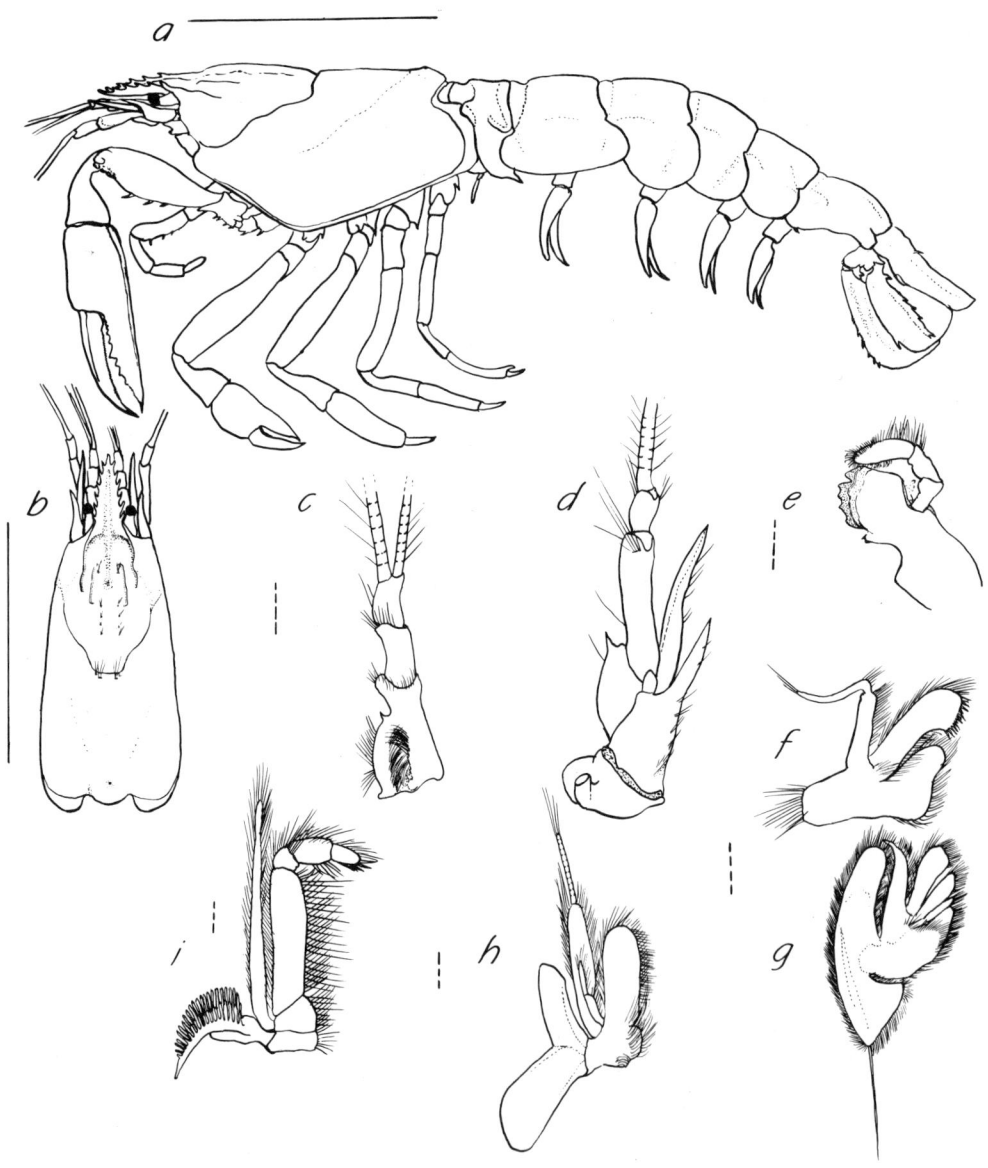

Fig. 175. *Axius serratus*: *a*, whole animal from left side; *b*, carapace in dorsal aspect; *c*, antennule; *d*, antenna; *e*, mandible; *f*, maxillule; *g*, maxilla; *h*, first maxilliped; *i*, second maxilliped; solid line = 10 mm, broken line = 1 mm.

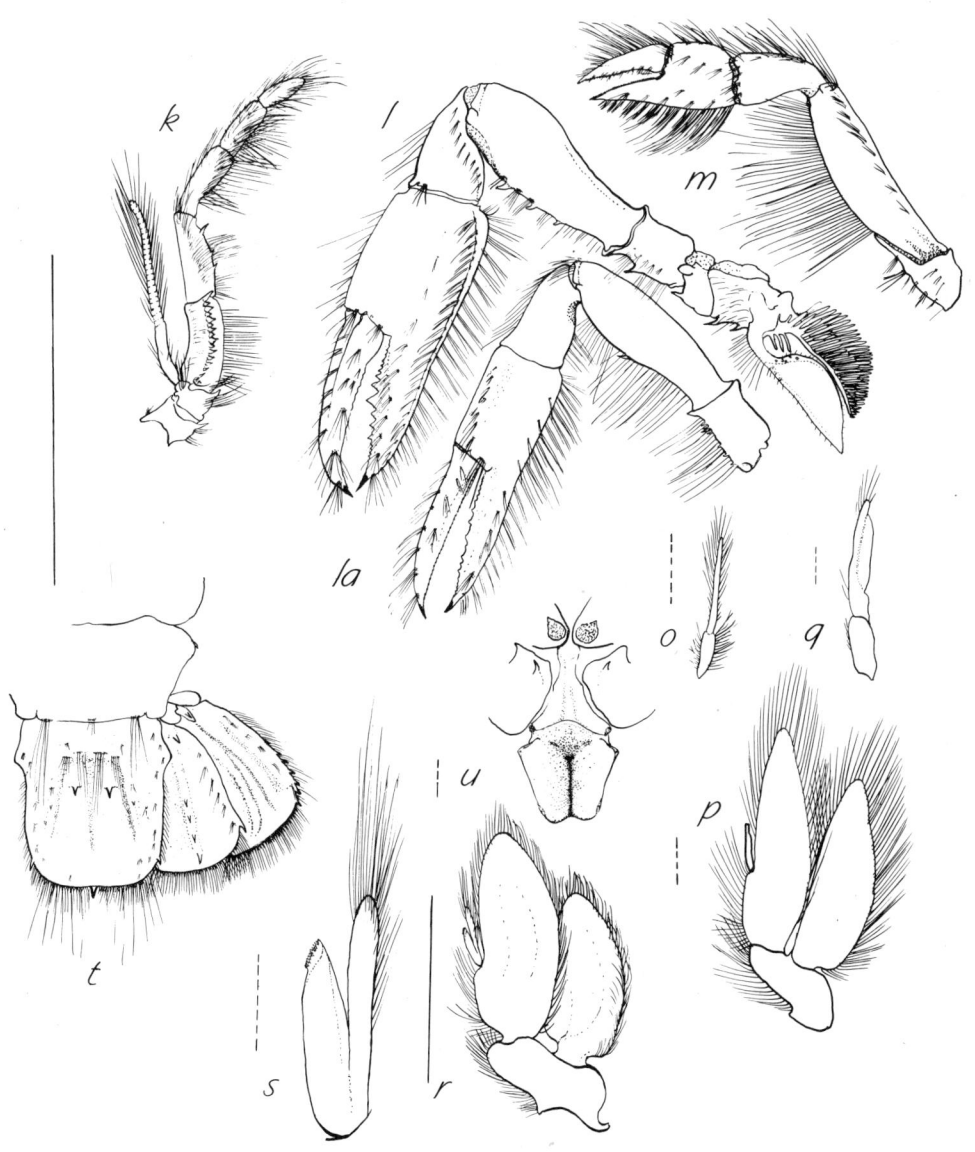

Fig. 176. *Axius serratus*: *k*, third maxilliped; *l*, first pereopod left; *la*, first pereopod right; *m*, second pereopod; *o*, F first pleopod; *p*, F second pleopod; *q*, M first pleopod; *r*, M second pleopod; *s*, appendix masculina; *t*, telson; solid line = 10 mm, broken line = 1 mm.

Maxilla (g) endites unequally bilobed, distal narrow expanding toward tips, proximal with one lobe very narrow, other subtriangular; endopod tapering, curved near tip, longer than anterior lobe of scaphognathite, latter subrectilinear, rounded at tip, posterior lobe shorter, triangular, with long whip at apex.

Maxilliped I (h) distal endite subrectilinear, long, with rounded end, proximal short, with thickened edge; endopod short with two segments, much shorter than laminate part of exopod, the latter with annulate lash; epipod bilobed, posterior lobe larger.

Maxilliped II (i) leglike, slightly compressed; distal segment longer than wide; exopod almost as long as endopod, slender; epipod with large podobranch.

Maxilliped III (k) leglike, stout, seven segments; merus with three lateral spines the distal one large; ischium with inner row of 16 strong fixed spines, corneous-tipped; basis and coxa with inner spine. Epipod large, expanded distally, with podobranch.

Pereopods: I (l) very large, chelate (left or right larger), heavy chela with sagittate teeth on propodal finger, smaller teeth on dactyl; propodus longer than merus, the latter with 3 spines on flexor surface; thin chela with small sagittate teeth on finger and many short corneous spines on dactyl, one spine on merus; epipod blade-shaped, with podobranch. II (m) chelate, palm somewhat inflated, dactyl over-reaching finger, cutting edges serrate; epipod as in I. III–V decreasing in length and stoutness: III with inflated propodus, V with pseudo-chela (dactyl opposed to distal extension of propodus), III and IV with epipod and podobranch as in I.

Pleopods: female I (o) very much reduced, short protopod and very slender endopod; female II (p) large wide subequal endopod and exopod with slender appendix interna. Male I (q) very small compared with II, slender, endopod more than twice as long as protopod, slightly constricted near centre, with apparent groove longitudinally, tip extended and with a few setae. Male II (r) large, wide, endopod slightly longer than exopod, appendix interna with distal patch of hooks, thicker but shorter than appendix masculina (s), the latter with long fine apical and inner lateral setae.

RANGE OF DISTRIBUTION

Nova Scotia to Long Island Sound. Depths 1–320 m (Williams 1984).
Records of distribution in the area of reference are in Fig. 177.

BIOLOGY

Lengths to 30 mm cl in males and 33 mm cl in females.

A burrower in mud and sand, this species constructs burrows to more than 2.5 m with entrances of 3 cm diameter, and in one area of observation at the Strait of Canso the density of such burrows was 9 per square metre (Pemberton et al 1976).

An ovigerous female was taken in October, and larvae have been observed during August to September at depths of 23–30 m.

A predator reported was the flounder *Glyptocephalus cynoglossus*.

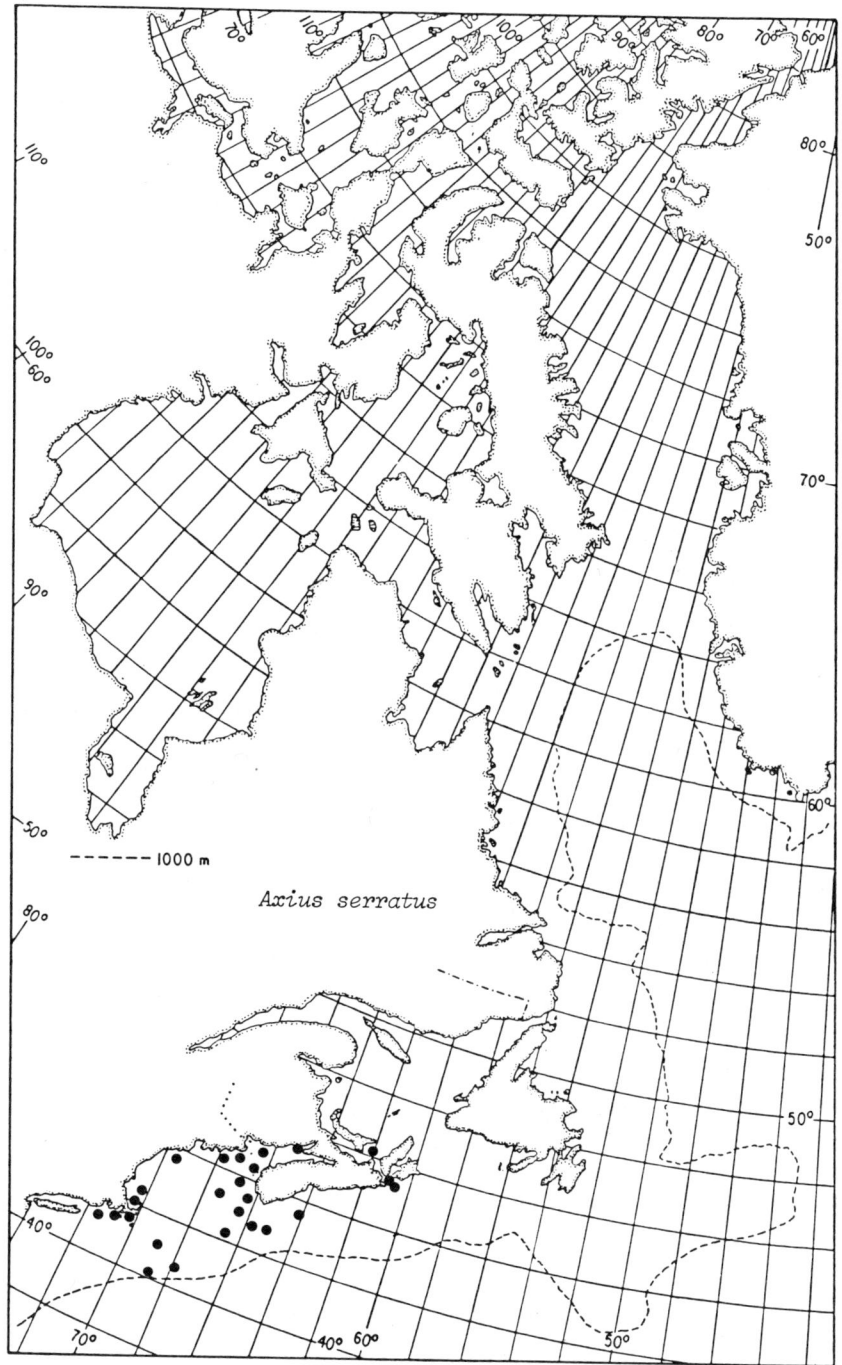

Fig. 177. *Axius serratus*, distribution records in the area of reference.

Genus *Calocaris* Bell, 1853
de Saint Laurent 1972: 354; Rathbun 1929: 25;
Zariquiey Alvarez 1968: 225.

Eyes reduced, without pigment; carapace with median carina from base of rostrum to posterior margin; carapace shorter than abdomen; rostrum triangular and hollow, its edges forming toothed carinae on anterior gastric area; prominent cervical groove; a suture on outer branch of uropod; first legs large with large subequal chelae, the latter with fingers longer than palm; telson longer than wide; hermaphroditic.

Calocaris templemani Squires, 1965
Squires 1965c: 2, figs. 1, 2A,B, CtC, CtD; 3Ct;
4; 5Ct; 6. Williams 1974: 17, figs. 47A,B (key)
Calocaris macandreae, Whiteaves 1901: 257;
Rathbun 1929: 25, fig.33.
(Figures 178 and 179)

DISTINGUISHING CHARACTERISTICS

Rostrum triangular, edges strongly dentate, forming carinae on anterior carapace; eyes without pigment, flat in front; prominent cervical groove; transverse suture on outer branch of uropod; first legs large with large subequal chelae; antennal thorn very small.

DESCRIPTION

Integument firm but flexible. covered with tiny roundish scales. Colour pale pink.

Carapace with anterior converging spinous carinae, meeting at tip (a turned up spine) of descending rostrum, with about nine spines on each carina, four of which are on the carapace; postorbital spine and smooth anterior margin, and a second margin behind it with two small teeth, a frontal notch below them and a rounded pterygostomian area without spine; median dorsal carina stronger anterior to cervical groove, forming a slight crest on proximal half of rostrum; cervical groove deepest middorsally sweeps obliquely forward becoming faint near anterior notch; posteriorly carapace is deeply notched on each side of median ridge.

Abdomen depressed dorsoventrally, longer than carapace, all pleura and terga rounded; tergum of first somite forming a lappet on each side to interlock with posterior edge of carapace; lateral oblique ridge above articulation point on each pleuron of somites 1–5; pleura of 6th proximally expanded laterally.

Telson (t) longer than wide, with lateral "shoulders" proximally bearing two spines followed by series of 3 marginal spines; dorsolaterally two diverging ridges bear three to five small spines each; terminally rounded without spines. Branches of uropod wide, about as long as telson, outer branch with transverse suture.

Eyes without pigment, flattened obliquely on anterior face, stalk not visible.

Antennule (c) 1st article longer than other two together, 2nd longer than 3rd; no stylocerite, may be coalesced to 1st; strong outer spine on proximal third.

Antenna (d) middle segment of peduncle longer than other two combined, 1st segment with inner distolateral fixed spine; 2nd segment of basal article with very short distal thorn and a strong outer spine, lateral inner expansion of basal segment with 4 teeth.

Mandibles (e) slightly different: left incisor with entire slightly rounded cutting edge with corner tooth, the right with a tooth at centre of edge; flat oval molar behind incisor; palp three-segmented.

Maxillule (f) distal endite narrow slightly expanded at tip, proximal wider subtriangular, point blunt; endopod with sinuous lash.

Maxilla (g) proximal endite with lobes almost equal, narrow, with curved row of setae at base; distal endite unequally bilobed, wide, rounded; endopod long , tapered,

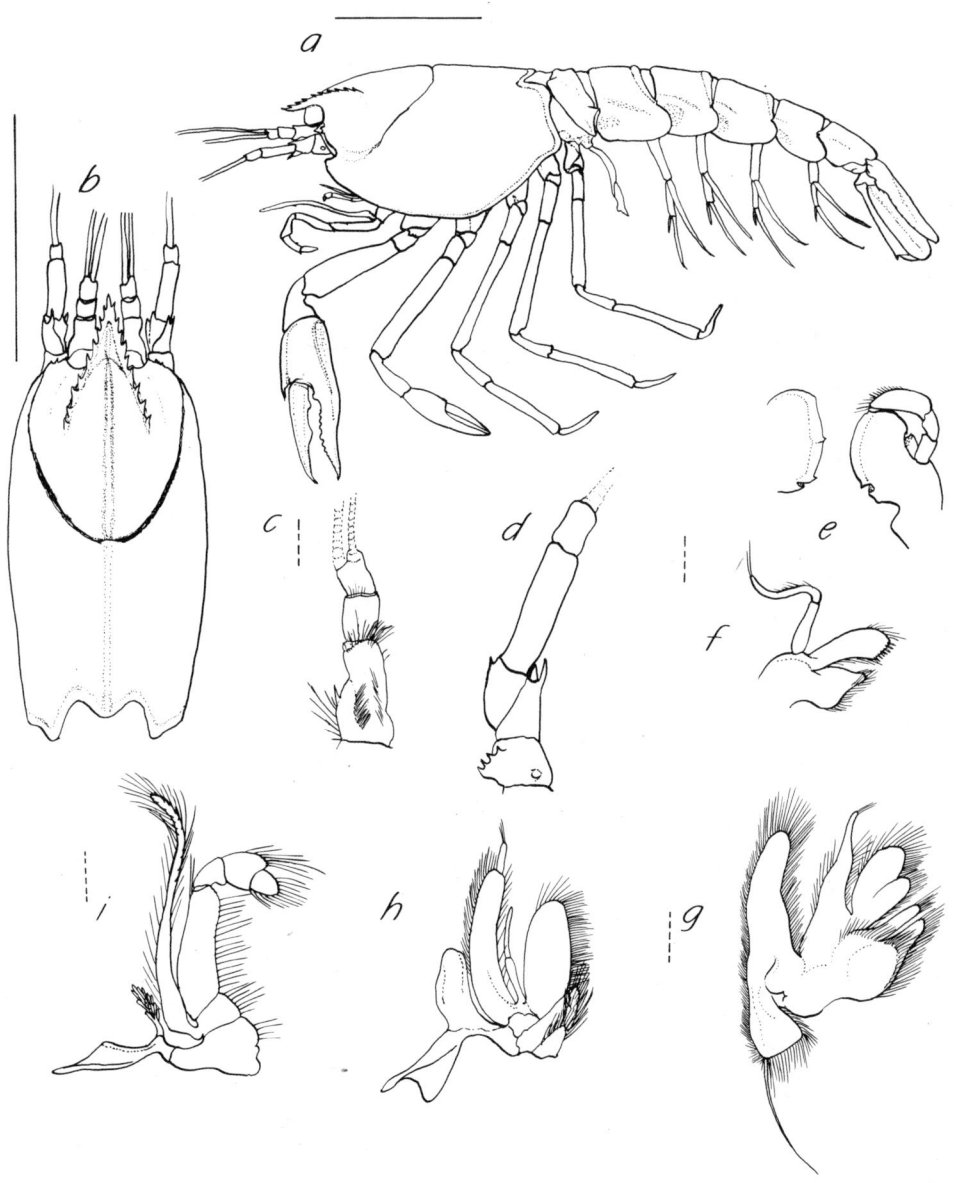

Fig. 178. *Calocaris templemani*: *a*, whole animal from left side; *b*, carapace in dorsal aspect; *c*, antennule; *d*, antenna; *e*, mandible; *f*, maxillule; *g*, maxilla; *h*, first maxilliped; *i*, second maxilliped; solid line = 10 mm, broken line = 1 mm.

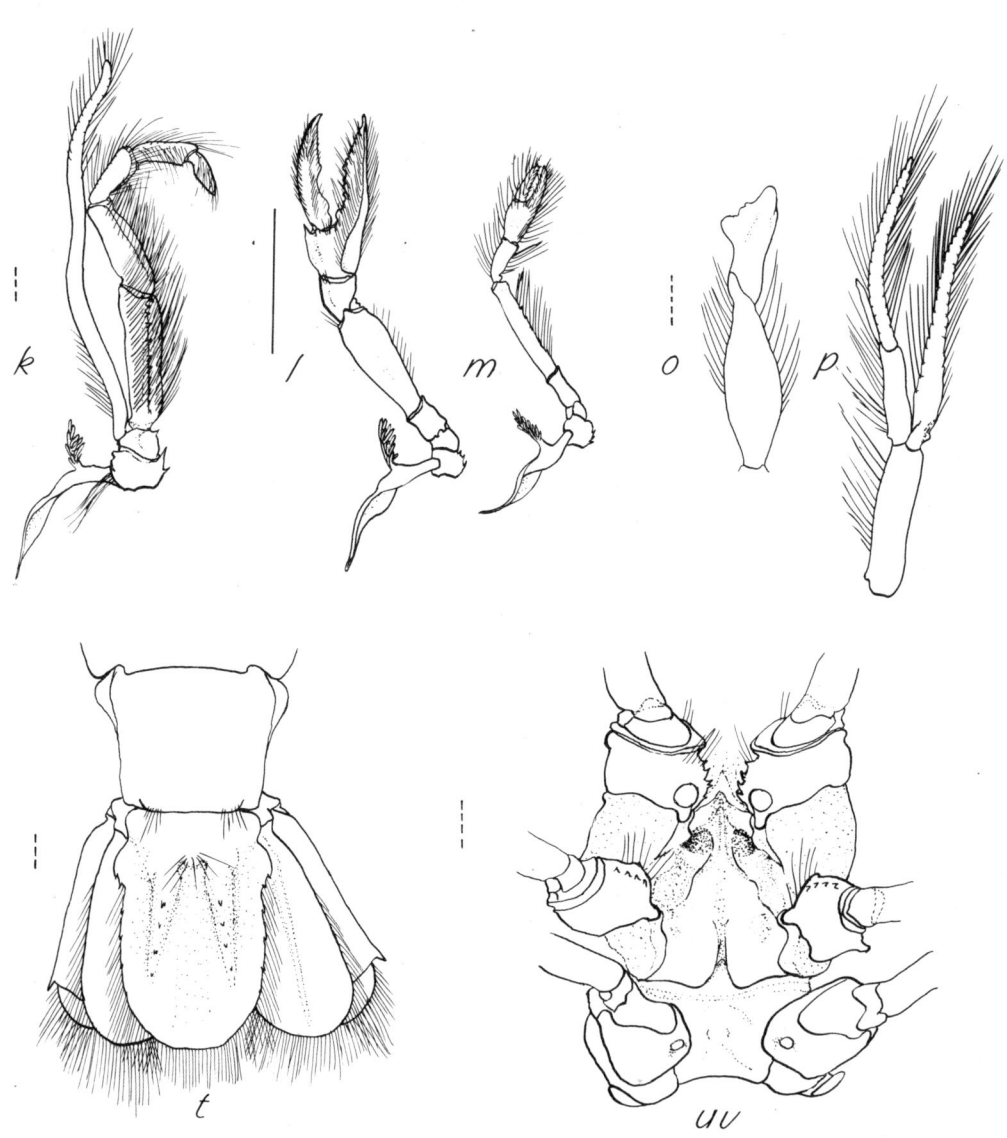

Fig. 179. *Calocaris templemani*: k, third maxilliped; l, first pereopod; m, second pereopod; o first pleopod; p, second pleopod; t, telson, uv, 3rd–5th sternal plates; solid line = 10 mm; broken line = 1 mm.

sinuous, exceeding anterior lobe of scaphognathite, the latter much longer than the posterior lobe which is axe-shaped with long whip at apex.

Maxilliped I (h) distal endite subrectilinear, large; proximal thickened, short, pyrimidal; endopod in two segments, narrow, much shorter than long narrow exopod, the latter with short narrow apical lash; epipod large bilobed, anterior lobe rounded, posterior with thickened supporting edge.

Maxilliped II (i) endopod leglike, distal segment slightly longer than wide; exopod long and slender; epipod elongate, with podobranch.

Maxilliped III (k) endopod leglike, moderately robust, all segments except dactyl triangular in cross-section, dactyl shovel-shaped; ischium with full-length inner row of strong teeth; basis with distolateral sharp tooth, coxa with distolateral and lateral sharp tooth; exopod slender and equal in length to endopod; epipod elongate, with podobranch.

Pereopods: I (l) large, heavy, chelate, fingers longer than palm, dactyl with gap between two large teeth, fixed finger with about five rounded teeth; propodus carinate; coxa with 3 sharp teeth; epipod elongate, with podobranch. II (m) more slender and shorter than I, chelate; coxa with 3 lateral sharp teeth; epipod with podobranch. III–IV slender, non-chelate, slightly longer than II, dactyl one-third propodus, inner edge of coxa with 4 sharp teeth; epipod as in II, with podobranch. V slightly shorter than others, no teeth on coxa, no epipod.

Pleopods: I (o) two-segmented, distal spatulate, with thin wavy edge, modified for spermatophore transferral. II (p) endopod two-segmented, slender, longer than exopod, with long slender appendix interna with apical hooks.

Thelycum (uv) sternal plate of 4th legs forming heart-shaped receptacle with central recess posteriorly; central ridge with anterior apex and a round depression on each side between 3rd legs. Female openings on coxae of 3rd legs, male openings on coxae of 5th legs.

RANGE OF DISTRIBUTION

From Gulf of St. Lawrence and Hermitage Bay on the south coast of Newfoundland to the Gulf of Maine and SE of Cape Lookout, North Carolina. Depths 200–700 m (Williams 1984).

Records of distribution in the area of reference are in Fig. 180.

BIOLOGY

Length to 14 mm cl in holotype. Hermaphroditic.

Williams (1984) suggests that because of similarity of habit, it is possible that life history would be similar to that of *Calocaris macandreae* Bell studied by Buchanan (1963) off Northumberland, U. K. In the latter species eggs were carried for 8–9 months, and the hatched larvae were not pelagic (Bull 1933). At an age of 3 years it functioned as a male although the ovary was present, by age 4 years the testes atrophied but the vas deferens remained filled with sperm; afterwards eggs were laid at 5 years, 7 years and possibly at 9 years. After 5 years moulting was once annually and possibly an age of 10 years was attained (Buchanan 1963).

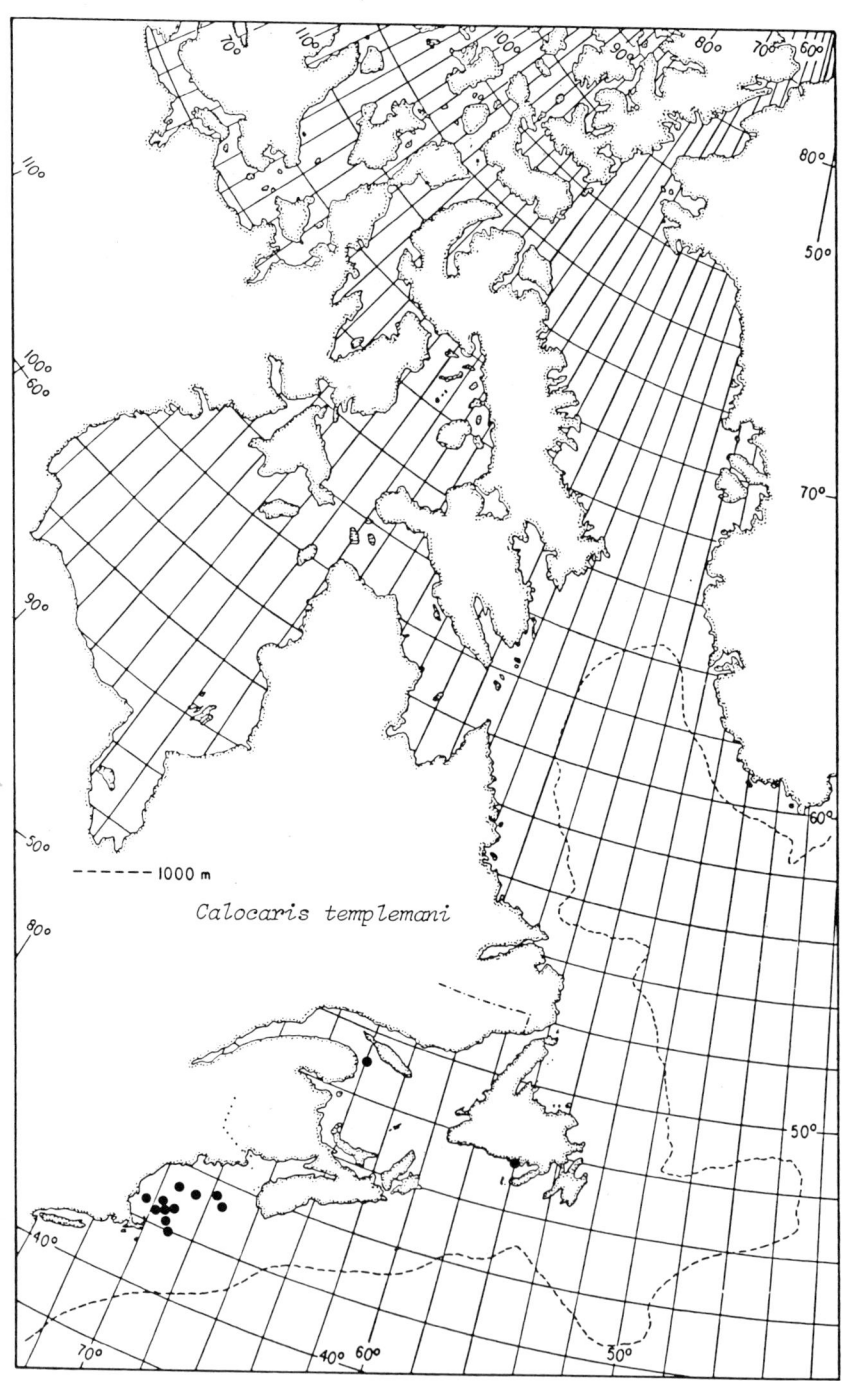

Fig. 180. *Calocaris templemani*, distribution records in the area of reference.

Family CALLIANASSIDAE Dana, 1852
Williams 1984: 180; Zariquiey Alvarez 1968: 228.

Rostrum very short, without spines; carapace divided by the cervical groove into long gastric area and short cardiac area; longitudinal linea thalassinica present in our species; cornea of eye small subapical on stalk; third maxilliped with merus and ischium expanded very wide to serve as operculum in our species, and without exopod; first two pairs of legs chelate, the first larger and unequal; fifth legs subcheliform.

Key to species of the Family Callianassidae in area of reference

1 Eyestalks with pointed tips; telson with terminal central spine ....
.... *Callianassa atlantica*

Eyestalks with rounded tips; telson without terminal spine ....
.... *Callianassa biformis*

Genus *Callianassa* Leach, 1814
Biffar 1971: 648; Makarov 1962: 61;
Manning & Felder 1986: 437;
Saint Laurent, de 1973; 514;
Williams 1984: 180;
Zariquiey Alvarez 1968: 228.

Rostrum very short; carapace divided by cervical groove into long anterior gastric area and short cardiac area; linea thalassinica present; cornea of eye small, circular, subapical; first legs unequal, large, with large chelae, second chelate, equal, somewhat smaller than first; abdomen much longer than carapace.

The subgenus *Callichirus* Stimpson, 1866, has endopod of uropod very narrow, and other differences from *Callianassa* (Manning and Felder 1986).

Callianassa atlantica Rathbun, 1926
Williams 1984: 180, fig. 125.
(Figures 181 and 182)

DISTINGUISHING CHARACTERISTICS

Eyestalks with pointed tips turning outward and upward, cornea subapical, on outer edge above; antennular peduncle longer than its flagella, also than peduncle of antenna; prominent cervical groove set far back and surrounding gastric area, leaving short cardiac area posteriorly; outer branch of uropod wide, of two fused parts, the outer one narrower; a small spine at centre of posterior edge of telson; third maxilliped with exopod longer than longest segment of endopod.

DESCRIPTION

Integument smooth, thin, wrinkling easily. Colour not available.

Carapace about one-third length of abdomen, pronounced continuous sulcus forms a large oval area or cap over most of carapace (gastric area) leaving a short cardiac area behind posterior cervical groove; rostrum short, curving downward, separating orbits, postorbital points not as far forward as rostrum; frontal margin of carapace notched below orbit where linea thalassinica begins and extends to posterior edge horizontally; ventral margin straight but expanded ventro-posteriorly; hepatic protuberance surrounded by a V-shaped groove laterally on anterior third of carapace, the posterior arm of which joins the linea thalassinica.

Abdomen long, depressed, wider than deep; first somite saddle-shaped, second somite longest; all pleura and terga rounded.

Telson (t) shorter than wide, with rounded boss on centre of proximal third; a small terminal spine (lost in some specimens); branches of uropod longer than telson, inner thin ovate, outer very wide, subtriangular, laminate dorso-anterior part fused with larger ventro-posterior part, both with setal fringe posteriorly.

Eyestalk wide at base, flattish but subtriangular in cross-section, tapering and curving outward and upward at tips anterior to the cornea which is small and round on outer edge where stalk begins to taper inward.

Antennule (c) 3rd article longer than 1st and 2nd together, and longer than flagella, all with many fine setae ventro-laterally; no stylocerite; opening of otocyst heavily setose.

Antenna (d) distal two articles each longer than 1st; one short blunt horn and small projecting lobe distally on 2nd article of peduncle.

Mandible (e) incisor with about 9 strong saw teeth along straight edge followed by curved edge with several small denticles; molar narrow and short behind incisor with corner tooth followed by a large saw tooth. Palp large, three-segmented.

Maxillule (f) distal endite narrow at base expanding to wide straight edge; proximal wider, subtriangular; endopod with short, oval, recurved distal segment.

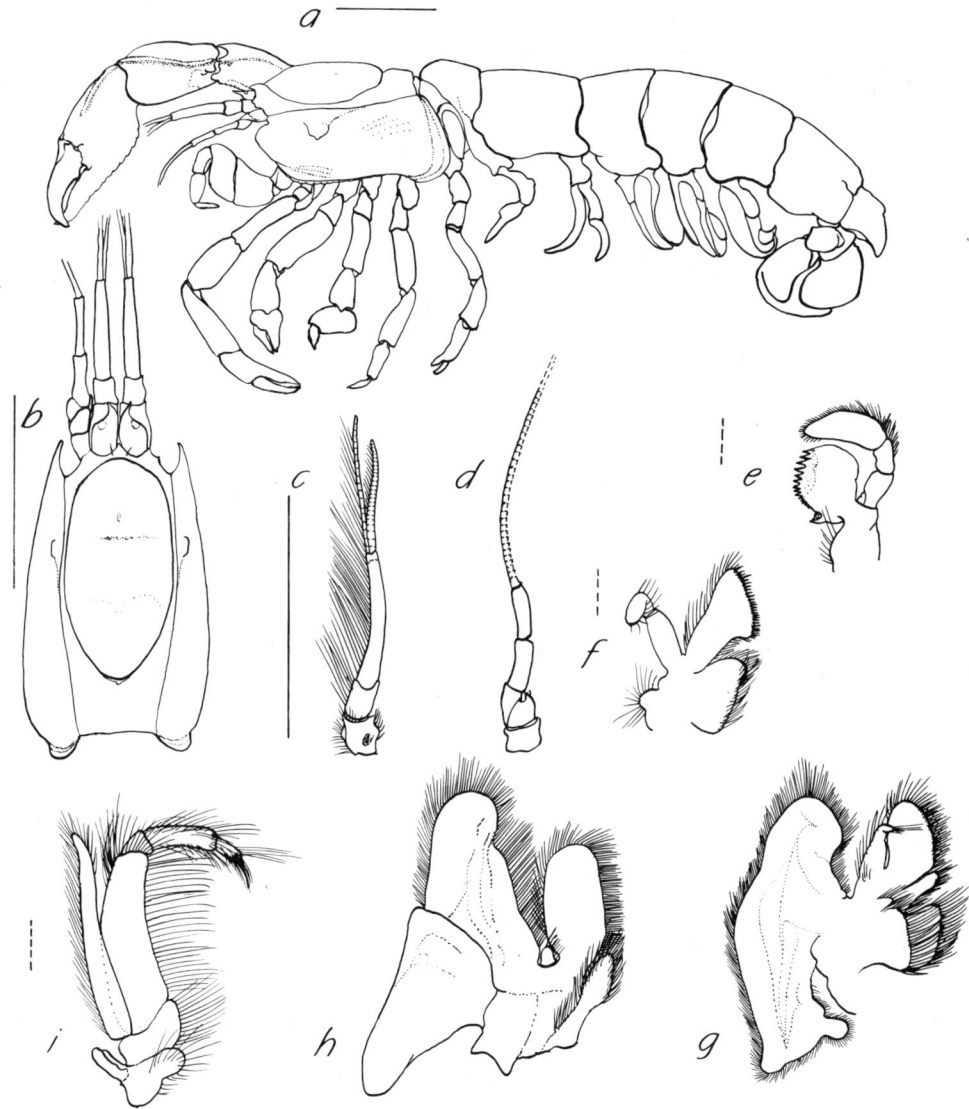

Fig. 181. *Callianassa atlantica*: *a*, whole animal from left side; *b*, carapace in dorsal aspect; *c*, antennule; *d*, antenna; *e*, mandible; *f*, maxillule; *g*, maxilla; *h*, first maxilliped; *i*, second maxilliped; solid line = 10 mm, broken line = 1 mm.

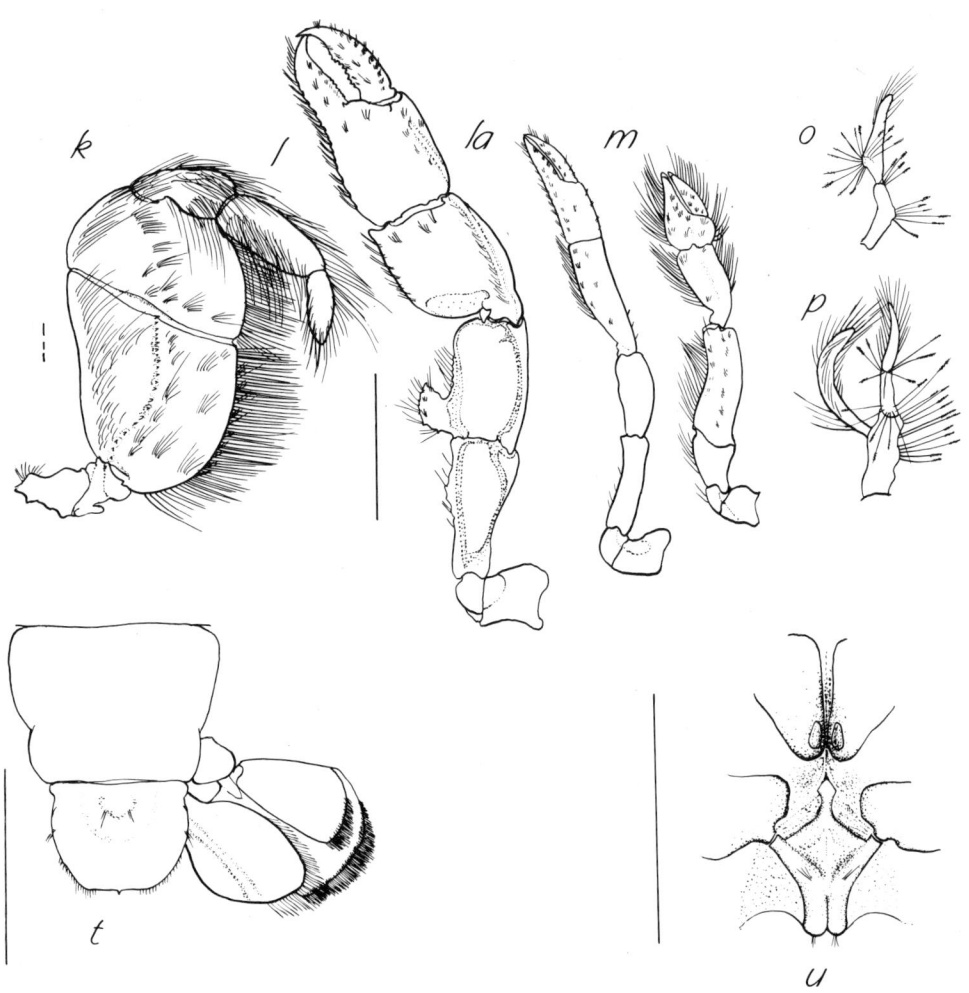

Fig. 182. *Callianassa atlantica*: *k*, third maxilliped; *l*, first pereopod right; *la*, first pereopod left; *m*, second pereopod; *o*, F first pleopod; *p*, F second pleopod; *t*, telson; *u*, thelycum; solid line = 10 mm, broken line = 1 mm.

Maxilla (g) distal endite with large distal lobe and narrow central lobe; proximal endite unilobed, wide, with row of long setae at base; endopod short with tip turning inward; short anterior lobe of scaphognathite skewed inward, shorter than irregularly shaped posterior lobe with projecting rounded inner corner.

Maxilliped I (h) distal endite large subrectilinear; proximal pointed obliquely, thickened and heavily fringed; endopod very short, rounded; exopod foliaceous, wide at base; epipod also foliaceous, large, foot-shaped in outline.

Maxilliped II (i) leglike but small, distal segments narrow, longer than wide; exopod blade-like, longer than longest segment of endopod; epipod small.

Maxilliped III (k) with seven segments; ischium and merus greatly expanded, operculiform, ischium with sinuous row of small teeth centrally (ischium flat triangular in cross-section).

Pereopods: I (l) unequal, chelate, the right (left) much the larger; propodus and carpus wide and compressed, and carinate above, carpus greatly narrowing to join merus; fingers with small teeth but strong curved tips; merus with proximal lateral wing with small teeth along its edge; smaller left (la) much more slender and shorter. II (m) shorter than I, moderately robust, chelate, palm slightly inflated. III about as stout as II, non-chelate but propodus wider than long with stout pointed dactyl. IV more slender and longer than III, dactyl moderate. V slightly more slender and shorter than IV, pseudochelate, the dactyl flattened and opposable to flattened distal expansion of propodus.

Pleopods: female I (o) uniramous; protopod and endopod both with flexure and an "elbow" with attached long bottle-brush setae; female II (p) biramous; endopod two-segmented with long bottle-brush setae at joint and distally on protopod, exopod sickle-shaped; III–V pleopods large thick and fleshy; endopod two-segmented wide, overlapping wide curved exopod; small appendix interna on distal part of endopod. No male first and second pleopods.

Thelycum (u) sternal plate of 4th legs modified to form diamond shaped receptacle, the posterior angle a fleshy protuberance divided at centre by deep groove, and anterior angle a short neck with small diamond-shaped head; centrally intersecting grooves form another diamond with central depression.

RANGE OF DISTRIBUTION

Bass River, Nova Scotia, to Georgia and Franklin County, Florida. Depths shoreline to 38 m, mostly in saline estuarine conditions in mud or muddy sand (Williams 1984).

Records of distribution in area of reference are in Fig. 183.

BIOLOGY

Lengths of body to 59 mm in males and 68 mm in females; 10–15 mm cl in females examined.

Ovigerous females have been taken in July, and larve from July to September (Sandifer 1973).

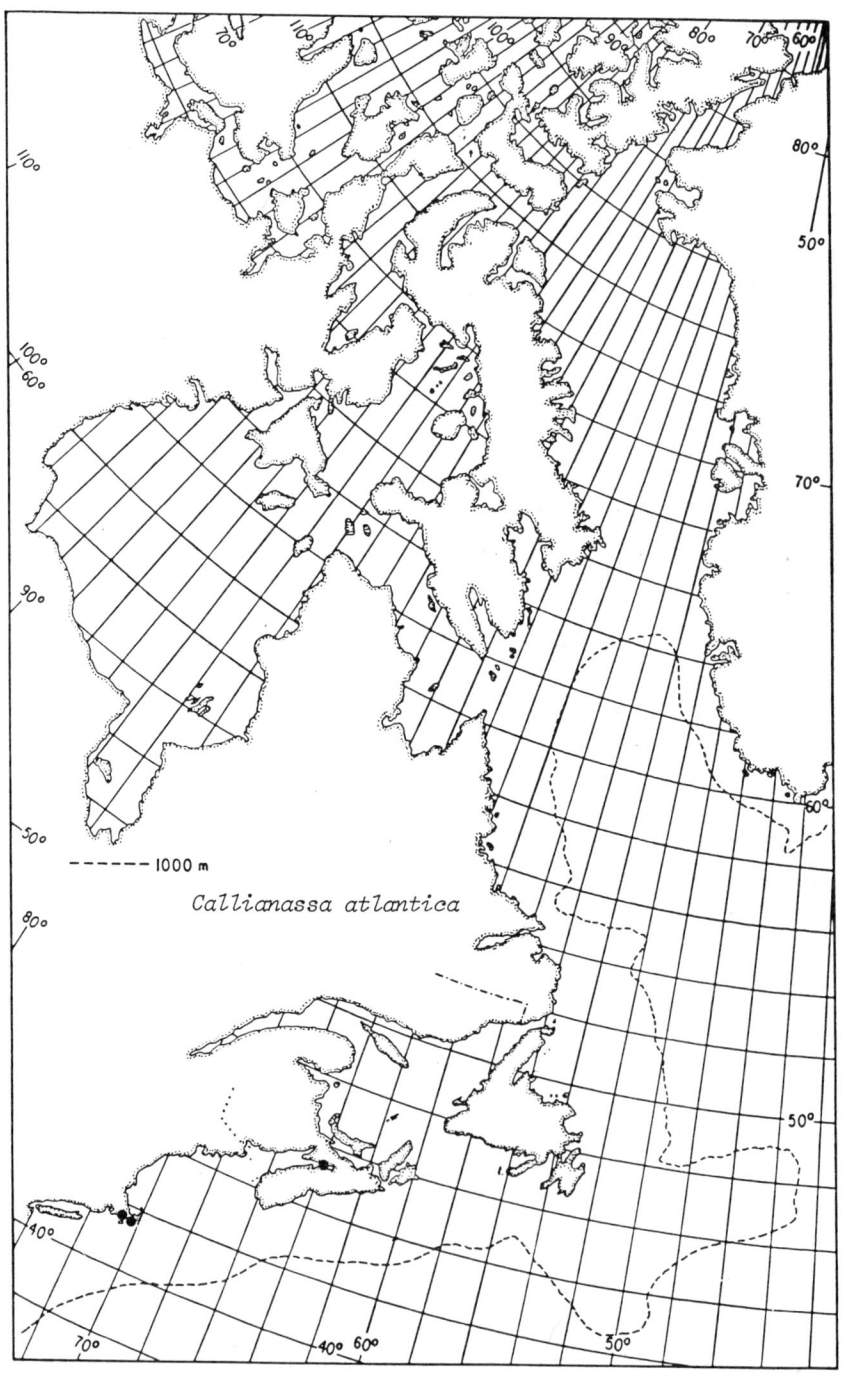

Fig. 183. *Callianassa atlantica*, distribution records in the area of reference.

Callianassa biformis Biffar, 1971
Williams 1984: 182, fig. 126.
(Figures 184 and 185)

Distinguishing Characteristics

A distinctly smaller species than *C. atlantica*; tips of eyestalks rounded; peduncle of antenna longer than that of the antennule; cervical groove on carapace prominent, continued forward to form an oval cap on carapace (groove faint anteriorly); short horizontal groove at ventro-anterior edge of carapace; first legs large, chelate, unequal, two different shapes for major chela in males; dividing line longitudinally on outer branch of uropod to show fusion of lobes missing or very faint; exopod of 3rd maxilliped shorter than longest segment.

Description

Integument thin, smooth, almost membranous.

Carapace with very short pointed rostrum dividing shallow orbits which have no postorbital points as in *C. atlantica*; cervical groove prominent, set far back to form a large gastric "cap" and narrow cardiac area posteriorly, groove becoming faint anteriorly; anterior notch with lower projection as far forward as rostrum, upper edge of notch with linea thalassinica extending to posterior margin; ventro-anterior edge has short horizontal groove reaching level of lateral V-shaped projection in hepatic area; ventral edge has narrow skirt with a few indentations anteriorly and slight expansion at posterior angle.

Abdomen about three times carapace length, 2nd somite longest; pleura and terga rounded.

Telson (t) shorter than uropods, slightly tapered, subtruncate with a few spinules at lateral corners, end convex with setal fringe only; exopod or outer branch of uropod showing fusion of parts by marginal overlay with extra fringe of setae.

Eyestalks flattened, tapering, rounded at tip, cornea small, round, subapical.

Antennule (c) 3rd article longer than 1st and 2nd combined, but shorter than flagella.

Antenna (d) peduncle longer than antennular peduncle; 2nd proximal segment with inconspicuous antennal thorns.

Mandible (e) incisor wide, squarish, with a few denticles along cutting edge; molar narrow lying obliquely inside incisor, with sharp inner corner tooth; palp larger than incisor, with 3 segments.

Maxillule (f) distal endite expanded distally foot-shaped; proximal much shorter, strongly pointed; endopod very small with recurved apex.

Maxilla (g) distal endite with wide triangular lobe and narrow lobe; proximal the same in reverse order, smaller, at base of lobes a curved row of long setae; endopod close to distal endite, with tip turned inward; anterior lobe of scaphognathite wide, squarish, about as long as subtriangular posterior lobe.

Maxilliped I (h) distal endite subrectilinear, long, proximal with conspicuous double edge; endopod very short; exopod wide, with distolateral, pointed projection with long setae; epipod large subtriangular.

Maxilliped II (i) leglike but small, distal segment longer than wide; thin exopod shorter than longest segment of endopod; epipod small.

Maxilliped III (k) operculiform, merus and ischium greatly widened, the latter flattened triangular in cross-section.

Pereopods: I (l) chelate, unequal and sexually dimorphic, of the two forms one with stronger curved dactyl in males; two forms also in females, the stronger with dactyl larger and serrate with curved tip, a shorter wider carpus and the merus with a proximo-lateral wing-like expansion with serrate edges and a sharp forward tip; in the weaker, a slightly longer narower carpus, less rounded proximally and with a simple curved spine ventro-laterally on merus. II (m) chelate, wide propodus and carpus, shorter than I. III about

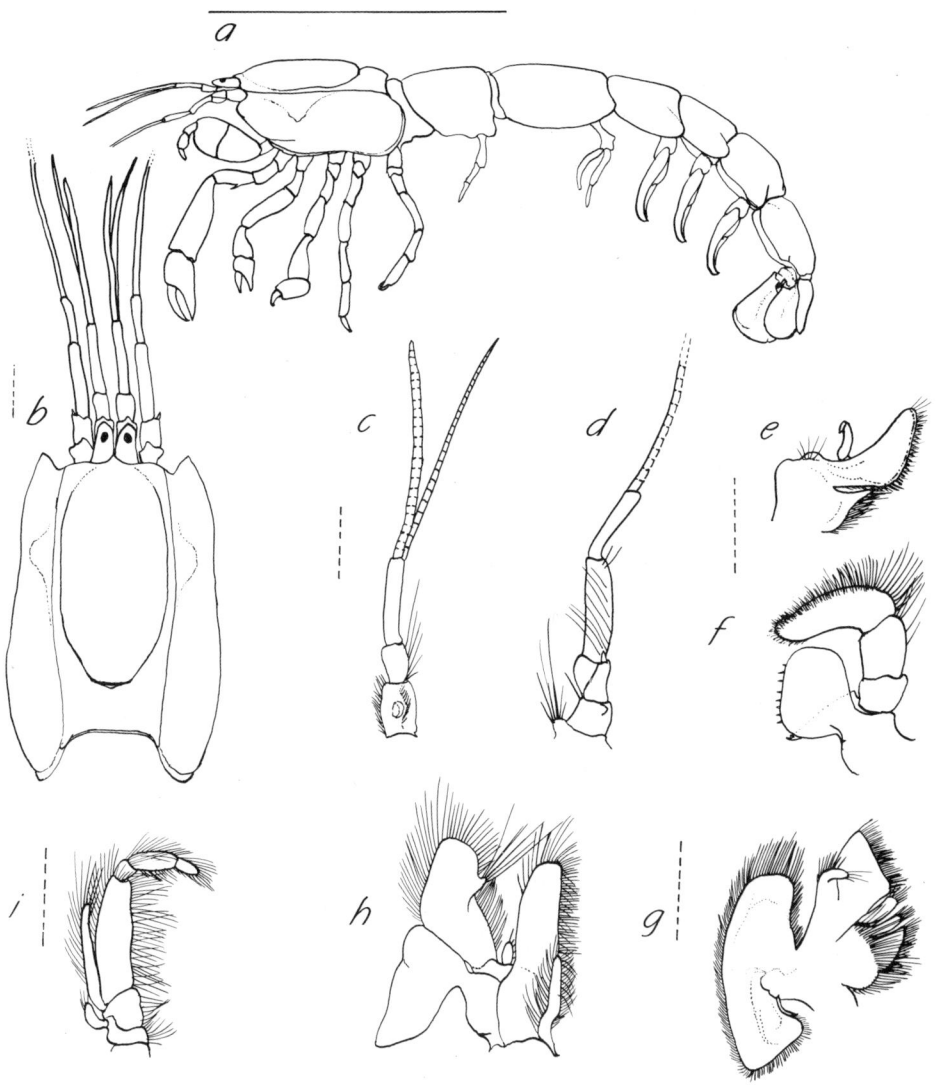

Fig. 184. *Callianassa biformis*: *a*, whole animal from left side; *b*, carapace in dorsal aspect; *c*, antennule; *d*, antenna; *e*, mandible; *f*, maxillule; *g*, maxilla; *h*, first maxilliped; *i*, second maxilliped; solid line = 10 mm, broken line = 1 mm.

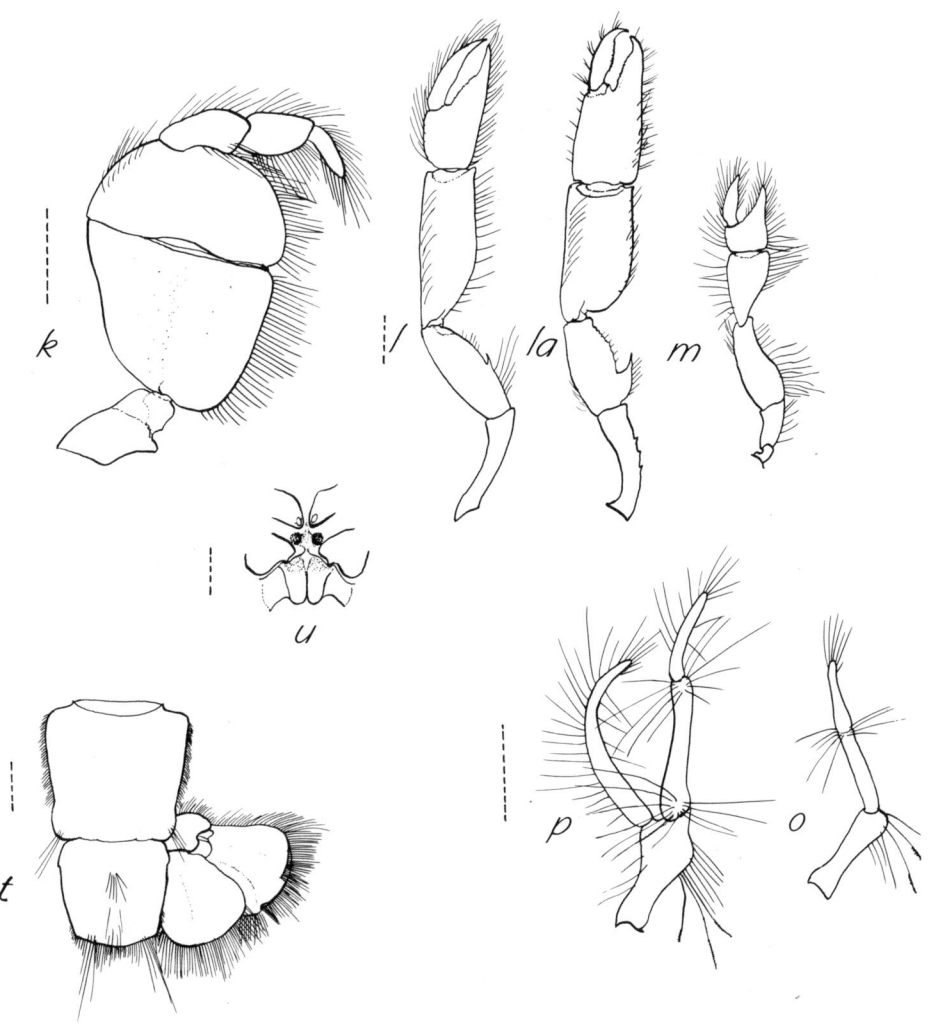

Fig. 185. *Callianassa biformis*: *k*, third maxilliped; *l*, first pereopod right; *la*, first pereopod left; *m*, second pereopod; *o*, F first pleopod; *p*, F second pleopod; *t*, telson; *u*, thelycum; broken line = 1 mm.

as long and stout as II, propodus wider than long with short curved dactyl at one corner. IV more slender than III. V slightly shorter than IV, propodus and dactyl modified to form a pseudo-chela with many bristles.

Pleopods: female I (o) uniramous, endopod with two segments, protopod thicker; female II (p) two-segmented endopod longer than sickle-shaped exopod, protopod expanded distally; III–V pleopods thick, fleshy, endopod and curved exopod overlapping each other, a short appendix interna on endopod. Male pleopods I and II non-existant.

Thelycum (u) sternal plate with median separation and rounded posteriorly, pointed anteriorly with rounded depression on each side of sternum of 4th legs.

RANGE OF DISTRIBUTION

Bass River and Yarmouth, Nova Scotia, and Nantucket Sound to Franklin County, Florida. Depths intertidal to 10 m mostly in saltwater estuaries in muddy sand (Williams 1984).

Records of distribution in the area of reference are in Fig. 186.

BIOLOGY

Lengths of body to 32 mm in males and 27 mm in females; 3.5 to 5.0 mm cl in females examined, USNM #372233.

Ovigerous females were taken in July and larvae were present during July to September (Williams 1984). Eggs 0.4 mm.

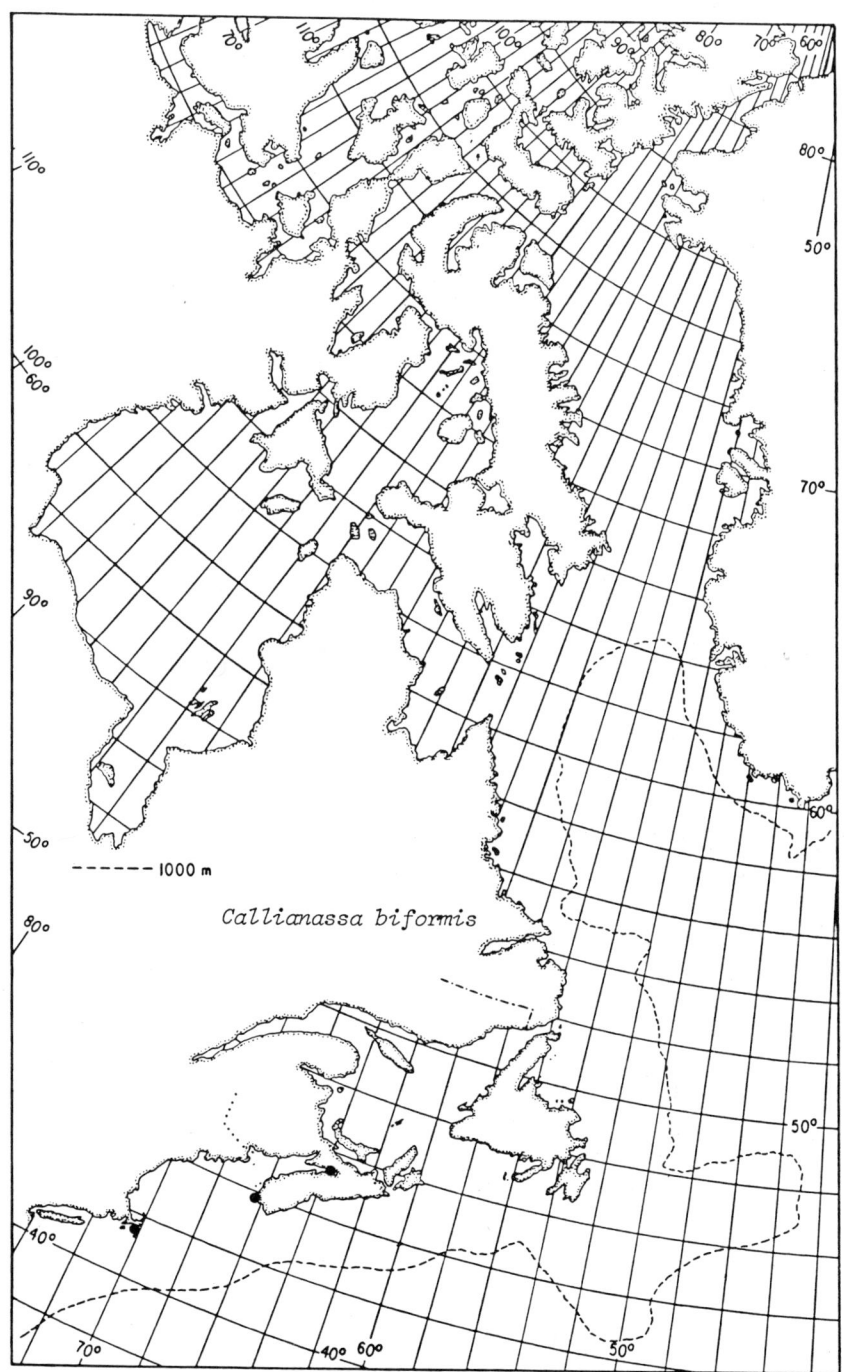

Fig. 186. *Callianassa biformis*, distribution records in the area of reference.

Family POLYCHELIDAE Wood-Mason, 1874
Burukovskii 1983: 134. Zariquiey Alvarez 1968: 205.

Carapace flattened dorsally, with spines along lateral edges, and middorsal carina from rostrum to posterior margin which is concave; eyes reduced, no cornea; pereopods I–IV chelate, I much longer than others, dactyl on outer side; abdomen shorter than carapace and with complex carination; telson ends in a sharp point.

Key to species of Polychelidae in area of reference

1 Carapace with well-defined pattern of spines along whole median carina; a double ridge dorsally on 6th abdominal somite 2

   Carapace median carina with only a few spines anteriorly but rest of carina with indefinite granulation; no ridge dorsally on 6th abdominal somite .... *Polycheles granulatus*

2 Posterior margin of carapace with two spines only at midline; edges of ridges on 6th abdominal somite smooth; .... *Stereomastis sculpta*

   Posterior margin of carapace with a few spines on each side in addition to two at midline; edges of ridges on 6th abdominal somite jagged .... *Stereomastis nana*

Genus *Polycheles* Heller, 1862
Zariquiey Alvarez 1968: 208.

Carapace wide, flattened, with sharp spiny edges laterally, the spines in three groups of 6–10 + 3 + 11–15; a low median carina with few spines anteriorly but many irregular small granules over the rest of its length; no spines on posterior margin; 1st to 5th abdominal somites with low dorsal ridge but no ridge on 6th.

Polycheles granulatus Faxon, 1893
Selbie 1914: pl. III, figs. 1–11;
Zariquiey Alvarez 1968: 210.
(Figures 187 and 188)

Distinguishing Characteristics

Large flattened carapace with edging of small sharp spines, longer than the variously carinated abdomen; rostrum of two small spines at anterior end of median carina with few spines and many granules reaching posterior margin; cervical groove at middle of carapace separating edge spines into three groups, the middle group of 3, anterior 7–10 and posterior 11–15; posterior margin without spines; large epipod on third maxilliped; abdomen with carina on somites 1–5 but none on 6th; very long first legs; first four legs chelate.

Description

Integument rough with innumerable small tubercles and occasional spines and setae. Colour pale reddish.

Carapace somewhat depressed and flattened above, with a double rostral spine followed on a low median carina by four single spines to about half the gastro-frontal area in front of the cervical groove and afterwards uneven rows of small spines to the posterior edge of the carapace; a pre-orbital and a post-orbital spine on edges of the orbits which is closed off about halfway down in place of eyes; a larger spine is at anterior edge, the last in a series of about 8 (7–10), 3 and 14 (11–15) lateral spines; an oval depression antero-lateral to cervical groove; a sinuous faint spinous carina from posterior corner over the branchial area. No marginal spines posteriorly. Area of carapace beneath lateral spinous margin canted inward and covered with rough setae.

Abdomen with marked separation of terga and pleura, the tergum of all but 6th somite with short middorsal ridge (1st to 4th with forward directed spine) and triangular divisions longitudinally, 6th without ridge or division; pleura with blunt point ventrally and with short, curved central carina.

Telson (t) wide, tapering to very sharp point, dorso-lateral carinae converging and meeting posteriorly; proximal triangular plateau with posterior fixed spine. Branches of uropod stiff, wide, with supporting rib, outer with distolateral tooth.

Eyes lacking; orbital notch with unpigmented closure where eye would be.

Antennule (c) 1st article apparently fused laterally with wide stylocerite with long attenuate point exceeding other two articles, 2nd and 3rd subequal, lateral edge of stylocerite with 6 small uneven teeth; dorsal flagellum much thicker and longer than short slender ventral one.

Antenna (d) scale calcified, rounded at tip, almost as long as peduncle, the 1st segment of which has a distolateral strong spine.

Mandible (e) without molar; incisor wide with two major toothed divisions and a large tooth at upper corner, making the edge W-shaped. Palp with 3 segments, central longest.

Maxillule (f) endites narrow, the distal with strong apical tooth; endopod very small, with tuft of setae.

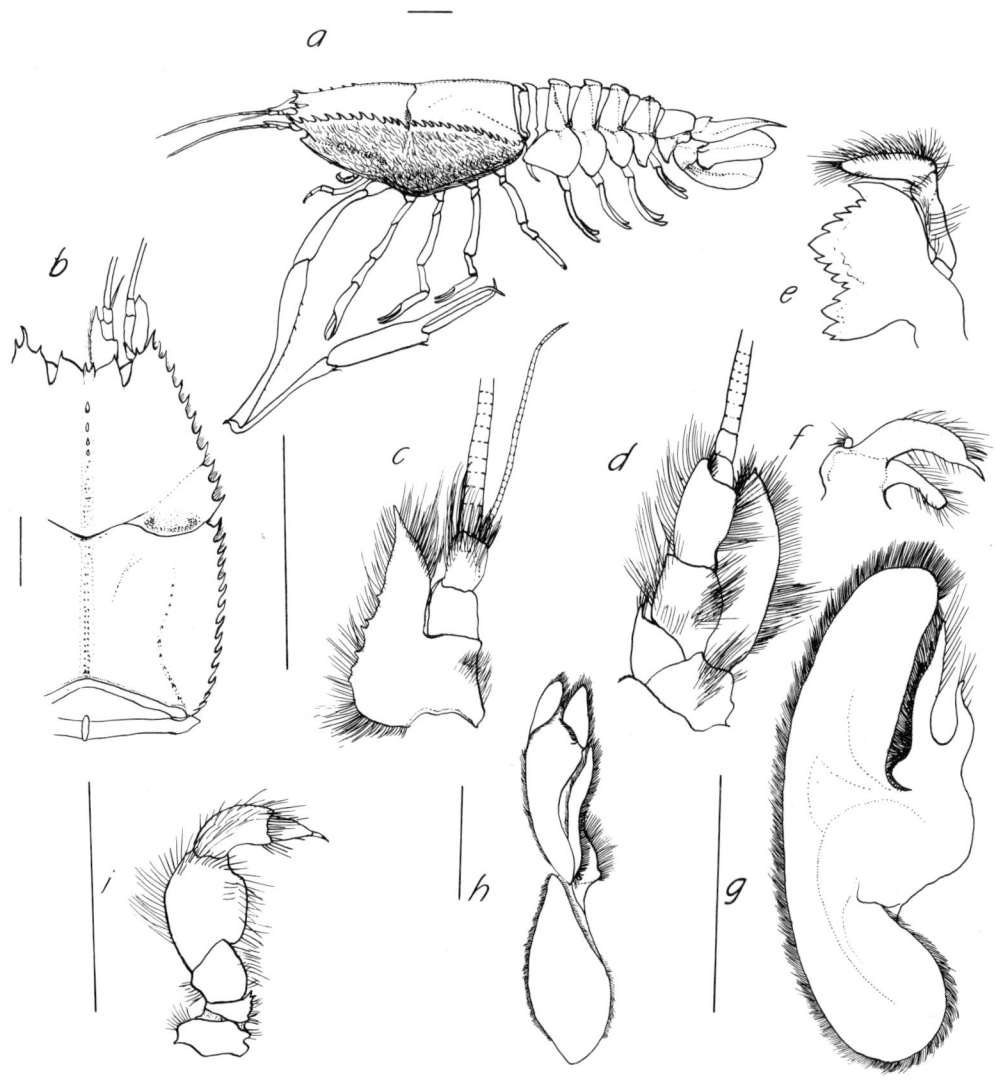

Fig. 187. *Polycheles granulatus*: *a*, whole animal from left side; *b*, carapace in dorsal aspect; *c*, antennule; *d*, antenna; *e*, mandible; *f*, maxillule; *g*, maxilla; *h*, first maxilliped; *i*, second maxilliped; solid line = 10 mm.

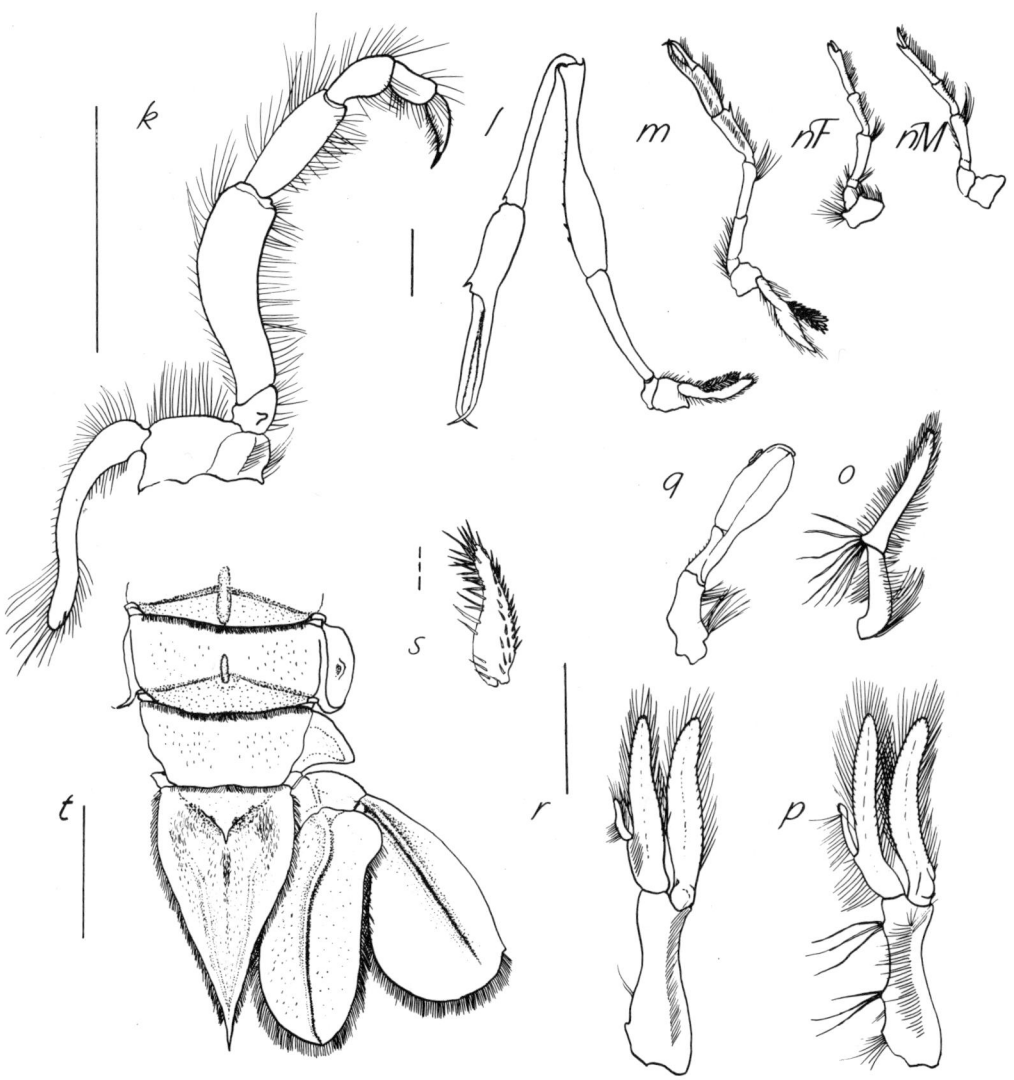

Fig. 188. *Polycheles granulatus*: k, third maxilliped; l, first pereopod; m, second pereopod; nF, F fifth pereopod; nM, M fifth pereopod; o, F first pleopod; p, F second pleopod; q, M first pleopod; r, M second pleopod; s, appendix masculina; t, telson; solid line = 10 mm, broken line = 1 mm.

Maxilla (g) endites reduced, the distal single-lobed and sinuous, slender; endopod long, sinuous; anterior lobe of scaphognathite subrectilinier, longer than rounded posterior lobe.

Maxilliped I (h) endites very much reduced; endopod slender, fusiform; exopod foliaceous, distal lobe curled in on itself; epipod about as large as exopod, wide, fusiform.

Maxilliped II (i) leglike, slightly compressed, distal segment with strong apical spine, middle segment wide and thin.

Maxilliped III (k) leglike, with 7 segments, distal segment with strong apical spine, coxa wide with long ribbon-like epipod extending into gill chamber.

Pereopods: I (l) chelate, much longer than others, chela with long slender fingers with curved tips, palm slightly inflated, a sharp tooth distolaterally on palm; carpus with distal tooth; merus with row of spines on ventral edge and distal tooth; epipod large, with podobranch; II (m) chelate, less than half length of I, slender fingers with curved tips not as long as palm, distolateral spine on carpus; III and IV chelate, about equal in length to II but chelae long, narrow and slightly curved; V (n) different in males and females, female chelate with short dactyl, male not chelate with longer dactyl possibly opposable to distal extension of propodus in part.

Pleopods: female I (o) uniramous, endopod narrow, longer than protopod, with long proximal egg-bearing setae in ovigerous female; female II (p) narrow exopod and endopod equal, the latter with appendix interna with distolateral patch of hooks; male I (q) semi-rigid, spatulate with distolateral rounded patch and distal curved area with hooks, a central longitudinal join in endopod; male II (r) exopod and endopod equal, the latter with appendix interna with distolateral hooks and appendix masculina (s) slightly shorter, curved fusiform with many moderate spines.

RANGE OF DISTRIBUTION

Western Atlantic off the southwest slope of the Grand Banks of Newfoundland and the Nova Scotian Shelf; eastern Atlantic SW of Ireland, the Azores, Cape Verde Islands, Madeira and the Canaries; also in the Pacific, the Gulf of Panama and Hawaii; and the Indian Ocean off Colombo, Sri Lanka (Zariquiey Alvarez 1968).

Records of distribution in the area of reference are in Fig. 189.

BIOLOGY

Lengths of males and females to 50 mm cl.

Females taken in May and November were ovigerous (eggs 0.8 mm average). The one in November also had large ova in the ovaries so would probably spawn after the eggs carried had hatched.

Stomach contents were fish bones and sponge spicules (Squires 1965a).

The decapod species present in catches with this species was *Stereomastis sculpta* (Squires 1965a).

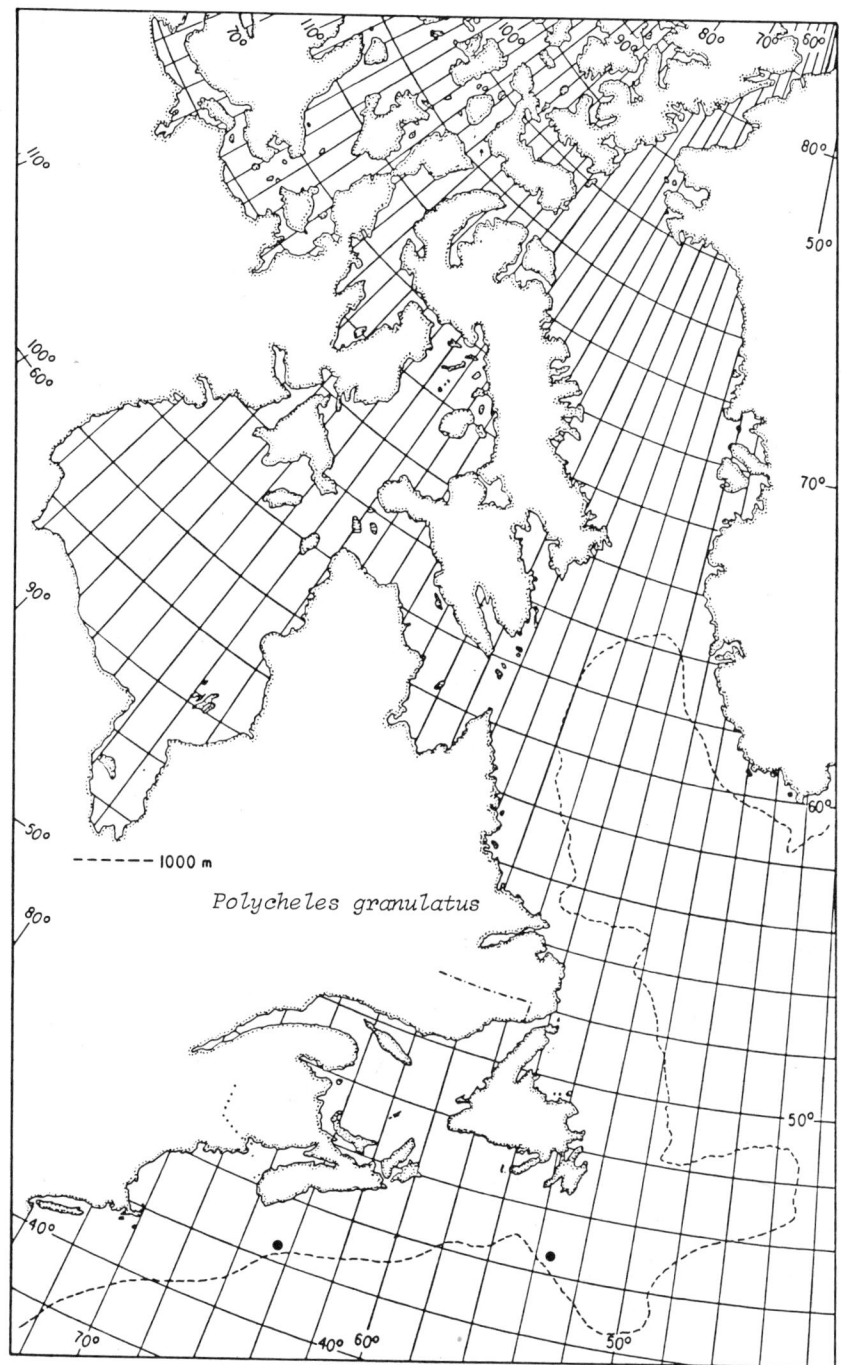

Fig. 189. *Polycheles granulatus*, distribution records in the area of reference.

Genus *Stereomastis* Bate, 1888
Zariquiey Alvarez 1968: 210.

Each lateral margin of carapace bears less than 20 spines; the median dorsal carina has a regular pattern of single and double spines (2, 1, 2, 1, 2, 2, 2) including 2 rostral and 2 at the posterior margin; epipod of third maxilliped is rudimentary; double ridge middorsally on 6th somite of abdomen.

Stereomastis sculpta (Smith, 1880)
Sivertsen and Holthuis 1956: 41;
Zariquiey Alvarez 1968: 210.
Polycheles sculptus, Selbie 1914: 18, pl. 2, figs. 1-9.
(Figures 190 and 191)

Distinguishing Characteristics

Carapace large, depressed, flattened dorsally, with margin fringed with small sharp spines, not more than 20 on each side; no eyes but obital notches closed; two spines at rostrum and a regular pattern of spines along median carina to posterior edge, the latter with two spines only at midline; 6th somite of abdomen with two smooth dorsal ridges; first legs as long as or slightly longer than body.

Description

Integument rigid, sparingly setose and minutely tuberculate. Colour pink on whitish background.

Carapace depressed and flattened dorsally, with strong lateral ridge along each side armed with sharp fixed spines in three groups of about 6, 3 and 7 spines from front to back; cervical groove divides the carapace into approximately two halves; median carina begins at rostrum with two spines and has 1, 2 and 1 on anterior half, and 2 spines behind groove, 2 at the posterior margin and 2 between them on the posterior half; a branchial carina with 4 spines is above the lateral carina and a ventro-lateral below it with 10 spines both beginning at the posterior margin; a series of 5 gastric spines begin behind the orbit and approach a spine on the cervical groove; a preorbital but no postorbital spine; a small spine on rudimentary eyestalk closing the orbit.

Abdomen slightly longer than carapace; terga of 1st to 5th somites with a short median ridge and anterior spine, the 6th with a double ridge and median sulcus; pleura are separated from terga by ridges and sulci, curved ridges on 2nd to 5th pleura.

Telson (t) wide, subtriangular, tapering to sharp angle; two dorsolateral carinae almost meet a median carina near tip; proximally an irregular ridge and hump with a posterior spine middorsally; branches of uropod wide, each with a supporting rib, the outer without a distolateral spine.

Eye rudimentary, stalk fused with anterior carapace (in orbit), a blunt spine at centre of anterior edge.

Antennule (c) 1st article with 3 lateral spines, apparently fused with stylocerite which greatly exceeds other two articles tapering to long attenuate point; ventromesial flagellum short, dorsolateral slightly longer than carapace.

Antenna (d) scale tapering to long apical spine; 2nd segment of peduncle with distal inner spine; flagellum about equal to carapace length.

Mandible (e) incisor only, compressed and slightly curved with W-shaped cutting edge of 3 major teeth with many accessory teeth along edges; palp large with 3 segments.

Maxillule (f) distal endite long and narrow with distal curved spine; proximal small narrow; no endopod.

Maxilla (g) endites reduced; endopod slender, not as long as narrow anterior lobe of scaphognathite with almost straight inner edge and length greater than wide, rounded posterior lobe.

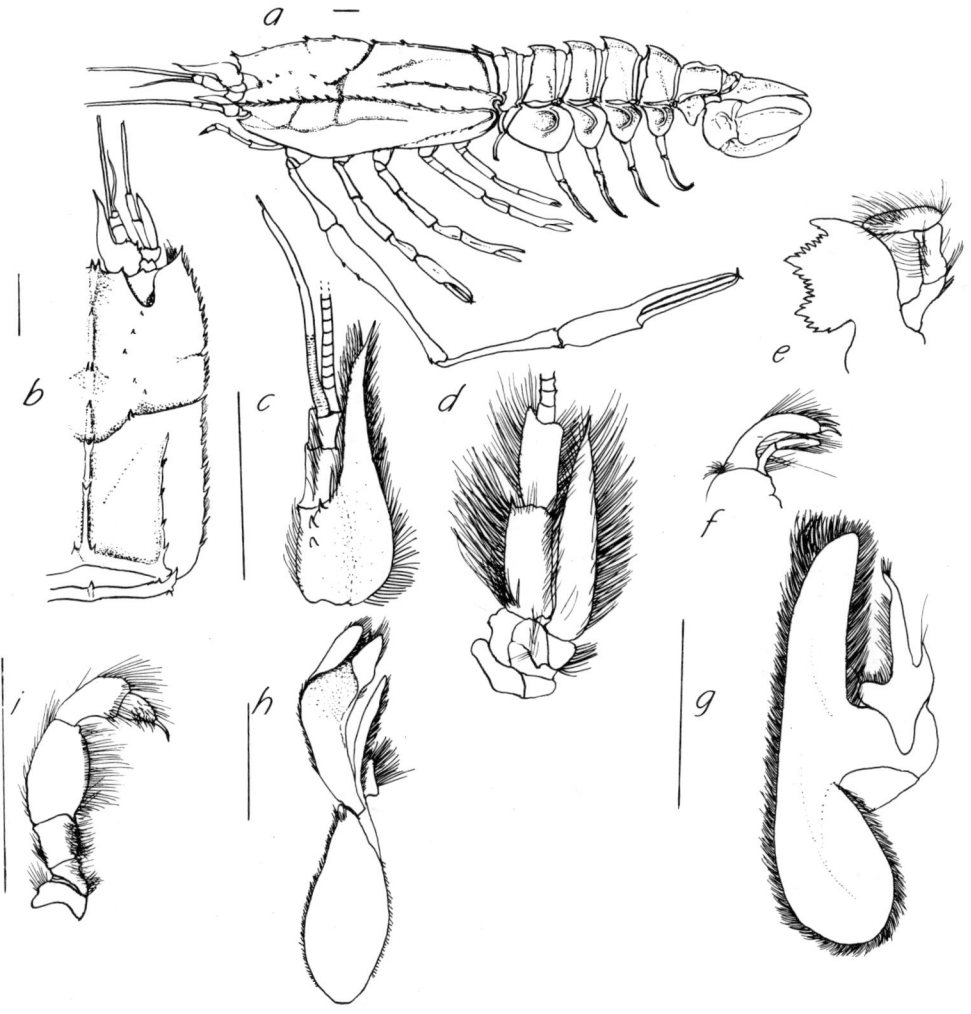

Fig. 190. *Stereomastis sculpta*: *a*, whole animal from left side; *b*, carapace in dorsal aspect; *c*, antennule; *d*, antenna; *e*, mandible; *f*, maxillule; *g*, maxilla; *h*, first maxilliped; *i*, second maxilliped; solid line = 10 mm.

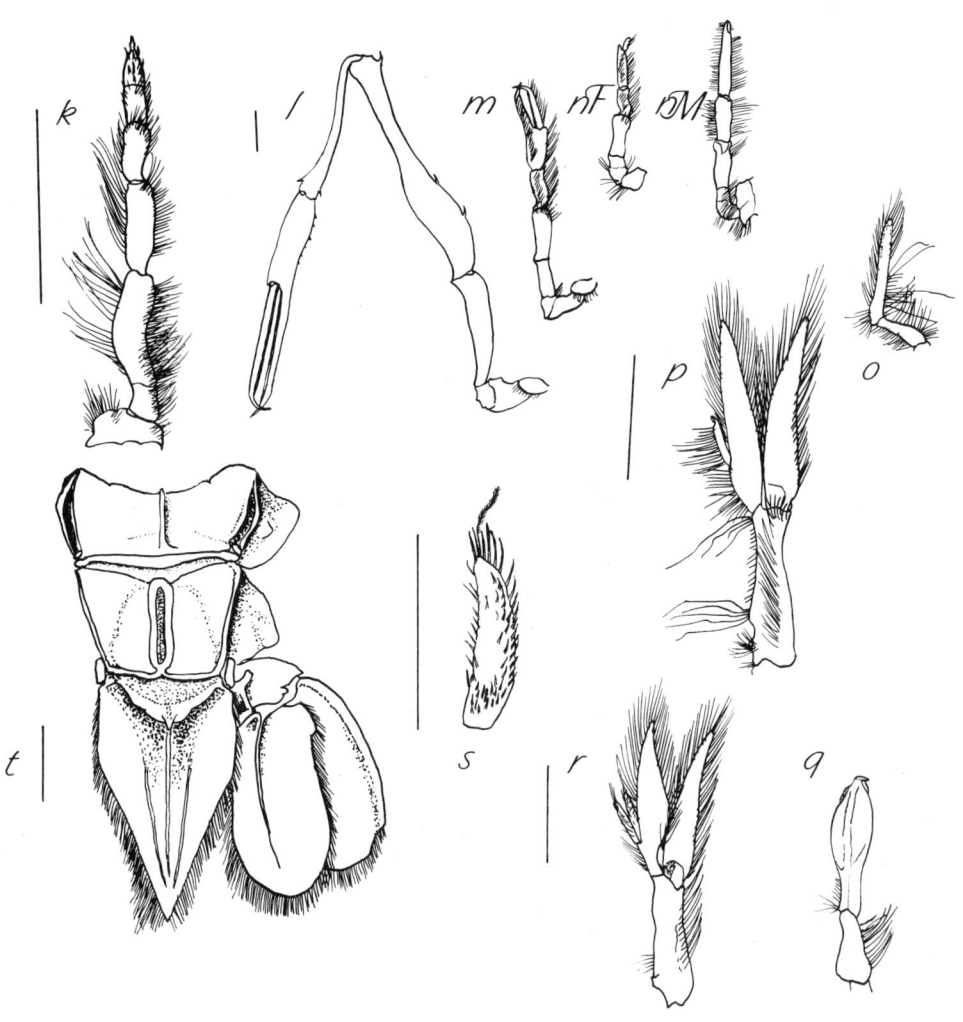

Fig. 191. *Stereomastis sculpta*; k, third maxilliped; l, first pereopod; m, second pereopod; nF, M fifth peropod; nM, F fifth pereopod; o, F first pleopod; p, F second pleopod; q, M first pleopod; r, M second pleopod; s, appendix masculina; t, telson; solid line = 10 mm.

Maxilliped I (h) endites reduced to a small triangular lobe; endopod very slender, triangular in cross-section, shorter than exopod, the latter foliaceous and folding in on itself distally; epipod large, foliaceous, ovate.

Maxilliped II (i) slightly compressed, leglike but small, distal segment small but with long curved apical spine.

Maxilliped III (k) stoutish, distal three segments short, with apical strong spine and spinous setae. No epipod.

Pereopods: I (l) slightly longer than body length, chelate, slender fingers longer than palm with curved sharp tips; merus longest compressed and expanded proximally with 2 lateral spines and a distal spine on dorsal edge; rounded epipod with small filaments of podobranch; II (m) to V (n) much shorter than I and decreasing in size, all chelate in female (V with very small chela) and in male the dactyl is longer than distal projection of propodus to which it may be opposable; all except V have rounded epipod and podobranch.

Pleopods: female I (o) endopod only, longer than protopod; left with short thick appendix interna with apical hooks, lacking in right. Female II (p) endopod and exopod about equal, appendix interna with apical pad of hooks. Male I (q) modified endopod, compressed, spatulate, with thick circular disc at apex and a projection below it with hooks; male II with exopod and endopod equal, appendix interna with subapical pad of hooks and slightly shorter than appendix masculina (s) the latter with about six moderate spines and a plumose seta apically and many small spines laterally.

RANGE OF DISTRIBUTION

In the western Atlantic from south of Iceland and in Davis Strait to the Caribbean; in the eastern Atlantic from Iceland to the Cape Verde Islands and Angola. Also in the Indian Ocean from the Arabian Sea to Malaysia. Depths 230–4 000 m (Sivertsen and Holthuis 1956).

Records of distribution in the area of reference are in Fig. 192.

BIOLOGY

Lengths to 70 mm cl in males and females.

Most females taken were ovigerous in May, October and November and about half of these already had large ova in the ovaries suggesting at least annual spawning. Egg diameter 0.8 mm.

Stomach contents were remains of fish, euphausiids, polychaetes and foraminiferans (Squires 1965a).

Species also present in catches were *Nematocarcinus cursor* and *Polycheles granulatus* (Squires 1965a).

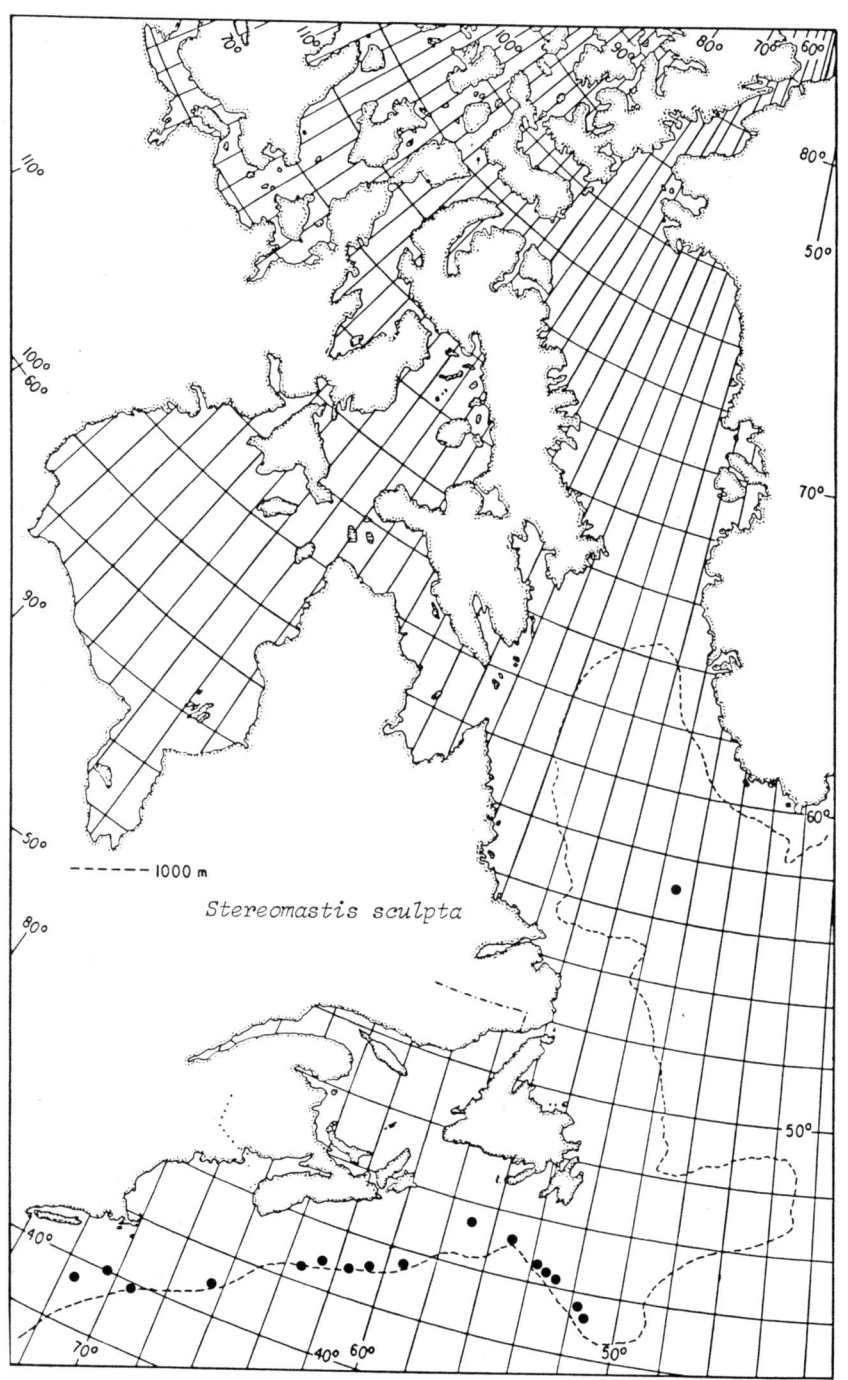

Fig. 192. *Stereomastis sculpta*, distribution records in the area of reference.

Family PAGURIDAE Latreille, 1803
McLaughlin 1974: 37; Rathbun 1929: 26; Williams 1984: 207; Zariquiey Alvarez 1968: 243.

Hermit crabs. Carapace in front of the cervical groove (the shield) calcified, posterior carapace soft, membraneous; eye scales usually triangular; maxillipeds widely separated at base, ischium of 3rd maxilliped with crista dentata; first pereopods with large chelae, right larger than left; abdomen asymmetrical, usually twisted spirally; tail fan adapted to hold body in hollow objects.

Only one genus in the area of reference.

Genus *Pagurus* Fabricius, 1775
McLaughlin 1974: 37; Williams 1984: 208.

Anterior shield well calcified, posterior soft not calcified; eye scales triangular; third maxillipeds widely separated at base by sternum, ischium with crista dentata; right chela of first legs much larger than left; fourth legs subchelate with rasp on propodus; males with paired sexual openings, vas deferens not protruding; abdomen well developed, uropods asymetrical; pleopods not paired.

Five species commonly in the area of reference.

Key to species of *Pagurus* in area of reference
(From Williams 1984)

1	Palm of left hand or chela with upper surface horizontal and not divided into two facets separated by a ridge	2
	Palm of left hand with upper surface divided into two facets by a ridge	3
2	Both chelae long and narrow	*Pagurus longicarpus*
	Both chelae stout and wide	*Pagurus acadianus*
3	Palm of left chela with single row of strong spines slanted inward; anterior part of sternite of 3rd legs subrectangular with few short setae	*Pagurus pubescens*
	Palm of left chela with double row of spines not slanted inward	4
4	Chelae with sharp spines forming double ridge on palm of small and many setae on upper surfaces of both; anterior part of sternite of 3rd legs semicircular with many and long setae	*Pagurus arcuatus*
	Chela with spines not forming ridges on palm and not strongly setose on upper surfaces; anterior part of sternite of 3rd legs a more pointed semicircle and with tufts of setae	*Pagurus politus*

Pagurus acadianus Benedict, 1901
Rathbun 1929: 26: fig. 34; Williams 1984: 209: fig. 147.
(Figures 193 and 194)

DISTINGUISHING CHARACTERISTICS

Large chelipeds stout and wide, rounded; upper surface of both chelae with broad longitudinal red-orange stripe; anterior shield of carapace about as wide as long, about equal to soft part of carapace in midline; eyescale concave on upper surface with terminal acuminate spine; anterior part of sternum of 3rd legs very narrow, slightly curved.

DESCRIPTION

Integument smooth, with occasional pits and tufts of setae, calcified on anterior half, soft and membraneous on posterior half. Colour of upper surface of both chelae a wide longitudinal red-orange stripe on reddish tinge.

Anterior carapacial shield about as wide as long, about equal in length to posterior soft part, both separated by cervical groove; rostrum obtuse but with a sharp point, lateral projections less than rostrum but each with a calcareous spine at tip; sternite (z) of 3rd legs with wide low anterior part fringed with long setae, the posterior parts squarish.

Abdomen sack-like, curved and slightly twisted posteriorly, somewhat longer than carapace; pleopods on one side only (three in males, four in females); ventrally across anterior abdomen is an asymmetrical ridge partly calcified.

Telson (t) posterior lobes with terminal margin widely notched with three major and two small spines on each side of notch. Left uropod much larger than right and with corresponding rasp areas.

Eye moderate with cornea slightly larger than stalk, the latter with slight middle constriction; eye scale with oval anterior lobe tapering to acuminate spine.

Antennule (c) slender 3rd article longer than each of 2nd and 1st; stylocerite short and fused with 1st article or rudimentary; flagella shorter than 3rd article.

Antenna (d) acicle with small spine at tip, exceeded by 5th segment of peduncle; terminal spines on 2nd and 3rd segments; flagellum about as long as carapace.

Mandible (e) incisor with slightly curved entire cutting edge, separated from molar by a notch; palp two-segmented.

Maxillule (f) distal endite expanded toward straight edge; proximal slightly wider with straight edge also; endopod with very short terminal lobes.

Maxilla (g) both endites unequally bilobed, the proximal with one very narrow lobe; endopod tapered to slender tip, slightly longer than anterior lobe of scaphognathite, posterior lobe about equal in length to but wider than posterior.

Maxilliped I (h) distal endite large, proximal small; endopod slender, shorter than exopod, the latter with lash or flagellum.

Maxilliped II (i) slightly compressed, dactyl longer than wide, propodus expanded; exopod long with moderate curved flagellum.

Maxilliped III (k) robust, ischium with crista dentata of about 14 small teeth with a distolateral accessory tooth; exopod about as long as ischium plus merus and with flagellum.

Pereopods: on (a) and (b) I left smaller than right, both covered with many tubercles or short rounded spines, outer face of propodus rounded and not divided into facets but there is a short central ridge of larger tubercles proximally; tips of right are calcareous but of left corneous and spooned. II and III longer than others: dactyl much longer than propodus, with longitudinal sulcus, tip corneous and sharp and edges with serrate corneous spines; IV short and stout, pseudochelate, a distal extension of propodus opposable to longer curved, sharp dactyl, distolateral rasp on propodus. V short and slender, articulated upward, short rounded dactyl opposed to extension of propodus forming small pseudochela, a distolateral rasp on propodus.

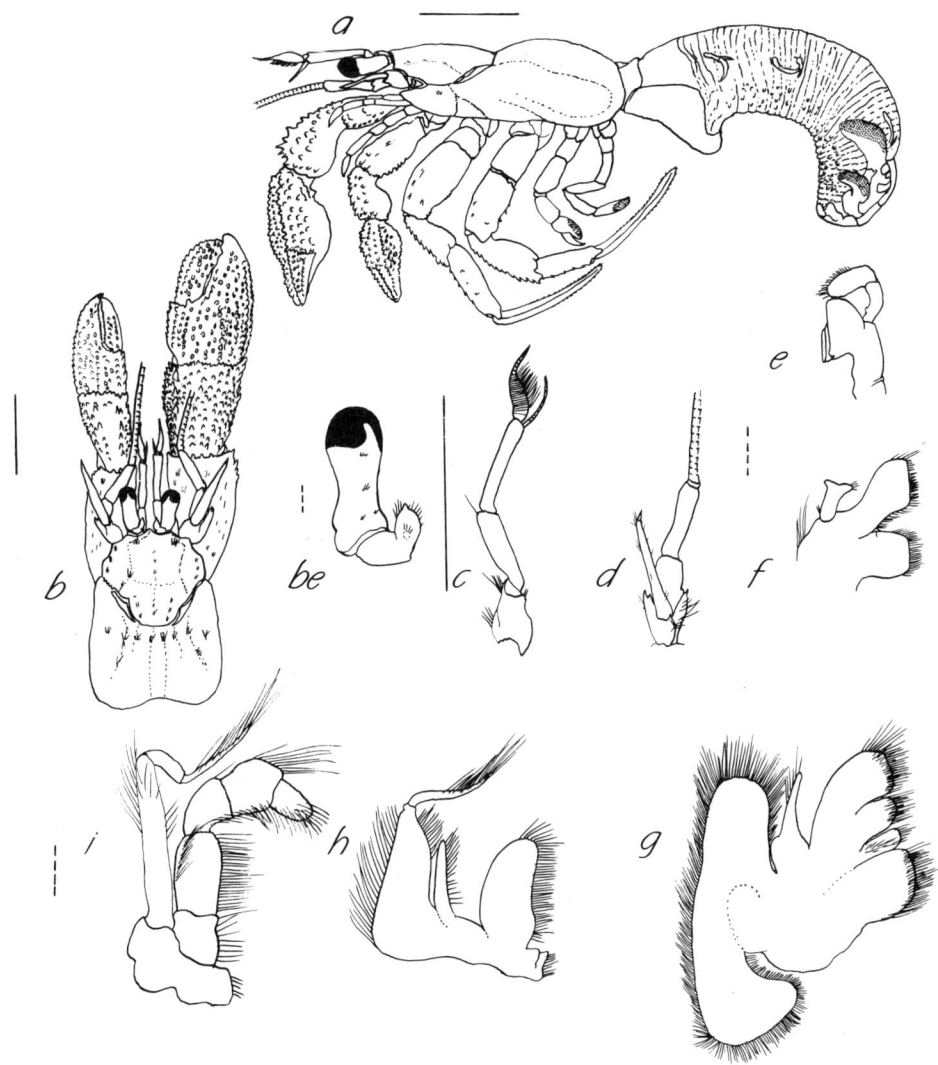

Fig. 193. *Pagurus acadianus*: *a*, whole animal from left side; *b*, carapace in dorsal aspect; *be*, eyescale; *c*, antennule; *d*, antenna; *e*, mandible; *f*, maxillule; *g*, maxilla; *h*, first maxilliped; *i*, second maxilliped; solid line = 10 mm, broken line = 1 mm.

Fig. 194. *Pagurus acadianus*: *k*, third maxilliped; *z*, sternal plates of fourth legs; *o*, F first pleopod; *p*, F fourth pleopod; *q*, M first pleopod; *t*, telson; solid line = 10 mm.

Pleopods: female I (o) (egg-bearing) exopod longer than endopod, with slight distolateral expansion with many long setae and long distal setae, protopod with long distolateral setae; endopod club-shaped with distal setae. Female II similar to I but endopod is longer than exopod. Female III similar to II but smaller. Female IV (p) non-eggbearing, with short spike-like endopod and flat narrow exopod fringed with a single row of setae. Male I (q) exopod is 2.7 times length of endopod, the latter slender and tapering, the former flat, moderately wide and curved with fringing straight setae. Male II similar to I. Male III (r) much smaller than others, exopod longer than slender and short endopod.

RANGE OF DISTRIBUTION

From the Straits of Belle Isle and Notre Dame Bay, Newfoundland, and Gulf of St. Lawrence to Chesapeake Bay, in the western Atlantic only. Depths from 0 to 485 m. (Squires 1966; Williams 1984).

Records of distribution in the area of reference are in Fig. 195.

BIOLOGY

Lengths of anterior shield to 18 mm in males and 13 mm in females.

In marking and recapture studies low recaptures indicated considerable movement in and out of limited study area. Preference for shells of *Buccinum* was demonstrated (Williams 1984).

Ovigerous females were taken in Cape Cod Bay in January and March (Williams 1984).

Stomach contents included fish offal, crustacean fragments and mussel shell (Squires 1965a).

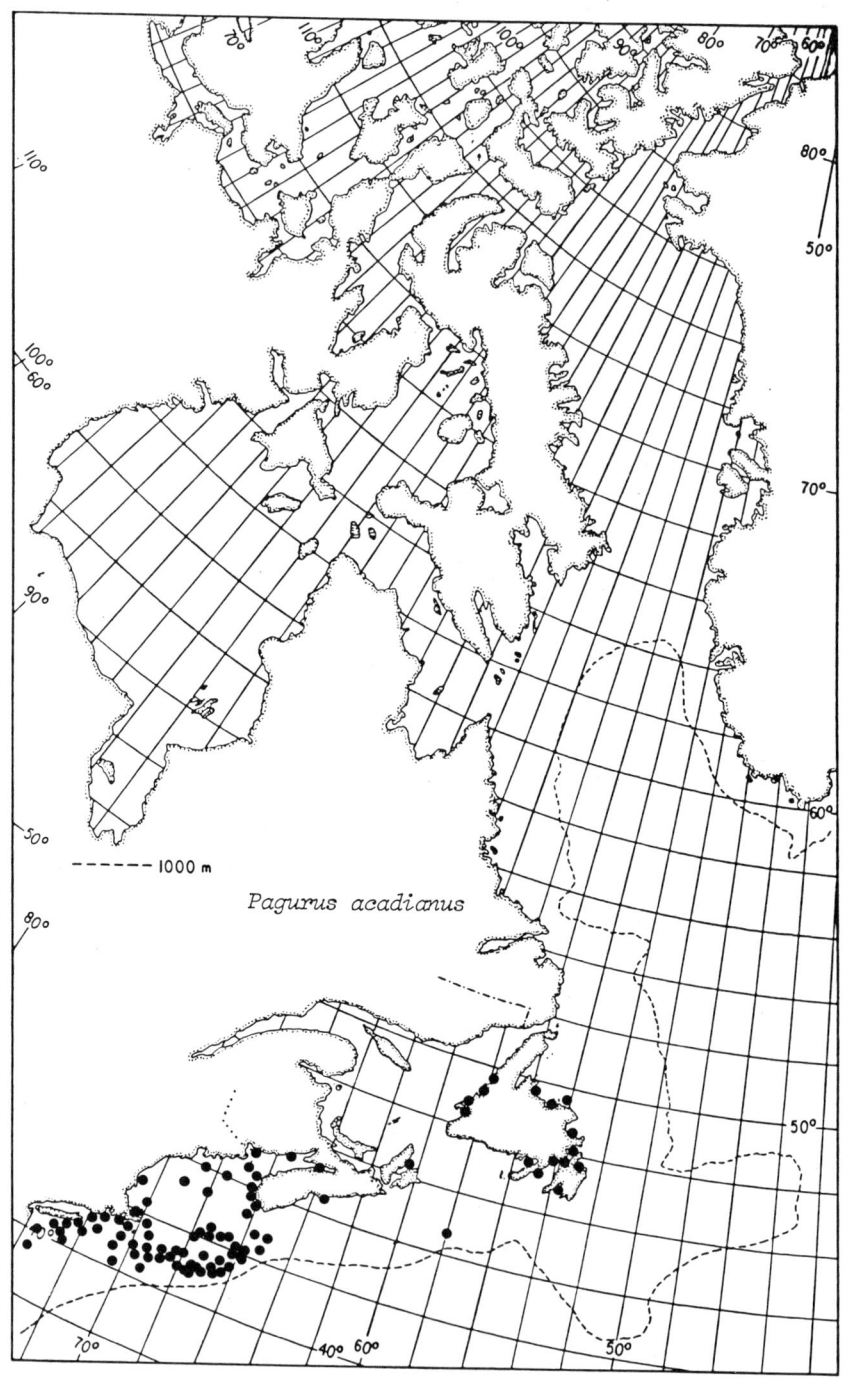

Fig. 195. *Pagurus acadianus*, distribution records in the area of reference.

Pagurus arcuatus Squires, 1964
Pagurus bankensis Nesis, 1964: 667, fig. 4;
Pagurus pubescens Rathbun, 1929: 28, fig. 37.
Pagurus arcuatus Williams, 1984: 212, fig. 149.
(Figures 196 and 197)

DISTINGUISHING CHARACTERISTICS

Left or small 1st pereopod with propodus divided dorsally into two facets, the outer convex or slightly rounded and narrow for its length, the inner with two rows of spines at the ridge separating the facets, many setae; anterior part of sternite of 3rd legs semicircular, almost as high as wide and with long setal fringe; narrow "V" dorsally on middle of right chela with no spines between sides.

DESCRIPTION

Integument smooth but with rugosities and pits with clumps of setae. Colour pink on dull white with yellowish setae.

Carapace shield with calcification, slightly longer than wide; rostrum rounded, with tuft of long setae; lateral projections with laterally directed spine; rough area behind rostrum. Sternite (z) of 3rd legs with anterior part almost semicircular, almost as long as wide, fringed with very long setae.

Abdomen slightly longer than total carapace, anterior muscular opercular part moderate; three pleopods in male, four in female, unpaired.

Telson (t) fairly wide cleft at posterior margin of posterior lobes with strong curved spine at corners and 5 equal spines on left and 5 similar on right side of cleft.

Eye moderate, slightly wider than long stalk, the latter with few setal tufts dorsally, not reaching end of antennal peduncle. Eye scales narrowish, with two tufts of setae on slightly hollow dorsal surface, and subapical sharp spine.

Antennule (c) from ventral aspect stylocerite appears to be fused with 1st article and about half length of article; 1st and 2nd articles equal, together slightly longer than 3rd.

Antenna (d) basal segment with outer spine and inner fan of 4 spines; 2nd segment with distal projection and two spines and a strong outer spine, attached acicle as long as 4/5 of 5th segment of peduncle; acicle with apical spine and many long lateral setae; 3rd segment (1st of peduncle) with inner distal spine.

Mandible (e) heavily calcified incisor edge not quite straight with both ends cut to lower level; heavy molar oval in shape at surface lies behind incisor; palp with 3 segments.

Maxillule (f) distal endite expanded to spinous edge, narrower than proximal endite; endopod with distal lobes spread, the outer slightly larger and with tip turned in.

Maxilla (g) endites unequally bilobed, proximal with one very narrow lobe; endopod rounded at base with straight taper to sharp point, leaning towards endite but slightly exceeding anterior lobe of scaphognathite, the latter rounded and about equal to wider posterior axe-shaped posterior lobe.

Maxilliped I (h) distal endite large, proximal smaller but greatly thickened; twisted endopod with pointed apex and long setal fringe, shorter than exopod, latter with short lash with very short annulated tip.

Maxilliped II (i) proximal segments triangular in cross-section, slightly compressed; two-parted exopod with long lash.

Maxilliped III (k) robust, leglike, proximal three segments triangular in cross-section; ischium with crista dentata of 23 corneous spines and an accessory large spine; distal and lateral tooth on merus, dactyl twice as long as wide.

Pereopods: I (l) chelate, very unequal, smaller with ridge of double row of spines separating palm into two facets, outer narrow and convex; two rows of spines also on carpus, and distal edge dorsally with three spines. Dorsal face of large chela (la) with long "V" of spines. II and III long and moderately slender, dorsal surfaces unarmed,

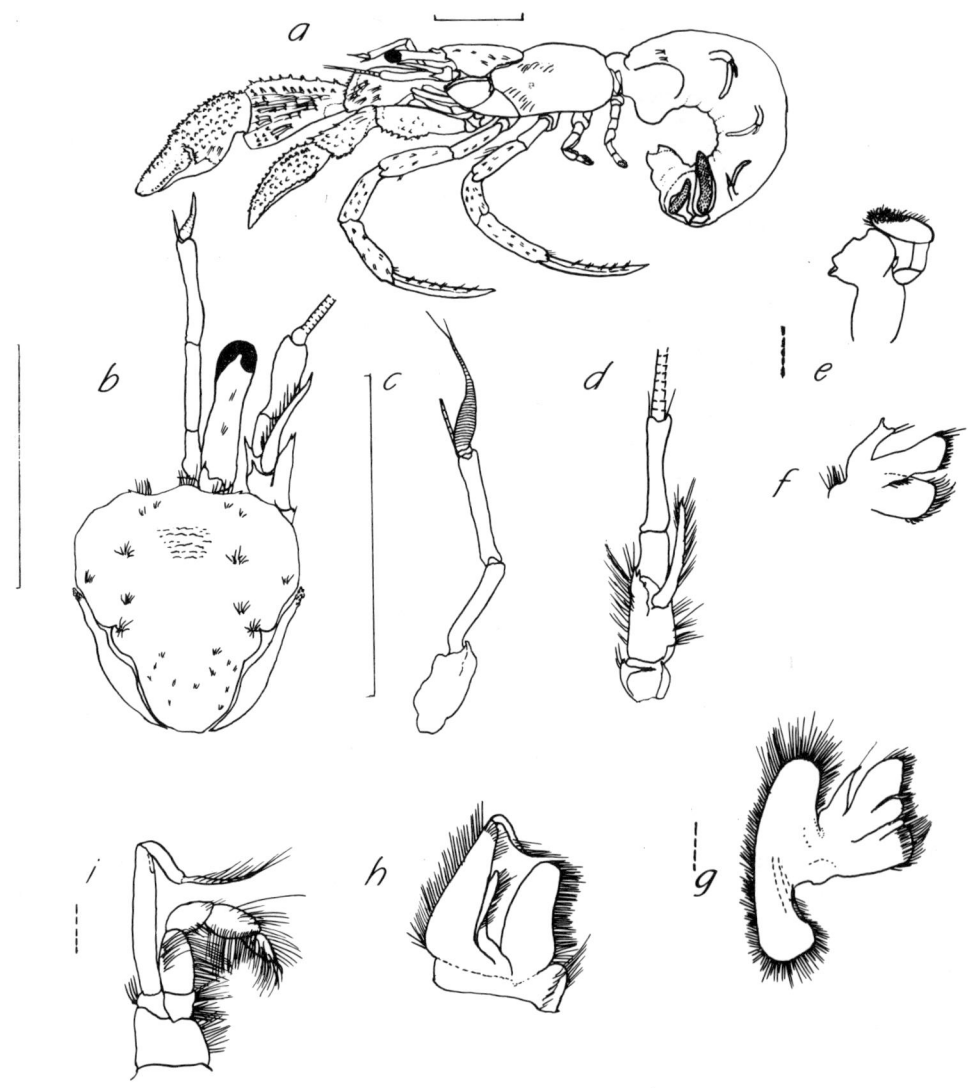

Fig. 196. *Pagurus arcuatus*: *a*, whole animal from left side; *b*, anterior carapace in dorsal aspect; *c*, antennule; *d*, antenna; *e*, mandible; *f*, maxillule; *g*, maxilla; *h*, first maxilliped; *i*, second maxilliped; solid line = 10 mm, broken line = 1 mm.

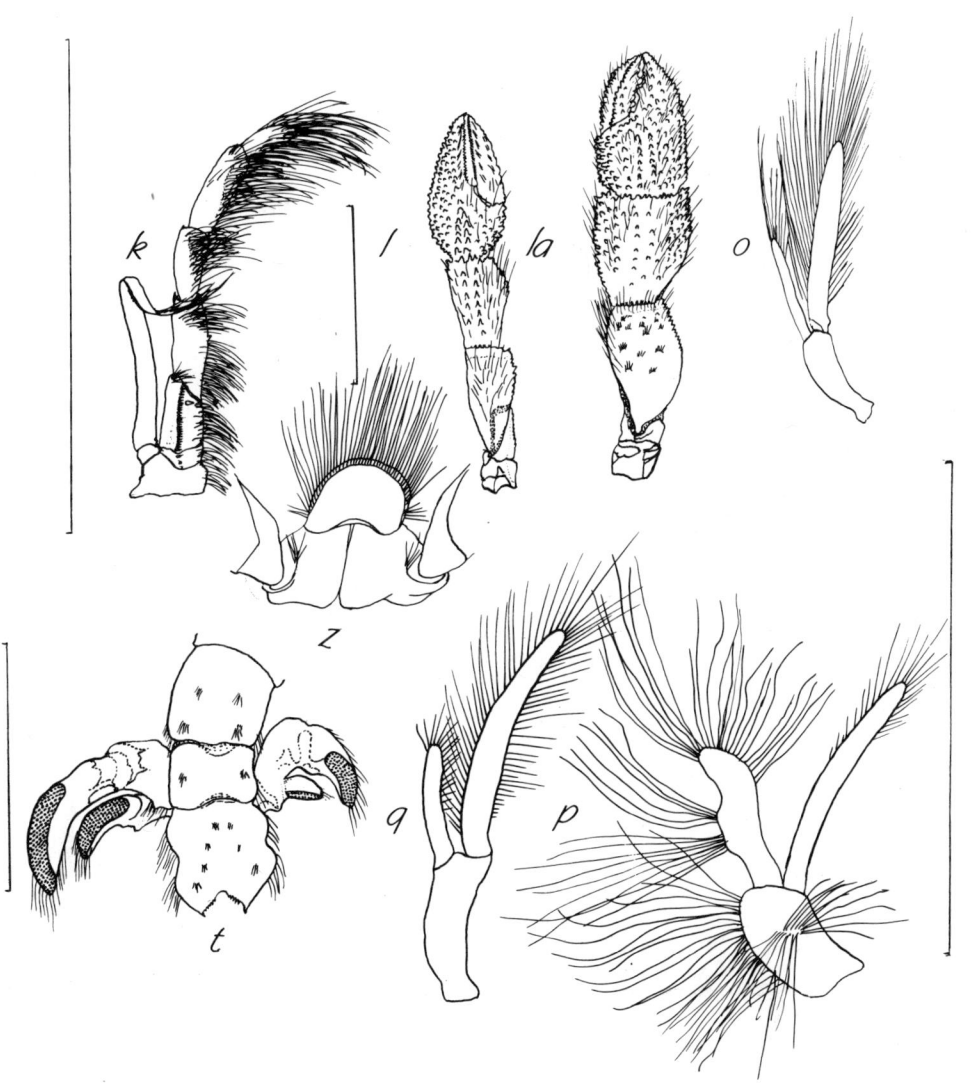

Fig. 197. *Pagurus arcuatus*: *k*, third maxilliped; *l*, first pereopod right; *la*, first pereopod left; *o*, F first pleopod; *p*, F fourth pleopod; *q*, M first pleopod; *t*, telson; *z*, sternal plates of fourth legs; solid line = 10 mm.

dactyl longer than propodus, with longitudinal sulcus and setal tufts and sharp point. IV much shorter than III, dactyl a curved claw, longer than distal extension of propodus with which it is partly opposable, end of propodus with rasp. V short, pseudo-chelate, dactyl rounded, opposable with extended tip of propodus.

Pleopods: female I (p) endopod with lateral expansion and long silky setae, II and III similar; IV (o) small with slender endopod and long narrow exopod with plain setae, attached near uropods (all on left side of abdomen). Male I (q) endopod slender, exopod more than twice as long, slightly curved, with plain setae; male II and III similar, no IV in males.

RANGE OF DISTRIBUTION

In the western Atlantic only from Greenland to off Virginia Capes. Depths from 0 to 270 m (Williams 1974b).

Records of occurrences in the area of reference are in Fig. 198.

BIOLOGY

Lengths of anterior shield 15 mm in males and 8 mm in females.

Females were first mature at a shield length of 5 mm.

Only 1.4% of specimens examined were infested with the parasitic barnacle, *Peltogaster paguri* (Squires 1965a).

Most specimens were in shells of *Buccinum* sp. but a few in shells of *Colus* sp., *Neptunea* sp. and *Lunatia* sp.

Podoceropsis sp., a commensal isopod, males and females together were frequently found in the shells with this species.

Stomach contents included phytobenthos, foraminiferans, small bivalve shells, crustacean fragments, small gastropods, hydroids and ophiuroids (Squires 1965a).

Occasionally present in catches with this species was *Pagurus pubescens* but more commonly *Spirontocaris spinus*, *Lebbeus polaris*, *L. groenlandicus* and *Pandalus montagui* (Squires 1965a).

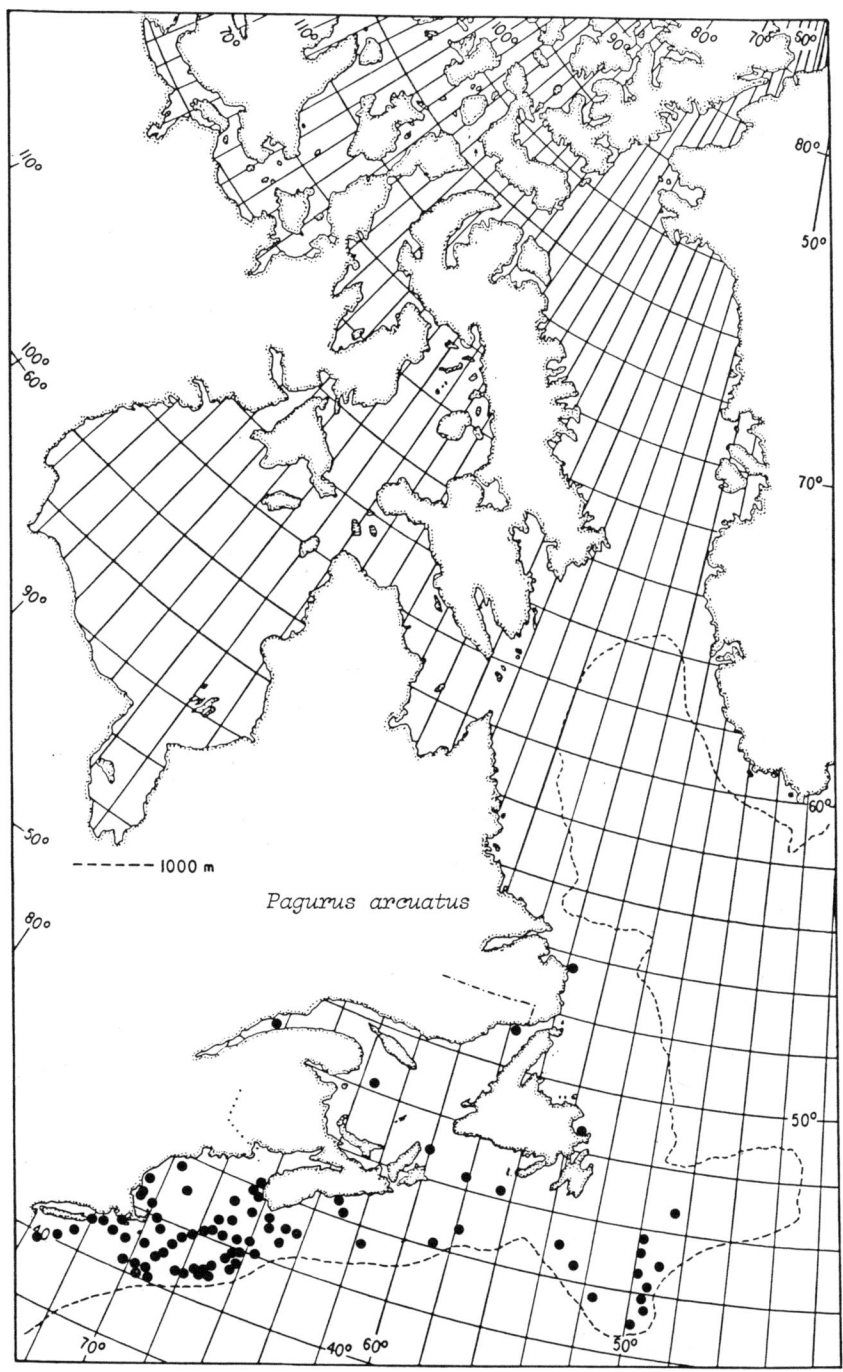

Fig. 198. *Pagurus arcuatus*, distribution records in the area of reference.

Pagurus longicarpus Say, 1817
Provenzano 1959: 394, fig. 13;
Rathbun 1929: 27, fig. 35;
Williams 1984: 216, fig. 154.
(Figures 199 and 200)

DISTINGUISHING CHARACTERISTICS

Right cheliped longer and stouter than left, longer than wide, appearing smooth and without setae but covered with small spinules, two rows of spines slightly larger than the rest with a few small spines between them on right palm, palm long, fingers short, tips somewhat hooked; small chela similar in form, subcylindrical, but tips not hooked; rostrum about even with lateral projections which have a laterally directed spine; eye scales large anteriorly, slightly concave above with one tuft of setae, also a subapical spine. Telson with posterior lobes much smaller than anterior ones, and with wide cleft at posterior margin. Anterior part of sternum of 3rd legs much wider than long and with long fringing setae.

DESCRIPTION

Integument smooth but with occasional pits and clumps of setae. Colour reddish brown on body with fingers whitish, lighter in colour at the southern part of its range, often with a darker longitudinal stripe on larger claw (Williams 1984).

Shield about as wide as long, soft part of carapace slightly longer; rostrum obtuse, about even with lateral projections which have laterally directed spine.

Sternal plate of 3rd legs (z) with anterior part moderately rounded, about 2.7 times as wide as long, centrally notched posteriorly, with long forward setae attached nearly at middle horizontally.

Abdomen somewhat longer than total carapace length when straightened; part of muscular anterior part tongue-like.

Telson (t) posterior lobes small, about equal, widely forked (divided) at posterior margin, with a pair of moderate spines at each corner and two groups of 5 and 4 small spines along each inner edge.

Eyestalks stout, length about 0.6 shield length, cornea large globular, exceeded by acicle and antennular and antennal peduncles; eye scales oval, hollowed dorsally, with subterminal spine and one dorsal tuft of setae.

Antennule (c) 1st article stout with inner spine distolaterally, 2nd only slightly shorter than 3rd, each longer than 1st; flagella short.

Antenna (d) outer projection of 2nd segment with inner edge sagittate; acicle sinuous, tapered to sharp point.

Mandible (e) almost straight incisor edge with strong cusp at each end; palp 3-segmented.

Maxillule (f) distal endite expanded to spinous edge, slightly longer than wider proximal endite; inner distal lobe of endopod pointed and with strong short seta.

Maxilla (g) endites bilobed, distal almost equal, proximal with one moderately narrow; endopod thin, slightly twisted, longer than anterior lobe of scaphognathite, anterior about equal to wider subtriangular posterior lobe.

Maxilliped I (h) distal endite slightly longer than thick edged proximal; endopod thin and curved setose on one side, shorter than exopod, the latter with long lash.

Maxilliped II (i) leglike, propodus expanded, dactyl twice as long as wide; exopod two-segmented tapered with long lash.

Maxilliped III (k) leglike, basis with distolateral spine opposite end of crista dentata on ischium, the latter with an accessory spine and 12 spines in a slight curve; exopod with two segments and long lash.

Pereopods: I (l) chelate, very unequal, larger right covered with small tubercles, almost no setae, two rows of spinules on propodus continued on carpus, the same on

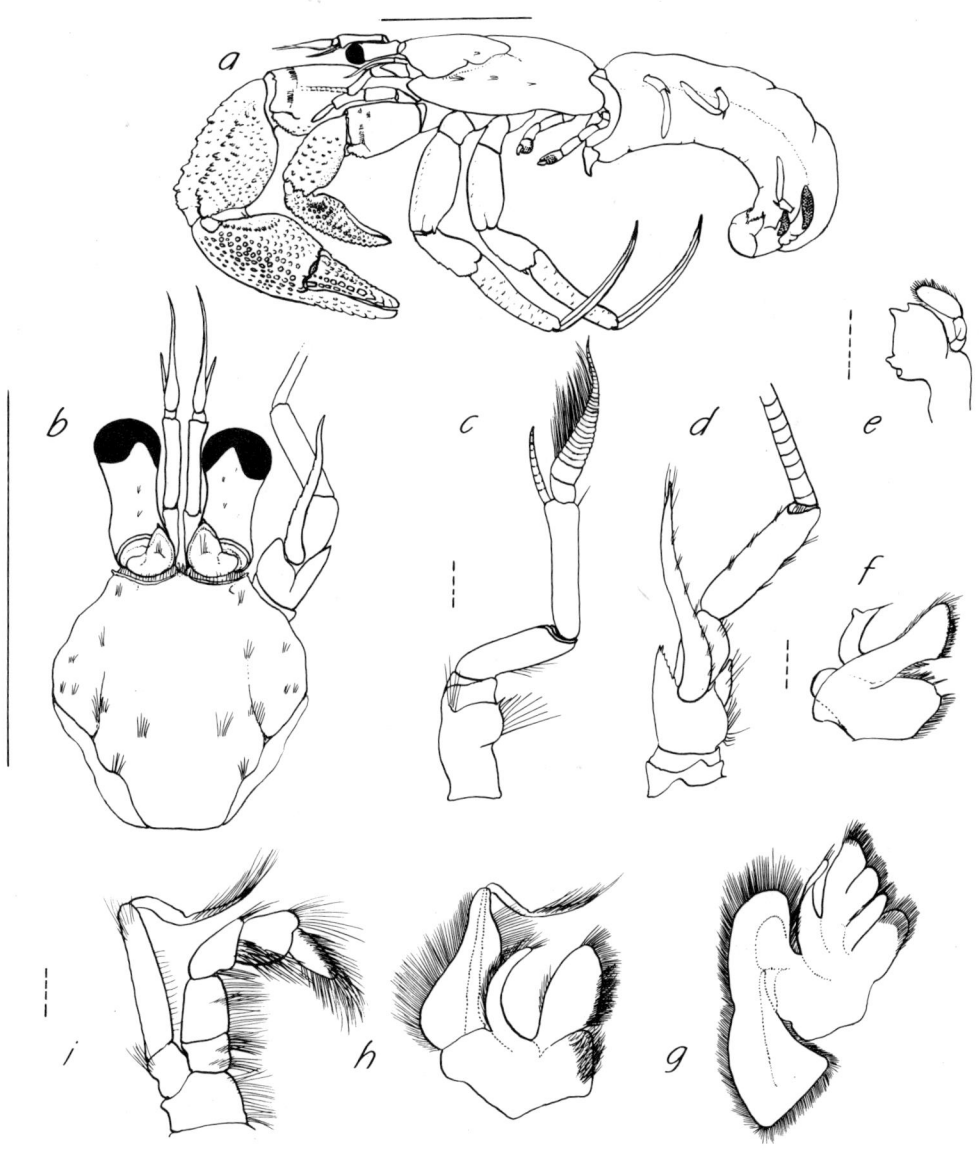

Fig. 199. *Pagurus longicarpus*: *a*, whole animal from left side; *b*, anterior carapace in dorsal aspect; *c*, antennule; *d*, antenna; *e*, mandible; *f*, maxillule; *g*, maxilla; *h*, first maxilliped; *i*, second maxilliped; solid line = 10 mm, broken line = 1 mm.

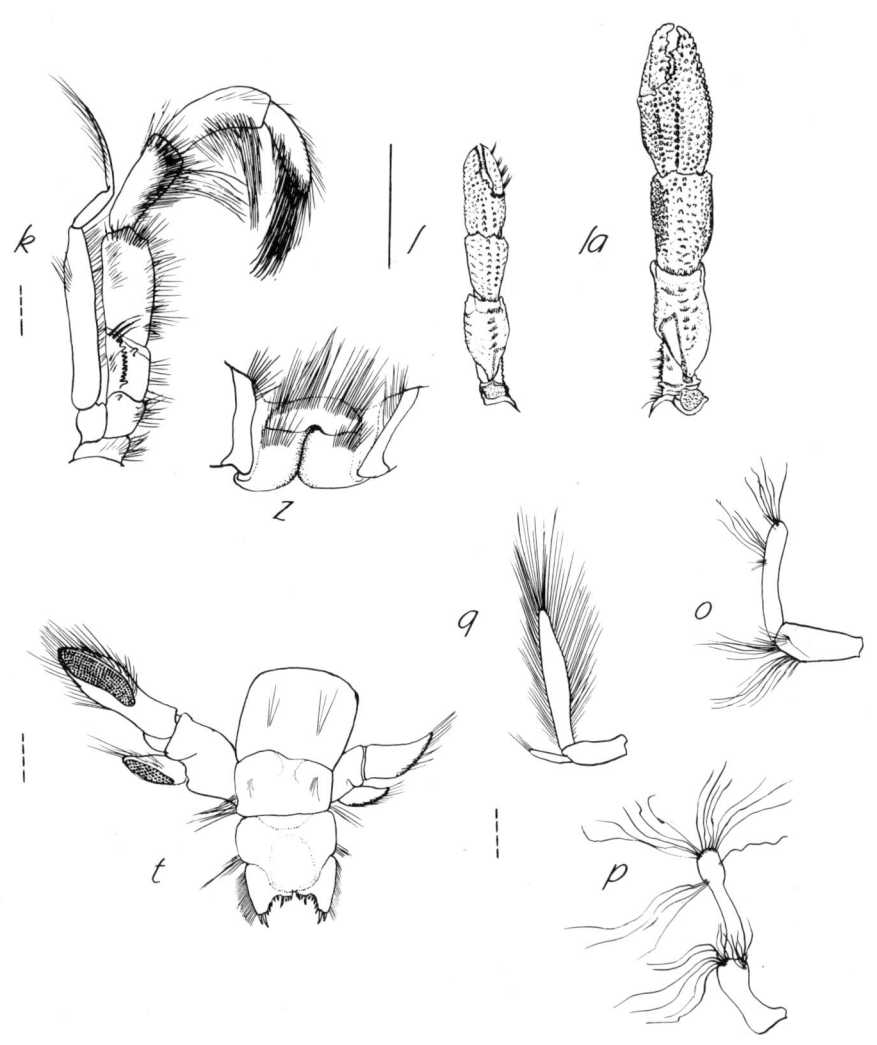

Fig. 200. *Pagurus longicarpus*: k, third maxilliped; l, first pereopod right; la, first pereopod left; o, F first pleopod; p, F second pleopod; q, M first pleopod; t, telson; z, sternal plates of fourth legs; solid line = 10 mm, broken line = 1 mm.

smaller; large chela with calcareous curved tips, small with chitinous tips to fingers. II and III (a) very long and slender, dactyls longer than propodus, with longitudinal sulcus. IV very short, dactyl curved, sharp, slightly longer than extension of propodus to which it is opposed in part. V almost as short as IV, articulated upwards, dactyl rounded and opposed to part of propodus forming a pseudochela.

Pleopods: female I (o) uniramous, endopod slightly longer than protopod, both with clumps of long setae; female II (p) uniramous endopod with rounded tip with long setae, also long setae distally on expanded tip of protopod; III similar to II; IV biramous with short slender endopod and long thin exopod. Male I (q) similar to IV of female and similar to other two (male with three only).

RANGE OF DISTRIBUTION

Minas Basin and Chignecto Bay to Florida and coast of Texas; depths 0–200 m (Williams 1984).

Records of occurrence in the area of reference are in Fig. 201.

BIOLOGY

Lengths of shield to 7 mm in males and 5 mm in females.

One of the most plentiful of hermit crabs in its range, it has been the subject of a great deal of research. The following facts have been summarized from Williams (1984). It has seasonal migratory and burrowing responses, moving into deeper water in winter. The breeding season is from May to September in the northern part of its range; extending into autumn and into winter according to distance farther south. It may mature in about four months after settling from planktonic larval stages and reach full size in about eight months. Females carried from 260 to 4 000 eggs.

Its diet is of diatoms, detritus and algae, also including scavanged materials and sand.

Low rates of parasitization by a larval acanthocephalan and an entoniscid isopod (both internal) have been observed (Williams 1984).

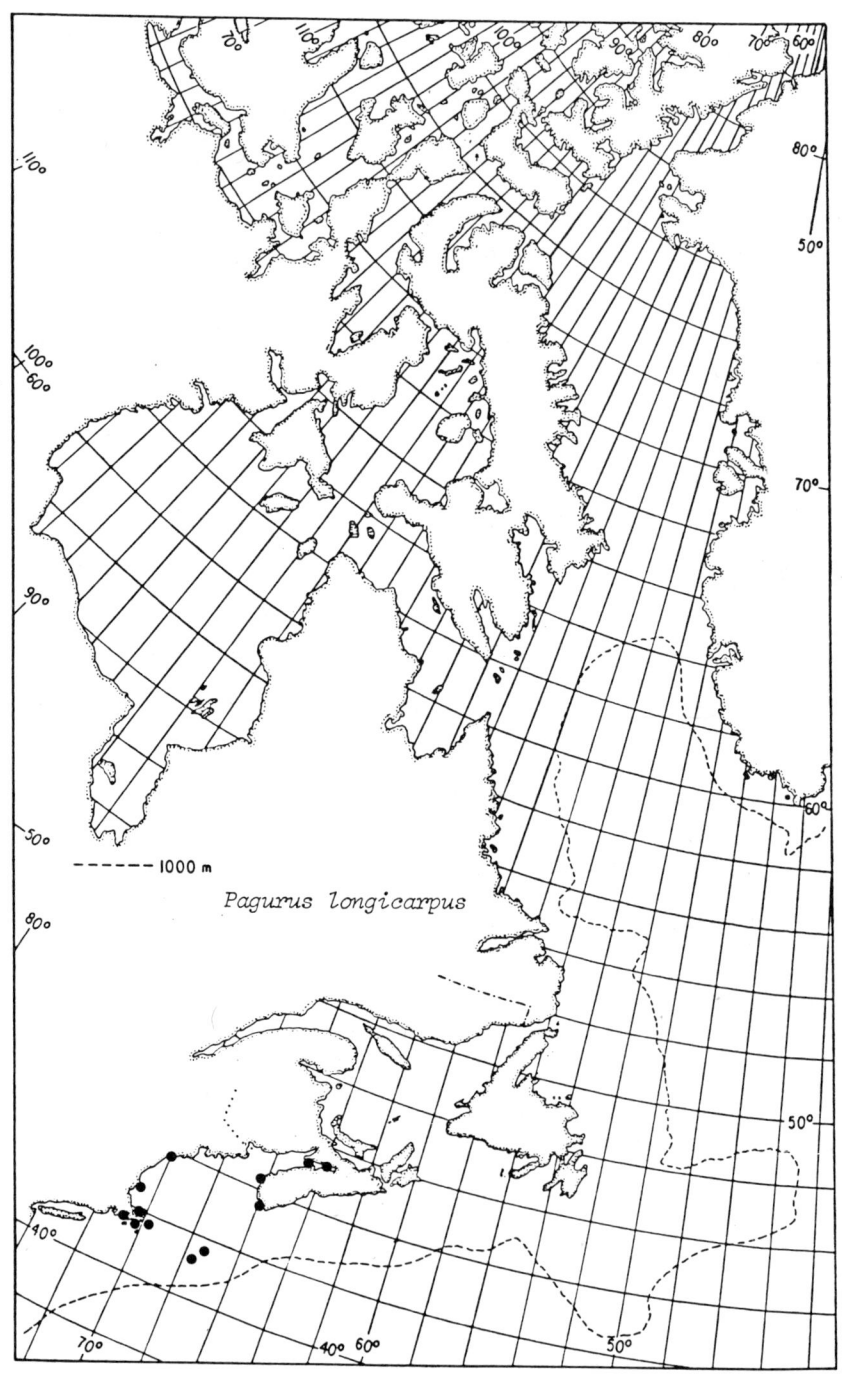

Fig. 201. *Pagurus longicarpus*, distribution records in the area of reference.

Pagurus politus (Smith, 1881)
Williams 1984: 219, fig. 156.
(Figures 202 and 203)

DISTINGUISHING CHARACTERISTICS

Chelae with moderately sharp spines not forming ridges and not conspicuously setose; anterior part of sternite of 3rd legs higher (longer) than wide, somewhat skewed and stretched semicircle; shield of carapace wider than long; acicle exceeding antennal peduncle; posterior lobes of telson greatly unequal with spines outside of small cleft; carpus and propodus of 2nd and 3rd legs dorsally armed with spines.

DESCRIPTION

Integument of anterior shield and legs strongly calcified, smooth on shield with a few pits and setal clumps in longitudinal rows, spinose on legs with few setae. Colour pale orange with tips of chelae and legs white.

Rostrum rounded, exceeded by lateral projections, each with a strong forward spine.

Anterior shield of carapace wider than long, posterior carapace lightly calcified; linea anomurica and sulcus cardiobranchialis medianly parallel are clearly marked (McLaughlin 1974).

Sternite (z) of 3rd legs with anterior part longer than wide, a stretched semicircle skewed to left of animal, equal in height to posterior part; tufts of setae around and inside edge.

Abdomen strongly rounded, anterior portion expanded, opercular lip present; dorsally an oval cap at join with thorax.

Telson (t) with unequal posterior lobes, cleft small with larger spine at tip of larger lobe and 3 spines at inner edge, moderate spine at tip of smaller lobe with 2 spines at inner edge, outside the cleft four spines of unequal size at edge of each lobe.

Eyes large, cornea globular, stalks tapered with longitudinal row of setal tufts; eye scales roundish, dorsally concave, with subapical spine.

Antennule (c) with stylocerite fused to 1st article and with terminal spine about half as long as article; 3rd article slightly longer than second and longer than short flagellum.

Antenna (d) distal projection of 2nd segment about half length of acicle, the latter sinuous, with sharp apical spine, reaching beyond peduncle.

Maxillule (f) distal endite expanded to spinous edge, proximal shorter but wider, subtriangular; endopod with terminal lobes unequal, the larger with terminal spine.

Maxillule (f) distal endite expanded to spinous edge, proximal shorter but wider, subtriangular; endopod with terminal lobes unequal, the larger with terminal spine.

Maxilla (g) endites unequally bilobed, proximal with one narrow lobe and semicircular row of setae behind margin; endopod tapered, close to endite, tip about even with tip of anterior lobe of scaphognathite which slopes inward and is not as long as subtriangular wide posterior lobe.

Maxilliped I (h) distal endite with slightly concave edge, larger than proximal thickened endite; endopod twisted, not as long as exopod, the latter with lash.

Maxilliped II (i) endopod leglike but small, distal segment short but longer than wide; exopod stout with long lash.

Maxilliped III (k) ischium with sinuous crista dentata of about 11 stout spines and without accessory spine.

Pereopods: I (l) chelate, unequal, dorsal face rounded not divided into facets, with moderate spines and few setae, fingers long with sharp tips not conspicuously hooked. II and III (a) long, carpus and propodus armed with strong spines, dactyl longer than carpus and propodus together, slender and sharp and with lateral sulcus. IV short, with hooked dactyl longer than extension of propodus with which it is opposable. V short, articulated upward, with rounded dactyl opposable to part of propodus forming pseudochela.

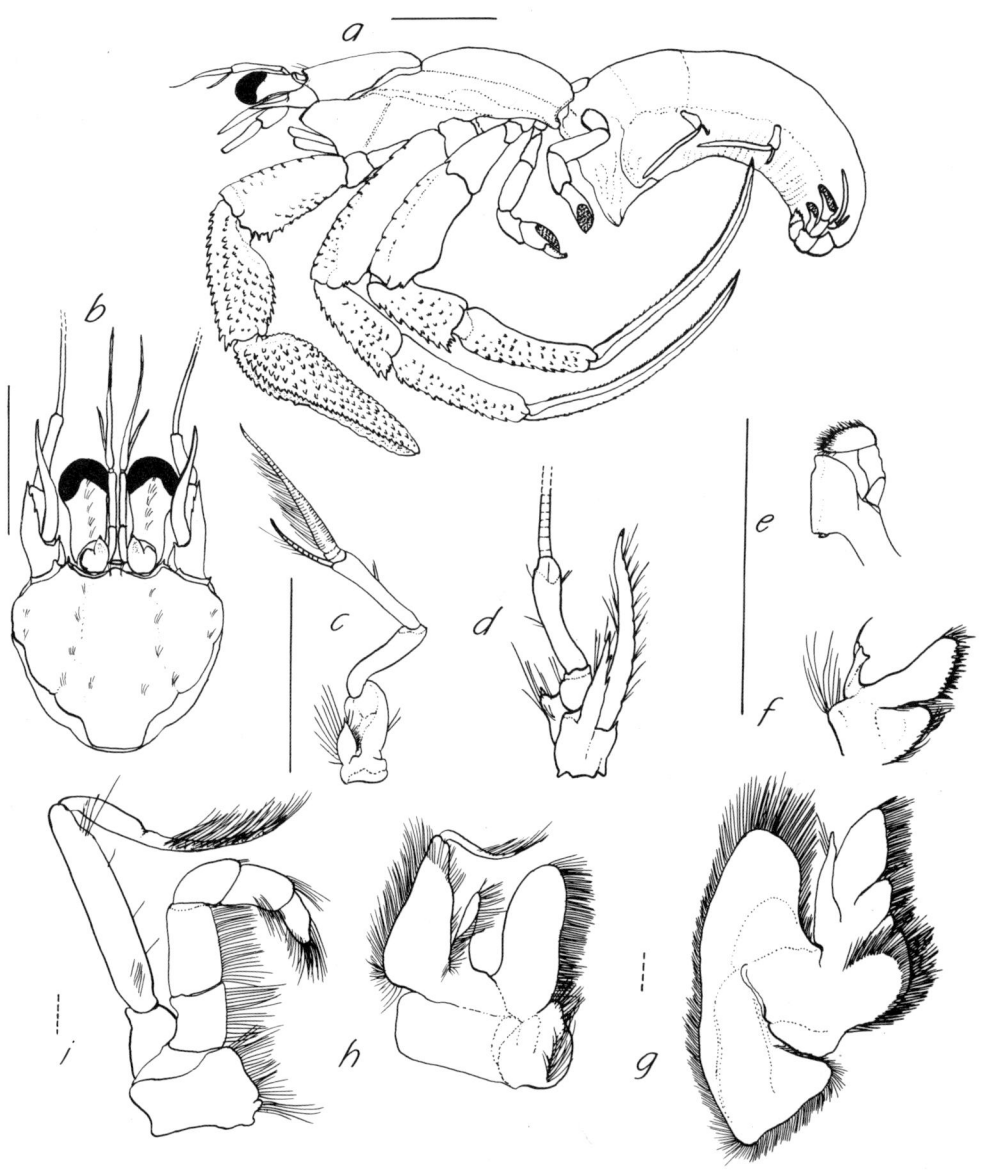

Fig. 202. *Pagurus politus*: *a*, whole animal from left side; *b*, anterior carapace in dorsal aspect; *c*, antennule; *d*, antenna; *e*, mandible; *f*, maxillule; *g*, maxilla; *h*, first maxilliped; *i*, second maxilliped; solid line = 10 mm, broken line = 1 mm.

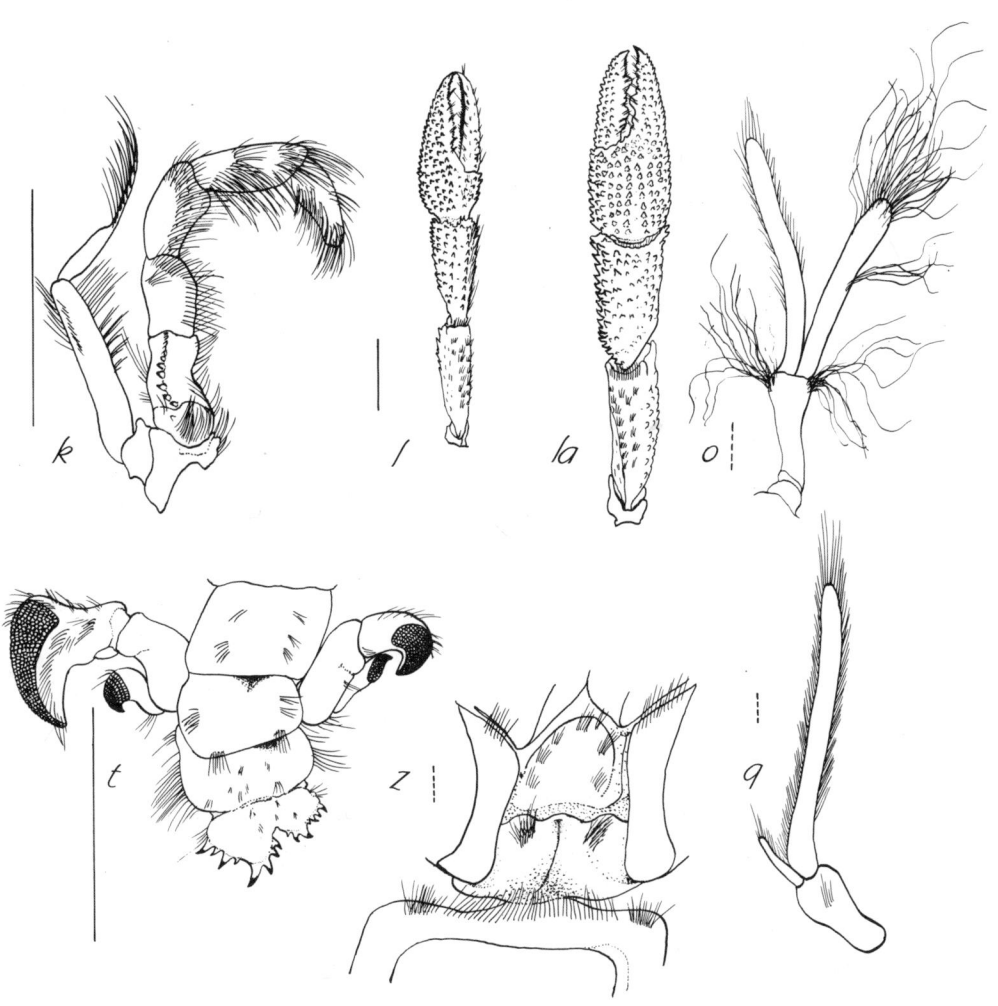

Fig. 203. *Pagurus politus*: *k*, third maxilliped; *l*, first pereopod right; *la*, first pereopod left; *o*, F first pleopod; *q*, M first pleopod; *t*, telson; *z*, sternal plates of 4th legs; solid line = 10 mm, broken line = 1 mm.

Pleopods: female I (o) protopod about two-thirds length of subequal endopod and exopod and with distolateral tufts of long thread-like (egg-bearing) setae, endopod distally with profuse long threadlike setae, exopod with fringe of short fine setae; female II and III similar to I but with exopod longer than endopod, the latter notched near apex and with distal expansion with profuse long threadlike setae; IV smaller and similar to male. Male I (q) endopod extremely short and slender, exopod long and ribbon-like, other two similar.

RANGE OF DISTRIBUTION

Nova Scotia to off Tortugas, Florida (Williams 1974). Essentially an offshore species on continental shelf; depths 30–1 170 m (Williams 1984).

Records of occurrence in the area of reference are in Fig. 204.

BIOLOGY

Anterior shield lengths to 13 mm in males and 6 mm in females. Records indicate year-round breeding (Williams 1984).

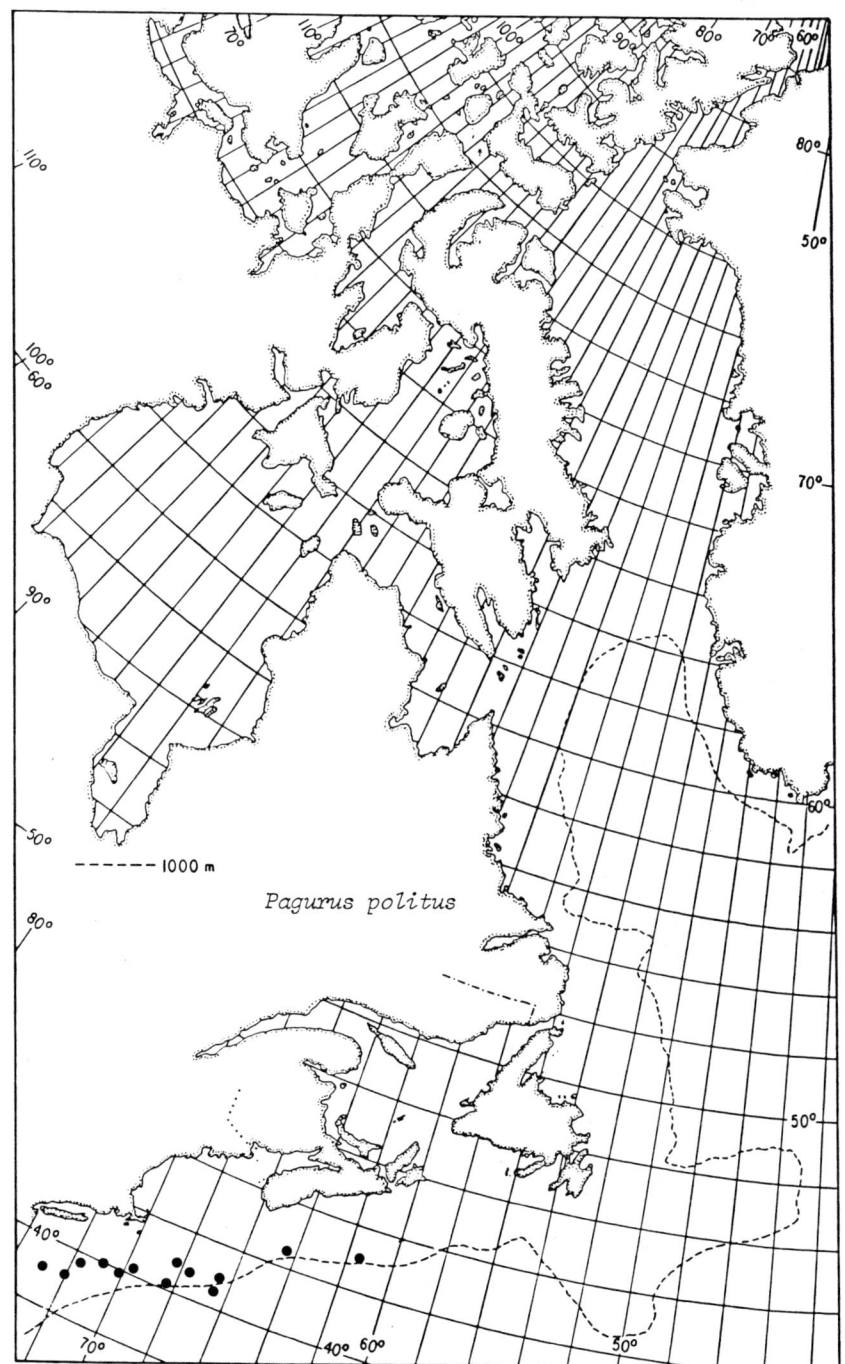

Fig. 204. *Pagurus politus*, distribution records in the area of reference.

Pagurus pubescens Krøyer, 1838
McLaughlin 1973: 564, figs. 1B, 2B, 3B;
1974, figs. 61-63; Squires 1964a: 355, figs. 1A, 2A, 6.
Williams 1984: 222, fig. 158.
Pagurus kroyeri, Rathbun 1929: 27, fig. 36.
(Figures 205 and 206)

DISTINGUISHING CHARACTERISTICS

Left first chela with strong single spinous ridge dividing the hand into two facets, the outer flat or slightly concave, the inner narrow and without spines; anterior part of sternite of 3rd legs about three times as wide as high and with moderate setal fringe; posterior lobes of telson slightly different in size, shallowly cleft with largest spine of each lobe at corners, two spines inside and one outside cleft.

DESCRIPTION

Integument heavily calcified anteriorly, smooth with pits and clumps of setae; posterior part leathery, shiny but finely pebbled surface. Colour reddish with light brownish legs.

Carapace shield slightly longer than wide, short median row of setal tufts. Rostrum obtuse, slightly rounded and with setal fringe, about even with lateral projections, each with a forward spine.

Sternite (z) of third legs with anterior part about three times as wide as long and with moderate setal fringe.

Abdomen about equal in length to whole carapace.

Telson (t) with posterior lobes slightly different, shallowly cleft with largest spine of each lobe at corners, two spines inside and one outside cleft on each side.

Eye moderate, stalk with longitudinal row of setal tufts; eye scales moderate, ovate, with subapical spine.

Antennule (c) 1st article short, the partly fused stylocerite almost as long; 2nd and 3rd articles subequal; flagella short.

Antenna (d) acicle exceeded by peduncle.

Mandible (e) incisor edge almost straight, slight notch at each corner; palp in 3 segments.

Maxillule (f) distal endite slightly expanded toward spinous edge, longer but narrower than pointed proximal endite. Endopod with one terminal lobe larger and setose.

Maxilla (g) endites unequally bilobed; endopod tapered from wide base leaning toward endite, slightly exceeding anterior lobe of scaphognathite, latter slightly tapering and about equal to axe-shaped posterior lobe.

Maxilliped I (h) distal endite larger than thickened proximal; endopod slightly twisted, shorter than exopod, the latter with moderate lash.

Maxilliped II (i) leglike but small; distal segment longer than wide, propodus and carpus inflated; exopod longer than endopod, with long lash.

Maxilliped III (k) leglike, basis with three spines, ischium with crista dentata with about 15 small spines, a large accessory spine; also spine on middle inside and distally outside of merus.

Pereopods: I (l) chelate, very unequal; left with propodus divided by a single row of strong spines into two facets, the outer flat or slightly concave with small rounded spines or tubercles, the inner sulcate with few or no spines; the right chela large with calcareous teeth on fingers, two rows of spines on dactyl. II and III (a) very long, carpus and propodus setose but unarmed, dactyl longer than propodus, sharp, with lateral sulcus. IV short, dactyl curved, beak-like, longer than distal process of propodus with which it is opposable in part. V short, articulated upward, dactyl rounded, opposable to distal part of propodus forming pseudochela.

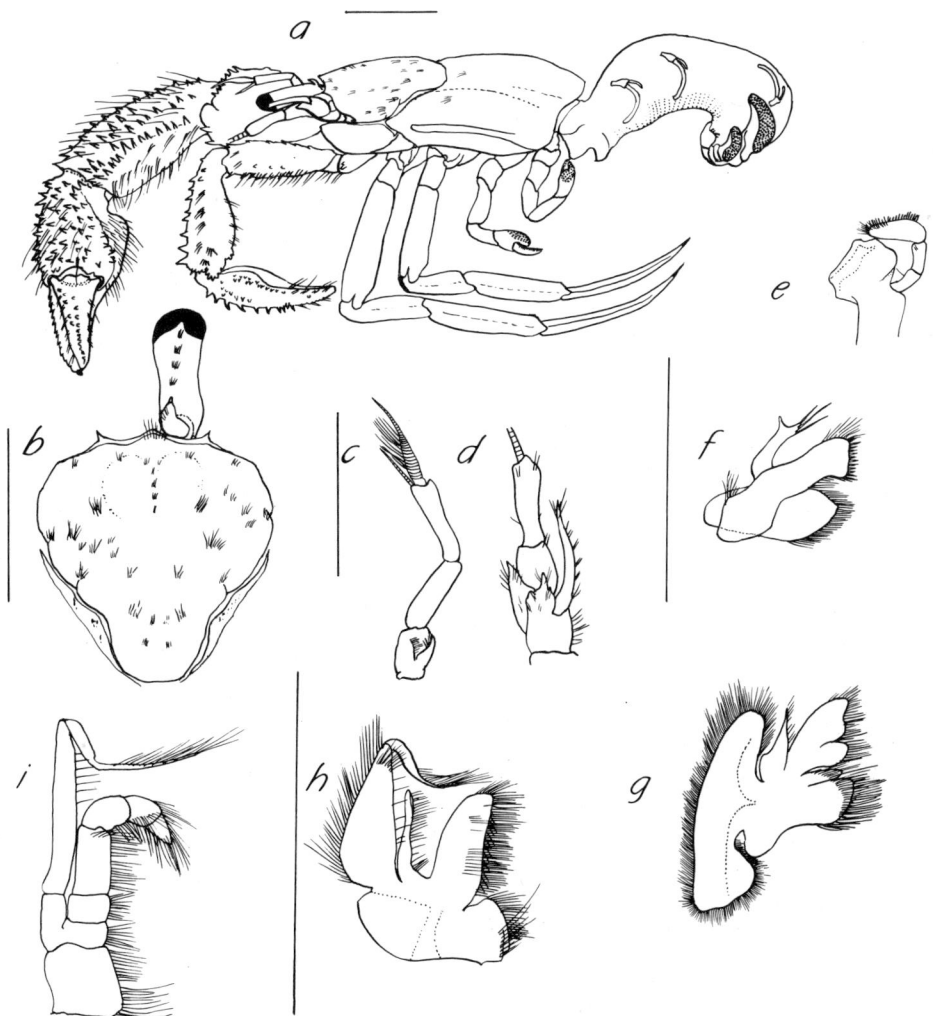

Fig. 205. *Pagurus pubescens*: *a*, whole animal from left side; *b*, anterior carapace in dorsal aspect; *c*, antennule; *d*, antenna; *e*, mandible; *f*, maxillule; *g*, antenna; *h*, first maxilliped; *i*, second maxilliped; solid line = 10 mm.

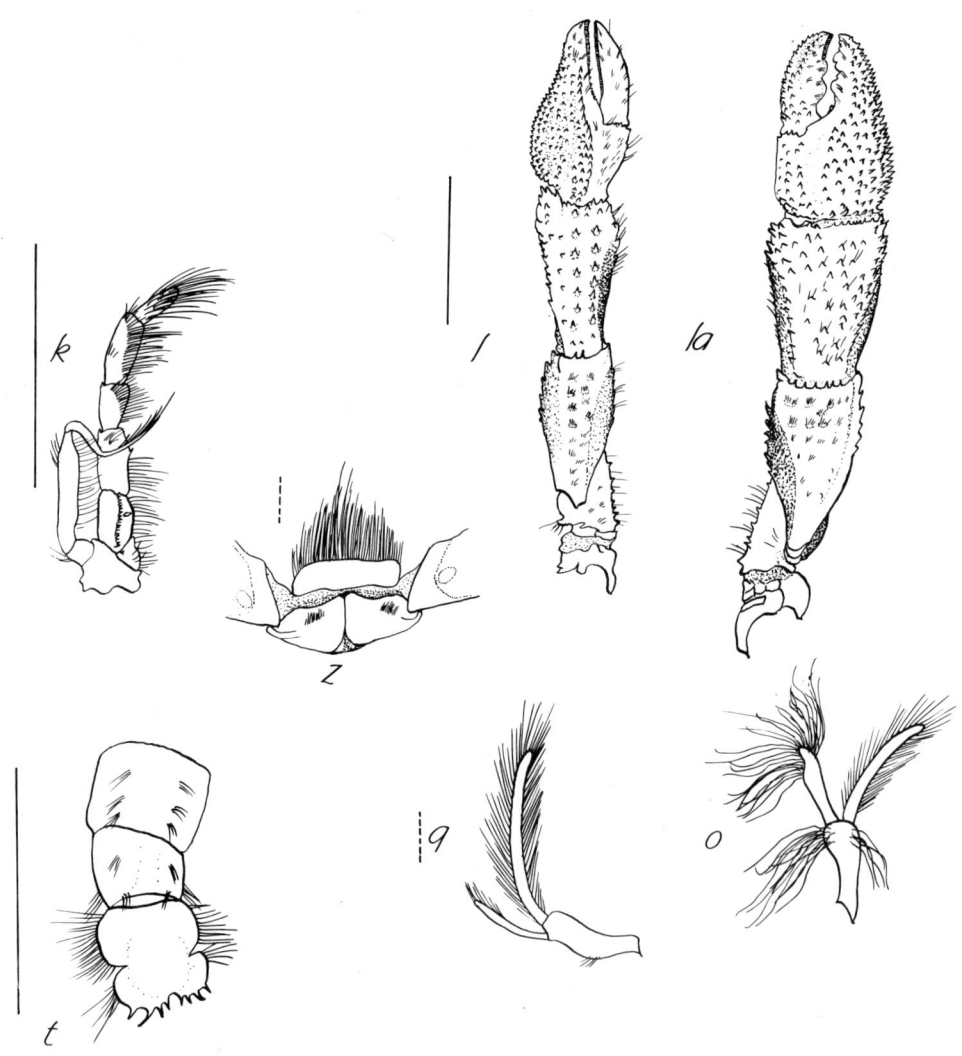

Fig. 206. *Pagurus pubescens*: *k*, third maxilliped; *l*, first pereopod right; *la*, first pereopod left; *o*, F first pleopod; *q*, M first pleopod; *t*, telson; *z*, sternal plates of fourth legs; solid line = 10 mm, broken line = 1 mm.

Pleopods: female I (o) protopod stout, expanded distally and with long threadlike setae, endopod slightly shorter than exopod but more robust with moderate subapical expansion and many threadlike (eggbearing) setae, exopod narrow with short setal fringe; female II and III similar to I; female IV with short endopod and long narrow exopod and no eggbearing setae. Male I (q) endopod slender much shorter than ribbon-like exopod, similar to other two pleopods present in males.

RANGE OF DISTRIBUTION

In the western Atlantic from Hudson Bay, Foxe Basin and West Greenland (73° N) to off Cape Hatteras; eastern Atlantic from East Greenland to 70° N, Iceland, Spitzbergen, Novaya Zemlya and Barents Sea, Faroes, and the British Isles. Depths 4–550 m; in deeper water in southern part of its range. Temperatures in the north from -1.6 to $4.6°$ C (Williams 1984).

Records of distribution in the area of reference are in Fig. 207.

BIOLOGY

Lengths of anterior shield to 14 mm in males and 10 mm in females.

Females were first ovigerous at 4 mm shield length in Foxe Basin, and 75% (47 females examined) were potentially ovigerous in autumn indicating annual spawning in that area.

Infestation with the parasite *Peltogaster paguri* occurred in 15% of specimens examined from off Labrador but did not appear in specimens from other areas.

Gastropod shells used were mostly *Buccinum* sp. Stomach contents were phytobenthos, foraminiferans, crustacean fragments (amphipods and ostracods) and fragments of bivalves, ophiurans, polychaetes and hydroids (Squires 1965a).

Decapod crustacean species most commonly present with this species in catches were *Eualus gaimardi gaimardi*, *Lebbeus polaris*, *Spirontocaris spinus*, *Spirontocaris phippsi* and *Sabinea septemcarinata* (Squires 1965a).

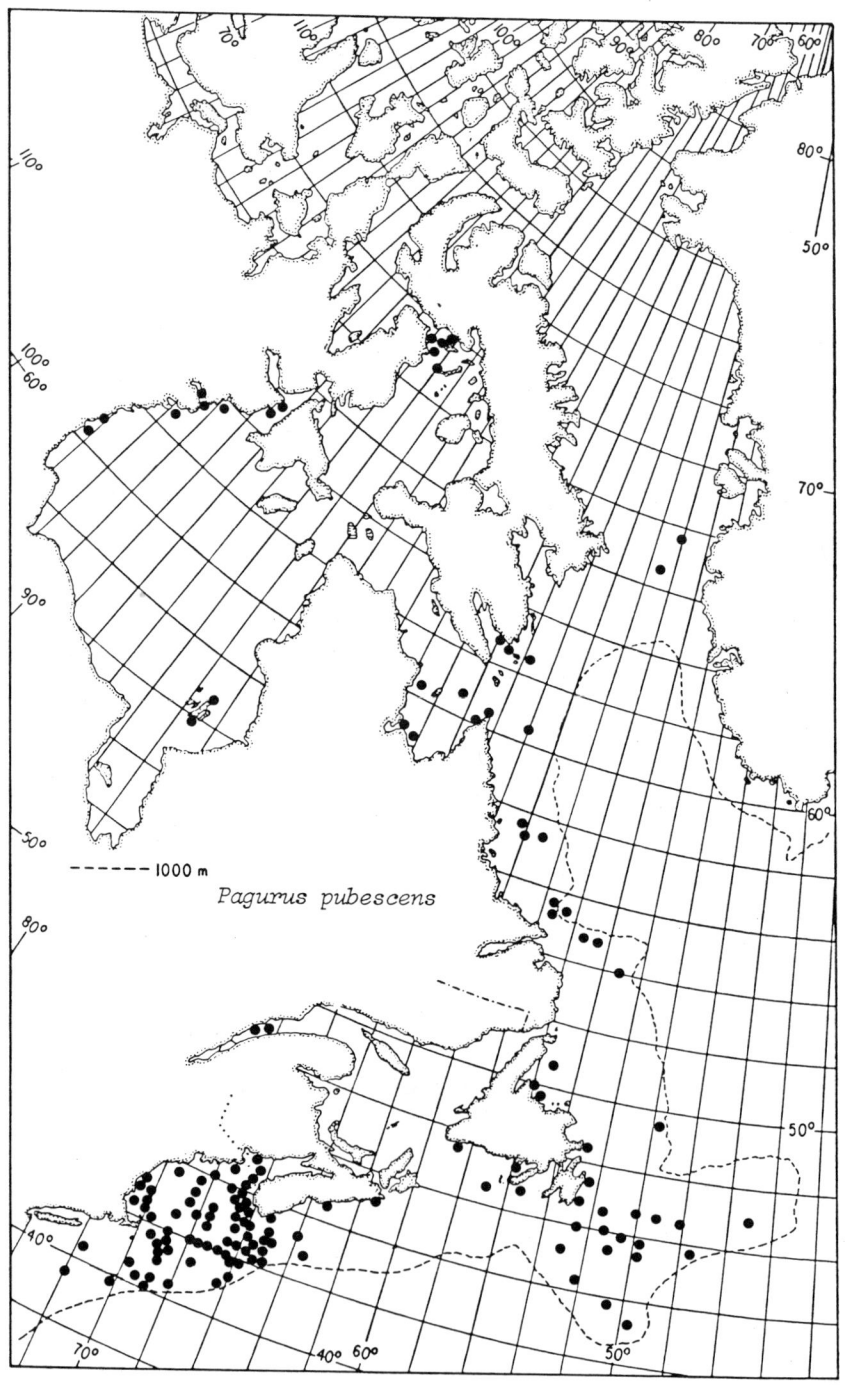

Fig. 207. *Pagurus pubescens*, distribution records in the area of reference.

Family LITHODIDAE Samouelle, 1819
Allen 1967: 60 (Key); Makarov 1962: 221;
Rathbun 1929: 29; Williams 1984: 230.

Crablike; abdomen calcified, showing segmentation, bent under the thorax. No uropods; also no pleopods in males but in females a pair of rudimentary pleopods on first somite and uniramous pleopods on 2nd to 5th somites. First pair of legs chelate, unequal; 2nd to 4th pair of legs similar; 5th pair very small and folded posteriorly under the carapace, making it seem that the animal has only four pairs of legs; legs and carapace with strong spines.

Key to species of Lithodidae in area of reference
(after Allen 1967)

1 Tip of rostrum bifid; second abdominal segment fused from median and lateral plates to form one piece; fairly long spines on carapace .... *Lithodes maja*

   Tip of rostrum simple; second abdominal segment composed of five pieces separated by sutures, the outermost indistinct, thus appearing to be three only; extremely long spines on carapace .... *Neolithodes grimaldii*

Genus *Lithodes* Latreille, 1806
Bouvier 1896: 10; Makarov 1962: 249;
Rathbun 1929: 29; Williams 1984: 230.

The 2nd abdominal segment is fused from central, lateral and marginal elements to form a single almost vertical plate, unseparated by sutures, posteriorly below carapace; the carapace is pear-shaped in outline and fringed with strong spines turned upward, the dorsal surface covered with tubercles and small spines; the rostrum is bifurcate at its tip and at about half its length a dorsolateral strong pair of spines is in front of the eyes; the antennal acicle is rudimentary.

Lithodes maja (Linnaeus, 1758)
Makarov 1962: 252, figs. 98–100;
Williams 1984: 230, fig. 166.
Lithodes maia, Rathbun 1929: 29, fig. 39.
"Stone crab" — "Crabe de roche"
(Figures 208 and 209, Plates 8 and 9)

DISTINGUISHING CHARACTERISTICS

Rostrum stout and bifurcate at the apex, slightly descending, with a lateral pair of strong divergent spines near proximal third (ahead of eyes); carapace pear-shaped in outline with a fringe of strong upturned spines along edge, dorsal surface covered with short spines and tubercles; abdomen turned under carapace formed of heavily calcified plates symmetrically arranged in males but strongly asymmetrical in females; first legs unequal, chelate, heavily calcified and spinose and shorter than the long spiny 2nd to 4th legs; 5th legs small and tucked under the carapace posteriorly.

DESCRIPTION

Integument heavily calcified and spiny. Colour pale purplish to red but occasionally brownish-purple.

Rostrum strong, slightly ascending and tapered to bifurcate tip and with stout divergent spines dorsolaterally just ahead of the eyes, three small spines at base and a large ventral spine curving forward.

The carapace is somewhat pear-shaped in outline; the rostrum, slightly ascending, is continued on the carapace as a rounded median ridge armed with 8 major and several smaller spines, reaching past half the carapace where it is interrupted by a transverse short groove and white depressed areas; posterior to it is another rounded cardiac ridge with 3 pairs of spines becoming attenuate before reaching the posterior edge; major spines around the edge of the carapace are 6 anterolaterals and 7 posterolaterals.

Abdomen with plates of 2nd somite fused forming a curved spiny plate at posterior carapace, other somites also fused and flattened, uncalcified ventrally, and pressed against thoracic sternites; somites are shown by calcareous plates arranged symmetrically in males, the 3rd, 4th and 5th decreasing in size and the 6th with telson at the end; in females these plates are strongly asymmetrical, the left three very large.

Telson is without uropods, semicircular in outline, with a few surface spines, at the centre posteriorly in males and to the right of centre in females.

Eyes small, branching outward on each side of rostrum.

Antennule (c) articles about equal in length, the 1st. stoutish; flagella short.

Antenna (d) 1st article with distolateral short spine, 2nd with a strong outer spine and small rudimentary acicle; others unequal the distal longest; flagellum only about half cl.

Mandible (e) incisor heavily calcified, edge almost straight but with a long anterior notch and a cleft where joined to molar. Palp with three segments.

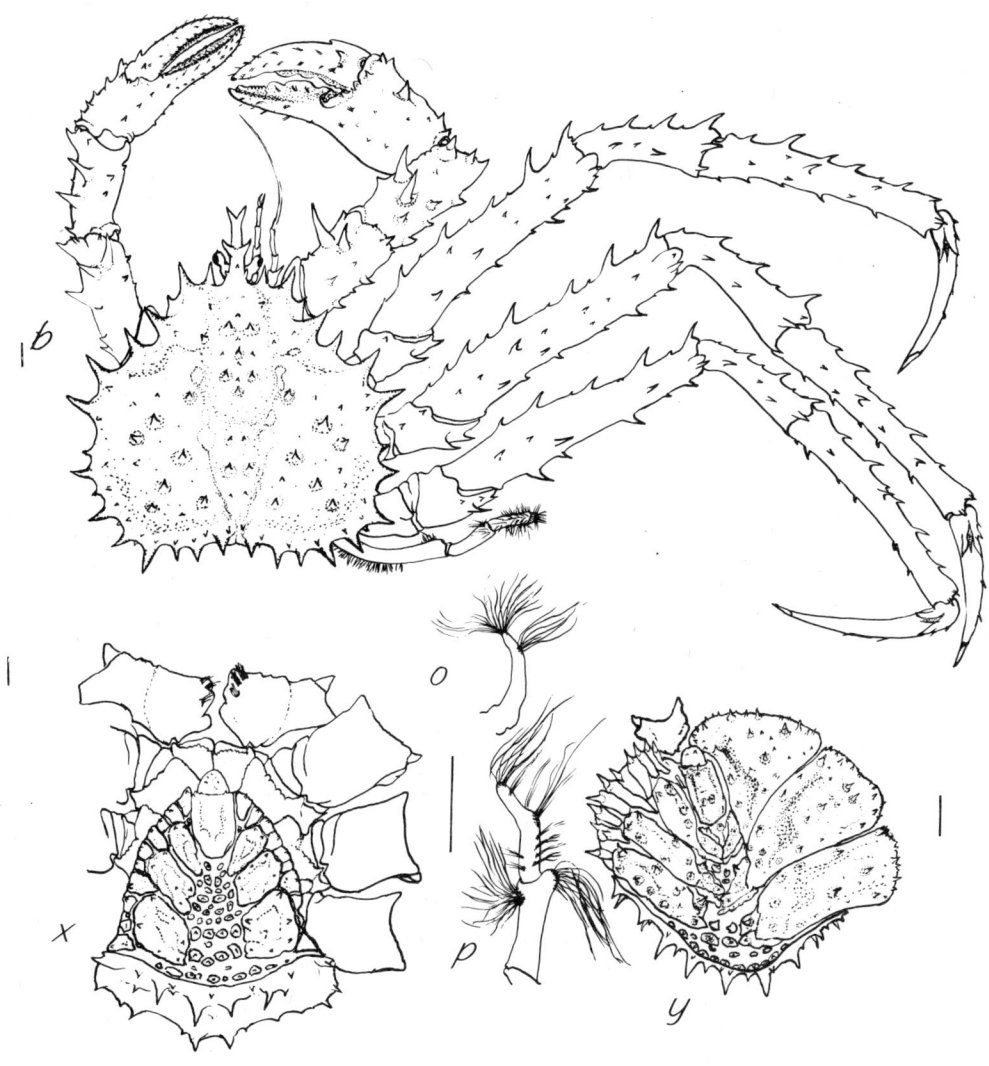

Fig. 208. *Lithodes maja*: *b*, carapace, etc., in dorsal aspect; *o*, F first pleopod; *p*, F second pleopod; *y*, F abdomen: *x*, M abdomen; solid line = 10 mm.

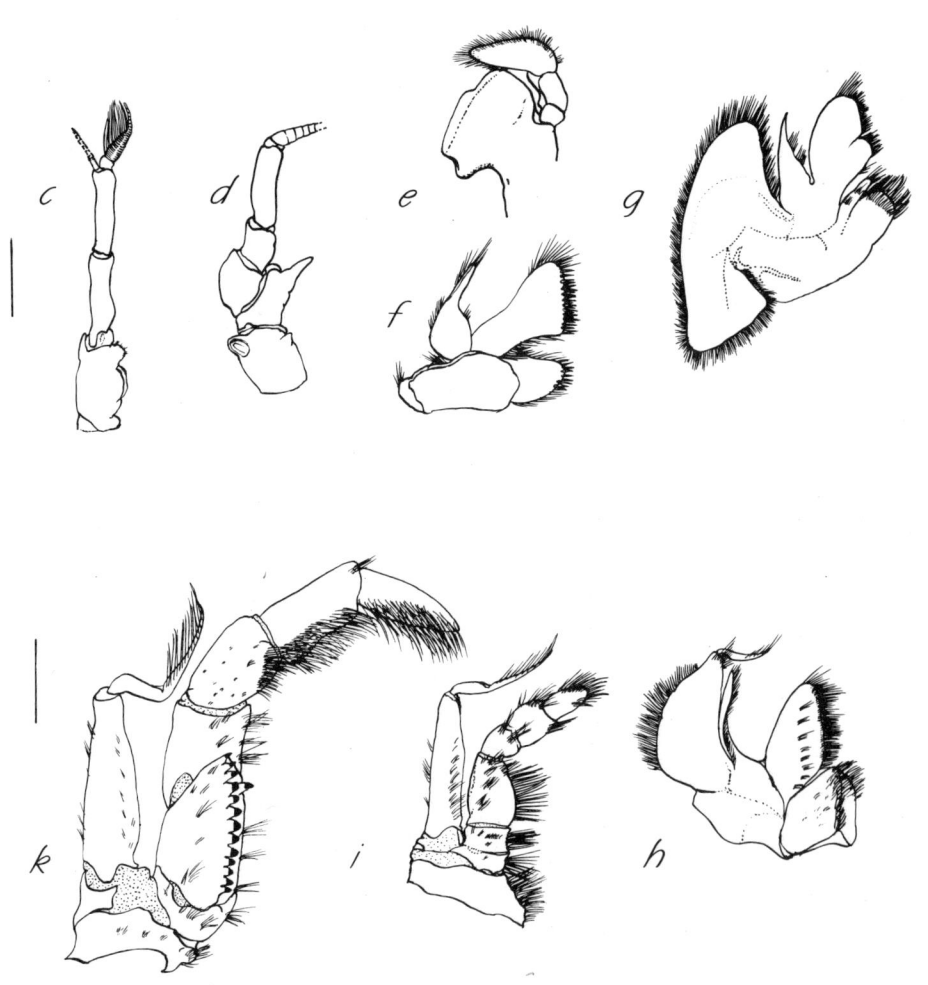

Fig. 209. *Lithodes maja*: *c*, antennule; *d*, antenna; *e*, mandible; *f*, maxillule; *g*, maxilla; *h*, first maxilliped; *i*, second maxilliped; *k*, third maxilliped; solid line = 10 mm.

Maxillule (f) distal endite widely expanded to spinous edge; proximal slightly pointed with sharp spines distally; endopod with narrow neck from rounded base.

Maxilla (g) endites unequally bilobed, the proximal with one lobe very narrow; endopod tapered to sharp point, exceeding slightly tapered anterior lobe of scaphognathite, the latter about equal to subtriangular posterior lobe.

Maxilliped I (h) distal endite subtriangular, slightly larger than thickened proximal endite; endopod thin, twisted, not as long as lamellate exopod, the latter with short lash.

Maxilliped II (i) small, stoutish, distal segment with spinous setae; exopod tapered, with moderate lash.

Maxilliped III (k) stout, ischium with crista dentata of 13 corneous-tipped large spines and large accessory spine; exopod tapered, with moderate lash, about as long as ischium and merus together.

Pereopods: I (a) chelate, unequal, shorter than II–IV, the larger with calcareous rounded teeth on both fingers, the smaller with fingers longer than palm and fine serrate teeth, both with black chitinous spooned tips; leg segments armed with strong sharp spines; II–IV long and spiny, dactyl very sharp, with proximal outer circle of sharp spurs and a couple of marginal spines. V very short and slender, articulated to fold under the posterior edge of carapace, distal segments heavily setose, the dactyl forming a cap-like pseudo-chela on end of propodus.

Pleopods: female I (o) paired, both a single slightly curved unsegmented appendage with apical and subapical tufts of long wavy setae; female II (p) uniramous, endopod with distal and several lateral bunches of long wavy setae, protopod with distal expansion and two bunches of long wavy setae; female III–V similar to II, all unpaired. No pleopods in male.

RANGE OF DISTRIBUTION

West Greenland to Sandy Hook, New Jersey, in the western Atlantic; East Greenland, Iceland and Spitzbergen to the British Isles and the Netherlands in the eastern Atlantic. Depths 65–790 m (Williams 1984).

Records of distribution in the area of reference are in Fig. 210.

BIOLOGY

Lengths to 105 mm cl in males and 81 mm cl in females in specimens examined.

Females were first mature at 37 mm cl. Egg diameters averaged 2.1 mm. Condition of ova indicated that about 60% would be ovigerous in autumn.

Stomach contents were phytobenthos, crustacean fragments, polychaetes, foraminiferans, sponge spicules, ophiuroids, small bivalve and gastropod shells, hydroids, a chiton and a small scallop (Squires 1965a).

This species was frequently taken with *Pandalus borealis*. Decapod crustacean species in other catches with this species were *Lebbeus polaris*, *Sabinea sarsi*, *Pandalus propinquus* and *Chionoecetes opilio* (Squires 1965a).

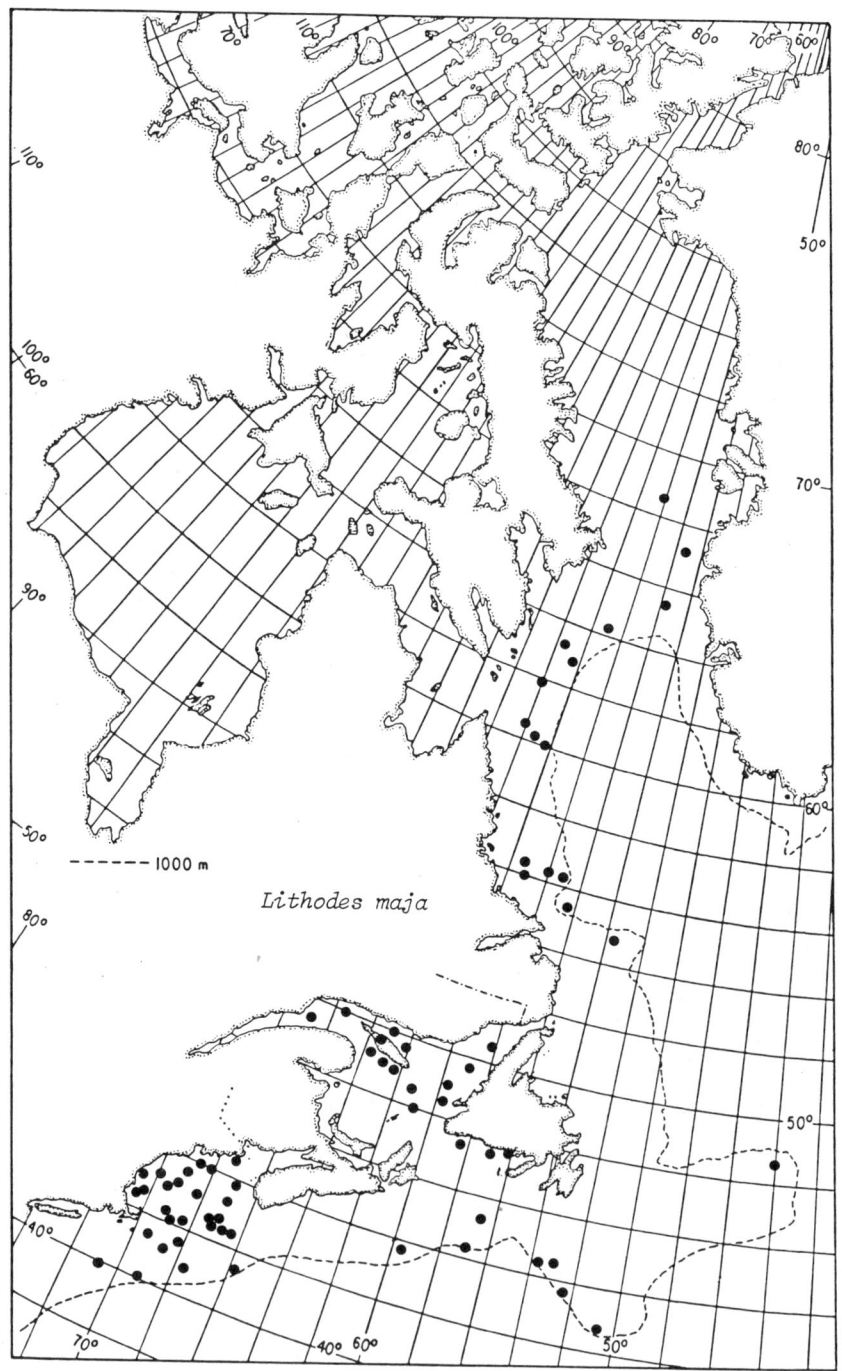

Fig. 210. *Lithodes maja*, distribution records in the area of reference.

Genus *Neolithodes* A. Milne-Edwards & Bouvier, 1894
Allen 1967: 60 (key); Bouvier 1896: 8.

Plates of second abdominal somite incompletely fused and separated by sutures into five pieces only three of which are clearly demarked; acicle small; spines of carapace extremely long; rostrum a single long spine, with two almost equally long spines at its base.

Neolithodes grimaldii (A. Milne-Edwards & Bouvier, 1894)
A. Milne-Edwards & Bouvier 1894: 62, pl. 3, figs. 1-6;
Sivertsen and Holthuis 1956: 48.
(Figures 211 and 212, Plates 10 and 11)

DISTINGUISHING CHARACTERISTICS

Crablike, with long legs and covered with very long spines; rostrum a single strong spine with divergent spines at its base almost as long; only first legs chelate, the right larger than the left; fifth legs hidden so it appears that only four pairs of legs are present; abdomen under carapace with no pleopods in males and surface not divided into plates, in females the opposite is true with plates arranged asymmetrically.

DESCRIPTION

Integument greatly calcified, smooth and shiny but covered with mostly large sharp spines, few small. Colour intense red throughout, but the large claw has white crushing teeth.

Rostrum a single large sharp spine with two diverging spines almost as large at its base rising at an angle of about 30 degrees over the eyes.

Carapace is almost round in outline but the front part projects so that it is somewhat pear-shaped; dorsally dominated by two groups of 4 major spines in raised turret-like gastric and cardiac areas and a group of 3 pairs of major spines behind them near posterior margin (intestinal area); branchial areas have about 7 major spines each and a few smaller ones; hepatic areas with one large spine; anteriorly are two moderate frontal spines, one on each side of rostral group.

Abdomen in males with rough leathery surface with numerous calcareous spinous tubercles and bosses and the 6th somite with telson at the centre of posterior margin or slightly skewed to the left; in the female the 3rd–5th somites are marked by three large surface plates increasing in size posteriorly on the left, but reduced to indistinct divisions on right with the 6th and telson skewed far to the right. 1st and second somites united in curved plate under posterior carapace with very long spines and indistinct sutures.

Telson somewhat pointed apically but smaller than 6th somital plate; located just left of centre in males but far to the right in females.

Eyes with short stalks, close to each other under the spines at base of rostrum, cornea larger than stalk diameter.

Antennule (c) 1st article stout with longitudinal cleft on distal two-thirds, 2nd and 3rd subequal, more slender than 1st; flagella short.

Antenna (d) dorsolateral distal angle of second segment greatly produced; acicle small, reduced; fifth segment equal to 3rd and 4th together; flagellum a little more than half cl.

Mandible (e) incisor edge long, almost straight, bevelled; molar separated by a groove from incisor and surface oval in outline; palp three-segmented.

Maxillule (f) distal endite widely expanded to spinous edge; proximal short, rounded; endopod forming neck distally from wide base and with indistinct apical lobes, one with brush of setae.

Maxilla (g) endites unequally bilobed, the proximal smaller, with submarginal rows of setae; endopod sharply tapered from wide rounded base, about even with anterior lobe

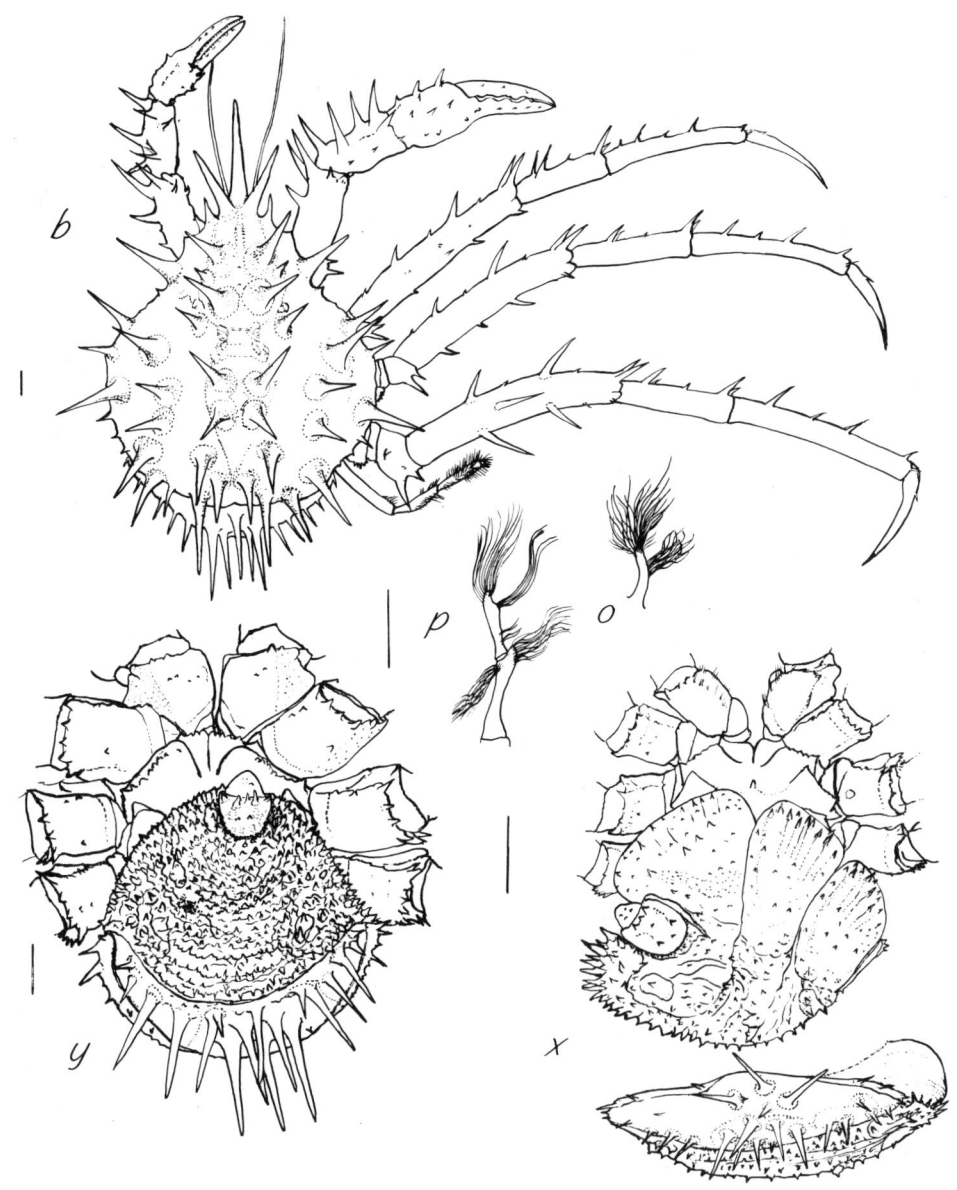

Fig. 211. *Neolithodes grimaldii*: *b*, carapace, etc., in dorsal aspect; *o*, F first pleopod; *p*, F second pleopod; *x*, female abdomen; *y*, male abdomen; solid line = 10 mm.

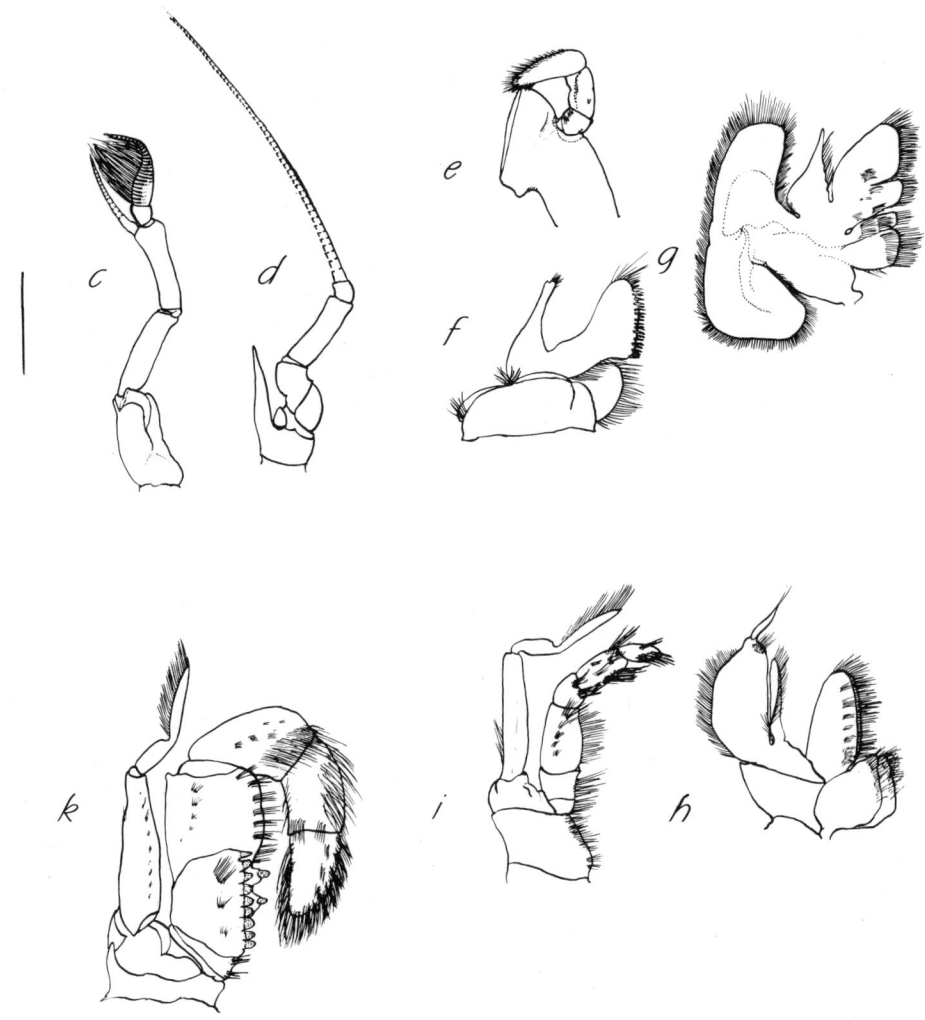

Fig. 212. *Neolithodes grimaldii*: c, antennule; d, antenna; e, mandible; f, maxillule; g, maxilla; h, first maxilliped; i, second maxilliped; k, third maxilliped; solid line = 10 mm.

of scaphognathite, the latter about equal in length to but narrower than the foot-shaped (apex squared off but rounded at corners) posterior lobe.

Maxilliped I (h) distal endite with submarginal row of setal tufts, longer than thickened (triangular in cross-section) proximal endite; endopod slender, twisted, not as long as lamellate wide exopod, the latter with thickened rounded inner edge and short, rigid, non-annulated lash.

Maxilliped II (i) leglike but small; papilla at base of exopod, the latter with rigid non-annulated lash.

Maxilliped III (k) very stout, leglike, ischium with crista dentata of 9 strong corneous-tipped spines and two accessory spines; exopod tapered, with rigid non-annulated flagellum or lash.

Pereopods: I chelate, unequal, larger on right, both with fingers longer than slighly inflated palm; on right teeth white calcareous, rounded, but tips sharp corneous; on left edges corneous with spooned tips; both with very strong spines on carpus and merus; II–IV much longer than I, merus longest segment with a few major spines including two large distal above, ischium short with one major and one minor spine, carpus with distally a large and a small spine and propodus with one small distally, dactyl shorter than propodus, sharp with one small proximal spine; V very short and slender , articulated to fit under carapace, dactyl domed on one side and flat on the other to form a pot-lid pseudo-chela with the propodus, both with strong brush setae.

Pleopods: female I (o) single segment, rudimentary, with two clumps of long wavy setae; female II (p) with endopod and protopod, both with clumps of long wavy setae (III–V on somites 3–5 similar to II). Male with no pleopods.

RANGE OF DISTRIBUTION

In the western Atlantic from Greenland to off North Carolina; in the eastern Atlantic from Iceland and the British Isles to the Azores and the Cape Verde Islands. Depths 800–2 000 m (Sivertsen and Holthuis 1956).

Records of occurrences in the area of reference are in Fig. 213.

BIOLOGY

Lengths to 133 mm cl in male and 124 mm cl in female in specimens examined (Squires 1965a).

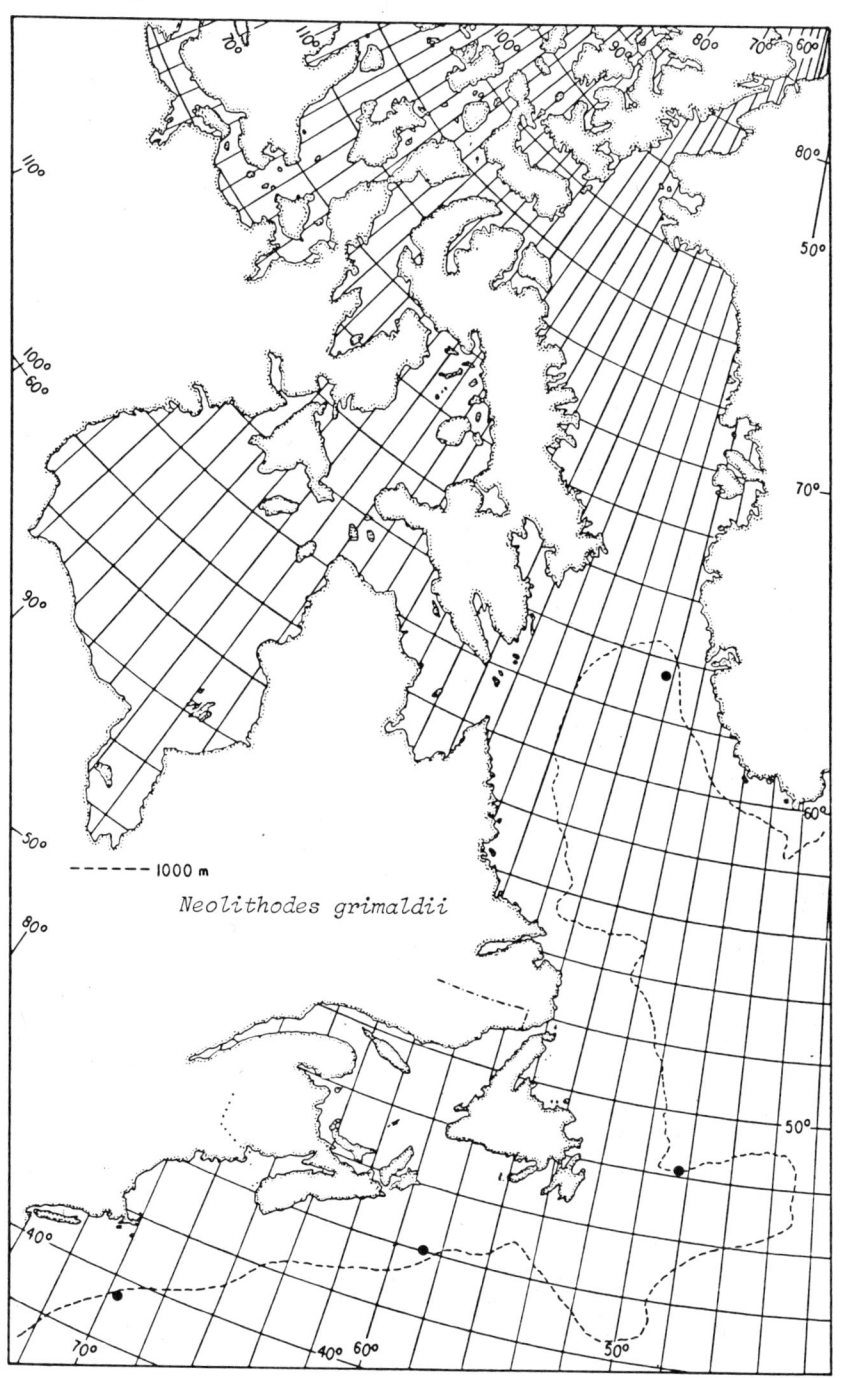

Fig. 213. *Neolithodes grimaldii*, distribution records in the area of reference.

Family PARAPAGURIDAE Smith, 1882
McLaughlin 1974: 371; de Saint Laurent 1972:

Males with paired pleopods on 1st and 2nd abdominal segments, female pleopods unpaired; males with paired gonopores, females with single gonopore on left coxa; telson entire, without pronounced median cleft on posterior margin; ischium of 3rd maxilliped with crista dentata without accessory tooth; maxillipeds widely separated at base; chelipeds unequal, right much larger than left, left chela and carpus compressed.

Only one genus in the family.

Genus *Parapagurus* Smith, 1879
McLaughlin 1974: 371; de Saint Laurent 1972b:
Smith 1879: 50; Rathbun 1929: 28;
Zariquiey Alvarez 1968: 251.

The genus has the same characteristics as the family. The chela are greatly unequal, the smaller one compressed; the anterior edge of the carapace is straighter than in *Pagurus*; antennules are much longer than the eyes and exceed the antennal peduncle; the eyes are small and eye scales long and narrow; in the male the first and second pairs of pleopods are calcified and spatulate.

Only one species is present in the area of reference.

Parapagurus pilosimanus Smith, 1879
Smith 1879: 51; de Saint Laurent 1972b:
McLaughlin 1974: 371-378; Rathbun 1929:
28, fig.38; Zariquiey Alvarez 1968: 252.
(Figures 214 and 215)

DISTINGUISHING CHARACTERISTICS

Chelipeds covered with short setae, chelae greatly unequal, the smaller compressed; eyes small, the antennules long, much exceeding peduncles of antenna and eyes; acicle longer than antennal peduncle; the front edge of the anterior shield of carapace is almost straight and the shield is squarish, not tapered posteriorly; shield wider than long; posterior margin of telson with only a small median notch and many small fringing spines, none inside the notch; first pleopods paired in males, a small rudimentary budlike appendage on right in females.

DESCRIPTION

Integument smooth on shield except for a few pits with clumps of setae, on chelipeds heavily calcified and with numerous pits with clumps of short setae. Colour pale dull orange red, darker at tips of legs.

Front of anterior shield of carapace has a low rounded rostrum with no lateral projections; posteriorly the shield is as wide as in front making it quadrate although slightly rounded posteriorly; antero-laterally in hepatic area is a pronounced small depression.

Sternite (z) of 3rd legs anterior part with a central rounded boss with long radiating setae, posterior part rounded and with a central fissure; female gonopore on left coxa only.

Abdomen robust, with pronounced torque; paired appendages calcified on first and second somites in male, uniramous.

Telson (t) not divided in four lobes as in *Pagurus* but is quadrate although rounded posteriorly and with a small median notch and a fringe of small spines on rounded edge, about 15 on the right and 25 on the left.

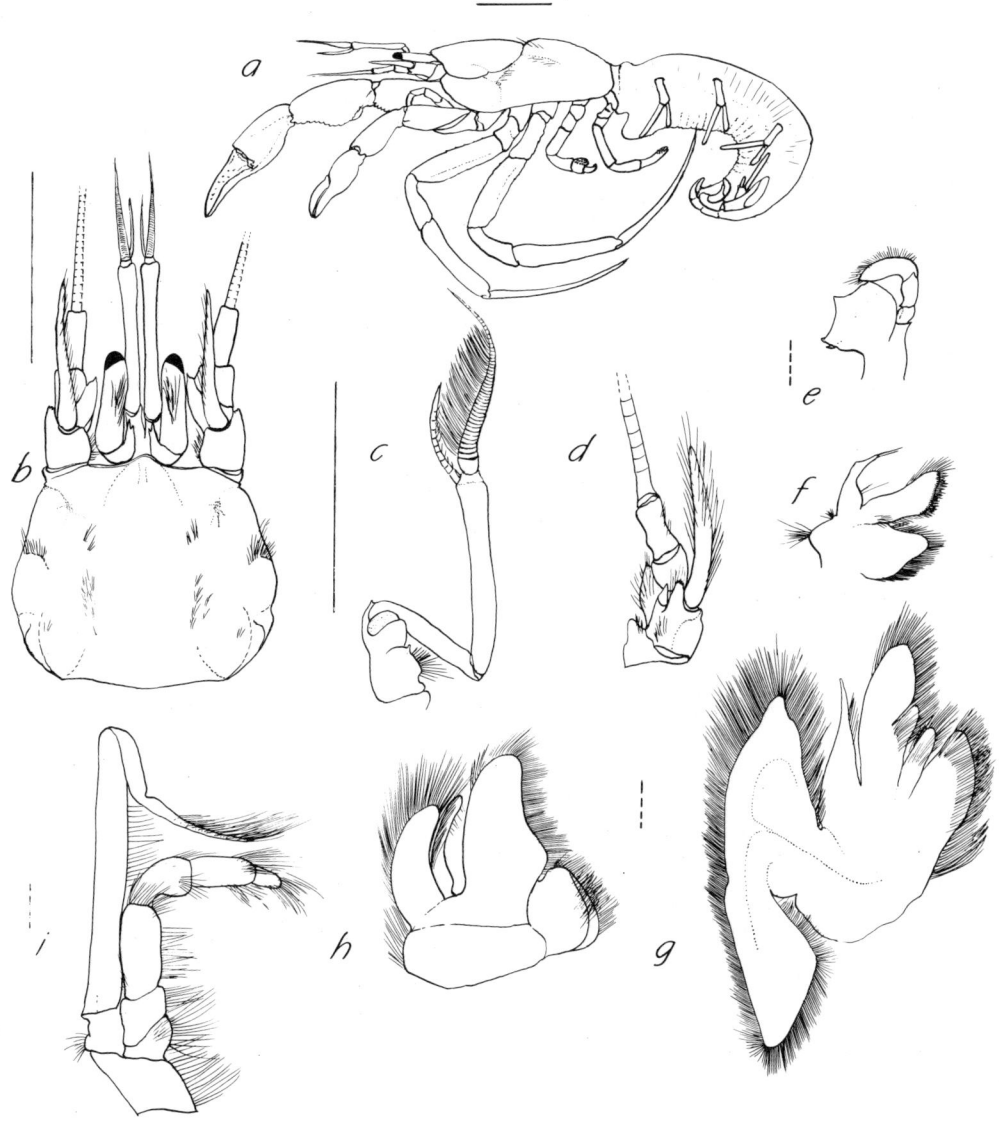

Fig. 214. *Parapagurus pilosimanus*: *a*, whole animal from left side; *b*, anterior carapace in dorsal aspect; *c*, antennule; *d*, antenna; *e*, mandible; *f*, maxillule; *g*, maxilla; *h*, first maxilliped; *i*, second maxilliped; solid line = 10 mm, broken line = 1 mm.

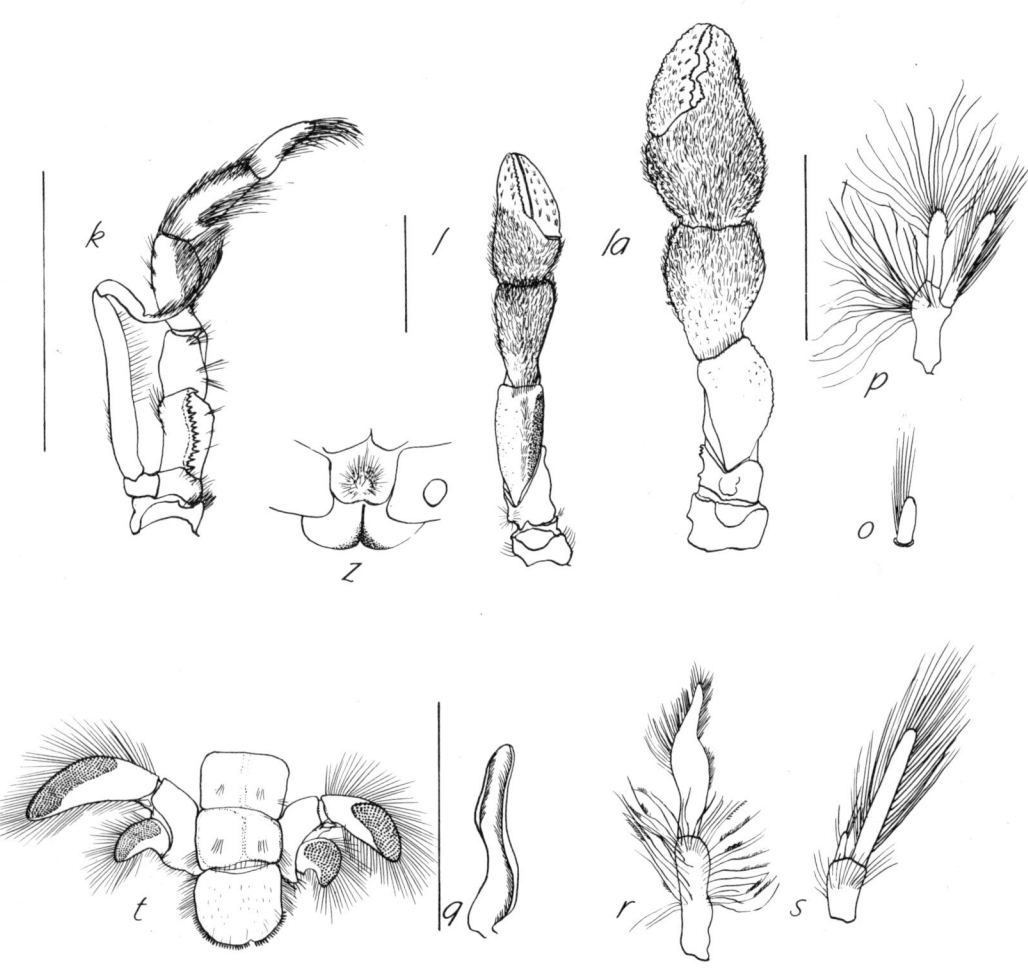

Fig. 215. *Parapagurus pilosimanus*: *k*, third maxilliped; *l*, first pereopod right; *la*, first pereopod left; *o*, F first pleopod; *p*, F second pleopod; *q*, M first pleopod; *r*, 2nd pleopod; *s*, third pleopod; *t*, telson; *z*, sternite of fourth legs; solid line = 10 mm.

Eyes with small cornea not as large in diameter as stalk; eye scales with double pointed slender apex.

Antennule (c) 1st article very short, robust; 3rd article very long, longer than other two combined and about equal to longer flagellum.

Antenna (d) acicle exceeding peduncle; dorsolateral distal angle of segment 2 short with four small apical spines, supernumerary segment (McLaughlin 1972) spinelike, other segments short.

Mandible (e) sharp-edged incisor sloped to strong central point, small cusps at each corner and behind lower corner with corneous tips; molar narrow. Palp 3-segmented.

Maxillule (f) distal endite slightly expanded toward spinous edge; proximal subtriangular with even setal fringe; endopod with one apical lobe reduced.

Maxilla (g) endites unequally bilobed, both with one narrow lobe; endopod with long attenuate point longer than anterior lobe of scaphognathite, latter pointed rather than rounded and slightly longer than posterior subtriangular lobe.

Maxilliped I (h) distal endite large, with concave edge, proximal thickened shorter; endopod not twisted, curved, longer than exopod, the latter without lash.

Maxilliped II (i) leglike, exopod longer than endopod and with long lash.

Maxilliped III (k) crista dentata with alternating small and large spines on ischium, no accessory spine; moderate slender exopod longer than proximal 3 segments of endopod.

Pereopods: I (l) chelate, unequal, larger right with low rounded spines along edges and calcareous teeth (corneous at tip of dactyl), smaller with calcareous teeth on propodus and corneous edge on dactyl, upper surface including carpus covered with moderate curled setae. II and III long unarmed with very long dactyl, narrow and sharp, with longitudinal groove. IV short, dactyl projecting and hooked, opposable to distally extended propodus forming pseudo-chela; V slightly longer than IV, dactyl and propodus modified to form small pseudo-chela.

Pleopods: female I, left (p) endopod shorter than exopod, with long wavy apical setae as on dilated tip of protopod, right (o) very small, female II–III similar to (p), IV different with very small endopod and long exopod. Male I (q) sinuous slightly hollowed with marginal fold and fringe of short setae, pair close together near sternite of 5th legs; male II (r) uniramous partly calcified; male III (s) and IV with very short endopod and long narrow exopod.

RANGE OF DISTRIBUTION

From Grand Banks of Newfoundland to the Gulf of Mexico in the western Atlantic (Rathbun 1929); in the Sargasso Sea and from Ireland and the Bay of Biscay to Sierra Leone off West Africa in the eastern Atlantic. In depths of 500–4 000 m (Zariquiey Alvarez 1968). Pacific records are of other subspecies (McLaughlin 1974; Saint Laurent, de 1972).

Records of occurrence in the area of reference are in Fig. 216.

BIOLOGY

Anterior shield length in male 13 mm and female 12 mm in specimens examined.

A rhizocephalan parasite *Angulosaccus tenuis* has been reported for related subspecies (McLaughlin 1974).

The specimen examined by Smith (1879) was in a gastropod shell that had been overgrown by an actinoid polyp which extended far beyond the shell.

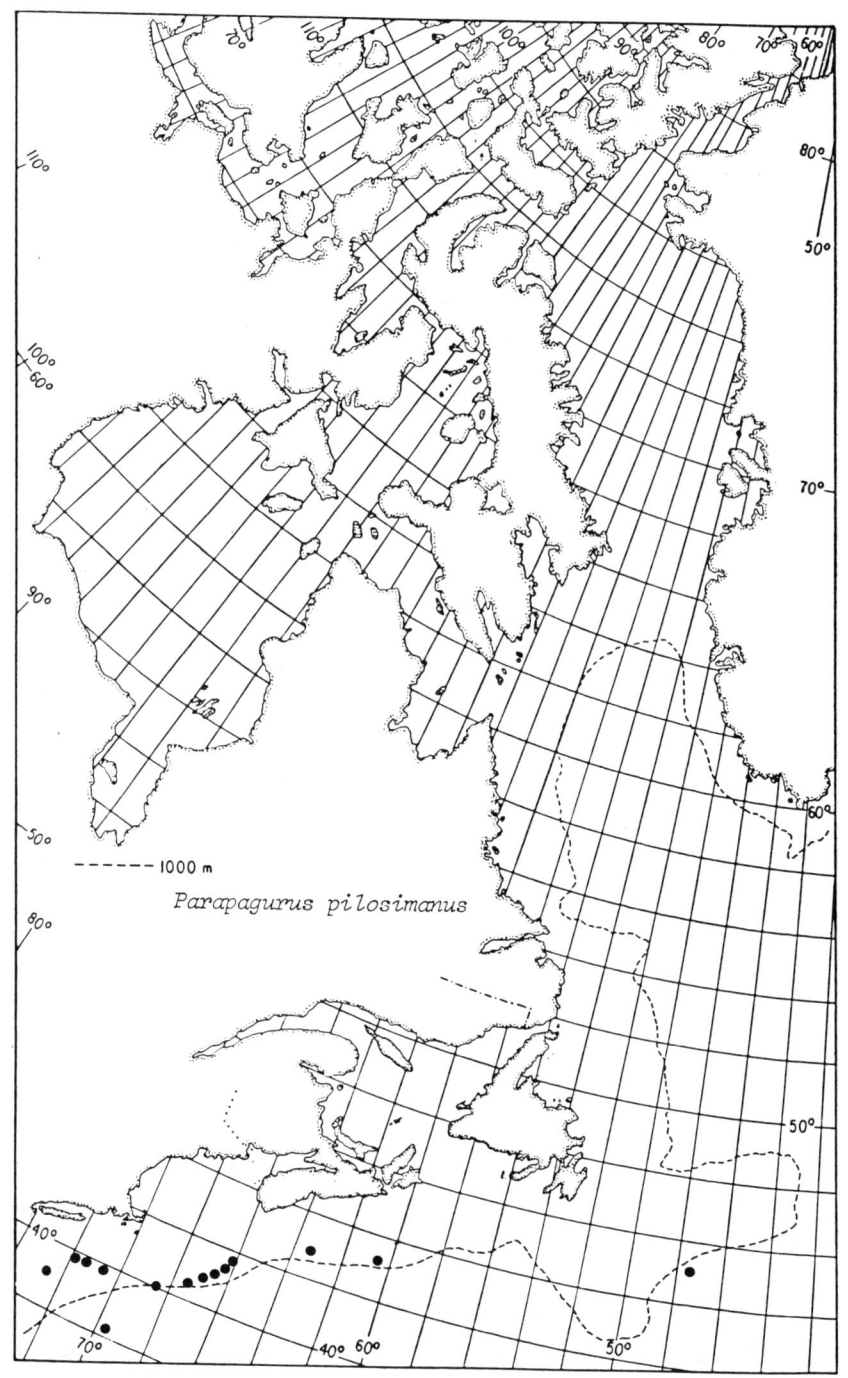

Fig. 216. *Parapagurus pilosimanus*, distribution records in the area of reference.

Family GALATHEIDAE Samouelle, 1819

Makarov 1962: 79; Rathbun 1929: 24;
Williams 1984: 231; Zariquiey Alvarez 1968: 268.

Body shrimplike, carapace elongate, with well-defined areas, sometimes ornamented with transverse ciliated ridges; rostrum prominent, spiniform, projecting beyond the eyes; first pereopods elongate, slender, chelate; abdomen depressed with bend at fourth rather than the third somite as in shrimp; sexually modified pleopods on 1st and 2nd somites in males, not present on first somite in females.

Key to species of Family Galatheidae in area of reference

1   Eyes opaque, non-pigmented and without facets; exopod of 1st maxilliped without flagellum .... *Munidopsis curvirostra*

   Eyes pigmented and with facets; exopod of 1st maxilliped with flagellum 2

2   Posterior margin of carapace without spines 3

   Posterior margin of carapace with spines .... *Munida tenuimana*

3   First article of antennule with outer spine smaller than inner .... *Munida iris iris*

   First article of antennule with distolateral (outer) spine larger than distomesial (inner) spine .... *Munida valida*

Genus *Munida* Leach, 1820
Makarov 1962: 90; Zariquiey Alvarez 1952: 148, figs. 1 and 2.

The rostrum is a long slender spine on each side of which is a somewhat shorter supraorbital spine; carapace and abdomen with transverse ridges bearing anteriorly short dense setae; cervical groove well marked at about the middle of the carapace; eyes are large and well pigmented; 1st pereopods are very long, slender and spinulose; exopod of 1st maxilliped with flagellum.

Munida iris iris A. Milne-Edwards, 1880
Williams 1984: 233, fig. 168.
(Figures 217 and 218)

DISTINGUISHING CHARACTERISTICS

Rostral spine with small serrate spines, about twice as long as supraorbital spines, the latter not extending to distal edge of cornea and almost parallel to the rostrum; 1st article of antennular peduncle with mesial spine longer than distolateral spine; carapace with no spines at posterior edge; 2nd abdominal somite with only one pair of small spinules on anterior margin; pubescence and fringes of transverse ridges of carapace, abdomen and legs iridescent.

DESCRIPTION

Integument rigid, smooth except for ridges or transverse striae with dense anterior fringes of short setae which are iridescent. Colour pinkish with setal iridescence.

Carapace almost as wide as long; rostral spine with low sagittate spines, long (0.5 cl) forming short ridge on carapace dividing the epigastric region, the latter with a moderate spine followed by three small epigastric spines; supraorbital spines about half rostrum, only slightly diveregent on each side of rostrum (almost parallel with it); large external orbital spine with a small lateral behind it, followed by 3 anterior branchial and 2 posterior branchial spines at lateral edge; 2 inner anterior branchial spines near cervical groove and 3 post-cervical spines; posterior cardiac region with median saddle of interrupted striae but no spines; no spines on posterior border.

Sternite of 3rd maxilliped with left and right oval halves with an anterior spine. Sternal plates (z) of legs also calcified and widening posteriorly, of the first legs with three anterior spines, and of legs II and III with lateral projections the latter with 3 teeth anteriorly, IV without teeth, all plates symmetrical with faint midventral separation. Sternum of V separated from others as in all anomurans.

Abdomen with 2 small anterior spines on tergum of 2nd somite (1st somite partly hidden under the carapace); no spines on 3rd somite; pleura of 2nd to 5th pointed ventrally; terga with transverse striae and no spines; 6th with a median join and a triangular median posterior area, ventrolaterally an acuminate spine.

Telson (t) scaly in appearance with inbricated striae in semicircles or minor arcs, and five major defined areas: the anterior or anal area with a proximal faint median separation, a wide lateral at each side with outer straight edge and small posterior spines, a central area with faint separation from laterals, and a terminal subtriangular area with faint median separation and small midmarginal notch and long setal fringe but no spines. Branches of uropod are shorter than telson, the inner with straight sagittate edge and the outer (shorter) with setose straight edge, both subtriangular in outline.

Eye large, cornea ovate, with short stalk.

Antennule (c) 1st article with 3 sharp inner spines, one distal shorter than long distal outer spine, 2nd article slender and about equal to 3rd, the latter expanded distally with very short ventromesial flagellum and slightly longer dorsolateral flagellum.

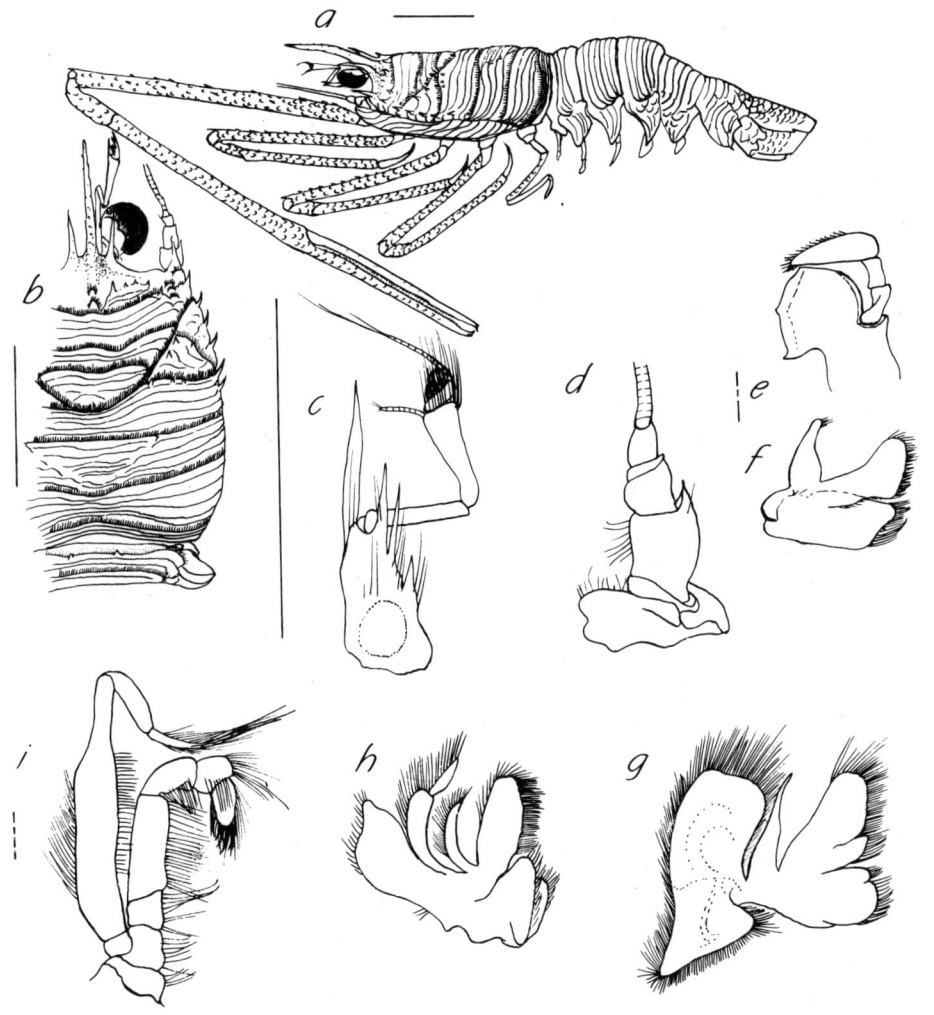

Fig. 217. *Munida iris iris*: *a*, whole animal from left side; *b*, carapace in dorsal aspect; *c*, antennule; *d*, antenna; *e*, mandible; *f*, maxillule; *g*, maxilla; *h*, first maxilliped; *i*, second maxilliped; solid line = 10 mm, broken line = 1 mm.

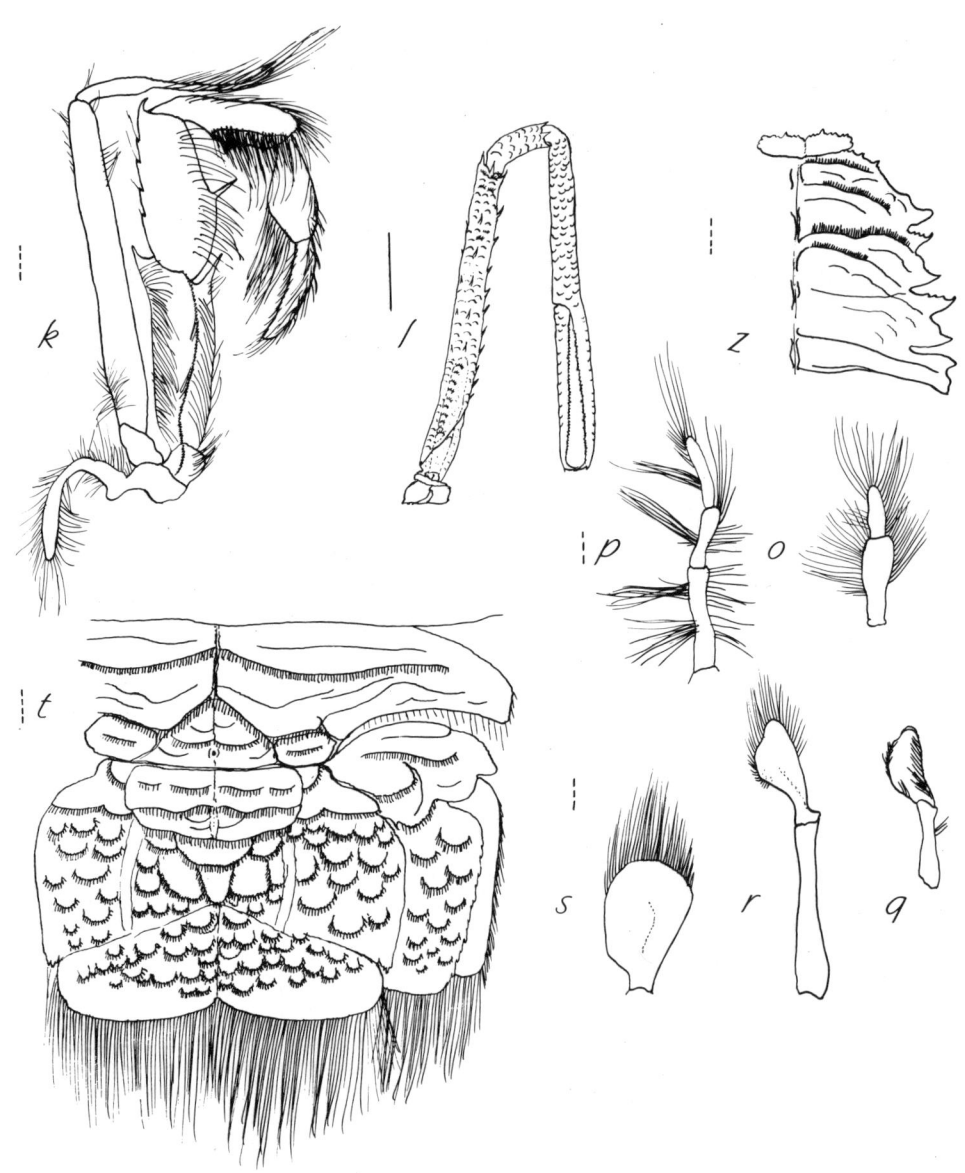

Fig. 218. *Munida iris iris*: *k*, third maxilliped; *l*, first pereopod; *z*, sternum; *o*, F second pleopod; *p*, F third pleopod; *q*, M first pleopod; *r*, M second pleopod; *s*, M third pleopod; *t*, telson; solid line = 10 mm, broken line = 1 mm.

Antenna (d) basal segment or synopod with squamous lobe and narrower lobe surrounding edge of 1st article; 1st article with distal spine outside about as long as other two together.

Mandible (e) heavily calcified, slightly curved edge with tooth at centre and sharp tooth at lower corner, separated by a cleft from short molar lying inside the incisor; palp with three segments.

Maxillule (f) distal endite slightly expanded to spinose edge, about as wide as slightly pointed proximal endite; endopod short, not clearly bifid.

Maxilla (g) proximal endite scarcely bilobed, a small short lobe at distal corner to wide rounded lobe; distal endite unequally bilobed; endopod slender with long attenuate point about equal to anterior lobe of scaphognathite, the latter wider and longer than subtriangular posterior lobe.

Maxilliped I (h) distal endite thin, narrower than thickened proximal endite; endopod slender, curved, slightly shorter than narrow exopod, the latter with short lash; epipod larger than exopod and slightly pointed.

Maxilliped II (i) leglike, distal segment longer than wide, penultimate slightly inflated; exopod longer than endopod, with long lash.

Maxilliped III (k) leglike, with seven segments, dactyl narrow; merus with distal spine and 4 low outer spines and one large inner spine on proximal third; ischium with large inner and outer distal spine and crista dentata of numerous small teeth continued on basis. Epipod long, club-like, setose.

Pereopods: I (l) chelate, much longer than others, covered with scale-like striae, subcylindrical, fingers long and slender but shorter than palm, with cutting edge of small serrate teeth with distal tooth and curved tips crossing over each other, merus with series of spines and strong distal spines. II–IV long and slender, scaly, merus with distal spines and series of lateral spines. V much smaller than others, with a small pseudo-chela and many brush setae terminally.

Pleopods: female II (o) uniramous, endopod with one segment, many long setae; female III (p) endopod with two segments and tufts of long wavy setae, setae also on protopod; male I (q) uniramous, endopod expanded, spatulate, about as long as protopod; male II (r) modified as in I for transferral of spermatophore, endopod expanded distally, spatulate, protopod much longer; other pleopods (s) foliaceous, wide and uniramous with apical setae.

RANGE OF DISTRIBUTION

From off Nova Scotia to the Gulf of Mexico and through the Caribbean Islands to the Amazon River, Brazil (Williams 1984). Depths 43–613 m (Wenner and Read 1982).

Records of occurrence in the area of reference are in Fig. 219.

BIOLOGY

Lengths of carapace including rostrum 47 mm in males and 45 mm in females.

Breeding season as shown by occurrence of ovigerous females is long, perhaps year round in parts of its range. Females bear an average of 7 900 eggs each. Males have larger chelae according to their size than females after maturity.

Parasitization by a bopyrid isopod (*Anuropodione carolinensis*) averaged about 10% off North Carolina (Williams and Brown 1972).

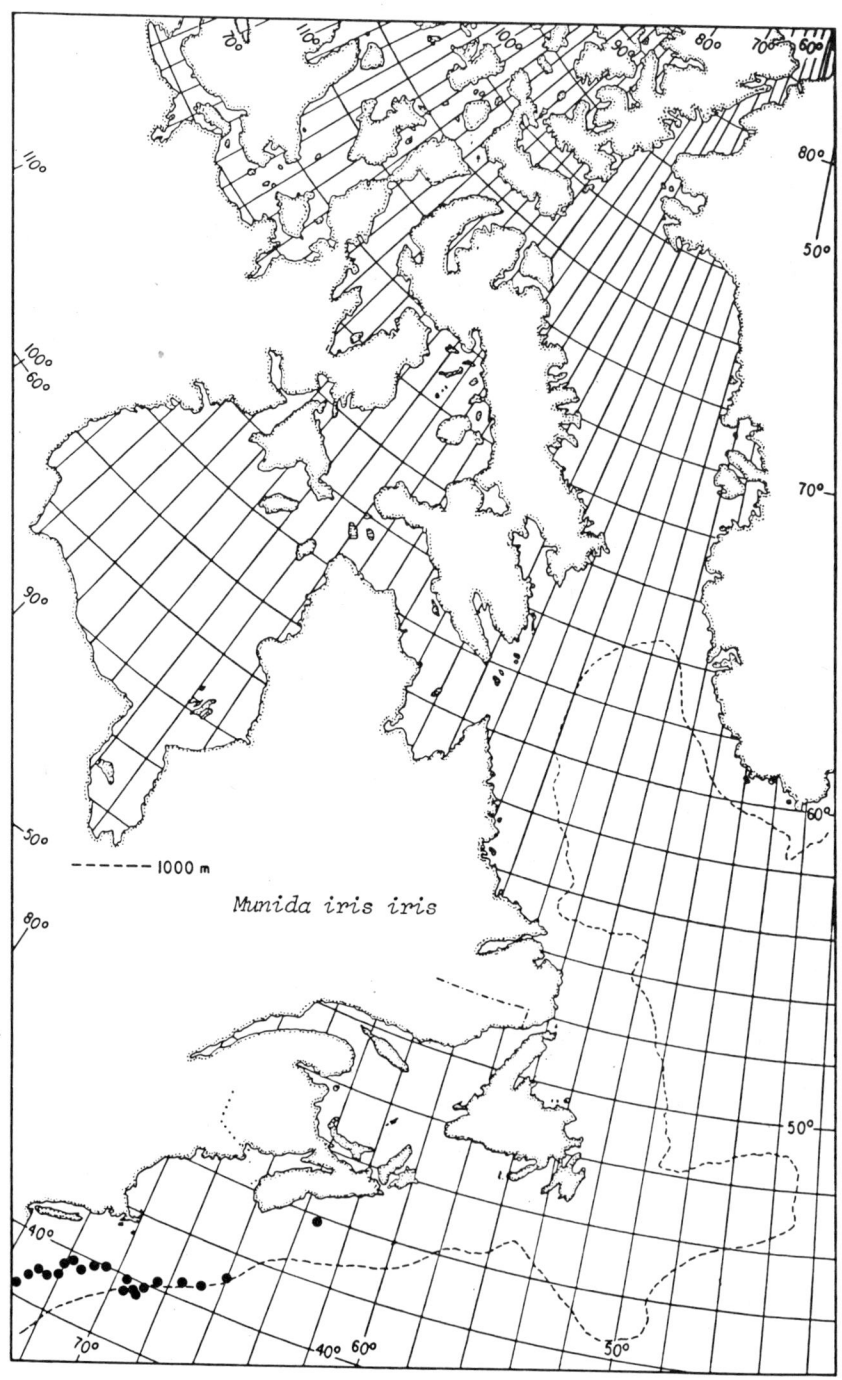

Fig. 219. *Munida iris iris*, distribution records in the area of reference.

Munida tenuimana G. O. Sars, 1871
Selbie 1914: pl. XI, figs. 15, 16;
Zariquiey Alvarez 1952: 197, fig. 6.
(Figures 220 and 221)

DISTINGUISHING CHARACTERISTICS

Carapace with about 6 spines on posterior margin; supraorbital spines about half as long as rostral spine and almost parallel with it; lateral margin with about 6 spines; two spines in epigastric and 4 in metagastric areas; two spines behind cervical groove; no middorsal spines. Abdominal somites 2–4 with spines along anterior margins (first somite is hidden under carapace).

DESCRIPTION

Integument calcified, covered with many transverse striations with short forward directed setae. Colour pale greyish-pink.

Rostrum a single spine or stylet almost half as long as carapace with fine granules laterally and scattered short setae dorsally, flanked by supraorbital spines about half as long and lying parallel to it; supraorbitals trending upward and rostrum slightly descending below level of carapace but turning up slightly at tip.

Carapace depressed, almost as wide as long, with many transverse striations or low ridges with forward-directed short setae; just behind and in line with supraorbitals are two pairs of spines in gastric area and a small spine in hepatic area; pronounced cervical groove behind which are two widely spaced spines near anterior corners of well-defined cardiac area (cardiac groove); lateral spines following external orbital and hepatic are two large anterior branchial and two posterior branchial in decreasing size; posterior margin with pronounced ridge and 6 stout spines directed forward.

Ventrally the sternites form a broad shield-shaped area (in reverse (z)) with four clear divisions for legs 1–4, with a midventral line (clearer posteriorly) and an anterior cross-piece at the third maxillipeds (fifth sternite separate).

Abdomen wide, depressed; first somite narrow, partly hidden under carapace; strong forward spines on somites 2–4 as follows: 6 (4–8), 4 and 2, respectively. Pleura pointed ventrally.

Telson (t) wider than long; surface with imbricate markings (scales) with setal fringes; with symmetrical divisions distally rounded on each side of anal tube and faint midline proximally, and without spines; uropodal branches with squarish outer corners, both shorter than telson, protopod with short spine.

Eye large, cornea globular, with very short thick stalk.

Antennule (c) 1st article broad, slightly compressed; inner edge grooved with a long distal spine and a slightly shorter outer spine; outer edge also has two long lateral spines on distal half; 2nd article shorter and more slender than 1st but longer than 3rd, the latter expanded slightly toward tip, flagella short.

Antenna (d) 1st article with long distal spine at each side; 2nd with long distal inner spine; 3rd (the shortest) with outer moderate distal spine.

Mandible (e) incisor calcified, edge forming a low peak with small tooth slightly off-centre and a lower extended corner with very small teeth; molar short with oval irregular surface behind incisor; palp with three segments.

Maxillule (f) distal endite curved and expanded distally to spinous edge; proximal wide with slightly concave edge; endopod slender.

Maxilla (g) endites large unequally bilobed, proximal with submarginal curved row of long setae; endopod tapering to sharp point, exceeding wide anterior lobe of scaphognathite, the latter longer than short triangular posterior lobe, a shallow notch separating them at outer edge.

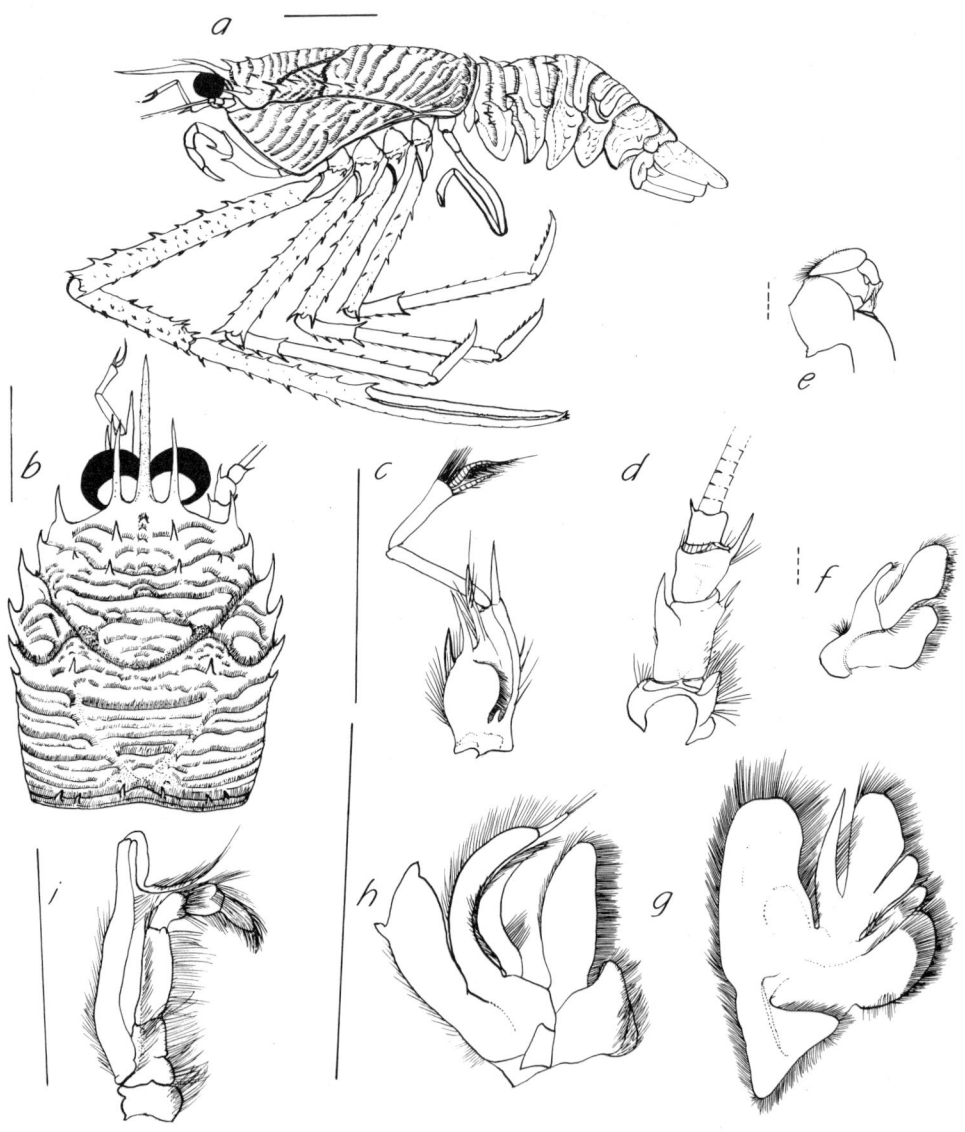

Fig. 220. *Munida tenuimana*: *a*, whole animal from left side; *b*, carapace in dorsal aspect; *c*, antennule; *d*, antenna; *e*, mandible; *f*, maxillule; *g*, maxilla; *h*, first maxilliped; *i*, second maxilliped; solid line = 10 mm, broken line = 1 mm.

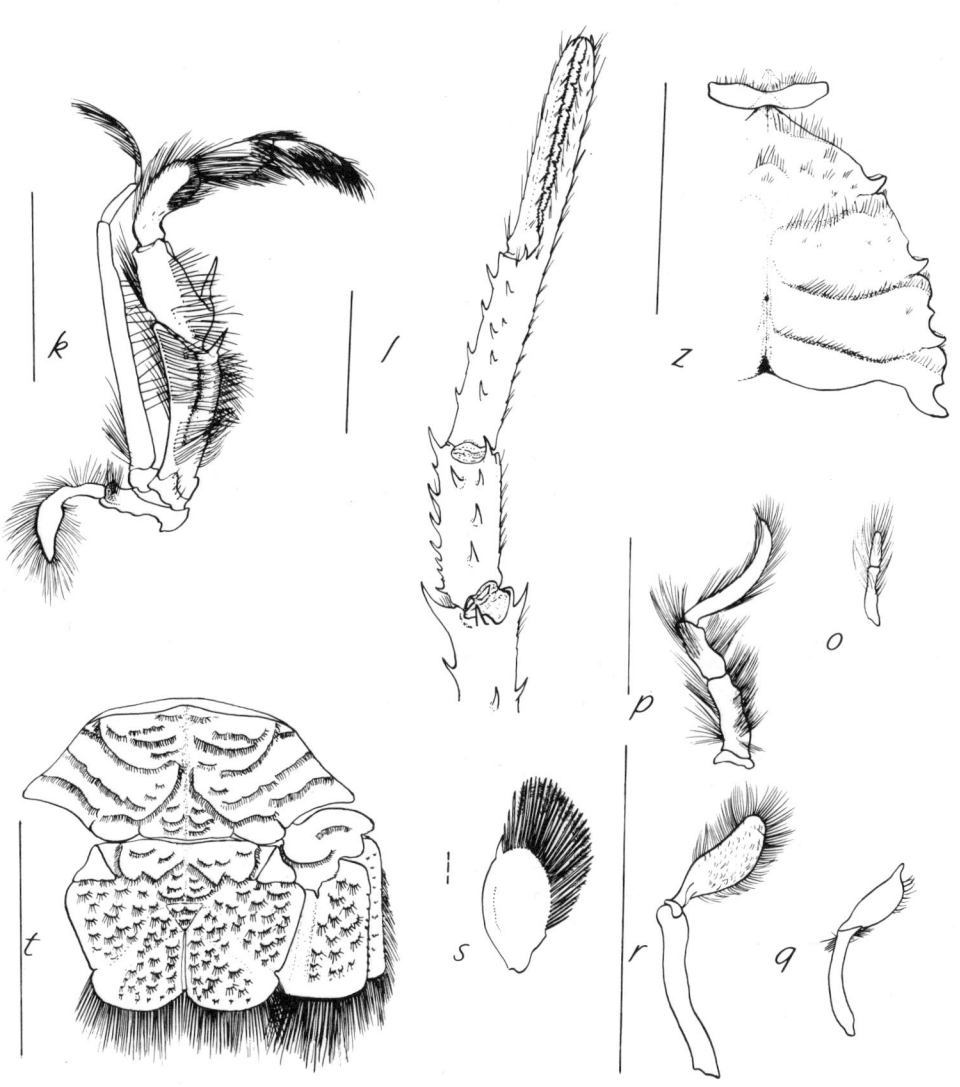

Fig. 221. *Munida tenuimana*: k, third maxilliped; l, first pereopod; z, sternum; o, F second pleopod; p, third pleopod; q, M first pleopod; r, M second pleopod; s, M third pleopod; t, telson; solid line = 10 mm, broken line = 1 mm.

Maxilliped I (h) distal endite long (wide) stepped in from shorter thickened proximal endite; endopod slender sickle-shaped, almost as long as narrow exopod, the latter with short one-piece flagellum or lash; epipod wider than exopod with small distal projections.

Maxilliped II (i) leglike, slightly compressed. Exopod long, with lash.

Maxilliped III (k) leglike, ischium with crista dentata of many very small spines, also strong distal spine; merus with very strong spine on proximal third; exopod slender, long, exceeding ischium and merus, with lash. Epipod long, dilated toward tip but pointed.

Pereopods: I (l) chelate, very long and slender, fingers longer than palm, ischio-merus, carpus and propodus with series of strong fixed spines, propodal finger terminating in spine as well as curved tip; II–IV long but shorter than I, series of fixed spines on ischio-merus to propodus, dactyl with moveable spines on flexor edge; V small and slender, with pseudo-chela, modified for grooming.

Pleopods: female I not present, II (o) small, endopod smaller than protopod; female III (p) endopod with two segments the distal sickle-shaped; male I (q) protopod slender, endopod shorter, expanded but pointed; male II (r) endopod spatulate, setose, shorter than protopod; male III–V (s) short and rounded with distal appendix and fan of plumose setae.

RANGE OF DISTRIBUTION

In the western Atlantic from Davis Strait to the Grand Banks, and eastern Atlantic from Iceland and Norway to British Isles and in the Mediterranean (Selbie 1914). Depths 440–650 m and temperatures 3.5–4.4° C (Squires 1965a).

Records of occurrences in the area of reference are in Fig. 222.

BIOLOGY

Lengths of males to 31 mm cl and females 24 mm cl. Stomachs of most specimens examined were empty but sponge spicules and a pycnogonid were found in one stomach (Squires 1965a)

Lebbeus polaris, *Pasiphaea tarda*, *Pandalus propinquus* and *Pontophilus norvegicus* were present in catches with this species (Squires 1965a).

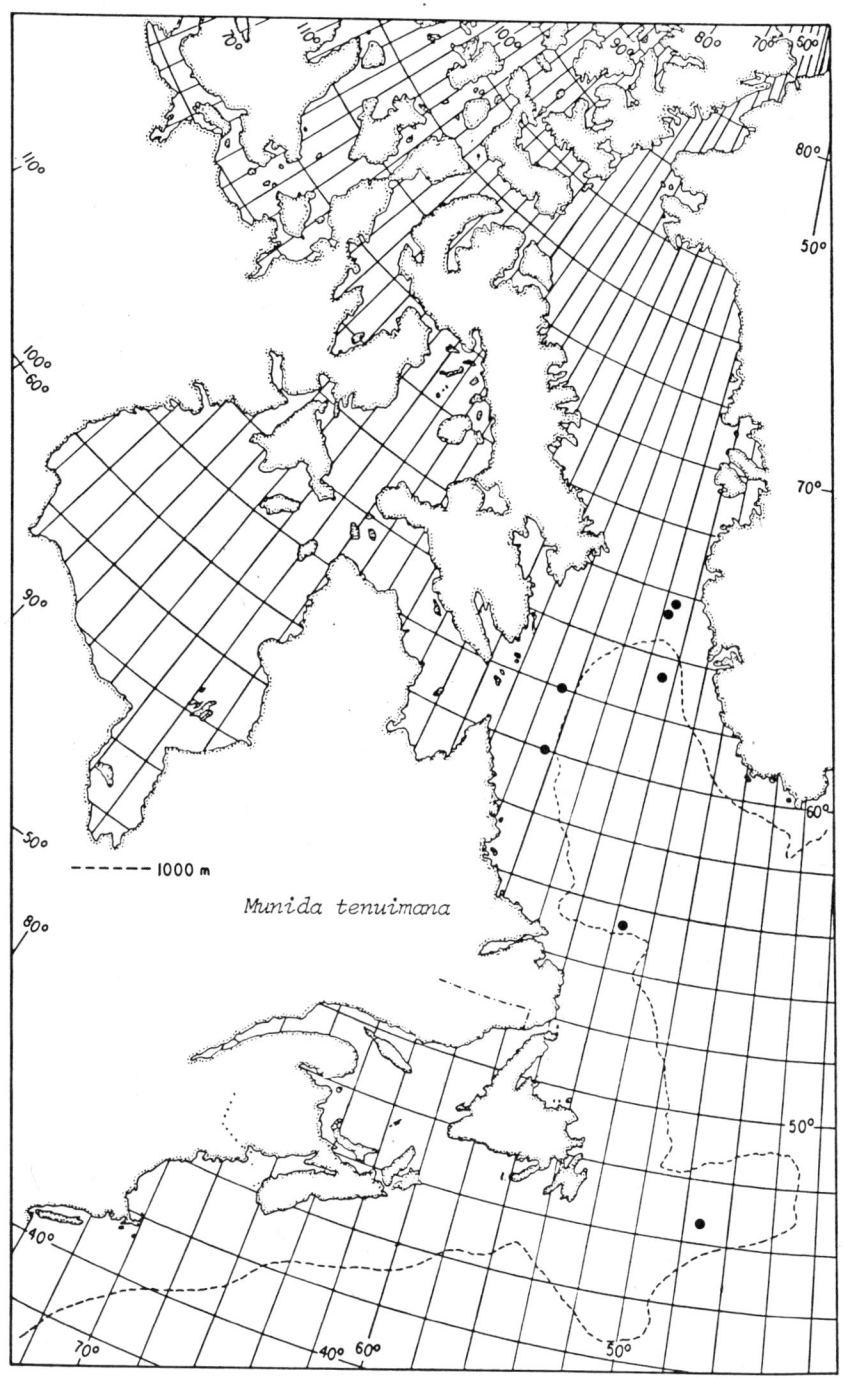

Fig. 222. *Munida tenuimana*, distribution records in the area of reference.

Munida valida Smith, 1883
Williams 1984: 237, figs. 172, 173.
(Figures 223 and 224)

DISTINGUISHING CHARACTERISTICS

Rostral spine more than twice as long as outwardly divergent supraorbital spines, the latter exceeding cornea of eyes, all three spines covered with fine granules, seemingly smooth; small spines in midline behind rostrum in epigastric area; no spines on posterior margin of carapace; 2nd and 3rd somites of abdomen (1st narrow and hidden under carapace) with anterior marginal spines; 1st article of antennule with outer distal spine longer than inner distal spine.

DESCRIPTION

Integument calcified, covered with numerous short setae, also transverse striations or low ridges with anterior short setal fringes. Colour greyish pink.

Rostrum a slender median spine about 0.4 cl about on level with dorsal carapace but tip slightly ascending, supraorbital spine on each side ascending and divergent slightly less than half rostrum but exceeding cornea of eye.

Carapace depressed, width about 0.8 cl; pronounced cervical groove with lateral branch separating anterior branchial (with three lateral spines) from posterior branchial area (with two lateral anterior spines); laterally ahead of cervical groove is large spine with accessory basal spine and anterior to it the large postorbital spine; behind and in line with supraorbital spines are two pairs of gastric spines the anterior pair much the larger; in the median line between the large pair are two small spines; a small spine in hepatic area lateral to smaller pair; behind cervial groove a widely separate pair of moderate spines and no spines posterior to them.

Ventrally sternal plates widening and decreasing in length posteriorly (z), separated by faint midventral line, with rounded depression at centre of sternum of fifth legs; plate between third maxillipeds narrow, with edges almost parallel but slightly curved.

Abdomen depressed, 1st somite narrow partly hidden, 2nd somite with about 11 stout spines along anterior edge directed forward and 3rd with about 6 spines; all terga except 6th with transverse striations fringed with short setae, 6th with ridges arcuate or scale-like and setal fringes directed posteriorly; pleura rounded except 2nd which is slightly pointed, all with scale-like striations.

Telson (t) wider than long, slightly narrowing distally; symmetrically divided at centre by anal tube but no clearly defined areas; shallowly notched at middle of posterior edge only slightly rounded. Uropodal branches squarish at outer corners both considerably shorter than telson, spiny at outer edges; protopod with posterior curved spine.

Eyes large, cornea globular, much larger than short stalk.

Antennule (c) 1st article robust, longer than much more slender 2nd or 3rd, with long sharp ascending dorsal spine on distal two-thirds and outer small lateral spine near its base, also two distal spines, the outer longer than the inner.

Antenna (d) 1st article of peduncle with distolateral spine at each side.

Mandible (e) heavily calcified, incisor with small tooth at centre of curved edge and larger tooth at lower corner; palp with three segments.

Maxillule (f) distal endite widely expanded to spinous edge, wider than thin pointed proximal endite; endopod short.

Maxilla (g) endites large, unequally bilobed; endopod with attenuate sharp apex, slightly exceeded by somewhat pointed anterior lobe of scaphognathite, the latter almost twice as long as axe-shaped posterior lobe.

Maxilliped I (h) distal endite large, set back from edge of thickened proximal endite; endopod slender, sickle-shaped, slightly longer than narrow exopod, the latter with short one-piece lash; epipod large foliaceous, tulip-shaped in outline.

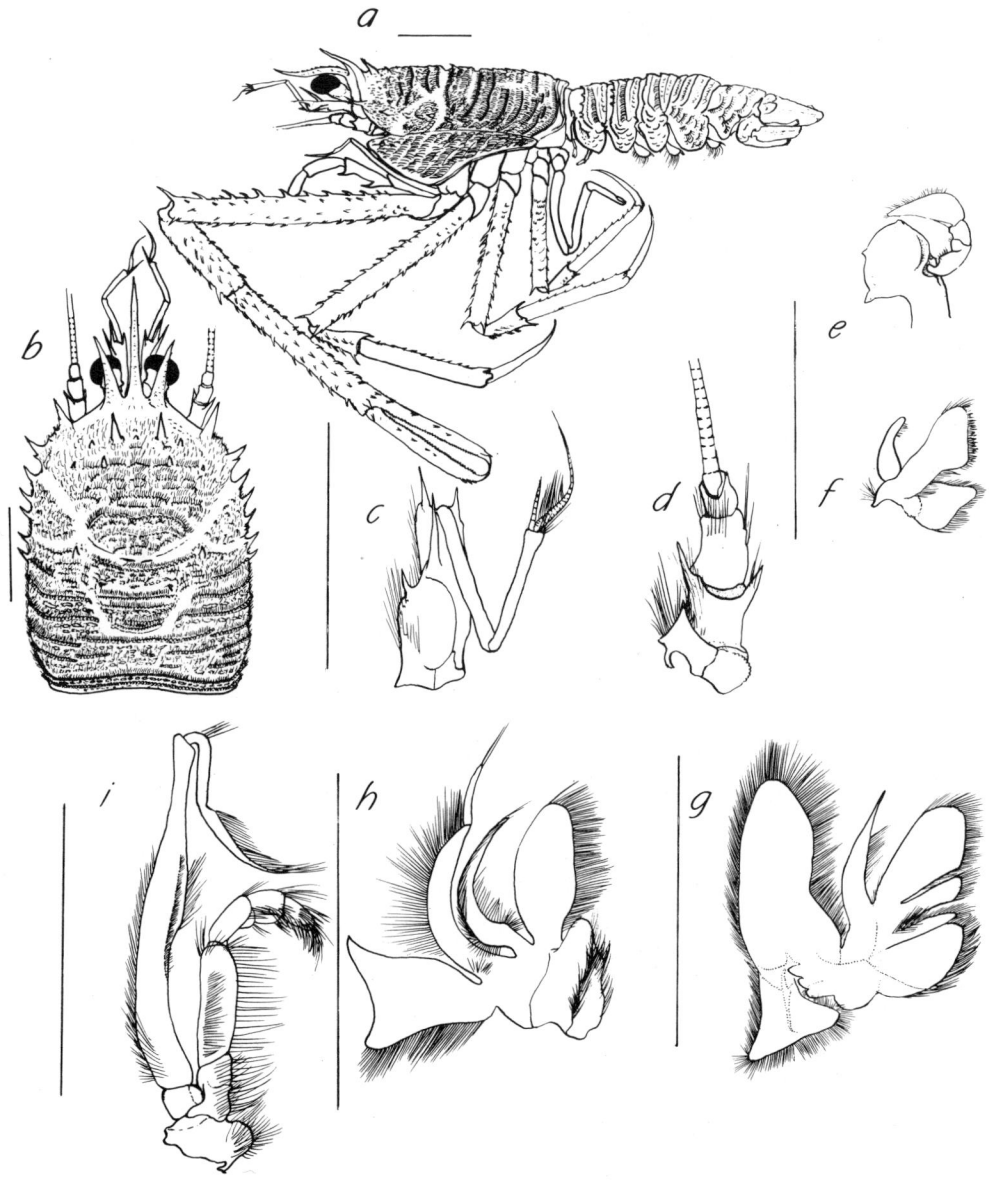

Fig. 223. *Munida valida*: *a* whole animal from left side; *b*, carapace in dorsal aspect; *c*, antennule; *d*, antenna; *e*, mandible; *f*, maxillule; *g*, maxilla; *h*, first maxilliped; *i*, second maxilliped; solid line = 10 mm.

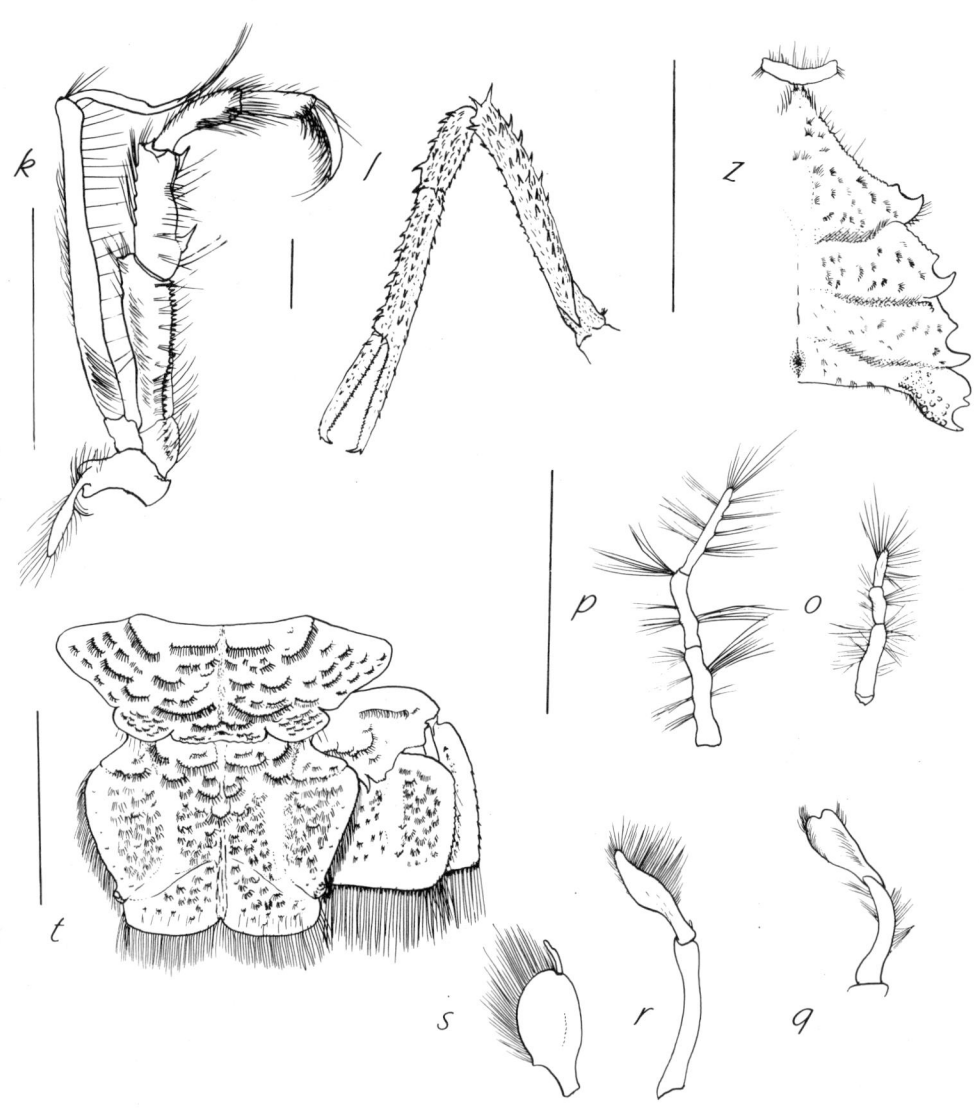

Fig. 224. *Munida valida*: k, third maxilliped; l, first pereopod; z, sternum; o, F second pleopod; p, third pleopod; q, M first pleopod; r, M second pleopod; s, M third pleopod; t, telson; solid line = 10 mm.

Maxilliped II (i) leg-like, slightly compressed, exopod much longer than endopod, with long lash.

Maxilliped III (k) leg-like, dactyl long and curved; merus with inner distal and proximal spine, ischium with crista dentata of many small spines (few also in line on basis) and distal outer blunt spine; exopod slender, longer than ischium plus merus, with long lash. Epipod moderate, club-shaped.

Pereopods: I (l) chelate, very long, fingers shorter than palm, series of strong fixed spines on merus, carpus and propodus, latter with apical spine as well as curved tip and dactyl with three apical spines as well as curved tip; II–IV in decreasing length and slenderness, series of fixed spines on merus, carpus and propodus (fewer on the latter) and strong distal spines on merus and carpus; dactyl long and sharp, without lateral spines. V much more slender and shorter than others, without spines, but with pseudo-chela for grooming.

Pleopods: female I not present, II (o) uniramous, endopod in two parts, with long setal bunches; III (p) uniramous, endopod in two parts, proximal curved distally, with many tufts of setae; male I (q) endopod spatulate, shorter than II (r), the latter also with spatulate endopod which could possibly fit with I to form a tube for transferral of spermatophore. III–V (s) wide, short, with small apical appendix and fringe of plumose setae.

RANGE OF DISTRIBUTION

Off Nova Scotia to Gulf of Mexico and Caribbean to Golfo de Morrosquillo, Colombia (Williams 1984, in part). Depths 90–1823 m (Wenner and Boesch 1979).

Records of occurrence in the area of reference are in Fig. 225.

BIOLOGY

Lengths of carapace including rostrum in male 54 mm and female 49 mm.

Ovigerous females have been taken almost year round suggesting continuous breeding (Williams 1984).

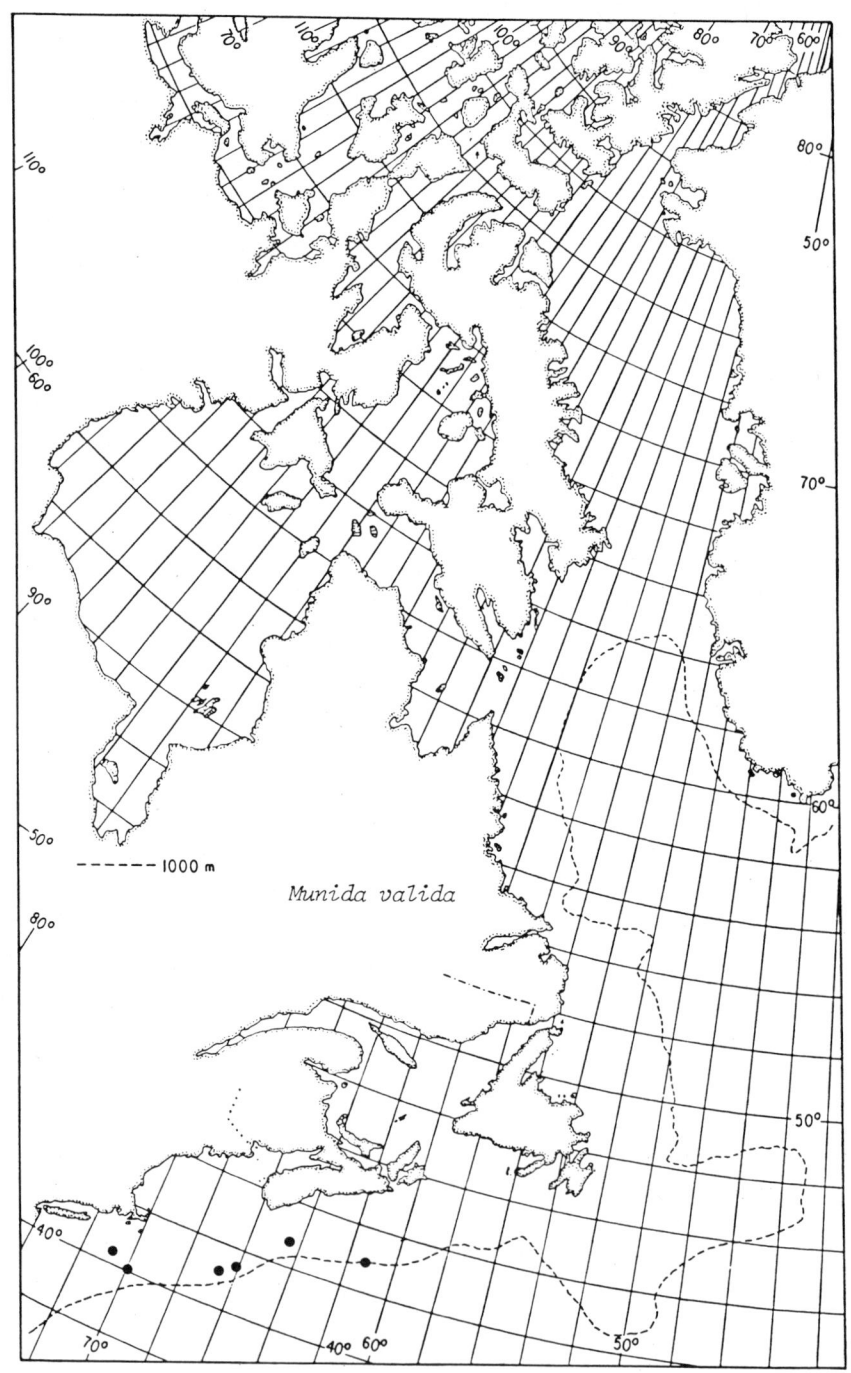

Fig. 225. *Munida valida*, distribution records in the area of reference.

Genus *Munidopsis* Whiteaves, 1874
Makarov 1962: 96; Rathbun 1929: 24.

Rostrum spiniform, sometimes with small but never long supraorbital spines; eyes without pigment or facets, and eyestalks may be free or fused with the orbit; carapace depressed, rectangular in outline, integument strongly calcified; first legs chelate about as long as or slightly longer than others.

Munidopsis curvirostra Whiteaves, 1874
Rathbun 1929: 24, fig. 31;
Whiteaves 1901: 257, fig. 1, 1a.
(Figures 226 and 227)

DISTINGUISHING CHARACTERISTICS

Rostrum a single long and stoutish spine curving upward, one-half to two-thirds carapace length; eyes without pigment or facets, eyestalks free; transverse ridge on middle of cardiac area with strong forward-directed spine; carapace heavily calcified, 2 middorsal spines behind base of rostrum; a middorsal spine on anterior edge of somites 2–4 of the abdomen; exopod of 1st maxilliped without flagellum.

DESCRIPTION

Integument rigid, opaque, rough, with ridges or striations and short setae. Colour greyish white, eyes yellowish.

Rostrum like a curved spike, tapering gradually to a fine point, slightly compressed dorso-ventrally, lateral ridges with fine spinules and confluent with orbit, also covered with fine setae, curved strongly upward.

Carapace squarish but longer than wide, surface covered with short setae and small transverse irregular ridges; in rounded gastric area is a pair of spines behind rostrum and ahead of two in midline; also anterolateral spine on a small mound with two small accessory spines behind anterior edge; cervical groove is followed by depressed area and a transverse cardiac ridge with median spine directed forward.

Sternal plates (z) of 1st to 4th legs fused but with clear divisions, increasing moderately in width posteriorly; 5th narrower, separate; plate of 3rd maxilliped small, with pointed ends laterally.

Abdomen heavily calcified, 2nd to 4th tergites with transverse ridge and forward-directed median spine; pleura extended ventrally but with rounded tips.

Telson (t) broadly shield-shaped and symmetrical on each side of anal tube, with five pairs of matching pieces of varying sizes and a wide proximal triangular centrepiece; terminally a shallow central notch and long setae; uropodal branches with smooth straight outer edges and slightly rounded corners, both shorter than telson.

Eye rounded, with free stalk, but unpigmented and without corneal facets.

Antennule (c) 1st article very stout and longer than other two combined; with dorsolateral spine longer than distomesial spines, the inner with lateral teeth, also spines on inner edge; 2nd and 3rd articles subequal; flagella short.

Antenna (d) peduncle short, articles decreasing distally, each with small distolateral spines.

Mandible (e) heavily calcified, incisor with very small tooth at centre of curved edge and small spine at lower corner; palp with three segments.

Maxillule (f) distal endite expanded to straight spinous edge; proximal with almost parallel sides and slightly concave edge and a short narrow accessory lobe; endopod slender, curved.

Maxilla (g) endites large, unequally bilobed; endopod tapering to sharp point, slightly exceeding anterior rounded lobe of scaphognathite, the latter much longer than triangular posterior lobe, both separated by a notch in outer edge, posterior with an inner fold line.

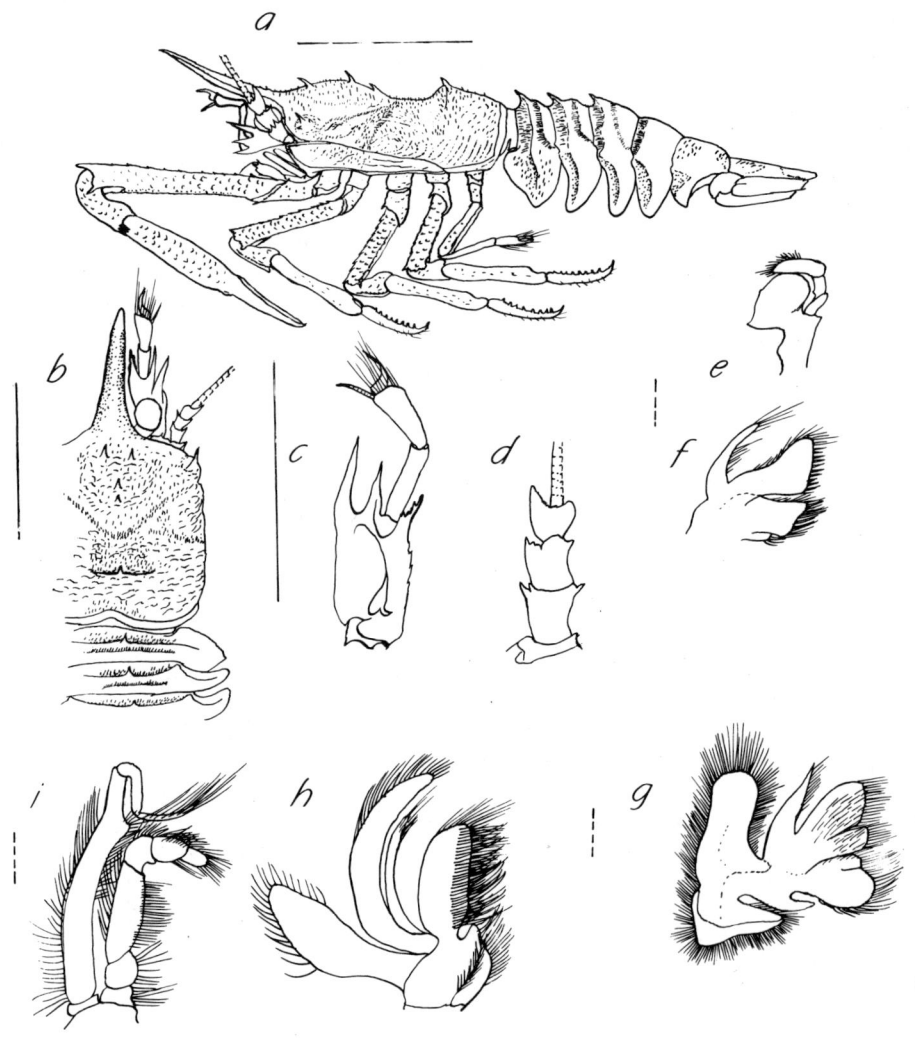

Fig. 226. *Munidopsis curvirostra*: *a*, whole animal from left side; *b*, carapace in dorsal aspect; *c*, antennule; *d*, antenna; *e*, mandible; *f*, maxillule; *g*, maxilla; *h*, first maxilliped; *i*, second maxilliped; solid line = 10 mm, broken line = 1 mm.

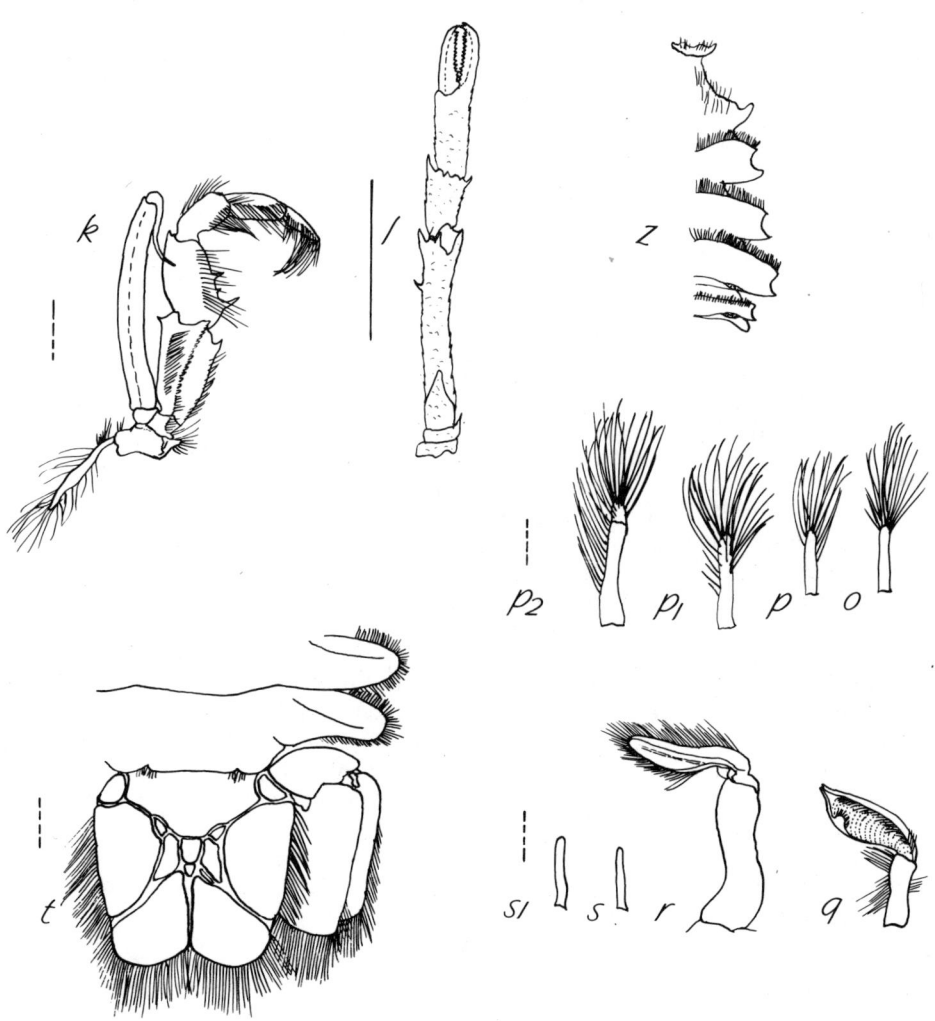

Fig. 227. *Munidopsis curvirostra*: k, third maxilliped; l, first pereopod; z, sternum; o, F second pleopod; p, F third pleopod; p1, F fourth pleopod; p2, F fifth pleopod; q, M first pleopod; r, M second pleopod; s, M third pleopod; s1, fourth pleopod; t, telson; solid line = 10 mm; broken line = 1 mm.

Maxilliped I (h) distal endite large, pedunculate and set back from thickened edge of proximal endite; endopod slender with apical setae only, shorter than narrow curved exopod, the latter without lash or flagellum; epipod large pedunculate.

Maxilliped II (i) leglike with seven segments, distal with apical strong curved spines; exopod long with distolateral projection and long lash.

Maxilliped III (k) leglike, with seven segments, ischium with crista dentata of many small spines and distolateral spine on each side; merus with two large spines and small one between them on inner edge and short distal tooth at outer edge; all segments triangular in cross-section; exopod slender, longer than merus and ischium combined, and with short lash. Epipod moderately long and narrow, somewhat club-shaped, setose.

Pereopods: I (l) chelate, longer and more robust than others, fingers with sagittate teeth, propodus with double spines at apex between which apical spine of dactyl fits; strong distal spines on carpus and merus, and a strong inner spine on distal third of merus; II–IV similar, with distal spine on merus and a row of about 8 strong teeth on flexor edge of long dactyls; V slender and short, with pseudo-chela (fingers flattened and oval) and strong setae on dactyl and propodus, the merus appears hollow along one side into which the carpus can fit.

Pleopods: female I not present, II (o) one short segment only with apical setae, III and IV similar but different in size; V (p2) slightly larger with additional short apical segment; male I not present, II (q) spatulate, endopod foliaceous folded inward, longer than protopod; III (r) protopod long and stout, endopod spatulate also folded inward: II and III together able to form a tube; IV and V (s) small.

RANGE OF DISTRIBUTION

Davis Strait to North Carolina in the western Atlantic and Iceland and the Britsh Isles to northwest Africa in the eastern Atlantic. Depths 75–2 325 m (Selbie 1914).

Records of occurrence in the area of reference are shown in Fig. 228.

BIOLOGY

Lengths to 12 mm cl in males and 15 mm cl in females.

Most females were ovigerous in March and one out of two in August; all ovigerous and large non-ovigerous females had large ova in the ovaries, and were first ovigerous at 9 mm cl. Males were first mature (carrying spermatophores) at 8 mm cl.

Stomach contents were phytobenthos, foraminiferans, crustaceans, polychaetes and small bivalves (Squires 1965a).

Decapod species also present in the catches with this species were *Acanthephyra pelagica*, *Pandalus borealis*, *Pandalus propinquus*, *Lebbeus polaris* and *Sabinea sarsi* (Squires 1965a).

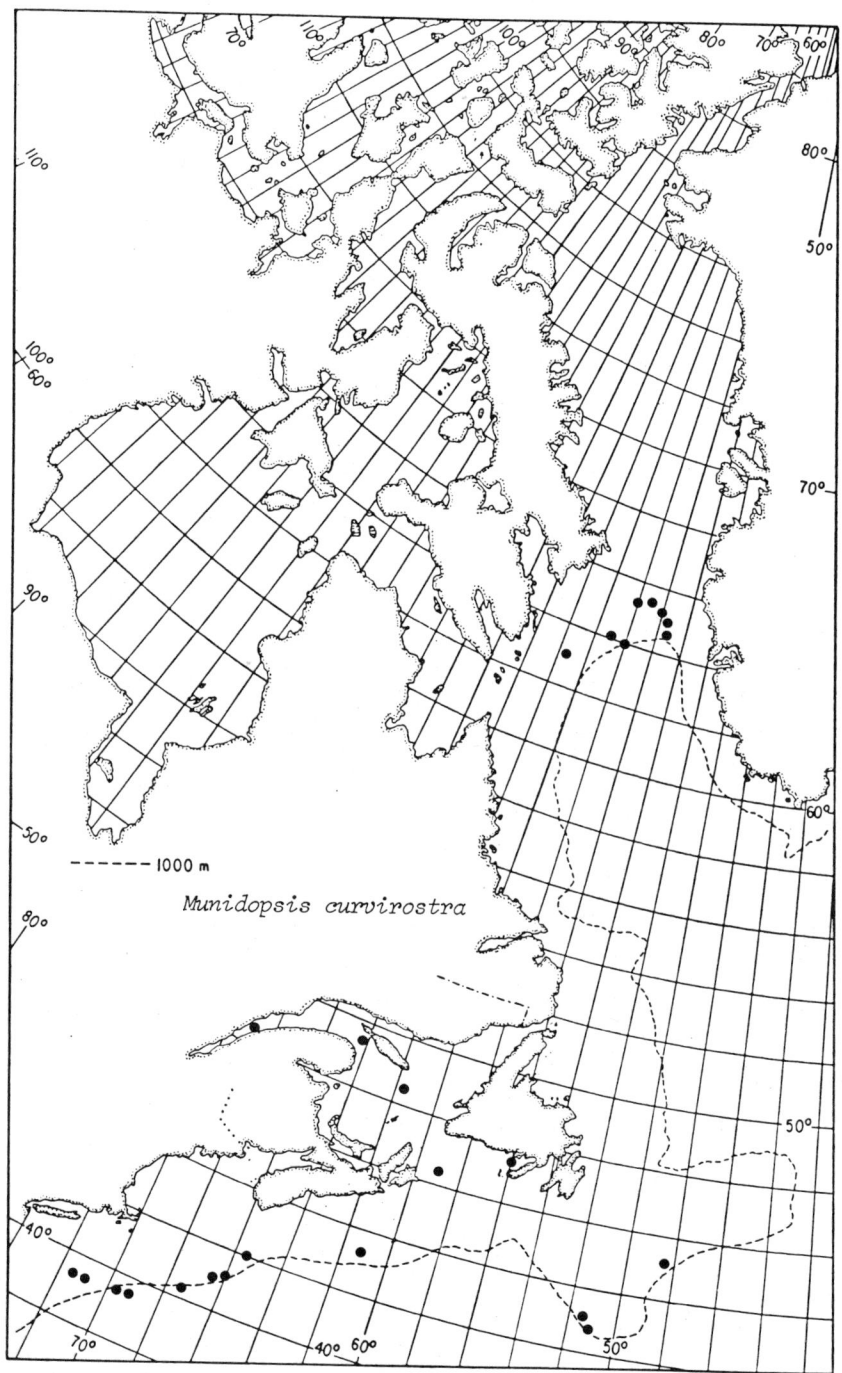

Fig. 228. *Munidopsis curvirostra*, distribution records in the area of reference.

Family MAJIDAE Samouelle, 1819
Rathbun 1929: 35; Williams 1984: 291.

Fore part of body narrow, forming a distinct rostrum; carapace subtriangular or oval; second article of antenna fused with epistome; orbits slightly defined by a projection at anterior edge of carapace; carapace with short hooked setae; chelipeds shorter or equal in length to walking legs, very mobile at base; spider crabs.

Key to species of Majidae in area of reference
(Modified from Williams 1984 and Rathbun 1929)

1 Walking legs much longer than chelipeds, compressed ...... *Chionoecetes opilio*

   Walking legs only slightly longer than chelipeds, roundish 2

2 Carapace with few setae, rostrum showing line of division 3

   Carapace densely pubescent, rostrum without centre line ..... *Libinia emarginata*

3 Carapace subtriangular; basal articles of antenna smooth and triangular in shape .... *Hyas araneus*

   Carapace lyrate; basal articles of antenna with tubercles along lateral edge and quadrilateral in shape 4

4 Rostrum moderately long (4.5 to 6.4 times rostrum length = carapace length); anterior lateral (hepatic) expansion of carapace almost as wide as posterior lateral (branchial) expansion .... *Hyas coarctatus coarctatus*

   Rostrum short and wide (7.1 to 9.3 times rostrum length = carapace length); anterior lateral expansion of carapace appreciably exceeded by posterior lateral expansion .... *Hyas coarctatus alutaceus*

Genus *Chionoecetes* Kroyer, 1838
Garth 1958: 148; Rathbun 1925: 232.

Carapace about as wide as long, depressed or partially inflated, tuberculate; rostrum short, divided into two flat, triangular horns; orbits large, shallow, open above; large postocular tooth directed forward; basal articles of antenna narrow, tapered distally and ending in a spine; cheliped shorter than first three walking legs, the latter compressed; abdomen of seven segments.

Chionoecetes opilio (O. Fabricius, 1788)
Rathbun 1925: 233, figs 84, 85; 1929: 36, fig. 49;
Williams 1984: 307, figs. 242, 245a.
"Snow crab" — "Crabe de neige"
Figures 229 and 230

DISTINGUISHING CHARACTERISTICS

Carapace about as wide as long, covered with tubercles and hooked setae; rostrum short and wide with a gap between flat, pointed horns; shallow open orbits; basal article of antenna long, tapering distally and ending in a blunt spine; first three walking legs compressed, with long merus, much longer than chelipeds; brownish in colour above and pale yellowish or buff below.

DESCRIPTION

Integument moderately calcified, covered with tubercles and hooked setae. Colour brownish with faint greenish tint above and pale buff below.

Rostrum short, flattish but hollow below, in two parts with points separated anteriorly.

Carapace about as wide as long; hepatic region short, slightly inflated; below it the pterygostomian region is also inflated with rows of short spines, some sharp and some with double tips; gastric area wider in front slightly inflated; arcuate grooves anterior to and posterior to raised central cardiac area with two larger tubercles; branchial areas wide, tuberculate. Orbit shallow, dorsal fissure with triangular spine, ventral fissure wider. Postero-lateral border with parallel grooves and edge rows of granules widely interrupted at intestinal region. Sternal plates all fused with lateral notches for articulation with pereopods.

Abdomen greatly compressed dorso-ventrally and tucked under the carapace; with rounded median ridge; much narrower in the male (y) than in the female (x) but only slightly tapered.

Eye with short thick stalk and large sub-spherical cornea.

Antennule (c) 1st article thick and wide, fused with endostome and immobile, 2nd and 3rd short, retractile into socket, flagella short.

Antenna (d) basal article formed of 2nd and 3rd articles fused, narrow, tapering to distal and lateral blunt spine and with tubercles along outer edge; fused with epistome; 4th and 5th articles diminishing in size and very short; flagellum also short, scarcely exceeding rostrum.

Mandible (e) incisor is a wide, curved, heavily calcified blade pointed at the middle and with the pointed lower corner projecting as a large tooth; molar surface behind incisor; palp in three segments.

Maxillule (f) proximal endite thick, narrow, strongly curved and pointed, with bunch of terminal spines; distal endite expanded widely to spinous edge; endopod foliaceous, in two segments, the distal with inward curved edge.

Maxilla (g) endites bilobed, the proximal with one very narrow lobe, the distal with short equal lobes; endopod with very wide rounded base and short slender tip from its centre; scaphognathite wide, with pointed anterior lobe and broad triangular posterior lobe slightly longer.

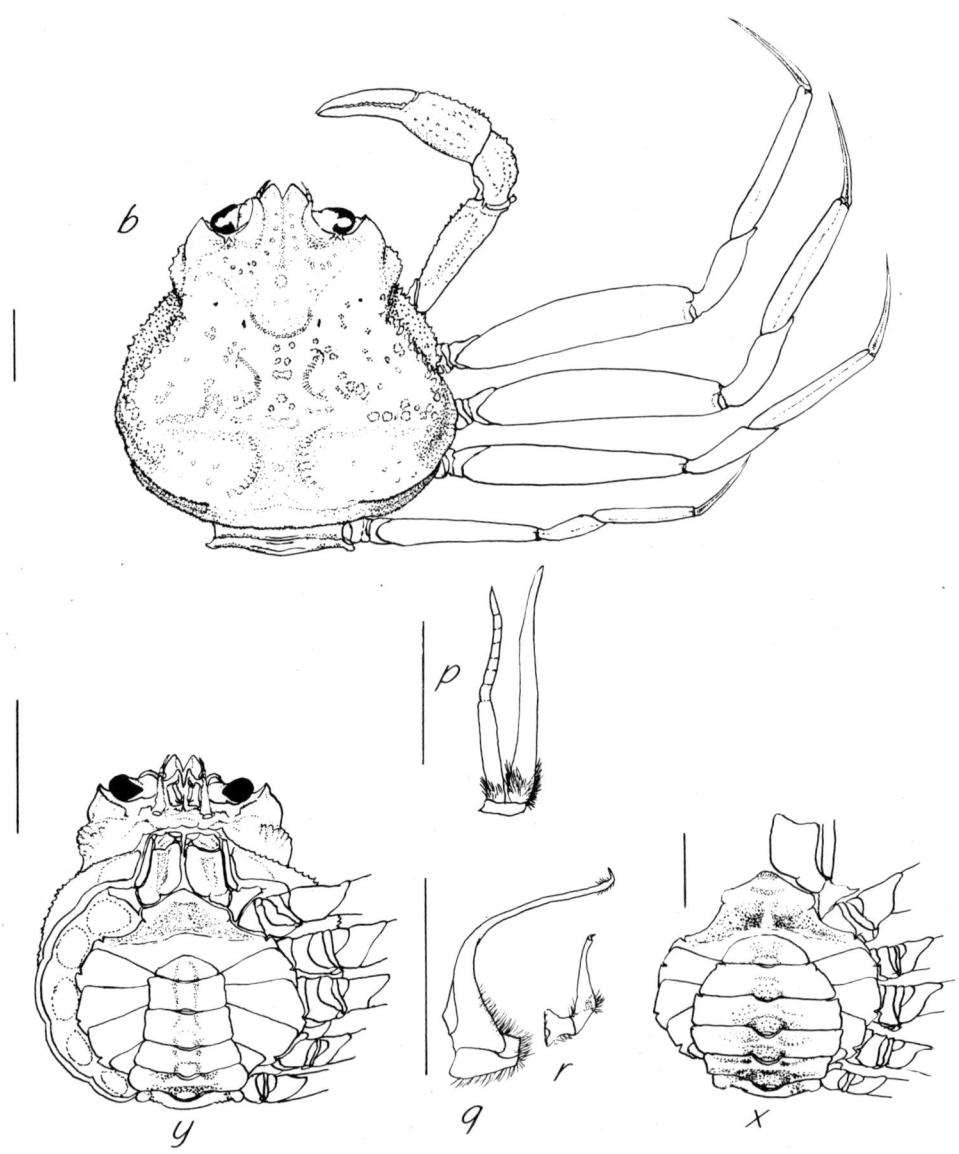

Fig. 229. *Chionoecetes opilio*: *b*, carapace in dorsal aspect; *p*, F second pleopod; *q*, M first pleopod; *r*, M second pleopod; *x*, F abdomen; *y*, M abdomen; solid line = 10 mm.

Fig. 230. *Chionoecetes opilio*: c, antennule; d, antenna; e, mandible; f, maxillule; g, maxilla; h, first maxilliped; i, second maxilliped; k, third maxilliped; broken line = 1 mm.

Maxilliped I (h) distal endite flat and curved wider than proximal thick endite; endopod or central lobe foliaceous, sinuous and with distal part in opposite plane, exceeding narrower exopod, the latter with small lash; epipod very large, wide and foliaceous proximally and with a long ribbonlike setose extension into the gill chamber.

Maxilliped II (i) endopod pediform, stout, distal segment with strong spines; exopod compresed but rigid, tapering and with short lash; epipod long, ribbonlike but somewhat rigid and with long lateral setae; podobranch.

Maxilliped III (k) operculiform, endopod with greatly expanded ischium (covering the buccal cavity and other mouth parts), merus short but wide with carpus inserted laterally; exopod long, slightly tapering and with very short lash; epipod long, ribbonlike but rigid and setose; podobranch.

Pereopods: I chelate, much shorter than walking legs, fingers slender, sharp with fine serrate teeth, palm inflated, with granulate or spiny ridges. II–IV long, with long compressed merus, dactyl long, thin and very sharp, with median groove. V shorter and more slender than II–IV.

Pleopods: female II (p) endopod slightly shorter than exopod and with end part annulate, exopod slightly tapering and bent near apex; all 4 pairs similar, setose in breeding dress. male I (q) tapering to long curved neck from wider base, neck grooved and terminating in a narrow modified arrowhead with several short retrorse lateral setae; male II (r) much shorter than I, flask-shaped and tapering to distal translucent tip with one small and one large point.

RANGE OF DISTRIBUTION

In the western Atlantic from Greenland to the Gulf of Maine (Rathbun 1924, recognized a different subspecies *elongatus* from the Pacific, which could be a gemminate species of *opilio*, in view of the latter not appearing in collections from the Canadian Arctic, and the distributions of both therefore should be separate. Records from Arctic Alaska, Siberia and the North Pacific therefore would probably be for the species *C. elongatus* and not for *C. opilio*). Depths 20–310 m but mostly from 70–280 m (Elner 1985).

Records of distribution in the area of reference are in Fig. 231.

BIOLOGY

Carapace widths (cw) of males reach 150 mm and of females 90 mm.

Maturity takes place at a minimum size of 51 mm cw in males and 41 mm cw in females (Watson 1970). Apparently competition during the mating period results in most successes for large males mating with females, however in observations of a natural population it was seen that small males (as small as 89 mm cw) also took part in mating to the same extent as larger males (Taylor et al. 1985). These observations also showed that females mated in the hard shell condition.

Reproductive potential in this species is very high with about 100% of the females carrying eggs each year (Taylor et al 1985).

Stomach contents of samples of small specimens were phytobenthos, foraminiferans, crustacean fragments (including shrimps, crabs, amphipods, copepods, isopods, cumaceans and ostracods), bivalves, brittle stars, polychaetes, gastropods, chitons and hydroids (Squires 1965a).

A common predator is the cod (*Gadus morhua*).

Decapod species taken in catches with this species were mostly *Pagurus pubescens*, *Sabinea septemcarinata*, *Pandalus borealis*, *Pandalus montagui*, *Hyas araneus* and *Hyas coarctatus*.

FISHERY

Areas fished include the south and western Gulf of St. Lawrence, around Cape Breton, east and southeast coasts of Newfoundland and southern Labrador.

Fishing for this species began in the early 1960s with by-catch trawling for groundfish off Gaspé (Brunel 1963). The trap fishery began in the Gulf of St. Lawrence in 1966 and in Trinity Bay, Newfoundland, in 1968. Expansion of the fishery was rapid during the 1970s and a peak in landings was reached in 1982 and 1983 with landings of 47 000 and 37 000 metric tonnes, respectively (Elner 1985; Taylor and O'Keefe 1987).

The fishery is regulated by a minimum size limit of 95 mm cw and limited entry. This allows the escape of all females (which have a maximum size of less than 95 mm cw) so that maximum production of eggs is always assured. However, the large males are taken in the fishery, and there is concern that if most of the large males were removed from an area, a reduction in numbers of viable eggs might occur (Conan and Comeau 1986). Observations have shown, however, that males less than regulation minimum size can mate successfully with smaller females (Ennis et al. 1988).

A constraint to the fishery is the relatively long period of recovery from a moult (lasting up to 3 months: Watson 1970) during which the crabs are not commercially acceptable.

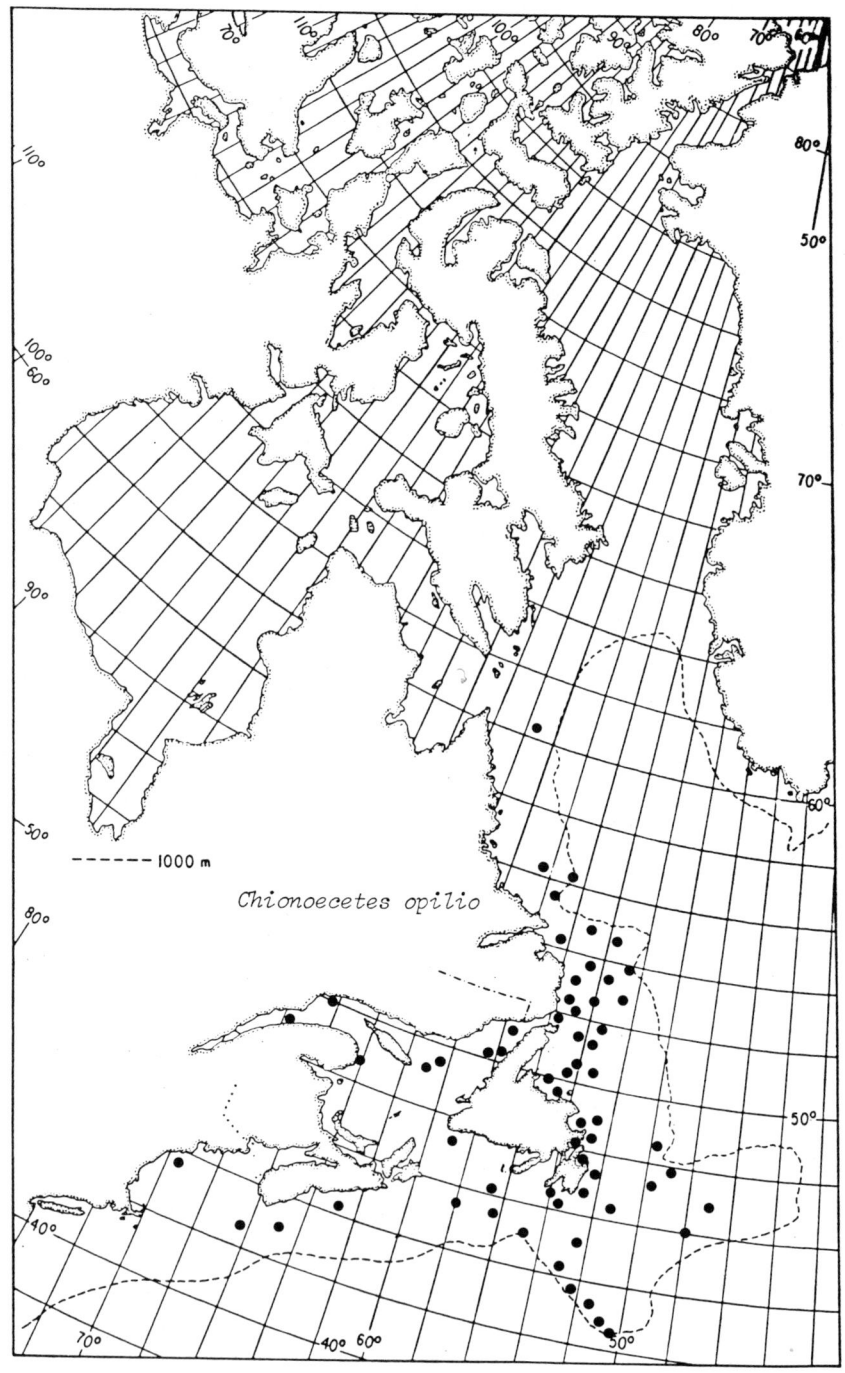

Fig. 231. *Chionoecetes opilio*, distribution records in the area of reference.

Genus *Hyas* Leach, 1814
Ingle 1983: 132; Rathbun 1925: 252.

Carapace subtriangular to lyrate, tuberculate and pubescent; rostrum triangular, flat, bifid, horns separated by narrow slit; a tubercle on middle of anterior surface of eyestalk; basal articles of antenna fused, longer than wide; chelipeds stout, chelae compressed; walking legs subcylindrical, only slightly longer than chelipeds.

Hyas araneus (Linnaeus, 1758)
Rathbun 1925: 253, figs. 92, 93;
1929: 37, fig. 50;
Williams 1984: 309, figs. 243, 245b.
"Toad crab" — "Crabe bufo"
(Figures 232 and 233)

DISTINGUISHING CHARACTERISTICS

Carapace longer than wide, subtriangular, lateral margins converging anteriorly; rostrum pointed anteriorly, flat, divided by median slit; basal antennal article triangular, smooth, longer than wide; first walking legs only slightly longer than chelipeds.

DESCRIPTION

Integument heavily calcified, rigid, tuberculate, dull, rough, with many hooked setae. Colour dull reddish brown dorsally and pale off-white ventrally.

Carapace roughly shield-shaped, dorsally inflated at the middle with many tubercles and hooked setae especially just behind rostrum; rounded posteriorly with two lateral tuberculate ridges curving inward and forming wide grooves near the margin, disappearing toward the centre; hepatic region slightly expanded with lateral tuberculate ridges extended posteriorly; orbital fissures open above and below; postorbital tooth cupped, acute; orbits open, eyes not concealed when retracted.

Rostrum flat, divided into two triangular parts by median slit and pointed anteriorly.

Abdomen wide, flattened, narrowed somewhat in male and widest at 3rd and 4th somites (7 in all with the telson the last somite), in female widest at the 5th.

Eyes small, cornea smaller than stalk, distal tubercle with setae and a small lateral tubercle on stalk directed antero-ventrally.

Antennule (c) 1st article large, rounded, containing otocyst; 2nd and 3rd stout, short; flagella short.

Antenna (d) basal article (2nd and 3rd fused) large, subtriangular in shape, smooth; 4th inflated at one side and 5th cylindrical and smaller; flagellum short.

Mandible (e) incisor heavily calcified, forming a sharp cutting blade only slightly curved and with a projecting tooth-like lower corner; palp with three segments; molar a shelf behind incisor.

Maxillule (f) proximal endite narrow, pointed; distal expanded distally, wide; endopod in 2 segments, the distal with a distolateral projection.

Maxilla (g) endites unequally bilobed, distal larger, exceeding endopod; endopod slender, papillate, from inner edge of wide base; anterior lobe of scaphognathite pointed, with irregular inner edge, longer than subtriangular posterior lobe.

Maxilliped I (h) distal endite squarish, not clearly separated from projecting thickened proximal endite; endopod or central lobe sinuous, foliaceous, twisted, almost as long as exopod, the latter with slender flagellum or lash; epipod foliaceous wide at base and tapering to long ribbon-like and setose blade.

Maxilliped II (i) pediform, distal segment with strong spinous setae; exopod tapered, with lash; epipod long ribbonlike with podobranch.

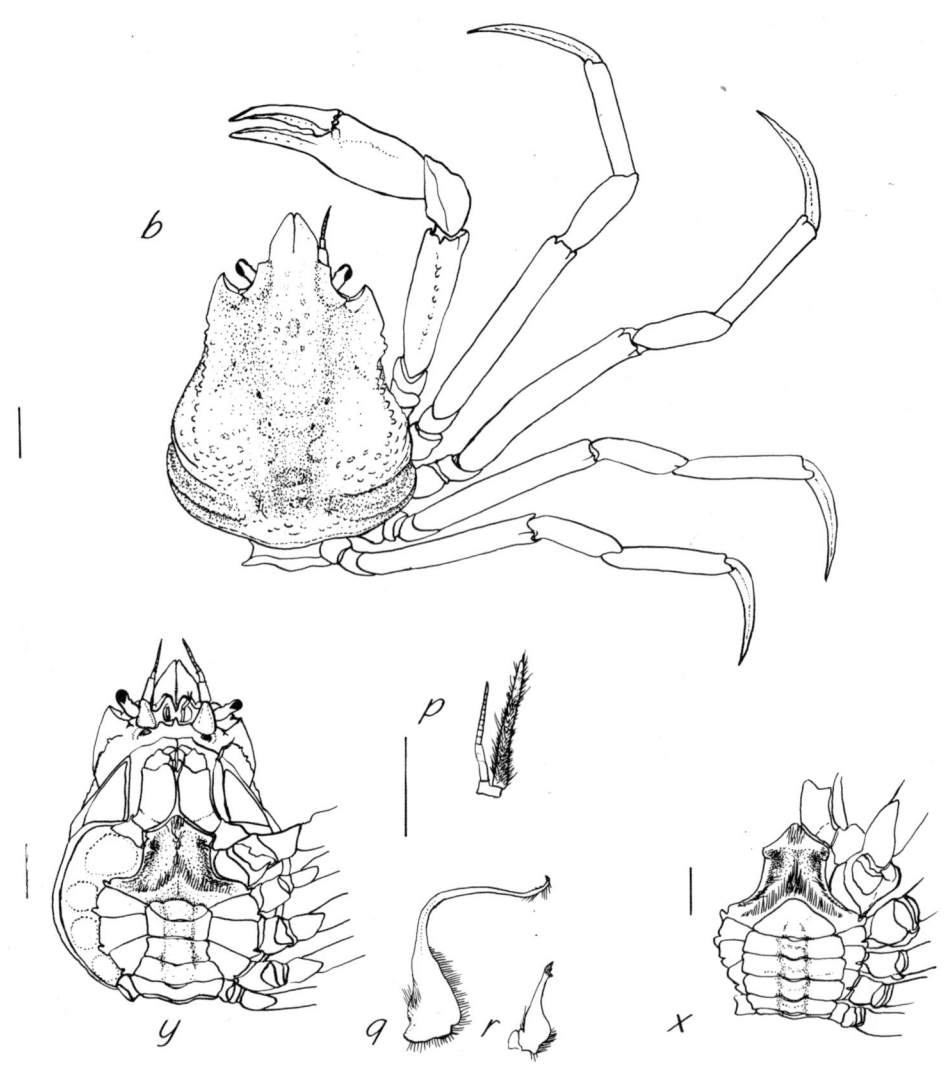

Fig. 232. *Hyas araneus*: b, carapace in dorsal aspect; p, F second pleopod; q, M first pleopod; r, M second pleopod; x, F abdomen; y, M in ventral aspect; solid line = 10 mm.

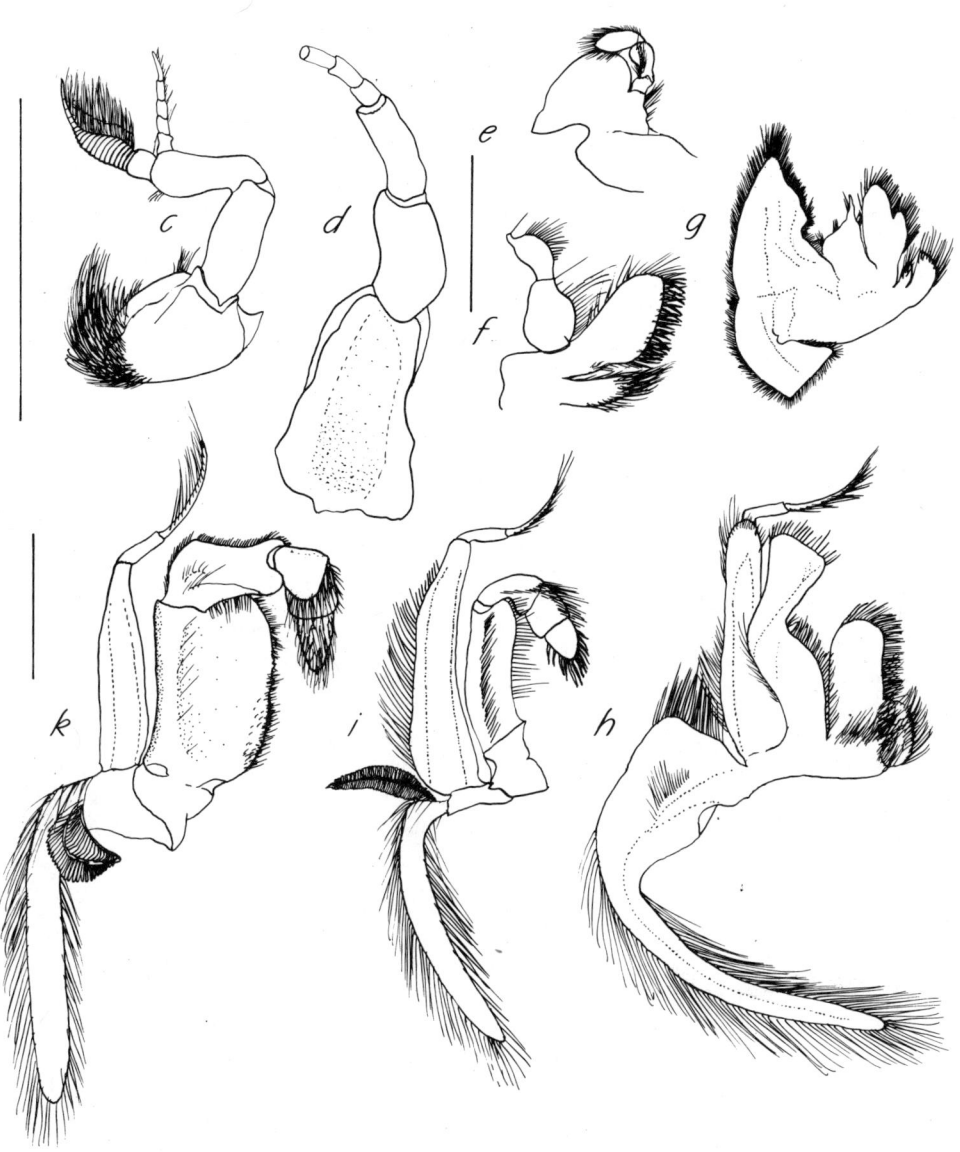

Fig. 233. *Hyas araneus*: c, antennule; d, antenna; e, mandible; f, maxillule; g, maxilla; h, first maxilliped; i, second maxilliped; k, third maxilliped; solid line = 10 mm.

Maxilliped III (k) operculiform, ischium very large, inner edge with a few small teeth, merus short, wide, with lateral attachment of carpus; exopod large, rigid, with lash; epipod long, ribbonlike but fairly rigid, with podobranch.

Pereopods: I chelate, fingers narrow and about as long as palm, the latter only slightly inflated; II the longest, more slender than I, subcylindrical, with long, curved and sharp dactyl; III–V similar but slightly decreasing in size posteriorly.

Pleopods: female II (p) endopod slender, distal part annulate, shorter than tapering setose exopod; other three pairs similar; male I (q) with long tapering neck from wide stout base, neck grooved and with slightly expanded tip with lateral retrorse setae or filaments; male II (r) flask shaped, much shorter than I, with neck and expanded tip (both combined form a means of transferral of spermatophores).

RANGE OF DISTRIBUTION

In the western Atlantic from Greenland to Rhode Island; in the eastern Atlantic from the Kara Sea and Iceland to the British Isles and France. Depths 0–360 m (Williams 1984); 4–730 m (Squires 1965a).

Records of distribution in the area of reference are in Fig. 234.

BIOLOGY

Lengths to 95 mm cl (75 mm cw) in males and 81 mm cl (64 mm cw) in females (Williams 1984).

Ovigerous females in spring had large ova in ovaries and would lay eggs after hatching larvae, indicating annual spawning. Eggs were carried about 10 months. Egg diameter 0.7 mm.

This species was not taken in the Arctic where *Hyas coarctatus* was common. But it occurred inshore in Labrador with water temperatures of -1.3 to $6.3°$ C, and offshore (35–730 m) with temperatures of $0.8–3.6°$ C. It was also present in the Gulf of St. Lawrence deep-water community of *Pandalus borealis* where positive temperatures prevail throughout the year, indicating a preference for warmer waters than *Hyas coarctatus*. Other species taken frequently with this species were *Pandalus montagui*, *Spirontocaris spinus*, *Pagurus pubescens*, *Lebbeus polaris* and *Sabinea septemcarinata* (Squires 1965a).

Stomach contents were phytobenthos, crustacean fragments (mostly gammarid amphipods and ostracods), small crabs and hermit crabs, polychaetes, foraminiferans, brittle stars, bivalves, gastropods, chitons and sea urchins. Also fish remains in inshore areas in Labrador (Squires 1965a).

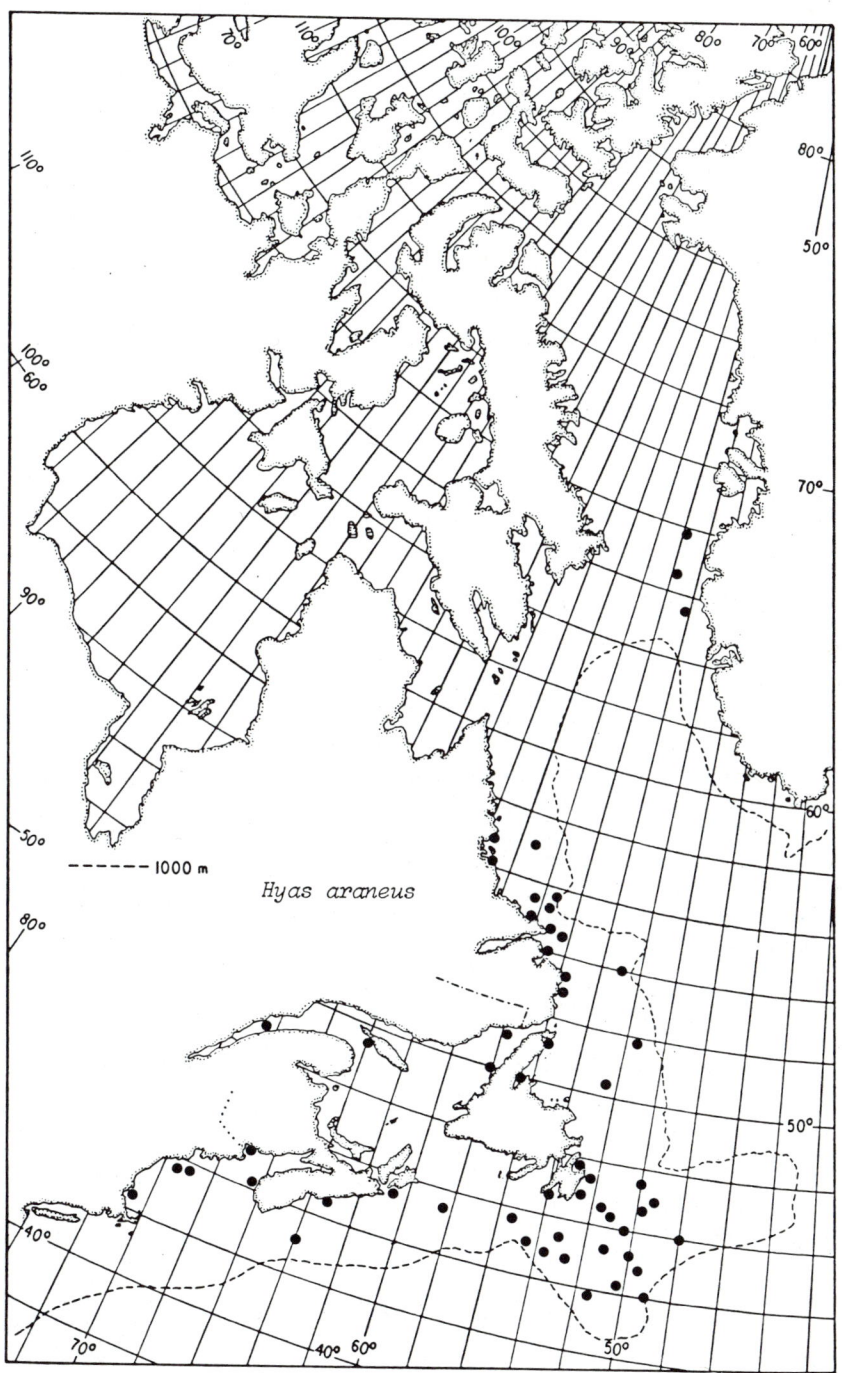

Fig. 234. *Hyas araneus*, distribution records in the area of reference.

Hyas coarctatus Leach, 1815
Rathbun 1925: 258, figs. 90, 93, pls. 94-97;
1929: 37, figs. 51, 52;
Williams 1984: 309, figs. 244, 245c.
(Figures 235 and 236)

DISTINGUISHING CHARACTERISTICS

Carapace lyrate, hepatic region dilated laterally, forming with postorbital region a winglike expansion anteriorly, its posterior angle rounded; basal antennal article with sides slightly converging anteriorly but not forming a triangle and with lateral and terminal tubercles or blunt spines; rostral horns flat, triangular, separated by a median slit; tubercles of lateral margin extend forward only as far as posterior hepatic region. Spine on eyestalk more prominent than in *H. araneus*. The form *alutaceus* differs from the typical *coarctatus* in its wider carapace, more rounded posteriorly, with fewer and smaller tubercles or ridges, and the rostrum is shorter and wider (cl 7.1 to 9.3 times rostrum length compared with 4.5 to 6.4 times rostrum length in *H. coarctatus*).

DESCRIPTION

Integument heavily calcified and rigid, tuberculate and with many hooked short setae. Colour dull reddish dorsally and light buff ventrally.

Carapace lyrate, almost as wide anteriorly as posteriorly because of lateral wings of hepatic and postorbital region; orbits shallow, with dorsal and ventral fissure; gastric area inflated, with rows of tubercles extending anteriorly to rostrum, one larger tubercle in the metagastric area; branchial areas each with a roughly V-shaped formation of tubercles, the arms of the V pointing anteriorly; posterolateral border with pronounced narrow groove ending at the intestinal area and a wider metabranchial groove above it ending at the raised cardiac area.

Rostrum flattened but hollowed underneath, and subtriangular, separated into halves by median slit; edges with low tubercles.

Abdomen flattened, pressed closely to fused plates of sternum; narrow in male, widest at 3rd somite; much wider and longer in female, widest at 4th somite; telson only slightly narrower than other somites.

Eyes not completely hidden when retracted in open orbit; cornea covering about two-thirds tip of stalk, the latter with a blunt spine at about the middle directed antero-ventrally.

Antennule (c) almost completely retractible in cavity posterior to rostrum; 1st article subspherical, 2nd longer than 3rd and with spinules along edge.

Antenna (d) basal article large, sides almost parallel with tubercles along edges and small blunt spines distally; next article (4th) dilated at outer edge with tubercles and small spines; distal article subcylindrical, small.

Mandible (e) incisor heavily calcified with sharp curved cutting edge, a pointed tooth-like lower corner; palp with 3 segments; molar a shelf behind incisor.

Maxillule (f) proximal endite slightly flattened, and curved, narrow and pointed; distal endite slightly curved and expanded to spinous edge; endopod in two parts, the distal truncate and turned in on mandible.

Maxilla (g) proximal endite with one very small narrow lobe, the other squarish; distal endite with subequal lobes separated only at the tips; endopod with wide rounded base and very short slender, spinelike tip; anterior lobe of scaphognathite pointed at apex, longer than axe-shaped posterior lobe.

Maxilliped I (h) distal endite flat and wider than short thick protruding proximal lobe; endopod foliaceous, sinuous, distally in a second plane forming rounded blade, almost as long as exopod, the latter narrow and with short lash; epipod very wide at base extended into a long ribbon with many long setae.

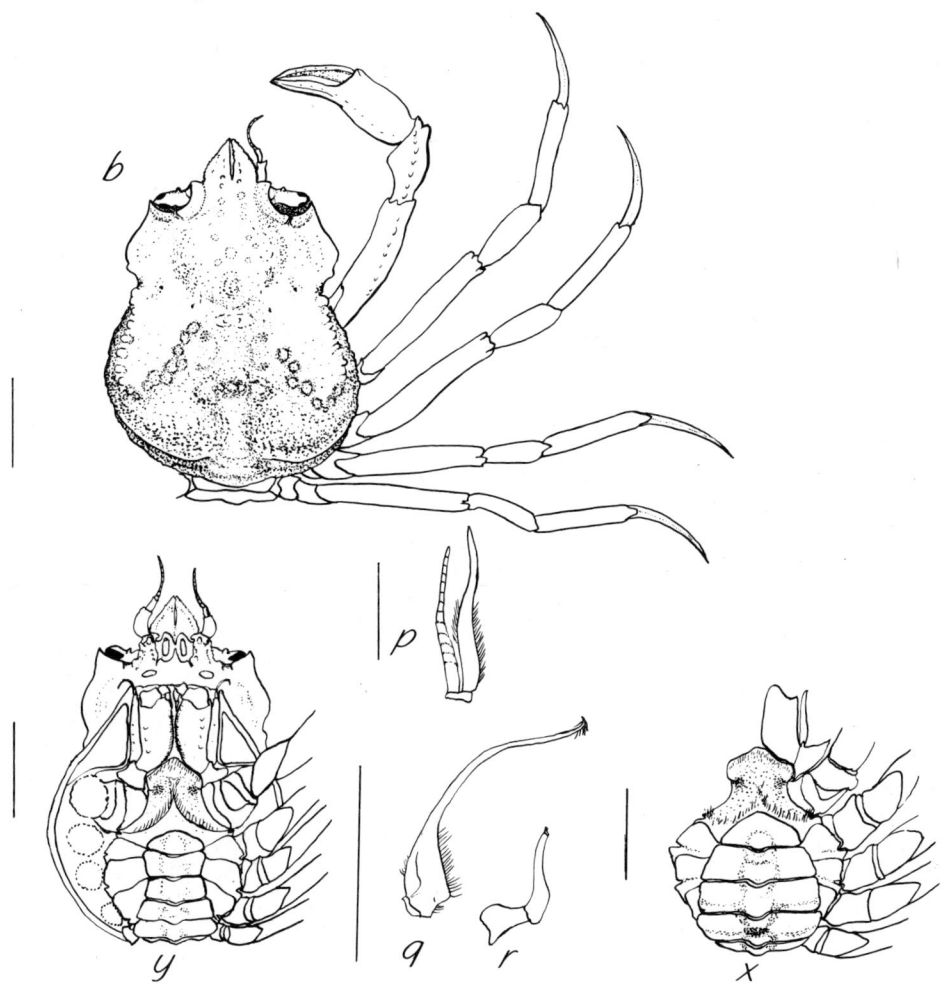

Fig. 235. *Hyas coarctatus*: *b*, carapace in dorsal aspect; *p*, F second pleopod; *q*, M first pleopod; *r*, M second pleopod; *x*, Female abdomen; *y*, Male in ventral aspect; solid line = 10 mm.

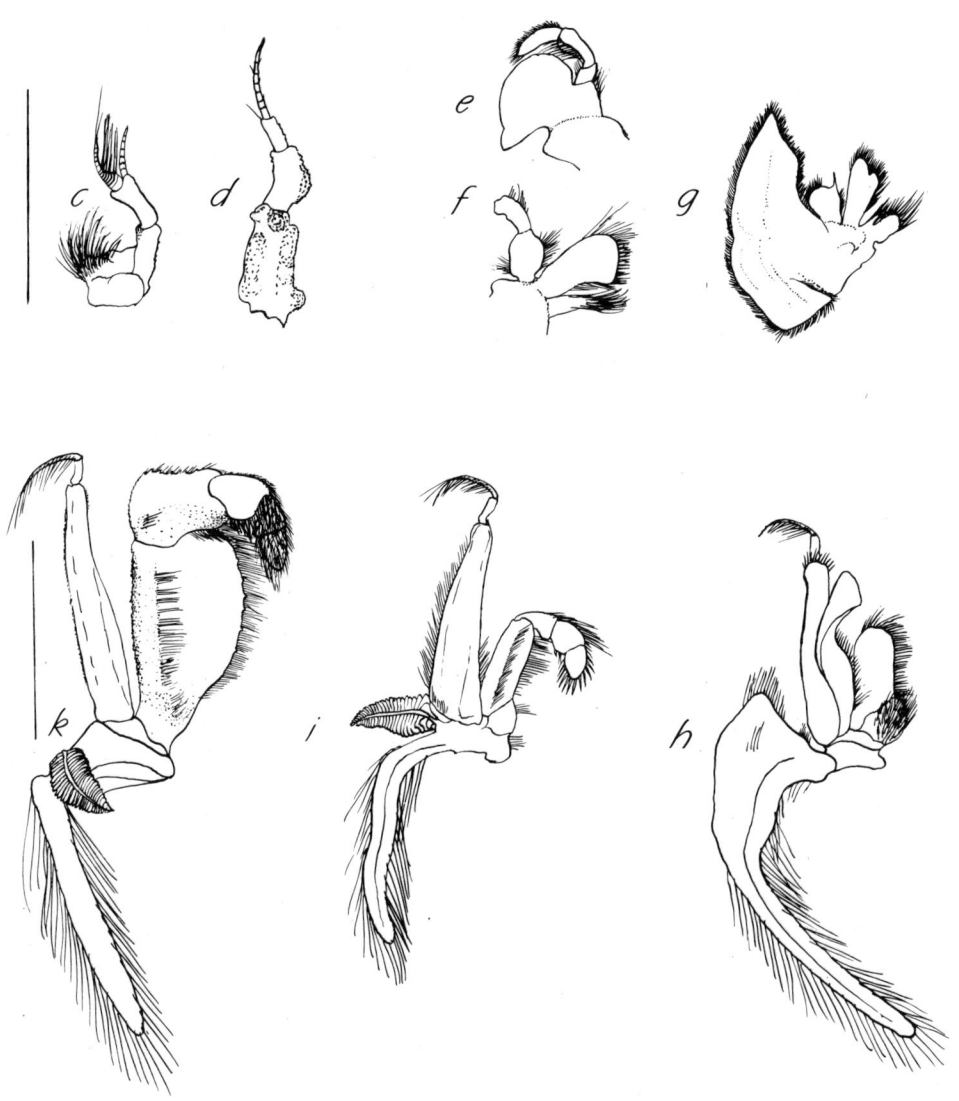

Fig. 236. *Hyas coarctatus*: c, antennule; d, antenna; e, mandible; f, maxillule; g, maxilla; h, first maxilliped; i, second maxilliped; k, third maxilliped; solid line = 10 mm.

Maxilliped II (i) endopod compressed distally, exceeded by tapering exopod with short apical lash; epipod long ribbonlike, flexible but semi-rigid, with podobranch.

Maxilliped III (k) operculiform, ischium greatly expanded and long, with evenly spaced spines on inner edge; merus much shorter, with row of spines at the top and insertion of carpus laterally; exopod large, rigid, with short lash; epipod long narrow, semi-rigid, with podobranch.

Pereopods: I chelate, stouter but shorter than others, chela with fingers shorter than palm; II–V non-chelate, longer than I but decreasing slightly in size posteriorly, dactyl long and sharp.

Pleopods: female II (p) biramous, endopod shorter and annulate distally; other pairs similar to II. Male I (q) with very long grooved neck from wide base, tip arrow-head-like and curved with several short retrorse setae or filaments; male II (r) short, flask-shaped with small distal translucent area with a few short setae.

RANGE OF DISTRIBUTION

In the western Atlantic from Hudson Bay and Greenland to Cape Hatteras, and in the eastern Atlantic from the Murman Sea to Iceland and the British Isles. Depths 0–550 m.

On the basis of carapace length compared with rostrum length (carapace length = 7.1 to 9.3 times rostrum length in *Hyas coarctatus alutaceus*) specimens that I have examined from Hudson Bay and Ungava Bay to Newfoundland do not belong to the form *alutaceus*. I believe therefore that this form does not occur in the Atlantic.

Records of distribution in the area of reference are shown in Fig. 237.

BIOLOGY

Lengths to 87 mm cl in males and to 49 mm cl in females.

Mature females taken from May to November from Hudson Strait, Labrador and Newfoundland were mostly carrying eggs and most also had large ova in the ovaries, indicating that spawning would occur after eggs hatched each year. Females were first mature at 22 mm cl. Those from Hudson Bay appeared to have lower reproductive potential, first maturing at 27 mm cl, and only 29% were ovigerous or ready to lay eggs, possibly indicating environmental stress (Squires 1967a).

Decapod species present in catches with this species were *Pandalus montagui, Lebbeus polaris, Spirontocaris spinus, Chionoecetes opilio, Hyas araneus* and *Pagurus pubescens* (Squires 1965a).

Stomach contents were phytobenthos, crustacean fragments (including shrimps, gammarid amphipods, euphausiids and copepods), foraminiferans, bivalves, brittle stars, hydroids, polychaetes and sponge spicules (Squires 1965a).

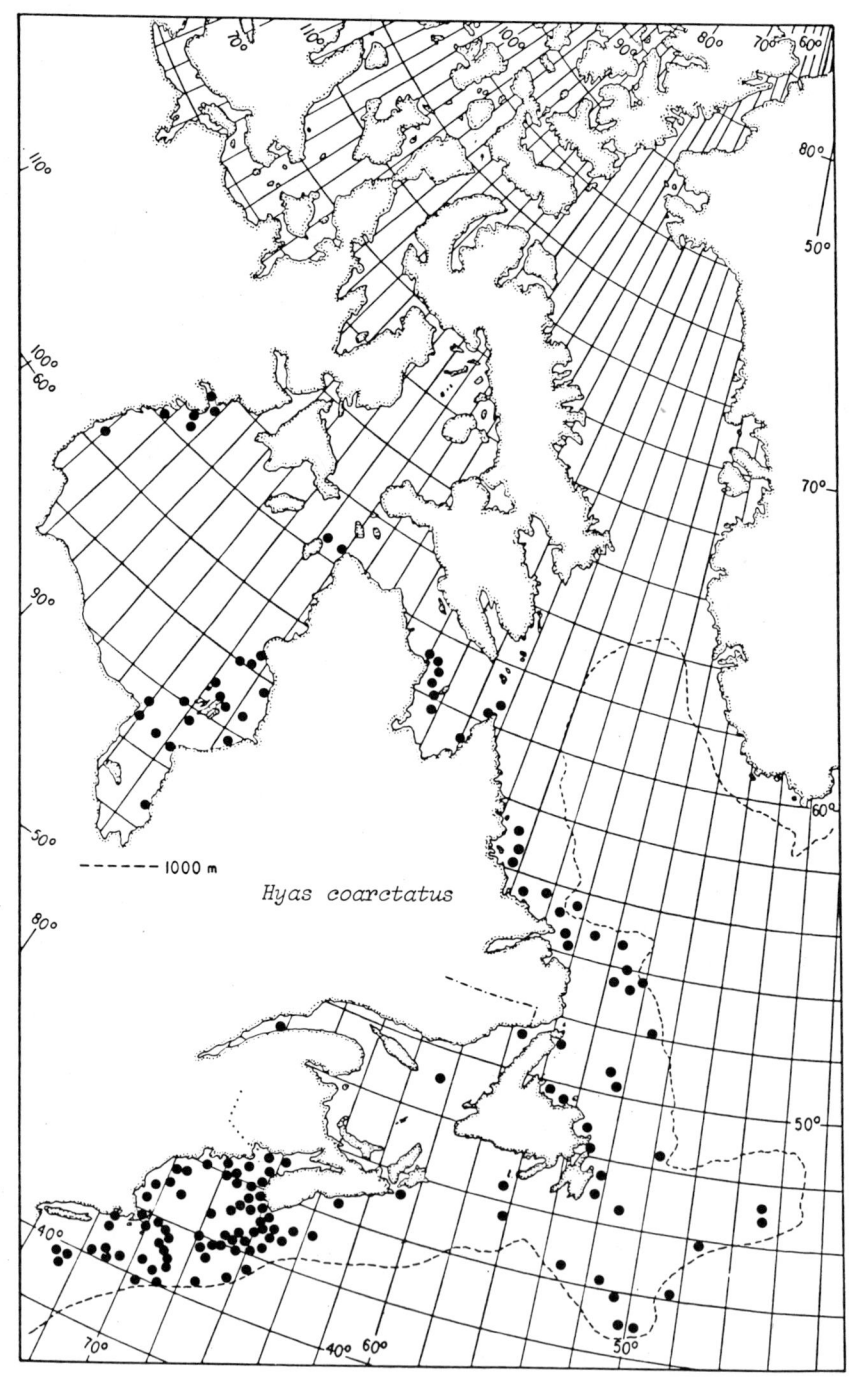

Fig. 237. *Hyas coarctatus*, distribution records in the area of reference.

Genus *Libinia* Leach, 1815
Rathbun 1925: 311; 1929: 38;
Williams 1984: 316.

Carapace behind rostrum almost circular, covered with spines and tubercles and hooked setae; rostrum small, bifid at apex; orbits small, subcircular; chelipeds stout, palm much longer than fingers. Walking legs not much longer than chelipeds, cylindrical.

Libinia emarginata Leach, 1815
Rathbun 1925: 311, figs. 103, 104, pls. 110-113;
1929: 38, fig. 53; Williams 1984: 318, figs. 253, 259h.
(Figures 238 and 239)

DISTINGUISHING CHARACTERISTICS

Carapace subcircular and with short blunt spines around the edge, inside the edge posteriorly; gastric and cardiac areas well-defined with surrounding grooves and well-rounded with median spines or tubercles; covered with dense short setae; rostrum short, tip divided into two short truncate tips, with median hollow; orbits with prominent preorbital spine, eyes small; chela with short fingers and long palm, granulate; legs covered with short setae and no spines.

DESCRIPTION

Integument heavily calcified and rigid, covered with short dense pile of setae and blunt spines or tubercles. Colour dark brown above and dirty yellowish below.

Carapace orbicular, with produced rostrum shallowly bifurcate and with truncate tips, shallowly excavate in median line; major embossing in gastric and cardiac areas with pronounced grooves; raised orbital area with preorbital blunt spine; about 9 major median blunt spines: 4 gastric, 1 urogastric, 2 cardiac and 2 intestinal; at each side about 4 major branchial spines at lateral edge in front, and 3 along emarginate edge behind, with one posteriorly; other blunt spines or tubercles in less well-defined clumps; 2 strong subhepatic spines at each side; supra- and infra-orbital fissures.

In females a thick fine setal semicircular row is on sternal plates along anterior edge of wide ovate abdomen. Male abdomen is narrow but widened at the 2nd to 4th somites.

Eye stalk short, setose, cornea wider, sublunate.

Antennule (c) 1st article subspherical, setose; 2nd and 3rd subequal, slender, flagella short.

Antenna (d) basal article massive, covered with dense setae, with stout blunt outer spine distally; next article twice size of distal article; flagellum short.

Mandible (e) incisor edge only slightly curved on left, with small central tooth on right, corner only slightly modified; palp of 3 segments.

Maxillule (f) proximal endite tapered to slender tip; distal expanded to receding edge; endopod stout with inflated inner edge and curved distal segment.

Maxilla (g) proximal endite reduced to small rounded projections; distal small, unequally bilobed; endopod slender and pointed at tip from very wide base; anterior lobe of scaphognathite subtruncate with slight distolateral concavity at outer edge, slightly longer than axe-shaped wide posterior lobe.

Maxilliped I (h) distal endite wide rounded, larger than thick-edged proximal endite; endopod wide foliaceous, twisted into second plane and inverted triangle shape distally; exopod pointed apically, with long lash; epipod extremely wide at base, tapering to long ribbonlike tail with long lateral setae.

Maxilliped II (i) endopod pediform, somewhat compressed; longer than tapered exopod, the latter with long flagellum; epipod ribbonlike, with long podobranch.

Maxilliped III (k) operculiform; ischium wide inner edge with series of rounded teeth; merus wide and almost as long as ischium, carpus attached at inner distal angle;

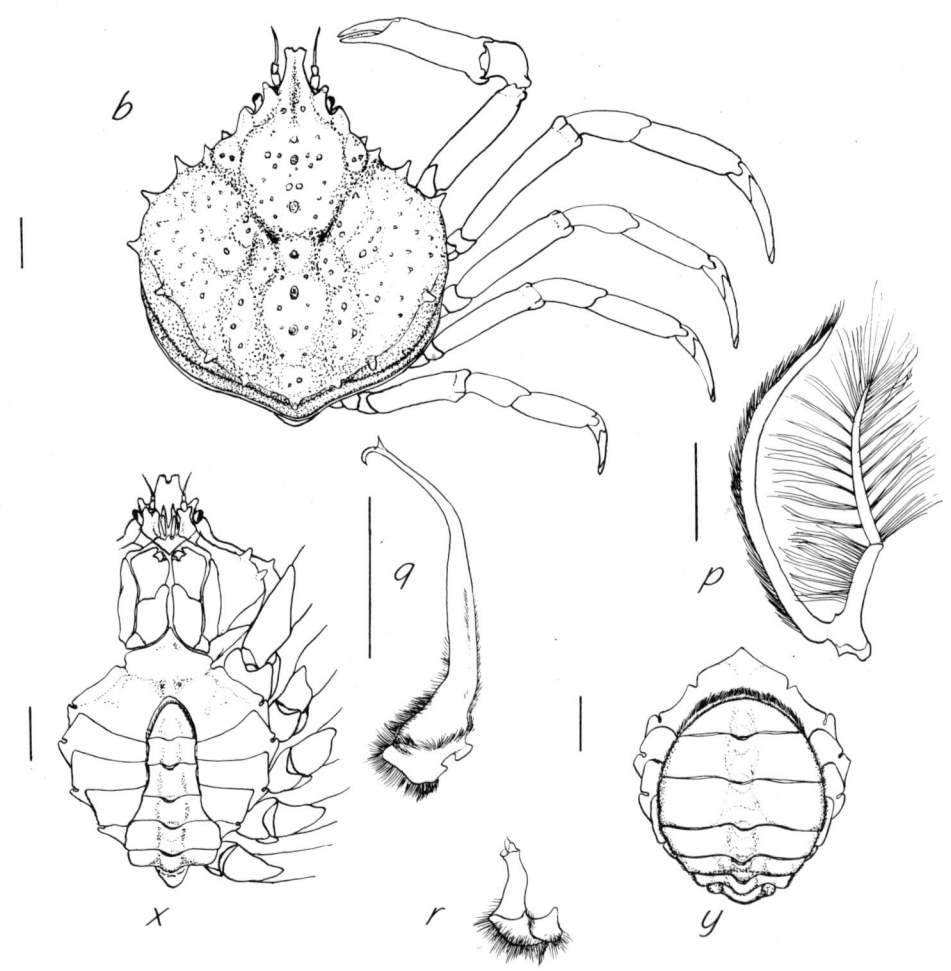

Fig. 238. *Libinia emarginata*: *b*, carapace in dorsal aspect; *p*, F second pleopod; *q*, M first pleopod; *r*, M second pleopod; *y*, F abdomen; *x*, M in ventral aspect; solid line = 10 mm.

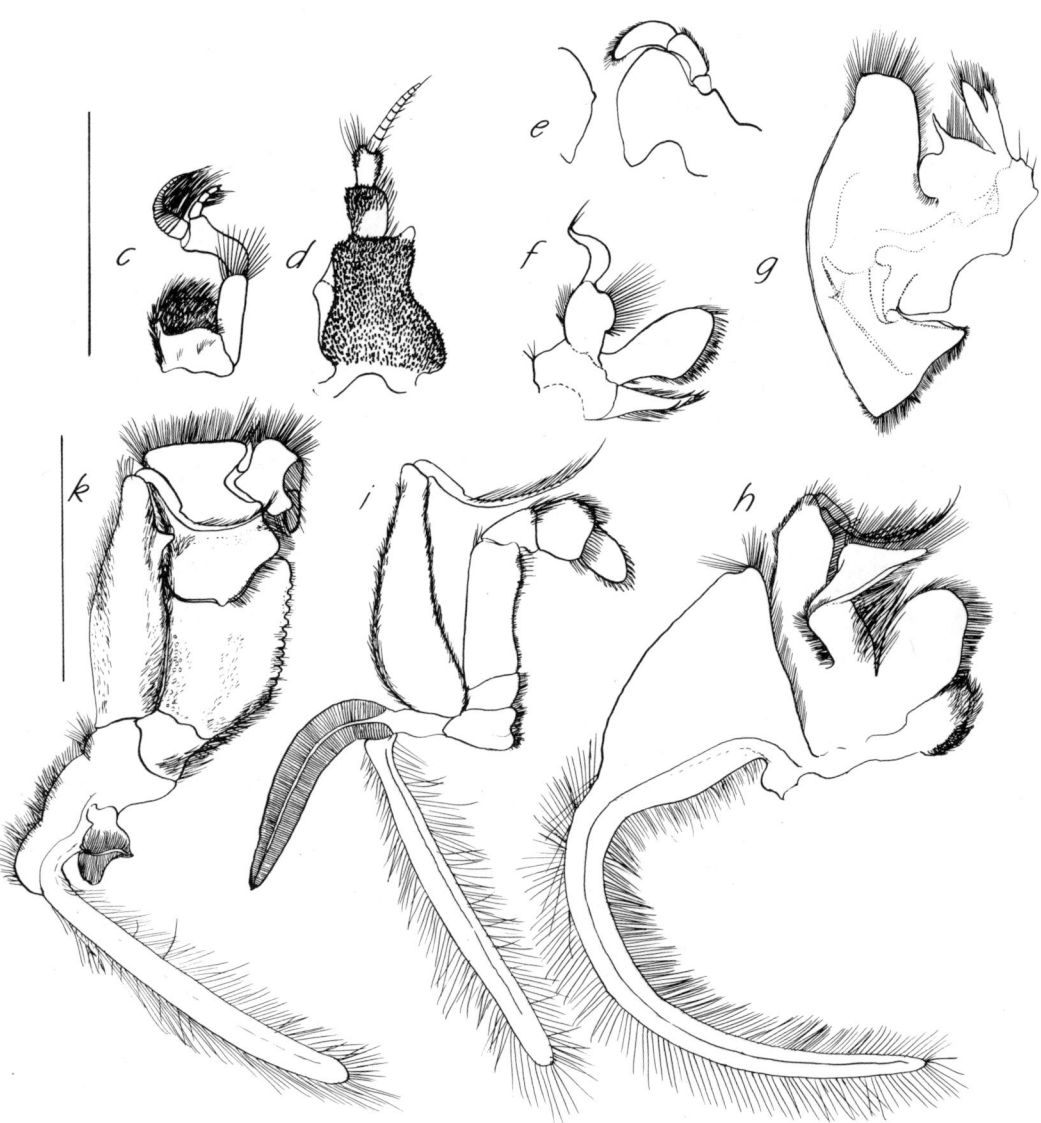

Fig. 239. *Libinia emarginata*: c, antennule; d, antenna; e, mandible; f, maxillule; g, maxilla; h, first maxilliped; i, second maxilliped; k, third maxilliped; solid line = 10 mm.

exopod large, rigid, with inner distolateral tooth; epipod long narrow, with short podobranch.

Pereopods: I chelate, stouter and shorter than others, fingers less than one-half palm; II–V non-chelate, in decreasing length posteriorly, dactyl stout, chitinous-tipped, blunt.

Pleopods: female II (p) endopod in two segments, with tufts of long setae; III–V similar to II. Male I (q), stout, long, tapering distally to narrow neck and apex with strongly curved tip on one side and distolateral fin on the other; male II (r) short, stout, tip with curved lip and central projection.

RANGE OF DISTRIBUTION

Prince Edward Island and Nova Scotia to western Gulf of Mexico. Depths 0–49 m, occasionally to 124 m (Williams 1984).

Records of distribution in the area of reference are in Fig. 240.

BIOLOGY

Lengths of males to 124 mm (width 124 mm) and of females 69 mm (width 66 mm).

Ovigerous females were ready to spawn after hatching of larvae. When first laid, eggs are orange-red in colour but become brown before hatching. Larval development is abbreviated with two zoeal stages and a megalopa reached in about 3 weeks at 20° C and 30‰ salinity (Williams 1984).

Mating of the female occurs in the hard shell condition.

Feeding on sea stars (*Asterias* sp.) has been observed (Williams 1984). It is also said to scavange on fish remains (M. J. Dadswell, personal communication).

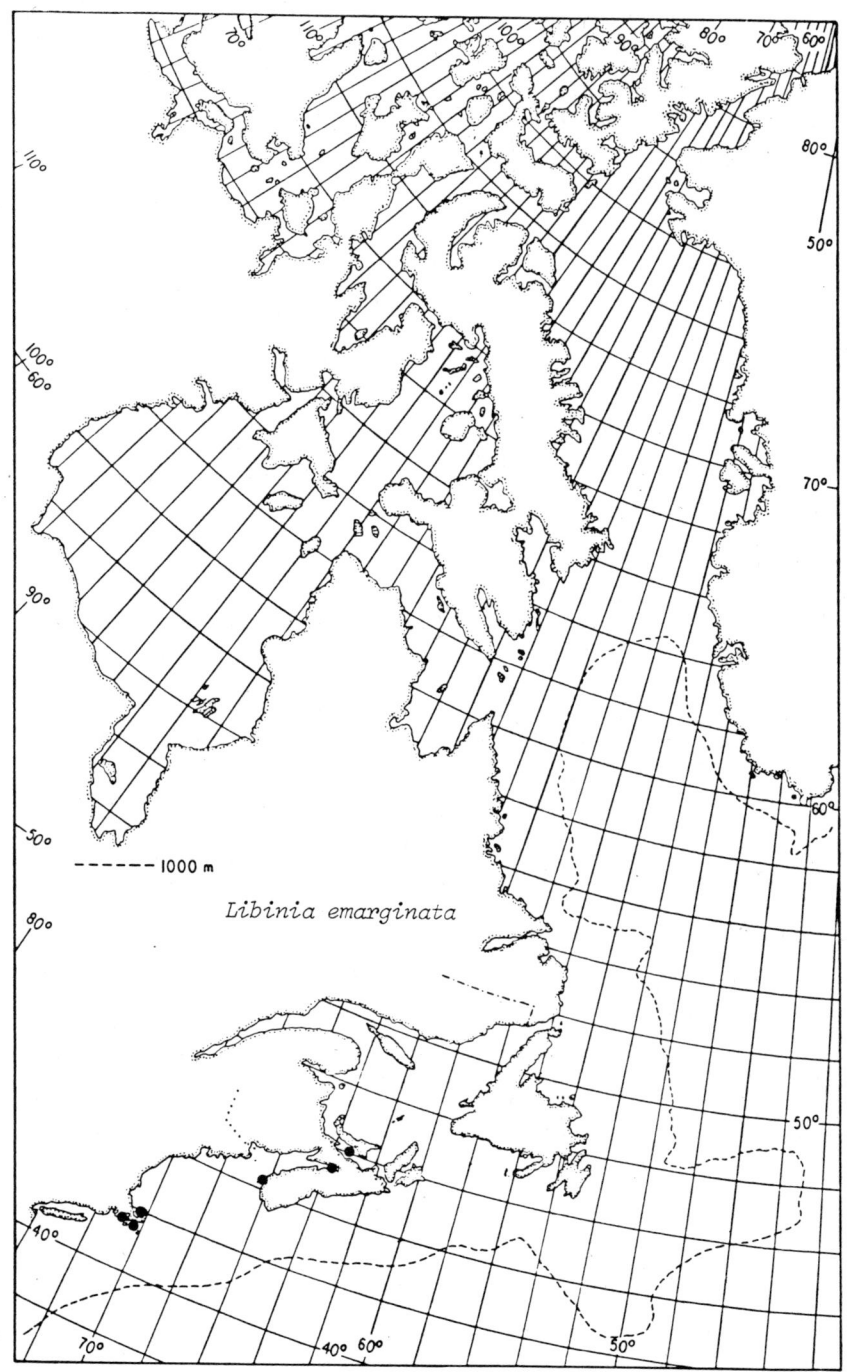

Fig. 240. *Libinia emarginata*, distribution records in the area of reference.

Family CANCRIDAE Latreille, 1803
Nations 1975: 30; Rathbun 1929: 32; Williams 1984: 351.

Carapace broadly oval or hexagonal, the front with several teeth one of which is median; antennules folded obliquely or almost vertical (lengthwise); antennal flagellum setose.

Key to species of the Family Cancridae in the area of reference

1   Anterolateral teeth of carapace with denticulate margins; upper surface of chelae denticulate .... *Cancer (Metacarcinus) borealis*

   Anterolateral teeth of carapace with granulate margins; upper surface of chelae granulate .... *Cancer (Cancer) irroratus*

Genus *Cancer* Linnaeus, 1758
Nations 1975: 30; Rathbun 1929: 32;
1930: 182; Williams 1984: 351.

Carapace ovate, wider than long; anterolateral margin convex, divided into many teeth; front between eyes narrow with few teeth; orbits small with two fissures in both upper and lower margins; eyestalks short; chelipeds subequal.

Cancer (Metacarcinus) borealis Stimpson, 1859
Nations 1975: 45, figs. 4, 38–2, 38–3;
Williams 351, fig. 287.
Cancer borealis, Rathbun 1929: 33, fig. 44; 1930: 182, fig. 30.
"Jonah crab" — "Crabe Jonah"
(Figures 241 and 242)

DISTINGUISHING CHARACTERISTICS

Antero-lateral margins more strongly arched than in *irroratus*, teeth denticulate; upper surface of chela denticulate; frontal centre tooth larger than in *irroratus*. Crabs large, robust.

DESCRIPTION

Integument heavily calcified, rigid and covered with small round even tubercles or granules. Colour reddish above and pale yellow or buff below.

Carapace front (between eyes) with 3 teeth, the centre one longest and slightly depressed; anterolateral margin widely curved, almost semicircular, with 9 crenulate (denticulate) teeth, and between each pair a short shallow fissure; posterolaterals two only, the second not clearly marked; slightly inflated in gastric and cardiac areas and generally convex dorsally; posteriorly the border has an emarginate even tuberculate ridge ending at intestinal area. Sub-branchial area under edge of carapace heavily setose.

Abdomen narrowish, elongate, forming in outline a dome with spire in female, the penultimate somite largest, the telson triangular with sides slightly concave; in male the 3rd to 5th somites fused, very slightly tapered, telson triangular.

Eye with short setose stalk and large cornea, completely retractable in orbit; orbit with two closed fissures above and two open below.

Antennule (c) first article subspherical (exposed part triangular), 2nd and 3rd slender and subequal; folding slightly obliquely (almost vertically) in socket.

Antenna (d) basal article massive, granulate, spiny distolaterally; other two segments tiny by comparison.

Mandible (e) heavily calcified, slightly curved cutting edge, smooth in left but with central tooth in right, lower corner projecting (more pronounced in left); molar short and with rounded boss at one end; 3-segmented palp setose.

Maxillule (f) proximal endite narrow tapered to point with curved spines; distal endite expanded greatly from narrow base; endopod wide foliaceous, proximally rounded, distal part narrower, curved in a second plane with single apical spine.

Maxilla (g) proximal endite bilobed, unequal, both narrow and widely separated; distal endite with both lobes short from long base; endopod tip curved spinelike from very wide base; anterior lobe of scaphognathite very wide, slightly pointed, about as long as axe-shaped posterior lobe.

Maxilliped I (h) distal endite flat and curved with straight inner edge, much larger than thick, projecting proximal endite; endopod foliaceous expanding slightly toward flattened top twisted into second plane; exopod longer than endopod, with long lash; epipod very wide at origin but tapering to long ribbonlike heavily setose tail.

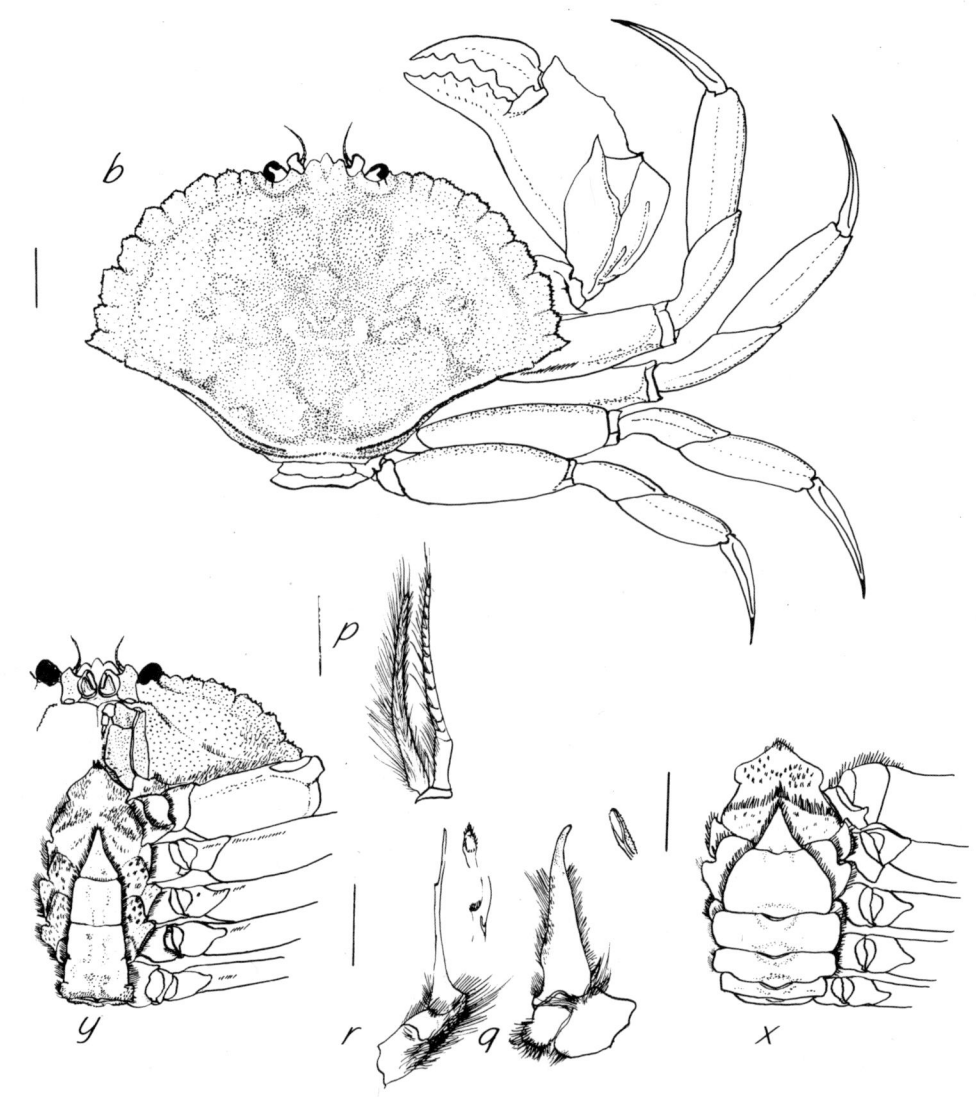

Fig. 241. *Cancer borealis*: b, carapace in dorsal aspect; p, F second pleopod; q, M first pleopod; r, M second pleopod; x, Female abdomen; y, M in ventral aspect; solid line = 10 mm.

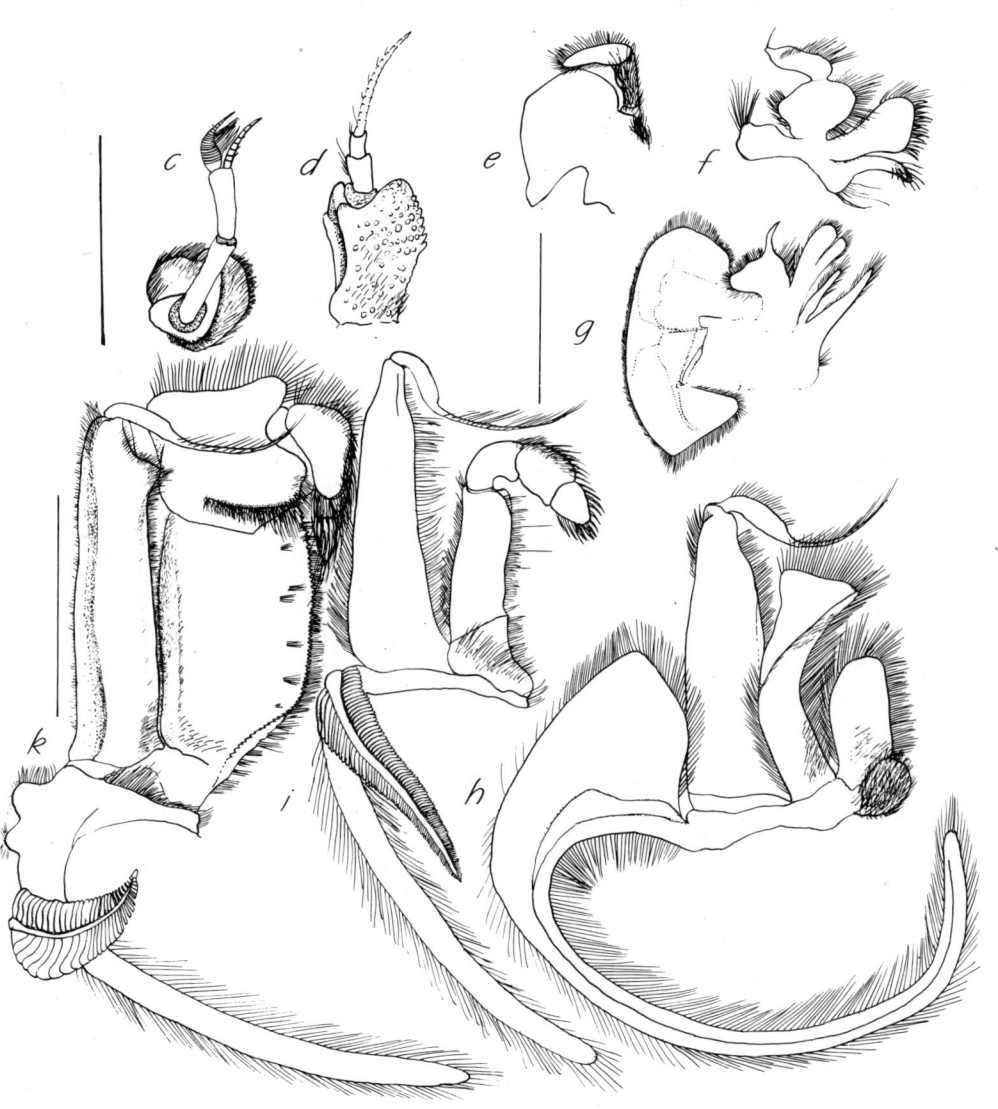

Fig. 242. *Cancer borealis*: *c*, antennule; *d*, antenna; *e*, mandible; *f*, maxillule; *g*, maxilla; *h*, first maxilliped; *i*, second maxilliped; *k*, third maxilliped; solid line = 10 mm.

Maxilliped II (i) pediform, distal segment with strong spines; exopod tapered, about as long as endopod, with long lash; epipod long ribbonlike, with podobranch.

Maxilliped III (k) operculiform, ischium wide and long, with sagittate inner edge; merus wide but only about half as long as ischium; exopod stout, rigid, with large distolateral inner projection and slender lash; epipod long, ribbonlike, semirigid, with podobranch.

Pereopods: I chelate, longer than II, propodus with 4 granulate ridges, dorsally somewhat denticulate, dactyl with one; carpus with ridges forming triangular area distally; II–V progressively shorter posteriorly but almost equal, compressed, dactyls stout but sharp.

Pleopods: female II (p) biramous, styliform, endopod with two segments the distal annulated, slightly longer than setose exopod; female III–V similar to II; male I (q) stouter and longer than II, tapering, grooved to open tip, distal area with many retrorse denticles, many plumose setae laterally; male II (r) slender, hollow tip drawn out from lateral shelf with spines, tip open with fringe of denticles.

RANGE OF DISTRIBUTION

Grand Banks to south of Tortugas, Florida. Depths 0–800 m (Williams 1984).
Records of occurrence in area of reference are in Fig. 243.

BIOLOGY

Lengths to 180 mm cw in males and 152 mm cw in females. Up to 180 mm cw and 0.4 kg (Elner 1985).

Females were first mature at about 21 mm cw, carrying 4 000 eggs, but in later development at 88 mm cw they carry about 330 000 eggs. New eggs are orange-red from November to May and are brown and ripe from March to June, but individuals carry eggs for about 5 months. Mating is during soft-shell condition of the female (Williams 1984).

Growth increments in females are less than in males after first maturity, reaching 100 mm cw in 8 years after 14 moults, while males reach 130 mm cw in 6–7 years after 13–14 moults (Williams 1984)

FISHERY

Intermittent landings as by-catch by lobster fishermen have occurred since the 1960s in the Bay of Fundy and off southern Nova Scotia. In 1983 landings amounted to 200 tons, including 90 tons from a directed pilot fishery on the Scotian Shelf (Elner 1985).

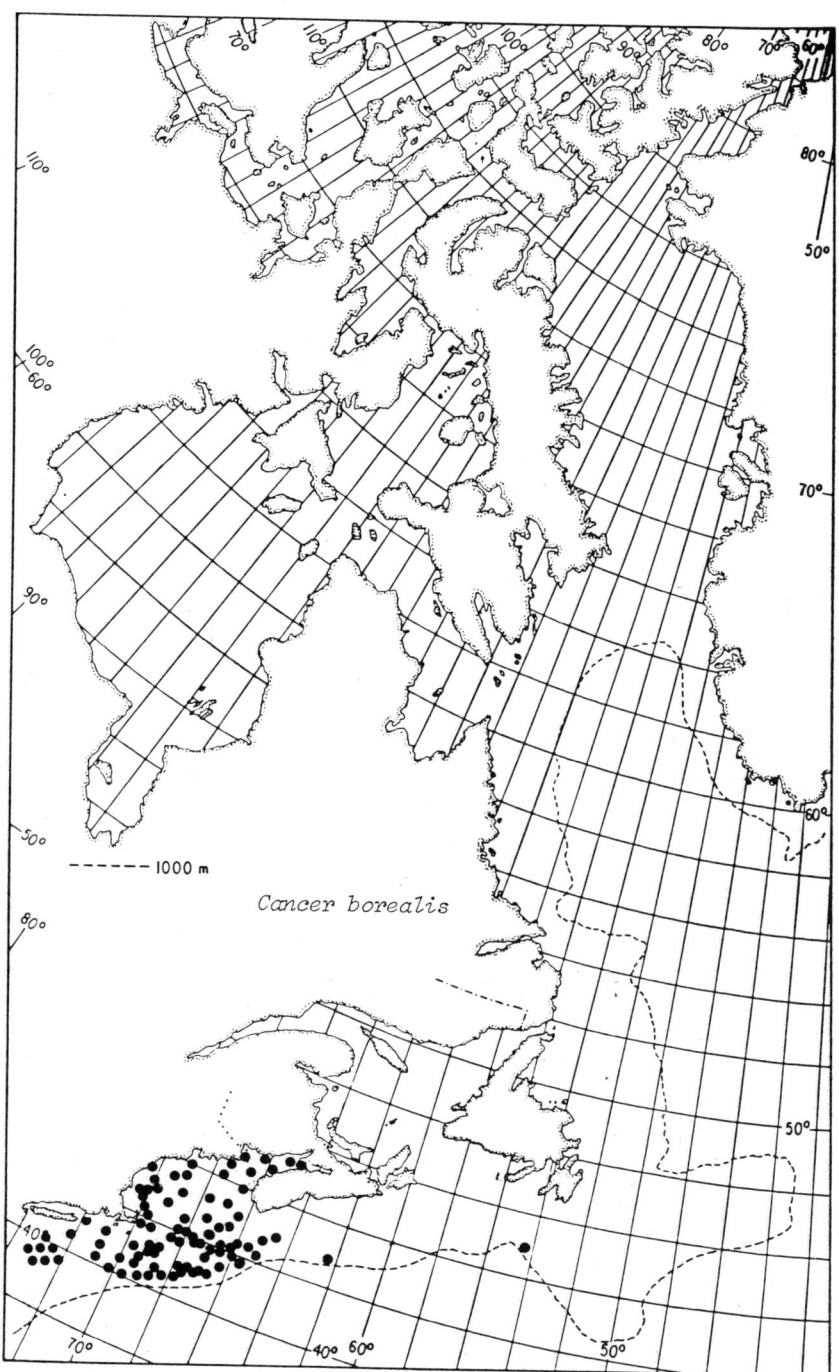

Fig. 243. *Cancer borealis*, distribution records in the area of reference.

Cancer (Cancer) irroratus Say, 1917
Nations 1975: 45, figs. 4, 42–3, 42–4;
Rathbun 1930: 180, fig. 29, pl. 85, fig 1;
1929: 32, fig. 43; Bigford 1979: 1, fig. 1;
Williams 1984, fig. 288.
"Rock crab" — "Crabe de roche"
(Figures 244 and 245)

DISTINGUISHING CHARACTERISTICS

Anterolateral teeth of carapace with entire or granulate edges; frontal outline wider as an arc of a circle rather than a semicircle; chelipeds granulate rather than denticulate as in *C. borealis*; basal article of antenna smooth or granulate and not denticulate.

DESCRIPTION

Integument heavily calcified, somewhat brittle, lightly granulate to smooth. Colour variations considerable in shallow areas with seaweed and rocks from brick red in females to dark brown in males. Pale yellow or buff ventrally.

Carapace with nine anterolateral teeth with even edges, granulate only; second posterolateral tooth not clearly marked in small specimens examined; gastric area slightly convex with three rounded areas at each side, median clear inverted V medially behind small rostral point; cardiac area slightly inflated, separated from gastric area by clear markings; posteriorly an emarginate granulate edge is interrupted in intestinal area. Sternal area with longer setae and generally more setose than in *C. borealis*.

Abdomen elongate in male with 3rd to 5th somites fused; 6th largest in female, dome-like with triangular telson.

Eyes on short thick stalks, rounded cornea, retractible in small rounded orbits; two closed fissures dorsally and two open ventrally.

Antennule (c) area of rounded 1st article exposed subtriangular; 2nd article longer than 3rd; folding slightly obliquely in socket.

Antenna (d) basal article massive, smoothish, small granules distolaterally; distal segments tiny.

Mandible (e) edge of incisor slightly curved, lower corner only slightly pointed; palp with 3 segments.

Maxillule (f) proximal endite tapered from rounded base; distal expanded from narrow base; endopod with basal segment as wide as long, inner edge expanded, distal segment slender curved, apically pointed and setose.

Maxilla (g) distal endite unequally bilobed, lobes short; proximal endite with long narrow separate lobes; endopod short and pointed from very wide base; anterior lobe of scaphognathite stubby, but body divided equally with short axe-shaped posterior lobe.

Maxilliped I (h) flat distal endite much larger than short thick proximal endite; endopod foliaceous, distally twisted into second plane and end slightly concave; exopod longer than endopod, forming distal neck, with long lash; epipod foliaceous, very wide at base and tapering to very long tail with many lateral setae.

Maxilliped II (i) pediform, distal segment with short strong spines; exopod long forming short distal neck, with long lash; epipod long, ribbonlike, with podobranch.

Maxilliped III (k) operculiform, ischium wide and long, with smooth inner edge; merus as wide but not as long as ischium; exopod not as long as both the former, massive, rigid, with distolateral inner projection; epipod long, ribbonlike but semirigid, with podobranch.

Pereopods: I chelate, carpus with ridges and reticulate pattern above, almost as long as II; II–V compressed, slightly decreasing in size posteriorly.

Pleopods: female II (p) endopod in two segments, the distal annulated, exopod not quite as long, sharply pointed and setose; female III–V similar. Male I (q) broad at base

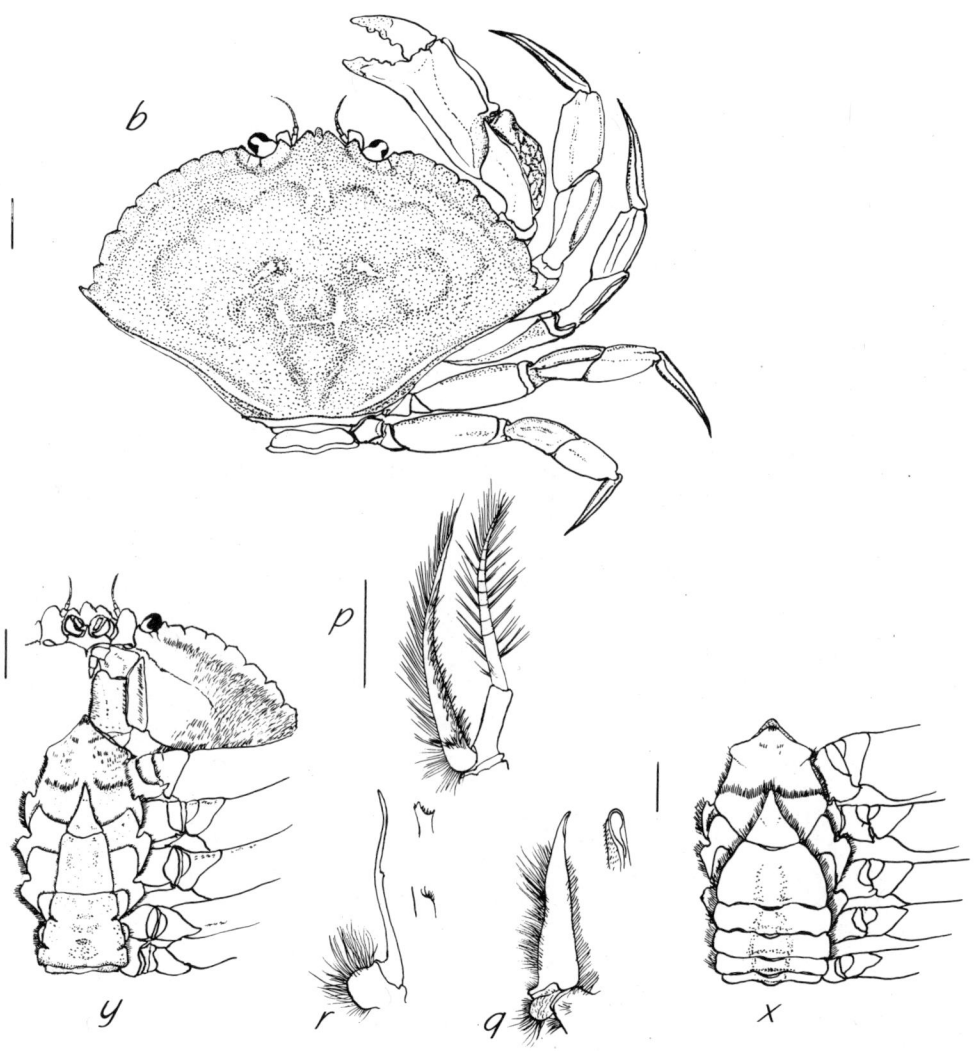

Fig. 244. *Cancer irroratus*: b, carapace in dorsal aspect; p, F second pleopod; q, M first pleopod; r, M second pleopod; x, Female abdomen; y, Male in ventral aspect; solid line = 10 mm.

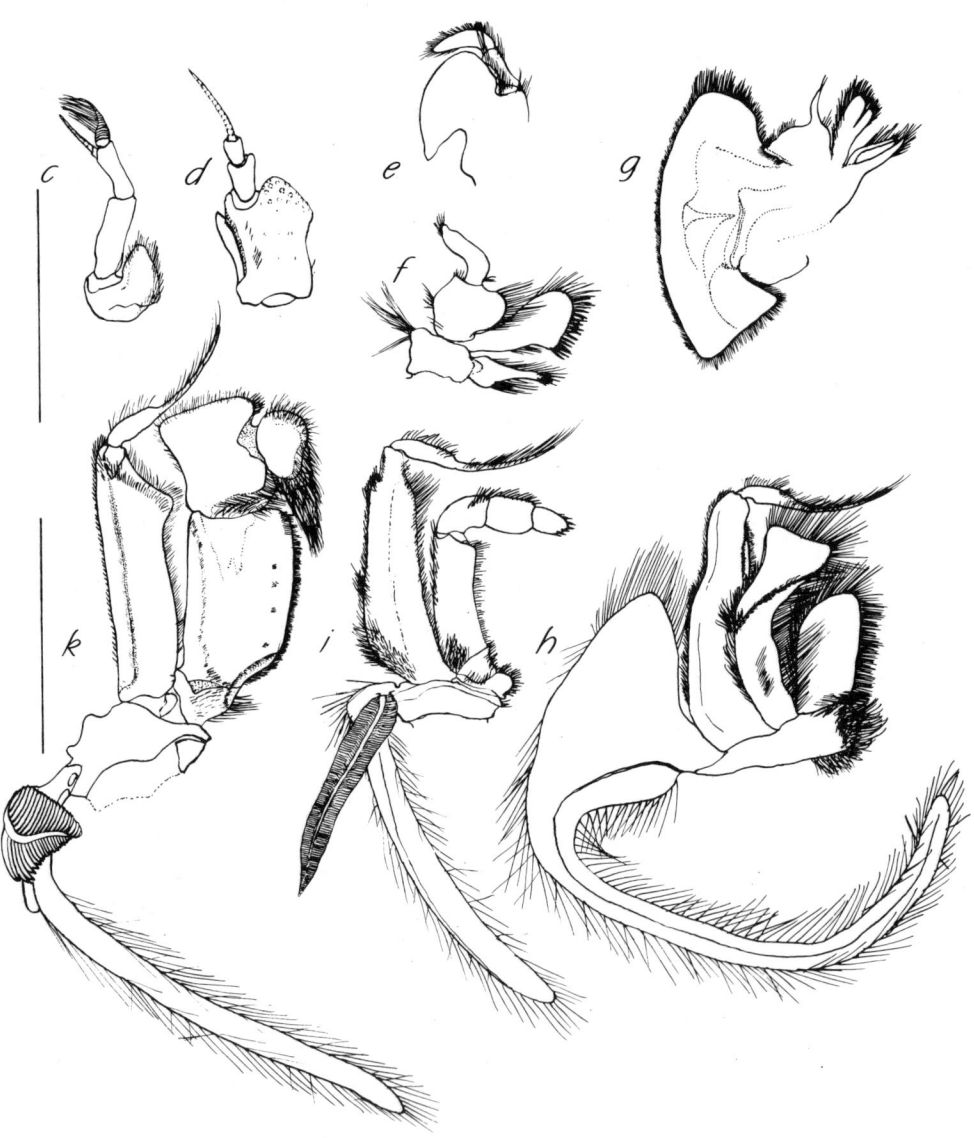

Fig. 245. *Cancer irroratus*: c, antennule; d, antenna; e, mandible; f, maxillule; g, maxilla; h, first maxilliped; i, second maxilliped; k, third maxilliped; solid line = 10 mm.

and tapering to hollow apex, the distal end of an almost closed groove, with short retrorse denticles distally, laterally setose; male II (r) very slender, with shoulder slightly more than halfway and long slightly curved neck appearing hollow, with an apical fringe of about 15 spinules.

RANGE OF DISTRIBUTION

Strait of Belle Isle in Gulf of St. Lawrence to Miami, Florida. Depths 0–575 m (Williams 1984).

Records of occurrence in the area of reference are in Fig. 246.

BIOLOGY

Size 82 mm cl, 125 mm cw, in males and 62 mm cl, 91 mm cw, in females. Up to 140 mm cw and 0.25 kg (Elner 1985).

Size at first maturity 55–60 mm cw in females and 70 mm cw in males from Northumberland Strait, but from populations farther south females were ovigerous at carapace widths of 14–25 mm at Rhode Island and 30 mm at Virginia. Bright orange-red eggs were carried from late fall to late spring and these turned brick red and pale grey brown before hatching in summer. Fecundity estimates for females 21–88 mm cw were 4 000 to 330 000 eggs. Mating takes place when the female is in the soft-shell condition (Bigford 1979).

Based on field collections from Rhode Island waters, age at growth stages were as follows: ages 1 and 2 years 14 mm and 40 mm cw in males and females, but afterwards males grew faster than females and were 80 mm at 4 years and 140 mm cw at 7 years, compared with 61 mm at 4 years and 89 mm cw at 7 years in females (Reilly and Saila 1978).

Predators were several species of inshore fishes and the lobster, *Homarus americanus* (Bigford 1979; Squires 1970).

Stomach contents observed were small *Littorina*, amphipods, *Crangon*, polychaetes and small bivalves (Squires 1965a).

FISHERY

By-catches of this species in the lobster fishery prior to 1973 were not likely to have exceeded 230 tons per year in Canada. However, an occasional directed fishery since that date has resulted in landings of about 500 tons in 1983, for example, and sustainable yields from the southern Gulf of St. Lawrence have been estimated at 1 360 to 2 270 tons. The fishery has not been developed to this extent, however, because of the small size, low meat yield and high processing costs of this species (Elner 1985). Approximately 45 crabs are required for a meat yield of 1 kilogram, using crabs of 89 mm cw or over (Bigford 1979).

Some landings of this species occur in the New England area from by-catches of other inshore fisheries, but they are not separated in statistics.

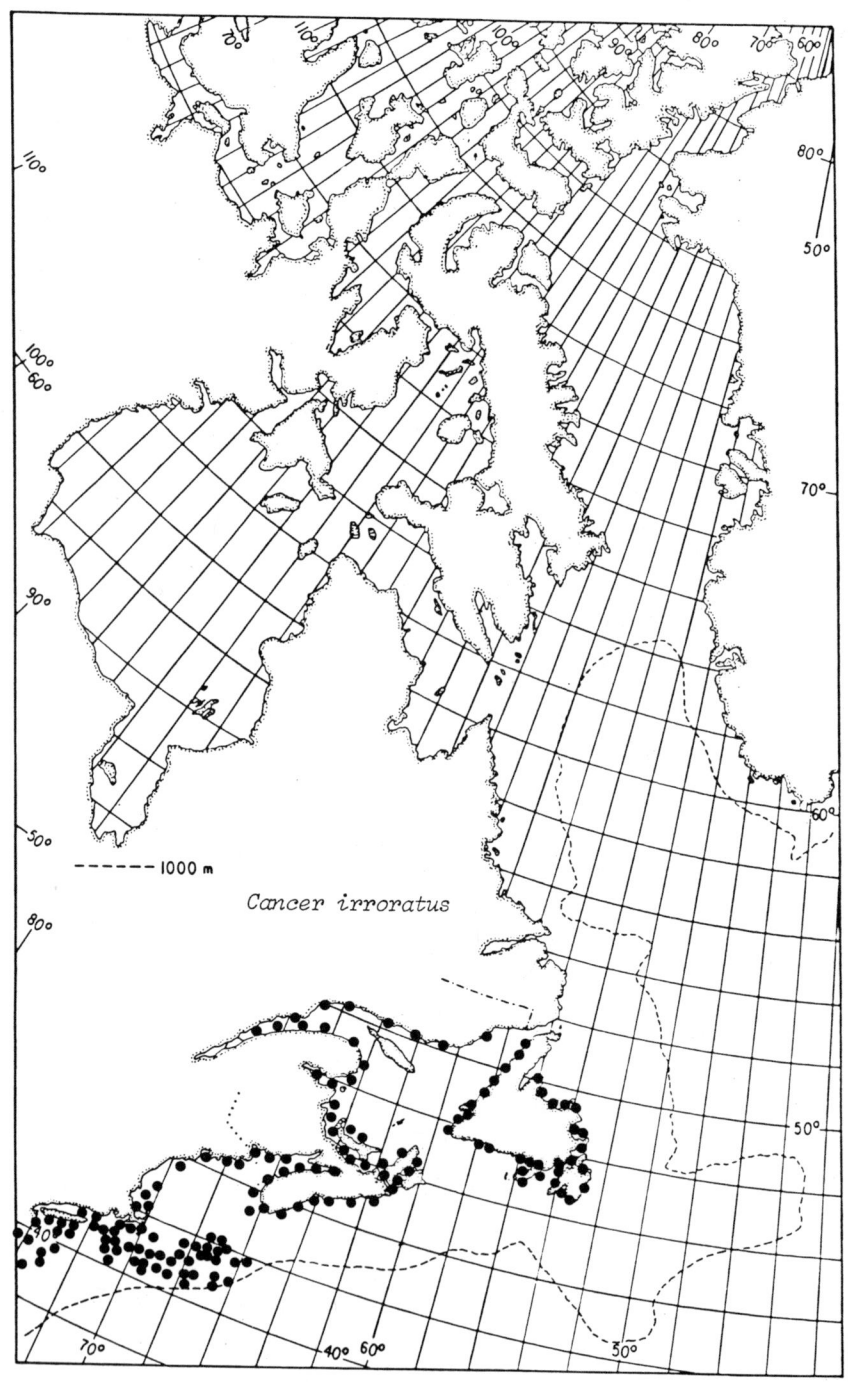

Fig. 246. *Cancer irroratus*, distribution records in the area of reference.

Family GERYONIDAE Colosi, 1924
Ingle 1980: 110.

Carapace hexagonal to trapezoidal, front wide, anterolateral margins with three to five spines; antennules fold back obliquely; fifth pair of pereopods longer than chelipeds; male genital openings coxal.

Genus *Geryon* Kroyer, 1835
Rathbun 1929: 34; 1937: 265;
ZariquieyAlvarez 1968: 388.

Carapace wider than long, hexagonal, antero-laterally arcuate with marginal spines, strongly convex from front to back, posteriorly truncate; front narrow, prominent with two median rostral spines; orbits large; buccal cavity widening anteriorly, not covered entirely by third maxillipeds; abdominal somites not fused.

Geryon quinquidens Smith, 1897
Rathbun 1929: 35, fig. 47;
1937: 271, pls. 85, 86.
"Deep-sea red crab" — "Crabe rouge de mer profonde"
(Figures 247 and 248)

Distinguishing Characteristics

Antero-lateral edge of carapace wide, with 5 spines at each side; carapace hexagonal; front narrow with two teeth forming rostrum and a less prominent tooth at each side; chelae large, walking legs much longer than chelipeds; colour dark red.

Description

Integument heavily calcified, smooth on anterolateral quadrants, finely granulate elsewhere. Colour dark red.

Carapace subhexagonal, convex, front wide, slightly descending, with 5 (3–5) anterolateral spines at each side, the posterior most prominent; two median rostral teeth protrude farther than the large orbital tooth which is in advance of the infraorbital and the strong suborbital; orbital fissure very short and faint dorsally, edge denticulate ventrally; proto-, meso- and metagastric areas are well defined by grooves and are slightly inflated; cardiac area also well-defined with cardiac markings and grooves; posterolateral margins converge slightly to subtruncate posterior margin, formed by three slightly concave intersecting lines, the longest at the centre. Carapace length about 76% of width (between tips of spines).

Sternum ovate: width about 0.8 times length; anterior portion with forward attenuate point and lateral slots for chelipeds, about equal to posterior portion with three imbricate plates visible (+ one under reflexed abdomen) and lateral slots for 2nd to 4th legs.

Abdomen of male subtriangular, with all somites free, 3rd widest; of female very wide, the 5th widest.

Eyes large, cornea subglobular.

Antennule (c) 1st article short, subspherical, triangular as seen; other articles about equal, folded in socket obliquely.

Antenna (d) not fused with epistome or orbit, basal segment not much larger than other two, the distal shortest and cylindrical; flagellum moderately long.

Mandible (e) cutting edge of incisor smoothly curved in left but with prominent central tooth in right. Palp with 3 segments.

Maxillule (f) distal endite expanded widely from narrow base; proximal endite with wide base but becoming very narrow; first segment of endopod wide with lateral expansion, second segment sinuous slender inserted at outer corner of first.

Maxilla (g) endites unequally bilobed; endopod slender from wide base; anterior lobe of scaphognathite subtriangular, about equal in length to wide axe-shaped posterior lobe.

Maxilliped I (h) distal endite flat and curved, slightly larger than thick setose proximal endite; endopod large foliaceous, twisted into other plane distally and exceeding the exopod, the latter with long lash; epipod very wide at base but tapering to long ribbonlike tail with many lateral setae.

Maxilliped II (i) endopod pediform, stout, merus with series of spinules at inner edge, dactyl with strong spines; exopod tapered, with long lash; epipod long, narrow, with podobranch.

Maxilliped III (k) operculiform, ischium wide, with inner edge irregular and with few small teeth, not much longer than merus, also wide with insertion of carpus at distal inner corner; exopod stout, rigid, not as long as merus and ischium combined, with inner distolateral expansion and long lash; epipod long and narrow, with podobranch.

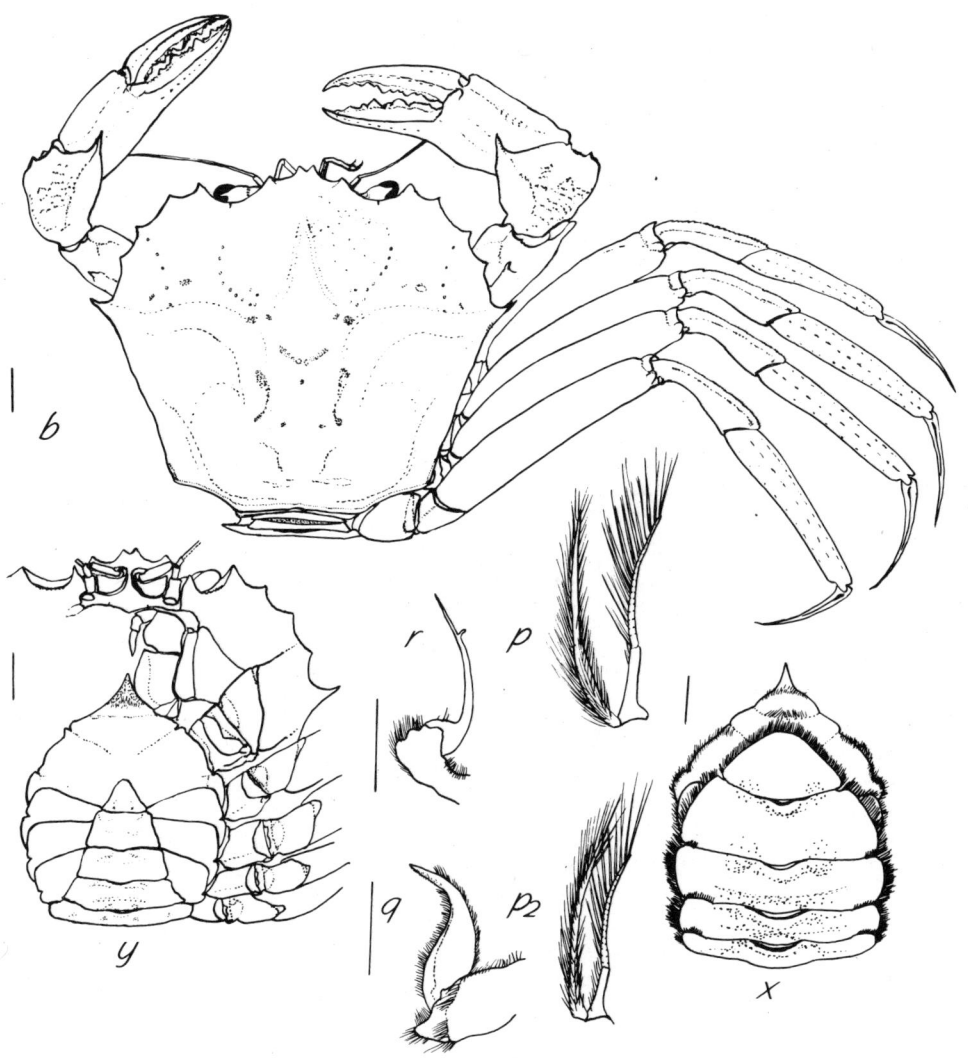

Fig. 247. *Geryon quinquidens*: b, carapace in dorsal aspect; p, F second pleopod; p2, F third pleopod; q, M first pleopod; r, M second pleopod; x, F abdomen; y, M in ventral aspect; solid line = 10 mm.

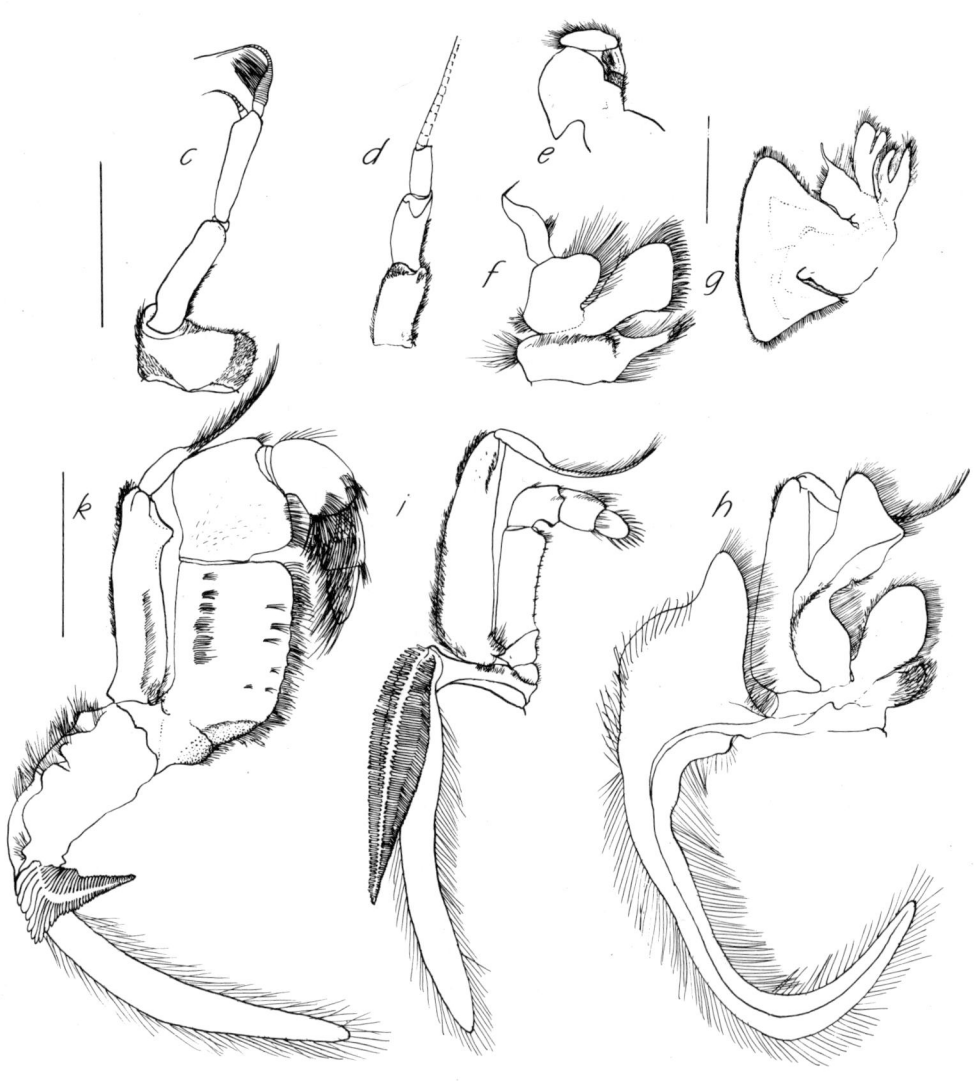

Fig. 248. *Geryon quinquidens*: *c*, antennule; *d*, antenna; *e*, mandible; *f*, maxillule; *g*, maxilla; *h*, first maxilliped; *i*, second maxilliped; *k*, third maxilliped; solid line = 10 mm.

Pereopods: I chelate, slightly unequal, the right larger, fingers longer than palm; carpus with strong inner distal spine, merus with dorsal spine. II–V very long, compressed, dactyl curved and sharp, outer edge of carpus denticulate.

Pleopods: female II (p) endopod with short proximal segment not annulated and long distal segment annulated and with long setae; exopod subequal to endopod, tapered and setose. Female III (p) to V similar to II. Male I (q) supported across 1st abdominal somite by yoke-shaped brace, robust and slightly curved and tapered distally, the apex hollow with retrorse spinules; male II (r) tapering from short, thick but compressed base to a long slender tip with a small tongue-like inner projection on distal fifth.

RANGE OF DISTRIBUTION

Western Atlantic only from south of Nova Scotia to South Carolina and possibly to Brazil. Depths 40–2 150 m (Schroeder 1959).

Records of occurence in the area of reference are in Fig. 249.

BIOLOGY

Width to 180 mm and weight 1.4 kg.

Females reached first maturity at 65–75 mm cl (80–91 mm cw) (Haefner 1977). In the annual reproductive cycle most females were ovigerous and ready to hatch larvae or had freshly extruded eggs in November, as shown by trawl surveys at Norfolk Canyon off Chesapeake Bight during a period of three years (Haefner 1978).

Most of the large females were captured in depths less than 600 m, while the larger males dominated the catch in deeper water (to 1300 m) (Haefner 1978).

FISHERY

Since 1973 annual catches from the continental shelf off New England at depths of 180–550 m have been from two fishing vessels landing at one plant (Elner 1985).

Exploratory fishing along the Scotian Shelf to the Fundian Channel gives an indication of a resource of 2 700 tons of large male crabs (Elner 1985).

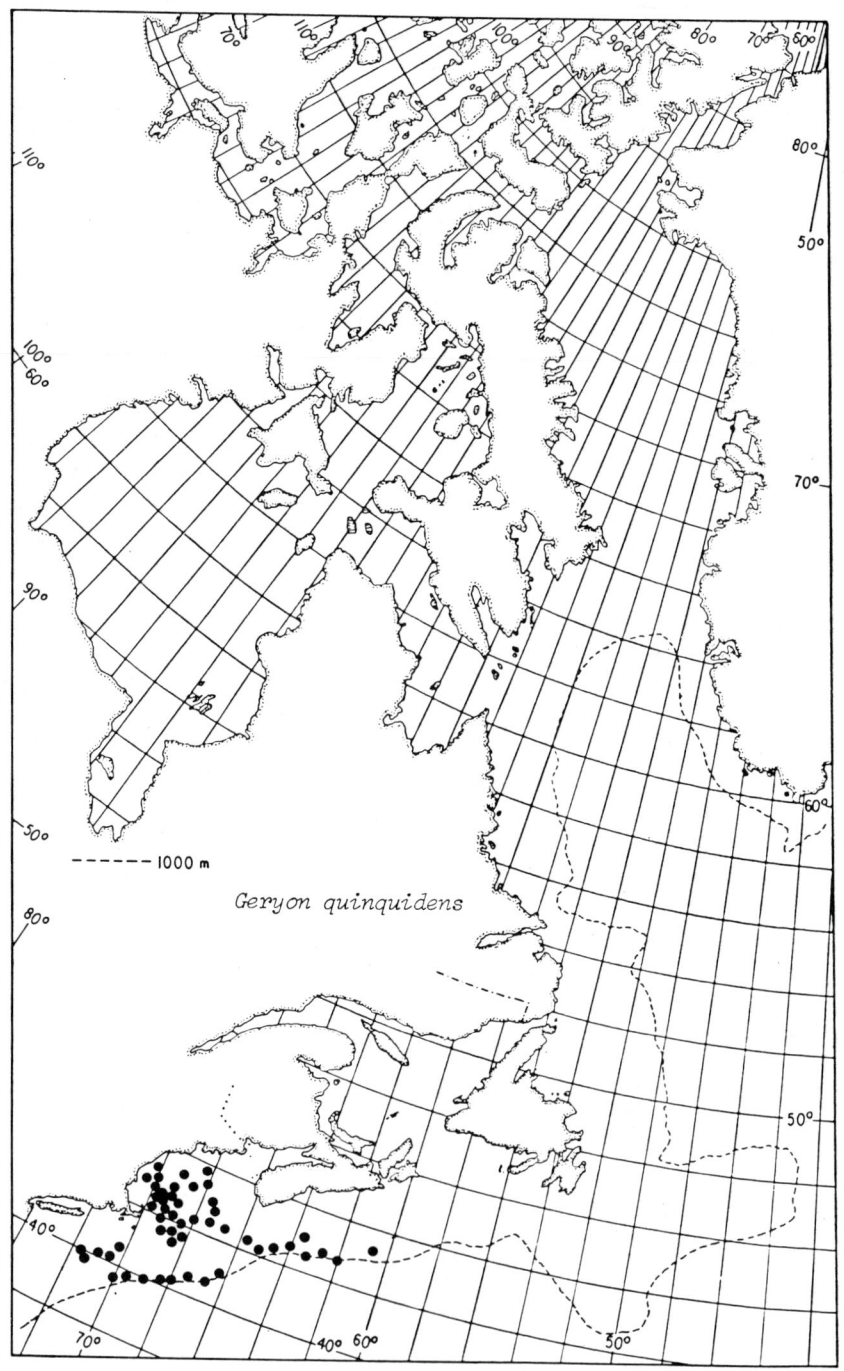

Fig. 249. *Geryon quinquidens*, distribution records in the area of reference.

Family PORTUNIDAE Rafinesque, 1815
Rathbun 1929: 30; 1930: 13; Williams 1984: 355.

Carapace much wider than long, greatest width marked by last antero-lateral spine; front dentate or lobate, anterolateral margin dentate; antennules folding obliquely or transversely; fifth legs with leaf-shaped or oval dactyl, dorsoventrally flattened, adapted for swimming.

Key to species of the family Portunidae in area of reference
(After Williams 1984)

1	Carapace with 5 teeth on anterolateral margin; lateral tooth equal to or not much larger than others	2
	Carapace with 9 anterolateral teeth; lateral tooth often much larger than others	3
2	Dactyl of 5th legs paddle-like	*Ovalipes ocellatus*
	Dactyl of 5th legs not paddle-like	*Carcinus maenas*
3	Carpus of cheliped with inner distal spine; abdomen of male triangular	*Portunus sayi*
	Carpus of cheliped without inner distal spine: abdomen of male inverted T-shaped	*Callinectes sapidus*

Genus *Callinectes* Stimpson, 1860
Rathbun 1929: 30; Williams 1984: 363.

Abdomen of male wide proximally and very narrow distally, roughly inverted T-shaped, third to fifth segments fused; no inner spine distally on carpus of chelipeds; anterolateral margins with nine spines the most posterior very large.

Callinectes sapidus Rathbun, 1896
Rathbun 1929: 31, fig. 41;
Williams 1984: 376, figs. 293g, 299.
"Blue crab" — "Crabe bleu"
(Figures 250 and 251)

DISTINGUISHING CHARACTERISTICS

Two frontal or inter-antennal teeth; width between tips of spines more than twice length of carapace; anterolateral teeth with sharp (or bluntish) acuminate tips directed outward; telson lanceolate, much longer than wide; propodus and carpus of cheliped with granulate ridges, width of chelae similar, propodal finger of major hand with lower margin decurved proximally.

DESCRIPTION

Integument heavily calcified, smooth or sparingly granulate. Colour greyish, bluish or brownish green of varying tints dorsally on carapace, spines with reddish tints, tubercles at articulation of legs orange and legs varying blue and white with traces of red or brownish green; palm of chelae of male blue and fingers blue on inner and white on outer surfaces tipped with red. Mature female with orange fingers tipped with purple. Ventrally off-white with tints of yellow and pink (Williams 1984).

Carapace with two frontal central teeth, the mesial slopes of which are longer than the lateral; nine anterolateral teeth, acuminate, the larger one at the lateral corner long and sharp-pointed, with a dorsal carina produced to the metagastric area.

Abdomen of male wide proximally, the 3rd to 5th somites fused and tapering shortly to join narrow 6th somite which is longer than 3rd–5th together, telson is narrow with attenuate tip. Female abdomen is triangular when immature but wide and ovate when mature, telson triangular narrow.

Eyestalk stout, cornea globular, brown and yellow striped, slightly narrower than stalk; dorsally two orbital fissures are closed.

Antennule (c) 1st article somewhat rounded and thick and of different colours; 2nd and 3rd articles very slender, retracted transversely into socket.

Antenna (d) basal segment short and wide and with a projecting wing; penultimate segment longer and stouter than distal segment; flagellum moderate, slender.

Mandibles (e) heavily calcified incisor edge slightly rounded forming a point at middle in left but a low tooth in right mandible; molar small, egg-shaped in outline; palp three-segmented.

Maxillule (f) distal endite narrow at base but widening toward curved edge; proximal endite thick, almost cylindrical but slender and pointed; endopod large, the proximal segment with inner lobe, the distal more slender, sinuous with attenuate tip.

Maxilla (g) endites unequally bilobed, the distal with expanded lobes at end of slender arm; proximal lobes separate at base, slender and long; slender pointed endopod from wide base; anterior lobe of scaphognathite wider than long, with irregular inner edge; posterior lobe subtriangular wide, with strong central rib.

Maxilliped I (h) distal endite flat and curved, larger than thick proximal endite; endopod widening and changing axis distally, inner distal corner lobate, shorter than exopod, the latter with short neck and long lash; epipod only moderately wide at base and tapering shortly to long slender ribbonlike blade.

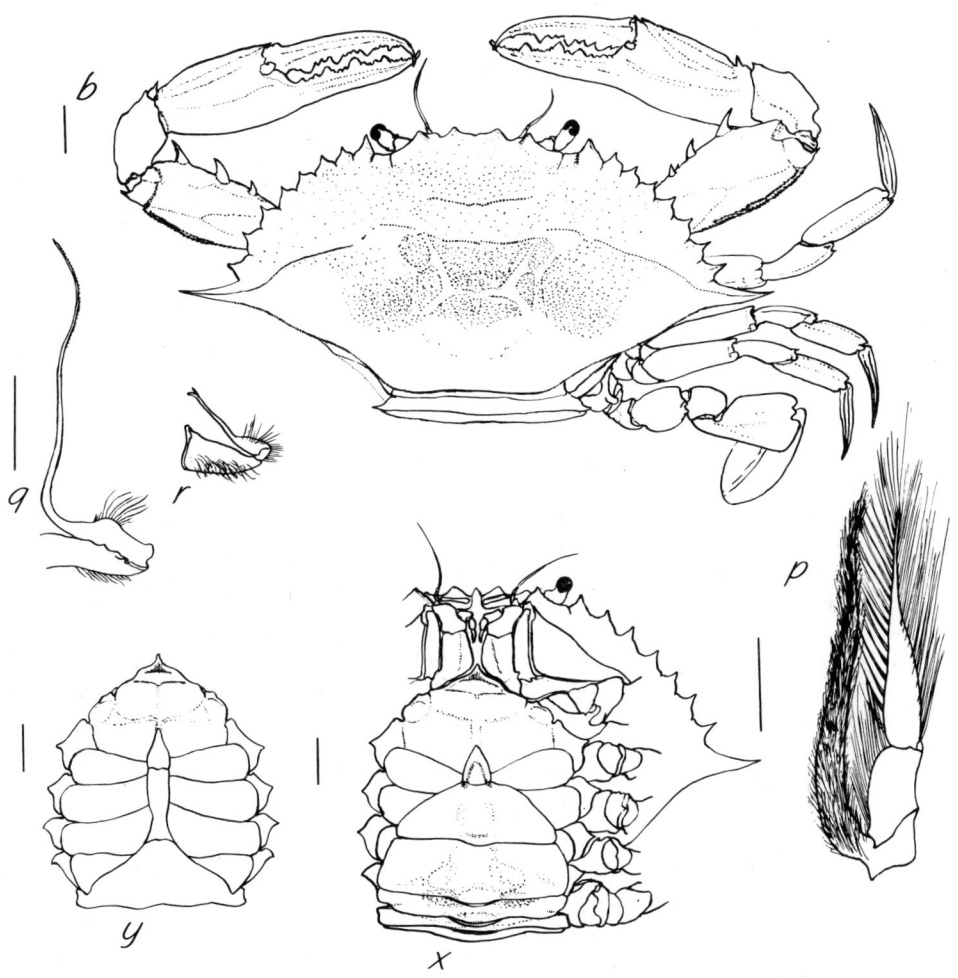

Fig. 250. *Callinectes sapidus*: *b*, carapace in dorsal aspect; *p*, F second pleopod; *q*, M first pleopod; *r*, M second pleopod; *x*, Female in ventral aspect; *y*, M abdomen; solid line = 10 mm.

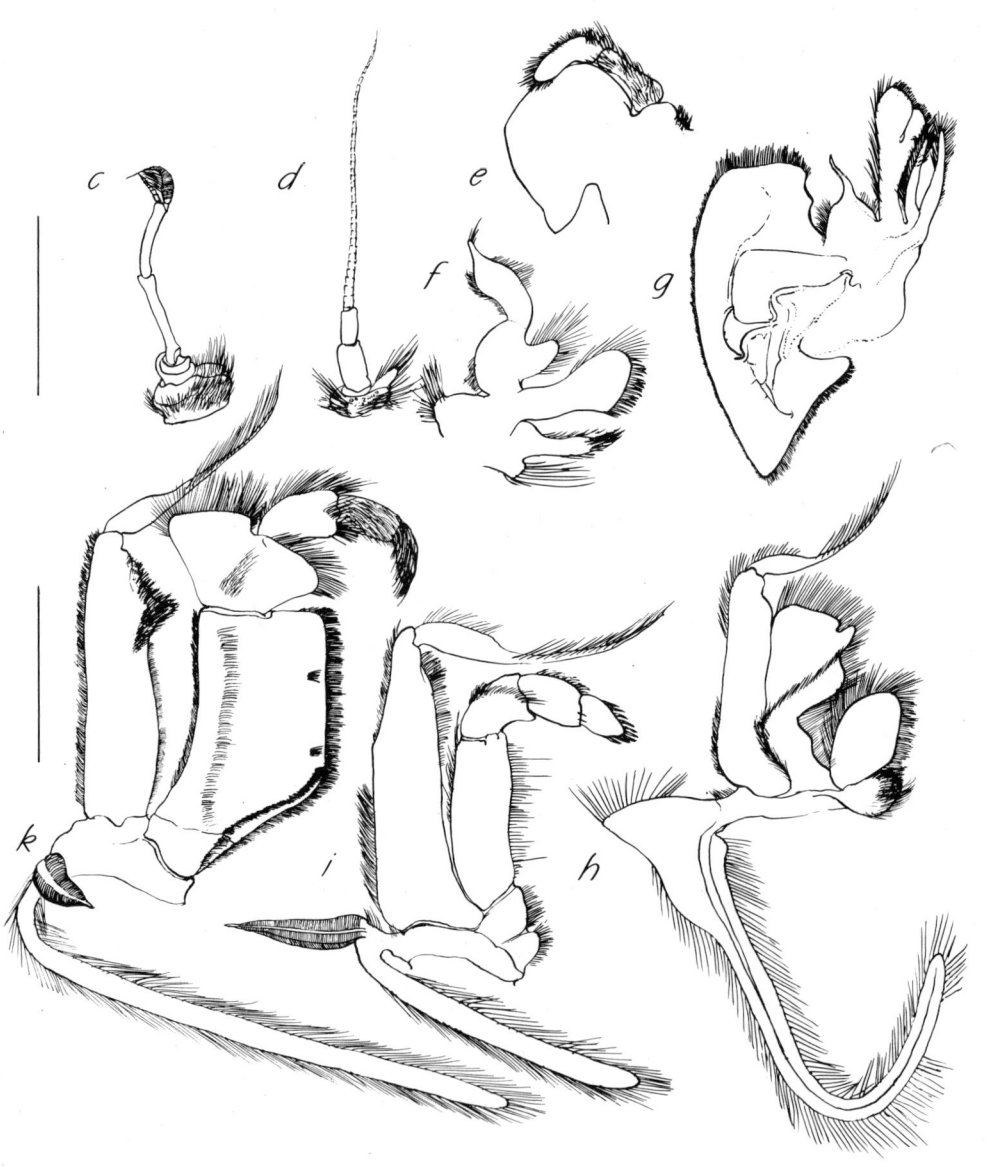

Fig. 251. *Callinectes sapidus*: *c*, antennule; *d*, antenna; *e*, mandible; *f*, maxillule; *g*, maxilla; *h*, first maxilliped; *i*, second maxilliped; *k*, third maxilliped; solid line = 10 mm.

Maxilliped II (i) endopod pediform, calcified; exopod slightly thicker but less calcified, with long lash; epipod not as long as that of I and III, with podobranch.

Maxilliped III (k) large, operculiform, heavily calcified, ischium wide and long, merus much shorter and narrow at join with ischium, inner edge pointed rather than straight and insertion of carpus at upper corner; exopod heavily calcified, with inner distolateral projection and long lash; epipod more than twice as long as that of II, with podobranch.

Pereopods: I chelate, very strong, with prominent ridges, fingers with strong irregular cutting teeth, about as long as palm, slightly unequal in shape but both large, merus with three very strong inner spines increasing in size distally; II–IV about equal, shorter than I, non-chelate, compressed; V more robust than IV with expanded and compressed propodus and dactyl, the latter ovate with central rib, paddle-like.

Pleopods: female II (p) endopod stout, in two segments, the distal with long attenuate tip; exopod sinuous heavily setose; female III–V similar to II. Male I (q) long, slender and curved, turning toward each other distally, with reflexed spinules on distal third outside, tip showing expansion of folds of groove which forms a half spiral around shaft as it ascends from bulbous base; male II (r) much shorter than I, tip unequally biramous, the thicker ramus with rosette of hooks.

Range of Distribution

Only occasionally north of Cape Cod in favourably warm periods as far as Nova Scotia and Maine. Otherwise from Cape Cod to Northern Argentina including the West Indies and Caribbean in the western Atlantic. Also in the eastern Atlantic where it has been inadvertently introduced and extends from the North Sea to the Black Sea and the Mediterranean. Depths 0–90 m (Williams 1984).

Records of occurrence in the area of reference are in Fig. 252.

Biology

Carapace widths including lateral spines to 209 mm in males (91 mm cl) and 204 mm in females (75 mm cl). Male crabs are significantly heavier than females for a given carapace width (Pullen and Trent 1970).

Although the species is tolerant of extremes it is mainly adapted to estuarine conditions. Its larvae are hatched near the mouths of estuaries and develop in oceanic water but very soon migrate into estuaries where they grow to maturity.

In Chesapeake Bay female crabs spawned in June of one year mature in about 14 months later and mate in the soft-shelled condition. Usually they spawn the following spring, and may spawn a second time during the summer, and a third time the following year. Under full tropical conditions spawning may be more frequent. The number of eggs at a spawning is estimated to be from 700 000 to 2 million. Eggs hatch in about 15 days at 26.1° C. (Williams 1984).

Three years is considered to be the normal maximum age.

Stomach contents observed include a variety of materials such as fishes, benthic invertebrates and plants (Williams 1984).

Fishery

The species is harvested throughout its range but the largest commercial catch is in the Chesapeake Bay area in the United States where it has been fished for more than 100 years. During the 1970s the average annual catch was about 65 000 tons with a landed value of about 20 million dollars (Williams 1984).

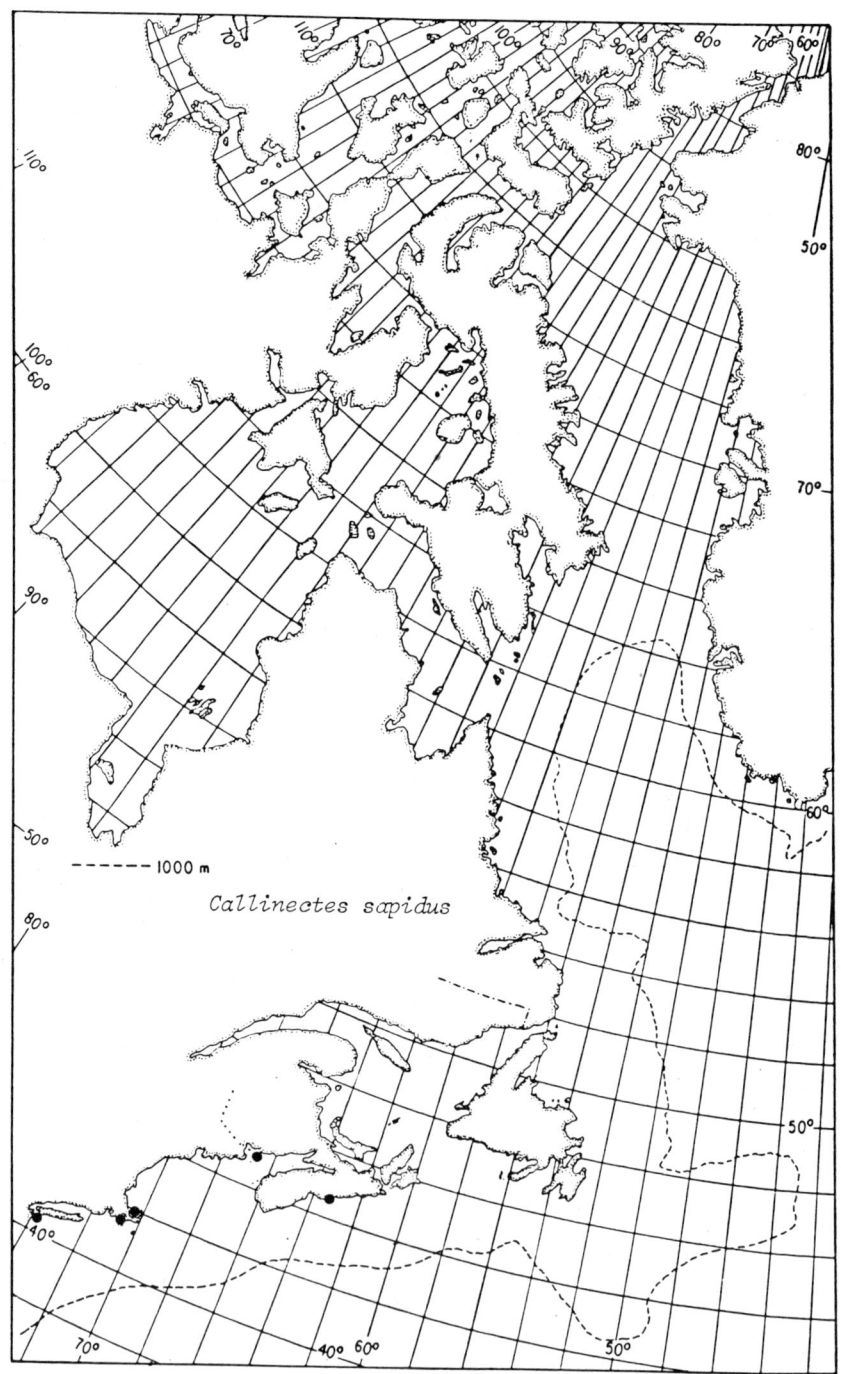

Fig. 252. *Callinectes sapidus*, distribution records in the area of reference.

Genus *Carcinus* Leach, 1814
Ingle 1980: 100; Williams 1984: 356.
Zariquiey Alvarez 1968: 353.

Anterolateral margins of carapace with 5 teeth, posterior lateral no larger than others; orbit with only one fissure dorsally; front of carapace only slightly projected with rounded rostral area; basal article of antenna fixed, longer than wide; 5th pereopods compressed but dactyl only slightly dilated not paddle-like.

Carcinus maenas (Linnaeus, 1758)
Ingle 1980: 100, fig. 44, 44a, pl. 12b;
1983: 106, fig. 29; Williams 1984: 356, fig. 289.
"Green crab" — "Crabe verde"
(Figures 253 and 254)

DISTINGUISHING CHARACTERISTICS

Carapace only slightly wider than long, with 5 antero-lateral teeth about the same in size; rostral area only slightly protruding, granulate; orbit with single fissure dorsally; chelae large, slightly unequal; distal inner spine on carpus; legs 2nd to 4th about equal but 5th more compressed and dactyl wider but not oval; abdomen of male triangular, somites 3–5 fused.

DESCRIPTION

Integument smooth to finely granulate. Colour dark greenish to brownish green or bluish, a row of lighter coloured spots following cervical groove on each side, yellowish white to darker colour ventrally.

Carapace with 5 anterolateral teeth about equal in size, width 1.3 times length; surface granulate with divisions well-defined; front projecting forward slightly with three divisions, the central slightly pointed; orbital margin with one dorsal fissure; posterior portion somewhat tapered laterally and longer than anterior portion.

Thoracic sterna wide, finely granulate anteriorly and smooth posteriorly.

Abdomen of male triangular, 3rd to 5th somites fused; in females somites are not fused, subtriangular but rounded laterally.

Eyes moderate, stalks short, cornea globular.

Antennule (c) 1st article wide but short, setose on exposed surface; other articles subequal, fitting in socket obliquely.

Antenna (d) basal segment fused to epistome at outer edge, much larger than other two, the 3rd only half length of 2nd.

Mandibles (e) similar but right has a central tooth at curved edge of incisor blade, lower corner pointed; palp with 3 segments.

Maxillule (f) distal endite very much expanded from narrow base, proximal narrow and shorter; endopod 1st segment with inner lobe or expansion, 2nd segment narrow curved outward.

Maxilla (g) both endites unequally bilobed, distal narrow, lobes short with long neck, proximal both lobes narrow widely separate; endopod narrow pointed from wide base; anterior lobe of scaphognathite wide, sloping inward, about equal to very wide pointed posterior lobe with slightly concave edge.

Maxilliped I (h) distal endite flat and curved, wider than thick proximal endite; endopod wide, expanded distally into second axis; exopod longer than endopod and with lash; epipod wide foliaceous and tapering to long narrow ribbon-like tail with setal fringe.

Maxilliped II (i) endopod pediform; exopod with lash; epipod with long end slightly expanded, with long podobranch.

Maxilliped III (k) endopod operculiform, ischium greatly expanded, with low sagittation along inner edge, merus shorter and narrower but squarish, carpus inserted at

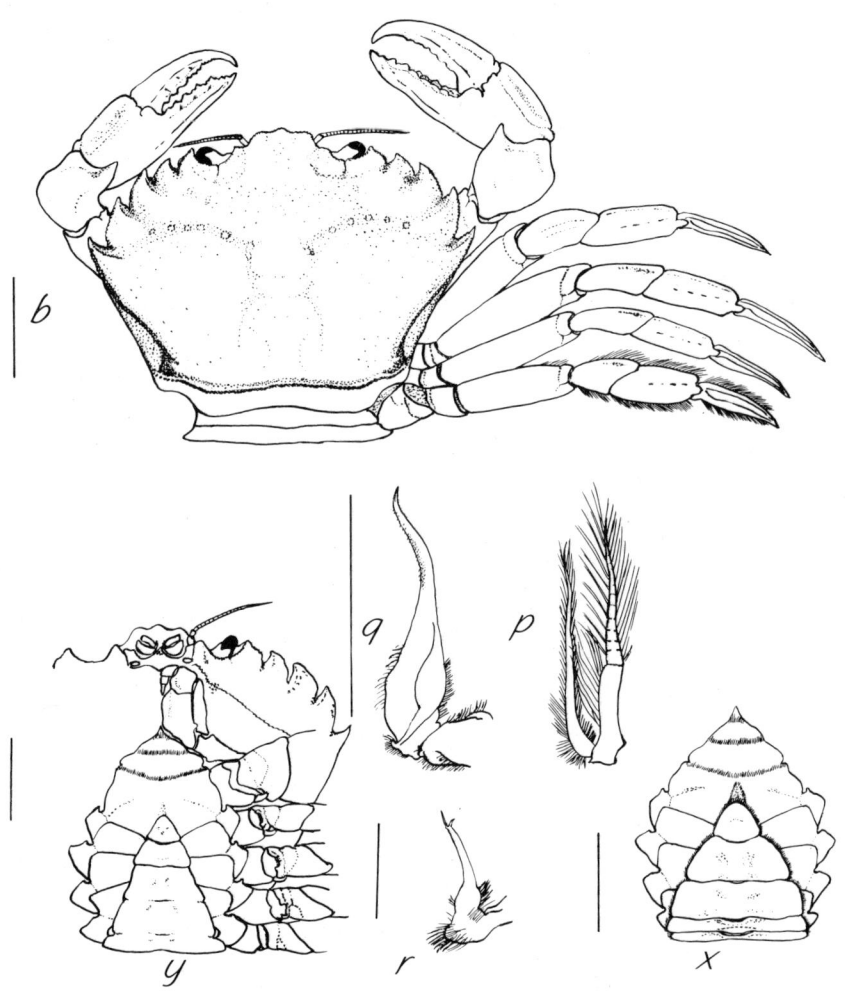

Fig. 253. *Carcinus maenas*: b, carapace in dorsal aspect; p, F second pleopod; q, M first pleopod; r, M second pleopod; x, F abdomen; y, Male in ventral aspect; solid line = 10 mm.

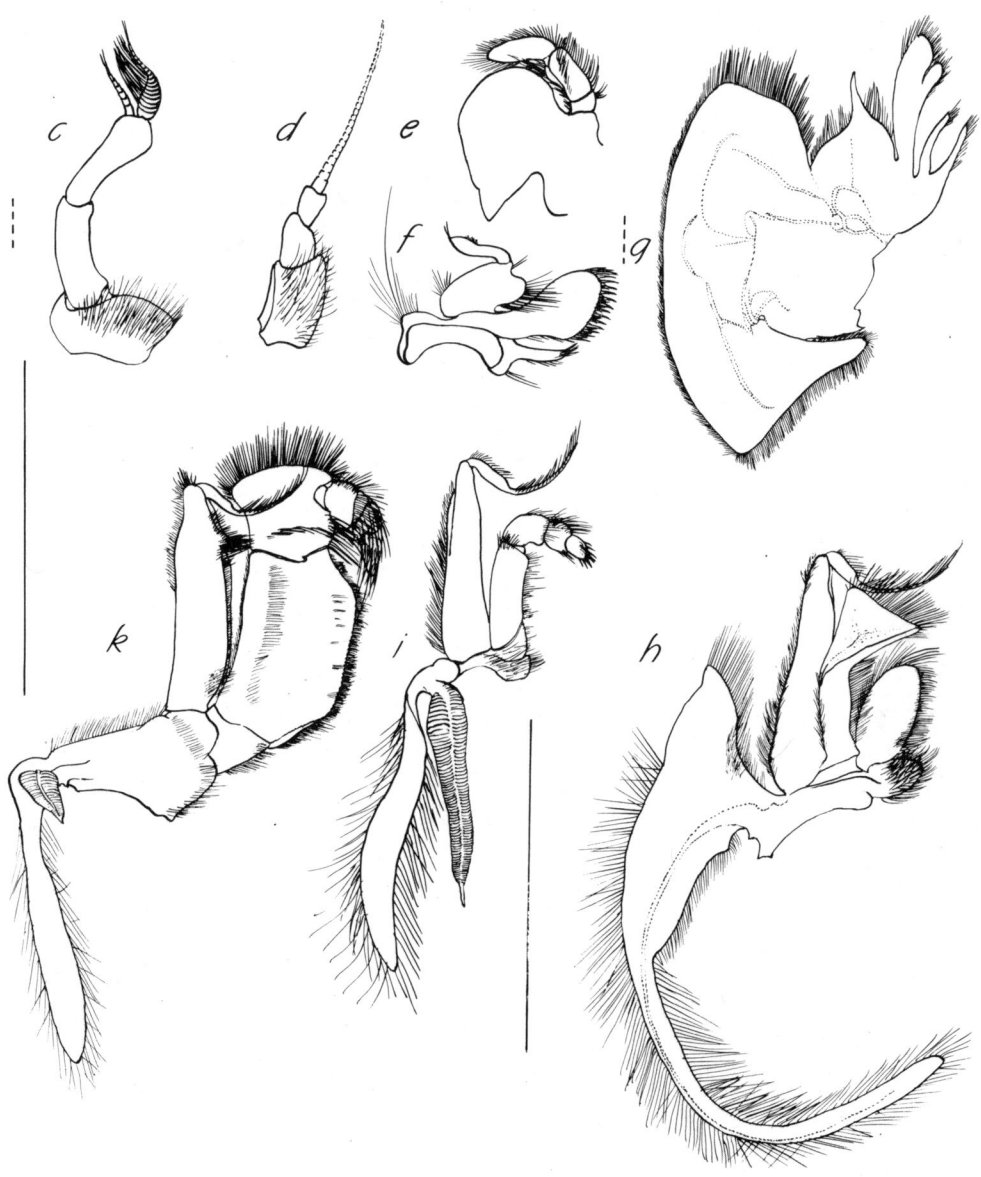

Fig. 254. *Carcinus maenas*: c, antennule; d, antenna; e, mandible; f, maxillule; g, maxilla; h, first maxilliped; i, second maxilliped; k, third maxilliped; solid line = 10 mm, broken line = 1 mm.

distal inner corner; exopod strongly calcified with distolateral projection and lash; epipod with long end slightly expanded distally, with podobranch.

Pereopods: I right chela larger than left and with two rows of crushing teeth on fingers proximally, left with single row of cutting teeth; carpus with strong inner tooth distally; II–IV slightly longer than cheliped, slightly compressed; V with carpus and propodus wider than in IV, compressed, dactyl lanceolate wider than in IV.

Pleopods: female II (p) endopod two-segmented the distal annulate, stouter and longer than tapered exopod; III–V similar to II; male I (q) tapering and sinuous from wide base to narrow grooved and hollow tip with retrorse denticles; male II (r) much shorter than I with slender neck and spike-like tip with short subapical lateral projection.

RANGE OF DISTRIBUTION

In the western Atlantic (to which it has been introduced from Europe) from Northumberland Strait and Cape Breton to Virginia; in the eastern Atlantic from Iceland and Norway through the North Sea to Mauritania, NW Africa. It has also been introduced to the Red Sea, Madagascar, India and Sri Lanka and Burma (Williams 1984). Its form *C. mediterraneus* occupies all the Mediterranean (Zariquiey Alvarez 1968). Depths 0–10 m (Elner 1985), rarely to 200 m (Christiansen 1969).

Records of occurrence in the area of reference are in Fig. 255.

BIOLOGY

Lengths to 60 mm cl (79 mm cw) in males and 60 mm cl (77 mm cw) in females (Williams 1984).

Sexual maturity may be reached in females at 19–20 mm cw. In Holland some spawn in autumn, hatch larvae in spring and spawn a second time in summer. Those hatched in spring reach about 20 mm cw in autumn, while those hatched in summer catch up with the others the following summer reaching about 30 mm cw and then moulting less frequently. Females are judged to live about 3 years (averaging 36, 42 and 50 mm cw) and males up to 5 years (averaging 36, 42, 48, 56 mm cw and over). A large female (46 mm cw) will produce 185 000 to 200 000 eggs in a spawning. A low temperature of $1.4°$ C in water of low salinity will kill eggs (Williams 1984).

Feeding is omniverous, but on bivalves when available, and the soft-shell clam in particular can be opened and eaten by a green crab equal to it in shell width (Ropes 1968).

FISHERY

Mostly because of its size, the green crab is of little commercial importance, although used as bait by sport fishermen where it is available (Elner 1985).

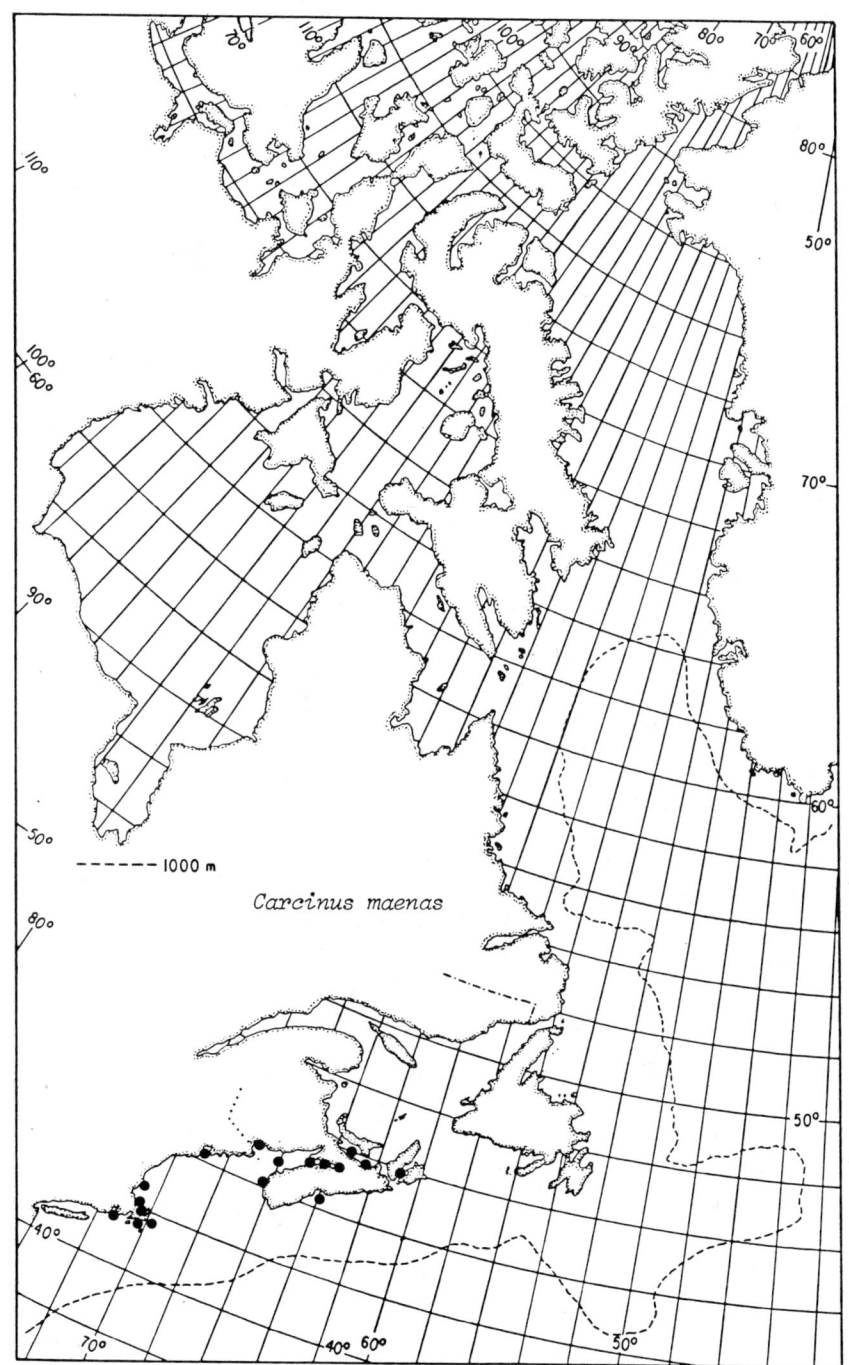

Fig. 255. *Carcinus maenas*, distribution records in the area of reference.

Genus *Ovalipes* Rathbun, 1898
Rathbun 1929: 31; 1930: 18;
Stephenson and Rees 1968: 215;
Williams 1984: 359.

Carapace somewhat narrow but wider than long, deep dorsoventrally; anterolateral margin curved, with 5 teeth about equal in size and spacing; front narrow, rostral tooth prominent, with a smaller tooth on each side; chelipeds subequal, carpus with outer and large inner spine distally; palm with conspicuous inner and outer dorsal carina continued on dactyl, inner surface dorsally with dense row of setae; 5th leg with large oval dactyl.

Ovalipes ocellatus (Herbst, 1799)
Rathbun 1929: 31, fig. 42; 1930: 19, fig. 5;
Stephenson and Rees 1968: 241, pls. 37B, 40D,
41C, 42I, figs. 1I, 2H, 3H, 4H;
Williams 1984: 359, fig. 290.
"Calico crab" — "Crabe calico"
(Figures 256 and 257)

DISTINGUISHING CHARACTERISTICS

Anterolateral margins of carapace curved and with 5 equal sharp teeth evenly spaced; frontal area between orbits narrow with three teeth, the centre one largest; carapace wider than long, convex, covered with small dots arranged in roundish clusters; chelae large, unequal, palm with inner setose ridge, continued on to moveable finger; 5th leg with large paddle-like oval dactyl.

DESCRIPTION

Integument smooth, shiny, finely granulate, thin, flexible. Colour yellowish grey, closely set with small annular spots of reddish purple, and with iridesence; chelipeds and legs light brownish tending to orange and bluish; paddles of 5th legs greenish yellow.

Rostrum flat, thin, pointed, with a faint rib, about 0.1 cl, flanked by similar but shorter pre-orbital spines; orbit wide with small notch dorsally, inner suborbital angle projecting as far forward as rostrum. Carapace wider than long with 5 sharp anterolateral teeth directed forward, narrowing posteriorly with emarginate granulate curving edge, rounded and narrow at posterior margin. Horseshoe shaped depression on each side of median line in metagastric area, between them and anterior are larger granulations in the midline but behind them are no granules to posterior edge of carapace. Ventrally at pterygostomian region is a long curved stridulating ridge made up of short striae and tubercles at the end; a short matching ridge is at the proximal end of the inner edge of the ischium of the cheliped.

Abdomen of male narrow, sides nearly parallel, second to fifth somites fused, sixth somite almost twice as long as 7th (telson), the latter subcircular; adult female abdomen suboval, occupying only half sternum.

Eyestalk slightly curved, protruding from orbit, cornea about as wide as stalk at apex.

Antennule (c) 1st article subspherical, inserted at front edge of carapace visible near rostrum, 2nd article stouter and longer than 3rd.

Antenna (d) basal article longer than wide, triangular in cross-section, attached near and, including other segments and flagellum, shorter than antennule.

Mandibles (e) heavily calcified, edge bevelled, curved evenly in left but with low tooth at centre in right, pointed at lower corner; small molar shelf behind incisor at lower corner; palp three-segmented.

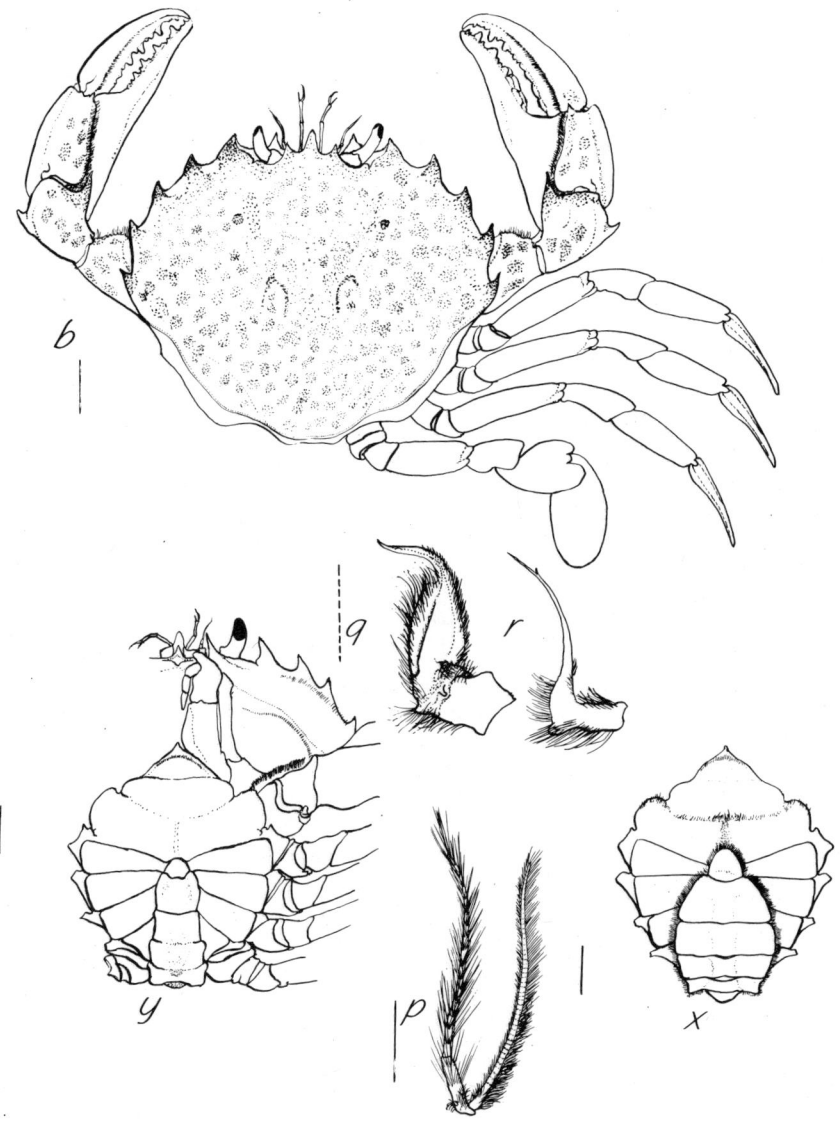

Fig. 256. *Ovalipes ocellatus*: b, carapace in dorsal aspect; p, F second pleopod; q, M first pleopod; r, M second pleopod; x, F abdomen; y, M in ventral aspect; solid line = 10 mm, broken line = 1 mm.

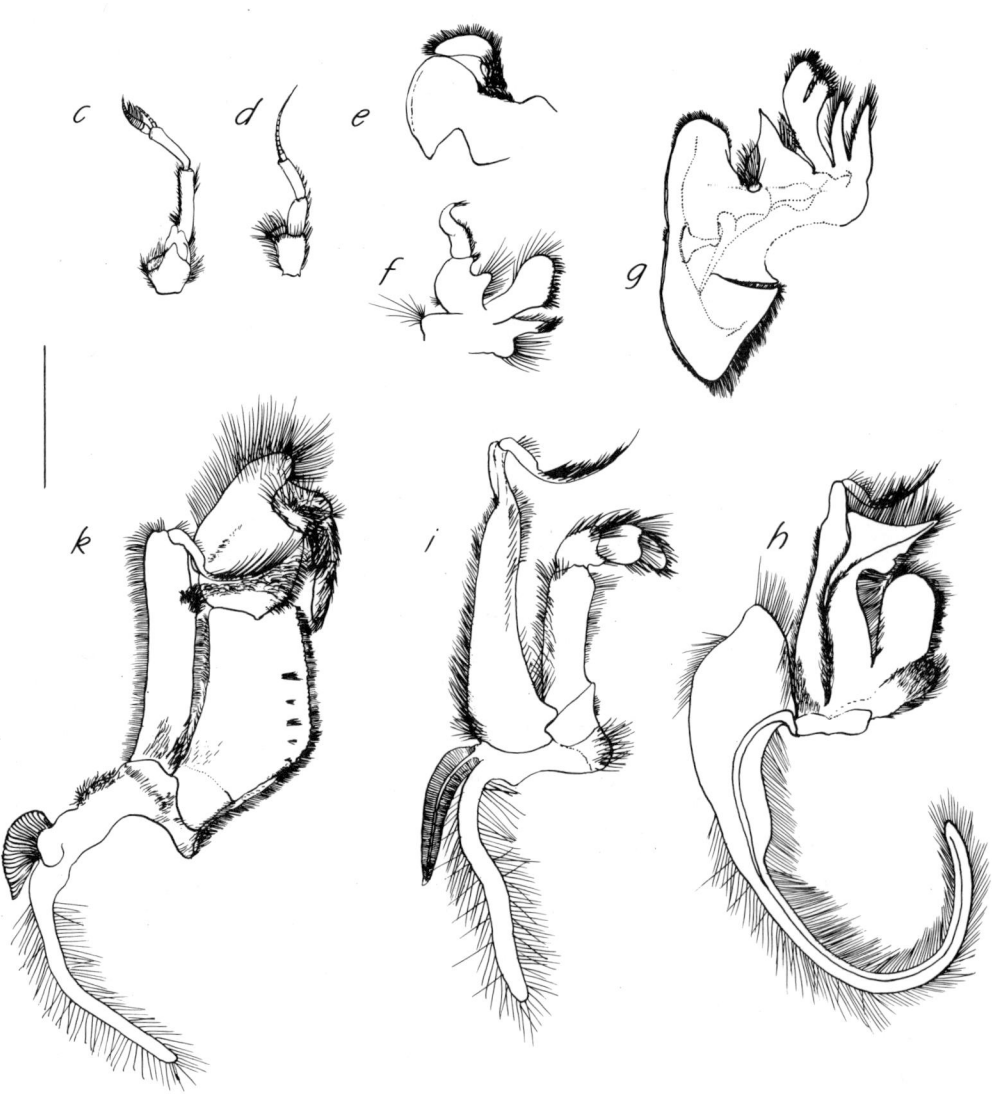

Fig. 257. *Ovalipes ocellatus*: *c*, antennule; *d*, antenna; *e*, mandible; *f*, maxillule; *g*, maxilla; *h*, first maxilliped; *i*, second maxilliped; *k*, third maxilliped; solid line = 10 mm.

Maxillule (f) distal endite greatly expanded from narrow base, proximal with wide base but narrow and pointed; endopod wide at base, 1st segment with lateral expansion, distal segment narrow, tapering and curved inward.

Maxilla (g) endites subequally bilobed, curving upward in line with endopod, distal with lobes short and neck long, proximal lobes separate, tapering; endopod subtriangular from wide base, slightly exceeding anterior lobe of scaphognathite, narrower and shorter than posterior subtriangular lobe; outer edge infolded with very short setal fringe.

Maxilliped I (h) distal endite flat, rounded apically, proximal shorter, thick, setose; palp foliaceous, wide, twisted into other axis apically with concave upper and irregular inner edge; exopod longer and with short lash; epipod foliaceous, wide at base and tapering into long tail.

Maxilliped II (i) endopod pediform, compressed; exopod long and tapering with pronounced "neck" distally and moderate lash; epipod moderately long, with podobranch.

Maxilliped III (k) operculiform, ischium wide and long with inner edge crenulate; merus not quite as long, extended as a lobe distally; exopod stout, rigid, distolateral projection small, setose; epipod moderate, narrow, with podobranch.

Pereopods: I large, chelate, subequal, with prominent ridges above on propodus, the inner ending in distal spine and continued on dactyl, chela with sharp-toothed fingers on left and some flat tipped spines on right, carpus with strong inner and small outer spine distolaterally; II–IV similar, non-chelate, compressed, dactyls with ridges and corneous tips; V stout shorter than others, propodus expanded and compressed, dactyl wide and thin, ovate, paddle-like.

Pleopods: female II (p) endopod slender, with short proximal segment and long annulate distal segment, exopod very slender, annulate; III–V similar. Male I (q) wide and stout for most of its length but tapering to narrow distal part, curved ventrolaterally, about one-third as long as proximal part; central cavity leading to tip. Male II (r) very slender but as long as I, with a short lateral projection on distal fifth.

RANGE OF DISTRIBUTION

Northumberland Strait, Prince Edward Island, to Georgia. Depths 0–95 m (Williams 1984).

Records of occurrence in the area of reference are in Fig. 258.

BIOLOGY

Carapace width to 87 mm in males and 60 mm in females.

Aparently favours nearshore and estuarine areas and moderate salinities. Females mate in the soft-shell condition. Ovigerous females have been observed with newly spawned eggs, orange in colour, in October and February, and with dark eggs in October and December. Larvae have been taken in plankton during June to October, most commonly in September.

Development of larvae to megalopa in the laboratory required 27 days at 20° C and 18 days at 25° C (Williams 1984).

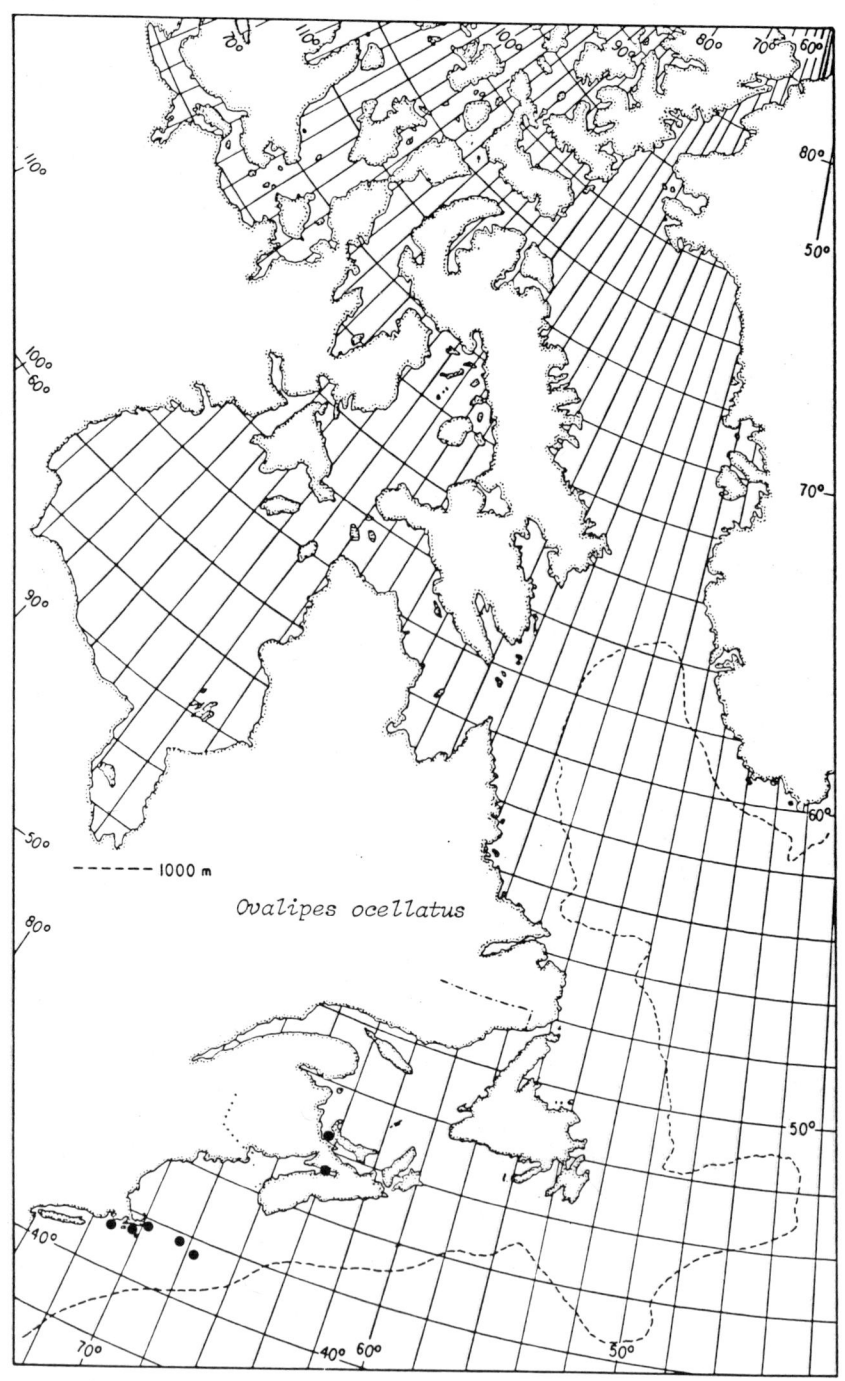

Fig. 258. *Ovalipes ocellatus*, distribution records in the area of reference.

Genus *Portunus* Weber, 1795
Rathbun 1929: 30; 1930: 33;
Stephenson and Campbell 1960: 75;
Williams 1984: 386.

Carapace transverse, widest at last anterolateral spine; front (between antennae) with 3–6 teeth; anterolateral margin arched, longer than posterolateral, with nine teeth; abdomen of male triangular; chelipeds longer and heavier than other legs, merus with spines, inner and outer angles of carpus with spines; palm triangular in cross-section, with ridges; 5th legs with expanded, compressed segments, including dactyl thin and ovate, paddle-like.

Portunus sayi (Gibbes, 1850)
Rathbun 1929: 30, fig. 40;
1930: 37, pl. 14, figs. 6, 7;
Williams 1984: 391, fig. 307.
"Gulf weed crab" — "Crabe de Sargassum"
(Figures 259 and 260)

DISTINGUISHING CHARACTERISTICS

Carapace high in the middle; anterolateral margins arched, longer than posterolateral margins, with nine teeth, the most lateral ones very large and sharp; frontal or rostral teeth four, the central ones smaller; two low transverse granulate ridges across the gastric area; merus of cheliped with 3 large inner spines, a strong spine at inner angle of carpus, a small outer spine; a spine on outer surface of palm near articulation with carpus.

DESCRIPTION

Integument smooth, shiny, finely granulate, thin. Colour mottled olive green or purplish and variegated yellow-brown with irregular white patches (Williams 1984).

Carapace very convex, high in the middle and sloping away on all sides, furrows shallow, two low ridges across gastric area curving backward, and a faint ridge from each lateral strong spine on to the branchial area; frontal or rostral teeth four, the center ones smaller, each equally advanced and the preorbitals slightly behind; anterolateral teeth nine, directed forward slightly, the edge arched and longer than posterolateral edge; two orbital fissures dorsally, one more open ventrally.

Sternum about as wide as long, plates rounded laterally.

Abdomen with sides slightly tapering after curving in from wide base in male, 3rd to 5th somites fused, 6th longest, telson small; abdomen very wide in female, no somites fused, telson small.

Eyes large, cornea globular, stalks stout, not enclosable in orbit.

Antennule (c) 1st article large, subspherical; 3rd more slender and shorter than 2nd, flagella very short.

Antenna (d) basal article subcylindrical, with distolateral projection; other articles about equal in length, distal more slender; flagellum slender, about one-third cl.

Mandibles (e) similar, incisor edge forming an obtuse angle at centre, lower corner a slender point; palp 3-segmented.

Maxillule (f) distal endite short and very wide distally; proximal wide at base but narrow and slightly curved to pointed tip; endopod with two segments, distal narrow and curved outward.

Maxilla (g) endites small, unequally bilobed, distal with lobes expanded at end of long narrow neck, proximal lobes separate narrow; endopod spike-like from wide base; anterior lobe of scaphognathite wide and rounded at corners, slightly shorter than posterior subtriangular lobe.

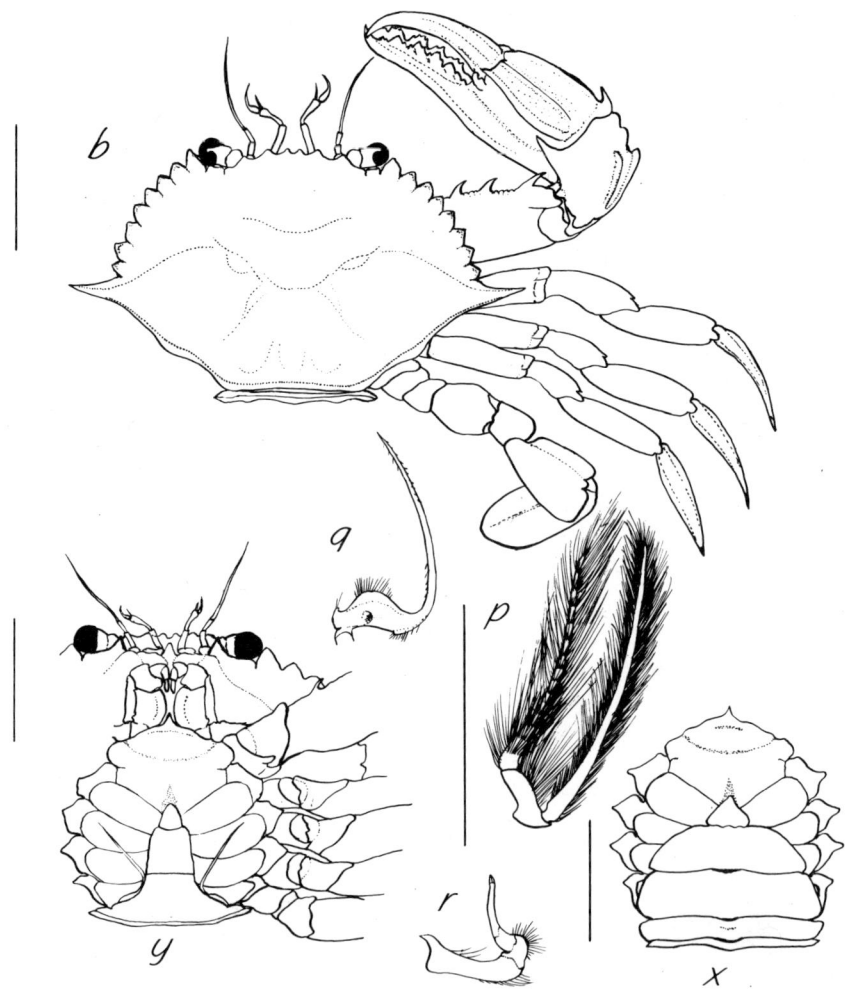

Fig. 259. *Portunus sayi*: *b*, carapace, etc., in dorsal aspect; *p*, F second pleopod; *q*, M first pleopod; *r*, M second pleopod; *x*, F abdomen; *y*, male in ventral aspect; solid line = 10 mm.

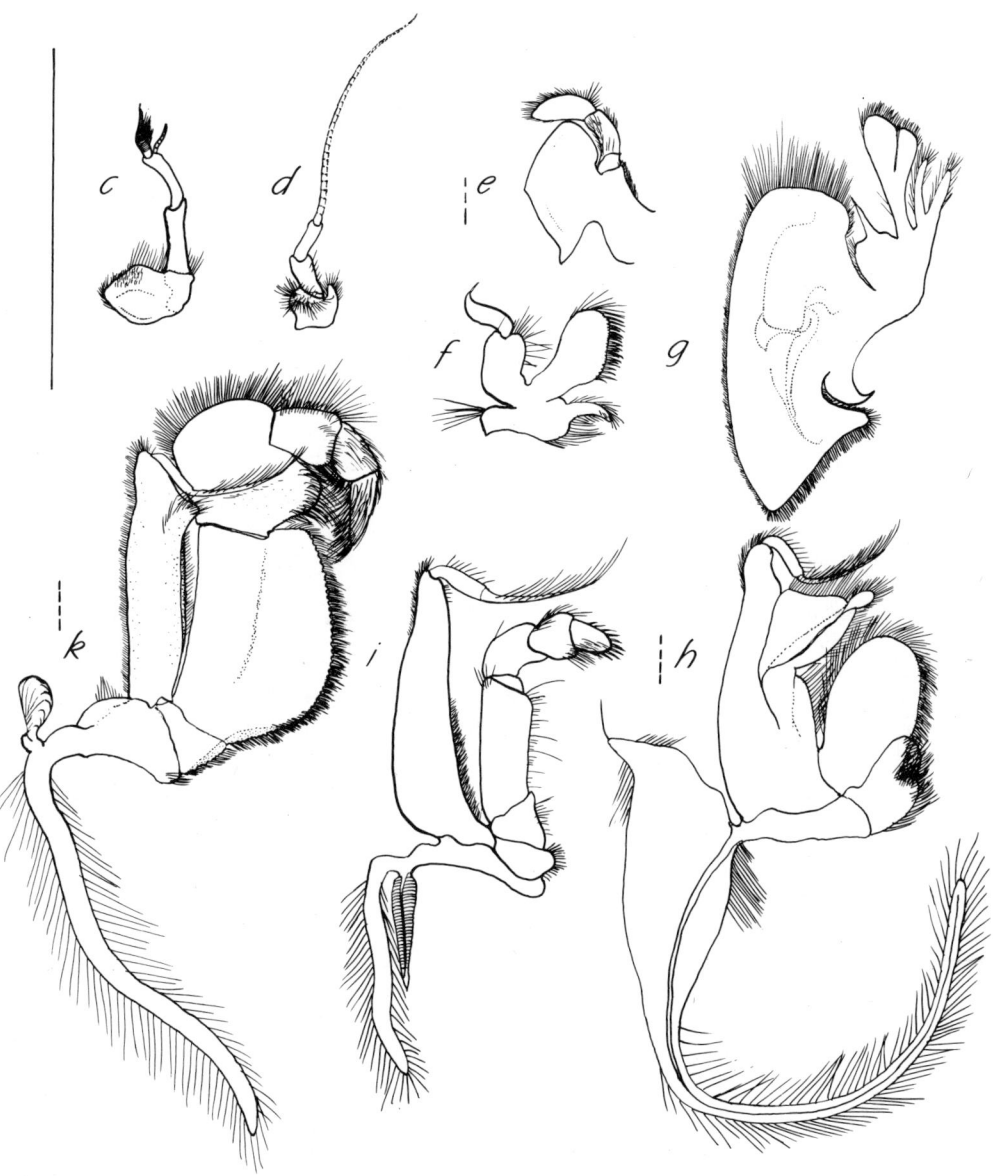

Fig. 260. *Portunus sayi*: *c*, antennule; *d*, antenna; *e*, mandible; *f*, maxillule; *g*, maxilla; *h*, first maxilliped; *i*, second maxilliped; *k*, third maxilliped; solid line = 10 mm, broken line = 1 mm.

Maxilliped I (h) distal endite wide, thin, ovate, proximal thick, bilobate; endopod partly coalesced with exopod, distally curved into separate axis and with small tongue-like lobe at inner corner; exopod exceeding endopod and with flagellum; epipod proximally wide and with pointed corner, tapering to very long tail.

Maxilliped II (i) endopod pediform, compressed; exopod with long lash; epipod small, narrow, with podobranch.

Maxilliped III (k) operculiform, ischium very wide, with crenulate inner edge; merus about half as long, carpus inserted at distal inner corner; exopod stout, rigid, with distolateral projection and lash; epipod moderately long, narrow, with podobranch.

Pereopods: I chelate, subequal, chela large, propodus with strong proximal spine on outer ridge, an inner ridge with distal spine and a middle ridge continued on dactyl; carpus with sharp inner spine and smaller outer spine distolaterally, merus with 4 fixed spines along inner edge. II–IV similar, compressed, dactyl lanceolate. V stout, shorter than IV, carpus expanded and compressed as well as propodus, and dactyl flattened, oval, paddle-like.

Pleopods: female II (p) proximal segment of endopod short, distal long, annulate; exopod a single segment, slender, faintly annulate; female III–V similar to II. Male I (q) long and slender, extending from beneath abdomen laterally, with short retrorse spinules near apex, opening into central canal at expanded base; male II (r) short, slender, with bifid tip.

RANGE OF DISTRIBUTION

North Atlantic Ocean from off the Grand Banks through Gulf of Mexico to the Guianas , the Canary Islands and Morocco — essentially the Gulf Stream and Sargasso Sea. It is pelagic among floating *Sargassum*, but may be carried into coastal areas by currents (Williams 1984).

Records of occurrence in the area of reference are in Fig. 261.

BIOLOGY

Size limits to 31 mm cl (61 mm cw) in males and 31 mm cl (64 mm cw) in females.

Ovigerous females are known from most months in the year (Williams 1984). Egg diameter is 0.2 mm.

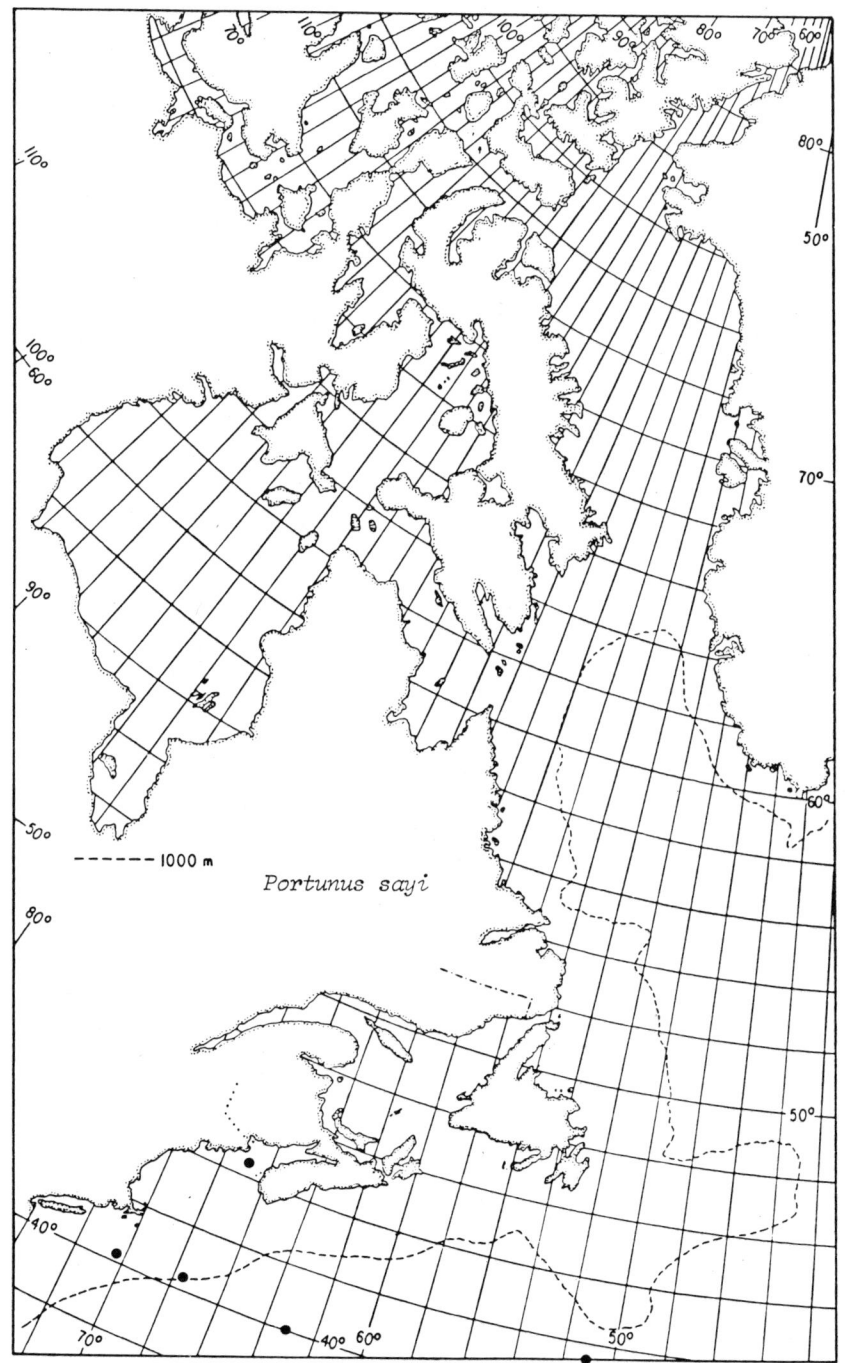

Fig. 261. *Portunus sayi*, distribution records in the area of reference.

Family XANTHIDAE MacLeay, 1838
Ingle 1980: 111; Rathbun 1929: 33; 1930: 233; Williams 1984: 395.

Carapace always nearly oval or hexagonal in outline; front wide, never produced as a rostrum; antennules folded transversely or obliquely; male gonopores on coxae of 5th legs.

Key to species of Xanthidae present in area of reference
(After Williams 1984)

1 Chelae with dark fingers, carpus with strong inner spine; frontal margin of carapace advanced, with simple edge .... *Neopanope sayi*

   Chelae without dark fingers, carpus without strong inner spine; frontal margin of carapace almost straight, with grooved edge, appearing double .... *Rhithropanopeus harrisi*

Genus *Neopanope* H. Milne-Edwards, 1800
Ingle 1980: 114; Rathbun 1929: 33; 1930: 366.

Carapace transversely oval, subhexagonal, high in the middle; antero-lateral margins with four teeth, the first two separated only by a shallow sinus, second lobiform; front advanced, arcuate; telson of male wider than long.

Neopanope sayi (Smith, 1869)
Ingle 1980: 114, fig. 61, pl. 20;
Williams 1984: 409, figs. 324, 331k.
Neopanope texana sayi, Rathbun 1929: 34, fig. 45;
1930: 369, pl. 168, figs. 3, 4.
(Figures 262 and 263)

DISTINGUISHING CHARACTERISTICS

Fingers of chelae dark-coloured or black; carapace oval or hexagonal, with four antero-lateral teeth (appearing as three) the last two more clearly defined with a dorsal ridge; front with small median notch; chelipeds unequal, more strikingly in males, fingers slightly longer than palm, carpus with a stout spine on inner distal margin; dactyls of walking legs stout, with terminal tooth.

DESCRIPTION

Integument granulate with a few setae especially along posterolateral margin. Colour a dark slaty bluish green, brown or buff, with dark reddish-brown speckles on yellowish background, or bluish purple on grey background, outer face of chelae yellowish grey, fingers dark or black, colour extending on to palm, tips light (Williams 1984).

Carapace subhexagonal, about three-quarters as long as wide, convex or somewhat inflated; three clear anterolateral teeth, pointed and somewhat upturned, but also a rounded flattish tooth anterior to them adjoining the orbit; posterior two teeth with a ridge extending on to carapace; front slightly rounded and with a median notch, and a median groove extending back to about level of orbit and branching into two over the gastric area; orbit with two dorsal fissures.

Abdomen of male wide at the base but converging at third to fifth fused somites, telson wider than long; female abdomen rounded, laterally with long setae.

Eyes large, covered by orbit when retracted.

Antennule (c) 1st article short and wide, 2nd and 3rd subequal, sloped obliquely in socket.

Antenna (d) basal article longer than wide, fused with epistome on one side, free side faintly serrate.

Mandibles (e) edge of incisor strongly curved, with central tooth on right only, lower corner only slightly produced, palp with three segments.

Maxillule (f) distal endite narrow at base but expanding to receding spinous edge; proximal short and narrow; endopod with two segments, proximal slightly expanded, distal very narrow and curved outward.

Maxilla (g) distal endite with unequal lobes at end of long neck, proximal with equal narrow lobes much shorter than distal endite; endopod a short curved tip from wide base; anterior lobe of scaphognathite wide but tapered, about as long as subtriangular posterior lobe.

Maxilliped I (h) proximal endite thick, pointed, heavily setose; distal endite flat curved, rounded apically; endopod wide foliaceous curved and twisted into second axis distally, about as long as exopod, the latter with slender lash; epipod moderately wide and prolonged at base but tapering to long ribbonlike part with many lateral setae.

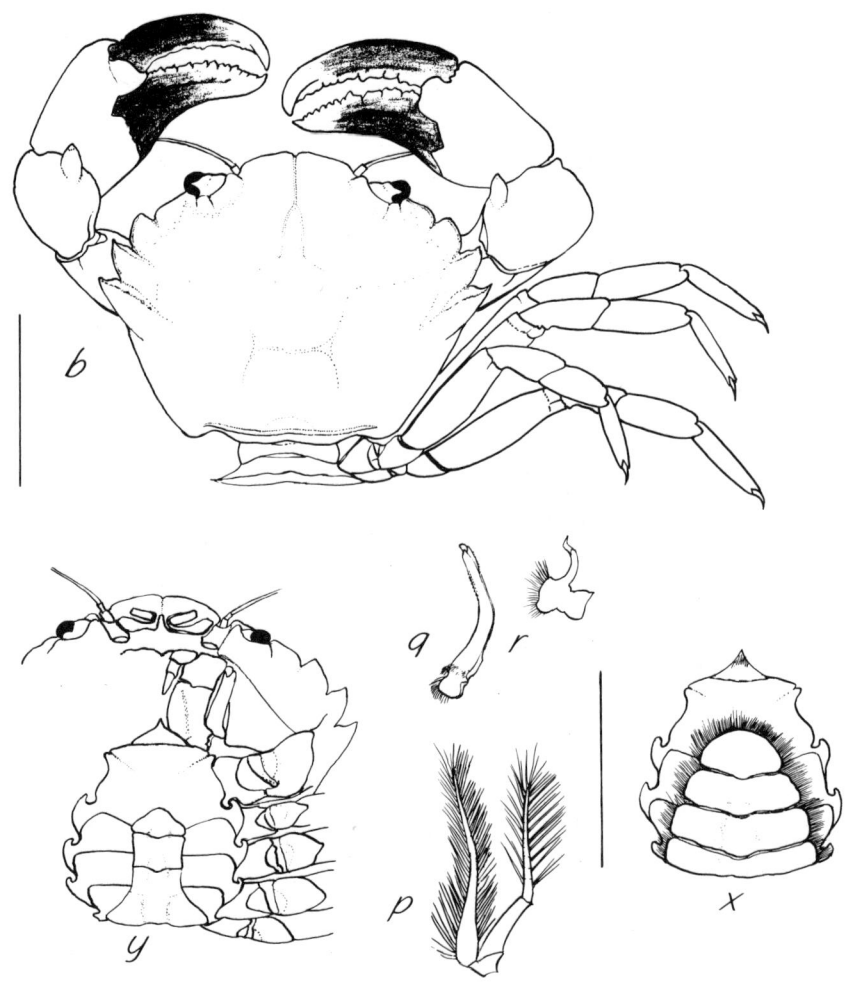

Fig. 262. *Neopanope sayi*: b, carapace, etc., in dorsal aspect; p, F second pleopod; q, M first pleopod; r, M second pleopod; x, F abdomen; y, male in ventral aspect; solid line = 10 mm.

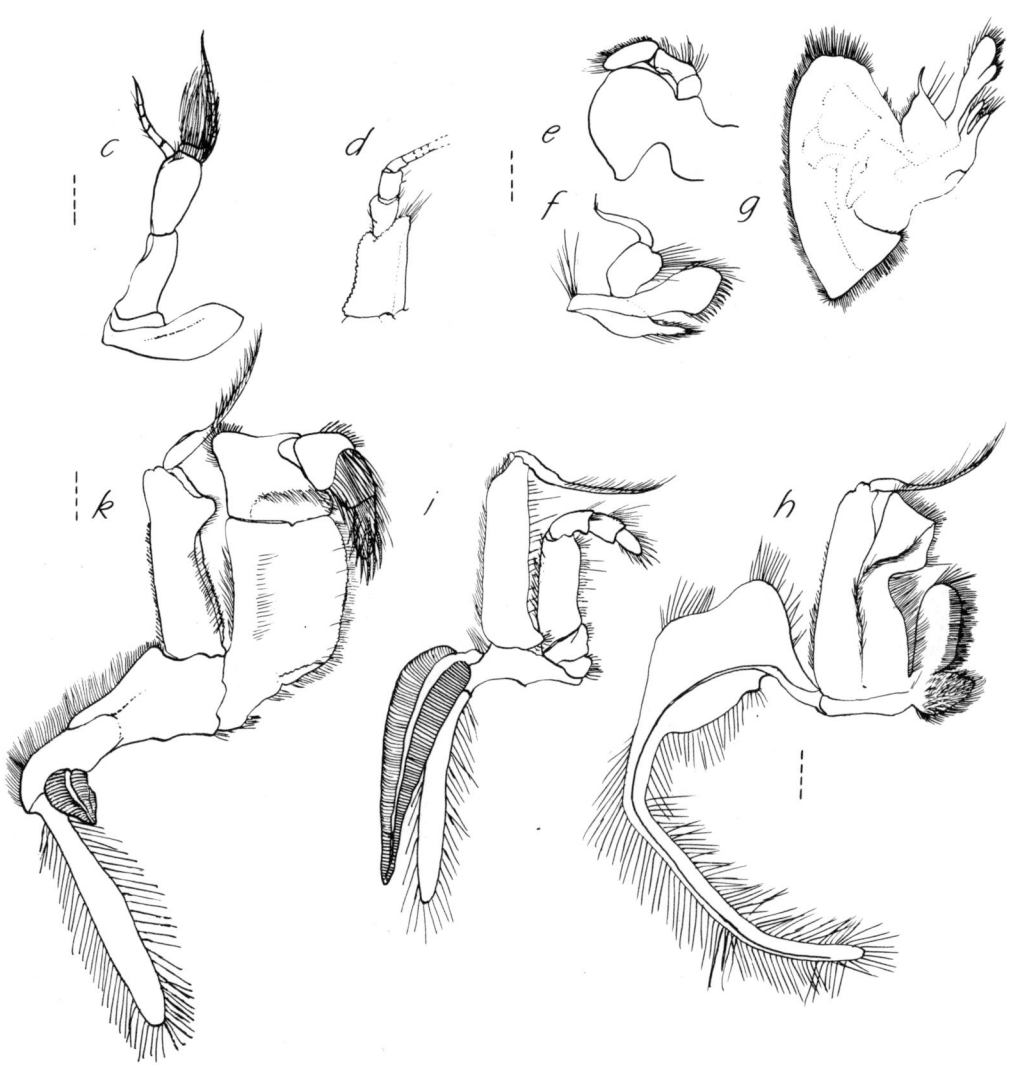

Fig. 263. *Neopanope sayi*: *c*, antennule; *d*, antenna; *e*, mandible; *f*, maxillule; *g*, maxilla; *h*, first maxilliped; *i*, second maxiliped; *k*, third maxillliped; broken line = 1 mm.

Maxilliped II (i) endopod pediform, compressed, smaller than exopod, the latter with flagellum; epipod with moderate narrow extension not much longer than large podobranch.

Maxilliped III (k) operculiform, ischium wide and long, inner edge crenulate, merus much shorter but as wide as ischium, carpus inserted at inner distal corner; exopod stout, rigid, with distolateral projection and short flagellum; epipod long partly expanded, with podobranch.

Pereopods: I chelipeds very large, slightly unequal , longer than walking legs, smooth, teeth of right larger, fingers dark grey but light at tips, dark colour extended to palm, carpus with stout spine at inner edge distally, merus with strong tooth on distal fourth. II–V similar, slightly compressed, dactyls stout with spinous tip.

Pleopods: female II (p) endopod two-segmented, distal part annulated, exopod simple tapered, slender, longer than endopod; female III–V similar to II; male I (q) stout bent at middle, stout short spines at inner edge of apex, retrorse spinules distolaterally; male II (r) much shorter than I, with wide base, curved neck and pointed apical cap.

RANGE OF DISTRIBUTION

Southern Gulf of St. Lawrence to the Florida Keys. Introduced inadvertently in shipping to the Bristol Channel, U.K. Depths 0–46 m (Williams 1984).

Records of occurrence in area of reference are in Fig. 264.

BIOLOGY

Size limits to 21 mm cl (30 mm cw) in males and 14 mm cl (20 mm cw) in females.

Females mate in the hard-shell condition. Reproduction takes place in temperatures of 24 to 30° C for the most part. Mature females spawned 10–16 days after moulting. Embryonic development was completed in 9–10 days in the breeding season. The interval between hatching one batch of eggs and spawning was about 3 days and the duration of a moult interval with one spawning was about 28 days, thus in the first breeding season the females moulted four or five times and spawned four to five times, and produced from 1 400 to 12 000 as they increased in size or 15 000–30 000 eggs in a season. Life span was presumed to be three summers (Schwartz 1978).

Larval development took 14 days at 30° C and 37 days at 21° C, and in the Chesapeake Bay area these crabs were mature the following summer.

Predation is on barnacles and small bivalves.

They are preyed upon by two species of hake in the Georgia area (Williams 1984).

Social behaviour studies show ritualized displays and stereotypic periodic movements which accompany reproductive activities. Agonistic encounters rarely result in combat or injury (Williams 1984).

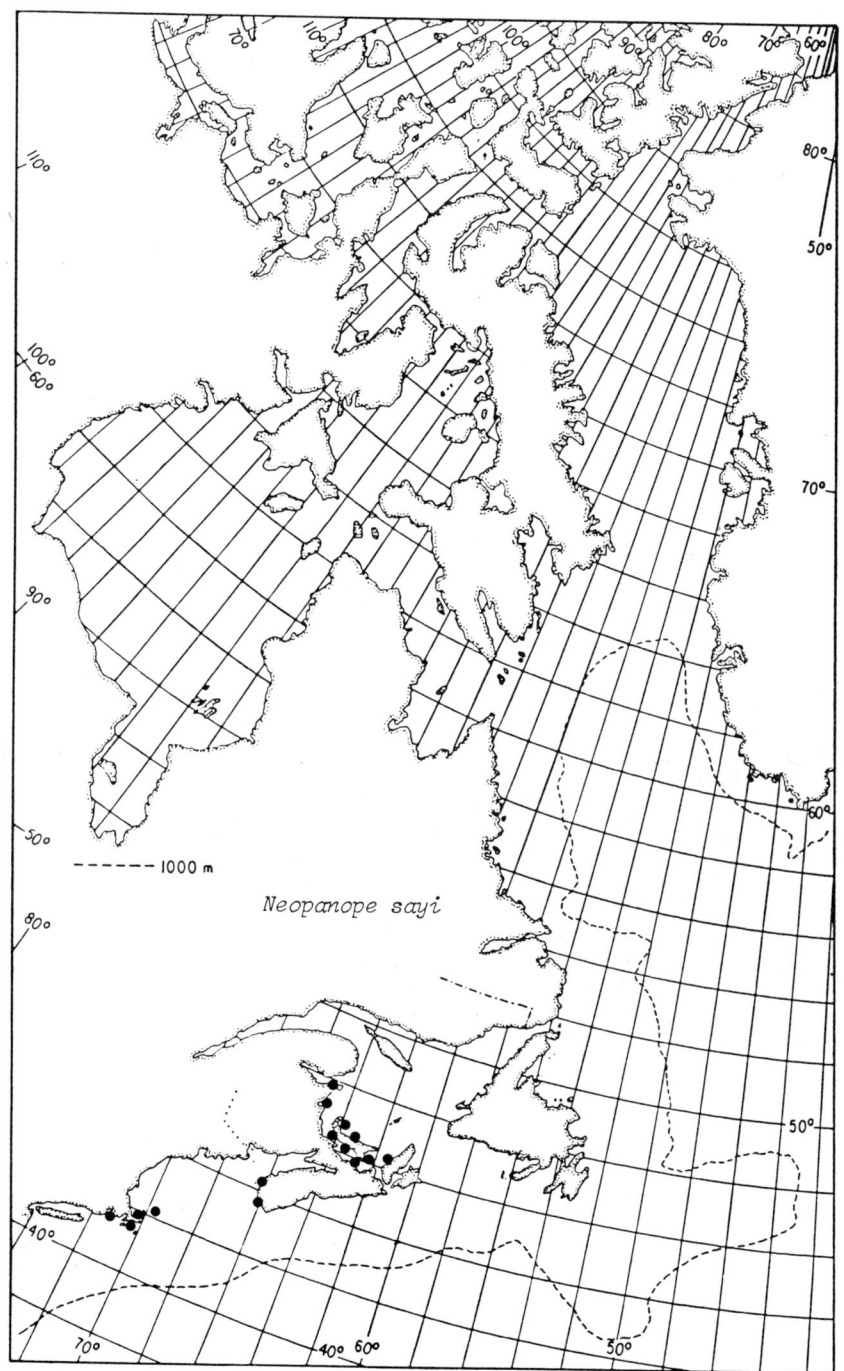

Fig. 264. *Neopanope sayi*, distribution records in the area of reference.

Genus *Rhithropanopeus* Rathbun, 1898
Ingle 1980: 113; Rathbun 1929: 34; 1930: 455.

Carapace subquadrate, wider than long, low transverse ridges dorsally across gastric area: 2 opposite pairs on mesogastric and two pairs in line across metagastric areas; front double-edged, almost straight, with very slight median notch; anterolateral margins with four teeth, the fourth (most posterior) smallest, bluntish; chelipeds very unequal, fingers shorter than palm in right, carpus without inner distolateral spine; telson of male longer than or equal to width, rounded distally.

Rhithropanopeus harrisi (Gould, 1841)
Rathbun 1929: 34, fig. 46; 1930: 456, pl. 183, figs. 7, 8;
Williams 1984: 401, figs. 316, 317, 331f.
(Figures 265 and 266)

DISTINGUISHING CHARACTERISTICS

Chelipeds large, unequal and dissimilar, fingers shorter than palm on right, no inner distolateral spine on carpus; carapace subquadrate, wider than long, front with faint median notch, almost straight and with a groove along edge giving appearance of a double edge; four blunt anterolateral teeth, the posterior smallest.

DESCRIPTION

Integument of preserved specimens examined appeared lightly calcified and covered with brownish layer that could be peeled off leaving dark iridescent, shiny coat, punctate and with scattered setae.

Carapace subquadrate, somewhat wider than long, anterolateral teeth low, bluntish, second and third largest, fourth small; two broken lines of granules across mesogastric area, and an irregular interrupted line of granules (in four parts) across metagastric area between the fourth lateral teeth; cardiac and intestinal areas lightly outlined, setose especially below in subhepatic subbranchial areas; front almost straight but with slight median notch, appearing double when seen from in front or below.

Abdomen narrow in males, sides of 6th somite parallel, 3rd to 5th somites fused, 3rd widest, telson longer than wide, rounded apically. Female abdomen wider than male, subtriangular in outline, 6th somite about as long as telson.

Eyes large, cornea globular, retractible into orbits.

Antennule (c) 1st article wide and short, 2nd and 3rd subequal, folding obliquely into socket.

Antenna (d) basal article not fused to epistome, wide, others very small, flagellum short with about 17 segments.

Mandibles (e) edge of incisor rounded, a central projection larger in right than in left; palp with 3 segments.

Maxillule (f) distal endite wide, with rounded spinous edge; proximal narrow, curved; endopod proximal segment expanded, distal narrow curved outward; exopod stubby.

Maxilla (g) distal endite unequally bilobed, lobes short; proximal lobes narrow, one pointed; endopod a short pointed tip from wide base; anterior lobe of scaphognathite wide tapered, about equal to pointed, wide posterior lobe.

Maxilliped I (h) distal endite squarish, proximal thick, protruding; endopod foliaceous, large, distally flared and with changed axis, exceeded by exopod with lash; epipod wide at base and tapered to moderately long tail part.

Maxilliped II (i) endopod pediform; exopod reaching as far as carpus and with long lash; epipod fusiform from narrow neck, with podobranch.

Maxiliped III (k) operculiform, ischium wide, squarish, slightly longer than square merus, the latter with carpus inserted at distal inner corner; epipod with rigid part long and a club-shaped tail part, with podobranch.

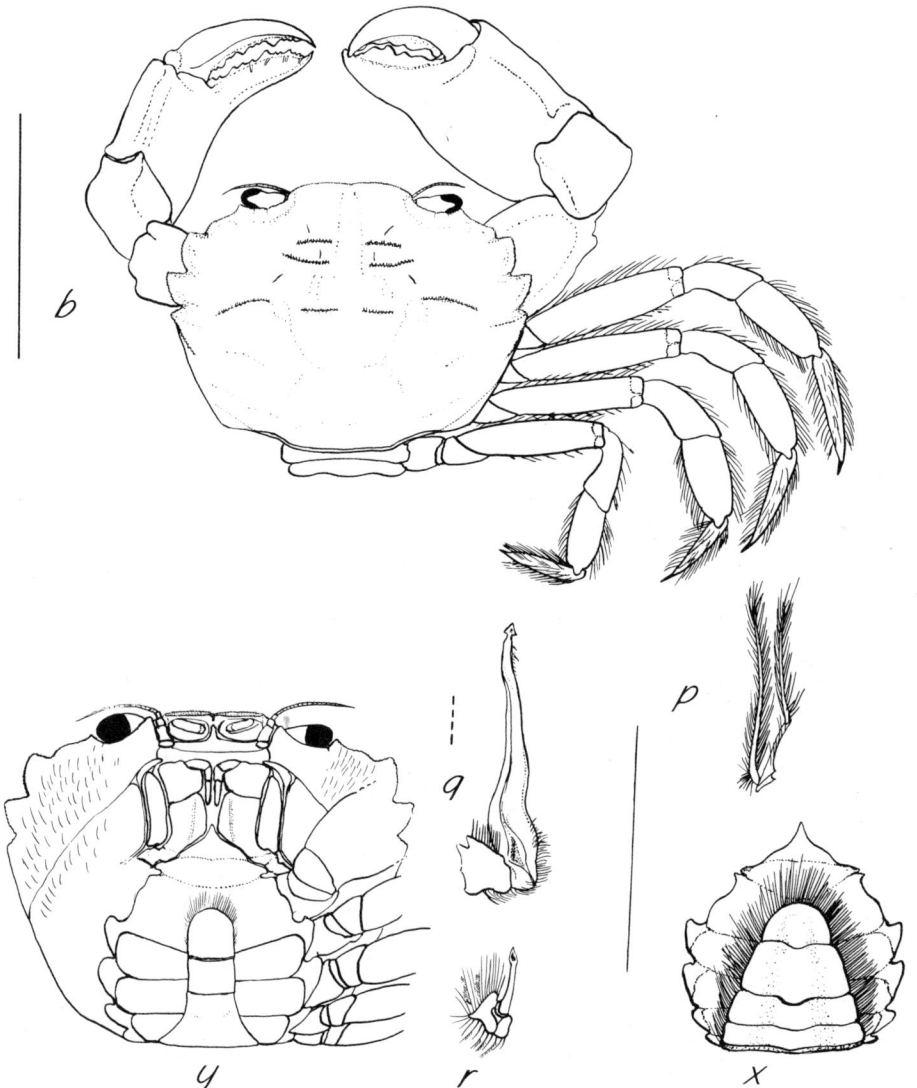

Fig. 265. *Rhithropanopeus harrisi*: b, carapace, etc, in dorsal aspect; p, F second pleopod; q, M first pleopod; r, M second pleopod; x, F abdomen; y, M in ventral aspect; solid line = 10 mm; broken line = 1 mm.

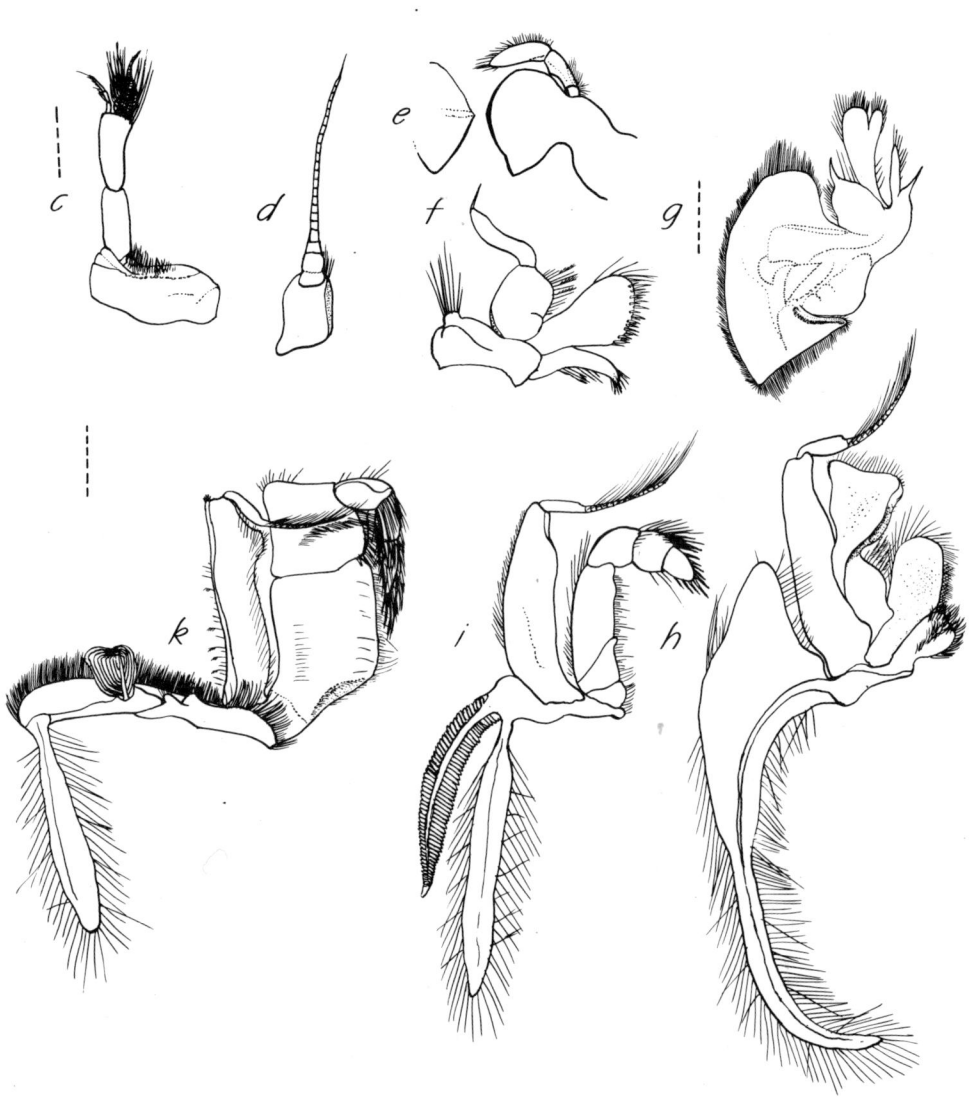

Fig. 266. *Rhithropanopeus harrisi*: *c*, antennule; *d*, antenna; *e*, mandible; *f*, maxillule; *g*, maxilla; *h*, first maxilliped; *i*, second maxilliped; *k*, third maxilliped; broken line = 1 mm.

Pereopods: I chelate, large, very unequal, right massive, with fingers shorter than palm, low granulate ridge above on propodus, teeth few, large and rounded; left slightly more slender, fingers about equal to palm, teeth several, sharp; carpus with blunt inner process. II–V similar, slightly decreasing in size posteriorly, slightly compressed, dactyl stout with strong spine at tip, strong setal fringes most profuse on dactyl.

Pleopods: female II (p) endopod in two segments proximal short, distal annulated, about equal to slender tapering exopod; female III–V similar to II. Male I (q) stout, tapering, sinuous, with groove originating in triangular cavity proximally and spiralling around side to near apex which is a small spade-shaped projection; II (r) much shorter than I, slender from wide base and with a flared and pointed concave apex.

RANGE OF DISTRIBUTION

In fresh to estuarine waters from southwestern Gulf of St. Lawrence to Veracruz, Mexico. It has been inadvertently introduced into Europe and to the west coast of the United States. Depths 0–37 m (Williams 1984).

Records of occurrences in the area of reference are in Fig. 267.

BIOLOGY

Size limits to 16 mm cl (21 mm cw) in males and 12 mm (16 mm cw) in females.

Osmotic regulation is highly developed; development of larval stages was best at 6–10‰, duration increasing with decreasing temperatures. Best survival to megalopa and first crab stages occurred at 20‰ salinity and 20–25° C temperatures.

Zoeae first appeared in plankton in May but were most common in July to September and very few in October. Concentrations of larvae were found (and retained) in mouths of estuaries near the bottom where net tidal flow would be zero.

First maturity of females was considered to be reached in the second summer at 6 mm cl. Adults continued to moult and grow reaching 10 mm cl from 5 mm cl in four moults in the Chesapeake Bay area.

Stomach contents indicated omniverous feeding on mangrove leaf detritus and crustaceans such as amphipods and harpacticoid copepods.

The species was preyed upon by two species of hake (*Urophycis* sp.) in a Georgia estuary and also by the white catfish (*Ictalurus* sp).

The above was reviewed in more detail by Williams (1984).

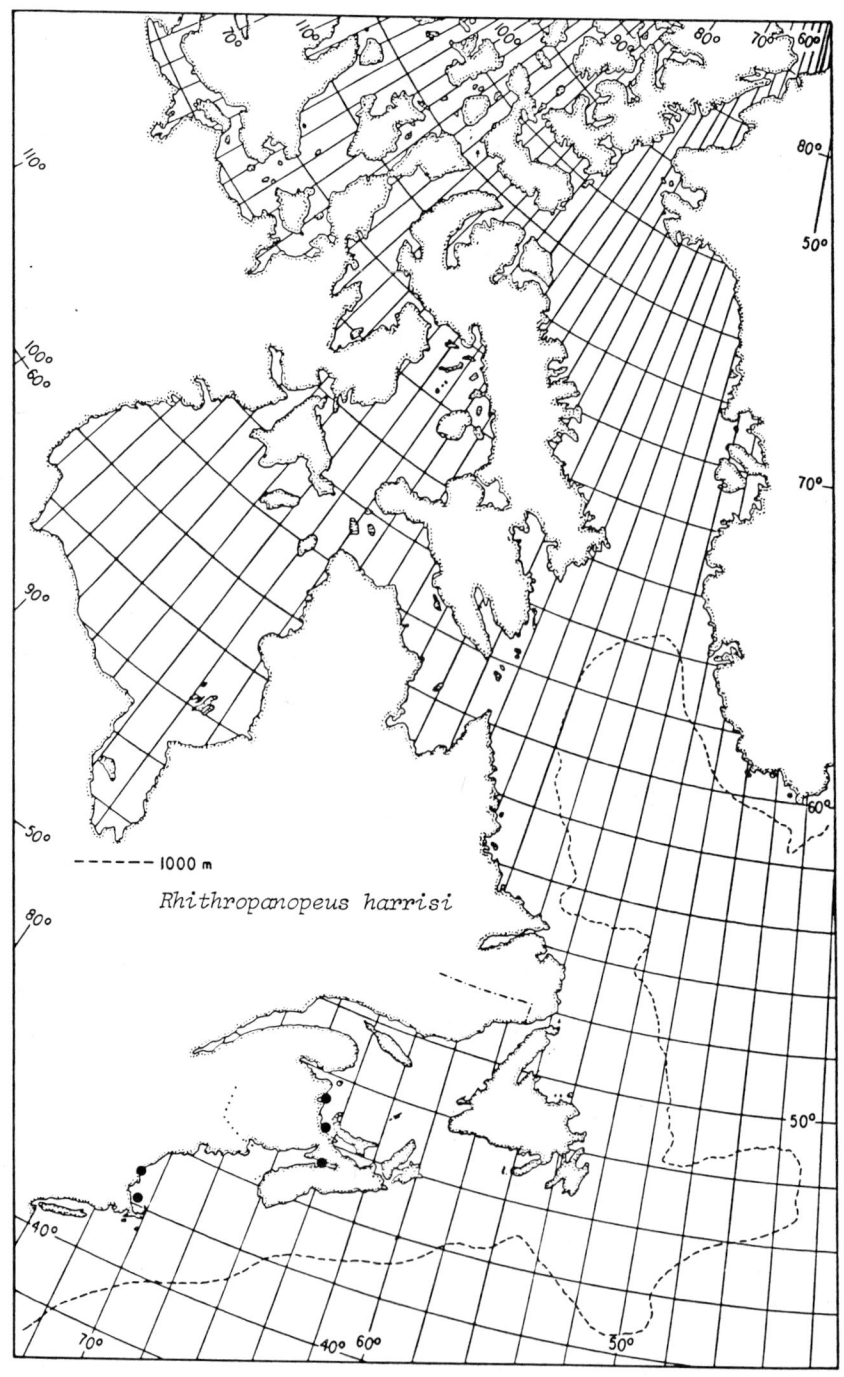

Fig. 267. *Rhithropanopaeus harrisi*, distribution records in the area of reference.

Family GRAPSIDAE MacLeay, 1838
Rathbun 1918: 224; 1929: 35; Williams 1984: 458.

Front quite wide; carapace with lateral margins either straight or slightly arched and orbits at or near anterolateral corners; buccal cavity square, generally a gap between the third maxillipeds; male openings sternal; carpus of third maxilliped inserted at middle of anterior edge of merus or at its anteroexternal angle.

Genus *Planes* Bowdich, 1825
Ingle 1980: 122; Rathbun 1918: 253; Williams 1984: 460.

Carapace almost square, front very wide with orbits at anterior corners; anterolateral margin with two teeth, the posterior one rudimentary; carpus of third maxilliped articulating near middle of anterior edge of merus; pereopods compressed and stout; abdominal segments free in both sexes, abdomen of male triangular.

Planes minutus (Linnaeus, 1758)
Chace 1951: 81, figs. 1a, 2a, d, g, j, k, l, 3a-h;
Ingle 1980: 122, fig. 73, pl. 24a;
Rathbun 1918: 253, pl. 63: Williams 1984: 460, fig. 369.
(Figures 268 and 269)

DISTINGUISHING CHARACTERISTICS

Carapace almost square, about as wide as long, front wide with faint median division, orbits large at anterior corners; surface convex dorsally and smooth with faint oblique markings at lateral margins; chelipeds large, heavy, equal, merus with spines distally and along inner edge, carpus with a blunt inner distal spine; legs compressed, stout, with fringes of sharp spines and setae; male with triangular abdomen and all somites free.

DESCRIPTION

Integument thin, lightly calcified, flexible, surface smooth to finely granulate. Colour irregularly mottled or blotched with light greenish yellow or pale yellow on darker olive green background; or reddish fawn colour blotched with dark brown and usually a white spot on each side or one large white spot on front of carapace (Williams 1984).

Carapace quadrate, squarish at front with faint central depression, about as wide as long, slightly inflated; orbits large, outer edge of orbit spinous with a basal notch or rudimentary spine; dorsal surface only lightly marked: anteriorly parallel median lines on gastric area, a transverse line delimiting metagastric area, oblique parallel lines on lateral branchial area, cardiac area faintly outlined; posteriorly truncate with short lateral grooves.

Abdomen of male wide triangular, somites free; female abdomen large covering almost all of sternum.

Eyes large, cornea globular.

Antennule (c) 1st article wide and short, subspherical; 3rd article small, the 2nd with lateral hollow for 3rd to fit into when folded obliquely into socket.

Antenna (d) basal article with distolateral expansion, other segments small; inserted in inner edge of orbit.

Mandibles (e) similar, incisor edge curved to form a small tooth at centre, a pointed tooth at lower corner, palp with 3 segments.

Maxillule (f) distal endite slightly curved, proximal ovate, pointed; endopod with moderately wide basal segment, distal segment with distolateral branch and apical expansion into two rounded leaflets.

Maxilla (g) distal endite with one very wide and one narrow lobe; proximal endite with unequal lobes and rounded setal fringe at base; endopod short and narrow from wide base; anterior lobe of scaphognathite pointed distally, slightly shorter than axe-shaped posterior lobe.

Maxilliped I (h) endites rounded, distal slightly larger than thick proximal; endopod distally flared and with changed axis, with inner lateral pointed projection; exopod tapered, with lash; epipod wide, foliaceous at base but tapering to narrow ribbonlike terminal part.

Maxilliped II (i) endopod pediform, dactyl with long curved spines; exopod about as long as endopod, with lash; epipod moderately long, narrow, with podobranch.

Maxilliped III (k) operculiform but not covering completely the buccal cavity, ischium wide, merus as wide as but only about half as long as ischium, insertion of carpus near middle of anterior edge of merus; exopod rigid, without distolateral projection, with lash; epipod with rigid part longer than terminal flexible narrow part, with podobranch.

Pereopods: I (l) massively chelate, equal, propodus inflated, fixed finger curved downward, dactyl with matching curve, both with bluntish teeth; inner edge of ischium and merus serrate, continued as toothed crest on merus distally, carpus with inner truncate toothed projection. II–V compressed, wide, merus ventrally hollowed distally to take

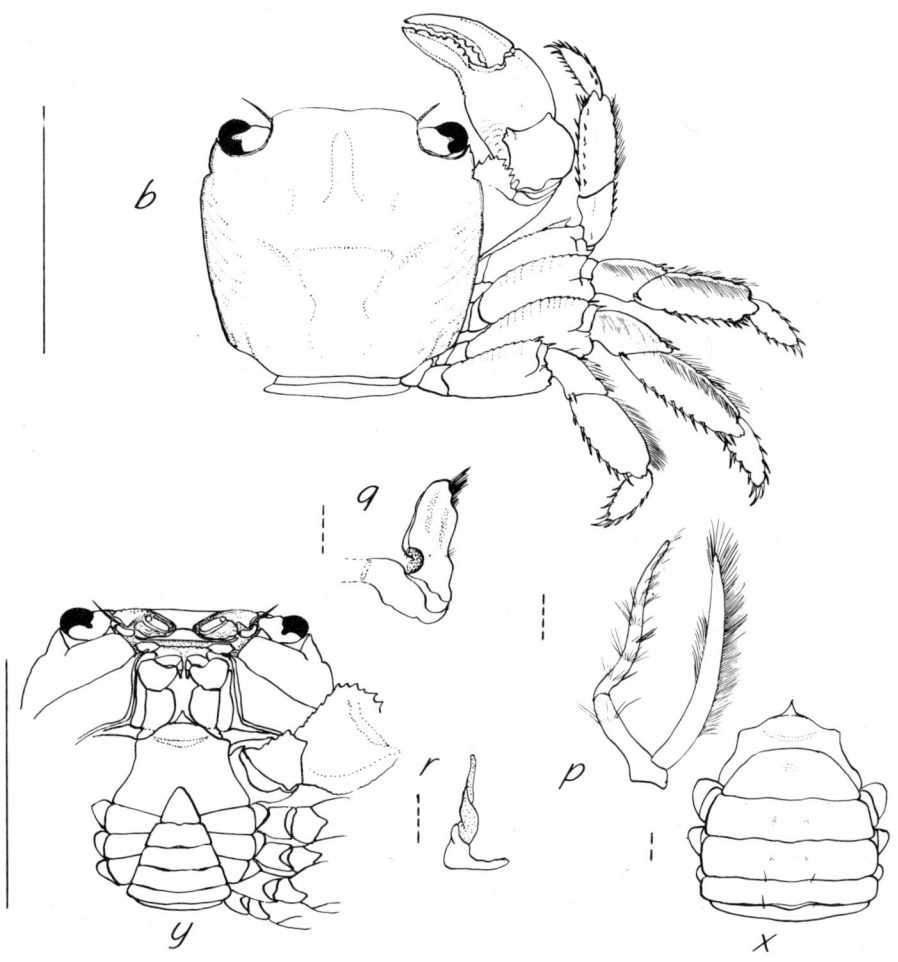

Fig. 268. *Planes minutus*: *b*, Carapace, etc, in dorsal aspect; *p*, F second pleopod; *q*, M first pleopod; *r*, M second pleopod; *x*, F abdomen; *y*, M in ventral aspect; solid line = 10 mm, broken line = 1 mm.

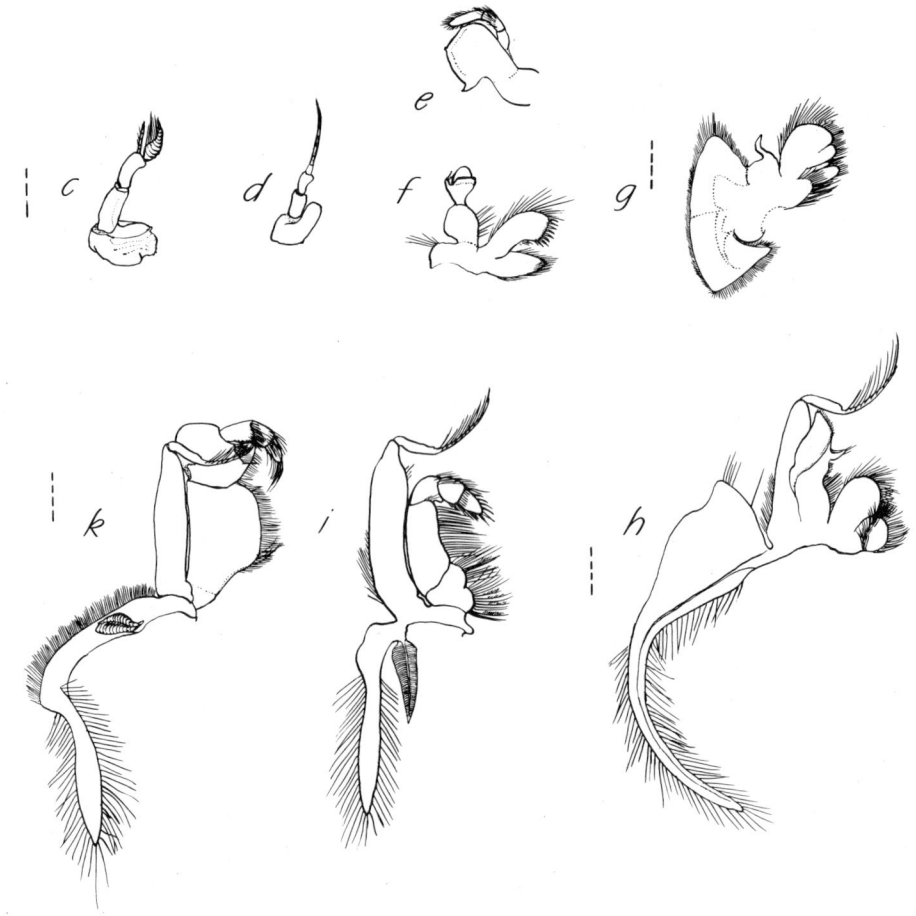

Fig. 269. *Planes minutus*: *c*, antennule; *d*, antenna; *e*, mandible; *f*, maxillule; *g*, maxilla; *h*, first maxilliped; *i*, second maxilliped; *k*, third maxilliped; broken line = 1 mm.

flexed carpus, propodus and dactyl with terminal and lateral strong sharp spines, and carpus and merus with distal strong spines; merus dorsally serrate with short transverse grooves.

Pleopods: female II (p) endopod two-segmented, distal segment the longer, sinuous, with setal tufts, exopod slender, tapered; female III–V similar to II. Male I (q) stout, short, with external groove and distal dense bunch of short setae, both joined by transverse bar across abdomen; male II (r) short and tapering with spiral groove and pointed apex.

RANGE OF DISTRIBUTION

North Atlantic from off the Grand Banks of Newfoundland to the southern North Sea and south to 11° N (essentially the Sargasso Sea and parts of the Caribbean), also including the Mediterranean Sea but not the Gulf of Mexico (Williams 1984).

Records of occurrences in the area of reference are in Fig. 270.

BIOLOGY

Size limits to 19 mm cl (19 mm cw) in males.

Ovigerous females have been observed throughout the year.

Food in stomachs examined comprised fragments of crustaceans, probably shrimps (Squires 1965a).

This species is more abundant on *Sargassum* in the Sargasso Sea than elsewhere, but throughout its range depends upon floating debris or organisms to which it clings (Chace 1951).

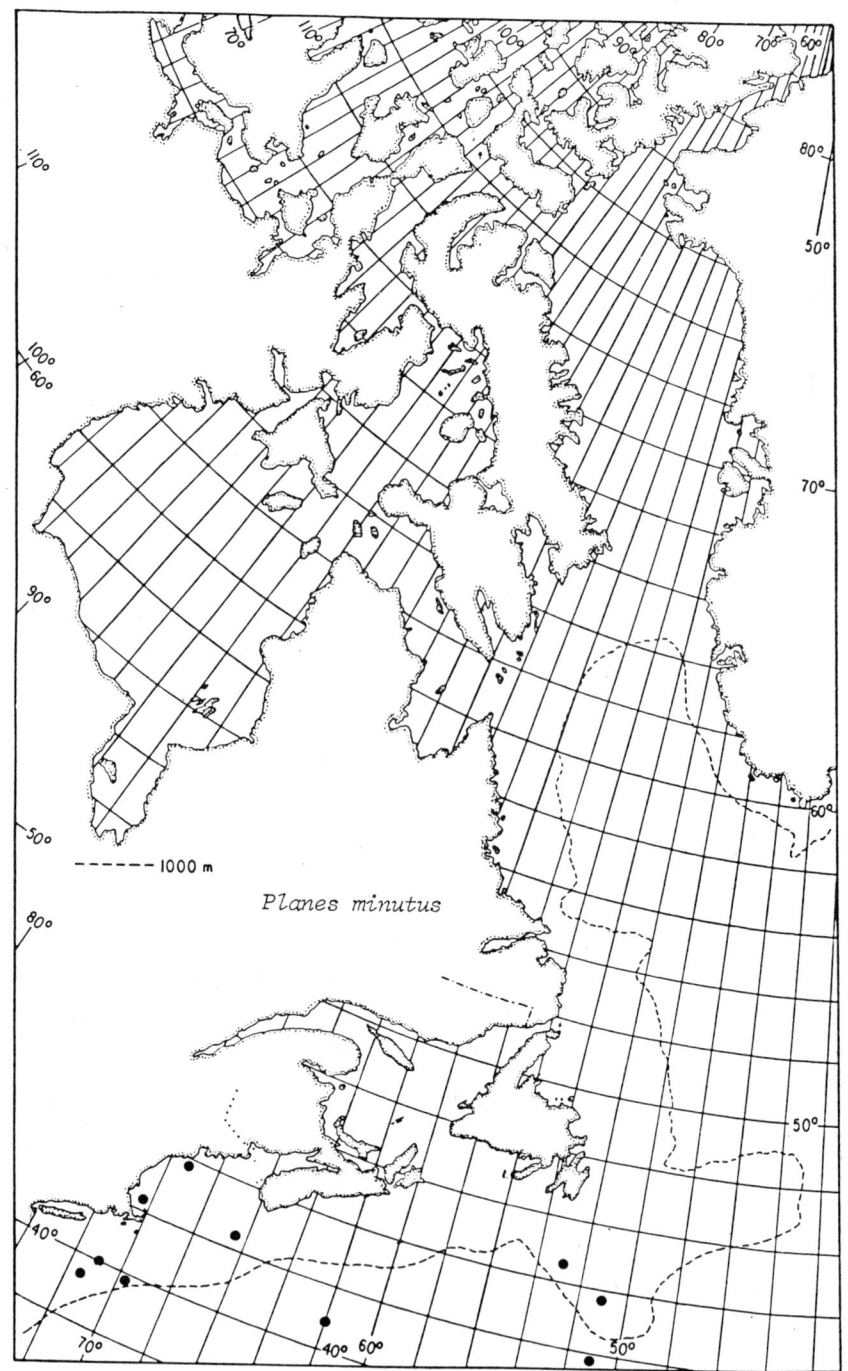

Fig. 270. *Planes minutus*, distribution records in the area of reference.

Extralimital Species

The following species have not been included because in the main they are either found more deeply than 1 000 m or belong farther south than 42° N latitude.

Acanthephyra acutifrons
Acanthephyra eximia
Ephyrina figueirai
Glyphocrangon sculpta

Hymenopenaeus laevis
Parapenaeus longirostris
Stereomastis nana

GLOSSARY

Abdomen — the flexible part of the body behind the carapace, the "tail" or hind part in shrimps and lobsters, held under the carapace in crabs.
Abdominal somite — a division or segment of the abdomen.
Accessory tooth — tooth distolateral to crista dentata.
Acicle — antennal scale reduced to a spine, as in hermit crabs.
Aesthetasc — a sensory seta as on the antenna.
Annulations — evenly spaced rings on an appendage such as on a flagellum of the antenna.
Antenna — outer sensory appendage lateral to the antennule at front of body: usually with one flagellum.
Antennal scale — flat blade forming exopod of antenna, sometimes reduced to a spine.
Antennal spine — spine at front of carapace just below orbit, behind antenna.
Antennule — inner sensory appendage at front of body: usually with two flagella.
Anterior — front or towards the front: opposite to posterior.
Anterolateral teeth — teeth at the front edge at the side in crabs, between the orbit and the extreme lateral spine.
Apical — at the tip.
Appendix interna — small branch on inner edge of endopod of pleopod; usually tipped with minute hooks.
Appendix masculina — small branch on inner edge of endopod of second pleopod of male between appendix interna and edge.
Arthrobranch — gill attached at the membrane between a leg and the body wall.
Ascending — sloping upward.

Basicerite — second segment of basal article of antenna, bearing scale.
Basis — second segment of a leg (pereopod).
Bifurcate — forming two branches.
Biunguiculate — with two strong spines at the tip of a leg (dactyl).
Biramus — with two branches.
Branchia — gill.
Branchial region — part of carapace over the gills.
Branchiocardiac groove — groove separating the branchial and cardiac regions on the carapace.
Branchiostegal spine — spine on or near anterior edge of carapace below the antennal spine or branchiostegal groove.
Brush setae — tuft of short stiff setae on terminal segments of legs used for grooming.
Buccal cavity — cavity in which the mouthparts are situated in crabs.

Carapace — chitinous case or shell covering the head and thorax or fore part of the body.
Cardiac region — part of the carapace over the heart, with the gastric region in front and branchial regions at the sides.
Caridean lobe — lamellate expansion of exopod of 1st maxilliped.
Carina — a ridge.
Carpus — fifth segment of a leg, forming the "knee" bend.
Central lobe — endopod in 1st maxilliped.
Cervical groove — groove or complex of grooves across carapace at about the middle and turning forward obliquely at each side.
Chela — pincer or claw formed by dactyl and forward part of propodus both usually with occluding rows of teeth.
Chelipeds — legs bearing chelae.
Chromatophore — contractile cell containing colour in integument.

Cincinnuli — grappling hooks distally on appendix interna, etc.
Compressed — flattened from side to side.
Cornea — outer faceted coating of the eye.
Coxa — first or proximal segment of leg.
Crista dentata — serrated median border or row of strong spines on ischium of third maxilliped.

Dactyl — (dactylus) terminal or seventh segment of leg; moveable finger of chela.
Depressed — flattened from top to bottom or dorsoventrally.
Descending — sloping downwards.
Diaeresis — transverse groove on exopod of uropod.
Distal — towards the tip or away from point of attachment.
Distolateral — outer side towards the tip.
Dorsal — back or upper surface.
Dorsolateral — upper surface and at the side.

Endites — medial branches on protopod of maxillule, maxilla and 1st maxilliped.
Endopod — inner branch of biramous appendage; central lobe of maxilla and 1st maxilliped.
Endostome — part of epistome forming palate in brachyuran crabs.
Epigastric spine — usually last spine in rostral-carapacial series in gastric region.
Epipod — a process attached to coxa of maxilliped or leg, sometimes with terminal hook for grooming setobranch.
Epistome — sternal plate lying between the labrum and antennae of crabs, the anterior part of the buccal frame.
Exopod — the outer branch of a biramous appendage: attached to basis of leg or protopod of pleopod.
Eyescale — a short flat pointed process at base of eyestalk in hermit crabs.
Eyestalk — a stalk at the end of which is the cornea of the eye.

Finger — one of the parts of a chela; the moveable part or dactyl occludes with the fixed finger or extension of the propodus.
Flagellum — the annulated, flexible or whiplike part of an antenna or antennule or maxilliped.
Flexor — the edge towards which a limb is flexed or bent.
Front — frontal part of carapace, between orbits in crabs.

Gastric region — median part of carapace over the stomach, in front of cervical groove.
Gill cavity — space under carapace containing gills or branchia.
Gonopores — openings for extrusion of sexual products on coxa of 3rd legs in females and 5th legs in males.

Hepatic region — small subtriangular area of carapace between the gastric and branchial regions.
Hepatic spine — spine at anterior edge of hepatic region.

Incisor — sharp-edged or cutting (or holding) part of mandible.
Infraorbital spine — spine on lower margin of orbit.
Ischium — third segment from attachment of leg.

Keel — a longitudinal ventral carina.
Keelson — extension to keel.
Key — arrangement of statements of characters of species giving alternatives to help in identification.

Lamella — thin flat blade.
Lamina — expanded part of lamellate appendage.
Larva — form of young decapod after hatching from egg, different from adult.
Lash — flagellum or distal part of exopod of maxilliped.
Lateral — at the side.
Linea anomurica — longitudinal groove or uncalcified line on carapace in anomurans such as hermit crabs.
Linea thalassinica — longitudinal groove or uncalcified line on carapace in anomalans such as *Callianassa*.
Luminescent — producing light.

Mandibles — mouthpart with heavy jaws lying behind other mouthparts.
Maxilla — the second pair of mouthparts following the maxillule.
Maxilliped — one of three pairs of mouthparts following the maxilla.
Maxillule — the first pair of mouthparts following the mandibles.
Median — the middle line.
Merus — the middle or fourth segment of a leg.
Mesial — referring to part nearest the midline.
Molar — thick part of mandible with cusps for grinding or holding.
Movable spine — spine in a socket allowing movement.

Ocellus — little eye or pigment spot near cornea.
Ocular acicle — eye scale in hermit crabs.
Ommatidia — visual unit of compound eye covered by cornea.
Orbit — cavity at front of carapace containing the eye.
Orbital region — narrow area around orbits.
Organs of Pesta — small luminescent areas on hepatopancreas in *Sergestes*.
Ovigerous — carrying eggs on pleopods.

Palate — roof of buccal cavity in crabs.
Palm — part of chela behind the fingers.
Palp — appendage attached to mandible.
Peduncle — proximal part of appendage to which organs are attached.
Pereopod — one of five pairs of legs.
Petasma — complex genital structure attached to inner edge of protopod of first pair of pleopods in male penaeid shrimps.
Photophore — luminescent organ.
Pleon — abdomen.
Pleonite — abdominal somite.
Pleopod — one of five pairs of abdominal appendages: swimmerets.
Pleurobranch — gill attached to wall of thorax above attachment of legs.
Pleuron — ventrolateral part of shell covering somites of abdomen below point of articulation.
Plumose seta — seta with lateral setae or plumes like a feather.
Podobranch — gill attached to an epipod or coxa of appendage.
Posterior — towards the hind part.
Posteroventral — lower part to the rear.
Prehensile — for grasping.
Propodus — sixth segment of leg.
Prosartema — a thin ciliated lobe from dorsal surface of first article of antennule, extending forward over the eye; in some Penaeidea only.
Protopod — proximal segment (fused coxa and basis) of appendage such as pleopod.
Proximal — towards the point of attachment.

Pseudochela — chela-like modification of dactyl occluding with extension of propodus of 5th pereopod.
Pterygostomian spine — spine at anterolateral edge of carapace.

Ramus — branch.
Rasp — rows of rough chitinous elements on surface of terminal segments of uropods or pereopods in hermit crabs.
Retrorse — bending backward from tip.
Rostral formula — number of dorsal spines on rostrum plus those on carapace over number of ventral spines.
Rostrum — projection of front of carapace between orbits.

Scale — antennal scale.
Scaphocerite — antennal scale.
Scaphognathite — "bailer"; exopod of maxilla functioning as a propellor to force water through gill cavity.
Serrate — saw-toothed or with fixed spines along edge.
Seta — slender bristlelike organ, more or less rigid, from surface of shell.
Setobranch — tuft of long threadlike setae on coxa of some legs of some species of decapods, trailing into gill cavity, possibly discouraging settling of fouling organisms. Cleaned by action of claw-bearing epipods (Bauer 1981).
Setose — bearing many setae.
Shield — anterior part of carapace of hermit crab.
Sinus — any space or cavity; may refer to concavity at edge of shell.
Somite — in body segmentation a part or section.
Spermatophore — aggregation of sperm cells in various forms extruded from the gonopore of the male decapod.
Spine — a sharp needle-like projection.
Sternal plates — partly fused ventral plates between legs.
Sternum — ventral part of thorax between legs.
Stylocerite — lateral lobe often with forward spine on 1st article of antennule.
Subchela — a chela formed by the dactyl folding back on front of propodus as in Crangonidae.
Subequal — almost but not quite equal.
Subtriangular — shaped almost like a triangle.
Suborbital spine — spine on lower rim of orbit.
Sulcus — a groove.
Supraorbital spine — a spine near front of carapace above the eye.
Suture — line or seam at junction of plates or around base of spine.
Sympod — protopod.

Tail fan — telson and uropod at each side of it.
Telson — last or seventh somite of abdomen bearing anal opening.
Tergum — dorsal plate of each abdominal somite, between pleura.
Thelycum — genital structure (seminal receptacle) formed from last two thoracic sternal plates in female Penaeidea.
Truncate — as if cut off.
Tubercle — small rounded protuberance.

Uropod — one of a pair of appendages attached to 6th abdominal somite, each with an inner and outer branch (endopod and exopod).

Vas deferens — tubule for passage of spermatozoa.

References

ALLEN, J. A. 1959. On the biology of *Pandalus borealis* Kroyer, with reference to a population off the Northumberland coast. J. Mar. Biol. Assoc. U. K. 38: 189-220.

_____ 1962. Observations on *Spirontocaris* from Northumberland waters. Crustaceana 3(3): 227–238.

_____ 1963. Observations on the biology of *Pandalus montagui* (Crustacea: Decapoda). J. Mar. Biol. Assoc. U.K. 43: 665–682.

_____ 1966. The dynamics and interrelationships of mixed populations of Caridea found off the north-east coast of England, p. 45–66. *In* H. Barnes [ed.] Some contemporary studies in marine science. Hafner Publishing Co., New York, N.Y. 716 p.

_____ 1967. The fauna of the Clyde Sea area. Crustacea: Euphausiacea and Decapoda. Scottish Mar. Biol. Assoc., Millport, 116 p.

ANTHONY, V. C. 1980. Review of lobster mortality estimates in the United States. Can. Tech. Rep. Fish. Aquat. Sci. 932: 17–35.

APOLLONIO, S. 1969. Breeding and fecundity of the glass shrimp, *Pasiphaea multidentata* (Decapoda, Caridea) in the Gulf of Maine. J. Fish. Res. Board. Can. 26: 1969–1983.

BATE, C. S. 1888. Report on the Crustacea Macrura collected by H. M. S. *Challenger* during the years 1873–76. Report on the scientific results of the voyage of H.M.S. *Challenger*, Zoology, 24, xc + 942 pages, Pls. 1–150.

BAUER, R. T. 1981. Grooming behaviour and morphology in the decapod Crustacea. J. Crustacean Biol. 1: 153–173.

BERNIER, L., AND L. POIRIER. 1981. Évaluation sommaire de possibilités d'exploitation commerciale du stock de crevette de roche, *Sclerocrangon boreas*, des iles de Mingan. Québec Direction générale des pêches maritimes, Cahier d'Information 94: 1–43.

BIGFORD, T. E. 1979 Synopsis of biological data on the rock crab *Cancer irroratus* Say. NOAA Tech. Rep. NMFS Circ. 426: 26 p.

BLACKER, R. W. 1957. Benthic animals as indicators of hydrographic conditions in Svalbard waters. Fish. Invest., Ser. 2. 20: 1–49.

BIFFAR, T. A. 1971. The genus *Callianassa* (Crustacea, Thalassinidea) in South Florida, with keys to the Western Atlantic species. Bull. Mar. Sci. 21: 637–715.

BOONE, L. 1930. Crustacea: Anomura, Macrura, Schizopoda, Isopoda, Amphipoda, Mysidacea, Cirrepedia and Copepoda. Bull. Vanderbilt Mar. Mus., 3: 1–221.

BOSCHMA, H. 1972. On the occurrence of *Carcinus maenas* L. and its parasite *Sacculina carcini* Thompson in Burma with notes on the transport of crabs to new localities. Zool. Meded. 47: 145–155.

BOUSFIELD, E. L. 1956. Studies on the shore Crustacea collected in eastern Nova Scotia and Newfoundland, 1954. Nat. Mus. Can. Bull. 142: 127–152.

_____ 1958. Littoral marine arthropods and mollusks collected in western Nova Scotia, 1956. Proc. N.S. Inst. Sci. 24: 303–325.

_____ 1962. Studies on littoral arthropods from the Bay of Fundy region. Nat. Mus. Can. Bull. 183: 42–62.

BOUSFIELD, E. L., AND D. R. LAUBITZ. 1972. Station lists and new distributional records of littoral marine invertebrates of the Canadian Atlantic and New England regions. Nat. Mus. Nat. Sci. Publ. Biol. Oceanogr. 5: 51 p.

BOUSFIELD, E. L., AND A. H. LEIM. 1960. The fauna of Minas Basin and Minas Channel. Nat. Mus. Can. Bull. 166: 1–30.

BOUVIER, E. L. 1896. Sur la classification des Lithodinés et sur leur distribution dans les océans. Ann. Sci. Nat. Zool. Palaeontol., Ser. 8, 1: 1–46.

Bowman, T. E. 1967. The planktonic shrimp *Lucifer chacei* sp. nov., (Sergestidae: Luciferinae), the Pacific twin of the Atlantic *Lucifer faxoni*. Pac. Sci. 21: 266–271.

Bowman, T. E., and J. C. McCain. 1967. Distribution of the planktonic shrimp *Lucifer* in the western North Atlantic. Bull. Mar. Sci. 17: 660–671.

Brian, A. 1941. I crostacei eduli del mercato de Genova (Decapoda Natantia), 51 p. (Genova, Laboratorio di Biologia Marina del Mare Ligure).

―――― 1942. I crostacei eduli del mercato de Genova (Decapoda Natantia). Boll. Pesca Piscic. Idribiol. 18: 25–60.

Brothers, G. 1971. Shrimp fishing gear experiments — Newfoundland. Can. Fish. Rep. 17: 155–168.

Brown, F. A. 1939. The coloration and color changes of the Gulf-weed shrimp, *Latreutes fucorum*. Am. Nat. 73: 564–568.

Bruce, A. J. 1974. On *Lysmata grabhami* (Gordon), a widely distributed tropical hippolytid shrimp (Decapoda, Caridea). Crustaceana 27: 107–109, 1 pl.

Brunel, P. 1963. Troisième série d'observations sur la biologie et la biométrie du crabe-araignée *Chionoecetes opilio* (Fabr.). Rapp. ann. 1962, Sta. Biol. mar. Grande-Rivière: 92–100.

Buchanan, J. B. 1963. The biology of *Calocaris macandreae* Bell (Crustacea: Thalassinidea). J. Mar. Biol. Assoc. U.K. 43: 729–747.

Bull, H. O. 1933. The newly hatched larvae of *Calocaris macandreae* Bell. Rep. Dove Mar. Lab. 3rd. Ser. 1: 48–50.

Burkenroad, M. D. 1934. Littoral Penaeidea chiefly from the Bingham Oceanographic collection. With a revision of *Penaeopsis* and description of two new genera and eleven new American species. Bull. Bingham Oceanogr. Coll. 4: 1–109.

―――― 1936. The Aristaeinae, Solenocerinae and pelagic Penaeinae of the Bingham Oceanographic collection. Bull. Bingham Oceanogr. Coll. 5: 1–151.

―――― 1963. The evolution of the Eucarida, (Crustacea, Eumalacostraca) in relation to the fossil record. Tulane Stud. in Geol. 2: 2–17.

―――― 1983. Natural classification of the Dendrobranchiata with a key to recent genera. Crustacean Issues 1: 279–290.

Burukovskii, R. N. 1983. Key to shrimps and lobsters. Russian Translations Series 5: i–xi, 174 p., Balkema, Rotterdam.

Butler, T. H. 1980. Shrimps of the Pacific coast of Canada. Can Bull. Fish. Aquat. Sci. 202: 280 p, 1–8 pls.

Caddy, J. F. 1979. The influence of variations in the seasonal temperature regime on survival of larval stages of the American lobster (*Homarus americanus*) in the southern Gulf of St. Lawrence. Rapp. P.-V. Reun. Cons. Int. Explor. Mer 175: 204–216.

Campbell, A., and R. K. Mohn. 1983. Definition of American lobster stocks for the Canadian Maritimes by analysis of fishery landing trends. Trans. Am. Fish. Soc. 112: 744–759.

Chace, F. A., Jr. 1940. Plankton of the Bermuda oceanographic expedition. XI. The bathypelagic caridean Crustacea. Zoologica 25: 117–209.

―――― 1942. Reports on the scientific results of the Atlantis expeditions to the West Indies under the joint auspices of the University of Havana and Harvard University. The Anomuran Crustacea. I. Galatheidae. Torreia, Havana 11: 1–106.

―――― 1951. The oceanic crabs of the generaa *Planes* and *Pachygraspus*. Proc. U.S. Nat. Mus. 101: 65–103.

―――― 1972. The shrimps of the Smithsonian–Bredin Caribbean Expeditions with a summary of the West Indian shallow-water species (Crustacea: Decapoda: Natantia). Smithson. Contr. Zool. 98: i–x, 1–179.

―――― 1985. The caridean shrimps (Crustacea: Decapoda) of the ALBATROSS Philippine Expedition, 1907–1910, Part 3: families Thalassocarididae and Pandalidae. Smithson. Contr. Zool. 411: i–iv, 1–143.

————— 1986. The caridean shrimps (Crustacea: Decapoda) of the ALBATROSS Philippine Expedition, 1907–1910, Part 4: Families Oplophoridae and Nematocarcinidae. Smithson. Contr. Zool. 432: 82 p.

CHACE, F. A., Jr., AND H. H. HOBBS, Jr. 1969. The freshwater and terrestrial decapod crustaceans of the West Indies with special reference to Dominica. Bredin–Archbold–Smithsonian Biological Survey of Dominica. U.S. Nat. Mus. Bull. 292: i–v, 258 p., 5 pls.

CHRISTIANSEN, M. E. 1969. Crustacea Decapoda Brachyura. Marine Invertebrates of Scandinavia. Universitetsforlaget, Oslo: 143 p.

CONAN, G. Y., AND M. COMEAU. 1986. Functional maturity and terminal molt of male snow crab (*Chionoecetes opilio*). Can. J. Fish. Aquat. Sci. 43: 1710–1719.

CORRIVEAULT, G. W., AND J. L. TREMBLAY. 1948. Contribution à la biologie du homard (*Homarus americanus* Milne-Edwards) dans la Baie-des-Chaleurs et la golfe Saint-Laurent. Contr. Stat. biol. Saint-Laurent, P. Q., Canada 19: 1–222.

COUTURE, R. 1968. Écologie d'*Argis dentata* Rathbun (Rapport préliminaire des travaux en cours, été 1967). Rapp. Ann. Sta. biol. Mar. Grande-Rivière: 57–59.

————— 1971. Shrimp fishing in the province of Quebec. Can. Fish. Rep. 17: 31–44.

COUTURE, R., AND G. FILTEAU. 1971. Âge, croissance et mortalitié d'*Argis dentata* (Crustacea: Decapoda) dans le sud-ouest du golfe Saint-Laurent. Le Naturaliste Canadien, 98: 837–850.

COUTURE, R., AND P. TRUDEL. 1968. Les crevettes des eaux côtière du Québec. Le Naturaliste Canadien 95: 857–885.

COUTURE, R., AND P. TRUDEL. 1969a. Biologie et écologie de *Pandalus montagui* Leach (Decapoda, Natantia). I. Distribution et migrations à Grande-Rivière (Gaspé) Québec. Le Naturaliste Canadien 96: 283–299.

————— 1969b. Biologie et écologie de *Pandalus montagui* Leach (Decapoda: Natantia). II. Âge, croissance et reproduction. Le Naturaliste Canadien 96: 301–315.

CROSNIER, A., AND J. FOREST. 1973. Les crevettes profondes de l'Atlantique oriental tropical. Faune Tropicale XIX. ORSTOM, Paris, 409 p.

DADSWELL, M. J. 1979. A review of the decline in lobster (*Homarus americanus*) landings in Chedabucto Bay between 1956 and 1977 with an hypothesis for a possible effect by the Canso Causeway on the recruitment mechanism of eastern Nova Scotia lobster stocks. Fish. Mar. Serv. Tech. Rep. 834: 114–144.

DAVANT, P. 1963. Clave para el identificacion de los camarones marinos y de rio. Instituto Oceanografico, Univ. de Oriente, Cumana, Venezuela, Cuadernos Oceanograficos 1: 1–113.

DOW, R. L. 1969. Cyclic and geographic trends in seawater temperature and abundance of the American lobster. Science 164: 1060–1063.

————— 1977. Relationship of sea surface temperature to American and European lobster landings. J. Cons., Cons. Int. Explor. Mer. 37: 186–190.

————— 1978. Effects of sea surface temperature cycles on landings of American, European and Norway lobsters. J. Cons., Cons. Int. Explor. Mer. 38: 271–272.

DOUGLIS, M. B. 1946. Some evidence of a dominance–subdominance relationship among lobster, *Homarus americanus*. Anat. Rec. 96: 57.

DRINKWATER, K. F., AND R. W. TRITES. 1988. Overview of environmental conditions in the Northwest Atlantic in 1986. NAFO Sci. Counc. Studies, 12: 43–55.

DUNBAR, M. J. 1951. Eastern Arctic waters. Bull. Fish. Res. Bd. Can. 88: 131 p.

————— 1960. The evolution of stability: natural selection at the level of the ecosystem. *In*: T. W. M. Cameron [ed.]. Evolution, its science and doctrine. Royal Society of Canada, Studia Varia 4: 98–109.

ELNER, R. W. 1985. Crabs of the Atlantic coast of Canada. DFO Underwater World Factsheet UW/43:: 8 p.

ENNIS, G. P. 1983. The effect of wind direction on the abundance and distribution of decapod crustacean larvae in a Newfoundland nearshore area. Can. Tech. Rep. Fish. Aquat. Sci. 1138: iv + 19 p.

    1984a. Territorial behaviour of the American lobster, *Homarus americanus*. Trans. Am. Fish. Soc. 113: 330–335.

    1984b. Small-scale seasonal movements of the American lobster *Homarus americanus*. Trans. Am. Fish. Soc. 113: 336–338.

    1986a. Swimming ability of larval American lobsters, *Homarus americanus*, in flowing water. Can. J. Fish. Aquat. Sci. 43: 2177–2183.

    1986b. Stock definition, recruitment variability, and larval recruitment processes in the American lobster, *Homarus americanus*: a review. Can. J. Fish. Aquat. Sci. 43: 2072–2084.

    1988. Changes in size composition of male crabs (*Chionoecetes opilio*) participating in the annual breeding migration in Bonne Bay Newfoundalnd. CAFSAC Res. Doc. 88/2: 14 p.

ENNIS, G. P., P. W. COLLINS, AND G. DAWE. 1982. Fisheries and population biology of lobsters (*Homarus americanus*) at Comfort Cove, Newfoundland. Can. Tech. Rep. Fish. Aquat. Sci. 1116: iv + 45 p.

FARMER, A. S. 1974. The functional morphology of the mouthparts and pereopods of *Nephrops norvegicus* (L.) (Decapoda: Nephropidae). J. Nat. Hist. 8: 121–142.

FLEMING, L. E. 1969. Use of male external genitalic details as taxonomic characters in some species of *Palaemonetes* (Decapoda: Palaemonidae). Proc. Biol. Soc. Washington 82: 443–452.

FONTAINE, B. 1977. Fixation d'une ponte de gasteropode sur des crevettes de la famille des Crangonidees. Rev. Trav. Inst. Pêches Mar. 41: 301–307.

FOREST, J. 1965. Campagnes du "Professeur Lacaze-duthiers" aux Beleares: Juin 1953 et Août 1954, Crustacés Decapodes. Vie et Milieu, séries B, 16: 325–413, 6 pls.

FOXTON, P. 1969. The morphology of the antennal flagellum of certain of the Penaeidea (Decapoda: Natantia). Crustaceana 16: 33–42.

    1970. The vertical distribution of pelagic decapods (Crustacea, Natantia) collected on the SUND cruise 1965. II. The Penaeidea and general discussion. J. Mar. Biol. Assoc. 50: 961–1000.

    1972. Further evidence of the taxonomic importance of the organs of Pesta in the genus *Sergestes* (Natantia: Penaeidea). Crustaceana 22: 181–189.

FRECHET, J. 1971. Shrimp exploration off the Canadian Atlantic coast. Can. Fish. Rep., 17: 45–50.

FRÉCHETTE, J., G. W. CORRIVEAULT, AND R. COUTURE. 1970. Hermaphrodisme protérandrique chez une crevette de la famille des Crangonides, *Argis dentata* Rathbun. Le Naturaliste Canadien 97: 805–822.

FROST, N. 1936. Decapod larvae from Newfoundland waters. Nfld. Dept. Nat. Resources, Division of Fishery Research. Reports: Faunistic Series 1, Res. Bull. 3: 11–24.

GARTH, J. S. 1958. Brachyura of the Pacific coast of America. Oxyrhyncha. Allen Hancock Pacific Expedition 21: i–xii, 1–499.

GENTHE, H. C. 1969. The reproductive biology of *Sergestes similis* (Decapoda, Natantia). Mar. Biol. 2: 203–217.

GREVE, L. 1963. The genera *Spirontocaris*, *Lebbeus*, *Eualus* and *Thoralus* in Norwegian waters (Crustacea, Decapoda). Sarsia 11: 29–42.

GURNEY, R. 1936. Notes on some decapod crustaceans of Bermuda. II. The species of *Hippolyte* and their larvae. Proc. Zool. Soc. London, 106: 25–32, 5 pls.

HACHEY, H. B. 1961. Oceanography and Canadian Atlantic waters. Bull. Fish. Res. Board. Can. 134: 1–120.

HAEFNER, P. A., Jr. 1977. Reproductive biology of the female deep-sea red crab, *Geryon quinquidens*, from Chesapeake Bight. Fish. Bull. 75: 91–102.

1978. Seasonal aspects of the biology, distribution and relative abundance of the deep sea crab *Geryon quinquidens* Smith in the vicinity of the Norfolk Canyon, western North Atlantic. Proc. National Shellfish. Assoc. 68: 49–62.

1979. Comparative review of the biology of North Atlantic caridean shrimps (*Crangon*) with emphasis on *C. septemspinosa*. Bull. Biol. Soc. Washington 3: 1–40.

HANSEN, H. J. 1908. Crustacea Malacostraca I. Danish Ingolf Exped. 3: 1–120.

HARTNOLL, R. G. 1968. The female reproductive organs of *Lucifer* (Decapoda, Sergestidae). Crustaceana 15: 263–271.

HAYASHI, K. -I. 1977. Studies of the hippolytid shrimps from Japan — IV. The genus *Spirontocaris* Bate. J. Shimonoseki Univ. Fish. 25: 155–186.

HAYNES, E. B. 1973. Description of prezoeae and Stage I zoeae of *Chionoecetes bairdi* and *C. opilio*. (Oxyrhyncha, Oregoniinae). Fish. Bull. 71: 769–775.

1976. Description of zoeae of coonstripe shrimp, *Pandalus hypsinotus*, reared in the laboratory. Fish. Bull. 74: 323–342.

1978a. Description of larvae of the humpy shrimp, *Pandalus goniurus*, reared in situ in Kachemak Bay, Alaska. Fish. Bull. 76: 235–248.

1978b. Description of larvae of a hippolytid shrimp, *Lebbeus groenlandicus*, reared in situ in Kachemak Bay, Alaska. Fish. Bull. 76: 457–465.

1979. Description of larvae of the northern shrimp, *Pandalus borealis*, reared in situ in Kachemak Bay, Alaska. Fish. Bull. 77: 157–173.

HAYNES, E. B., AND R. L. WIGLEY. 1969. Biology of the northern shrimp, *Pandalus borealis*, in the Gulf of Maine. Trans. Am. Fish. Soc. 98: 60–76.

HEEGAARD, P. E. 1941. The zoology of East Greenland. Decapod crustaceans. Medd. Gronl. 126: 1–72.

HEDGECOCK, D., K. Nelson, J. Simons, and R. Schleser. 1977. Genic similarity of American and European species of the lobster *Homarus*. Biol. Bull. 152: 41–50.

HERBST, G. N., A. B. WILLIAMS, AND B. B. BOOTHE, Jr. 1979. Reassessment of northern geographic limits for decapod crustacean species in the Carolinian Province, USA; some major range extensions. Proc. Biol. Soc. Washington 91: 989–998.

HINSCH, G. W. 1972. Some factors controlling reproduction in the spider crab, *Libinia emarginata*. Biol. Bull. 143: 358–366.

1988. Morphology of the reproductive tract and seasonality of reproduction in the golden crab *Geryon fenneri* from the eastern Gulf of Mexico. J. Crustacean Biol. 8: 254–261.

HOFSTEN, N. von. 1916. Die decapoden Crustaceen des Eisfjords. Zool. Ergebn. Schwed. Exped. nach Spitzbergen, 1908. Kungl. Svenska Vet. Akad. Handl. 54: 1–108.

HOLTHUIS, L. B. 1947. The Decapoda of the Siboga Expedition. Part IX. The Hippolytidae and Rhynchocinetidae collected by the Siboga Expeditions and Snellius Expeditions with remarks on other species. Siboga-Expeditie, Monographie 39a[8]: 100 p.

1949. Note on the species of *Palaemonetes* (Crustacea, Decapoda) found in the United States of America. Proc. K. Ned. Akad. Wet. 52: 85–95.

1952. A general revision of the Palaemonidae (Crustacea, Decapoda, Natantia) of the Americas. II. The Subfamily Palaemoninae. Allan Hancock Found. Publ. Occas. Pap. 12: 1–396, 55 pls.

1955. The recent genera of caridean and stenopodidean shrimps (Class Crustacea, Order Decapoda, Supersection Natantia) with keys for their determination. Zool. Verh. 26: 157 p.

1971. The Atlantic shrimps of the deep-sea genus *Glyphocrangon* A. Milne-Edwards, 1881. Bull. Mar. Sci. 21: 267–373, figs. 1–15.

1974. Biological results of the University of Miami Deep-sea expeditions. 106.

The lobsters of the superfamily Nephropidea of the Atlantic Ocean (Crustacea: Decapoda). Bull. Mar. Sci. 24: 723–884.

———. 1980. FAO species catalogue. Shrimps and prawns of the world. An annotated catalogue of species of interest to fisheries. FAO Fish. Synop. 125, Vol. 1: i–xviii, 271 p.

HOLTHUIS, L. B., AND A. J. PROVENZANO, Jr. 1970. New distribution records for species of *Macrobrachium* with notes on the distribution of the genus in Florida (Decapoda, Palaemonidae). Crustaceana 19: 211–213.

HORSTED, S. A., AND E. SMIDT. 1956. The deep sea prawn (*Pandalus borealis*) in Greenland waters. Medd. Dan. Fisk. Havunders. N. S. 1: 1–118.

INGLE, R. W. 1980. British crabs. Oxford University Press. 222 p.

———. 1983. Shallow water crabs. Synopsis of the British Fauna (New Series). 206 p.

KEMP, S. 1910. The Decapoda Natantia of the coasts of Ireland. Fisheries Branch, Ireland, Sci. Invest. 1908 1: 1–190, 23 pls.

LAUZIER, L. M., AND R. W. TRITES. 1958. The deep waters in the Laurentian Channel. J. Fish. Res. Board. Can. 15: 1247–1257.

LEGARE, J. E. H. 1971. Canadian shrimp catching methods and gear. Can. Fish. Rep. 17: 117–125.

LEIM, A. H. 1921. A new species of *Spirontocaris* with notes on other species from the Atlantic coast. Trans. R. Can. Inst. 29, 13: 133–145, pls. 2–6.

LITTLE, G. 1968. Induced winter breeding and larval development in the shrimp *Palaemonetes pugio* Holthuis (Caridea, Palaemonidae). Crustaceana, Supp. 2, Studies on decapod larval development: 19–26.

MAKAROV, R. R. 1968. On the larval development of the genus *Sclerocrangon* G. O. Sars (Caridea, Crangonidae). Crustaceana Supp. 2: 27–37.

MAKAROV, V. V. 1935. Beschriebung neuer Dekapoden-formen aus den Fernen Ostens. Zool. Anz. 109: 319–325.

———. 1941. The decapod Crustacea of the Behring and Chuckchi Seas, p. 111–163. *In* K. M. Derjugin and S. A. Sernov [eds.] Investigations of the Far East Seas of the USSR, Academy of Sciences of the USSR, Moscow and Leningrad.

———. 1962 [1938]. Fauna of the USSR, Crustacea. 10. Anomura. Zool. Inst. Akad. Nauk SSSR, 278 p. (Translated by F. D. Por for the National Science Foundation and Smithsonian Institution).

MAN, J. G. de 1920. Decapoda of the Siboga Expedition. IV. Families: Pasiphaeidae, Stylodactylidae, Hoplophoridae, Nematocarcinidae, Thalassocaridae, Pandalidae, Psaliopodidae, Gnathophyllidae, Processidae, Crangonidae and Glyphocrangonidae. Siboga-Expeditie, Monographie 39a^3: 1–318.

———. 1925. The Decapoda of the Siboga Expedition. VI. The Axiidae of the Siboga Expedition. Siboga-Expeditie 39a^5, 127 p.

MANNING, R. B., AND D. L. FELDER. 1986. The status of the callianassid genus *Callichirus* Stimpson, 1866 (Crustacea, Decapoda, Thalassinidea). Proc. Biol. Soc. Washington 99: 437–443.

MARKHAM, J. C. 1977. Distribution and systematic review of the bopyrid isopod *Probopyrinella latreuticola* (Gissler, 1882). Crustaceana 33: 189–197.

MCLAUGHLIN, P. A. 1973. Remarks on the presumed example of hybridization between two species of *Pagurus* (Crustacea: Decapoda: Paguridae). J. Mar. Biol. Assoc. U.K. 53: 563–568.

———. 1974. The hermit crabs (Crustacea, Decapoda, Paguridea) of northwestern North America. Zool. Verh. 130: 396 p., 1 pl.

———. 1980. Comparative morphology of recent Crustacea. W. H. Freeman Co., San Francisco, CA.

MILNE-EDWARDS, A., ET E. L. BOUVIER. 1894. Crustacées decapodes provenant des campagnes du yacht l'Hirondelle (1886–1888). I. Brachyures et Anomures. Res. Camp. Sci. Prince de Monaco 7: 1–112.

MILNE, D. S. 1968. *Sergestes similis* Hansen and *S. consobrinus* n. sp. (Decapoda) from the northeastern Pacific. Crustaceana 14: 21–34.

MISTAKIDIS, M. N. 1957. The biology of *Pandalus montagui* Leach. Fish. Invest., Ser. 2, 21: i–iv, 1–52, 8 pls.

MORGAN, S. G., J. W. GOY, AND J. D. COSTLOW, Jr. 1988. Effect of density, sex ratio, and refractory period on spawning of the mud crab *Rhithropanopeus harrisii* in the laboratory. J. Crustacean Biol. 8: 245–249.

NATIONS, J. D. 1975. The genus *Cancer* (Crustacea: Brachyura) systematics biogeography and fossil record. Nat. Hist. Mus. Los Angeles County Sci. Bull. 23: 1–104.

――― 1979. The genus *Cancer* and its distribution in time and space. Bull. Biol. Soc. Washington 3: 153–187.

NESIS, K. N. 1964. Systematic and zoogeographic position of two benthic invertebrates in the north-western Atlantic Ocean. Zool. Zh. 43: 662–670 (Russian with English summary).

OHLIN, A. 1895. Bidrag til kannedomen om malakostrakfaunan i Baffin Bay och Smith Sound. Acta Univ. Lund. Bd. 31 (K. Fysiogr. sallsk. handl. N. F. Bd. 6).

PACKARD, A. S. 1866–69. Observations on the glacial phenomena of Labrador and Maine, with a view of the recent fauna of Labrador. Mem. Boston Soc. Nat. Hist. Vol. 1: 210–303.

PARSONS, D. G., AND J. FRÉCHETTE. 1989. Fisheries for northern shrimp (*Pandalus borealis*) in the northwest Atlantic from Greenland to the Gulf of Maine, p. 63–85. *In* J.F. Caddy [ed.] Marine invertebrate fisheries: their assessment and management. Wiley, New York, NY.

PARSONS, D. G., AND R. A. KHAN. 1986. Microsporidiosis in the northern shrimp, *Pandalus borealis*. J. Inverteb. Pathol. 47: 74–81.

PARSONS, D. G., G. R. LILLY, AND G. J. CHAPUT. 1986. Age and growth of northern shrimp *Pandalus borealis* off northeastern Newfoundland and southern Labrador. Trans. Am. Fish. Soc. 115: 872–881.

PARSONS, G. D., AND G. E. TUCKER. 1986. Fecundity of northern shrimp, *Pandalus borealis* (Crustacea, Decapoda), in areas of the northwest Atlantic. Fish. Bull. 84: 549–558.

PARSONS, D. G., P. J. VEITCH, AND G. E. TUCKER. 1983. Distribution, abundance and some biolgoical characteristics of the striped pink shrimp (*Pandalus montagui*) in the eastern Hudson Strait and Ungava Bay. CAFSAC Res. Doc. 83/11: 28 p.

PEARCY, W. G., AND C. A. FORSS. 1966. Depth distribution of oceanic shrimps (Decapoda, Natantia) off Oregon. J. Fish. Res. Board. Can. 23: 1135–1143.

PEMBERTON, G. S., M. J. RISK, AND D. E. BUCKLEY. 1976. Supershrimp: deep bioturbation in the Strait of Canso, Nova Scotia. Science 192: 790–791.

PEQUEGNAT, L. H. 1970. Deep-sea caridean shrimps with descriptions of six new species, p. 59–123. *In* W. E. Pequegnat and F. A. Chace, Jr. [ed.] Texas A & M University Oceanographic Studies, 1 (4), Contributions on the biology of the Gulf of Mexico.

PEREZ FARFANTE, I. 1977. American solenocerid shrimps of the genera *Hymenopenaeus, Haliporoides, Pleoticus, Hadropenaeus* new genus, and *Mesopenaeus* new genus. Fish. Bull. 75: 261–346.

PEREZ FARFANTE, I., AND H. R. BULLIS, Jr. 1973. Western Atlantic shrimps of the genus *Solenocera* with descriptions of a new species (Crustacea: Decapoda: Penaeidae). Smithsonian Contr. Zool. 153: 1–33.

PETRIE, B., S. AKENHEAD, J. LAZIER, AND J. LODER. 1988. The cold intermediate layer on the Labrador and Northeast Newfoundland shelves 1978–86. NAFO Sci. Counc. Stud. 12: 21–25.

Pfeffer, G. 1886. Mollusken, Krebse und Echinodermen von Cumberland-Sund nach der Ausbeute der deutschen Nord Expeditionen 1882 und 1883. Jahrb. Hamburgischen Wiss. Anstalt. 3: 25–49.

Pike, R. B. 1954. Notes on the growth and biology of the prawn *Spirontocaris lilljeborgi* (Danielssen). J. Mar. Biol. Assoc. U.K., 33: 739–747.

Prefontaine, G. et P. Brunel. 1962. Listes d'invertèbres marine recueillis dans l'estuaire du Saint-Laurent de 1929 et 1934. Le Naturaliste Canadien 89: 237–263.

Price, K. S. 1962. Biology of the sand shrimp, *Crangon septemspinosa*, in the shore zone of the Delaware Bay region. Chesapeake Sci. 3: 244–255.

Pringle, J. D., and R. E. Duggan. 1984. Latent lobster fishing effort along Nova Scotia's Atlantic coast. CAFSAC Res. Doc. 84/56: 23 p.

Provenzano, A. J., Jr. 1959. The shallow-water hermit crabs of Florida. Bull. Mar. Sci. 9: 349–420.

Pullen, E. J., and W. L. Trent. 1970. Carapace width-total weight relation of blue crab from Galveston Bay, Texas. Trans. Am. Fish. Soc. 99: 795–798.

Rasmussen, B. 1953. On the geographical variation in growth and sexual development of the deep-sea prawn (*Pandalus borealis* Kr.). Rep. Norw. Fish. Mar. Invest. 10: 1–160.

1967. Note on growth and protandric hermaphroditism in the deep sea prawn, *Pandalus borealis*. Proc. Symp. Crustacea, Mar. Biol. Assoc. India 2: 701–706.

Rathbun, M. J. 1896. The genus *Callinectes*. Proc. U. S. Nat. Mus., 18: 349–375.

1902. Description of new decapod crustaceans from the west coast of North America. Proc. U. S. Nat. Mus. 24: 885–905.

1913. List of Crustacea on the Labrador coast. *In* Labrador, the country and its people, by W. T. Grenfell et al. Appendix VI: 506–513.

1918. The grapsoid crabs of America. U. S. Nat. Mus. Bull. 97: i–xxii, 461 p., 161 pls.

1919. Part A: Decapod crustaceans. Vol. VII: Crustacea. Report of the Canadian Arctic Expedition 1913–1918, VII: 1A–14A.

1925. The spider crabs of America. U. S. Nat. Mus. Bull. 129: i–x, 613 p., 283 pls.

1929. Canadian Atlantic Fauna. 10. Arthropoda, 10m. Decapoda. Biological Board Canada, Atlantic Biological Station, St. Andrews, N.B., 38 p.

1930. The cancroid crabs of America of the families Euryalidae, Portunidae, Atelicyclidae, Cancridae and Xanthidae. U. S. Nat. Mus. Bull. 152: i–xvi, 609 p, 230 pls.

Rebach, S. 1974. Burying behaviour in relation to substrate and temperature in the hermit crab, *Pagurus longicarpus*. Ecology 55: 195–198.

Reilly, P. N., and S. B. Saila. 1978. Biology and ecology of the rock crab, *Cancer irroratus* Say, 1817, in southern New England waters (Decapoda, Brachura). Crustaceana 34: 121–140.

Roberts, M. M., Jr. 1968. Functional morphology of the mouthparts of the hermit crabs, *Pagurus longicarpus* and *P. pollicaris*. Chesapeake Sci. 9: 9–20.

Ropes, J. W. 1968. The feeding habits of the green crab *Carcinus maenas* (L.) Fish. Bull. 67: 183–203.

Ross, J. C. 1835. Marine invertebrate animals, p. i–c. *In* "Narrative of a second voyage in search of a residence in the Arctic during 1829–33, including the reports of J. C. Ross and the discovery of the north magnetic pole". A. W. Webster. London.

Rouse, W. L. 1970. Littoral Crustacea from southwest Florida. Q. J. Florida Acad. Sci. 32: 127–152.

Sabine, E. 1824. Invertebrate animals, p. ccxix–ccxxix. *In* W. E. Parry, Journal of a voyage for the discovery of a north-west passage from the Atlantic to the Pacific;

performed in the years 1819–20 in H. M. Ships Hecla and Griper ... Appendix X. John Murray, London.

SAINT LAURENT, M. de. 1972a. Un thalassinide nouveau de Golfe de Gascogne, *Calastacus laevis* sp. nov. Remarques sur le genre *Calastacus* Faxon (Crustacea, Decapoda, Axiidae). Bull. Mus. Nat. Hist. Nat., Ser. 3, 29: 347–356.

    1972b. Sur la famille des Parapaguridae Smith, 1882. Description de *Typhlopagurus foresti* gen. nov., sp. nov., et de quinze especes ou sous-especes nouvelles de *Parapagurus* Smith (Crustacea, Decapoda). Bijdr. Dierk. 42: 97–123.

    1973. Sur la systématique et la phylogénie des Thalissinidea: définition des familles des Callianassidae et des Upogebiidae et diagnose de cinq genres nouveaux (Crustacea Decapoda). C.R. Hebd. Sëances Acad. Sci., 290, series D (19): 513–516.

SANDIFER, P. A. 1973. Mud shrimp (*Callianassa*) larvae (Crustacea, Decapoda, Callianassidae) from Virginia plankton. Chesapeake Sci. 14: 149–159.

SARS, G. O. 1885. Crustacea I. Norwegian North Atlantic Expedition, 1876–78. Grondahl & Son, Christiana. 280 pages, 21 plates.

    1912. On the genera *Cryptocheles* and *Bythocaris* with description of the type species of each genus. Archiv for Mathematik og Naturvidenskab. B. XXXII, 5: 1–19, Pls. I, II.

SCARRETT, D. J. 1964. Abundance and distribution of lobster larvae (*Homarus americanus*) in Northumberland Strait. J. Fish. Res. Board Can. 21: 661–680.

SCARRETT, D. J., AND R. LOWE. 1972. Biology of rock crab (*Cancer irroratus*) in Northumberland Strait. J. Fish. Res. Board Can. 29: 161–166.

SCHMITT, W. L. 1935. Mud shrimps of the Atlantic coast of North America. Smithsonian Misc. Coll. 93: 1–21, 3 pls.

SCHRAM, F. R. 1986. Crustacea. Oxford University Press, New York, NY. 606 p.

SCHROEDER, W. C. 1959. The lobster *Homarus americanus*, and the red crab *Geryon quinquidens*, in the offshore waters of the western North Atlantic. Deep-Sea Res. 5: 266–282.

SCOTT, W. B., AND M. G. SCOTT. 1988. Atlantic fishes of Canada. Can. Bull. Fish. Aquat. Sci. 219: 731 p.

SELBIE, C. M. 1914. The Decapoda Reptantia of the coasts of Ireland. Part I. Palinura, Astacura and Anomura (except Paguridea). Fish. Ireland Sci. Invest.: 116 p.

    1921. The Decapoda Reptantia of the coasts of Ireland. Part II. Fish. Ireland Sci. Invest. 1: 1–68, 9 pls.

SHUMWAY, S. E., H. C. PERKINS, D. F. SCHICK, AND A. P. STICKNEY. 1985. Synopsis of biolgoical data on the pink shrimp *Pandalus borealis* Kroyer. FAO Fish. Synop. 144; NOAA Tech. Rep., NMFS Circ. 30: 57 p.

SIMPSON, A. C., B. R. HOWELL, AND P. J. WARREN. 1970. Synopsis of biological data on the shrimp *Pandalus montagui* Leach, 1814. FAO Fish. Rep. 57: 1225–1249.

SIVERTSEN, E., AND L. B. HOLTHUIS. 1956. Crustacea Decapoda (The Penaeidea and Stenopodidea excepted). Rep. Sci. Res. "Michael Sars" North Atlantic Deep-Sea Exped., 1910, V: 1–54.

SMALDON, G. 1979. British coastal shrimps and prawns. Synopsis of the British Fauna (New Series) 15: 126 p.

SMITH, S. I. 1879. The stalk-eyed crustaceans of the Atlantic coast of North America north of Cape Cod. Trans. Connecticut Acad. Arts Sci. 5: 27–136, 4 pls.

    1882. Report on the Crustacea. Part I. Decapoda. Reports on the results of dredging under the supervision of Alexander Agassiz on the east coast of the United States during the summer of 1880 by the US Coast Survey Steamer "Blake", Commander J. R. Bartlett USN. Bull. Mus. Comp. Zool. Harvard 10: 1–108, 16 pls.

    1884a. List of the Crustacea dredged on the coast of Labrador by the Expedition ... of W. A. Stearns, in 1882. Proc. U. S. Nat. Mus. 6.

1884b. List of Crustacea. Appendix iv. *In* Report of Progress of the Geological Survey of Canada for 1882–83–84 by Robert Bell.

1884c. XV. Report on the decapod Crustacea of the ALBATROSS dredgings off the east coast of the United States in 1883: 345–424. *In* U. S. Commission of Fish and Fisheries. Part X. Report of the Commission for 1882: i–xcii, 1101 p.

1885. On some genera and species of Penaeidae mostly from recent dredgings of the United States Fish Commission. Proc. U. S. Nat. Mus. 8: 170–190.

1886. Report of the decapod Crustacea of the "Albatross" dredgings off the east coast of the United States during the summer and autumn of 1884, p. 605–705 (22 plates). *In* United States Commission of Fish and Fisheries, Part XIII. Report of the Commissioner for 1885, cxii + 1108 p.

SMITH, P. C., B. D. PETRIE, AND C. R. MANN. 1978. Circulation, variability, and dynamics of the Scotian Shelf and Slope. J. Fish. Res. Board. Can. 35: 1067–1083.

SMITH, P. C., AND H. SANDSTROM. 1988. Physical processes at the shelf edge in the Northwest Atlantic. J. Northw. Atl. Fish. Sci. 8: 25–31.

SQUIRES, H. J. 1957. Decapod Crustacea of the CALANUS expeditions to Ungava Bay, 1947–50. Can. J. Zool. 35: 463–494.

1961. Shrimp survey in the Newfoundland fishing area, 1957 and 1958. Bull. Fish. Res. Board. Can. 129: 29 p.

1962. Decapod Crustacea of the CALANUS expeditions in Frobisher Bay, Baffin Island, 1951. J. Fish. Res. Board. Can. 19: 677–686.

1964a. *Pagurus pubescens* and a proposed new name for a closely related species in the northwest Atlantic (Decapoda, Anomura). J. Fish. Res. Board. Can. 21: 355–365.

1964b. Neotype of *Argis lar* compared with *Argis dentata* (Crustacea,Decapoda). J. Fish. Res. Board. Can. 21: 461–467.

1965a. Decapod crustaceans of Newfoundland, Labrador and the Canadian eastern Arctic. Fish. Res. Board. Can. Rep. (Biol.) 8l0: 212 p.

1965b. Larvae and megalopa of *Argis dentata* (Crustacea, Decapoda) from Ungava Bay. J. Fish. Res. Board. Can. 22: 69–82.

1965c. A new species of *Calocaris* (Crustacea, Decapoda, Thalassinidea) from the northwest Atlantic. J. Fish. Res. Board. Can. 22: 1–11.

l966. Distribution of decapod Crustacea in the northwest Atlantic. Amer. Geogr. Soc. Serial Atlas of Marine Environ. 12: 4 p., 4 pls.

1967a. Decapod Crustacea from CALANUS expeditions in Hudson Bay in 1953, 1954 and 1958–61. J. Fish. Res. Board. Can. 24: 1873–1903.

1967b. Some aspects of adaptation in decapod crustaceans of the northwest Atlantic. Mar. Biol. Assoc. India, Proc. Symp. Crustacea III: 987–995.

1968a. Decapod Crustacea from the Queen Elizabeth and nearby islands in 1962. J. Fish. Res. Board. Can. 25: 347–362.

1968b. Relation of temperature to growth and self-propagation in *Pandalus borealis* from Newfoundland. FAO Fish. Rep. 57: 243–250.

1969. Decapod Crustacea of the Beaufort Sea and arctic waters eastward to Cambridge Bay, 1960-65. J. Fish. Res. Board. Can. 26: 1899–1918.

1970. Lobster (*Homarus americanus*) fishery and ecology in Port au Port Bay, Newfoundland, 1961–65. Proc. National Shellfish. Assoc. 60: 22–39.

1973. El reproductivo de los crustaceos decapodos. Bol. Museo del Mar (Univ. de Bogota) 5: 3–7.

SQUIRES, H. J., G. P. ENNIS, AND G. E. TUCKER. 1973. Lobsters of the northwest coast of Newfoundland, 1964–67. Proc. National Shellfish. Assoc. 64: 16–27.

SQUIRES, H. J., AND A. J. G. FIGUEIRA. 1974. Shrimps and shrimp-like anomurans (Crustacea, Decapoda) from southeastern Alaska and Prince William Sound. Nat. Mus. Can. Publ. Biol. Oceanogr. 6: 23 p.

SQUIRES, H. J., G. E. TUCKER, AND G. P. ENNIS. 1971. Lobster (*Homarus americanus*) in Bay of Islands, Newfoundland, 1963–65. Fish. Res. Board. MS Rep. (Biol.) 1151: 58 p.

SQUIRES, H. J., AND G. RIVEROS C. 1978. Fishery biology of spiny lobster (*Panulirus argus*) of the Guajira Peninsula of Colombia, South America, 1969–70. Proc. Nat. Shellfish. Assoc. 68: 63–74.

STEPHENSEN, K. 1935. Crustacea Decapoda. The GODTHAAB Expedition, 1928. Medd. om Gronl. 80: 1–94.

STEPHENSON, W., AND B. CAMPBELL. 1960. The Australian portunids (Crustacea, Portunidae). IV. Remaining genera. Aus. J. Mar. Freshw. Res. 11: 73–122, 6 pls.

STEPHENSON, W., AND M. REES. 1968. A revision of the genus *Ovalipes* Rathbun, 1898 (Crustacea, Decapoda, Portunidae). Records of the Australian Museum, Sydney 27: 213–261, pls. 35–42.

STIMPSON, W. 1854. Synopsis of the marine Invertebrata of Grand Manan: or the region about the mouth of the Bay of Fundy. Smithson. Contr. (1853), 6: iv + 67 p. 3 pls.

1866. Descriptions of new genera and species of macrurous Crustacea from the coasts of North America. Proc. Chicago Acad. Sci. 1: 46–48.

SUTCLIFFE, W. H., Jr. 1973. Correlation between seasonal river discharge and local landings of American lobster (*Homarus americanus*) and Atlantic halibut (*Hippoglossus hippoglossus*) in the Gulf of St. Lawrence. J. Fish. Res. Board. Can. 30: 856–859.

SWARTZ, R. C. 1978. Reproductive and molt cycles in the xanthid crab *Neopanope sayi* (Smith, 1869). Crustaceana 34: 15–32.

TAYLOR, D. M., R. G. HOOPER, AND G. P. ENNIS. 1985. Biological aspects of snow crab, *Chionoecetes opilio*, in Bonne Bay, Newfoundland (Canada). Fish. Bull. 83: 707–7ll.

TAYLOR, D. M., AND P. G. O'KEEFE. 1986. Analysis of the snow crab (*Chionoecetes opilio*) fishery in Newfoundland for 1985. CAFSAC Res. Doc. 86/57: 24 p.

1987. Analysis of the snow crab (*Chionoecetes opilio*) fishery in Newfoundland for 1986. CAFSAC Res. Doc. 87/57: 26 p.

TEMPLEMAN, W. 1935. Local differences in the body proportions of the lobster, *Homarus americanus*. J. Biol. Board. Can. 1: 213–226.

1936a. Local differences in the life history of the lobster (*Homarus americanus*) on the coast of the Maritime provinces of Canada. J. Biol. Board. Can. 2: 41–88.

1936b. The influence of temperature, salinity, light and food conditions on the survival and growth of the larvae of the lobster (*Homarus americanus*). J. Biol. Board. Can. 2: 485–497.

1937. Habits and distribution of larval lobsters (*Homarus americanus*). J. Biol. Board. Can. 3: 343–347.

1939. Investigation into the life history of the lobster (*Homarus americanus*) on the west coast of Newfoundland, 1938. Nfld. Dep. Nat. Resources Bull. (Fish.) 7: 52 p.

1944. Abdominal width and sexual maturity of female lobsters on the Canadian Atlantic coast. J. Fish. Res. Board. Can 6: 281–290.

1975. Comparison of temperatures in July–August hydrographic sections of the eastern Newfoundland area in 1972 and 1973 with those from 1951 to 1971. ICNAF Spec. Publ. 10: 17–39.

TEMPLEMAN, W., AND S. N. TIBBO. 1945. Lobster investigations in Newfoundland, 1938 to 1941. Nfld. Dep. Nat. Resourc. Bull. (Fish.) 16: 98 p.

TRITES, R. W., AND K. F. DRINKWATER. 1984. Overview of environmental conditions in the Northwest Atlantic in 1982. NAFO Sci. Counc. Stud., 7: 7–25.

1985. Overview of environmental conditions in the Northwest Atlantic in 1983. NAFO Sci. Counc. Studies, 8: 7–20.

1986. Overview of environmental conditions in the Northwest Atlantic in 1984. NAFO Sci. Counc. Studies, 10: 21–34.

TUCK, L. M., AND H. J. SQUIRES. 1955. Food and feeding habits of Brunnich's Murre (*Uria lomvia lomvia*) on Akpatok Island. J. Fish. Res. Board. Can. 12: 781–792.

TUROBOYSKI, K. 1973. Biology and ecology of the crab *Rhithropanopeus harrisi* ssp. *tridentatus*. Mar. Biol. 23: 303–313.

UZMANN, J. R., AND E. B. HAYNES. 1968. A mycosis of the gills of the pandalid shrimp, *Dichelopandalus leptocerus*. J. Invertebr. Pathol. 12: 275–277.

VAN WINKLE, M. E., AND W. L. SCHMITT. 1936. Notes on the Crustacea, chiefly Natantia, collected by Captain Robert A. Bartlett in Arctic Seas. J. Wash. Acad. Sci. 26: 324–331.

VINOGRADOV, L. G. 1947. Decapod crustaceans of the Okhotsk Sea. Izv. Tikhookean. Nauchno Issled. Inst. Rybn. Khoz. Okeanogr. 25: 67–124. (Transl. from Russian by Fish. Res. Board. Transl. Ser. 477, 1964.)

WATSON, J. 1970. Maturity, mating and egg-laying in the spider crab, *Chionoecetes opilio*. J. Fish. Res. Board. Can. 27: 1607–1616.

WENNER, E. L., AND D. F. BOESCH. 1979. Distribution patterns of epibenthic decapod Crustacea along the shelf-slope coenocline, Middle Atlantic Bight, USA. Bull. Biol. Soc. Washington 3: 106–133.

WENNER, E. L., AND T. READ. 1982. Seasonal composition and abundance of decapod crustacean assemblages from the South Atlantic Bight, USA. Bull. Mar. Sci. 32: 181–206.

WHITEAVES, J. F. 1901. Catalogue of marine Invertebrata of eastern Canada. Geol. Surv. Can. 722: 1–272.

WICKSTEN, M. K., AND M. MENDEZ G. 1982. New records and new species of the genus *Lebbeus* (Caridea: Hippolytidae) in the eastern Pacific Ocean. Bull. Southern Calif. Acad. Sci. 81: 106–120.

WIGLEY, R. L. 1970. A tropical shrimp in the Bay of Fundy (Decapoda, Palaemonidae). Crustaceana 19: 107–108.

WIGLEY, R. L., R. B. THEROUX, AND H. E. MURRAY. 1975. Deep-sea red crab, *Geryon quinquidens*, survey off northeastern United States. Mar. Fish. Rev. 37: 1–21.

WILLIAMS, A. B. 1974a. Two new axiids (Decapoda: Thalassinidea: *Calocaris*) from North Carolina and the Straits of Florida. Proc. Biol. Soc. Washington 87: 451–464.

1974b. Marine flora and fauna of the northeastern United States Crustacea: Decapoda. NOAA Tech. Rep. NMFS Circ. 389: 50 p.

1984. Shrimps, lobsters and crabs of the Atlantic coast of the eastern United States, Maine to Florida. Smithsonian Institution Press, Washington, DC. 550 p.

WILLIAMS, A. B., AND W. S. BROWN. 1972. Notes on structure and parasitism of *Munida iris* A. Milne-Edwards (Decapoda, Galatheidae) from North Carolina, USA. Crustaceana 22: 303–308.

WILLIAMS, A. B., AND R. L. WIGLEY. 1977. Distribution of decapod Crustacea off northeastern United States based on specimens at the Northwest Fisheries Center, Woods Hole, Massachusetts. NOAA Tech. Rep. NMFS Circ. 407: i–iv, 44 p.

WILLIAMS, A. B., AND D. MCN. WILLIAMS. 1981. Carolinian records for American lobster, *Homarus americanus* and tropical swimming crab, *Callinectes boucourti*. Postulated means of dispersal. Fish. Bull. 79: 192–198.

WOLFF, T. 1978. Maximum size of lobsters (*Homarus*) (Decapoda, Nephropidae. Crustaceana 34: 1–14.

WOLLEBAEK, A. 1908. Remarks on decapod crustaceans of the North Atlantic and the Norwegian fjords. I and II. Bergens Museum Aarbog 12: 1–77, 8 pls.

YALDWIN, J. C. 1957. Deep water Crustacea of the genus *Sergestes* (Decapoda, Natantia) from Cook Strait, N. Z. Zool. Publ. Victoria Univ. Wellington 22: 1–27.

ZARENKOV, N. A. 1965. Reviziya rodov *Crangon* Fabricius i *Sclerocrangon* G. O. Sars (Decapoda, Crustacea). (Revision of the genera ... etc.). Zool. Zh. 44: 1761–1775.

ZARIQUIEY ALVAREZ, R. 1952. Estudio de las especias europeas del genero *Munida* Leach, 1818. Eos (Revista Espanola de Entomologia) 28: 143–231.

———. 1968. Crustaceos decapodos ibericos. Invest. Pesq. (Barcelona) 32: i–xi, 510 p.

Additional Reference:

BOWMAN, T. E., AND R. B. MANNING. 1972. Two arctic bathyal crustaceans: the shrimp *Bythocaris cryonesus* new species, and the amphipod *Eurythenes gryllus,* with in situ photographs from Ice Island T-3. Crustaceana 23: 187-201.

APPENDICES

Appendix 1. Summary of some distinctive features of mouthparts and antennae of selected families of decapods from the Atlantic coast of Canada.

TABLE 1.

Appendage	Families of decapods		
	ARISTEIDAE	SERGESTIDAE	OPLOPHORIDAE
Antennule	Stylocerite small, partly fused with 1st article.	Stylocerite small, partly fused with 1st article; ventro-mesial flagellum modified in males for grasping.	Stylocerite almost as long as 1st article (except in *Oplophorus*).
Antenna	Scale large; tip long and attenuate in male of *Plesiopenaeus*.	Scale large, about two-thirds as long as carapace.	Scale large; tapered strongly in *Acanthephyra*; greatly modified and spiny in *Oplophorus*; peduncle less than half scale.
Mandible	Palp very long and wide, 2-segmented; no gap between incisor and molar.	Palp very long, 2-segmented, proximal longer and wider than distal segment.	Palp moderate, 3-segmented; incisor not clearly separated from molar.
Maxillule	Distal larger than proximal endite; endopod long and curved.	Distal endite very large, rounded: proximal small and ovate.	Proximal almost as large as distal endite; endopod not bifid except in *Oplophorus*.
Maxilla	Distal and proximal endites unequally bilobed; endopod with hooked spines on neck in *Gennadas*.	Proximal endite unilobed.	Proximal endite unilobed with lamina behind and just along edge forming a large and small lobe. (Except in *Hymenodora*).
Scaphognathite	Anterior lobe longer than posterior, latter rounded.	Anterior lobe long and narrow, posterior rounded.	Anterior lobe long, wide, sloped on inner distolateral margin (except truncate in *Hymenodora*).
Maxilliped I	Endopod with 3 segments (plus one small distal in *Benthesicymus*); distal endite long and narrow.	Endopod 2-segmented longer than blade of exopod; distal endite long and narrow.	Endopod 3-segmented in most species; exopod with short lash from inner corner (except in *Hymenodora* and *Acanthephyra*); distal endite large with concave inner margin.
Maxilliped II	Dactyl longer than wide.	Dactyl longer than wide.	Dactyl inserted diagonally, wider than long.

TABLE 2.

Appendage	Families of decapods		
	NEMATOCARCINIDAE	PASIPHAEIDAE	PALAEMONIDAE
Antennule	Stylocerite with outer expansion at middle, about two-thirds as long as 1st article.	Stylocerite as long as 1st article, sharp.	Stylocerite shorter than half 1st article, sharp-pointed; except in *Leander*. Two branches in dorsolateral flagellum.
Antenna	Scale large; distolateral spine slightly curved; peduncle about one-quarter scale.	Distolateral spine exceeding scale. Peduncle about one-half scale (more in *Parapasiphaea*).	Scale wide, exceeds distolateral spine. Peduncle about one-half scale (less in *Leander*).
Mandible	Palp 3-segmented; incisor thin, with few apical teeth well separated from molar.	No palp in *Pasiphaea*; no molar, incisor saggitate.	No palp in *Palaemonetes*; incisor well separated from molar, with distal teeth.
Maxillule	Proximal wider than distal endite, latter curved. Endopod not clearly bifid.	Proximal endite rounded, small. Endopod simple in *Pasiphaea*.	Proximal endite pointed, narrow; endopod clearly bifid, curved distally.
Maxilla	Proximal endite with lamina along inner edge.	No endites.	No proximal endite.
Scaphognathite	Ant. lobe constricted at middle; post. lobe with long trailing setae at end and side.	Ant. lobe narrow, from wide base, longer than endopod.	Ant. lobe longer than posterior lobe and endopod.
Maxilliped I	Endopod 3-segmented, distal segment small; exopod with long lash.	No endites; exopod with reduced lobe, lash leaflike, short.	Endites short, squarish; endopod much shorter than exopod.
Maxilliped II	Dactyl wider than long.	Dactyl longer than wide; spine at tip.	Dactyl wider than long.

TABLE 3.

Appendage	Families of decapods		
	HIPPOLYTIDAE	PANDALIDAE	CRAGONIDAE
Antennule	Stylocerite long and sharp (except in *Hippolyte* and *Latreutes*) usually longer than 1st article.	Stylocerite truncate less than half 1st article.	Stylocerite wide and robust, pointed; usually more than half 1st article.
Antenna	Scale large, usually exceeded by distolateral spine (except in *Bythocaris Caridion* and *Latreutes*).	Scale moderately narrow and tapering, exceeded by spine except in *P. borealis*.	Scale wide, robust, spine may exceed blade.
Mandible	Palp 2-segmented in most species; incisor a thin blade with few distal teeth, separated from molar (except without palp in *Hippolyte*, and palp or incisor in *Latreutes*).	Palp 3-segmented; incisor a thin blade with few teeth, separated from molar.	No palp or incisor; molar with distal fangs.
Maxillule	Proximal endite much narrower than distal. Endopod bifid, simple in *Bythocaris* and *Latreutes*.	Proximal much narrower and smaller than distal endite. Endopod bifid.	Small proximal endite, (except in *Sclerocrangon*). Endopod simple.
Maxilla	Proximal endite one lobe or small unequal lobes.	Proximal endite a large squarish lobe and small pointed distal lobe.	No endites. Endopod short except in *Sclerocrangon*.
Scaphognathite	Ant. lobe longer than posterior, the latter extremely narrow in *Hippolyte*. Setal fringe short.	Ant. lobe shorter than posterior, the latter pointed and with long trailing setae at end and side (except in *Stylopandalus*).	Lobes about equal in length. Post. lobe with long trailing setae at end (except in *Crangon*). Post. lobe narrow and pointed in *Sclerocrangon*.
Maxilliped I	Endopod 2-segmented; endites large; exopod with lash longer than lobe (except in *Bythocaris*).	Endopod 3-segmented, (except in *Stylopandalus*), about as long as lobe of exopod (except in *Dichelopandalus*.	Endopod obscurely 2-segmented, almost as long as exopod. No endites.
Maxilliped II	Dactyl much wider than long (except in *Hippolyte* and *Latreutes*).	Dactyl much wider than long (except in *Stylopandalus*).	Dactyl wider than long inserted diagonally.

TABLE 4.

Appendage	Families of decapods		
	AXIIDAE & NEPHROPIDAE	POLYCHELIDAE	PAGURIDAE
Antennule	No stylocerite, spine present.	1st article with lateral expansion and long, sharp anterior extension.	No stylocerite (except in *Pagurus politus*); 1st article short, robust, others slender.
Antenna	Scale reduced to a thorn.	Scale modified, pointed, about as long as peduncle.	Scale reduced to spike like acicle.
Mandible	Palp 3-segmented; incisor heavily calcified, molar behind it, few or no blunt teeth.	Palp 3-segmented; incisor saggitate, large, no molar.	Palp 3-segmented; incisor heavily calcified, even edged, molar behind it.
Maxillule	Proximal endite wider than distal; endopod with a flagellum.	Both endites narrow, curved, distal one with strong apical spine; endopod small, none in *Stereomastis*.	Proximal endite larger than distal; endopod bifurcate at tip.
Maxilla	Endites unequally bilobed; endopod long, sinuous.	Endites reduced to one slender distal projection; endopod long, sinuous.	Endites unequally bilobed, large. Endopod long.
Scaphognathite	Post. lobe with long terminal whiplike seta; endopod longer than ant. lobe.	Ant. lobe long and narrow, post. lobe rounded, moderate.	Ant. lobe about as long as posterior lobe and endopod, rounded. Post. lobe subtriangular (Except in *P. arcuatus*).
Maxilliped I	Endopod 2-segmented, shorter than exopod. Endites large.	Endites reduced to small lobe; endopod slender; exopod expanded distally with recurved lobes.	Endites large; endopod shorter than exopod, latter with lash.
Maxilliped II	Dactyl longer than wide; epipod very long in *Homarus*.	Dactyl longer than wide and with strong apical spine.	Dactyl longer than wide.
Maxilliped III	Crista dentata on ischium without accessory tooth.	Leglike. Without crista dentata on ischium.	Crista dentata on ischium with accessory tooth (except in *Pagurus politus*).

TABLE 5.

Appendages	Families of decapods		
	LITHODIDAE	GALATHEIDAE	MAJIDAE
Antennule	No stylocerite; 1st article stout, others more slender.	No stylocerite; 1st article much stouter than others and with large sharp distal spines.	No stylocerite; 1st article short, almost spherical; others slender, folding obliquely in socket.
Antenna	No scale or acicle; basal article with strong spine.	No scale; articles with a few stout spines.	No scale; basal article fused with carapace, longer than wide.
Mandible	Palp 3-segmented; incisor heavily calcified, edge almost straight, molar behind it.	Palp 3-segmented; incisor heavily calcified, edge with central and corner tooth; molar behind it.	Palp 3-segmented; incisor heavily calcified, edge curved with blunt corner tooth; molar behind.
Maxillule	Distal endite larger and wider than proximal, latter rounded; endopod moderate, slightly curved.	Endites about equal; endopod with wide base and slender neck.	Distal endite much larger than narrow tapered proximal; endopod 2-segmented, stout.
Maxilla	Endites unequally bilobed, large.	Endites unequally bilobed, proximal with small distal lobe.	Endites unequally bilobed (proximal reduced in *Libinia*).
Scaphognathite	Lobes about equal, posterior subtriangular; endopod about as long as anterior.	Ant. lobe much larger than posterior, latter subtriangular; endopod as long as anterior.	Ant. lobe wider than long, subtriangular (except in *Libinia*); endopod short.
Maxilliped I	Endites large; endopod slender almost as long as exopod, latter with short lash. Epipod reduced.	Endites large; endopod slender about as long as exopod, latter with short lash. (except in *Munidopsis*) Epipod larger than exopod.	Endites large; endopod thick, subtruncate, about as long as exopod (with long lash). Epipod extremely long, ribbonlike.
Maxilliped II	Dactyl longer than wide.	Dactyl longer than wide.	Dactyl longer than wide. Epipod extremely long, ribbonlike.
Maxilliped III	Crista dentata with one accessory tooth in *Lithodes*, two in *Neolithodes*.	Crista dentata without accessory tooth.	Operculiform, rigid. Epipod extremely long, ribbonlike.

Appendix 2. Mythological References of Names of Some Families, Genera and Species in the Text
(From Graves, Robert. 1955. *The greek Myths*. Pelican A509, Vols. 1 and 2, Penguin Books, Harmondsworth, Middlesex)

Argus (*bright*) — origin of genus name *Argis*. Builder of the "*Argo*", the ship in which Jason and a company of heros (including Argus) sailed to Colchis to bring back the Golden Fleece.

Aristaeus (*the best*) — origin of family name Aristeidae. Aristaeus was the son of Appolo and Cyrene, a Naiad, grand-daughter of the river-god Peneius.

Benthesicyme (*wave of the deep*) — origin of the genus name *Benthesicymus*. The daughter of Poseidon (god of the sea) and the sea-goddess, Amphitrite.

Boreas (*the north wind*) — species name of *Sclerocrangon*. Son of Eos (the daughter of the Titans Hyperion and Theia) and Astraeus, also of Titan stock, and brother of the South and West winds.

Chione (*snow queen*) — origin of the genus name *Chionoecetes*. Daughter of Boreas and Oreithyia, the daughter of Erechtheus king of Athens; secretly loved by Poseidon for whom she had a son.

Cranae (*stony*) — presumed origin of the genus name *Crangon* and family name Crangonidae. Sister of Atthis (goddess of the rugged coast) and Cranaechme (rocky point) daughters of Cranaus (rocky), king of Athens at the time of the Deucalonian flood.

Enalus (*child of the sea*) — by presumed misspelling = *Eualus*. Rescued by a dolphin when he jumped into the sea to join his sweetheart, Phineis, who had been thrown into the sea chosen by lot to appease the sea-goddess, Amphitrite; Phineis was also rescued by the dolphin's mate.

Galatea (*milk-white*) — origin of the family name Galatheidae. An ivory statue brought to life by Aphrodite (Venus) who became the lover of Pygmalion by whom she had two children.

Geryon (*crane*) — genus name of the deep-sea red crab of the family Geryonidae. A son of Chrysior and Callirrhoea, a daughter of the Titan Oceanus; he was born with three heads, six hands and three bodies joined at the waist. He owned a herd of beautiful red cattle and was king of Tartessus in Spain.

Hippolyte (*of the stampeding horses*) — family name Hippolytidae and genus of shrimp, *Hippolyte*. Amazonian queen, one of the children of Ares (Mars) and the Naiad Harmonia.

Iris (*rainbow*) — species name of *Munida*. A goddess, messenger of Hera.

Maia (*grandmother*) — a former spelling of *maja*, a species of *Lithodes*. The mother of Hermes, messenger of the gods; his father was Zeus.

Palaemon (*wrestler*) — origin of the genus name *Palaemonetes* of the family Palaemonidae. Ino's son Melicertes by Athamas was deified by Zeus as the god Palaemon and sent to the Isthmus of Corinth riding on a dolphin's back. He became the patron of Isthmian games held every four years.

Panopeus (*all-viewing or full moon*) — origin of genus name *Neopanope*. and the genus *Rhithropanopaeus*. The only one of the allies of Amphitryon who broke his oath to Ares and Athene not to hide any of the spoils of war.

Pasiphaea (*she who shines for all- the moon*) — genus *Pasiphaea* in the family Pasiphaeidae. A daughter of the sun-god Helius and the nymph Crete who married Minos, a son of Zeus. She gave birth to the Minotaur, a monster with the body of a man and the head of a bull, as a result of vengence by Poseidon against Minos.

Xanthus (*yellow*) — family name, Xanthidae, of crab species. The name of one of the horses of Achilles, one of the two immortal horses given by Poseidon to Achilles' father, Peleus, at his wedding to the nereid Thetis.

INDEX OF SCIENTIFIC NAMES

Acanthephyra 65
   pelagica 65–69
   purpurea 70–74
Anomala 16 (key)
Argis 262
   dentata 262–266
Aristeidae 20
Astacidea 16 (key)
Axiidae 332 (key)
Axius 333
   serratus 333–337

Benthesicymus 21
   bartletti 21–25
Brachyura 17 (key)
Bythocaris 148
   gracilis 148–152
   payeri 153–157
   spinipleura 158–162

Callianassa 344
   atlantica 344–348
   biformis 349–353
Callianassidae 343 (key)
Callinectes 468
   sapidus 468–472
Calocaris 338
   templemani 338–342
Cancer 451
   borealis 451–455
   irroratus 456–460
Cancridae 450 (key)
Carcinus 473
   maenas 473–477
Caridea 15 (key)
Caridion 163
   gordoni 163–167
Chionoecetes 429
   opilio 429–434
Crago 267
Crangon 267
   septemspinosa 267–271
Crangonidae 261 (key)

Dendrobranchiata 15 (key)
Dichelopandalus 234
   leptocerus 234–238

Eualus 168
   fabricii 168–172
   gaimardi belcheri 173–177
   gaimardi gaimardi 178–182
   macilentus 183–187
   pusiolus 188–192
Eukyphida 15 (key)

Galatheidae 407 (key)
Gennadas 26
   elegans 26–30
   valens 31–35
Geryon 461
   quinquidens 462–466
Geryonidae 461
Grapsidae 499

Hippolyte 193
   coerulescens 193–197
Hipplolytidae 147 (key)
Homarus 326
   americanus 326–331
Hyas 435
   araneus 435–439
   coarctatus 440–444
   coarctatus alutaceus 440, 443
Hymenodora 75
   glacialis 75–79
Hymenopenaeus 36
   robustus 36

Latreutes 198
   fucorum 198–202
Leander 127
   tenuicornis 127–131
Lebbeus 203
   groenlandicus 203–207
   microceros 208–212
   polaris 213–217
   zebra 208
Libinia 445
   emarginata 445–449
Lithodes 392
   maja 392–396
Lithodidae 391 (key)
Lucifer 57
   faxoni 57–61
Luciferidae 57

Macrobrachium 132–136
Majidae 428 (key)
Metacrangon 272
   jacqueti agassizi 272–276
Munida 408
   iris iris 408–412
   tenuimana 413–417
   valida 418–422
Munidopsis 423
   curvirostra 423–427

Nematocarcinidae 100
Nematocarcinus 100
   cursor 100–104

rotundus 105–109
Neolithodes 397
  grimaldii 397–401
Neopanope 489
  sayi 489–493
Nephropidae 326
Notostomus 80
  elegans 80–84
  robustus 85–89

Oplophoridae 64 (key)
Oplophorus 90
  spinosus 90–94
Ovalipes 478
  ocellatus 478–482

Paguridae 365
Pagurus 365 (key)
  acadianus 366–370
  arcuatus 371–375
  longicarpus 376–380
  politus 381–385
  pubescens 386–390
Palaemonetes 137
  pugio 137–141
  vulgaris 142–146
Palaemonidae 126 (key)
Pandalidae 233 (key)
Pandalus 239
  borealis 239–244
  montagui 245–250
  propinquus 251–255
Parapaguridae 402
Parapagurus 402
  pilosimanus 402–406
Parapandalus 256
Parapasiphaea 111
  sulcatifrons 111–115
Pasiphaea 116
  multidentata 116–120
  tarda 121–125
Pasiphaeidae 110 (key)
Penaeidea 15
Planes 500
  minutus 500–504
Pleoticus 36
  robustus 36–40
Plesiopenaeus 41
  edwardsianus 41–45

Polycheles 355
  granulatus 355–359
Polychelidae 354 (key)
Pontophilus 277
  brevirostris 277–281
  norvegicus 282–286
Portunidae 467 (key)
Portunus 483
  sayi 483–487

Reptantia 16 (key)
Rhithropanopeus 494
  harrisi 494–498

Sabinea 287
  hystrix 287–291
  sarsi 292–296
  septemcarinata 297–301
Sclerocrangon 302
  boreas 302–306
  ferox 307–311
Sergestes 47
  arcticus 47–51
Sergestidae 46 (key)
Sergia 52
  robusta 52–56
Spirontocaris 218
  fabricii 168
  gaimardi belcheri 173
  gaimardi 178
  groenlandicus 203
  lilljeborgi 218–222
  macilenta 193
  microceros 208
  phippsi 223–227
  polaris 213
  pusiola 188
  spinus 228–232
  zebra 208
Stereomastis 360
  sculpta 360–364
Stylopandalus 256
  richardi 256–260
Systellaspis 95
  debilis 95–99

Thalassinidea 17 (key)

Xanthidae 488 (key)